LANDOLT-BÖRNSTEIN

Numerical Data and Functional Relationships
in Science and Technology

New Series

Editor in Chief: K.-H. Hellwege

Group V: Geophysics and Space Research

Volume 1

Physical Properties of Rocks

Subvolume b

M. Beblo · A. Berktold · U. Bleil · H. Gebrande · B. Grauert
U. Haack · V. Haak · H. Kern · H. Miller
N. Petersen · J. Pohl · F. Rummel · J.R. Schopper

Editor: G. Angenheister

Springer-Verlag Berlin · Heidelberg · New York 1982

LANDOLT-BÖRNSTEIN

Zahlenwerte und Funktionen aus Naturwissenschaften und Technik

Neue Serie

Gesamtherausgabe: K.-H. Hellwege

Gruppe V: Geophysik und Weltraumforschung

Band 1

Physikalische Eigenschaften der Gesteine

Teilband b

M. Beblo · A. Berktold · U. Bleil · H. Gebrande · B. Grauert
U. Haack · V. Haak · H. Kern · H. Miller
N. Petersen · J. Pohl · F. Rummel · J. R. Schopper

Herausgeber: G. Angenheister

Springer-Verlag Berlin · Heidelberg · New York 1982

CIP-Kurztitelaufnahme der Deutschen Bibliothek

Zahlenwerte und Funktionen aus Naturwissenschaften und Technik/Landolt-Börnstein. – Berlin; Heidelberg; New York: Springer
Parallelt.: Numerical data and functional relationships in science and technology

NE: Landolt, Hans [Begr.]; PT. N.S. Gesamthrsg.: K.-H. Hellwege. Gruppe 5, Geophysik und Weltraumforschung. Bd. 1
Physikalische Eigenschaften der Gesteine/Hrsg.: G. Angenheister. Teilbd. b. M. Beblo … – 1982.

ISBN 3-540-11070-4 (Berlin, Heidelberg, New York)
ISBN 0-387-11070-4 (New York, Heidelberg, Berlin)

NE: Hellwege, Karl-Heinz [Hrsg.]; Angenheister, Gustav [Hrsg.]; Beblo, Martin [Mitverf.]

Typesetting, printing and bookbinding: Universitätsdruckerei H. Stürtz AG Würzburg

2163/3020 – 543210

Editor

G. Angenheister
Institut für Allgemeine und Angewandte Geophysik und Geophysikalisches
Observatorium, Ludwig-Maximilians-Universität, München, FRG

Contributors

M. Beblo
Institut für Allgemeine und Angewandte Geophysik und Geophysikalisches
Observatorium, Ludwig-Maximilians-Universität, München, FRG

A. Berktold
Institut für Allgemeine und Angewandte Geophysik und Geophysikalisches
Observatorium, Ludwig-Maximilians-Universität, München, FRG

U. Bleil
Institut für Geopyhsik der Ruhr-Universität, Bochum, FRG

H. Gebrande
Institut für Allgemeine und Angewandte Geophysik und Geophysikalisches
Observatorium, Ludwig-Maximilians-Universität, München, FRG

B. Grauert
Institut für Mineralogie der Westfälischen Wilhelms-Universität, Münster, FRG

U. Haack
Geochemisches Institut, Universität Göttingen, FRG

V. Haak
Institut für Geophysikalische Wissenschaften, Freie Universität, Berlin, FRG

H. Kern
Mineralogisch-Petrologisches Institut und Museum der Universität, Kiel, FRG

H. Miller
Institut für Allgemeine und Angewandte Geophysik und Geophysikalisches
Observatorium, Ludwig-Maximilians-Universität, München, FRG

N. Petersen
Institut für Allgemeine und Angewandte Geophysik und Geophysikalisches
Observatorium, Ludwig-Maximilians-Universität, München, FRG

J. Pohl
Institut für Allgemeine und Angewandte Geophysik und Geophysikalisches
Observatorium, Ludwig-Maximilians-Universität, München, FRG

F. Rummel
Institut für Geophysik der Ruhr-Universität, Bochum, FRG

J.R. Schopper
Institut für Geophysik, Technische Universität Clausthal, Clausthal-
Zellerfeld, FRG

Übersicht Band V/1

Survey of Volume V/1

Inhaltsverzeichnis

Contents

4 Thermische Eigenschaften   siehe Teilband V/1a

5 Elektrische Eigenschaften 239

3 Elasticity and inelasticity — Elastizität und Inelastizität

3.1 Elastic wave velocities and constants of elasticity of rocks and rock forming minerals — Geschwindigkeiten elastischer Wellen und Elastizitäts-Konstanten von Gesteinen und gesteinsbildenden Mineralen

3.1.1 Introduction — Einleitung

3.1.1.1 Notation, units and abbreviations — Bezeichnungen, Einheiten und Abkürzungen

The following notations are used throughout chapter 3. Some other notations are used in individual sections only and are explained there.

Die folgenden Bezeichnungen werden einheitlich im ganzen Kapitel 3 verwendet. Einige weitere Bezeichnungen werden nur in einzelnen Abschnitten verwendet und sind dort erläutert.

P	[MPa]	pressure — Druck
P_{ext}	[MPa]	external (confining) pressure — äußerer Druck (Umschließungsdruck)
P_{pore}	[MPa]	pore fluid pressure — Druck der Porenflüssigkeit
T	[°C, K]	temperature — Temperatur
ϱ	[kg m^{-3}]	density — Dichte
v	[m s^{-1}]	velocity — Geschwindigkeit
v_p, v_P	[m s^{-1}]	compressional wave velocity, longitudinal wave velocity, P wave velocity — Kompressionswellen-Geschwindigkeit, Geschwindigkeit longitudinaler Wellen, P-Wellen-Geschwindigkeit
v_s, v_S	[m s^{-1}]	shear wave velocity, transverse wave velocity, S wave velocity — Scherwellen-Geschwindigkeit, Geschwindigkeit transversaler Wellen, S-Wellen-Geschwindigkeit
K	[GPa]	bulk molulus — Kompressionsmodul
G ($=\mu$)	[GPa]	shear modulus — Scherungsmodul
E	[GPa]	Young's modulus — E-Modul
σ		Poisson's ratio — Poisson-Zahl
β ($=1/K$)	[TPa^{-1}]	compressibility — Kompressibilität
ϕ	[vol %]	porosity (volume %) — Porosität (Volumen-%)
A	[%]	velocity anisotropy — Geschwindigkeits-Anisotropie $A=100(v_{max}-v_{min})/\bar{v}$
$\bar{m}$		mean atomic weight — mittleres Atomgewicht
C_{mn}	[GPa]	elastic stiffnesses — Elastizitätsmoduln (m, n = 1···6)
S_{mn}	[TPa^{-1}]	elastic compliances — Elastizitätskoeffizienten (m, n = 1···6)
$\perp$		orthogonal to foliation or bedding — senkrecht zur Schichtung oder Schieferung
$\parallel$		parallel to foliation or bedding — parallel zur Schichtung oder Schieferung

Superscript symbols denote: Hochgestellte Symbole bedeuten:

$^-$	arithmetic mean — arithmetische Mittelung
V	Voigt's average — Mittelung nach Voigt
R	Reuss's average — Mittelung nach Reuss
VRH	Voigt-Reuss-Hill average — Voigt-Reuss-Hill-Mittelung

The following abbreviations are used in the column "Notes":

In der Spalte „Notes" werden folgende Abkürzungen verwendet:

$\bar{3}$D(3S)	mean values of 3 directions, measured on 3 different samples — Werte sind gemittelt über 3 Richtungen, gemessen an 3 verschiedenen Proben
$\bar{3}$oD(1S)	mean values of 3 orthogonal directions, measured on the same sample — Werte sind gemittelt über 3 orthogonale Richtungen, gemessen an der gleichen Probe
1D($\bar{4}$S)	mean values of 4 different samples, measured in the same direction — Werte sind gemittelt über 4 verschiedene Proben, gemessen in der gleichen Richtung
	Other combinations are to be understood accordingly! — Andere Kombinationen sind analog zu verstehen!
dry	dry sample — trockene Probe
wet	wet sample — feuchte Probe

satw   water-saturated sample — wassergesättigte Probe
satb   brine-saturated sample — salzwassergesättigte Probe
?   questionable datum — Angabe unsicher

The modal composition of rocks is characterized by the following abbreviations:

Zur Charakterisierung des modalen Stoffbestands von Gesteinen werden folgende Abkürzungen verwendet:

ab	albite	fo	forsterite	opx	orthopyroxene
am	amphibole	fsp	feldspar	or	orthoclase
an	anorthite	gar	garnet	phy	phyllosilicates
ap	apatite	gl	glass	plg	plagioclase
au	augite	hbl	hornblende	pump	pumpellyite
bi	biotite	hy	hypersthene	px	pyroxene
br	bronzite	kfsp	alkali feldspar	qu	quartz
ca	calcite	kel	kelyphite	sau	saussuritization
car	carbonate	ma	magnetite	sil	sillimanite
chl	chlorite	mi	mica	serp	serpentine
chr	chromite	mu	muscovite	sph	sphene
cpx	clinopyroxene	ne	nepheline	sym	symplectite
do	dolomite	ol	olivine	tit	titanite
en	enstatite	om	omphacite	zeo	zeolite
ep	epidote	op	opaque	zo	zoisite
fa	fayalite				

Prefixed numbers specify the content (vol%) of the respective minerals in the rock. An annexed capital symbol characterizes the composition of solid solution series, the subscript indicating the mol% fraction (weight% as exceptions) of the respective component. For example:

Vorangestellte Zahlen geben den Gehalt des betreffenden Minerals im Gestein in vol% an. Nachgestellte großgeschriebene Symbole mit Index charakterisieren die Zusammensetzung von Mischkristallen, wobei der Index den Anteil der betreffenden Komponente in mol% (ausnahmsweise auch in Gewichts%) angibt. Z.B. bedeuten:

35 serp   stands for 35 vol% serpentine
30 plg/An$_{20}$ stands for 30 vol% plagioclase
               with 20 mol% anorthite.

35 serp   35 vol% Serpentin im Gestein
30 plg/An$_{20}$ 30 vol% Plagioklas mit 20 mol% Anorthit.

The modal analyses have been simplified in some cases. For more detailed informations the reader should refer to the quoted original literature.

Values given in other units have been converted into SI-units by means of the following identities:

Der modale Stoffbestand ist z.T. in vereinfachter Form angegeben. Genauere Informationen sind der angegebenen Originalliteratur zu entnehmen.

Zur Umrechnung von Meßwerten in SI-Einheiten wurden folgende Relationen verwendet:

$$1 \text{ m} = 39.37 \text{ in} = 3.2808 \text{ ft}$$
$$1 \text{ MPa} = 10 \text{ bar} = 10^7 \text{ dyn/cm}^2 = 9.869 \text{ atm} = 145.04 \text{ psi (lbf/in}^2)$$

3.1.1.2 General remarks — Allgemeine Bemerkungen

By seismic methods the velocities of seismic waves and their variation in the Earth's interior can be determined. To interpret these results in terms of composition and physical state, comparative laboratory investigations of rocks and minerals under controlled conditions are required. This chapter summarizes results of such *laboratory* measurements. Results of *in-situ* measurements are presented in Volume V/2.

Mit seismischen Methoden können die Geschwindigkeiten seismischer Wellen und deren Variation im Erdinnern bestimmt werden. Um hieraus auf den Stoffbestand und den physikalischen Zustand im Erdinnern schließen zu können, sind Vergleichsuntersuchungen an Gesteinsproben und Mineralen unter kontrollierten Bedingungen im Laboratorium erforderlich. Der vorliegende Abschnitt enthält Ergebnisse derartiger *Laboruntersuchungen*. Die Ergebnisse seismischer *in-situ*-Messungen werden im Band V/2 dargestellt.

Seismic *in-situ* measurements are carried out in the frequency range from about 1···150 Hz, laboratory investigations, however, generally in the ultrasonic range of some $10^4 \cdots 10^7$ Hz. Fortunately, the elastic

Bei seismischen *in-situ*-Messungen liegen die Frequenzen etwa zwischen 1 und 150 Hz, bei Laboruntersuchungen dagegen im allgemeinen im Ultraschallbereich zwischen etwa 10^4 und 10^7 Hz. Zum Glück

properties of rocks are practically *frequency independent* [Ku54, Pe61]. (The theoretical possibility of dispersion in partially saturated high porosity rocks [Ge61, Do77] seems less important in practice.) If the wavelength is greater than about 3 times the diameter of the grains and the sample diameter is at least 5 times the wavelength, the velocities of ultrasonic elastic waves in rock samples can be considered equal to the velocities of seismic body waves in large rock bodies [An68]. However, systematic differences between statically (zero frequency) and dynamically determined elastic properties of rocks exist, especially at zero pressure (see 3.1.3.1.8). If not explicitly stated otherwise all elastic constants in this chapter have been dynamically measured.

Several *measuring methods* are suited · for the determination of elastic properties of rocks and minerals and have been summarized among others by [Si65] and [An68]. More recently, the method of Brillouin scattering has become available for minerals [An69a, We75]. Nowadays seismic wave velocities in rock samples are measured almost exclusively by different variations of the ultrasonic pulse-transmission technique [Bi60, Si64a, Ke75]. These methods are readily adaptable to measurements at high temperatures and pressures. The accuracy as quoted by most authors is 0.5 to 2%; the relative precision within a single series of measurement may be much better [Ch74], which is important for the determination of pressure and temperature derivatives (see 3.1.3 and 3.1.4).

The elastic properties of rocks depend on a large number of *lithologic parameters* (such as chemical and mineralogical composition, rock texture, porosity, pore geometry, pore content), on *physical parameters* (such as temperature, confining pressure, internal stresses and strains) and eventually on the *sample history* or the experimental conditions (see Fig. 2). The above-mentioned parameters are often known insufficiently only and their influence may therefore be difficult to assess. It is hoped that the following tables and figures will not only be useful as reference for individual values which are widespread in the literature, but will also facilitate the understanding of the different factors influencing seismic velocities.

Older tables on elastic wave velocities and elastic constants of rocks may be found in [Bi60, Pr66] and [An68]. Furthermore, extensive tables on elastic properties of crystals [He66, He69, Si71, He79] are available.

Whereas the literature up to 1966 has been almost completely covered in [An68], the number of published

sind die elastischen Eigenschaften von Gesteinen praktisch *frequenzunabhängig* [Ku54, Pe61]. (Die theoretische Möglichkeit einer Dispersion in partiell gesättigten hoch-porösen Gesteinen [Ge61, Do77] scheint für die Praxis weniger bedeutend.) Sofern die verwendeten Wellenlängen mindestens 3mal größer als die Korndurchmesser und die Probendurchmesser mindestens 5mal größer als die Wellenlängen sind, können die Geschwindigkeiten von Ultraschallwellen in Gesteinsproben mit denen seismischer Raumwellen in ausgedehnten Gesteinskörpern gleichgesetzt werden [An68]. Systematische Unterschiede bestehen allerdings, besonders bei Normaldruck, zwischen den statisch (Frequenz Null) und den dynamisch bestimmten elastischen Eigenschaften von Gesteinen (vgl. 3.1.3.1.8). Sofern nicht anders vermerkt, sind die in diesem Abschnitt angegebenen elastischen Konstanten dynamisch bestimmt worden.

Zur Bestimmung der elastischen Eigenschaften von Gesteinen und Mineralen sind verschiedene *Meßmethoden* geeignet, die u.a. in [Si65] und [An68] zusammenfassend beschrieben sind. Für Minerale wird neuerdings zusätzlich das Verfahren der Brillouin-Streuung verwendet [An69a, We75]. Die seismischen Geschwindigkeiten in Gesteinsproben werden heute fast ausschließlich nach verschiedenen Varianten des Ultraschall-Impuls-Verfahrens [Bi60, Si64a, Ke75] bestimmt. Es eignet sich insbesondere auch zu Messungen bei erhöhten Drucken und Temperaturen. Die absolute Genauigkeit des Verfahrens wird von den meisten Autoren mit 0.5···2% angegeben; die relative Genauigkeit innerhalb einzelner Meßserien kann wesentlich besser sein [Ch74], was für die Bestimmung der Druck- und Temperaturabhängigkeit von Bedeutung ist (vgl. 3.1.3 und 3.1.4).

Die elastischen Eigenschaften von Gesteinen werden durch eine Vielzahl von *lithologischen Parametern* (wie z.B. chemische und mineralogische Zusammensetzung, Gesteinsgefüge, Porosität, Porengeometrie, Porenfüllung) und *physikalischen Parametern* (wie Temperatur, externer Druck, interner Spannungs- und Deformationszustand) beeinflußt und hängen u.U. auch von der *Vorgeschichte* (vgl. Fig. 2), bzw. den experimentellen Bedingungen ab. Die genannten Parameter sind oft nur unzureichend bekannt und ihr Einfluß ist daher nur schwer zu überblicken. Es ist zu hoffen, daß die nachfolgenden Tabellen und Abbildungen nicht nur das Auffinden konkreter, in der Literatur weit verstreuter Meßwerte ermöglichen, sondern auch den Überblick über die verschiedenen Einflußfaktoren erleichtern können.

Tabellen älteren Datums über Geschwindigkeiten elastischer Wellen und elastische Konstanten von Gesteinen sind in [Bi60, Pr66] und [An68] zu finden. Darüber hinaus existieren umfangreiche Tabellen über die elastischen Eigenschaften von Kristallen [He66, He69, Si71, He79].

Während in [An68] die Literatur bis etwa 1966 nahezu vollständig erfaßt war, ist die Zahl der publi-

measurements – especially at elevated pressures – has increased in the meantime to such an extent that a complete presentation seems neither possible nor meaningful.

Velocity determinations on rocks of the same type at atmospheric pressure generally show large scatter and the significance of single values may be questioned. Therefore mainly mean values and standard deviations for different groups of rocks are presented in 3.1.2.1. Series of measurements on individual rock samples, however, have been compiled in 3.1.3 for varying pressures and in subsection 3.1.4 for varying pressures *and* temperatures.

zierten Messungen – insbesondere auch bei erhöhtem Druck inzwischen so sehr angewachsen, daß eine vollständige Wiedergabe weder möglich noch sinnvoll scheint.

Bei Normaldruck streuen die Schallgeschwindigkeiten in gleichartigen Gesteinen meist sehr stark, so daß ein einzelner Meßwert nur wenig Aussagekraft hat. Im Abschnitt 3.1.2.1 sind daher vorwiegend Mittelwerte und Streubereiche für verschiedene Gesteinsgruppen angegeben. Meßserien an einzelnen Gesteinsproben sind dagegen im Abschnitt 3.1.3 für variablen Druck und im Abschnitt 3.1.4 für variablen Druck *und* variable Temperatur zusammengestellt.

3.1.1.3 Formulae and theoretical relations — Formeln und theoretische Zusammenhänge

Many rocks behave *elastically isotropic*. The elasticity of isotropic bodies is fully described by two elastic constants. Different pairs of constants, however, are in use in the literature. They are summarized in Table 1 together with their connecting identities.

Viele Gesteine verhalten sich *elastisch isotrop*. Die Elastizität isotroper Körper wird durch zwei elastische Konstanten vollständig beschrieben. Es sind allerdings in der Literatur unterschiedliche elastische Konstanten gebräuchlich. Sie sind in Tab. 1 mit den zugehörigen Verknüpfungen zusammengefaßt.

Table 1. Elastic constants of isotropic media and their connecting identities – Elastische Konstanten isotroper Medien und ihre Verknüpfungen.

K	bulk modulus (Kompressionsmodul)
G	shear modulus (Scherungsmodul)
λ	1. Lamé's constant (1. Lamé'sche Konstante)
$\mu(=G)$	2. Lamé's constant (2. Lamé'sche Konstante)
E	Young's modulus (Elastizitätsmodul)
σ	Poisson's ratio (Poisson-Zahl)

	K, G	λ, μ	E, σ
K	K	$\lambda + \frac{2}{3}\mu$	$\dfrac{E}{3(1-2\sigma)}$
G, μ	G	μ	$\dfrac{E}{2+2\sigma}$
λ	$K - \frac{2}{3}G$	λ	$\dfrac{E\sigma}{(1+\sigma)(1-2\sigma)}$
E	$\dfrac{9KG}{3K+G}$	$\mu\dfrac{3\lambda+2\mu}{\lambda+\mu}$	E
σ	$\dfrac{3K-2G}{2(3K+G)}$	$\dfrac{\lambda}{2(\lambda+\mu)}$	σ

In geophysics, the constants K and G are mainly used. The wave velocities v_p and v_s depend upon them and on density ϱ according to Eqs. (1a) and (1b).

In der Geophysik wird meist das Konstanten-Paar K und G verwendet. Die Geschwindigkeiten v_p und v_s hängen hiervon und von der Dichte ϱ gemäß Gl. (1a) und (1b) ab.

$$v_\mathrm{p} = \left(\frac{K+4G/3}{\varrho}\right)^{\frac{1}{2}} \tag{1a}$$

$$v_\mathrm{s} = \left(\frac{G}{\varrho}\right)^{\frac{1}{2}} \tag{1b}$$

By using the relations listed in Table 1 the wave velocities can be expressed by any other pair of elastic constants.

Durch Verwendung der Verknüpfungen aus Tab. 1 lassen sich die Wellengeschwindigkeiten auch durch andere elastische Konstanten ausdrücken.

Poisson's ratio and the ratio of wave velocities are unambiguously connected by:

Die Poisson-Zahl und das Verhältnis der Wellengeschwindigkeiten sind eindeutig miteinander verknüpft gemäß:

$$(v_p/v_s)^2 = \frac{2-2\sigma}{1-2\sigma} \tag{2a}$$

$$\sigma = \frac{1}{2}\frac{(v_p/v_s)^2-2}{(v_p/v_s)^2-1} \tag{2b}$$

This relation is shown graphically in Fig. 1.

Der Zusammenhang ist in Fig. 1 dargestellt.

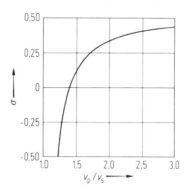

Fig. 1. Interrelation between the v_p/v_s-ratio and Poisson's ratio σ for isotropic media.

Another quantity often used is the seismic parameter φ as defined by

Eine weitere in der Seismik häufig verwendete Größe ist der Seismische Parameter φ, der wie folgt definiert ist:

$$\varphi = \frac{K}{\varrho} = v_p^2 - \frac{4}{3}v_s^2 \tag{3}$$

Because of the small thermal conductivity of rocks (see Chapter 4.1) elastic waves propagate practically adiabatically. The dynamically measured elastic constants are therefore constants at constant entropy and are distinguished by the subscript S from the elastic constants at constant temperature (subscript T). With thermodynamics it follows that:

Wegen der geringen Wärmeleitfähigkeit der Gesteine (vgl. Kap. 4.1) erfolgt die Ausbreitung elastischer Wellen praktisch adiabatisch. Die dynamisch gemessenen elastischen Konstanten werden daher auch als Konstanten bei konstanter Entropie bezeichnet und durch den Index S von den elastischen Konstanten bei konstanter Temperatur (Index T) unterschieden. Theoretisch gilt der Zusammenhang:

$$K_S = K_T(1 + \alpha^2 K_T T^*/\varrho c_v), \tag{4a}$$

$$G_S = G_T \tag{4b}$$

α, T^* and c_v are the coefficient of thermal volume expansion, absolute temperature and specific heat capacity at constant volume, respectively. The difference between isentropic and isothermal bulk modulus is about $0.5 \cdots 1.3\%$ for rocks at room temperature and can therefore be neglected for most applications. In the following, the dynamically measured elastic constants will be written without the subscript S.

Dabei ist α der thermische Volumen-Ausdehnungskoeffizient, T^* die absolute Temperatur und c_v die spezifische Wärme bei konstantem Volumen. Der Unterschied zwischen dem isentropen und dem isothermen Kompressionsmodul beträgt für Gesteine bei Normaltemperatur etwa $0.5 \cdots 1.3\%$ und ist daher für die meisten Anwendungen zu vernachlässigen. Im weiteren Verlauf dieses Kapitels werden die dynamisch bestimmten Konstanten ohne den Index S geschrieben.

All minerals and many sedimentary and metamorphic rocks behave *elastically anisotropic*. This also applies for the wave propagation. The elastic stiffnesses C_{ijkl} of anisotropic media are defined by the generalized Hooke's law:

Alle Minerale und auch viele sedimentäre und metamorphe Gesteine verhalten sich *elastisch anisotrop* und dies gilt daher auch für die Ausbreitungsgeschwindigkeiten. Die Elastizitätsmoduln eines anisotropen Mediums sind durch das allgemeine Hookesche Gesetz

$$\tau_{ij} = C_{ijkl}\varepsilon_{kl} \qquad (i, j, k, l = 1, 2, 3) \tag{5}$$

where τ_{ij} and ε_{kl} are stress and deformation tensors, respectively, and the summation convention applies

definiert, wobei τ_{ij} und ε_{kl} Spannungs- und Deformationstensor bedeuten und nach der Summenkonven-

(products with the same subscript appearing twice are summed up). The stiffnesses C_{ijkl} form a fourth-rank tensor with a maximum of 21 independent components. As usual (see [Ny57, He61, Tr69]) the contracted notation of Voigt [Vo28]

tion über zweifach in einem Produkt erscheinende Indizes summiert wird. Die C_{ijkl} bilden einen Tensor 4. Stufe mit maximal 21 unabhängigen Komponenten. In den folgenden Tabellen wird, wie allgemein üblich, die auf Voigt [Vo28] zurückgehende verkürzte Schreibweise

$$\tau_m = C_{mn}\varepsilon_n \qquad (n, m = 1\cdots 6) \tag{6}$$

is used in the following tables. Stresses and strains are written as column matrices and the stiffnesses as a 6×6 matrix. The indices n and m are obtained from the ij and kl by the index transformation:

verwendet (vgl. [Ny57, He61, Tr69]). Die Spannungen und Deformationen sind hier in Form von Spaltenmatrizen und die Elastizitätsmoduln als Matrix der Ordnung sechs geschrieben. Die n und m gehen aus den ij und kl durch die Indextransformation

ij, kl	11	22	33	23 32	13 31	12 21
m, n	1	2	3	4	5	6

The shear strains are transformed according to

hervor. Die Scherdeformationen werden dabei gemäß

$$\varepsilon_n = 2\varepsilon_{kl} \qquad (n > 3);$$

all other matrix elements are numerically identical to the tensor components in Eq. (5). Alternatively to Eq. (6) the strains may be expressed in terms of stresses by the equation

transformiert; die übrigen Matrix-Elemente unterscheiden sich dagegen numerisch nicht von den Tensorkomponenten in Gl. (5). Die Auflösung von Gl. (6) nach den Deformationen ergibt:

$$\varepsilon_m = S_{mn}\tau_n \qquad (n, m = 1\cdots 6) \tag{7}$$

where the constants S_{mn} are the so-called compliances. The C_{mn} and S_{mn} form symmetrical 6×6 matrices with 21 independent elements in the general case of a triclinic material. Symmetry properties of minerals or rocks lead to a reduction in the number of independent constants. Schemes for the different symmetry classes are shown in section 3.1.2.5, Table 6.

Die hierbei auftretenden Konstanten S_{mn} werden als Elastizitätskoeffizienten bezeichnet. Die C_{mn} und S_{mn} bilden symmetrische 6×6 Matrizen mit maximal 21 unabhängigen Elementen (triklines System). Symmetrie-Eigenschaften der Minerale oder Gesteine führen zu einer Verminderung der Zahl der unabhängigen Konstanten. Die Schemata sind jeweils bei den einzelnen Symmetrieklassen in 3.1.2.5, Tab. 6 angegeben.

Application of the matrix notation for anisotropic media to the isotropic case gives the C_{mn}-matrix:

Die Anwendung der Matrix-Schreibweise für anisotrope Medien auf den isotropen Fall ergibt die C_{mn}-Matrix:

$$\begin{pmatrix} C_{11} & C_{12} & C_{12} & \cdot & \cdot & \cdot \\ C_{12} & C_{11} & C_{12} & \cdot & \cdot & \cdot \\ C_{12} & C_{12} & C_{11} & \cdot & \cdot & \cdot \\ \cdot & \cdot & \cdot & C_{44} & \cdot & \cdot \\ \cdot & \cdot & \cdot & \cdot & C_{44} & \cdot \\ \cdot & \cdot & \cdot & \cdot & \cdot & C_{44} \end{pmatrix} \quad \text{with} \quad \begin{array}{l} C_{11} = K + \frac{4}{3}G \\ C_{12} = K - \frac{2}{3}G \\ C_{44} = G \\ \cdot \quad = 0 \end{array}$$

If no preferred orientation of mineral grains and their crystallographic axes exists, polycrystalline aggregates of anisotropic minerals may behave isotropically. The elastic constants of monomineral isotropic aggregates can be estimated from the single crystal properties according to the averaging schemes of Voigt [Vo28] and Reuss [Re29]:

Sofern keine bevorzugte Orientierung von Mineralkörnern und ihrer kristallographischen Achsen vorhanden ist, verhalten sich auch polykristalline Mineralaggregate isotrop. Die elastischen Konstanten monomineralischer isotroper Aggregate können nach den Mittelungsverfahren von Voigt [Vo28] und Reuss [Re29] abgeschätzt werden:

$$K^V = \frac{A + 2B}{3}$$

$$3A = C_{11} + C_{22} + C_{33} \tag{8a}$$
$$3B = C_{23} + C_{31} + C_{12}$$

$$G^V = \frac{A - B + 3C}{5}$$

$$3C = C_{44} + C_{55} + C_{66} \tag{8b}$$

Gebrande

$$K^R = \frac{1}{3(a+2b)}$$

$$3a = S_{11} + S_{22} + S_{33} \tag{9a}$$

$$3b = S_{23} + S_{31} + S_{12}$$

$$G^R = \frac{5}{4a - 4b + 3c}$$

$$3c = S_{44} + S_{55} + S_{66} \tag{9b}$$

The superscripts V and R characterize the bulk and shear moduli as defined by Voigt and Reuss. The Voigt- and Reuss-moduli arise from the assumption of homogeneous strain and homogeneous stress, respectively, throughout the aggregate. Hill [Hi52] has shown that the Voigt- and Reuss-moduli are upper and lower bounds of the effective moduli of an isotropic aggregate and has suggested using their arithmetic means as better approximations. These mean values are widely known as Voigt-Reuss-Hill (VRH) averages. They are tabulated for rock forming minerals in 3.1.2.5, Table 9.

A generalization of the Voigt and Reuss averaging schemes for aggregates composed of different crystalline components has been given by Hill [Hi63] (see also [Wa76]). Hashin and Shtrikman [Ha62, Ha62a] have shown that the Voigt-Reuss bounds could be improved, and they have derived closer bounds for aggregates of cubic minerals. Corresponding expressions for hexagonal, trigonal and tetragonal minerals are given in [Pe65] and [Me66]. For perfectly disordered crystals Kröner [Kr67] has developed an averaging method giving unambiguous values of effective aggregate moduli. Generally the differences to the VRH-values are small. Even with best available data on single crystals and aggregates a superiority of Kröner's averages over the VRH-averages cannot be clearly proven [Th72]. For most geophysical applications the accuracy of the VRH-averages seems to be adequate.

Theories on the elasticity of heterogeneous materials are comprehensively summarized in [Wa76].

If crystallographic axes or grain shapes of minerals are not distributed statistically isotropically, the rock becomes macroscopically anisotropic. Anisotropy can also be produced by preferred orientations of pores or microcracks. For each direction in an anisotropic medium, three independent elastic waves may be propagated, their velocities v being given by the solutions of the equation

Die Indizes V und R kennzeichnen die nach Voigt bzw. Reuss bestimmten Kompressions- und Schermoduln. Die Voigt- und Reuss-Moduln ergeben sich aus den Annahmen homogener Deformation bzw. homogener Spannung im Aggregat. Hill [Hi52] hat gezeigt, daß die Voigt- und Reuss-Moduln obere und untere Schranken für die effektiven Moduln eines isotropen Aggregats darstellen und hat ihre arithmetischen Mittelwerte als bessere Näherungen vorgeschlagen. Sie werden meist als Voigt-Reuss-Hill-Mittelwerte bezeichnet (VRH). Sie sind im 3.1.2.5, Tab. 9 für gesteinsbildende Minerale tabelliert.

Eine Verallgemeinerung der Voigt- und Reuss'schen Mittelungsverfahren für Aggregate aus mehreren kristallinen Komponenten ist von Hill [Hi63] angegeben worden (vgl. auch [Wa76]). Hashin und Shtrikman [Ha62, Ha62a] haben gezeigt, daß die Voigt-Reuss-Schranken verschärft werden können und haben entsprechende Ausdrücke für Aggregate aus kubischen Mineralen abgeleitet. Entsprechende Schranken für hexagonale, trigonale und tetragonale Minerale sind in [Pe65] und [Me66] angegeben. Für vollständig ungeordnete Kristallite hat Kröner [Kr67] eine Mittelungsmethode entwickelt, die eindeutige Werte für die effektiven Moduln des Aggregats liefert. Die Unterschiede zu den VRH-Mittelwerten sind in der Regel klein. Selbst mit den besten verfügbaren Daten für Einkristalle und Aggregate läßt sich eine Überlegenheit der Krönerschen Mittelung nicht eindeutig nachweisen [Th72]. Für die meisten geophysikalischen Anwendungen dürfte die Genauigkeit der VRH-Mittelwerte ausreichen.

Ein umfassender Überblick über Theorien zur Elastizität heterogener Materialien wird in [Wa76] gegeben.

Sofern die kristallographischen Achsen und die Kornformen der Minerale in einem Gestein nicht statistisch isotrop verteilt sind, verhält es sich makroskopisch anisotrop. Eine bevorzugte Orientierung von Poren und Mikrorissen kann ebenfalls Anisotropie bedingen. In jeder Richtung eines anisotropen Mediums können sich drei unabhängige elastische Wellen ausbreiten. Ihre Geschwindigkeiten v sind durch die Lösungen der Gleichung

$$\begin{vmatrix} \Gamma_{11} - \varrho v^2 & \Gamma_{12} & \Gamma_{13} \\ \Gamma_{12} & \Gamma_{22} - \varrho v^2 & \Gamma_{23} \\ \Gamma_{13} & \Gamma_{23} & \Gamma_{33} - \varrho v^2 \end{vmatrix} = 0, \tag{10}$$

where the Γ_{ik} are the Christoffel stiffnesses given by

bestimmt, wobei Γ_{ik} die durch

$$\Gamma_{ik} = a_j a_l C_{ijkl} \tag{11}$$

and the a_i (i = 1, 2, 3) are the direction cosines of the normal to the wave front [He61]. One wave has quasi-

gegebenen Christoffel-Moduln und die a_i (i = 1, 2, 3) die Richtungskosines der Wellennormalen sind [He61].

longitudinal and the other two have quasi-transversal polarization. Their displacements are mutually orthogonal, but in general they are not parallel and orthogonal to the wave normal. There are, however, specific directions depending on the symmetry properties of the medium for which pure longitudinal and transverse waves can be propagated. Even in these directions the velocities of transverse waves with different polarizations are different in general (see Table 7).

In most cases, sound-velocity measurements on anisotropic rock samples are carried out in such specific directions. Many rocks behave transversely isotropic (hexagonal symmetry). In this case pure longitudinal and transverse waves may be propagated parallel and orthogonal to the hexagonal axis (z-axis). Thereby the velocities of the transverse waves propagating orthogonal to the z-axis are different for polarization directions parallel and orthogonal to the z-axis. In media with hexagonal or lower symmetry, velocity measurements of P- and S-waves along the three principal axes are not sufficient for the determination of the complete set of stiffnesses; for this purpose, additional measurements oblique to the principal axes have to be carried out. Only few rock samples have been studied accordingly, e.g. in [Ch71] and [Ba72]. Using velocity measurements parallel and orthogonal to the principal axis of a transversely isotropic medium only 4 of its 5 stiffnesses can be determined; C_{13} remains undetermined.

In most cases transverse isotropy of rocks is due to foliation or bedding. An additional lineation parallel to the foliation produces orthorhombic anisotropy. The superposition of different foliations may give rise to monoclinic or even triclinic anisotropy. Generalized Voigt-Reuss schemes for the calculation of elastic constants and wave velocities of anisotropic rocks from single crystal data and petrofabric measurements have been described in [Kl68, Cr71, Ba72] and [Ba72a].

Eine Welle ist quasi-longitudinal, die anderen beiden sind quasi-transversal polarisiert. Ihre Verrückungen sind paarweise zueinander senkrecht, aber im allgemeinen sind sie nicht parallel und senkrecht zur Wellennormalen. Es gibt allerdings spezielle, von den Symmetrieeigenschaften des Mediums abhängige Richtungen, in denen sich rein longitudinale und transversale Wellen ausbreiten können. Auch in diesen Richtungen sind die Ausbreitungsgeschwindigkeiten der Transversalwellen mit unterschiedlicher Polarisationsrichtung im allgemeinen verschieden (vgl. Tab. 7).

Schallgeschwindigkeitsmessungen an anisotropen Gesteinsproben werden meist in solchen bevorzugten Richtungen durchgeführt. Weit verbreitet sind transversal-isotrope Gesteine (hexagonale Symmetrie). Reine Longitudinal- und Transversalwellen können sich darin parallel und senkrecht zur hexagonalen Achse (z-Achse) ausbreiten. Die Geschwindigkeiten der Transversalwellen senkrecht zur z-Achse sind dabei für die Polarisationsrichtungen senkrecht und parallel zur z-Achse verschieden. Geschwindigkeitsmessungen von P- und S-Wellen in Richtung der Hauptachsen reichen zur vollständigen Bestimmung der Elastizitätsmoduln von Medien mit hexagonaler oder geringerer Symmetrie nicht aus; hierfür sind zusätzlich Messungen schräg zu den Hauptachsen erforderlich. Nur wenige Gesteinsproben sind bisher entsprechend untersucht worden, z.B. in [Ch71] und [Ba72]. Von den 5 unabhängigen Elastizitätsmoduln eines transversal-isotropen Gesteins können durch Geschwindigkeitsmessungen in den Hauptachsenrichtungen nur 4 bestimmt werden; C_{13} bleibt unbestimmt.

Transversal-Isotropie von Gesteinen wird meist durch Schieferung oder Schichtung verursacht. Eine zusätzliche lineare Gefügeregelung parallel zur Schieferung ergibt rhombische Symmetrie. Die Überlagerung mehrerer Schieferungen führt im allgemeinen zu monokliner oder trikliner Anisotropie. Verallgemeinerte Voigt-Reuss-Methoden zur Berechnung der elastischen Konstanten und Wellengeschwindigkeiten anisotroper Gesteine aus Einkristall-Daten und Gefügemessungen sind in [Kl68, Cr71, Ba72] und [Ba72a] beschrieben.

3.1.2 Elastic wave velocities and constants of elasticity at normal conditions — Geschwindigkeiten elastischer Wellen und Elastizitäts-Konstanten bei Normalbedingungen

3.1.2.1 Rocks — Gesteine

Elastic wave velocities measured at atmospheric pressure on different samples of the same rock type generally show large scatter, even when made by a single method. They depend not only on petrological composition and rock fabric, but are influenced to a large degree by the presence, size and orientation of pores and cracks, and by the presence or absence of interstitial fluids. Without detailed information about these factors elastic wave velocities in a rock of known

Die Geschwindigkeiten elastischer Wellen bei Normaldruck in verschiedenen Proben desselben Gesteins-Typs streuen im allgemeinen stark, auch wenn sie mit der gleichen Methode bestimmt werden. Sie hängen nicht nur von der petrologischen Zusammensetzung und dem Gesteinsgefüge ab, sondern werden darüber hinaus in hohem Maße durch die Existenz, Größe und Orientierung von Poren und Rissen und durch das Vorhandensein oder Fehlen einer Poren-

composition can hardly be predicted. Fig. 2 clearly demonstrates that wave velocities may not only depend on the actual conditions but also on the sample's history. By heating a rock sample at atmospheric pressure new microcracks may be formed due to inhomogeneous and anisotropic thermal expansion of mineral grains and an irreversible velocity decrease remains after cooling to the original temperature [Id37]. Probably this process can also take place under natural conditions; only by the simultaneous application of rather high confining pressure can it be avoided (see 3.1.4, Fig. 1).

flüssigkeit beeinflußt. Ohne genaue Kenntnisse über diese Faktoren lassen sich die seismischen Geschwindigkeiten in einem Gestein bekannter Zusammensetzung kaum voraussagen. Fig. 2 zeigt klar, daß die Wellengeschwindigkeiten nicht nur von den momentanen Zustandsbedingungen sondern auch von der Geschichte der Probe abhängen können. Wegen der uneinheitlichen und anisotropen thermischen Dehnung von Mineralkörnern können durch Erhitzen einer Gesteinsprobe bei Normaldruck neue Mikrorisse entstehen, was zu einer irreversiblen Geschwindigkeitsabnahme nach Abkühlung auf die Ausgangstemperatur führt [Id37]. Sehr wahrscheinlich kann dieser Prozeß auch unter natürlichen Bedingungen ablaufen; nur durch gleichzeitiges Aufbringen eines relativ großen Umschließungsdrucks kann er verhindert werden (vgl. 3.1.4, Fig. 1).

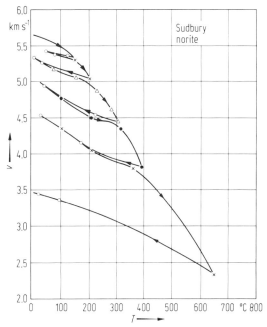

Fig. 2. The effect of thermal cycling at atmospheric pressure on the elastic wave velocity of rocks according to [Id37]; the measured velocity is the rod velocity.

Since single velocity values are of questionable significance, only mean values and ranges of velocity variation in the most common rock types at atmospheric pressure are presented in Table 2. A Gauss standard distribution law has been assumed for the calculation of the 80%-confidence limits. Mean values and 80%-confidence limits for the different rock types are also shown graphically in Figs. 3 to 6. It is quite evident from these figures that the rock type cannot be predicted unambiguously from a measured pair of v_p and v_s velocities, even if the rock class (plutonic, volcanic, sedimentary or metamorphic) is known. Under in-situ conditions the overlapping ranges may be even broader, because rock samples for ultrasonic velocity measurements are generally selected with regard to maximum homogeneity and minimum defects such as void or cracks.

Da die Aussagekraft einzelner Geschwindigkeitswerte fragwürdig ist, werden für Normaldruck in Tabelle 2 nur Mittelwerte und Variationsbereiche der Geschwindigkeiten für die wichtigsten Gesteinstypen angegeben. Die 80%-Vertrauens-Grenzen wurden unter der Annahme einer Gaußschen Normal-Verteilung berechnet. Mittelwerte und 80%-Vertrauens-Grenzen sind auch graphisch in den Fig. 3···6 dargestellt. Aus diesen Abbildungen ist klar, daß aus einem gemessenen Paar von v_p- und v_s-Geschwindigkeiten nicht eindeutig auf den Gesteinstyp geschlossen werden kann, selbst wenn die Gesteinsklasse (plutonisch, vulkanisch, sedimentär oder metamorph) bekannt ist. Unter in-situ-Bedingungen dürften die Überlappungsbereiche sogar noch breiter sein, da Gesteinsproben für Geschwindigkeitsbestimmungen im allgemeinen auf maximale Homogenität und minimale Störungen durch Hohlräume und Risse hin ausgesucht werden.

Table 2. Means ($\bar\varrho$, $\bar v_p$, $\bar v_s$) and standard deviation ($s(\varrho)$, $s(v_p)$, $s(v_s)$) of density and elastic wave velocities for rocks at normal temperature and pressure: a Gauss standard distribution has been assumed for the calculation of the 80% confidence limits. See Figs. 3···6 and 13···15.

Material	Number of samples	$\bar\varrho$ kg m⁻³	$s(\varrho)$ kg m⁻³	$\bar v_p$ m s⁻¹	$s(v_p)$ m s⁻¹	80% confid. limits of v_p m s⁻¹	$\bar v_s$ m s⁻¹	$s(v_s)$ m s⁻¹	80% confid. limits of v_s m s⁻¹	Ref.
Plutonic rocks										
Granite, USSR	>100			5060	330	4630···5480				Vo75
Granite	31	2705	85	5010	650	4180···5858	2980	290	2610···3350	Table 11, An68
Diorite	6	2823	105	5600	600	4830···6370	3220	200	2960···3470	Table 11, An68
Gabbro, USSR	75			6420	350	5970···6870				Vo75
Gabbro + norite	10	2991	65	6460	370	5990···6930	3500	150	3310···3690	Table 11, Hu50
Pyroxenite	15	3180	120	7010	610	6230···7790	3830	370	3350···4300	Table 11, An68, As76
Ultrabasic rocks, USSR	80			7310	280	6950···7670				Vo75
Peridotites	16	3265	50	7480	700	6580···8370	3860	390	3360···4360	Table 11, An68,
	7	3245	50							Ke78, Ke81
Volcanic rocks										
Acid volcanics, dry	6	2395	190	4500	1150	3000···6000	2670	650	1840···3500	Wo63
Andesite + trachyte, dry	15	2636	160	5050	840	3970···6130	2970	410	2440···3500	Wo63, Hu57, Ma68
Tholeiites, satb	70	2684	160	4990	620	4190···5780	2600	400	2080···3110	Ko80
Alkalic basalts, satb	25	2726	130	5170	470	4560···5770	2720	270	2370···3070	Ko80
Oceanic basalts, satb	189	2785	130	5480	480	4860···6090	3100	210	2830···3370	Ha79
	65									Ha79
Basalts, dry, low porosity	29	2840	140	5410	580	4670···6150	3060	350	2610···3510	Wo63, Ma68, As76, Ke79a
Nephelinite, dry	9	2846	110	5730	470	5130···6330	3210	260	2870···3550	Ma68
Diabase, dry	17	2880	95	6030	610	5240···6810	3550	130	3380···3720	Table 12, Pr62, Wo63,
	6	2950	50							Hu68, As76
Sedimentary rocks										
Coal		1200···1500				(1100···2800)				La78
Clay		1500···2200				(1800···2400)				La78
Rock salt		2200		4000						Dr74
Gypsum		2250		5450			2100			Dr74
Sand, satw						(1500···1600)			($v_p/v_s = 3\cdots10$)	Mo60

(continued)

Gebrande

Table 2 (continued)

Material	Number of samples	$\bar{\varrho}$ kg m⁻³	$s(\varrho)$ kg m⁻³	$\bar{v}_p$ m s⁻¹	$s(v_p)$ m s⁻¹	80% confid. limits of v_p m s⁻¹	$\bar{v}_s$ m s⁻¹	$s(v_s)$ m s⁻¹	80% confid. limits of v_s m s⁻¹	Ref.
Sandstones	21	2385	210	3630	650	2800⋯4460	2260	520	1590⋯2930	An68, Hu68, Jo80
	14	2320	210							An68, Jo80
Limestones	26	2658	90	5630	580	4890⋯6370	3120	450	2540⋯3700	Table 13, Pe62, Pe68, Jo74, Ma76
Silt- and claystones ‖	13	2683	70	4730	640	3910⋯5550				Hu68
⊥	13	2683	70	4100	990	2830⋯5370				Hu68
Marls ‖	8	2688	40	5680	400	5160⋯6200				Hu68
⊥	8	2688	40	5410	490	4780⋯6040				Hu68
Graywackes	13	2717	70	5970	250		3680	150	(3480⋯3870)	Pr62, Bi60, Hu68
	4	2704	30							Pr62
Slates ‖	15	2755	50	5910	430					Hu68, An68, Bi60
⊥	15	2755	50	5150	790					
Dolomites	10	2794	80	5710	620	4920⋯6500	3320	330	2900⋯3740	Table 13, Hu51, Ma76
Anhydrites		2900		5500						Dr74
Metamorphic rocks										
Serpentinites	23	2620	160	5240	770	4250⋯6230	2780	550	2070⋯3490	Table 14, As76
Quartzites	10	2636	10	4010	1210	2460⋯5560	2560	600	1790⋯3330	Table 14, Jo74
Anisotropic gneisses ‖	62	2668	50	4020	690	3130⋯4900	2460	350	2010⋯2910	Jo74, Ma76, An68
⊥	56	2668	50	2750	710	1840⋯3660	2050	370	1580⋯2520	
Marbles	57	2728	65	5730	540	5040⋯6420	3190	370	2710⋯3660	Hu51, Ku54, Wo63, St73, Jo74, Ma76
Schists, dry ‖	25	2741	70	5220	860	4120⋯6320	3210	410	2680⋯3740	Jo74, Ma76
⊥	20	2761	90	4080	1060	2760⋯5440	2710	590	1950⋯3470	
Quasi-isotropic gneisses	8	2852	130	5090	940	3880⋯6300	3120	560	2400⋯3840	Table 14, An68, Jo74
"Crystalline shales", USSR	50			5820	600	5050⋯6590				Vo75
Metagabbros	7	2877	90	6610	220	6320⋯6890	3590	60	3510⋯3670	Table 14, Ch78
Granulites	11	2932	160	5930	440	5360⋯6500	3340	180	3100⋯3570	Table 14
Eclogites	20	3410	100	7270	580	6530⋯8010	4060	390	3560⋯4560	Table 14, Ba78

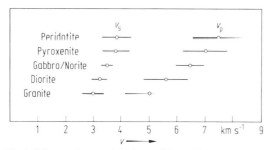

Fig. 3. Mean values and 80%-confidence limits of v_p and v_s velocities in plutonic rocks at normal temperature and pressure; data from Table 2.

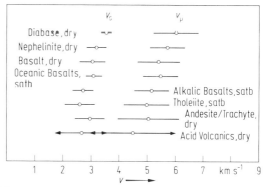

Fig. 4. Mean values and 80%-confidence limits of v_p and v_s velocities in volcanic rocks at normal temperature and pressure; data from Table 2.

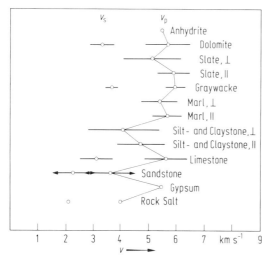

Fig. 5. Mean values (circles) and 80%-confidence limits of v_p and v_s velocities in sedimentary rocks at normal temperature and pressure; data from Table 2. If mean values only are shown, the confidence limits are unknown; if one information only is given, if refers to the v_p velocity. Rocks are arranged from bottom to top in the order of increasing density.

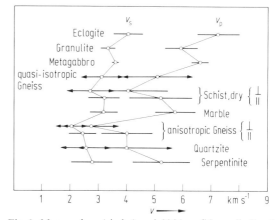

Fig. 6. Mean values (circles) and 80%-confidence limits of v_p and v_s velocities in metamorphic rocks at normal temperature and pressure; data from Table 2. Rocks are arranged from bottom to top in the order of increasing density.

As can be seen from Table 2 the number of samples is not large enough in all cases to give statistically reliable results on mean values and standard deviations. For some rock types, especially of sedimentary origin, only relatively few shear wave velocity determinations have been published.

Wie aus Tabelle 2 zu ersehen, ist die Zahl der Proben nicht in allen Fällen groß genug, um statistisch zuverlässige Berechnungen von Mittelwerten und Standardabweichungen zu ermöglichen. Für einige Gesteine, insbesondere Sedimente, sind bisher nur relativ wenige Bestimmungen der S-Wellen-Geschwindigkeiten publiziert.

3.1.2.2 Influence of porosity and pore content — Einfluß von Porosität und Porenfüllung

Porosity of rocks is extensively described in section 2.1. Some aspects which are important for seismic wave velocities can be illustrated by Table 3 and Fig. 7···11. In Table 3 [Ha70] and Fig. 7 [Bo80] it can be seen that compressional wave velocities in water-saturated unconsolidated sediments with porosities greater than 50% are only slightly different from water sound velocity. With decreasing porosities velocities strongly increase. The relation between P-wave velocity and porosity is often approximated by one of the

Die Porosität von Gesteinen wird ausführlich in Kapitel 2.1 behandelt. Einige Aspekte, die für die seismischen Wellengeschwindigkeiten von Bedeutung sind, können durch Tab. 3 und die Abbildungen 7 bis 11 verdeutlicht werden. Aus Tab. 3 nach [Ha70] und Fig. 7 nach [Bo80] ist zu ersehen, daß die P-Wellengeschwindigkeiten in wassergesättigten Lockersedimenten für Porositäten größer als 50% nur wenig von der Schallgeschwindigkeit in Wasser verschieden sind. Mit kleiner werdenden Porositäten

following Eqs. (12)···(14) introduced by Wood [Wo41], Wyllie et al. [Wy56] and Nafe and Drake [Na57], respectively:

nehmen die Geschwindigkeiten stark zu. Der Zusammenhang zwischen P-Wellengeschwindigkeit und Porosität wird oft näherungsweise durch eine der folgenden Gl. (12)···(14) beschrieben, die auf Wood [Wo41], Wyllie et al. [Wy56] sowie auf Nafe und Drake [Na57] zurückgehen:

$$v = \left\{ \frac{K_{mat} K_{por}}{[(K_{mat} - K_{por})\, \phi + K_{por}]\, \varrho_{tot}} \right\}^{\frac{1}{2}} \tag{12}$$

$$1/v = \phi/v_{por} + (1 - \phi)/v_{mat} \tag{13}$$

$$v^2 = \phi\, v_{por}^2 \left[1 + \frac{(1 - \phi)\, \varrho_{por}}{\varrho_{tot}} \right] + \frac{\varrho_{mat}}{\varrho_{tot}} (1 - \phi)^n\, v_{mat}^2 \tag{14}$$

The suffixes mat, por and tot refer to the properties of the solid component, the pore content and the total (bulk) material. Furthermore

Die Indizes mat, por und tot beziehen sich jeweils auf die Eigenschaften der festen Komponente, der Porenfüllung und des Gesamtgesteins. Es gilt ferner

$$\varrho_{tot} = \phi\, \varrho_{por} + (1 - \phi)\, \varrho_{mat} \tag{15}$$

applies. Eq. (12) theoretically applies to suspensions without rigidity. Since real unconsolidated sediments still have a certain rigidity, Eq. (12) gives a lower bound for P-wave velocities at a given porosity, as can be seen from Fig. 7. Eq. (13) corresponds to the velocity orthogonal to layering through a medium composed of alternating layers of pore fluid and solid material with thicknesses in the ratio ϕ: $(1 - \phi)$. Expression (13) is often termed "time average" equation. Since, in contrast to P-wave propagation through a real sediment, no lateral dilatation of the solid grains is possible in the underlying theoretical model Eq. (13) gives some kind of upper bound for observed velocities at a given porosity (see Fig. 7). Expression (14) is a semi-empirical equation with a free matching parameter n. For the extremes $\phi = 1$ and $\phi = 0$ all three equations give the velocities in the pore fluid and in the solid material.

Gl. (12) entspricht theoretisch dem Fall einer Suspension ohne Scherfestigkeit. Da reale Lockersedimente doch eine gewisse Scherfestigkeit besitzen, bildet Gl. (12) – wie z.B. aus Fig. 7 zu ersehen – eine untere Grenze für die P-Wellengeschwindigkeiten bei gegebener Porosität. Gl. (13) entspricht der Ausbreitungsgeschwindigkeit senkrecht zur Schichtung durch ein Medium, welches aus alternierenden Schichten aus Porenflüssigkeit und Festmaterial aufgebaut ist, deren Mächtigkeiten im Verhältnis ϕ: $(1 - \phi)$ zueinander stehen. Die Gl. (13) wird häufig auch als Zeitmittelungs-Formel bezeichnet. Da beim theoretisch zugrundeliegenden Ausbreitungsvorgang, im Gegensatz zur P-Wellenausbreitung in einem realen Sediment, keine Querdehnung der festen Partikel möglich ist, bildet Gl. (13) näherungsweise eine obere Schranke für praktisch beobachtete Geschwindigkeiten bei gegebener Porosität (vgl. Fig. 7). Gl. 14 beschreibt einen semiempirischen Zusammenhang und enthält zur Anpassung einen freien Parameter n. Alle drei Gleichungen liefern für die Grenzfläche $\phi = 1$ und $\phi = 0$ die Geschwindigkeiten in der Porenflüssigkeit und in der festen Materie.

In water-saturated hard rocks, velocities decrease systematically with increasing porosity as is to be seen from Fig. 9a and 9b. Generally the v_p/v_s ratio in such rocks is greater than in rocks without porosity and it is especially greater than for dry porous rocks. The influence of pore content on P-wave velocities is clearly reflected in Fig. 8 and 10 which show measurements with varying degree of saturation under otherwise constant conditions. Fig. 10 makes clear that propagation velocities in different directions may be influenced in different proportions by the pore content and therefore the degree of anisotropy may be changed.

Die Fig. 9a und 9b lassen erkennen, daß auch in wassergesättigten Festgesteinen die Geschwindigkeiten systematisch mit steigender Porosität abnehmen. Das v_p/v_s-Verhältnis ist dabei in der Regel größer als im Festgestein ohne Porosität und insbesondere größer als bei gleicher Porosität ohne Porenfüllung. Der Einfluß der Porenfüllung auf die P-Wellengeschwindigkeit ist besonders deutlich in den Meßreihen der Fig. 8 und 10 zu erkennen, während derer sich die Porenfüllung bei sonst gleichbleibenden Bedingungen änderte. Fig. 10 macht deutlich, daß die Ausbreitungsgeschwindigkeiten in unterschiedlichen Richtungen durch die Porenfüllung verschieden stark beeinflußt werden können, so daß dadurch auch eine vorhandene Anisotropie verändert werden kann.

The effect of freezing pore water on P-wave velocities through a sandstone is shown in Fig. 11.

Der Einfluß des Gefrierens des Porenwassers auf die P-Wellen-Geschwindigkeit in einem Sandstein ist in Fig. 11 dargestellt.

Gebrande

Table 3. Properties of unconsolidated marine sediments at 23 °C and atmospheric pressure [Ha70]. Density: ϱ_{tot}, saturated bulk density. Porosity: ϕ, measured salt free. Velocity: v_p, compressional wave velocity. Ratio: ratio of velocity in sediment to velocity in sea water with salinity of sediment pore water. Avg: mean value. SE: standard error of the mean.

Environment sediment type	No. of samples	Grain diameter Mean mm	Median mm	Sand %	Silt %	Clay %	Mean grain density g cm⁻³	Density g cm⁻³ Avg.	SE	Porosity % Avg.	SE	Velocity m s⁻¹ Avg.	SE	Ratio Avg.	SE
Continental Terrace (Shelf and Slope) Environment															
Sand															
Coarse	2	0.530	0.520	100.0			2.71	2.03		38.6		1836		1.201	
Fine	9	0.153	0.171	88.1	6.3	7.1	2.70	1.98	0.024	43.9	1.29	1742	10	1.139	0.006
Very fine	3	0.090	0.094	83.9	13.0	2.9	2.74	1.91		47.4		1711		1.121	
Silty sand	11	0.073	0.126	65.0	21.6	13.4	2.71	1.83	0.025	52.8	1.55	1677	9	1.096	0.006
Sandy silt	6	0.036	0.051	34.5	51.2	14.3	2.75	1.56		68.3		1552		1.015	
Sand-silt-clay	17	0.018	0.041	32.6	41.2	26.1	2.71	1.58	0.030	67.5	1.66	1578	9	1.032	0.006
Clayey silt	40	0.006	0.011	6.1	59.2	34.8	2.71*)	1.43	0.016	75.0	0.87	1535	3	1.004	0.002
Silty clay	17	0.003	0.004	5.3	41.5	53.6	2.69	1.42	0.013	76.0	0.74	1519	3	0.994	0.002
Abyssal Plain (Turbidite)															
Sandy silt	1	0.017	0.017	19.4	65.0	15.6	2.46	1.65		56.6		1622		1.061	
Silt	1	0.016	0.018	7.2	79.5	13.3	2.47	1.60	0.029	60.6	1.53	1634	2	1.069	0.001
Clayey silt	15	0.005	0.006	7.6	50.3	42.1	2.61	1.38		78.6		1535		1.003	
Silty clay	35	0.002	0.003	2.9	36.1	61.3	2.55	1.24	0.010	85.8	0.49	1521	2	0.994	0.001
Clay	2	0.001	0.001	0.1	20.3	79.6	2.67	1.26		85.8		1505		0.985	
Abyssal Hill (Pelagic)															
Clayey silt	3	0.0035	0.0053	3.3	50.0	46.7	2.58	1.41		76.4		1531		1.000	
Silty clay	32	0.0026	0.0023	2.6	32.9	65.2	2.71	1.37	0.014	79.4	0.77	1507	2	0.985	0.C01
Clay	6	0.0015	0.0013	0.6	20.7	78.9	2.76	1.42	0.023	77.5	1.35	1491	1.4	0.975	0.C01

*) Five samples.

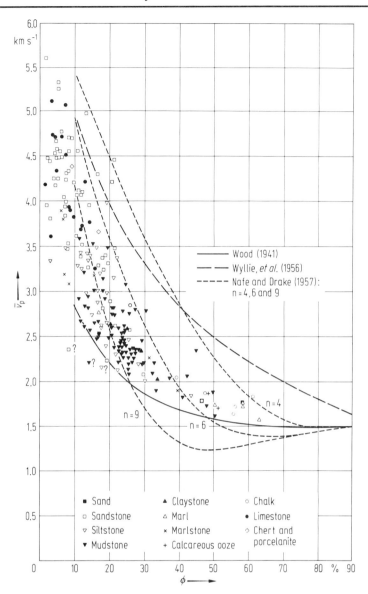

Fig. 7. Average of the horizontal and vertical compressional wave velocity $\bar{v}_p$ in unconsolidated marine sediments vs. porosity, ϕ. Included are equations of Wood (1941), Wyllie et al. (1957), and Nafe and Drake (1957), assuming a limestone matrix $(2.72\ \mathrm{g\ cm^{-3}},\ 6.45\ \mathrm{km\ s^{-1}})$ with sea water $(1.025\ \mathrm{g\ cm^{-3}},\ 1.53\ \mathrm{km\ s^{-1}})$ in the pores of the sedimentary rock; simplified after [Bo80].

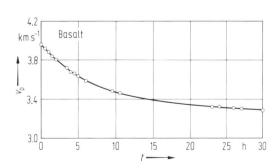

Fig. 8. Compressional wave velocity v_p vs. time t for a sample of water-saturated basalt allowed to air dry at atmospheric pressure [Ch75a].

Gebrande

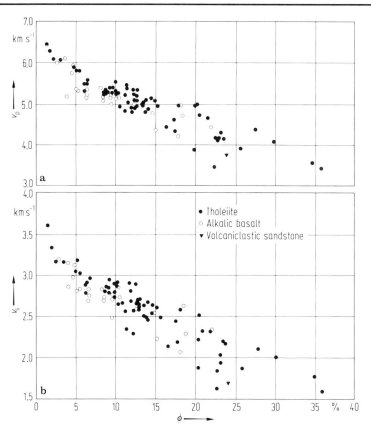

Fig. 9. Porosity dependence of (a) compressional and (b) shear wave velocities in brine-saturated oceanic rocks (Deep Sea Drilling Project Leg 55) [Ko80].

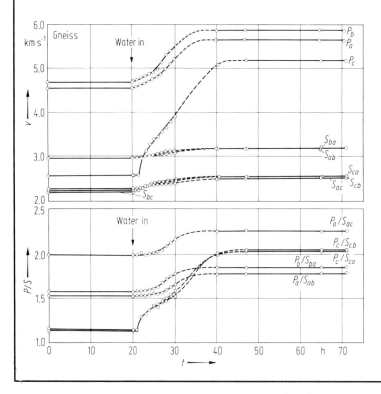

Fig. 10. Velocities of elastic waves in an Alpine gneiss as a function of saturation time with water as the fluid medium. The specimen was immersed in water after it was subjected to vacuum for 20 hours. The subscript c indicates the direction perpendicular to the fabric of the rock and thus perpendicular to the microcracks. The subscripts a and b indicate directions parallel to the fabric of the rock. P_a is the compressional wave propagating in the a direction; S_{ab} is the shear wave propagating in the a direction but polarized in the b direction, etc.; and P_a/S_{ab} is the ratio of the two respective velocities [Wa75].

Gebrande

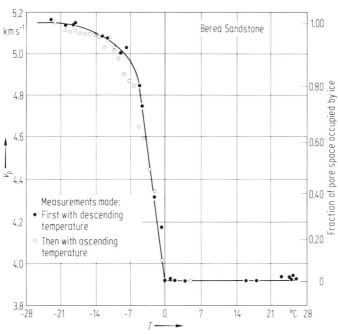

Fig. 11. Compressional wave velocity v_p vs. temperature T in Berea Sandstone saturated with distilled water (porosity 18.1%). Rock matrix is under uniaxial pressure of 31.7 MPa and pore fluid is under atmospheric pressure [Ti68].

3.1.2.3 Velocity-density relations — Geschwindigkeits-Dichte-Beziehungen

Velocity-density relations are useful for the combined interpretation of gravimetric and seismic measurements. They allow the estimation of seismic velocities from densities or vice versa. Generally seismic velocities increase with increasing density. This may be surprising at first glance because of the position of density in the denominator of Eqs. (1a, b). It can be understood, however, if for instance porosity is eliminated from Eqs. (12)···(14) by means of Eq. (15).

In Fig. 12, P-wave velocity in water-saturated marine sediments is plotted against density. Due to the close relation between density and porosity (Eq. 15) the general trend is very similar to Fig. 7. The velocities shown are mean values for vertical and horizontal wave propagation. Many marine sediments show clear anisotropy with greater velocities parallel to bedding. Anisotropy may range up to 30% and is particularly pronounced for P-wave velocities from about 2.0 to 4.2 km s⁻¹ [Bo80]. In less or more consolidated sediments anisotropy seems to be less significant.

Velocity-density relations for plutonites, volcanites and quasi-isotropic metamorphites based on data from Table 2 are shown in Figs. 13···15. In spite of the large scatter of single values within different rock groups mean values of velocities and densities cor-

Geschwindigkeits-Dichte-Beziehungen sind nützlich für die kombinierte Auswertung gravimetrischer und seismischer Messungen. Sie ermöglichen die Abschätzung seismischer Geschwindigkeiten aus Dichten oder umgekehrt. Im allgemeinen nehmen die seismischen Geschwindigkeiten mit zunehmender Dichte zu. Dies mag zunächst überraschen, da die Dichte in den Gln. (1a, b) im Nenner steht. Es ergibt sich aber z.B. aus den Gln. (12)···(14), wenn dort mittels Gl. (15) die Porosität eliminiert wird.

In Fig. 12 ist die P-Wellen-Geschwindigkeit für wassergesättigte marine Sedimente über der Dichte aufgetragen. Wegen des engen Zusammenhangs zwischen Dichte und Porosität (Gl.15) ergibt sich ein ähnlicher Verlauf wie in Fig. 7. Bei den dargestellten Geschwindigkeiten handelt es sich um Mittelwerte für vertikale und horizontale Wellenausbreitung. Viele marine Sedimente zeigen eine deutliche Anisotropie mit größeren Geschwindigkeiten parallel zur Schichtung. Die Anisotropie kann bis zu 30% betragen und ist besonders ausgeprägt für P-Wellen-Geschwindigkeiten im Bereich von etwa 2.0 bis 4.2 km s⁻¹ [Bo80]. In schwächer oder stärker verfestigen Sedimenten scheint die Anisotropie weniger signifikant.

Geschwindigkeits-Dichte-Beziehungen für Plutonite, Vulkanite und quasi-isotrope Metamorphite auf der Basis der Daten der Tabelle 2 sind in den Fig. 13··· 15 dargestellt. Trotz der großen Streuung der Einzelwerte innerhalb der Gesteinsgruppen sind die Mittel-

relate clearly. The following Eqs. (16a)···(18b) have been obtained by regression analysis:

werte der Geschwindigkeiten und Dichten deutlich miteinander korreliert. Durch Ausgleichsrechnung wurden die Gln. (16a)···(18b) erhalten:

Plutonites (Plutonite):	$\bar{v}_p = (-6.73 + 4.36\,\bar{\varrho} \pm 0.03)\,\text{km s}^{-1}$	(16a)
	$\bar{v}_s = (-1.48 + 1.66\,\bar{\varrho} \pm 0.06)\,\text{km s}^{-1}$	(16b)
Volcanites (Vulkanite):	$\bar{v}_p = (-2.37 + 2.81\,\bar{\varrho} \pm 0.18)\,\text{km s}^{-1}$	(17a)
	$\bar{v}_s = (-1.02 + 1.46\,\bar{\varrho} \pm 0.22)\,\text{km s}^{-1}$	(17b)
Metamorphites (Metamorphite):	$\bar{v}_p = (-6.93 + 4.41\,\bar{\varrho} \pm 0.37)\,\text{km s}^{-1}$	(18a)
	$\bar{v}_s = (-1.62 + 1.70\,\bar{\varrho} \pm 0.22)\,\text{km s}^{-1}$	(18b)

The indicated errors are standard deviations; densities should be inserted in (g cm^{-3}). Apart from the different standard deviations the parameters of regression lines for plutonites and quasi-isotropic metamorphites are

Die angegebenen Fehlergrenzen sind Standardabweichungen; die Dichte ist jeweils in g cm^{-3} einzusetzen. Von den verschiedenen Standardabweichungen abgesehen unterscheiden sich die Parameter der Aus-

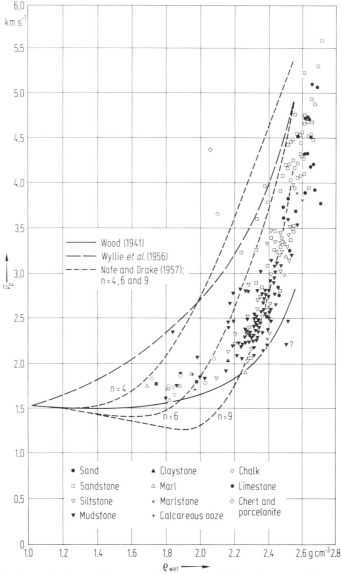

Fig. 12. Average of the horizontal and vertical compressional wave velocity $\bar{v}_p$ in unconsolidated marine sediments vs. wet-bulk density, ϱ_{wet}. Included are equations of Wood (1941), Wyllie et al. (1956), and Nafe and Drake (1957), assuming a limestone $(2.72\,\text{g cm}^{-3}, 6.45\,\text{km s}^{-1})$ matrix with sea water $(1.025\,\text{g cm}^{-3}, 1.53\,\text{km s}^{-1})$ in the pores of the sedimentary rock; simplified after [Bo80].

very similar. For practical purposes both rock types need not be distinguished from each other. It should be noted that the regression lines (16a)···(18b) connect *mean* values. If seismic velocities are to be estimated from the density of an *individual* sample, the standard deviation for the respective rock type (as given in Table 2) must be added to the standard deviation of the regression line.

gleichsgeraden für Plutonite und quasi-isotrope Metamorphite nur wenig. In der Praxis braucht dazwischen nicht unterschieden zu werden. Es ist zu beachten, daß es sich um Ausgleichsgeraden für *Mittelwerte* handelt. Wenn aus der Dichte einer *einzelnen* Probe eine seismische Geschwindigkeit abgeschätzt werden soll, ist zur Standardabweichung der Ausgleichsgeraden die Standardabweichung für den jeweiligen Gesteinstyp (siehe Tab. 2) zu addieren.

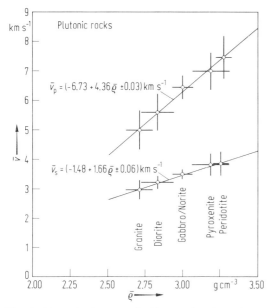

Fig. 13. Mean elastic wave velocities $\bar{v}$ as functions of mean density $\bar{\varrho}$ for plutonic rocks at normal temperature and pressure. Data is from Table 2. The length of the error bars corresponds to the standard deviations of density and velocities for the different rock groups. Solid lines are least-squares best-fit lines.

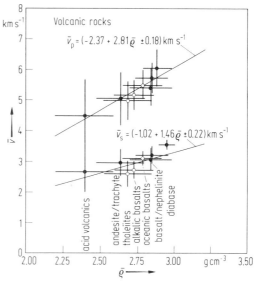

Fig. 14. Mean elastic wave velocities $\bar{v}$ as functions of mean density $\bar{\varrho}$ for dry (full circles) and brine-saturated (open circles) volcanic rocks at normal temperature and pressure. Data is from Table 2. The length of the error bars corresponds to the standard deviation of density and velocities for different rock groups. Solid lines are least-squares best-fit lines.

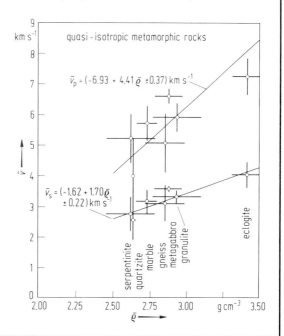

Fig. 15. Mean elastic wave velocities $\bar{v}$ as functions of mean density $\bar{\varrho}$ for quasi-isotropic metamorphic rocks at normal temperature and pressure. Data is from Table 2. The length of the error bars corresponds to the standard deviation of density and velocities for the different rock groups. Solid lines are least-squares best-fit lines and have similar parameters as for plutonic rocks (Fig. 13).

For the estimation of densities from seismic velocities or vice versa the "Woollard curve" [Wo59] and the "Nafe-Drake curve" (in [Ta59]) have been extensively used in the literature. They are shown in Figs. 16 and 17. For the construction of the "Woollard curve" some velocity determinations at elevated pressures have also been taken into account. The data base of the "Nafe-Drake curve" is unknown.

For velocity-density relations at elevated pressures see 3.1.3.1.7.

Für die Abschätzung von Dichten aus Geschwindigkeiten und umgekehrt sind in der Literatur häufig die „Woollard-Kurve" [Wo59] und die „Nafe-Drake-Kurve" (in [Ta59]) verwendet worden, die in den Fig. 16 und 17 dargestellt sind. In der „Woollard-Kurve" sind z.T. auch Geschwindigkeitsbestimmungen bei höheren Drucken berücksichtigt. Die Datenbasis für die „Nafe-Drake-Kurve" ist unbekannt.

Bezüglich Geschwindigkeits-Dichte-Beziehungen bei höheren Drucken wird auf 3.1.3.1.7 verwiesen.

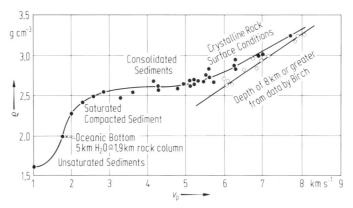

Fig. 16. Relation of density ϱ to compressional wave velocity v_p in rocks; quoted from Woollard [Wo59]. The source material includes laboratory measurements at normal and elevated pressures as well as field measurements.

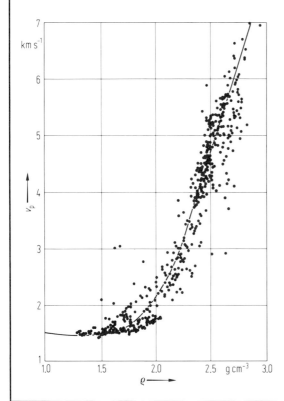

◄

Fig. 17. A plot of compressional wave velocity v_p vs. bulk density ϱ [Gr65] according to Nafe and Drake in [Ta59]. The solid line is the so-called Nafe-Drake curve which is often used in the literature for estimating densities from seismic velocities or vice versa. Data base for this curve is somewhat unclear.

3.1.2.4 Various materials of geophysical interest — Diverse Materialien von geophysikalischem Interesse

This section includes in Table 4 some data for solid materials which are sometimes used for model-seismic investigations or are of general interest for comparison purposes. P-wave velocities in gases and fluids, some of which constitute pore fillings of natural rocks, are found in Table 5 as well as in Figs. 18···20.

Dieser Abschnitt enthält in Tab. 4 einige Daten für feste Materialien, die bisweilen für modell-seismische Untersuchungen verwendet werden oder die allgemein für Vergleichszwecke von Interesse sind. P-Wellen-Geschwindigkeiten in Gasen und Flüssig-keiten, die z.T. als Porenfüllungen natürlicher Gesteine vorkommen, sind in Tab. 5 sowie den Fig. 18···20 zu finden.

Table 4. Elastic wave velocities and constants of elasticity of miscellaneous materials at normal temperature and pressure.

Material	ϱ kg m^{-3}	v_p m s^{-1}	v_s m s^{-1}	v_p/v_s	K GPa	G GPa	E GPa	σ	Ref.
Ice	see chapter 8, section 8.4								
Plexiglass	1180	2670	1121	2.38	6.4	1.5	4.1	0.39	Be54
Obsidian, Lake County, Oregon, USA	2350	5820	3570	1.63	40	30	72	0.20	Wo63
Porcelain	2410	5340	3120	1.71	37.4	23.5	58.2	0.24	Be54
Concrete		3560		1.65				0.21	Mo60
Crown-glass	2500	5660	3420	1.66	41.1	29.2	70.9	0.21	Be54
Quartz-glass	2600	5570	3515	1.59	37.8	32.1	75.1	0.17	Be54
Aluminum	2700	6260	3080	2.03	72	26	69	0.34	Be54
Flint-glass	3600	4260	2560	1.66	33.9	23.6	57.4	0.22	Be54
Flint-glass	4600	3760	2220	1.69	34.8	22.7	55.9	0.23	Be54
Iron	7800	5850	3230	1.81	158	81	208	0.28	Be54
Brass	8100	4430	2123	2.89	110	37	99	0.35	Be54
Copper	8800	5240	2640	1.99	160	61	163	0.33	Be54
Lead	11400	2160	700	3.09	45.7	5.6	16.1	0.44	Be54

Table 5. Sound velocity in some gases and fluids.

Material	T °C	v m s^{-1}	dv/dT m s^{-1}°C^{-1}	Ref.
CO_2	0	259	0.4	Be54
Air	0	331	0.59	Be54
Methane	0	430		Be54
Steam	134	494		Be54
Hydrogen	0	1284	2.2	Be54
Petroleum	34	1295		Be54
Petroleum		1326··· 1395		Pr66
Mercury	20	1451	−0.46	Be54
Water	25	1497 (see Fig. 18)		Be54, Sch67
Salt water	see Figs. 19, 20			Wy56, Ha71

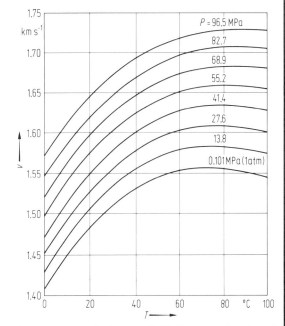

Fig. 18. Sound velocity v in distilled water as function of temperature T and pressure P [Sch67].

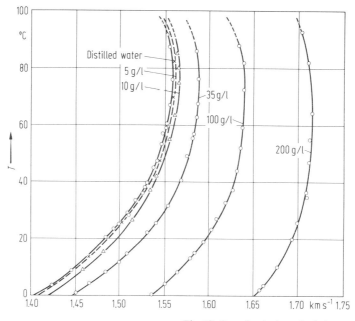

Fig. 19. Sound velocity v in brine as function of temperature T and NaCl concentration [Wy56].

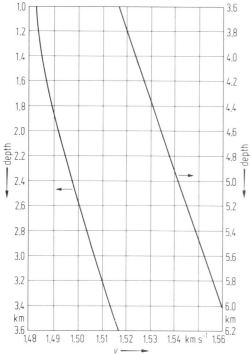

Fig. 20. Water sound velocity v vs. depth, Central Pacific [Ha71].

3.1.2.5 Rock forming minerals — Gesteinsbildende Minerale

The elastic properties of rocks are largely determined by the properties of the minerals of which they are composed. The elasticity of minerals is described by the elastic stiffnesses C_{mn} or the compliances S_{mn} (see section 3.1.1.3). In Tables 8a···8e both sets of constants have been listed for the most important rock-forming minerals, subdivided according to their crystal systems. The schemes of C_{mn}-matrices for the

Die elastischen Eigenschaften der Gesteine werden weitgehend durch die Eigenschaften der Minerale bestimmt, aus denen sie aufgebaut sind. Die Elastizität von Mineralen wird durch die Elastizitätsmoduln C_{mn} oder die Elastizitätskoeffizienten S_{mn} (vgl. 3.1.1.3) beschrieben. In den Tab. 8a bis 8e sind beide Sätze von Konstanten für die wichtigsten gesteinsbildenden Minerale, gegliedert nach Kristall-Klassen, zusammenge-

crystal systems of Tables 8a···8e are shown in Table 6. From this data and Eq. (10) elastic wave velocities can be calculated for arbitrary directions of wave propagation. For specific directions, in which pure longitudinal or transverse waves can be propagated, simple relations between velocities, densities and stiffnesses are obtained; they are given in Table 7.

If axis directions and grain shapes are statistically isotropically distributed, the effective bulk and shear moduli of a mineral aggregate can be estimated by the Voigt and Reuss mean values, which are given in Table 9; they are upper and lower bounds of the effective isotropic moduli [Hi52]. Table 9 includes P- and S-wave velocities v_p^{VRH} and v_s^{VRH} (Voigt-Reuss-Hill averages) as calculated from the mean values of the bounds as well as the pertinent Poisson's ratio σ^{VRH}. These mean values correspond to polycrystalline aggregates with zero porosity; measurements on real aggregates generally give lower velocities and seem not well suited for the exptrapolation to high pressures [Sp70, Sp72, So79].

The Reuss moduli apply to the case of homogeneous stress in the aggregate. Therefore preference may be given to them as compared to the VRH-values for the calculation of mineral densities in the Earth's interior because under conditions prevailing there persistently, relaxation of small-scale stress inhomogeneities is to be expected [Ku69b, Th72].

Because of their relation to elastic properties (see 3.1.3.1.7) mean atomic weights of rock forming minerals are compiled in Table 10, p. 34.

stellt. Die Schemata der C_{mn}-Matrizen für die Kristall-Klassen der Tab. 8a bis 8e sind in Tab. 6 angegeben. Aus diesen Angaben können nach Gl. (10) die Geschwindigkeiten elastischer Wellen für beliebige Ausbreitungsrichtungen errechnet werden. Für spezielle Richtungen, in denen sich reine Longitudinal- oder Transversalwellen ausbreiten können, ergeben sich einfache Zusammenhänge zwischen Geschwindigkeiten, Dichte und Elastizitätsmoduln; sie sind in Tab. 7 zusammengestellt.

Wenn die Achsenrichtungen und Kornformen eines Minerals in einem Mineralaggregat statistisch isotrop verteilt sind, können seine effektiven Kompressions- und Scherungsmoduln durch die Voigt'schen und Reuss'schen Mittelwerte abgeschätzt werden, die in Tab. 9 angegeben sind; sie stellen obere und untere Schranken für die effektiven isotropen Moduln dar [Hi52]. Ferner enthält Tab. 9 die aus den Mittelwerten beider Schranken errechneten P- und S-Wellengeschwindigkeiten v_p^{VRH} und v_s^{VRH} (Voigt-Reuss-Hill-Mittelwerte) sowie die zugehörige Querkontraktionszahl σ^{VRH}. Diese Mittelwerte entsprechen polykristallinen Aggregaten mit Porosität Null; Messungen an realen Aggregaten ergeben im allgemeinen kleinere Geschwindigkeiten und scheinen für Extrapolationen auf hohe Drucke wenig geeignet [Sp70, Sp72, So79].

Die Reuss'schen Moduln entsprechen dem Fall homogener Spannungsverteilung im Aggregat. Für die Berechnung von Mineraldichten im Erdinnern sind sie daher den VRH-Werten vorzuziehen, da unter den dortigen persistenten Bedingungen mit einem Abbau von kleinräumigen Spannungsinhomogenitäten durch inelastische Prozesse zu rechnen ist [Ku69b, Th72].

Wegen ihres Zusammenhanges mit den elastischen Eigenschaften (vgl. 3.1.3.1.7) sind in Tab. 10, S. 34, die mittleren Atomgewichte gesteinsbildender Minerale zusammengestellt.

Table 6. Stiffness matrices for crystal systems of Tables 8a···8e. The same schemes apply to the compliance matrices with the modifications:
a) $S_{46} = -2 S_{15}$ and $S_{56} = 2 S_{14}$ in the trigonal system,
b) $S_{66} = 2(S_{11} - S_{12})$ in the hexagonal and trigonal systems.

Cubic system:
(3 constants)

$$\begin{pmatrix} C_{11} & C_{12} & C_{12} & \cdot & \cdot & \cdot \\ C_{12} & C_{11} & C_{12} & \cdot & \cdot & \cdot \\ C_{12} & C_{12} & C_{11} & \cdot & \cdot & \cdot \\ \cdot & \cdot & \cdot & C_{44} & \cdot & \cdot \\ \cdot & \cdot & \cdot & \cdot & C_{44} & \cdot \\ \cdot & \cdot & \cdot & \cdot & \cdot & C_{44} \end{pmatrix}$$

Hexagonal system:
(5 constants)

$$\begin{pmatrix} C_{11} & C_{12} & C_{13} & \cdot & \cdot & \cdot \\ C_{12} & C_{22} & C_{13} & \cdot & \cdot & \cdot \\ C_{13} & C_{13} & C_{33} & \cdot & \cdot & \cdot \\ \cdot & \cdot & \cdot & C_{44} & \cdot & \cdot \\ \cdot & \cdot & \cdot & \cdot & C_{44} & \cdot \\ \cdot & \cdot & \cdot & \cdot & \cdot & (C_{11} - C_{12})/2 \end{pmatrix}$$

(continued)

Table 6 (continued)

Trigonal system:
(7 constants; if mirror planes or 2-fold axes exist C_{15} is zero)

$$\begin{pmatrix} C_{11} & C_{12} & C_{13} & C_{14} & C_{15} & . \\ C_{12} & C_{11} & C_{13} & -C_{14} & -C_{15} & . \\ C_{13} & C_{13} & C_{33} & . & . & . \\ C_{14} & -C_{14} & . & C_{44} & . & -C_{15} \\ C_{15} & -C_{15} & . & . & C_{44} & C_{14} \\ . & . & . & -C_{15} & C_{14} & (C_{11}-C_{12})/2 \end{pmatrix}$$

Orthorhombic system:
(9 constants)

$$\begin{pmatrix} C_{11} & C_{12} & C_{13} & . & . & . \\ C_{12} & C_{22} & C_{23} & . & . & . \\ C_{13} & C_{23} & C_{33} & . & . & . \\ . & . & . & C_{44} & . & . \\ . & . & . & . & C_{55} & . \\ . & . & . & . & . & C_{66} \end{pmatrix}$$

Monoclinic system:
(13 constants)

$$\begin{pmatrix} C_{11} & C_{12} & C_{13} & . & C_{15} & . \\ C_{12} & C_{22} & C_{23} & . & C_{25} & . \\ C_{13} & C_{23} & C_{33} & . & C_{35} & . \\ . & . & . & C_{44} & . & C_{46} \\ C_{15} & C_{25} & C_{35} & . & C_{55} & . \\ . & . & . & C_{46} & . & C_{66} \end{pmatrix}$$

Table 7. Elastic wave velocities for specific directions of pure longitudinal or transverse wave propagation; after [Al61a] with supplements.

Direction of propagation	polarization	$\varrho v^2 =$ Cubic	Hexagonal*)	Trigonal*)**)	Orthorhombic	Monoclinic
[001]	[001]	C_{11}	C_{33}	C_{33}	C_{33}	
	[100]	C_{44}	C_{44}	C_{44}	C_{55}	
	[010]	C_{44}	C_{44}	C_{44}	C_{44}	C_{44}
[010]	[010]	C_{11}	C_{11}		C_{22}	C_{22}
	[001]	C_{44}	C_{44}		C_{44}	
	[100]	C_{44}	$(C_{11}-C_{12})/2$	$(C_{11}-C_{12})/2$	C_{66}	
[100]	[100]	C_{11}	C_{11}	C_{11}	C_{11}	
	[010]	C_{44}	$(C_{11}-C_{12})/2$		C_{66}	C_{66}
	[001]	C_{44}	C_{44}		C_{55}	
[011]	[011]	$(C_{11}+C_{12}+2C_{44})/2$				
	[01̄1]	$(C_{11}-C_{12})/2$				
	[100]	C_{44}	$(C_{66}+C_{44})/2$	$(C_{66}+C_{44}+2C_{14})/2$	$(C_{55}+C_{66})/2$	
[101]	[101]	$(C_{11}+C_{12}+2C_{44})/2$				
	[010]	C_{44}	$(C_{66}+C_{44})/2$		$(C_{44}+C_{66})/2$	$(C_{44}+C_{66}+2C_{46})/2$
	[1̄01]	$(C_{11}-C_{12})/2$				
[111]	[111]	$(C_{11}+2C_{12}+4C_{44})/3$				
	$\perp$[111]	$(C_{11}-C_{12}+C_{44})/3$				

*) $C_{66}=(C_{11}-C_{12})/2$.
**) Trigonal system with 6 constants.

Table 8a. Elastic constants C_{mn} and S_{mn} and derivatives for some rock forming minerals of *cubic* symmetry; in alphabetical order.

Material	ϱ [kg m^{-3}]	C [GPa] S [(TPa)$^{-1}$] derivative[3])	Indices 11	44	12	Ref.
Calcium fluoride, CaF$_2$	3180	C	165	33.9	46	He79
		S	6.94	29.5	−1.53	
Chromite, FeCr$_2$O$_4$	4450	C	322	117	144	He79
		S	4.27	8.57	−1.32	
Diamond, C	3511	C	1040	550	170	He79
		S	1.01	1.83	−0.14	
Garnets:						
Pyrope, Mg$_3$Al$_2$Si$_3$O$_{12}$ [1])	3582	C	287.4	91.6	105.0	He79,
		S	4.325	10.92	−1.157	Is76
Almandine, Fe$_3$Al$_2$Si$_3$O$_{12}$ [1])	4318	C	310.1	92.9	115.1	He79,
		S	4.036	10.77	−1.093	Is76
Spessartine, Mn$_3$Al$_2$Si$_3$O$_{12}$ [1])	4190	C	302.0	95.9	107.5	He79,
		S	4.073	10.43	−1.069	Is76
Grossular, Ca$_3$Al$_2$Si$_3$O$_{12}$ [1])	3594	C	319.5	102.3	95.9	He79,
		S	3.634	9.78	−0.839	Is76

(Fe$_w$Mg$_x$Ca$_y$Mn$_z$)$_3$Al$_2$Si$_3$O$_{12}$

w	x	y	z[2])						

w x y z	ϱ	deriv.	11	44	12	Ref.
36 61 2 0	3730	C	292.2	91.6	106.2	He79,
		S	4.245	10.92	−1.132	Bo77
		dC/dP	6.78	1.43	3.72	
		dC/dT	−32.9	−7.5	−11.8	
52 0 1 46	4240	C	306.5	94.4	111.2	He79,
		S	4.05	10.6	−1.08	Is76
		dC/dP	6.69	1.26	3.54	
		dC/dT	−33.3	−9.1	−17.2	
76 21 3 0	4160	C	306.2	92.7	112.5	He79,
		S	4.08	10.8	−1.10	Bo77,
		dC/dP	7.48	1.31	4.41	So67
		dC/dT	−34.8	−10.4	−12.8	
46 0 0 54	4249	C	308.5	94.8	112.3	He79,
		S	4.02	10.5	−1.07	Wa74,
		dC/dP	7.15	1.30	3.85	Bo77
8 0 85 7	3617	C	317.7	101.1	98.3	He79,
		S	3.69	9.89	−0.87	Wa74
		dC/dP	5.1	0.52	3.83	
Magnetite, Fe$_3$O$_4$	5180	C	275	95.5	104	He79
		S	4.59	10.5	−1.26	

(continued)

[1]) Extrapolated.
[2]) mol-%.
[3]) dC/dP unitless; dC/dT in [MPa/°C].

Table 8a (continued)

Material	ϱ [kg m^{-3}]	C [GPa] S [(TPa)$^{-1}$] derivative[3])	Indices			Ref.
			11	44	12	
NaCl	2160?	C	49.1	12.7	14.0	He79
		S	23.3	78.6	−5.16	
NaCl		C	49.0	12.8	13.0	He79
		S	23.0	78	−4.8	
Periclase, MgO	3583	C	297.4	156.2	95.6	Sp70
		S	3.99	6.40	−0.97	
		dC/dP	8.7	1.09	1.42	
		dC/dT	−60.6	−10.3	7.4	
Pyrite, FeS$_2$	5010?	C	366	107	32	He79
		S	2.77	9.32	−0.22	
Spinels:						
MgO · Al$_2$O$_3$	3581	C	282	154	154	He79
		S	5.80	6.49	−2.05	
MgO · 2.61Al$_2$O$_3$	3619	C	299	158	154	He79,
		S	5.15	6.35	−1.75	Wa72
		dC/dP	4.90	0.85	3.90	
(Mg$_{75}$Fe$_{36}$)Al$_{1.90}$O$_4$ (Pleonaste)	3826	C	269.5	143.5	163.3	He79,
		S	6.84	6.97	−2.58	Wa72
		dC/dP	4.85	0.75	4.95	

[3]) dC/dP unitless; dC/dT in [MPa/°C].

Table 8b. Elastic constants C_{mn} and S_{mn} of some rock forming minerals of *hexagonal* symmetry in alphabetical order.

Material	ϱ [kg m^{-3}]	C [GPa] S [(TPa)$^{-1}$]	Indices					Ref.
			11	33	44	12	13	
Apatite, Ca$_{10}$(PO$_4$)$_6$F$_2$	3218	C	139	178	44.3	45	56	He79,
		S	8.68	6.96	22.9	−1.36	−2.10	Si71
Beryl, Be$_3$Al$_2$Si$_6$O$_{18}$	2660	C	290	257	67.7	107	83	He79,
		S	4.22	4.53	14.8	−1.28	−0.94	Si71
Biotite*), K(Mg, Fe)$_3$AlSi$_3$O$_{10}$(OH, F)$_2$	3050	C	186	54	5.8	32.4	11.6	Al61,
		S	5.6	18.9	172	−0.9	−1.1	He79
Ice, see chapter 8.4								
Muscovite*), KAl$_2$Si$_3$AlO$_{10}$(OH, F)$_2$	2790	C	178	54.9	12.2	42.4	14.5	Al61,
		S	6	18.9	81.9	−1.3	−1.2	He79
Nepheline, Na$_3$KAl$_4$Si$_4$O$_{16}$	2620	C	79	126	37.3	38.0	19.2	Ry62,
		S	17.0	8.5	27.0	−6.2	−1.5	He79
Phlogopite*), A KMg$_3$Si$_3$AlO$_{10}$(OH, F)$_2$	2800	C	179	51.7	5.6	32.4	25.8	Al61,
		S	6.1	22.0	179	−0.7	−2.7	He79
Phlogopite*), B	2820	C	178	51.0	6.5	30.2	15.2	Al61,
		S	5.9	20.5	154	−0.9	−1.5	He79
β-Quartz, SiO$_2$	2533	C	117	110	36.0	16	33	He79,
		S	9.41	10.6	27.7	−0.6	−2.6	Si71

*) Monoclinic quasi hexagonal.

Table 8c. Elastic constants C_{mn} and S_{mn} and derivatives for some rock forming minerals of *trigonal* symmetry.

Material	ϱ [kg m⁻³]	C [GPa] S [(TPa)⁻¹] derivative*)	Indices							Ref.
			11	33	44	12	13	14	15	
Alumina, Corundum, Ruby, Sapphire, Al₂O₃	3986	C	495	497	146	160	115	−23		He79, Si71
		S	2.38	2.19	7.03	−0.70	−0.38	0.49		
		dC/dP	6.16	5.07	2.24	3.22	3.68	0.15		
		dC/dT	−37	−42	−26	6?	−9?	2?		
Calcite, CaCO₃ (see also Fig. 22)	2712	C	144	84.0	33.5	53.9	51.1	−20.5		He79, Si71
		S	11.4	17.4	41.4	−4.0	−4.5	9.5		
		dC/dP	3.2	2.9	0.9	2.3	3.4	−1.3		
(300···500 °K)		dC/dT	−56?	−14	−61	−32	−24	8.2		
Haematite, Fe₂O₃	5240	C	242	228	85.3	54.9	15.7	−12.5		He79, Si71
		S	4.41	4.43	11.9	−1.02	−0.23	0.79		
Magnesite, MgCO₃	2980?	C	259	156	54.8	75.6	58.8	−19.0		He79
		S	4.67	7.41	19.7	−1.22	−1.30	2.04		
α-Quartz, SiO₂ (see also Fig. 21)	2650	C	86.6	106.1	57.8	6.7	12.6	−17.8		He79, Si71
		S	12.8	9.75	20.0	−1.74	−1.32	4.48		
		dC/dP	3.29	10.8	2.66	8.04	6.3	1.96		
		dC/dT	−4.2	−17.0	−10.2	−20	−6.9	−1.8		
Tourmaline	3050	C	277	163	64.0	64.5	31.9	−6.9		He79, Si71
		S	3.92	6.46	15.7	−0.83	−0.63	0.52		
Dolomite, CaMg(CO₃)₂ (7 constants)	2860	C	205	113	39.8	71.0	57.4	−19.5	13.7?	He79
		S	7.0	11.2	31.9	−2.4	−2.3	4.6	−3.3?	

*) dC/dP unitless, dC/dT in [MPa/°C].

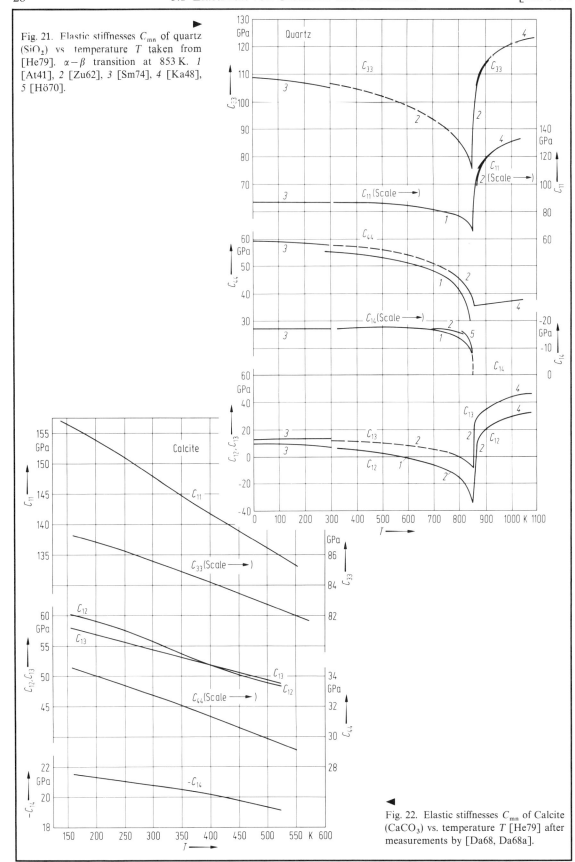

Fig. 21. Elastic stiffnesses C_{mn} of quartz (SiO$_2$) vs temperature T taken from [He79]. $\alpha - \beta$ transition at 853 K. *1* [At41], *2* [Zu62], *3* [Sm74], *4* [Ka48], *5* [Hö70].

Fig. 22. Elastic stiffnesses C_{mn} of Calcite (CaCO$_3$) vs. temperature T [He79] after measurements by [Da68, Da68a].

Table 8d. Elastic constants C_{mn} and S_{mn} and derivatives for some rock forming minerals of *orthorhombic* symmetry; in alphabetical order.

Material	ϱ [kg m⁻³]	C [GPa] S [(TPa)⁻¹] derivative*	Indices 11	22	33	44	55	66	12	13	23	Ref.
Andalusite, Al₂SiO₅	3145	C	233	289	380	99.5	87.8	112	81.4	116	97.7	Va78
		S	5.3	4.0	3.2	10.1	11.4	8.9	−1.0	−1.4	−0.7	
Anhydrite, CaSO₄	2970	C	93.8	185	112	32.5	26.5	9.26	16.5	15.2	31.7	He79,
		S	11.0	5.72	9.55	30.8	37.7	108	−0.76	−1.28	−1.52	Sch65
Aragonite, CaCO₃	2930	C	160	87.2	84.8	41.3	25.6	42.7	37.3	1.7	15.7	He79,
		S	6.95	13.2	12.2	24.2	39.0	23.4	−3.04	0.43	−2.38	Si71
Baryte, BaSO₄	4500	C	89.0	81.0	107	12.0	28.1	26.9	47.9	31.7	29.8	He79,
		S	17.2	18.8	10.8	83.7	35.5	37.2	−9.2	−2.5	−2.5	Si71
Bronzite, Bavaria $\phi = 1.8\%$	3380	C	188	158	208	70.0	59.2	54.4	69	60	56	He79,
		S	6.63	7.90	5.55	14.3	16.7	18.4	−2.43	−1.27	−1.42	Ry66
(Mg₈₄Fe₁₆)SiO₃	3335	C	230	165	206	83.1	76.4	78.5	70	57.3	50	Ku69a
		S	5.17	7.21	5.43	12.0	13.1	12.7	−1.9	−0.98	−1.21	
(Mg₈₀Fe₂₀)SiO₃	3354	C	228.6	160.5	210.4	81.8	75.5	77.6	71.0	54.8	46.0	Fr72
		S	5.25	7.45	5.23	12.2	13.3	12.9	−2.05	−0.91	−1.1	
		dC/dP	11.04	9.19	16.42	2.38	2.92	2.75	6.79	9.09	8.73	
(25···350 °C)		dC/dT	−35.2	−32.8	−51.6	−13.1	−13.8	−14.5	−21.2	−31.8	−10.7	
Enstatite, MgSiO₃	3200	C	225	178	214	77.6	75.9	81.6	72.4	54	53	We78
		S	5.3	6.7	5.2	12.9	13.2	12.3	−1.9	−0.9	1.2	
Natrolite, Na₂Al₂Si₃O₁₀·2H₂O	2250	C	71.6	63.2	137.8	19.6	24.8	42.3	26.1	29.7	29.7	Ry66
		S	17.2	19.2	8.44	51.1	41.2	23.6	−5.68	−2.43	−2.98	
Olivine, (Mg, Fe)₂SiO₄: Forsterite, Mg₂SiO₄	3224	C	328.4	199.8	235.3	65.9	81.2	80.9	63.9	68.8	73.8	Ku69, He79
		S	3.35	5.86	4.97	15.19	12.32	12.36	−0.80	−0.73	−1.60	
		dC/dP	8.47	6.56	6.57	2.12	1.66	2.37	4.67	4.84	4.11	
(25 °C; 0···200 MPa)		dC/dT	−33.1	−28.1	−28.3	−13.0	−13.2	−15.1	−10.4	−8.2	−4.6	
	3221.7	C	329.1	200.5	236.3	67.2	81.4	81.1	66.3	68.4	72.8	Gr69, He79
		S	3.36	5.83	4.92	14.87	12.28	12.32	−0.85	−0.71	−1.55	
(25 °C; 0···1 GPa)		dC/dP	8.32	5.93	6.21	2.12	1.65	2.32	4.30	4.23	3.53	
(200···400 °C; 0.1 MPa)		dC/dT	−38.9	−31.1	−26.9	−13.0	−14.4	−16.3	−11.7	−8.7	−9.2	

(continued)

Table 8d (continued)

Material	ϱ [kg m⁻³]	C [GPa] S [(TPa)⁻¹] derivative*)	11	22	33	44	55	66	12	13	23	Ref.
			Indices									
Olivine (continued)												
$(Mg_{92.7}Fe_{7.2})_2SiO_4$	3311	C	323.7	197.6	235.1	64.6	78.7	79.0	66.4	71.6	75.6	Ku69
		S	3.44	5.99	5.02	15.48	12.72	12.65	−0.86	−0.77	−1.66	
(25 °C; 0···200 MPa)		dC/dP	7.98	6.37	6.38	2.17	1.64	2.31	4.74	4.48	3.76	
		dC/dT	−34.0	−28.5	−28.6	−12.8	−13.0	−15.7	−10.5	−9.4	−5.1	
$(Mg_{91.7}Fe_{8.3})_2SiO_4$	3324	C	324	198	249	66.7	81.0	79.3	59	79	78	Si71
		S	3.42	5.90	4.81	15.0	12.3	12.6	−0.68	−0.87	−1.63	
$(Mg_{91.3}Fe_{8.1})_2SiO_4$	3316	C	324	196	232	63.9	77.9	78.8	71.5	71.5	68.8	Oh76
$(Mg_{92.0}Fe_{7.5})_2SiO_4$	3299	C	319	192	238	63.8	78.3	79.7	59	76	72	Oh76
Sillimanite, Al_2SiO_5	3241	C	287	232	388	122	80.7	89.3	94.7	83.4	159	Va78
		S	4.1	6.5	3.6	8.2	12.4	11.9	−1.5	−0.3	−2.3	
Staurolite, $(Fe, Mg)_2(Al, Fe)_9O_6SiO_4(O, OH)_2$	3369	C	343	185	147	46	70	92	67	61	12.8	Si71
		S	3.37	5.82	7.35	21.7	14.3	10.9	−1.1	−1.3	−0.4	

*) dC/dP unitless; dC/dT in [MPa/°C].

Table 8e. Elastic constants C_{mn} and S_{mn} of some rock forming minerals of monoclinic symmetry; in alphabetical order.

Material	ϱ [kg m⁻³]	C [GPa] S [(TPa)⁻¹]	11	22	33	44	55	66	12	13	23	15	25	35	46	Ref.
			Indices													
Aegirite, $NaFeSi_2O_6$	3500	C	186	181	234	62.9	51.0	47.4	68.5	70.7	62.6	9.8	9.4	21.4	7.7	Al64
		S	6.7	6.7	5.2	16.2	20.4	21.5	−2.0	−1.5	−1.1	−0.3	−0.4	−1.7	−2.6	
Aegirite-Augite	3420	C	156	152	216	40.0	46.5	49.2	81.1	66.0	68.4	25.3	26.0	19.2	4.1	Si71, Al64
		S	9.5	10.0	5.6	25.2	24.5	20.5	−4.0	−1.4	−1.7	−2.3	−2.7	−0.6	−2.1	
Augite	3320	C	182	151	218	69.7	51.1	55.8	73.4	72.4	33.9	19.9	16.6	24.6	4.3	Si71, Al64
		S	7.7	8.4	5.5	14.4	21.3	18.0	−3.2	−1.9	−0.1	−1.0	−1.4	−1.8	−1.1	
Coesite, SiO_2	2920	C	161	230	232	67.8	73.3	58.8	82	103	36	−36	3	−39	10	He79, We77
		S	11.3	5.3	6.2	15.1	16.2	17.4	−3.5	−3.9	0.4	3.6	−1.7	1.4	−2.6	

(continued)

Table 8e (continued)

Material	Composition	ϱ [kg m⁻³]	C [GPa] / S [(TPa)⁻¹]	11	22	33	44	55	66	12	13	23	15	25	35	46	Ref.
Diallage		3300	C	154	150	211	63.9	62.2	52.3	56.9	37.4	30.5	14.6	14.2	11.9	−8.6	Al64, Si71
			S	7.8	7.9	5.0	16.0	16.7	19.6	−2.7	−0.9	−0.6	−1.0	−1.1	−0.6	2.6	
Diopside, CaMgSi₂O₆		3310	C[1]	204	175	238	67.5	58.8	70.5	84.4	88.3	48.2	19.3	19.6	33.6	11.3	Al64, Ba72a
			S	6.9	7.3	5.3	15.2	18.9	14.6	−2.8	−2.0	−0.2	−0.2	−1.4	−2.3	−2.4	
		3290	C	223	171	235	74	67	66	77	81	57	17	7	43	7.3	Le79
			S	5.7	7.2	5.6	13.7	16.9	15.4	−2.1	−1.4	−1.1	−0.4	0.5	−3.1	−1.5	
Epidote		3400	C	212	239	202	39.1	43.2	77.5	66.3	45.2	45.6	0.0	−8.2	−14.3	−3.4	Ry66
			S	5.32	4.71	5.43	25.7	23.9	12.9	−1.3	−0.93	−0.72	−0.56	−0.66	1.66	1.13	

Feldspars[2]

1. Alkali Feldspars (mol% Or Ab An)

Material	mol% Or Ab An	ϱ [kg m⁻³]	C / S	11	22	33	44	55	66	12	13	23	15	25	35	46	Ref.
Microcline	78.5 19.4 2.1	2560	C	62.5	172	124	14.3	22.3	37.4	42.8	35.8	24.1	−15.4	−14.3	−11.5	−2.8	Ry65
			S	25	7.1	9.7	71.0	54.5	27.1	−4.5	−5.3	0.1	11.6	1.4	1.4	5.3	
	78.5 19.4 2.1	2560	C	66.4	171	122	14.3	23.8	36.1	43.8	25.9	19.2	−3.3	−14.8	−13.1	−1.5	Al62
	75 22 –	2540	C	57.2	148	103	13.7	18.0	32.3	32.8	33.3	19.3	−12.4	−6.1	−11.2	−2.5	Ry65
			S	26.6	7.7	12.2	73.9	66.1	31.4	−4.5	−6.4	−0.1	12.8	−0.5	3.2	5.7	
	74 18.9 1.95	2570	C	61.9	158	100	14.1	20.3	36.0	43.4	36.8	21.8	−10.0	−1.8	−12.1	−2.3	Ry65
			S	26.0	7.9	13.1	71.6	55.5	28.0	−6.0	−7.3	0.3	7.9	−2.1	4.3	4.6	
	66.6 28.6 –	2540	C	58.4	147	98.8	12.4	18.5	34.3	33.3	34.0	21.6	−10.7	−4.3	−13.0	−3.0	Ry65
			S	25.1	7.8	13.1	82.5	62.8	29.8	−4.5	−6.5	−0.3	8.9	−1.0	5.4	7.2	
	64.9 26.6 3.6	2570	C	59.6	158	105	13.9	20.3	37.0	36.2	36.0	28.5	−11.8	−5.7	−12.9	−2.6	Ry65
			S	25.0	7.4	12.4	73.0	57.0	27.4	−4.3	−6.3	−0.6	9.4	−0.8	4.0	5.1	
	60.7 35.6 1.64	2570	C	59.6	157	120	13.6	22.6	34.2	34.4	28.0	21.6	−17.0	−5.9	−12.9	−1.8	Ry65
			S	24.4	7.4	9.6	74.0	57.5	29.4	−4.4	−3.5	−0.5	16.1	−1.8	2.8	3.9	
	53.5 34.6 9.15	2580	C	63.0	152	118	10.1	26.8	35.6	35.9	49.0	36.1	−12.9	−1.8	−18.1	−2.6	Ry65
			S	26.3	7.7	13.1	101	43.4	28.6	−4.1	−8.7	−1.0	6.5	−2.1	4.7	7.4	

2. Plagioclases (mol% An)

Material	mol% An	ϱ [kg m⁻³]	C / S	11	22	33	44	55	66	12	13	23	15	25	35	46	Ref.
Albite[3] NaAlSi₃O₈	0	2630	C	74	131	128	17.3	29.6	32.0	36.4	39.4	31.0	−6.6	−12.8	−20.0	−2.5	He79, Za74
			S	19.7	9.1	10.2	58.5	38.5	31.6	−4.0	−4.7	−0.6	−0.8	2.7	5.6	4.6	
	9	2610	C	74.9	138	129	17.2	30.3	31.1	36.3	37.6	32.6	−9.1	−10.4	−19.1	−1.3	Ry64
			S	17.2	8.5	9.8	58.4	36.9	32.3	−3.5	−3.9	−0.9	−1.5	−1.3	4.7	2.4	

(continued)

Table 8e (continued)

Material		ϱ [kg m⁻³]	C [GPa] / S [(TPa)⁻¹]	Indices 11	22	33	44	55	66	12	13	23	15	25	35	46	Ref.
Oligoclase	15…16	2640	C	80.6	163	124	17.7	27.4	36.2	41.7	53.8	37.4	16.1	17.1	−7.4	1.0	Al62
			S	23.5	7.42	13.7	56.6	50.9	27.7	−2.0	−10.5	−1.6	−15.4	−3.9	10.9	−1.6	
	24	2640	C	81.8	145	133	17.7	31.2	33.3	39.3	40.7	34.1	−9.0	−7.9	−18.5	−0.8	Ry64
			S	15.9	8.1	9.5	56.5	35.2	30.1	−3.3	−3.8	−1.0	1.5	−0.5	4.3	1.4	
	29	2640	C	84.5	151	133	18.5	31.4	34.3	41.7	40.9	33.0	−8.7	−6.9	−18.5	−1.1	Ry64
			S	15.5	7.8	9.5	54.1	34.9	29.2	−3.4	−3.7	−0.9	1.3	−0.3	4.4	1.7	
Labradorite	53	2680	C	97.0	163	141	19.6	33.0	37.0	50.7	44.2	37.0	−9.6	−5.1	−15.0	−1.6	Ry64
			S	13.8	7.4	8.3	51.3	32.2	27.2	−3.5	−3.2	−0.9	2.0	−0.3	2.9	2.2	
	56	2690	C	98.9	172	141	19.9	34.1	37.6	52.1	44.1	36.6	−8.1	−5.1	−19.1	−1.9	Ry64
			S	13.4	7.0	8.8	50.5	31.8	26.8	−3.3	−3.2	−0.8	0.9	−0.2	4.1	2.6	
	57…60	2680	C	101	158	151	21.4	33.5	37.0	61.7	48.0	26.0	−0.3	−8.0	9.6	−5.9	Al62, Si71
			S	14.9	8.4	8.0	48.7	31.0	28.1	−5.2	−3.8	0.1	0.0	1.9	−2.3	7.4	
Anorthite,³) CaAl₂Si₂O₈	100	2760	C	124	205	156	23.5	40.4	41.5	66	50	42	−19	−7	−18	−1	He79, Za74
			S	11.2	6.0	7.7	42.7	27.3	24.2	−3.0	−2.4	−0.7	−3.6	−0.7	2.2	1.2	
Gypsum, CaSO₄ · 2H₂O			C	78.6	62.7	72.6	9.1	26.4	10.4	41.0	26.8	24.2	−7.0	3.1	−17.4	−1.55	He79
			S	20.7	27.9	20.1	113	49.5	98	−13.1	−1.9	−7.2	5.8	−11.5	13.6	17.0	
			C	94.5	65.2	50.2	8.6	32.4	10.8	37.9	28.2	32.0	−11.0	6.9	−7.5	−1.1	He79
			S	15.4	29.5	32.8	117	38.2	93.5	−8.6	−2.2	−15.9	6.6	−12.8	10.2	12.0	
Hornblende		3124	C	116	160	192	57.4	31.8	36.8	45	61	66	4	−2	10	−6	He79, A.61
			S	10.9	7.7	6.9	17.7	32.2	27.7	−1.9	−2.8	−2.1	−0.7	1.5	−2.0	3.0	
		3153	C	130	188	198	61.1	38.7	45.0	61	59	61	10	−7	−41	−1	He79, A.61
			S	11.4	6.6	8.6	16.4	38.2	22.2	−2.6	−4.1	−1.1	−7.6	0.7	9.9	0.3	
Hyalophane, (Ba, K)Al₂Si₂O₈		2646	C	67.4	168	124	13.6	25.3	35.4	43	45	25.6	−12.8	−7.6	−15.8	−1.7	He79, Za74
			S	23.8	7.5	10.9	72.2	44.9	28.5	−4.9	−6.8	0.24	6.3	−0.11	3.4	3.6	
Spodumene, LiAlSi₂O₆		3107	C	245	199	287	70.1	62.8	70.7	88	64	69	−40	−26.7	−14.2	−7.1	Za74a
			S	5.30	6.38	3.88	14.4	18.1	14.3	−1.80	−0.62	−1.08	2.46	1.32	0.03	1.45	

¹) Uncorrect negative signs of the constants C_{15}, C_{25}, C_{35}, C_{46} have been changed [Ba72a].
²) All feldspars of this table are triclinic quasi monoclinic.
³) Extrapolated.

Table 9. Elastic properties of statistically isotropic polycrystalline aggregates, calculated from single crystal data of Tables 8a···8e; K^R, G^R are bulk and shear modulus averaged according to Reuss [Re29]; K^V, G^V bulk and shear modulus averaged according to Voigt [Vo28]; σ^{VRH}, v_p^{VRH}, v_s^{VRH} Poisson's ratio, compressional and shear wave velocities calculated from Voigt-Reuss-Hill [Hi52] averaged elastic constants.

Material	ϱ [kg m^{-3}]	K^R GPa	K^V GPa	G^R GPa	G^V GPa	v_p^{VRH} m s^{-1}	v_s^{VRH} m s^{-1}	σ^{VRH}
Cubic system								
Calcium fluoride	3180	86	86	41	44	6700	3660	0.28
Chromite	4450	204	204	104	106	8790	4860	0.28
Garnets:								
Pyrope	3582	165.8	165.8	91.4	91.4	8960	5050	0.27
Almandine	4318	180.1	180.1	94.7	94.7	8420	4680	0.28
Spessartine	4190	172.3	172.3	96.5	96.5	8470	4800	0.26
Grossular	3594	170.4	170.4	105.9	106.1	9310	5430	0.24
Magnetite	5180	161.0	161.0	91.1	91.5	7390	4200	0.26
NaCl	2160	25.7	25.7	14.3	14.6	4560	2590	0.26
Periclase	3583	162.9	162.9	128	134	9710	6050	0.18
Pyrite	5010	143	143	125	131	7920	5060	0.16
Spinel	3581	196	196	98	118	9750	5500	0.27
Pleonaste	3826	198	198	96	107	9250	5020	0.29
Hexagonal system								
Apatite	3218	76	86	47	40	6680	3830	0.26
Beryl	2660	150	154	81	83	9910	5540	0.27
Biotite	3050	42	60	13	42	5350	3000	0.27
Muscovite	2790	43	62	22	41	5810	3370	0.25
Nepheline	2620	42	49	30	33	5750	3450	0.22
Phlogopite	2800	46	64	12	39	5620	3000	0.30
β-Quartz	2533	56.1	56.4	40.9	42.0	6635	4043	0.20
Trigonal system								
Alumina, Corundum	3986	248	252	159	165	10810	6380	0.23
Calcite	2712	70	76	27	37	6540	3430	0.31
Haematite	5240	97	98	92	94	6510	4220	0.14
Magnesite	2980	110	118	64	72	8280	4780	0.25
α-Quartz	2650	37.6	38.1	41.0	47.6	6050	4090	0.08
Tourmaline	3050	99	108	80	95	8490	5340	0.17
Dolomite	2860	89	99	40	52	7370	4000	0.29
Orthorhombic system								
Andalusite	3145	159	166	98	100	9670	5610	0.25
Anhydrite	2970	52	58	23	36	5620	3140	0.27
Aragonite	2930	45	49	37	40	5790	3630	0.18
Baryte	4500	54.7	55.1	22.8	24.6	4350	2250	0.32
Bronzite, Bavaria	3380	101.6	102.7	60.2	61.3	7360	4240	0.25
$(Mg_{80}Fe_{20})SiO_3$	3354	101.9	104.8	73.8	75.5	7780	4720	0.21
Enstatite	3200?	107	108	75	76	8080	4860	0.22
Natrolite	2250	46.7	49.3	28.0	29.8	6110	3530	0.25
Olivines:								
Forsterite	3224	128.5	130.7	79.4	82.7	8570	5015	0.24
	3222	126.7	131.2	80.1	83.2	8590	5030	0.24
$(Mg_{92.7}Fe_{7.2})_2SiO_4$	3311	129.3	131.5	79.1	80.6	8420	4890	0.25
Sillimanite	3241	166	175	88	95	9520	5330	0.27
Staurolite	3369	91	106	69	77	7630	4660	0.20

(continued)

Gebrande

Table 9 (continued).

Material	ϱ [kg m^{-3}]	K^R GPa	K^V GPa	G^R GPa	G^V GPa	v_p^{VRH} m s^{-1}	v_s^{VRH} m s^{-1}	σ^{VRH}
Monoclinic system								
Aegirite	3500	106	112	56	59	7280	4050	0.28
Augite	3420	89	101	56	60	7100	4120	0.25
Coesite	2920	114	118	67	67	8240	4500	0.27
Diallage	3300	81	85	59	62	7030	4270	0.21
Diopside	3310	105	117	62	66	7700	4380	0.26
	3290	108	118	65	69	7860	4520	0.25
Epidote	3400	105	108	61	65	7430	4240	0.26
Feldspars:								
Or$_{78.5}$Ab$_{19.4}$An$_{2.1}$	2560	45	63	23	32	5930	3260	0.28
Or$_{75}$ Ab$_{22}$	2540	41	53	20	28	5570	3070	0.28
Or$_{66.6}$Ab$_{28.6}$	2540	43	54	20	27	5600	3050	0.29
Or$_{60.7}$Ab$_{35.6}$An$_{1.64}$	2570	41	56	22	31	5700	3210	0.27
Or$_{53.5}$Ab$_{34.6}$An$_{9.15}$	2580	51	64	20	29	5900	3060	0.32
Ab$_{100}$ An$_0$ [1])	2630	49	61	28	31	5940	3290	0.28
An$_9$	2610	53	62	27	31	6070	3340	0.28
An$_{15\cdots16}$	2640	61	70	23	32	6230	3240	0.32
An$_{24}$	2640	58	65	28	32	6220	3400	0.29
An$_{29}$	2640	60	67	29	34	6300	3450	0.29
An$_{53}$	2680	70	74	31	36	6600	3540	0.30
An$_{56}$	2690	69	75	32	37	6610	3570	0.29
An$_{57\cdots60}$	2680	74	76	31	37	6690	3550	0.30
An$_{100}$ [1])	2760	79	89	37	43	7050	3800	0.29
Hornblende	3124	84	90	41	45	6810	3720	0.29
	3153	91	98	41	51	7030	3820	0.29
Hyalophane	2646	52	65	23	31	5980	3200	0.30
Spodumene	3107	117	130	69	75	8410	4813	0.26

[1]) Extrapolated.

Table 10. Mean atomic weight of some rock forming minerals; according to [Bi61] with supplements.

	Ideal composition	$\bar{m}$
Magnesite	$MgCO_3$	16.87
Chrysoberyl	$BeAl_2O_4$	18.14
Dolomite	$CaMg(CO_3)_2$	18.44
Spodumene	$LiAlSi_2O_6$	18.60
Brucite	$Mg(OH)_2$	19.44
Serpentine	$Mg_3Si_2O_5(OH)_4$	19.80
Chlorite	$Mg_5Al_2Si_3O_{10}(OH)_8$	19.85
Talc	$Mg_3Si_4O_{10}(OH)_2$	19.96
Anthophyllite	$Mg_7Si_8O_{22}(OH)_2$	20.02
Calcite	$CaCO_3$	20.02
Quartz	SiO_2	20.03
Enstatite	$MgSiO_3$	20.08
Glaucophane	$Na_2Mg_3Al_2Si_8O_{22}(OH)_2$	20.09
Forsterite	Mg_2SiO_4	20.10
Pyrope	$Mg_3Al_2Si_3O_{12}$	20.15
Periclase	MgO	20.16
Albite	$NaAlSi_3O_8$	20.17
		(continued)

Gebrande

Table 10 (continued)

	Ideal composition	$\bar{m}$		Ideal composition	$\bar{m}$
Cordierite	$Mg_2Al_4Si_5O_{18}$	20.17	Chloritoid	$Fe_2Al_4Si_2O_{10}(OH)_4$	22.90
Jadeite	$NaAlSi_2O_6$	20.21	Wollastonite	$CaSiO_3$	23.23
Kyanite	Al_2SiO_5	20.25	Biotite	$KAlFe_3Si_3O_{10}(OH)_2$	23.27
Spinel	$MgAl_2O_4$	20.32	Spessartite	$Mn_3Al_2Si_3O_{12}$	24.75
Corundum	Al_2O_3	20.39	Almandite	$Fe_3Al_2Si_3O_{12}$	24.88
Phlogopite	$KMg_3AlSi_3O_{10}(OH)_2$	20.86	Andradite	$Ca_3Fe_2Si_3O_{12}$	25.40
Nepheline	$Na_3KAl_4Si_4O_{16}$	20.87	Calcium fluoride	CaF_2	26.03
Muscovite	$KAl_3Si_3O_{10}(OH)_2$	20.96	Ferrosilite	$FeSiO_3$	26.38
Orthoclase	$KAlSi_3O_8$	21.40	Rutile	TiO_2	26.63
Anorthite	$CaAl_2Si_2O_8$	21.40	Fayalite	Fe_2SiO_4	29.11
Zoisite	$Ca_2Al_3Si_3O_{12}(OH)$	21.63	Halite	$NaCl$	29.22
Diopside	$CaMgSi_2O_6$	21.65	Ilmenite	$FeTiO_3$	30.35
Leucite	$KAlSi_2O_6$	21.83	Haematite	Fe_2O_3	31.94
Grossular	$Ca_3Al_2Si_3O_{12}$	22.52	Magnetite	Fe_3O_4	33.08
Anhydrite	$CaSO_4$	22.69	Pyrite	FeS_2	39.99

3.1.3 Elastic wave velocities and constants of elasticity of rocks at room temperature and pressures up to 1 GPa — Geschwindigkeiten elastischer Wellen und Elastizitäts-Konstanten von Gesteinen bei Zimmertemperatur und Drucken bis 1 GPa

Measurements compiled in this section have been carried out exclusively by the method of ultrasonic pulse transmission (see e.g. [Bi60, Si64a, Ke75]). Pressure may be applied to the samples either by gaseous, fluid or solid pressure transmission. Most authors, following Birch [Bi60], use fluid pressure transmission. With this method the samples must be jacketed to exclude the pressure fluid from the pore spaces of the rocks. If the pore pressure equals the confining pressure, pores are kept open and seismic velocities increase much less with confining pressure than at zero pore pressure (compare Fig. 23). The same effect is observed with perfectly saturated jacketed samples. If building up of pore pressure is to be avoided in water-saturated jacketed samples, a mesh screen must be placed between sample and jacket to allow the drainage of pore fluid with increasing compression of the sample [Ch75a, Ch78]. Only such water-saturated samples (satw and satb, respectively) are included in Tables 11…14. Up to now most velocity determinations at elevated pressures have been carried out on dry samples. For these, it should be unimportant whether pressure is applied directly by the anvils of a triaxial press to cubic samples [Fi71, Ke75, Ke78, Ke81, Mü79] or whether it is transmitted by fluid pressure to jacketed cylindrical samples [Hu56, Bi60, Si64a, Ch65, Ka65, Ma74, Ch75, Kr76 and others].

Die in diesem Abschnitt zusammengestellten Messungen sind ausschließlich nach dem Ultraschall-Impuls-Verfahren mit Frequenzen von 1 bis 3 MHz durchgeführt worden (vgl. z.B. [Bi60, Si64a, Ke75]). Die Aufbringung des Drucks kann durch gasförmige, flüssige oder feste Druckübertragungsmittel auf die Probe erfolgen. Von den meisten Autoren wird nach dem Vorbild von Birch [Bi60] mit flüssigen Druckübertragungsmitteln gearbeitet. Die Probe muß dabei ummantelt werden, um das Eindringen des Druckübertragungsmittels in den Porenraum des Gesteins zu verhindern. Falls der Porendruck gleich dem Umschließungsdruck ist, bleiben die Poren geöffnet und die seismischen Geschwindigkeiten nehmen mit steigendem Umschließungsdruck weit weniger rasch zu als bei Porendruck Null (vgl. Fig. 23). Der gleiche Effekt ist bei vollständig wassergesättigten ummantelten Proben zu beobachten. Wenn bei ummantelten wassergesättigten Proben der Aufbau eines Porenwasserdrucks vermieden werden soll, muß die Probe vor Ummantelung mit einem Metallnetz umgeben werden, welches ein Entweichen der Porenflüssigkeit bei zunehmender Kompression der Probe ermöglicht [Ch75a, Ch78]. Nur solche wassergesättigten Proben (satw bzw. satb) wurden in die Tabellen 11…14 aufgenommen. Die meisten Geschwindigkeitsmessungen bei höheren Drucken sind bisher an trockenen Proben durchgeführt worden. Hierbei sollte es unerheblich sein, ob der Druck direkt durch die Stempel einer Triaxialpresse auf würfelförmige Proben [Fi71, Ke75, Ke78, Mü79] oder durch Flüssigkeitsdruck auf ummantelte zylindrische Proben [Hu56, Bi60, Si64a, Ch65, Ka65, Ma74, Ch75, Kr76 und andere] übertragen wird.

Gebrande

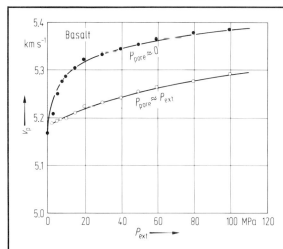

◄ Fig. 23. Compressional wave velocity v_p in a sea-water-saturated basalt for pore pressure conditions $P_{pore} \approx 0$ and $P_{pore} \approx P_{ext}$ as function of external (confining) pressure [Ha79].

The use of triaxial presses [Ke75, Ke78, Ke79, Ke81], however, offers the advantages of simultaneous velocity measurements in three mutually orthogonal directions and of direct control of length changes of the sample with increasing pressure (see also 3.1.4.2). If fluid pressure is used, anisotropy must generally be checked by separate measurements on different cylindrical cores taken from the same blocks of rock in three orthogonal directions. Thereby it cannot always be ascertained, whether velocity differences are due to anisotropy or to inhomogeneities of the samples. Length changes of the samples with increasing pressure are generally not directly measured when fluid pressure is used; for isotropic samples, however, they can be calculated from density (ϱ_0) at zero pressure and measured velocities by means of the following equation:

Die Verwendung von Triaxialpressen [Ke75, Ke78, Ke79, Ke81] bietet aber die Vorteile, daß simultan an einer Probe die Geschwindigkeiten in drei zueinander senkrechten Richtungen gemessen und daß Längen-, bzw. Dichteänderungen der Probe mit steigendem Druck direkt beobachtet werden können (vgl. auch 3.1.4.2). Bei Verwendung von Flüssigkeitsdruck muß die Anisotropie im allgemeinen durch getrennte Meßserien an drei verschiedenen zylindrischen Proben überprüft werden, die demselben Gesteinsblock in drei zueinander senkrechten Richtungen entnommen werden. Dabei ist nicht immer eindeutig festzustellen, ob Geschwindigkeitsunterschiede auf Anisotropie oder Inhomogenitäten der Proben zurückzuführen sind. Längenänderungen der Proben mit steigendem Druck können bei Flüssigkeitsdruck im allgemeinen nicht direkt gemessen werden; für isotrope Proben können sie aber aus den gemessenen Geschwindigkeiten v_p, v_s und der Dichte bei Normaldruck (ϱ_0) nach folgender Gleichung bestimmt werden:

$$\frac{L(P)}{L_0} = \left\{ 1 + \frac{1}{\varrho_0} \int_0^P (v_p^2 - \tfrac{4}{3}v_s^2)^{-1} \, dP \right\}^{-1/3} \tag{19}$$

L_0 and $L(P)$ are the length of the sample at normal and elevated pressure respectively. The pertinent density is given by:

L_0, $L(P)$ Länge der Probe bei Druck 0 und Druck P. Für die Dichtezunahme gilt entsprechend:

$$\varrho(P)/\varrho_0 = \{L_0/L(P)\}^3 \tag{20}$$

By means of Eq. (19) the measured velocities can be corrected iteratively. In most cases the corrections are less than 1 % for pressures up to 1 GPa. They may be important, however, for the calculation of pressure derivatives of velocities. The original data has been published partly corrected and partly uncorrected. For the sake of homogeneity the corrections have been carried out for all velocities in Tables 11···15, if not already done by the original authors. No later corrections for length changes have been applied to the data of section 3.1.3.2.

Nach Gl. (19) können die gemessenen Geschwindigkeiten iterativ korrigiert werden. In den meisten Fällen beträgt die Korrektur bis zu Drucken von 1 GPa weniger als 1 %. Sie kann aber für die Berechnung von Druck-Gradienten der Geschwindigkeiten wichtig sein. Die Original-Daten sind teilweise korrigiert oder unkorrigiert veröffentlicht. Aus Gründen der Einheitlichkeit wurde die Korrektur für alle Geschwindigkeiten der Tabellen 11···15 durchgeführt, sofern dies nicht bereits von den Erst-Autoren geschehen war. An den Daten des Abschnitts 3.1.3.2 wurde keine nachträgliche Korrektur auf Längenänderungen durchgeführt.

A typical example of the pressure dependence of seismic wave velocities (at zero pore pressure or $P_{pore} \ll P_{ext}$) is shown in Fig. 24. Within the first 0.1···0.2 GPa a relatively strong velocity increase is observed in most rocks. It is explained by a decrease of porosity with increasing pressure [Bi60 and many others]. At higher pressures the velocity increase declines and the velocities follow a flattened, approximately linear trend. It can be shown theoretically [Wa65], that pores with ratio α of minimum to maximum diameter are closed elastically at a pressure of about αE ($E = $ Young's modulus). Therefore, the velocity increase at small pressures can be explained by closure of very oblate pores ($\alpha < 1/100$) such as microcracks and loose grain contacts. More spherical pores cannot be closed elastically and may influence the velocities and elastic constants even at the highest pressures in Tables 11···15. Only if the total porosity is present as crack-porosity, such as seems to be the case in some plutonic and metamorphic rocks, the measured velocities at pressures above about 0.4 GPa may correspond to those in the compact rock matrix.

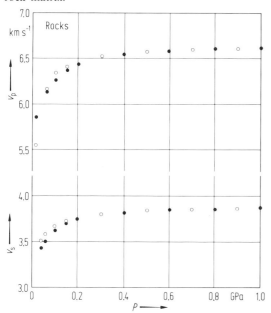

Below some 0.2 GPa some samples show a certain velocity hysteresis as for instance in Fig. 24. This may be explained by unreversible crack closure or by delayed adjustment of elastic equilibrium within the sample. In most cases lower velocities are measured with increasing and higher velocities with decreasing pressure respectively. Generally, mean values between both series of measurements are reported by the authors. In samples with high porosity the case may arise that pores collapse with increasing pressure and crack porosity thereby increases (compare [Jo80] and Fig. 32); such samples show a reversed hysteresis at low pressures.

Ein typisches Beispiel für die Druckabhängigkeit der seismischen Geschwindigkeiten (bei Porenflüssigkeitsdruck Null oder $P_{pore} \ll P_{ext}$) zeigt Fig. 24. Innerhalb der ersten 0.1···0.2 GPa wird in den meisten Gesteinen eine relativ starke Geschwindigkeitszunahme beobachtet, die auf eine Abnahme der Porosität mit zunehmendem Druck zurückgeführt wird [Bi60 und viele andere]. Bei höheren Drucken verringert sich die Zunahme, und die Geschwindigkeitskurven nehmen einen abgeflachten, annähernd linearen Verlauf. Theoretisch kann gezeigt werden [Wa65], daß Poren mit einem Verhältnis α von kleinstem zu größtem Durchmesser etwa bei einem Druck αE ($E = E$-Modul) elastisch geschlossen werden. Die Geschwindigkeitszunahme bei kleinen Drucken kann daher nur durch das Schließen sehr flacher Poren, das heißt von Mikrorissen und aufgelockerten Korngrenzen ($\alpha < 1/100$), erklärt werden. Mehr kugelförmige Poren können auch bei hohen Drucken nicht elastisch geschlossen werden und beeinflussen die Geschwindigkeiten und elastischen Konstanten auch bei den höchsten Drucken der Tabellen 11···15. Nur wenn die gesamte Porosität als Rißporosität vorliegt, wie dies bei Plutoniten und Metamorphiten teilweise der Fall zu sein scheint, entsprechen die bei Drucken oberhalb etwa 0.4 GPa gemessenen Geschwindigkeiten denen in der kompakten Gesteinsmatrix.

◄

Fig. 24. Typical dependence of elastic wave velocities in solid rocks on pressure. Except for most compact rocks, some hysteresis is usually observed at small pressures; full circles are measurements with increasing pressure; open circles, with decreasing pressure. Example is taken from [Ch70] for a partially water-saturated basalt of 3.7% porosity.

Unterhalb etwa 0.2 GPa wird bei manchen Proben, wie z.B. in Fig. 24, eine gewisse Hysterese der Geschwindigkeiten beobachtet. Sie kann durch irreversibles Schließen von Mikrorissen oder durch Verzögerungen in der Einstellung des elastischen Gleichgewichts der Probe erklärt werden. Meist sind die Geschwindigkeiten bei zunehmendem Druck etwas kleiner als die bei abnehmendem Druck. In der Regel geben die Autoren Mittelwerte zwischen beiden Meßreihen an. Bei Proben mit großer Porosität kann auch der Fall eintreten, daß bei zunehmendem Druck Poren zusammenbrechen und hierdurch die Rißporosität zunimmt (vgl. [Jo80] und Fig. 32); solche Proben zeigen bei niedrigen Drucken eine entgegengesetzte Hysterese.

If hysteresis effects are neglected, the dependence of seismic velocities on pressure in general can be represented within the accuracy of measurements by a relation of the following type:

Wenn von Effekten der Hysterese abgesehen wird, kann die Druckabhängigkeit der seismischen Geschwindigkeiten in der Regel im Rahmen der Meßgenauigkeit durch eine Beziehung der Form:

$$v(P) = (a + bP)(1 - c\,e^{-P/d}) \qquad (21)$$

The constants a and b determine the velocity in the state without cracks.

dargestellt werden; a und b bestimmen die Geschwindigkeit im Zustand ohne Risse.

3.1.3.1 Isotropic and quasi-isotropic rocks — Isotrope und quasi-isotrope Gesteine

For most isotropic and quasi-isotropic rocks a large number of velocity determinations at elevated pressures is available; only a part of these measurements and quantities derived from them can be presented in the following Tables 11···15. We use the term quasi-isotropic for rocks with an ascertained or suspected anisotropy $A < 5\%$ for both P- and S-waves (A_p, A_s). Some samples with larger anisotropies are included in Tables 11···15, if they could not be incorporated into Table 18 because of incomplete information (e.g. on polarization of shear waves). The A_p and A_s values given in the column "Notes" refer to the highest pressure of the respective series of measurements.

Für die meisten isotropen und quasi-isotropen Gesteine liegt eine große Zahl von Geschwindigkeitsbestimmungen bei höheren Drucken vor; nur ein Teil dieser Messungen und daraus abgeleitete Größen können in den folgenden Tabellen 11···15 wiedergegeben werden. Als quasi-isotrop werden hier Gesteine mit einer nachgewiesenen oder vermuteten Anisotropie $A < 5\%$ für P- und S-Wellen (A_p, A_s) bezeichnet. Die Tabellen 11···15 enthalten auch einige wenige Proben mit größerer Anisotropie, die aber wegen unvollständiger Angaben (z.B. über die Polarisationsrichtung der S-Wellen) nicht in die Tabelle 18 aufgenommen werden konnten. Die in der Spalte „Notes" vermerkten Werte A_p und A_s gelten jeweils beim größten Druck der Meßserie.

Fig. 25.

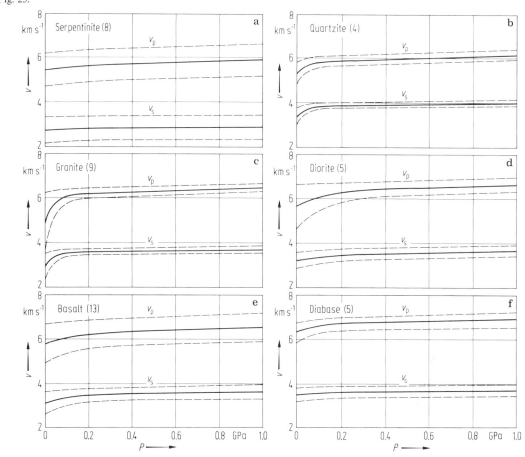

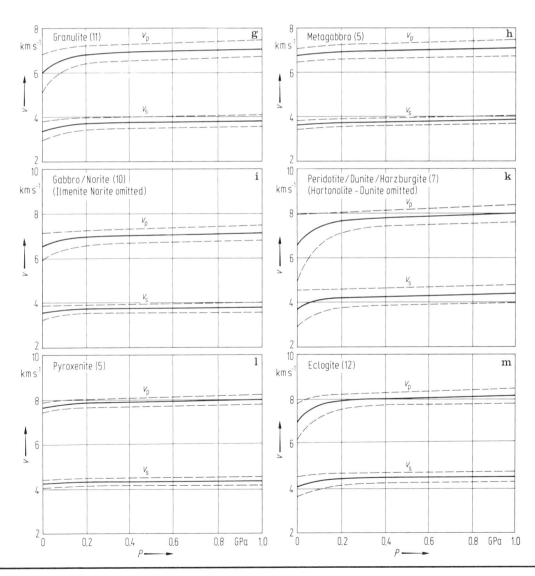

Fig. 25a···m. Mean values and 80% confidence limits of elastic wave velocities as functions of confining pressure for different rock types. Curves were calculated from data of Tables 11, 12 and 14; identical samples were used for the calculation of v_p and v_s curves; number of samples used is indicated in the figures in parentheses after the rock names.

Preferably petrographically well characterized samples, for which measurements along three mutually orthogonal directions have been carried out, were selected. Tables 11···15 contain only mean values of velocities. Only such samples are included for which both P- and S-wave velocities have been measured. For P-waves alone much more measurements are available. As far as information on the modal composition are given by the original authors, it has been included – sometimes slightly simplified – in the Tables; for the abbreviations, see section 3.1.1.1. If possible, information on mean atomic weight $\overline{m}$ and porosity ϕ is also given. *)

All velocities given in Tables 11···15 are corrected for length changes caused by compression of the samples. In some cases [Ke79, Ke79a, Ke81, Ke81a] length changes have been measured directly, otherwise they have been calculated by means of Eq. (19) and density variations by Eq. (20). The elastic constants K, G, E and σ have been calculated by means of Eqs. (1a), (1b) and the relations of Table 1. Rock samples are arranged within each table in order of increasing density.

Velocities are generally given to four figures; the accuracy of most measurements, however, may not be better than 1%. Some authors have originally given two or three figures only; in these cases the fourth figure arises with the application of compression correction. It was retained in order not to loose the relative precision of the original data.

For all series of v_p and v_s measurements of Tables 11···15 best-fitting curves of the type of Eq. (21) were calculated; generally a mean prediction error of less than 10 m/s was obtained. From the results 80% confidence limits and mean values of the velocities as functions of pressure were calculated for different rock types; they are shown graphically in Fig. 25a···m. They may be recommended to the reader with a more general interest in the pressure dependence of seismic velocities in different rocks. More detailed results for individual rock samples must be taken from Tables 11···15.

Es wurden bevorzugt solche Proben ausgewählt, für die Messungen in drei zueinander senkrechten Richtungen vorliegen und die petrographisch gut charakterisiert sind. Nur die Mittelwerte der Geschwindigkeiten sind in den Tabellen 11···15 enthalten. Nur solche Proben wurden aufgenommen, für die sowohl P- wie auch S-Geschwindigkeiten gemessen wurden. Für P-Wellen alleine liegen weit mehr Messungen vor. Soweit von den Primär-Autoren Angaben über den modalen Stoffbestand gemacht wurden, wurden sie – z.T. leicht vereinfacht – in die Tabellen übernommen; bzgl. der Abkürzungen s. 3.1.1.1. Soweit möglich wurden auch Angaben über das mittlere Atomgewicht $\overline{m}$ und die Porosität ϕ gemacht. *)

Alle Geschwindigkeitsangaben der Tabellen 11···15 sind auf Längenänderungen der Proben durch Kompression korrigiert. In einigen Fällen sind die Längenänderungen direkt gemessen [Ke79a, Ke81, Ke81a], in den anderen Fällen wurden sie nach Gl. (19) berechnet. Entsprechendes gilt für die Variation der Dichten mit dem Druck. Die Elastizitäts-Konstanten K, G, E und σ wurden nach den Gln. (1a), (1b) und den Gleichungen der Tabelle 1 berechnet. Die Gesteinsproben sind innerhalb der einzelnen Listen nach zunehmender Dichte angeordnet.

Die Geschwindigkeiten sind im allgemeinen auf vier Stellen angegeben; die meisten Messungen dürften dagegen eine absolute Genauigkeit von nur etwa 1% haben. Manche Autoren haben ihre Werte ursprünglich nur auf drei Stellen genau angegeben; die vierte Stelle ergab sich in diesen Fällen erst durch die Anwendung der Kompressions-Korrektur. Die vierte Stelle wurde beibehalten, um die relative Genauigkeit der Originaldaten nicht zu verlieren.

Zu sämtlichen v_p- und v_s-Meßreihen der Tabellen 11···15 wurden nach der Methode der kleinsten Quadrate optimal angepaßte Ausgleichskurven vom Typ der Gl. (21) berechnet; in der Regel ergibt sich dabei ein mittlerer Anpassungsfehler kleiner als 10 m/s. Aus den Ergebnissen wurden für verschiedene Gesteinstypen 80%-Vertrauens-Grenzen und Mittelwerte der Geschwindigkeiten als Funktionen des Drucks bestimmt; sie sind in den Fig. 25a···m graphisch dargestellt. Sie können dem Leser empfohlen werden, der sich ein generelles Bild über die Druckabhängigkeit der seismischen Geschwindigkeiten in verschiedenen Gesteinen verschaffen möchte. Detail-Ergebnisse für individuelle Gesteinsproben müssen den Tabellen 11···15 entnommen werden.

---

*) For the mean atomic weights of individual minerals, see Table 10, p. 34.

*) Bezüglich mittlerer Atomgewichte einzelner Minerale siehe Tab. 10, S. 34.

3.1.3.1.1 Plutonic rocks — Tiefengesteine

Table 11. Density, elastic wave velocities and elastic constants as function of pressure for isotropic and quasi-isotropic *plutonic* rocks at room temperature; densities and velocities are corrected for compression; abbreviations are explained in section 3.1.1.1.

Material	P MPa	ϱ kg m⁻³	v_p m s⁻¹	v_s m s⁻¹	v_p/v_s	K GPa	G GPa	E GPa	σ	Notes	Ref.
Granite, Westerly, USA	0	2614	4150	2686	1.545	19.8	18.8	43.0	0.14	$\bar{3}$oD(3S)	Si65a,
(35.4 or, 31.4 plg/An$_{17}$, 27.5 qu, 4.5 mi)	10	2615	4824	3016	1.60	29.1	23.8	56.1	0.18	dry	Bi61
$\bar{m}=20.9$	40	2617	5383	3251	1.66	39.0	27.7	67.1	0.21		
	70	2619	5569	3332	1.67	42.5	29.1	71.0	0.22		
	100	2621	5706	3382	1.69	45.4	30.0	73.7	0.23		
	200	2627	5827	3476	1.68	46.9	31.7	77.7	0.22		
	300	2632	5954	3531	1.69	49.6	32.8	80.6	0.23		
	500	2643	6064	3566	1.70	52.4	33.6	83.0	0.24	$A_p=0.8\%$	
	1000	2667	6208	3593	1.73	56.9	34.4	85.9	0.25	$A_s=3.1\%$	
Granite, Westerly, USA	0	2615								1D(1S)	Hu56
$\bar{m}=20.9$	20	2616	5900	3500	1.69	48.3	32.0	78.7	0.23	dry	
	50	2618	5980	3520	1.70	50.3	32.4	80.0	0.23		
	100	2620	6056	3548	1.71	52.1	32.9	81.7	0.24		
	150	2623	6084	3557	1.71	52.8	33.2	82.3	0.24		
	200	2625	6112	3565	1.71	53.6	33.4	82.9	0.24		
	300	2630	6128	3573	1.72	54.0	33.6	83.4	0.24		
	400	2635	6164	3581	1.72	55.1	33.8	84.2	0.25		
	500	2640	6201	3579	1.73	56.4	33.8	84.5	0.25		
Albitite, Sylmar, Pennsylvania, USA	0	2615	6070	3430	1.77	55.3	30.8	77.9	0.27	$\bar{3}$oD(6S)	Bi61,
$\bar{m}=20.2$	50	2617	6178	3539	1.75	56.2	32.8	82.3	0.26	dry	Si64a
	100	2620	6236	3568	1.75	57.4	33.3	83.8	0.26		
	200	2624	6303	3606	1.75	58.7	34.1	85.8	0.26		
	400	2633	6385	3642	1.75	60.8	34.9	87.9	0.26		
	600	2642	6428	3668	1.75	61.8	35.5	89.4	0.26	$A_p=2.6\%$	
	1000	2658	6484	3710	1.75	63.0	36.6	92.0	0.26	$A_s=1.1\%$	

(continued)

Table 11 (continued)

Material	P MPa	ϱ kg m⁻³	v_p m s⁻¹	v_s m s⁻¹	v_p/v_s	K GPa	G GPa	E GPa	σ	Notes	Ref.
Granite, Stone Mountain, Georgia, USA	0	2623	3640	2421	1.50	14.3	15.4	33.9	0.10	3oD(6S)	Si65a,
	10	2625	4336	3008	1.44	17.7	23.8	49.2	0.04	dry	Bi61
(38 plg/An₅, 32 or, 26 qu, 4 mu)	40	2628	5181	3287	1.58	32.7	28.4	66.1	0.16		
$\bar{m}=20.7$	70	2630	5524	3447	1.60	38.6	31.2	73.8	0.18		
	100	2632	5774	3535	1.63	43.9	32.9	79.0	0.20		
	200	2638	6029	3640	1.66	49.3	35.0	84.9	0.21		
	300	2643	6138	3705	1.66	51.2	36.3	88.0	0.21		
	500	2653	6235	3733	1.67	53.8	37.0	90.3	0.22		
	1000	2677	6368	3758	1.70	58.1	37.8	93.2	0.23		
Granite, Casco, Maine, USA	0	2626	3300	2320	1.42	9.8	14.1	28.6	0.01	1D(1S)	Nu69,
(45 or, 28 qu, 22 plg/An₁₅, 5 mi)	10	2627	5049	2790	1.81	39.7	20.4	52.4	0.28	dry	Br65b
$\phi=0.7\%$	20	2628	5689	3119	1.82	50.9	25.6	65.7	0.29	$A_p=0.8\%$	
(see Figs. 27, 28)	40	2629	5998	3439	1.74	53.1	31.1	78.0	0.26	$A_s=3.0\%$	
	70	2630	6297	3598	1.75	58.9	34.1	85.6	0.26		
	100	2632	6455	3657	1.77	62.7	35.2	89.0	0.26		
	200	2636	6542	3725	1.76	64.0	36.6	92.2	0.26		
	300	2640	6569	3753	1.75	64.3	37.2	93.5	0.26		
Granite, Quincy, Massachusetts, USA	0	2629	5919	3340	1.77	53.0	29.3	74.3	0.27	1D(1S)	Hu55,
(63 fsp, 26 qu, 10 am)	20	2630	6058	3429	1.77	55.3	30.9	78.2	0.26	dry?	Bi61
$\bar{m}=20.9$	50	2631	6206	3438	1.81	59.9	31.1	79.6	0.28		
	100	2634	6255	3447	1.81	61.4	31.3	80.3	0.28		
	150	2636	6293	3456	1.82	62.5	31.5	80.9	0.28		
	200	2638	6329	3464	1.83	63.6	31.7	81.6	0.29		
	300	2642	6366	3472	1.83	64.7	31.9	82.2	0.29		
	400	2646	6383	3491	1.83	64.9	32.3	83.1	0.29		
	500	2650									
Biotite Granite, Woodbury, Vermont, USA	0	2634	5769	3310	1.74	49.2	28.9	72.4	0.26	1D(1S)	Hu55
(40 qu, 20…30 kfsp, plg, bi, chl)	20	2635	6048	3459	1.75	54.4	31.5	79.3	0.26	dry	
	50	2637	6156	3558	1.73	55.5	33.4	83.5	0.25		
	100	2639	6194	3607	1.72	55.5	34.4	85.5	0.24		
	150	2641	6212	3626	1.71	55.7	34.8	86.3	0.24		
	200	2644	6249	3653	1.71	56.3	35.3	87.7	0.24		
	300	2648	6275	3661	1.71	57.0	35.6	88.3	0.24		
	400	2653	6291	3669	1.72	57.5	35.8	88.9	0.24		
	500	2658									

(continued)

Table 11 (continued)

Material	P MPa	ρ kg m⁻³	v_p m s⁻¹	v_s m s⁻¹	v_p/v_s	K GPa	G GPa	E GPa	σ	Notes	Ref.
Granite, Llano County, Texas, USA (50 kfsp, ≈25 qu, ≲25 plg)	0	2636	6139	3270	1.88	61.8	28.2	73.4	0.30	1D(1S) dry?	Hu56
	20	2637	6288	3349	1.88	64.9	29.6	77.1	0.30		
	50	2638	6337	3348	1.89	66.5	29.6	77.3	0.31		
	100	2640	6405	3377	1.90	68.2	30.1	78.8	0.31		
	150	2642	6424	3387	1.90	68.7	30.3	79.3	0.31		
	200	2644	6460	3375	1.91	70.3	30.2	79.2	0.31		
	300	2648	6487	3353	1.94	71.8	29.8	78.6	0.32		
	400	2652	6504	3362	1.94	72.3	30.0	79.1	0.32		
	500	2655	6521	3380	1.93	72.6	30.4	80.0	0.32		
	600	2659	6538	3379	1.94	73.3	30.4	80.1	0.32		
	700	2662									
Granite (Quartz Monzonite), Porterville, California, USA (34 qu, 40 plg/An₃₀, 23 or, 7 bi, 2 chl) $\bar{m}=21.2$	0	2648	5100	3160	1.61	33.6	26.4	62.8	0.19	3̄oD(6S) dry	Bi60, Si64a, Bi61
	100	2655	5945	3627	1.64	47.3	34.9	84.1	0.20		
	200	2660	6061	3704	1.64	49.0	36.5	87.7	0.20		
	400	2671	6202	3769	1.65	52.1	37.9	91.6	0.21	$A_p=1.6\%$	
	600	2681	6254	3794	1.65	53.4	38.6	93.3	0.21	$A_s=4.4\%$	
	1000	2700	6329	3835	1.65	55.2	39.7	96.1	0.21		
Trondhjemite, Trinity Complex, California, USA (40 plg (sau), 40 qu, 17 chl, 3 op)	0	2648	5999	3560	1.69	50.6	33.6	82.5	0.23	3̄oD satw	Ch78
	20	2649	6118	3579	1.71	53.9	33.9	84.2	0.24		
	40	2650	6198	3599	1.72	56.1	34.3	85.5	0.25		
	60	2651	6276	3628	1.73	57.9	34.9	87.2	0.25		
	100	2653	6363	3656	1.74	60.2	35.5	89.0	0.25		
	200	2657	6465	3712	1.74	62.5	36.7	92.1	0.25		
	400	2666	6538	3738	1.75	64.5	37.4	93.9	0.26		
	600	2674	6674	3790	1.76	68.3	38.6	97.5	0.26		
	1000	2690									
Granite, Barre, Vermont, USA $\bar{m}=20.8$	0	2660	5100	2790	1.83	41.6	20.7	53.3	0.29	3̄oD(6S) dry	Bi60, Si64a, Bi61
	50	2663	5858	3349	1.75	51.6	29.9	75.1	0.26		
	100	2665	6056	3478	1.74	54.8	32.2	80.8	0.25		
	200	2670	6142	3516	1.75	56.7	33.0	82.9	0.26		
	400	2680	6235	3631	1.72	57.1	35.3	87.9	0.24	$A_p=1.2\%$	
	600	2689	6297	3657	1.72	58.7	36.0	89.6	0.25	$A_s=1.6\%$	
	1000	2707	6353	3679	1.73	60.4	36.6	91.4	0.25		

(continued)

Gebrande

Table 11 (continued)

Material	P MPa	ϱ kg m^{-3}	v_p m s^{-1}	v_s m s^{-1}	v_p/v_s	K GPa	G GPa	E GPa	σ	Notes	Ref.
Quartz Diorite (Biotite Granodiorite), Senmaya, Japan (74 plg, 10 qu, 10 cpx, 6 bi) $\bar{m}=21.0$	0	2690	5950	3300	1.80	56.2	29.3	74.9	0.28	1D(1S)	Ka65, Ka70
	200	2699	6173	3446	1.79	60.1	32.1	81.7	0.27	dry ?	
	400	2708	6266	3552	1.76	60.8	34.2	86.3	0.26		
	600	2717	6339	3628	1.75	61.5	35.8	90.0	0.26		
	800	2726	6372	3664	1.74	61.9	36.6	91.7	0.25		
	1000	2735	6415	3680	1.74	63.2	37.0	92.9	0.26		
Anorthosite, Lake Placid, USA (100 plg/An_{49}) $\bar{m}=20.7$	0	2707	5632	3270	1.72	47.3	28.9	72.1	0.25	$\bar{3}oD$(3S)	Ma74
	50	2709	6406	3518	1.82	66.5	33.5	86.1	0.28	dry	
	100	2711	6578	3575	1.84	71.1	34.6	89.4	0.29		
	200	2715	6740	3624	1.86	75.8	35.7	92.5	0.30		
	400	2722	6832	3670	1.86	78.2	36.7	95.1	0.30		
	600	2729	6888	3694	1.87	79.8	37.2	96.7	0.30	$A_p=2.3\%$	
	800	2736	6922	3706	1.87	81.0	37.6	97.6	0.30	$A_s=1.4\%$	
	1000	2742	6950	3710	1.87	82.1	37.7	98.2	0.30		
Quartz Diorite (Hornblende Granodiorite), Kuji, Japan (57 plg, 19 qu, 16 bi, 8 hbl) $\bar{m}=21.0$	0	2710	5720	3180	1.80	52.1	27.4	70.0	0.28	1D(1S)	Ka65, Ka70
	200	2720	6113	3446	1.77	58.6	32.3	81.8	0.27	dry ?	
	400	2729	6246	3532	1.77	61.1	34.0	86.1	0.27		
	600	2738	6319	3548	1.78	63.4	34.5	87.5	0.27		
	800	2746	6372	3564	1.79	65.0	34.9	88.8	0.27		
	1000	2754	6445	3581	1.80	67.3	35.3	90.2	0.28		
Trondhjemite, Trinity Complex, California, USA (55 plg (sau), 44 qu, 5 chl)	0	2731	6109	3540	1.73	56.3	34.2	85.4	0.25	$\bar{3}oD$	Ch78
	20	2732	6209	3609	1.72	57.9	35.6	88.6	0.25	satw	
	40	2733	6288	3649	1.72	59.6	36.4	90.7	0.25		
	60	2734	6376	3698	1.72	61.3	37.4	93.3	0.25		
	100	2736	6483	3756	1.73	63.6	38.7	96.4	0.25		
	200	2740	6586	3832	1.72	65.4	40.4	100.4	0.24		
	400	2749	6649	3868	1.72	66.9	41.2	102.6	0.24		
	600	2757									
	1000	2773	6766	3940	1.72	69.5	43.0	107.0	0.24		
Anorthosite (Stillwater), Montana, USA (93 plg/An_{80}, 7 br)	0	2760	6500	3560	1.83	70.0	35.0	90.0	0.29	$\bar{3}oD$(6S)	Bi60, Si64a
	100	2764	6967	3688	1.89	84.0	37.6	98.1	0.31	dry	
	200	2767	7004	3717	1.88	84.5	38.2	100.0	0.30		
	400	2773	7039	3754	1.88	85.3	39.1	102.0	0.30		
	600	2780	7053	3761	1.88	85.9	39.3	102.3	0.30	$A_p=4\%$	
	1000	2793	7072	3795	1.86	86.1	40.2	104.4	0.30	$A_s=2.6\%$	

(continued)

Table 11 (continued)

Material	P MPa	ϱ kg m⁻³	v_p m s⁻¹	v_s m s⁻¹	v_p/v_s	K GPa	G GPa	E GPa	σ	Notes	Ref.
Quartz Diorite (Hornblende Granodiorite), Tejika, Japan (69 plg, 15 qu, 10 bi, 5 hbl) $\bar m=21.1$	0	2770	4560	2800	1.63	28.6	21.7	52.0	0.20	1D(1S) dry ?	Ka65, Ka70
	200	2782	6171	3175	1.94	68.5	28.1	74.1	0.32		
	400	2790	6445	3292	1.96	75.6	30.2	80.0	0.32		
	600	2797	6519	3339	1.95	77.3	31.2	82.5	0.32		
	800	2804	6573	3366	1.95	78.8	31.8	84.0	0.32		
	1000	2812	6617	3433	1.93	78.9	33.1	87.2	0.32		
Quartz Mangerite, Saranac Lake, New York, USA (46.3 kfsp, 17.9 plg, 12.6 opx, 7.4 cpx, 5.5 am, 2.3 ap, 2.1 op) $\bar m=21.7$	0	2826	6133	3550	1.73	58.8	35.6	88.9	0.25	$\bar3$oD(3S) dry	Ma74
	50	2828	6366	3730	1.71	62.1	39.3	97.5	0.24		
	100	2831	6429	3761	1.71	63.6	40.0	99.3	0.24		
	200	2835	6475	3783	1.71	64.8	40.6	100.7	0.24		
	400	2844	6522	3807	1.71	66.0	41.2	102.4	0.24		
	600	2852	6549	3817	1.72	66.9	41.6	103.3	0.24	$A_p=1.3\%$	
	800	2861	6574	3823	1.72	67.9	41.8	104.1	0.25	$A_s=0.9\%$	
	1000	2869	6594	3827	1.72	68.7	42.0	104.7	0.25		
Olivine Gabbro (Bytownite Gabbro), slightly altered, Duluth, Minnesota, USA (70 plg/An₈₀, ≈15 ol, ≈10 px)	0	2885	6449	3420	1.89	75.0	33.7	88.0	0.30	1D(1S) dry ?	Hu57
	20	2886	6609	3449	1.92	80.3	34.3	90.2	0.31		
	50	2887	6687	3469	1.93	82.8	34.8	91.5	0.32		
	100	2889	6716	3508	1.92	82.9	35.6	93.4	0.31		
	150	2890	6754	3517	1.92	84.2	35.8	94.0	0.31		
	200	2892	6772	3516	1.93	85.1	35.8	94.2	0.32		
	300	2896	6799	3524	1.93	86.0	36.0	94.8	0.32		
	400	2899	6816	3523	1.94	86.8	36.0	94.9	0.32		
	500	2902	6824	3532	1.93	87.0	36.2	95.5	0.32		
	600	2906									
Norite, Radautal, Germany (57 plg, 28 px, 8 hbl, 5 sau, 1 qu, 1 mi) $\bar m=21.67$	50	2906	6484	3639	1.78	70.9	38.5	97.8	0.27	1D(1S) dry	Ke81, Ke81a
	100	2906	6743	3761	1.79	77.3	41.1	104.8	0.27		
	200	2912	6859	3817	1.80	80.4	42.4	108.2	0.28		
	400	2921	6925	3842	1.80	82.6	43.1	110.2	0.28	$A_p=1.62\%$	
	600	2930	6974	3856	1.81	84.4	43.6	111.5	0.28		
Quartz Monzodiorite (Quartz Diorite), Dedham, Massachusetts, USA (13 qu, 6 or, 48 plg/An₂₀, 21 hbl, 5 chl)	0	2917	5500	3390	1.62	43.5	33.5	80.0	0.19	$\bar3$oD(6S) dry	Bi60, Si64a
	100	2922	6456	3688	1.75	68.8	39.7	100.0	0.26		
	200	2927	6523	3736	1.75	70.1	40.8	102.6	0.26		
	400	2935	6587	3772	1.75	71.6	41.8	104.9	0.26	$A_p=0.1\%$	
	600	2943	6630	3799	1.75	72.8	42.5	106.7	0.26	$A_s=0.3\%$	
	1000	2959	6678	3822	1.75	74.3	43.2	108.6	0.26		
(continued)											

Table 11 (continued)

Material	P MPa	ϱ kg m⁻³	v_p m s⁻¹	v_s m s⁻¹	v_p/v_s	K GPa	G GPa	E GPa	σ	Notes	Ref.
Hornblende Gabbro, origin unknown (40 plg/An₅₅, 35 hbl, 15 px, 10 bi)	0	2933								1D(1S)	Hu57
	20	2934	6599	3560	1.85	78.2	37.2	96.3	0.30	dry	
	50	2935	6669	3589	1.86	80.1	37.8	98.0	0.30		
	100	2937	6737	3648	1.85	81.2	39.1	101.1	0.29		
	150	2938	6776	3658	1.85	82.5	39.3	101.8	0.29		
	200	2940	6794	3687	1.84	82.4	40.0	103.2	0.29		
	300	2944	6832	3705	1.84	83.5	40.4	104.4	0.29		
	400	2947	6849	3704	1.85	84.3	40.4	104.6	0.29		
	500	2951	6866	3703	1.85	85.2	40.5	104.8	0.30		
	600	2954	6873	3711	1.85	85.3	40.7	105.3	0.29		
Diorite, Dando, Japan (41 plg, 46 hbl, 14 bi) $\bar{m}=21.8$	0	2943	6527	3380	1.93	80.5	33.6	88.5	0.32	3̄oD(3S)	Ka55
	200	2950	6677	3480	1.87	81.1	37.8	98.2	0.30	dry ?	Ka70
	400	2957	6772	3514	1.93	86.9	36.5	96.1	0.32		
	600	2964	6837	3552	1.93	88.7	37.4	98.3	0.32	$A_p=2.6\%$	
	800	2971	6869	3572	1.92	89.6	37.9	99.7	0.32	$A_s=1.0\%$	
	1000	2977	6926	3593	1.93	91.6	38.4	101.2	0.32		
Norite, Bushveld Complex, Transvaal, S. Africa (53 plg/An₆₀, 46 px, 1 am)	0	2981	6600	3560	1.85	79.5	37.8	97.8	0.30	3̄oD(6S)	Bi60,
	50	2983	7019	3809	1.84	89.2	43.3	111.8	0.29	dry	Si64a
	100	2984	7067	3839	1.84	90.4	44.0	113.5	0.29		
	200	2988	7105	3857	1.84	91.5	44.4	114.8	0.29		
	400	2994	7149	3884	1.84	92.8	45.2	116.6	0.29	$A_p=2\%$	
	600	3001	7184	3891	1.85	94.3	45.4	117.5	0.29	$A_s=1\%$	
	1000	3013	7254	3926	1.85	96.6	46.4	120.1	0.29		
Gabbro, San Marcos, California, USA (55 plg/An₆₀, 35 hbl)	0	2993								1D(1S)	Hu57
	20	2994	6689	3470	1.93	85.9	36.0	94.9	0.32		
	50	2995	6789	3479	1.95	89.7	36.3	95.8	0.32		
	100	2996	6877	3499	1.97	92.8	36.7	97.2	0.33		
	150	2998	6926	3498	1.98	95.0	36.7	97.5	0.33		
	200	3000	6945	3507	1.98	95.5	37.0	98.1	0.33		
	300	3003	6973	3506	1.99	96.8	36.9	98.2	0.33		
Norite, Radautal, Germany (58 plg, 31 opx, 4 cpx, 4 bi)	50	2998	6456	3655	1.77	71.6	40.1	101.3	0.26	3̄oD(1S)	Ke81a
	100	3001	6536	3676	1.78	74.1	40.6	102.9	0.27	dry	
	200	3009	6586	3690	1.79	75.9	41.0	104.2	0.27		
	400	3021	6634	3700	1.79	77.8	41.4	105.4	0.27	$A_p=0.9\%$	
	600	3030	6670	3699	1.80	79.5	41.5	106.0	0.28	$A_s=1.2\%$	

(continued)

Gebrande

Table 11 (continued)

Material	P MPa	ϱ kg m⁻³	v_p m s⁻¹	v_s m s⁻¹	v_p/v_s	K GPa	G GPa	E GPa	σ	Notes	Ref.
Gabbro, Canyon Mountain, Oregon, USA	0	3013								$\bar{5}$oD satw	Ch78
(40 plg (sau), 35 px, 15 serp, 5 am, 5 chl)	20	3014	6839	3670	1.86	86.9	40.6	105.3	0.30		
	40	3014	6879	3689	1.86	87.9	41.0	106.5	0.30		
	60	3015	6908	3719	1.86	88.3	41.7	108.1	0.30		
	100	3016	6967	3759	1.85	89.6	42.6	110.3	0.30		
	200	3020	7065	3817	1.85	92.1	44.0	113.9	0.29		
	400	3026	7170	3894	1.84	94.4	45.9	118.5	0.29		
$\bar{m}=21.5$	600	3033	7244	3932	1.84	96.7	46.9	121.1	0.29		
	1000	3045	7374	4006	1.84	100.4	48.9	126.1	0.29		
Gabbro, Papua, New Guinea	5	3030	7070	3760	1.88	94.3	42.8	111.6	0.30	$\bar{5}$oD(3S)	Kr76
(50 au, 46 plg/An₈₀, 3 hy)	50	3032	7250	3800	1.91	101.0	43.8	114.8	0.31	dry	
	100	3033	7330	3810	1.92	104.3	44.0	115.8	0.32		
	200	3036	7410	3830	1.94	107.3	44.5	117.4	0.32		
$\bar{m}=21.5$	300	3039	7450	3830	1.95	109.2	44.6	117.7	0.32		
	500	3044	7500	3840	1.95	111.4	44.9	118.7	0.32	$A_p=2.5\%$	
	1000	3058	7590	3860	1.97	115.4	45.6	120.8	0.33	$A_s=2.3\%$	
Hornblende Gabbro, Ayabe, Japan	0	3110	6600	3540	1.86	83.5	39.0	101.2	0.30	1D(1S)	Ka65,
(31 plg, 68hbl)	200	3117	6955	3647	1.91	95.5	41.5	108.7	0.31	dry ?	Ka70
	400	3124	7010	3745	1.87	95.1	43.8	113.9	0.30		
	600	3130	7045	3772	1.87	96.0	44.5	115.7	0.30		
$\bar{m}=21.8$	800	3137	7080	3809	1.86	96.5	45.5	118.0	0.30		
	1000	3143	7125	3836	1.86	97.9	46.3	119.9	0.30		
Harzburgite, Papua, New Guinea	5	3160	7240	3760	1.93	106.1	44.7	117.6	0.32	$\bar{5}$oD(3S)	Kr76
(59 ol, 21 opx, 16 serp)	50	3161	7390	3820	1.94	111.1	46.1	121.6	0.32	dry	
	100	3163	7440	3840	1.94	112.9	46.6	123.0	0.32		
	200	3166	7500	3860	1.94	115.2	47.2	124.5	0.32		
$\bar{m}=21.4$	300	3168	7550	3880	1.95	117.0	47.7	126.0	0.32		
	500	3174	7610	3910	1.95	119.1	48.5	128.2	0.32	$A_p=2.9\%$	
	1000	3187	7730	3950	1.96	124.1	49.7	131.6	0.32	$A_s=1.9\%$	
Peridotite, Higashiakaishi, Japan	0	3163	6927	3633	1.91	96.1	41.8	109.4	0.31	$\bar{5}$oD(3S)	Ka65,
(91 ol, 9 serp)	200	3170	7195	3814	1.89	102.6	46.1	120.3	0.31	dry	Ka70
	400	3176	7380	3895	1.90	108.7	48.2	125.9	0.31		
$\bar{m}=21.0$	600	3181	7476	3929	1.90	112.3	49.1	128.6	0.31		
	800	3187	7548	3983	1.90	114.1	50.6	132.2	0.31	$A_p=5.0\%$	
(continued)	1000	3193	7570	4021	1.88	114.1	51.6	134.6	0.30	$A_s=4.5\%$	

Gebrande

Table 11 (continued)

Material	P MPa	ϱ kg m⁻³	v_p m s⁻¹	v_s m s⁻¹	v_p/v_s	K GPa	G GPa	E GPa	σ	Notes	Ref.
Pyroxenite, Canyon Mountains, Oregon, USA (90 px, 7 serp, 2 ol, 1 am)	0	3209								$\bar{3}$oD satw	Ch78
	20	3210	7730	4220	1.83	115.6	57.2	147.2	0.29		
	40	3210	7749	4230	1.83	116.2	57.4	147.9	0.29		
	60	3211	7779	4239	1.84	117.3	57.7	148.7	0.29		
	100	3212	7808	4259	1.83	118.1	58.3	150.1	0.29		
	200	3214	7856	4288	1.83	119.6	59.1	152.2	0.29		
	400	3220	7901	4325	1.83	120.7	60.2	155.0	0.29		
	600	3225	7937	4343	1.83	122.1	60.8	156.5	0.29		
	1000	3238	8009	4377	1.83	124.9	62.0	159.7	0.29		
Dunite, Åheim, Norway (92 ol, 6 px, 2 chl) $\bar{m}=20.78$	50	3248	7786	4439	1.75	111.6	64.0	161.2	0.26	1D(1S) dry	Ke81, Ke81a
	100	3251	7937	4547	1.75	115.2	67.2	168.8	0.26		
	200	3257	8017	4592	1.75	117.8	68.7	172.5	0.26		
	400	3265	8117	4628	1.75	121.9	69.9	176.1	0.26	$A_p=4.8\%$	
	600	3278	8228	4662	1.77	126.9	71.2	180.0	0.26		
Pyroxenite-Harzburgite, Papua, New Guinea (62 opx, 35 ol) $\bar{m}=21.3$	5	3250	7750	4120	1.88	121.6	55.2	143.8	0.30	$\bar{3}$oD(3S) dry	Kr75
	50	3251	7870	4180	1.88	125.6	56.8	148.1	0.30		
	100	3253	7930	4200	1.89	128.1	57.4	149.8	0.31		
	200	3255	7980	4230	1.89	129.6	58.2	152.0	0.31		
	300	3258	8010	4240	1.89	131.0	58.6	153.0	0.31		
	500	3263	8080	4250	1.90	134.4	59.0	154.3	0.31	$A_p=2.2\%$	
	1000	3275	8170	4270	1.91	139.0	59.7	156.7	0.31	$A_s=2.6\%$	
Dunite, Webster, N.Carolina, USA (78 ol/Fo₉₀, 19(?) serp) $\bar{m}=21.0$	0	3254	7000	4010	1.75	89.7	52.3	131.4	0.26	$\bar{3}$oD(6S) dry	Bi6C, Bi61, Si64a
	100	3257	7537	4279	1.76	105.6	59.6	150.5	0.26		
	200	3260	7585	4297	1.77	107.3	60.2	152.2	0.26		
	400	3266	7640	4325	1.77	109.2	61.1	154.5	0.26		
	600	3272	7676	4352	1.76	110.2	62.0	156.6	0.26	$A_p=1.8\%$	
	1000	3284	7756	4387	1.77	113.3	63.2	159.9	0.27	$A_s=1.4\%$	
Dunite, Mt. Dun, New Zealand (97 ol/Fa₉, 3 serp) $\bar{m}=21.1$	0	3264	7500	4170	1.80	107.9	56.8	144.9	0.28	$\bar{3}$oD(6S) dry	B 60, Si64a, B 61
	50	3265	7689	4339	1.77	111.1	61.5	155.7	0.27		
	100	3267	7748	4369	1.77	113.0	62.4	158.0	0.27		
	200	3270	7795	4407	1.77	114.0	63.5	160.7	0.27		
	400	3276	7851	4445	1.77	115.6	64.7	163.6	0.26	$A_p=10\%$	
	600	3281	7906	4472	1.77	117.6	65.6	166.0	0.27	$A_s=4.6\%$	
	1000	3292	7977	4527	1.76	119.5	67.5	170.4	0.26		

(continued)

Table 11 (continued)

Material	P MPa	ϱ kg m⁻³	v_p m s⁻¹	v_s m s⁻¹	v_p/v_s	K GPa	G GPa	E GPa	σ	Notes	Ref.
Pyroxenite, Canyon Mountain, Oregon, USA (84 px, 15 ol, 1 serp)	0	3267								$\bar{3}oD$	Ch78
	20	3268	7740	4160	1.86	120.3	56.5	146.7	0.30	satw	
	40	3268	7769	4180	1.86	121.1	57.1	148.0	0.30		
	60	3269	7799	4199	1.86	122.0	57.6	149.4	0.30		
	100	3270	7828	4229	1.85	122.4	58.5	151.3	0.29		
	200	3272	7886	4258	1.85	124.4	59.3	153.6	0.29		
	400	3278	7951	4295	1.85	126.6	60.5	156.5	0.29		
	600	3283	7987	4303	1.86	128.4	60.8	157.5	0.30		
	1000	3293	8059	4319	1.87	132.0	61.4	159.5	0.30		
Bronzitite, Stillwater, Montana, USA (94 br, 4 hbl, 2 ol) $\bar{m}=21.2$	0	3283	7420	4480	1.66	92.9	65.9	159.9	0.21	$\bar{3}oD(6S)$	Bi60, Bi61, Si64a
	100	3286	7617	4558	1.67	99.6	68.3	166.8	0.22	dry	
	200	3290	7645	4577	1.67	100.4	68.9	168.2	0.22		
	400	3296	7710	4614	1.67	102.4	70.2	171.4	0.22	$A_p = 1.3\%$	
	600	3303	7735	4621	1.67	103.6	70.5	172.4	0.22	$A_s = 2.4\%$	
	1000	3315	7805	4645	1.68	106.6	71.5	175.3	0.23		
Peridotite, Kailua, Hawaii (58.3 ol/Fo₈₂, 39.5 cpx, 2.2 op)	0	3290								$\bar{3}oD(3S)$	Ch66
	10	3291	5400	3200	1.69	51.0	33.7	82.8	0.23	dry	
	50	3293	6598	3579	1.84	87.1	42.2	108.9	0.29		
	100	3294	7207	3918	1.84	103.7	50.6	130.5	0.29		
	200	3297	7624	4157	1.83	115.7	57.0	146.8	0.29		
	400	3303	7990	4344	1.84	127.7	62.3	161.0	0.29	$A_p = 3.3\%$	
	600	3308	8075	4422	1.83	129.5	64.7	166.3	0.29	$A_s = 7.3\%$	
	800	3313	8131	4450	1.83	131.6	65.6	168.7	0.29		
	1000	3318	8187	4477	1.83	133.7	66.5	171.2	0.29		
Peridotite, Kailua, Hawaii (70 ol/Fo₈₅, 27.2 cpx, 2.8 op)	0	3290								$\bar{3}oD(3S)$	Ch66
	10	3291	5400	3400	1.59	45.2	38.0	89.1	0.17	dry	
	50	3293	6438	3759	1.71	74.5	46.5	115.5	0.24		
	100	3295	7176	4078	1.76	96.6	54.8	138.3	0.26		
	200	3298	7794	4336	1.80	117.6	62.0	158.2	0.28		
	400	3303	8119	4514	1.80	128.0	67.3	171.8	0.28	$A_p = 4.0\%$	
	600	3308	8205	4581	1.79	130.1	69.4	176.9	0.27	$A_s = 3.0\%$	
	800	3314	8270	4629	1.79	132.0	71.0	180.6	0.27		
	1000	3319	8336	4667	1.79	134.2	72.3	183.8	0.27		

(continued)

Table 11 (continued)

Material	P MPa	ϱ kg m^{-3}	v_p m s^{-1}	v_s m s^{-1}	v_p/v_s	K GPa	G GPa	E GPa	σ	Notes	Ref.
Pyroxenite, Papua, New Guinea (66 opx, 28 cpx, 5 plg/An$_{80}$) $\bar{m}=21.1$	5	3340	7850	4290	1.83	123.9	61.5	158.2	0.29	3̄oD(3S)	Kr76
	50	3341	7970	4320	1.85	129.1	62.4	161.1	0.29	dry	
	100	3343	8020	4340	1.85	131.1	63.0	162.8	0.29		
	200	3345	8070	4360	1.85	133.1	63.6	164.6	0.29		
	300	3348	8100	4370	1.85	134.4	63.9	165.6	0.30		
	500	3353	8140	4400	1.85	135.6	64.9	167.9	0.29	$A_p=0.5\%$	
	1000	3365	8240	4410	1.87	141.2	65.4	170.1	0.30	$A_s=2.7\%$	
Ilmenite Norite, Tellnes, Norway (53 plg/An$_{40\ldots50}$, 10 opx, 29 ilm, 3 bi)	50	3440	6809	3673	1.85	97.6	46.4	120.2	0.30	3̄oD(1S)	Ke81a
	100	3445	6882	3715	1.85	99.8	47.5	123.1	0.29	dry	
	200	3453	6931	3734	1.86	101.7	48.1	124.7	0.30		
	400	3464	6980	3744	1.86	104.0	48.6	126.1	0.30	$A_p=3.45\%$	
	600	3475	7009	3750	1.87	105.6	48.9	127.0	0.30	$A_s=1.53\%$	
Hortonolite Dunite, Mooihoek, Transvaal, S.Africa (90 ol/Fa$_{56}$, 9 bowlingite) $\bar{m}=24.3$	10	3752	6700	3680	1.82	100.7	50.8	130.5	0.28	3̄oD(6S)	Ma70, Bi60, Bi61
	50	3754	7129	3759	1.90	120.0	53.1	138.7	0.31		
	100	3755	7158	3769	1.90	121.3	53.3	139.6	0.31		
	200	3758	7206	3798	1.90	122.9	54.2	141.8	0.31		
	400	3764	7262	3826	1.90	125.1	55.1	144.1	0.31		
	600	3770	7288	3854	1.89	125.6	56.0	146.3	0.31	$A_p=4.8\%$	
	1000	3782	7340	3890	1.89	127.5	57.2	149.3	0.31	$A_s=5.9\%$	
Hortonolite Dunite, Monroe, New York, USA	10	3934	7200	3680	1.96	132.9	53.3	141.0	0.32	3̄oD(3S)	Ma70
	50	3935	7239	3920	1.85	125.6	60.5	156.3	0.29		
	100	3937	7268	3949	1.84	126.1	61.4	158.5	0.29		
	200	3940	7296	3968	1.84	127.0	62.0	160.1	0.29		
	400	3946	7343	3986	1.84	129.2	62.7	161.9	0.29		
	600	3952	7379	4014	1.84	130.3	63.7	164.3	0.29	$A_p=2.9\%$	
	1000	3964	7441	4040	1.84	133.2	64.7	167.0	0.29	$A_s=7.9\%$	

3.1.3.1.2 Volcanic rocks — Ergußgesteine

Table 12. Density, elastic wave velocities and elastic constants as function of pressure for isotropic and quasi-isotropic *volcanic* rocks at room temperature; densities and velocities are corrected for compression; abbreviations are explained in section 3.1.1.1.

Material	P MPa	ϱ kg m^{-3}	v_p m s^{-1}	v_s m s^{-1}	v_p/v_s	K GPa	G GPa	E GPa	σ	Notes	Ref.
Trachyte, origin unknown (60 kfsp, 15···20 plg/An$_{60}$, 10···15 px, analc, ol)	0	2712								1D(1S) dry	Hu57
	20	2713	5409	3050	1.77	45.7	25.2	63.9	0.27		
	50	2715	5478	3079	1.78	47.2	25.7	65.3	0.27		
	100	2718	5546	3088	1.80	49.0	25.9	66.1	0.28		
	200	2723	5662	3096	1.83	52.5	26.1	67.2	0.29		
	300	2728	5719	3104	1.84	54.2	26.3	67.9	0.29		
	400	2733	5745	3102	1.85	55.1	26.3	68.1	0.29		
	500	2738	5762	3100	1.86	55.8	26.3	68.2	0.30		
Tholeiitic Basalt, Toftavatn, Faeroe (76 plg+cpx, 18 zeo+op)	50	2757	4986	2809	1.78	39.5	21.8	55.1	0.27	1D(1S) dry	Ke79a
	100	2763	5066	2889	1.75	40.2	23.1	58.1	0.26		
	200	2779	5198	2980	1.74	42.2	24.7	62.0	0.26		
	400	2805	5327	3020	1.76	45.5	25.6	64.6	0.26	$A_p < 2\%$	
	600	2834	5415	3043	1.78	48.1	26.2	66.6	0.27		
Basalt, East Pacific Rise (45 cpx, 40 plg, ol, op) $\phi \approx 3\%$	0	2823								$\bar{3}$oD(3S) satw	Ch72
	40	2825	5859	3119	1.88	60.3	27.5	71.6	0.30		
	100	2828	5922	3212	1.84	60.3	29.2	75.4	0.29		
	200	2832	5999	3290	1.82	61.1	30.7	78.8	0.29		
	400	2842	6082	3363	1.81	62.3	32.1	82.2	0.28		
	600	2851	6143	3387	1.81	64.0	32.7	83.8	0.28		
	800	2859	6166	3395	1.82	64.7	33.0	84.5	0.28	$A_p = 2.5\%$	
	1000	2868	6178	3401	1.82	65.2	33.2	85.1	0.28	$A_s = 3.9\%$	
Greenstone, Mid Atlantic Ridge $\phi \approx 4.3\%$	0	2838								$\bar{3}$oD(3S) wet	Ch70
	20	2839	5986	3200	1.87	63.0	29.1	75.6	0.30		
	60	2841	6038	3221	1.88	64.3	29.5	76.7	0.30		
	100	2842	6077	3239	1.88	65.2	29.8	77.6	0.30		
	200	2847	6152	3274	1.88	67.1	30.5	79.5	0.30		
	400	2855	6258	3320	1.89	69.9	31.5	82.1	0.30		
	600	2863	6321	3343	1.89	71.7	32.0	83.6	0.31		
	800	2871	6368	3358	1.90	73.3	32.4	84.7	0.31	$A_p = 2.1\%$	
	1000	2879	6401	3365	1.90	74.5	32.6	85.3	0.31	$A_s = 3.8\%$	

(continued)

Table 12 (continued)

Material	P MPa	ϱ kg m^{-3}	v_p m s^{-1}	v_s m s^{-1}	v_p/v_s	K GPa	G GPa	E GPa	σ	Notes	Ref.
Tholeiitic Basalt, Vagar, Faeroe	50	2855	6040	3430	1.76	59.4	33.6	84.8	0.26	1D(1S)	Ke79a
(88 plg + px + zeo, 12 ol)	100	2856	6166	3480	1.77	62.5	34.6	87.6	0.27	dry	Ke81a
$\bar{m} = 22.76$	200	2864	6280	3516	1.79	65.7	35.4	90.1	0.27		
	400	2881	6332	3530	1.79	67.6	35.9	91.5	0.28		
	600	2898	6354	3540	1.80	68.6	36.3	92.6	0.28	$A_p < 2\%$	
Diabase (altered), Paskenta,	0	2857	5949	3110	1.91	64.3	27.6	72.5	0.31	$\bar{3}$oD	Cɔ78
California, USA	20	2858	6029	3119	1.93	66.8	27.8	73.3	0.32	satw	
(40 ab, 35 cpx, 20 chl, ep, ca)	40	2859	6058	3129	1.94	67.6	28.0	73.8	0.32		
	60	2860	6147	3158	1.95	70.1	28.5	75.4	0.32		
	100	2861	6234	3207	1.94	72.1	29.5	77.8	0.32		
	200	2865	6348	3284	1.93	74.5	31.0	81.6	0.32		
	400	2873	6402	3321	1.93	75.7	31.8	83.6	0.32		
	600	2881	6461	3355	1.93	77.4	32.6	85.7	0.32		
	1000	2896									
Basalt (altered), Mid Atlantic Ridge	0	2859	5466	3170	1.73	47.2	28.7	71.6	0.25	$\bar{3}$oD(3S)	Ch70
(40 plg/An$_{68}$, 20 op, 10 px)	20	2860	5899	3358	1.76	56.6	32.3	81.3	0.26	wet	
$\phi \approx 3.7\%$	60	2862	6100	3443	1.77	61.3	34.0	86.0	0.27		
	100	2864	6345	3548	1.79	67.3	36.1	91.9	0.27		
	200	2869	6497	3627	1.79	71.0	37.9	96.4	0.27		
	400	2877	6554	3652	1.80	72.6	38.5	98.1	0.28		
	600	2885	6574	3659	1.80	73.4	38.7	98.8	0.28		
	800	2893	6576	3659	1.80	73.7	38.8	99.1	0.28	$A_p = 1\%$	
	1000	2901								$A_s = 0.7\%$	
Tholeiitic Basalt, Vidoy or Bordoy,	50	2863	6192	3517	1.76	62.6	35.4	89.4	0.26	1D(1S)	Ke79a
Faeroe	100	2864	6334	3576	1.77	66.1	36.6	92.7	0.27	dry	
(40 plg, 35 cpx, 13 ol, 8 zeo, 4 ilm)	200	2873	6452	3625	1.78	69.3	37.8	95.8	0.27		
	400	2888	6517	3645	1.79	71.5	38.4	97.6	0.27		
	600	2902	6551	3658	1.79	72.8	38.8	98.9	0.27	$A_p = 1 \cdots 2\%$	

(continued)

Table 12 (continued)

Material	P MPa	ϱ kg m⁻³	v_p m s⁻¹	v_s m s⁻¹	v_p/v_s	K GPa	G GPa	E GPa	σ	Notes	Ref.
Basalt, East Pacific Rise (45 cpx, 40 plg, ol, op)	0	2871	6289	3279	1.92	72.4	30.9	81.1	0.31	$\bar{3}$oD(3S)	Ch72
	40	2873	6419	3460	1.86	72.6	34.4	89.2	0.30	satw	
	100	2875	6541	3590	1.82	73.7	37.1	95.3	0.29		
	200	2879	6753	3667	1.84	79.9	38.8	100.2	0.29		
	400	2886	6700	3680	1.82	77.7	39.2	100.7	0.28		
	600	2894	6714	3692	1.82	78.0	39.5	101.5	0.28		
	800	2901	6715	3698	1.82	78.1	39.8	102.0	0.28	$A_p = 1\%$	
	1000	2909								$A_s = 1.3\%$	
Basalt, East Pacific Rise (45 cpx, 40 plg, ol, op)	0	2877	6039	3119	1.94	67.6	28.0	73.8	0.32	$\bar{3}$oD(3S)	Ch72
	40	2879	6244	3338	1.87	69.5	32.1	83.5	0.30	satw	
	100	2881	6427	3499	1.84	72.1	35.3	91.1	0.29		
	200	2885	6572	3610	1.82	74.7	37.7	96.8	0.28		
	400	2893	6633	3640	1.82	76.4	38.4	98.7	0.29		
	600	2901	6659	3650	1.82	77.3	38.7	99.6	0.29		
	800	2908	6664	3656	1.82	77.5	39.0	100.1	0.29	$A_p = 0.9\%$	
	1000	2916								$A_s = 1.5\%$	
Dolerite, Mid Atlantic Ridge (61 plg/An$_{36\ldots55}$, 29 px, 3 op) $\phi \approx 2.4\%$	0	2878	6122	3548	1.73	59.6	36.2	90.4	0.25	$\bar{3}$oD(3S)	Ch70
	20	2879	6200	3578	1.73	61.6	36.9	92.2	0.25	wet	
	60	2881	6237	3605	1.73	62.2	37.5	93.6	0.25		
	100	2883	6314	3658	1.73	63.6	38.6	96.4	0.25		
	200	2887	6413	3725	1.72	65.5	40.2	100.1	0.25		
	400	2896	6478	3760	1.72	67.1	41.1	102.4	0.25		
	600	2905	6512	3781	1.72	68.0	41.7	103.8	0.25		
	800	2914	6525	3793	1.72	68.4	42.0	104.7	0.25	$A_p = 0.5\%$	
	1000	2922								$A_s = 0.9\%$	
Diabase, Příbram, CSSR (49.7 am, 29.5 plg/An$_{60}$, 17.4 au, 3.4 op)	0	2879	6466	3489	1.85	73.6	35.0	90.7	0.30	1D(1S)	Pr62
	50	2881	6652	3546	1.88	79.2	36.2	94.3	0.30		
	100	2883	6716	3578	1.88	80.8	36.9	96.1	0.30		
	200	2886	6767	3597	1.88	82.4	37.3	97.3	0.30		
	300	2890	6799	3606	1.89	83.5	37.6	98.0	0.30		
	400	2893	6826	3608	1.89	84.6	37.7	98.4	0.31		

(continued)

Table 12 (continued)

Material	P MPa	ϱ kg m^{-3}	v_p m s^{-1}	v_s m s^{-1}	v_p/v_s	K GPa	G GPa	E GPa	σ	Notes	Ref.
Diabase, Příbram, CSSR (45.4 am, 34.2 plg/An$_{60}$, 16.6 au, 3.07 op)	0	2903	6509	3605	1.81	72.7	37.7	96.5	0.28	1D(1S)	Pr62
	50	2905	6686	3634	1.84	78.7	38.4	99.0	0.29		
	100	2907	6802	3653	1.86	82.8	38.8	100.7	0.30		
	200	2910	6876	3677	1.87	85.1	39.3	102.3	0.30		
	300	2914	6897	3690	1.87	85.7	39.7	103.1	0.30		
	400	2917	6909	3697	1.87	86.1	39.9	103.6	0.30		
Tholeiitic Basalt, Suduroy, Faeroe (39 px, 11 phy + gl, 13 op) $\bar m = 21.27$	50	2938	5788	3377	1.71	53.8	33.5	83.2	0.24	1D(1S) dry	Ke79a, Ke81a
	100	2940	5837	3396	1.72	55.0	33.9	84.4	0.24		
	200	2947	5914	3425	1.73	57.0	34.6	86.3	0.25		
	400	2965	6017	3454	1.74	60.2	35.4	88.7	0.25	$A_p = 1\cdots2\%$	
	600	2981	6107	3475	1.76	63.2	36.0	90.8	0.26		
Tholeiitic Basalt, Vagar, Faeroe (37 plg, 39 px, 11 phy, 8 op, 5 ol) $\bar m = 22.22$	50	2939	5705	3283	1.74	53.4	31.7	79.3	0.25	1D(1S) dry	Ke79a
	100	2943	5802	3332	1.74	55.5	32.7	81.9	0.25		
	200	2956	5935	3392	1.75	58.8	34.0	85.5	0.26		
	400	2981	6074	3439	1.77	63.0	35.3	89.1	0.26	$A_p < 2\%$	
	600	3006	6192	3475	1.78	66.9	36.3	92.2	0.27		
Basalt, East Pacific Rise (45 cpx, 40 plg, ol, op)	0	2945								$\bar 3$oD(3S) satw	Ch72
	40	2947	6329	3319	1.91	74.7	32.5	85.1	0.31		
	100	2949	6466	3490	1.85	75.4	35.9	93.0	0.29		
	200	2953	6590	3610	1.83	76.9	38.5	98.9	0.29		
	400	2960	6696	3693	1.81	78.9	40.4	103.5	0.28		
	600	2968	6738	3708	1.82	80.3	40.8	104.7	0.28		
	800	2975	6754	3715	1.82	81.0	41.1	105.4	0.28	$A_p = 1.0\%$	
	1000	2982	6758	3720	1.82	81.2	41.3	105.9	0.28	$A_s = 1.0\%$	
Basalt, East Pacific Rise (45 cpx, 40 plg, ol, op)	0	2953								$\bar 3$oD(3S) satw	C172
	40	2955	6369	3409	1.87	74.1	34.3	89.2	0.30		
	100	2957	6457	3488	1.85	75.3	36.0	93.1	0.29		
	200	2961	6551	3571	1.84	76.7	37.8	97.3	0.29		
	400	2968	6652	3655	1.82	78.5	39.6	101.8	0.28		
	600	2976	6692	3679	1.82	79.5	40.3	103.4	0.28		
	800	2983	6715	3686	1.82	80.5	40.5	104.1	0.28	$A_p = 1.1\%$	
	1000	2991	6725	3691	1.82	80.9	40.8	104.7	0.28	$A_s = 1.0\%$	

(continued)

Table 12 (continued)

Material	P MPa	ϱ kg m^{-3}	v_p m s^{-1}	v_s m s^{-1}	v_p/v_s	K GPa	G GPa	E GPa	σ	Notes	Ref.
Tholeiitic Basalt, Sandoy, Faeroe	50	2972	5861	3276	1.79	59.6	31.9	81.2	0.27	1D(1S)	Ke79a,
(64 plg, 21 px, 11 phy, 4 op)	100	2973	5913	3406	1.74	58.0	34.5	86.3	0.25	dry	Ke81a
	200	2983	5980	3444	1.74	59.5	35.4	88.6	0.25		
	400	3004	6070	3463	1.75	62.6	36.0	90.7	0.26		
	600	3022	6148	3480	1.77	65.4	36.6	92.5	0.26	$A_p < 2\%$	
Diabase, Centreville, Virginia, USA	0	2980	6140	3490	1.76	63.9	36.3	91.6	0.26	$\bar{3}$oD(6S)	Bi60,
(45 an, 45 px, 3 kfsp, 1.8 qu, 1.8 bi)	100	2984	6697	3678	1.82	80.0	40.4	103.7	0.28	dry	Si64a
$\bar{m} = 22.0$	200	2988	6754	3717	1.82	81.3	41.3	105.9	0.28		
	400	2995	6808	3744	1.82	82.9	42.0	107.7	0.28		
	600	3002	6843	3761	1.82	84.0	42.5	109.0	0.28	$A_p = 0.3\%$	
	1000	3016	6902	3785	1.82	86.1	43.2	111.0	0.29	$A_s = 1.0\%$	
Tholeiitic Basalt, Suduroy, Faeroe	50	2993	5778	3330	1.74	55.7	33.2	83.1	0.25	1D(1S)	Ke79a
(43 plg, 33 au, 17 phy, 7 op)	100	2994	5828	3361	1.73	56.6	33.8	84.6	0.25	dry	
	200	3006	5905	3383	1.75	58.9	34.4	86.4	0.26		
	400	3027	6024	3413	1.77	62.8	35.3	89.1	0.26		
	600	3049	6141	3449	1.78	66.6	36.3	92.1	0.27	$A_p < 2\%$	
Diabase, Frederick, Maryland, USA	0	3017	6650	3724	1.79	77.6	41.8	106.4	0.27	3oD(3S)	Si65a,
(49 au, 48 plg/An$_{67}$, 1 mi)	40	3019	6720	3764	1.79	79.3	42.8	108.8	0.27		Br65,
$\bar{m} = 22.0$, $\phi = 0.1\%$	70	3020	6756	3774	1.79	80.5	43.0	109.5	0.27		Bi61
	100	3021	6783	3781	1.79	81.4	43.2	110.1	0.28		
	200	3024	6828	3792	1.80	83.0	43.5	111.1	0.28		
	300	3028	6887	3807	1.81	85.1	43.9	112.3	0.28		
	500	3035	6922	3823	1.81	86.3	44.4	113.6	0.28	$A_p = 0.7\%$	
	1000	3052	7018	3843	1.83	90.2	45.1	115.9	0.29	$A_s = 1.4\%$	

Gebrande

3.1.3.1.3 Sedimentary rocks — Sedimentgesteine

In spite of the great practical significance in the connexion with hydro-carbon exploration only relatively few investigations on sedimentary rocks at elevated pressures have been published. The main reason for this may be that in this field preference is usually given to ultrasonic *in-situ*-measurements in boreholes (logging) rather than laboratory measurements. For the influence of porosity and pore content, see also 3.1.3.1.6.

Trotz der großen praktischen Bedeutung für die Kohlenwasser-stoff-Exploration sind nur relativ wenig Untersuchungen an Sedimentgesteinen bei höheren Drucken publiziert worden. Der Hauptgrund hierfür dürfte sein, daß man auf diesem Gebiet *in-situ*-Messungen mit Ultraschall-Verfahren in Bohrlöchern (logging) gewöhnlich gegenüber Labormessungen den Vorzug gibt. Bezüglich des Einflusses von Porosität und Porenfüllung siehe auch 3.1.3.1.6.

Table 13. Density, elastic wave velocities and elastic constants as function of pressure for isotropic and quasi-isotropic *sedimentary* rocks at room temperature; densities and velocities are corrected for compression; abbreviations are explained in section 3.1.1.

Material	P MPa	ϱ kg m^{-3}	v_p m s^{-1}	v_s m s^{-1}	$v_\mathrm{p}/v_\mathrm{s}$	K GPa	G GPa	E GPa	σ	Notes	Ref.
Sand, Ottawa, Illinois, USA pore content (nitrogen) at atmospheric pressure! ϕ decreases from 38.2% to 36.7% with increasing confining pressure	2.9	1638	869	575	1.51	0.52	0.54	1.20	0.11	dry	Do77
	3.6	1640	917	615	1.49	0.55	0.62	1.35	0.09		
	5.3	1645	1010	680	1.49	0.66	0.76	1.65	0.09		
	7.0	1648	1075	728	1.48	0.74	0.87	1.88	0.08		
	10.4	1653	1206	809	1.49	0.96	1.08	2.36	0.09		
	13.9	1657	1300	869	1.50	1.13	1.25	2.74	0.10		
	20.8	1664	1429	958	1.49	1.36	1.53	3.34	0.09		
	27.7	1671	1547	1034	1.50	1.62	1.79	3.92	0.10		
	34.6	1677	1655	1101	1.50	1.88	2.03	4.49	0.10		
Sand, Ottawa, Illinois, USA pore content (brine) at atmospheric pressure! ϕ decreases from 38.2% to 36.7% with increasing confining pressure	2.9	2056	1911							satb	Do77
	3.6	2058	1895								
	5.3	2060	1964	672	2.92	6.71	0.93	2.67	0.43		
	7.0	2062	1984	717	2.77	6.70	1.06	3.02	0.43		
	10.4	2065	2016	798	2.53	6.64	1.32	3.70	0.41		
	13.9	2068	2074	842	2.46	6.94	1.47	4.11	0.40		
	20.8	2072	2105	930	2.26	6.79	1.79	4.94	0.38		
	27.7	2076	2168	988	2.19	7.06	2.03	5.55	0.37		
	34.6	2079	2217	1023	2.17	7.32	2.18	5.94	0.37		
Limestone, Solnhofen, Germany (99 ca)	50	2607	5780	3124	1.85	53.2	25.4	65.8	0.29	$\bar{3}$oD(1S)	Ke81a
	100	2610	5807	3127	1.86	54.0	25.5	66.1	0.30	dry	
	200	2617	5834	3133	1.86	54.8	25.7	66.7	0.30		
	400	2632	5867	3163	1.86	55.5	26.3	68.2	0.30	$A_\mathrm{p}=2.6\%$	
	600	2654	5887	3138	1.88	57.1	26.1	68.0	0.30	$A_\mathrm{s}=1.7\%$	

(continued)

Table 13 (continued)

Material	P MPa	ϱ kg m^{-3}	v_p m s^{-1}	v_s m s^{-1}	v_p/v_s	K GPa	G GPa	E GPa	σ	Notes	Ref.
Limestone, Solnhofen, Germany	0	2663	5590	2990	1.87	51.5	23.8	61.9	0.30	1D(1S)	Nu69,
(99 ca)	10	2664	5620	3010	1.87	51.9	24.1	62.7	0.30	dry	Br65b
$\phi = 4.7\%$	20	2664	5629	3030	1.86	51.8	24.5	63.4	0.30		
	40	2665	5639	3049	1.85	51.7	24.8	64.1	0.29		
	70	2667	5657	3079	1.84	51.7	25.3	65.2	0.29		
	100	2668	5676	3088	1.84	52.0	25.4	65.6	0.29		
	200	2673	5713	3096	1.85	53.1	25.6	66.2	0.29		
	300	2678	5739	3114	1.84	53.6	26.0	67.1	0.29		
Limestone, Solnhofen, Germany	0	2710	5620	2980	1.89	53.5	24.1	62.8	0.30	1D(1S)	Nu69,
(99 ca)	10	2711	5640	3010	1.87	53.5	24.6	63.9	0.30	satw	Br65b
$\phi = 4.7\%$	20	2711	5669	3040	1.87	53.7	25.0	65.0	0.30		
	40	2712	5699	3059	1.86	54.2	25.4	65.9	0.30		
	70	2714	5658	3079	1.84	52.6	25.7	66.3	0.29		
	100	2715	5746	3088	1.86	55.1	25.9	67.2	0.30		
	200	2720	5773	3096	1.87	55.9	26.1	67.7	0.30		
	300	2725	5789	3104	1.87	56.3	26.3	68.2	0.30		
Limestone, Oak Hall,	0	2712	6299	3366	1.87	66.6	30.7	79.9	0.30	$\bar{3}$oD(3S)	Si65a,
Pennsylvania, USA	40	2714	6445	3378	1.91	71.4	31.0	81.2	0.31	dry	Br65
(99 ca)	70	2715	6480	3381	1.92	72.6	31.0	81.5	0.31		
$\phi = 0\%$	100	2716	6496	3384	1.92	73.1	31.1	81.7	0.31		
$\bar{m} = 20.02$	200	2720	6529	3388	1.93	74.3	31.2	82.2	0.32		
	300	2723	6562	3389	1.94	75.6	31.3	82.4	0.32	$A_p = 1\%$	
	500	2730	6596	3389	1.95	77.0	31.4	82.8	0.32	$A_s = 0.2\%$	
	1000	2748	6599	3386	1.95	77.7	31.5	83.3	0.32		
Magnesite, origin unknown	0	2765	6730	3910	1.72	68.9	42.3	105.3	0.25	1D(1S)	Si64,
(100% MgCO$_3$?)	50	2767	6878	3959	1.74	73.1	43.4	108.6	0.25	dry	Si64a
$\bar{m} = 16.87$	100	2769	6927	3998	1.73	73.8	44.3	110.7	0.25		
	200	2773	7014	4036	1.74	76.2	45.2	113.2	0.25		
	400	2780	7097	4093	1.73	77.9	46.6	116.5	0.25		
	600	2787	7161	4139	1.73	79.3	47.7	119.3	0.25		
	1000	2801	7289	4202	1.74	82.9	49.5	123.7	0.25		

(continued)

Table 13 (continued)

Material	P MPa	ϱ kg m^{-3}	v_{p} m s^{-1}	v_{s} m s^{-1}	$v_{\mathrm{p}}/v_{\mathrm{s}}$	K GPa	G GPa	E GPa	σ	Notes	Ref.
Magnesite, origin unknown $\bar{m}=16.87$	0	2838	7080	4230	1.67	74.6	50.8	124.2	0.22	1D(1S) dry	Si64, Si64a
	50	2840	7233	4269	1.69	79.6	51.8	127.6	0.23	dry	
	100	2842	7287	4288	1.70	81.2	52.3	129.1	0.24		
	200	2845	7354	4306	1.71	83.5	52.8	130.8	0.24		
	400	2852	7408	4333	1.71	85.1	53.5	132.8	0.24		
	600	2858	7452	4350	1.71	86.6	54.1	134.3	0.24		
	1000	2871	7551	4383	1.72	90.2	55.2	137.4	0.25		
Dolomite, Wetabuck, USA (99 do) $\phi=0.7\%$ $\bar{m}=18.44$	0	2867	5000	3700	1.35	19.3	39.2	70.2	0.11	1D(1S) dry	Nu69, Br65a
	10	2868	6399	3840	1.67	61.1	42.3	103.1	0.22	dry	
	20	2868	6689	3969	1.69	68.1	45.2	111.0	0.23		
	40	2869	6868	4059	1.69	72.3	47.3	116.4	0.23		
	70	2870	6917	4098	1.69	73.1	48.2	118.6	0.23		
	100	2871	6936	4118	1.68	73.2	48.7	119.6	0.23		
	200	2875	6993	4156	1.68	74.4	49.7	121.9	0.23		
	300	2879	7020	4184	1.68	74.7	50.4	123.4	0.23		
Dolomite, Wetabuck, USA (99 do) $\phi=0.7\%$ $\bar{m}=18.44$	0	2874	6400	3700	1.73	65.3	39.3	98.3	0.25	1D(1S) satw	Nu69, Br65a
	10	2874	6730	3830	1.76	74.0	42.2	106.3	0.26	satw	
	20	2875	6809	3930	1.73	74.1	44.4	111.0	0.25		
	40	2876	6889	4069	1.69	73.0	47.6	117.3	0.23		
	70	2877	6958	4149	1.68	73.2	49.5	121.2	0.22		
	100	2878	6987	4188	1.67	73.2	50.5	123.1	0.22		
	200	2882	7044	4256	1.66	73.4	52.2	126.6	0.21		
	300	2886	7080	4284	1.65	74.0	53.0	128.3	0.21		

3.1.3.1.4 Metamorphic rocks — Metamorphe Gesteine

Table 14. Density, elastic wave velocities and elastic constants as function of pressure for isotropic and quasi-isotropic *metamorphic* rocks at room temperature; densities and velocities are corrected for compression; abbreviations are explained in section 3.1.1.1.

Material	P MPa	ϱ kg m⁻³	v_p m s⁻¹	v_s m s⁻¹	v_p/v_s	K GPa	G GPa	E GPa	σ	Notes	Ref.
Serpentinite, Mt. Boardman, California, USA (95 serp, 5 op)	0	2513	4899	2440	2.01	40.4	15.0	40.0	0.34	$\overline{3}$oD	Ch78
	20	2514	4918	2449	2.01	40.7	15.1	40.3	0.34	satw	
	40	2515	4938	2459	2.01	41.1	15.2	40.6	0.34		
	60	2517	4956	2468	2.01	41.4	15.3	41.0	0.34		
	100	2519	5022	2476	2.03	43.0	15.5	41.5	0.34		
	200	2525	5134	2482	2.07	46.0	15.6	42.1	0.35		
	400	2536	5226	2489	2.10	48.5	15.8	42.7	0.35		
	600	2547	5382	2492	2.16	53.1	16.0	43.4	0.36		
	1000	2567									
Serpentinite, Paskenta, California, USA (96 serp, 4 op)	0	2517	4889	2400	2.04	40.9	14.5	38.9	0.34	$\overline{3}$oD	Ch78
	20	2518	4928	2419	2.04	41.5	14.7	39.6	0.34	satw	
	40	2519	4958	2429	2.04	42.1	14.9	39.9	0.34		
	60	2521	4996	2438	2.05	43.0	15.0	40.3	0.34		
	100	2523	5062	2456	2.06	44.5	15.3	41.1	0.35		
	200	2529	5174	2492	2.08	47.0	15.8	42.6	0.35		
	400	2540	5287	2529	2.09	49.5	16.3	44.1	0.35		
	600	2550	5452	2552	2.14	54.1	16.7	45.5	0.36		
	1000	2570									
Serpentinite, Canyon Mountain, Oregon, USA (94 serp, 6 op)	0	2550	4779	2359	2.03	39.3	14.2	38.0	0.34	$\overline{3}$oD	Ch78
	20	2551	4818	2379	2.03	40.0	14.4	38.7	0.34	satw	
	40	2553	4848	2389	2.03	40.6	14.6	39.0	0.34		
	60	2554	4896	2398	2.04	41.7	14.7	39.4	0.34		
	100	2556	4982	2426	2.05	43.5	15.1	40.6	0.35		
	200	2562	5094	2452	2.08	46.2	15.5	41.8	0.35		
	400	2574	5187	2469	2.10	48.5	15.8	42.6	0.35		
	600	2585	5342	2502	2.14	52.6	16.3	44.3	0.36		
	1000	2605									

(continued)

Table 14 (continued)

Material	P MPa	ϱ kg m^{-3}	v_p m s^{-1}	v_s m s^{-1}	v_p/v_s	K GPa	G GPa	E GPa	σ	Notes	Ref.
Serpentinite, Black Mountain, California, USA (90 serp, 4 opx, 3 ol, 3 op)	0	2623								B̄oD	Ch78
	20	2624	5509	2660	2.07	54.9	18.6	50.0	0.35	satw	
	40	2625	5519	2669	2.07	55.0	18.7	50.4	0.35		
	60	2626	5538	2669	2.08	55.6	18.7	50.5	0.35		
	100	2628	5557	2678	2.08	56.0	18.9	50.8	0.35		
	200	2632	5623	2697	2.09	57.7	19.1	51.7	0.35		
	400	2641	5697	2704	2.11	60.0	19.3	52.3	0.36		
	600	2650	5760	2711	2.13	62.0	19.4	52.9	0.36		
	1000	2667	5858	2715	2.16	65.3	20.0	53.6	0.36		
Quartzite 1463, origin unknown	50	2624	5415	3473	1.56	34.7	31.7	72.8	0.15	B̄oD(1S)	Ke81a
	100	2625	5655	3666	1.54	36.9	35.3	80.3	0.14		
	200	2634	5802	3767	1.54	38.8	37.4	84.9	0.14		
	400	2651	5907	3814	1.55	41.1	38.6	88.1	0.14	$A_p = 6.0\%$	
	500	2663	5942	3823	1.55	42.1	38.9	89.3	0.15	$A_s = 1.7\%$	
Quartzite, Rutland, Vermont, USA (91 qu, 7 or, 2 mi) $\phi = 0.3\%$	0	2627	5418	3182	1.70	41.7	26.6	65.6	0.24	B̄oD(3S)	Si65a, Bi65a
	10	2628	5631	3482	1.62	40.8	31.9	75.9	0.19	dry	
	40	2630	5806	3803	1.53	37.9	38.0	85.5	0.12		
	70	2632	5903	3886	1.52	38.7	39.7	88.8	0.12		
	100	2634	5959	3920	1.52	39.6	40.5	90.5	0.12		
	200	2640	6014	3960	1.52	40.3	41.4	92.5	0.12		
	300	2647	6071	3977	1.53	41.7	41.9	94.1	0.12		
	500	2659	6133	3987	1.54	43.7	42.3	95.9	0.13	$A_p = 0.4\%$	
	1000	2689	6236	3994	1.56	47.4	42.9	98.9	0.15	$A_s = 1.2\%$	
Quartzite, Clarendon Springs, Vermont, USA (95.6 qu, 3.2 ca, 1.1 fsp)	0	2630								B̄oD(3S)	Ch65, Ch66a
	10	2631	5499	3600	1.53	34.1	34.1	76.7	0.13	dry	
	20	2632	5699	3699	1.54	37.5	36.0	81.8	0.14		
	40	2633	5848	3799	1.54	39.4	38.0	86.2	0.14		
	60	2634	5937	3898	1.52	39.5	40.0	89.7	0.12		
	80	2636	5996	3917	1.53	40.8	40.4	91.2	0.13		
	100	2637	6045	3937	1.54	41.9	40.9	92.5	0.13		
	200	2643	6110	3973	1.54	43.0	41.7	94.6	0.13		
	400	2655	6170	4007	1.54	44.2	42.6	96.8	0.14		
	600	2667	6211	4021	1.55	45.4	43.1	98.3	0.14		
	800	2679	6232	4025	1.55	46.2	43.4	99.1	0.14	$A_p = 1\%$	
	1000	2690	6253	4039	1.55	46.6	43.9	100.2	0.14	$A_s = 1\%$	

(continued)

Gebrande

Table 14 (continued)

Material	P MPa	ϱ kg m⁻³	v_p m s⁻¹	v_s m s⁻¹	v_p/v_s	K GPa	G GPa	E GPa	σ	Notes	Ref.
Serpentinite, Black Mountain, California, USA (88 serp, 7 opx, 5 op)	0	2631								3̄oD satw	Ch78
	20	2632	5519	2560	2.16	57.2	17.2	47.0	0.36		
	40	2633	5529	2569	2.15	57.3	17.4	47.4	0.36		
	60	2634	5548	2579	2.15	57.7	17.5	47.7	0.36		
	100	2636	5567	2589	2.15	58.1	17.7	48.1	0.36		
	200	2640	5614	2607	2.15	59.3	17.9	48.9	0.36		
	400	2649	5687	2634	2.16	61.2	18.4	50.1	0.36		
	600	2657	5741	2651	2.17	62.7	18.7	51.0	0.36		
	1000	2674	5799	2666	2.18	64.6	19.0	52.0	0.37		
Serpentinite, Stonyford, California, USA (70 serp, 20 opx, 5 ol, 5 op)	0	2632								3̄oD satw	Ch78
	20	2633	5759	2800	2.06	59.8	20.6	55.5	0.35		
	40	2634	5829	2839	2.05	61.2	21.2	57.1	0.34		
	60	2635	5868	2869	2.05	61.8	21.7	58.2	0.34		
	100	2636	5917	2918	2.03	62.4	22.5	60.1	0.34		
	200	2641	5994	3027	1.98	62.6	24.2	64.3	0.33		
	400	2649	6087	3153	1.93	63.0	26.3	69.4	0.32		
	600	2657	6160	3180	1.94	65.0	26.9	70.8	0.32		
	1000	2673	6277	3223	1.95	68.3	27.8	73.4	0.32		
Quartzphyllite, S. Alta, Norway (48 qu, 23 mu, 21 or, 5 plg)	50	2665	5922	3768	1.57	43.0	37.8	87.8	0.16	3̄oD(1S) dry	Ke81a
	100	2665	5967	3789	1.58	43.9	38.3	88.9	0.16		
	200	2671	6013	3807	1.58	45.0	38.7	90.2	0.17	$A_p = 3.4\%$	
	400	2686	6061	3819	1.59	46.4	39.2	91.7	0.17	$A_s = 3.6\%$	
	600	2698	6100	3824	1.60	47.8	39.5	92.8	0.18		
Serpentinite, Stonyford, California, USA (95 serp, 5 op)	0	2665								3̄oD satw	Ch78
	20	2666	6439	3520	1.83	66.5	33.0	85.0	0.29		
	40	2667	6459	3529	1.83	67.0	33.2	85.5	0.29		
	60	2667	6468	3539	1.83	67.1	33.4	85.9	0.29		
	100	2669	6487	3548	1.83	67.5	33.6	86.5	0.29		
	200	2673	6534	3576	1.83	68.5	34.2	87.9	0.29		
	400	2681	6587	3583	1.84	70.4	34.4	88.8	0.29		
	600	2688	6621	3600	1.84	71.4	34.8	89.9	0.29		
	1000	2703	6658	3603	1.85	73.1	35.1	90.7	0.29		

(continued)

Table 14 (continued)

Material	P MPa	ϱ kg m⁻³	v_p m s⁻¹	v_s m s⁻¹	v_p/v_s	K GPa	G GPa	E GPa	σ	Notes	Ref.
Quartzite, Bayrischer Wald, Germany (100 qu) $\bar m = 20.03$	50	2692	5595	3760	1.49	33.5	38.1	82.8	0.09	1D(1S)	Ke81,
	100	2692	5662	3852	1.47	33.0	39.9	85.4	0.07	dry	Ke81a
	200	2699	5726	3923	1.46	33.1	41.5	87.9	0.06		
	400	2715	5805	3970	1.46	34.4	42.8	90.8	0.06	$A_p = 6.1\%$	
	500	2735	5883	3999	1.47	36.3	43.7	93.6	0.07		
Spilite, Black Mountain, California, USA (45 ab, 15 cpx, 40 chl, ep, pump, sph)	0	2704	4849	2779	1.75	35.7	20.9	52.5	0.26	$\bar{3}$oD	Ch78
	20	2706	5058	2859	1.77	39.8	22.1	56.0	0.27	satw	
	40	2707	5227	2918	1.79	43.2	23.1	58.8	0.27		
	60	2708	5426	2978	1.82	47.7	24.0	61.7	0.29		
	100	2711	5662	3085	1.84	52.6	25.9	66.6	0.29		
	200	2716	5844	3191	1.83	56.1	27.8	71.5	0.29		
	400	2726	5967	3237	1.84	59.2	28.7	74.1	0.29		
	600	2735	6153	3290	1.87	64.5	29.8	77.5	0.30		
	1000	2753									
Granulite, Tichborne, Ontario, Canada (63 plg, 21 qu, 7 mi, 4 px, 1 gar)	0	2712	5969	3249	1.84	58.5	28.7	73.9	0.29	$\bar{3}$oD(3S)	Ch75
	40	2714	6057	3308	1.83	60.0	29.7	76.5	0.29	dry	
	100	2717	6159	3387	1.82	61.6	31.2	80.1	0.28		
	200	2721	6299	3502	1.80	63.7	33.5	85.5	0.28		
	400	2730	6388	3565	1.79	65.4	34.8	88.6	0.27		
	600	2738	6435	3593	1.79	66.5	35.5	90.3	0.27	$A_p = 1\%$	
	800	2746	6449	3598	1.79	67.0	35.7	90.9	0.27	$A_s = 2\%$	
	1000	2755									
Spilite, Canyon Mountain, California, USA (40 ab, 40 ca, chl, pump, sph, qu, 15 cpx, 5 am)	0	2713	5739	3150	1.82	53.5	26.9	69.2	0.29	$\bar{3}$oD	Ch78
	20	2714	5769	3169	1.82	54.0	27.3	70.0	0.28	satw	
	40	2715	5808	3189	1.82	54.8	27.6	70.9	0.28		
	60	2716	5876	3208	1.83	56.6	28.0	72.0	0.29		
	100	2718	5943	3236	1.84	58.1	28.5	73.5	0.29		
	200	2723	6036	3262	1.85	60.8	29.1	75.2	0.29		
	400	2732	6109	3279	1.86	63.0	29.5	76.5	0.30		
	600	2741	6246	3312	1.89	67.2	30.2	78.9	0.30		
	1000	2757									

(continued)

Gebrande

Table 14 (continued)

Material	P MPa	ϱ kg m⁻³	v_p m s⁻¹	v_s m s⁻¹	v_p/v_s	K GPa	G GPa	E GPa	σ	Notes	Ref.
Serpentinite, California, USA	0	2714	5800	3120	1.86	56.1	26.4	68.5	0.30	$\overline{3}$oD(6S)	Bi60,
	100	2719	6017	3178	1.89	61.8	27.5	71.8	0.31	dry	Bi61,
	200	2723	6073	3196	1.90	63.3	27.8	72.8	0.31		Si64a
	400	2731	6137	3223	1.90	65.0	28.4	74.3	0.31		
	600	2740	6191	3230	1.92	66.9	28.6	75.1	0.31	$A_p = 1\%$	
	1000	2756	6278	3263	1.92	69.5	29.3	77.2	0.32	$A_s = 1\%$	
Granulite, Tupper Lake, New York, USA	0	2728	5793	3137	1.85	55.8	26.8	69.4	0.29	$\overline{3}$oD(3S)	Ma74
	50	2730	6304	3368	1.87	67.2	31.0	80.5	0.30	dry	
(40.7 kfsp, 39.9 plg, 7.9 qu, 3.4 opx,	100	2732	6365	3407	1.87	68.4	31.7	82.4	0.30		
3.1 cpx, 5.6 am, 2.6 op)	200	2736	6432	3455	1.86	69.6	32.7	84.7	0.30		
$\bar{m} = 21.3$	400	2744	6492	3488	1.86	71.1	33.4	86.6	0.30		
	600	2752	6536	3501	1.87	72.6	33.7	87.6	0.30		
	800	2759	6568	3505	1.87	73.8	33.9	88.2	0.30	$A_p = 2.3\%$	
	1000	2767	6592	3508	1.88	74.8	34.1	88.7	0.30	$A_s = 2.7\%$	
Spilite, Mt. Boardman, California, USA	0	2738	5659	3220	1.76	49.9	28.4	71.6	0.26	$\overline{3}$oD	Ch78
(55 chl, ca, ep, 45 ab)	20	2739	5718	3339	1.71	48.9	30.6	75.9	0.24	satw	
	40	2740	5768	3369	1.71	49.7	31.1	77.2	0.24		
	60	2741	5866	3398	1.73	52.2	31.7	79.0	0.25		
	100	2743	6022	3456	1.74	56.0	32.8	82.4	0.26		
	200	2749	6125	3521	1.74	57.9	34.2	85.7	0.25		
	400	2758	6178	3537	1.75	59.5	34.6	87.0	0.26		
	600	2768	6244	3559	1.75	61.5	35.3	88.9	0.26		
	1000	2786									
Charnockite, Tupper Lake, New York, USA	0	2739	5533	3033	1.82	50.3	25.2	64.8	0.29	$\overline{3}$oD(3S)	Ma74
	50	2741	6204	3409	1.82	63.0	31.9	81.8	0.28	dry	
(44.1 kfsp, 21 plg, 17.6 qu, 12.8 am,	100	2744	6275	3456	1.82	64.3	32.8	84.1	0.28		
3.5 opx)	200	2748	6352	3504	1.81	65.9	33.7	86.5	0.28		
$\bar{m} = 21.3$	400	2756	6435	3545	1.82	67.9	34.6	88.8	0.28		
	600	2764	6484	3563	1.82	69.4	35.1	90.1	0.28		
	800	2772	6517	3574	1.82	70.5	35.4	91.0	0.29	$A_p = 0.6\%$	
	1000	2780	6539	3576	1.83	71.5	35.6	91.5	0.29	$A_s = 0.6\%$	

(continued)

Table 14 (continued)

Material	P MPa	ϱ kg m⁻³	v_p m s⁻¹	v_s m s⁻¹	v_p/v_s	K GPa	G GPa	E GPa	σ	Notes	Ref
Gneiss, Karasjok, Norway (66 plg, 15 qu, 13 px, 3 mi) $\bar m=21.55$	50	2785	6201	3611	1.72	58.7	36.3	90.3	0.24	1D(1S)	Ke81, Ke81a
	100	2786	6344	3685	1.72	61.7	37.8	94.2	0.25	dry	
	200	2794	6443	3735	1.73	64.0	39.0	97.2	0.25		
	400	2804	6527	3763	1.74	66.5	39.7	99.3	0.25		
	600	2816	6579	3774	1.74	68.4	40.1	100.7	0.26		
Anorthosite Granulite, Willsboro, New York, USA (86.4 plg/An₄₁, 8.6 am, 4.4 cpx) $\bar m=20.9$	0	2785	6178	3518	1.76	60.3	34.5	86.9	0.26	$A_p=2.4\%$	Ma74
	50	2787	6618	3590	1.84	74.2	35.9	92.8	0.29	$\bar{3}$oD(3S)	
	100	2789	6749	3625	1.86	78.2	36.6	95.1	0.30	dry	
	200	2792	6832	3662	1.87	80.4	37.4	97.2	0.30		
	400	2799	6896	3702	1.86	82.0	38.4	99.5	0.30		
	600	2806	6943	3724	1.86	83.4	38.9	101.0	0.30		
	800	2813	6969	3739	1.86	84.2	39.3	102.1	0.30	$A_p=1\%$	
	1000	2820	6984	3747	1.86	84.8	39.6	102.8	0.30	$A_s=0.6\%$	
Metagabbro, Canyon Mountain, Oregon, USA (50 prehnite, ca, chl, 30 plg, 18 am, 2 px)	0	2816	6579	3589	1.83	73.5	36.3	93.5	0.29	$\bar{3}$oD	Ch78
	20	2816	6588	3599	1.83	73.6	36.5	94.0	0.29	satw	
	40	2817	6608	3609	1.83	74.1	36.7	94.5	0.29		
	60	2818	6637	3638	1.82	74.4	37.3	95.9	0.29		
	100	2820	6684	3677	1.82	75.3	38.2	97.9	0.28		
	200	2824	6718	3733	1.80	75.2	39.5	100.7	0.28		
	400	2831	6782	3770	1.80	76.8	40.3	103.0	0.28		
	600	2839	6910	3843	1.80	80.0	42.1	107.5	0.28		
	1000	2853									
Gneiss, Karasjok, Norway (40 plg, 33 qu, 24 hbl, 1 gar) $\bar m=21.65$	50	2822	5694	3526	1.62	44.7	35.1	83.4	0.19	1D(1S)	Ke81, Ke81a
	100	2823	6088	3753	1.62	51.6	39.8	94.9	0.19	dry	
	200	2831	6321	3889	1.63	56.0	42.8	102.4	0.20		
	400	2844	6441	3948	1.63	58.9	44.3	106.3	0.20		
	600	2858	6497	3967	1.64	60.7	45.0	108.2	0.20		
Granulite, Saranac Lake, New York, USA (43 kfsp, 36 plg, 11 px, 2 qu, 1 gar)	0	2830	6089	3239	1.88	65.4	29.7	77.4	0.30	$A_p=1.46\%$	Ch75
	40	2832	6397	3328	1.92	74.1	31.4	82.5	0.31	$\bar{3}$oD(3S)	
	100	2834	6536	3377	1.94	78.1	32.4	85.3	0.32	dry	
	200	2838	6644	3410	1.95	81.5	33.1	87.4	0.32		
	400	2845	6690	3425	1.95	83.0	33.5	88.5	0.32		
	600	2852	6719	3436	1.96	84.1	33.8	89.3	0.32	$A_p=1\%$	
	800	2859	6739	3443	1.96	84.8	34.0	89.9	0.32	$A_s=2\%$	
	1000	2865									

(continued)

Table 14 (continued)

Material	P MPa	ϱ kg m^{-3}	v_p m s^{-1}	v_s m s^{-1}	$v_\mathrm{p}/v_\mathrm{s}$	K GPa	G GPa	E GPa	σ	Notes	Ref.
Serpentinized Peridotite, Mt. Boardman, California, USA (65 ol, 23 px, 10 serp, 2 op)	0	2836								$\bar{3}$oD satw	Ch78
	20	2837	6069	3280	1.85	63.8	30.5	79.0	0.29		
	40	2838	6089	3289	1.85	64.3	30.7	79.5	0.29		
	60	2839	6108	3299	1.85	64.7	30.9	80.0	0.29		
	100	2840	6137	3308	1.86	65.5	31.1	80.5	0.30		
	200	2845	6184	3317	1.86	67.1	31.3	81.2	0.30		
	400	2853	6248	3333	1.87	69.1	31.7	82.5	0.30		
	600	2861	6291	3340	1.88	70.7	31.9	83.2	0.30		
	1000	2877	6349	3344	1.90	73.1	32.2	84.2	0.31		
Serpentinized Peridotite, Burro Mountain, California, USA (52.7 serp, 37.2 ol/Fo$_{94}$, 9.1 au/En$_{92}$)	0	2840								$\bar{3}$oD(3S) dry	Ch66
	10	2840	6000	3000	2.00	68.2	25.6	68.2	0.33		
	50	2842	6089	3039	2.00	70.4	26.3	70.0	0.33		
	100	2844	6137	3079	1.99	71.2	27.0	71.8	0.33		
	200	2848	6194	3087	2.01	73.1	27.1	72.5	0.34		
	400	2856	6268	3104	2.02	75.5	27.5	73.6	0.34	$A_\mathrm{p} = 3.3\%$	
	600	2863	6353	3122	2.04	78.4	27.9	74.8	0.34	$A_\mathrm{s} = 0.6\%$	
	800	2870	6407	3139	2.04	80.1	28.3	75.9	0.34		
	1000	2877	6472	3156	2.05	82.3	28.7	77.0	0.34		
Granulite, Saranac Lake, New York, USA (48 kfsp, 30 plg, 9 px, 4 hbl, 3 qu)	0	2845								$\bar{3}$oD(3S) dry	Ch75
	40	2847	6159	3309	1.86	66.4	31.2	80.9	0.30		
	100	2849	6397	3368	1.90	73.5	32.3	84.6	0.31		
	200	2853	6528	3424	1.91	77.0	33.4	87.6	0.31		
	400	2860	6638	3475	1.91	80.0	34.5	90.6	0.31		
	600	2867	6686	3492	1.92	81.5	35.0	91.8	0.31		
	800	2874	6714	3496	1.92	82.7	35.1	92.3	0.31		
	1000	2881	6731	3500	1.92	83.5	35.3	92.8	0.32		
Metagabbro, Canyon Mountain, Oregon, USA (50 ab, 40 am, 5 sp, 4 prehnite, 1 chl)	0	2871								$\bar{3}$oD satw	Ch78
	20	2872	6609	3640	1.82	74.7	38.0	97.6	0.28		
	40	2873	6629	3639	1.82	75.5	38.0	97.7	0.28		
	60	2873	6638	3649	1.82	75.6	38.3	98.2	0.28		
	100	2875	6657	3658	1.82	76.1	38.5	98.8	0.28		
	200	2879	6694	3677	1.82	77.1	38.9	99.9	0.28		
	400	2886	6738	3694	1.82	78.5	39.4	101.2	0.29		
	600	2893	6763	3710	1.82	79.2	39.8	102.3	0.29		
	1000	2908	6811	3744	1.82	80.5	40.8	104.6	0.28		

(continued)

Table 14 (continued)

Material	P MPa	ϱ kg m⁻³	v_p m s⁻¹	v_s m s⁻¹	v_p/v_s	K GPa	G GPa	E GPa	σ	Notes	Ref.
Metagabbro, Trinity Complex, California, USA (40 plg, 35 am, 23 ep)	0	2907								3oD	Ch78
	20	2908	6829	3540	1.93	87.0	36.4	95.9	0.32	satw	
	40	2908	6839	3559	1.92	86.9	36.8	96.9	0.31		
	60	2909	6858	3569	1.92	87.4	37.1	97.4	0.31		
	100	2910	6877	3599	1.91	87.4	37.7	98.9	0.31		
	200	2914	6925	3647	1.90	88.0	38.8	101.4	0.31		
	400	2920	6999	3704	1.89	89.6	40.1	104.6	0.31		
	600	2927	7054	3742	1.89	91.0	41.0	106.9	0.30		
	1000	2939	7144	3796	1.88	93.5	42.4	110.4	0.30		
Granulite, Adirondack Mountains, New York, USA (38 px, 32 plg, 30 kfsp)	0	2928								3oD(3S)	Ch75
	40	2929	6749	3679	1.83	80.6	39.7	102.2	0.29	dry	
	100	2932	6817	3728	1.83	81.9	40.8	104.9	0.29		
	200	2935	6888	3760	1.83	83.9	41.5	106.9	0.29		
	400	2942	6961	3788	1.84	86.3	42.2	108.9	0.29		
	600	2949	6995	3803	1.84	87.4	42.6	110.1	0.29		
	800	2956	7022	3814	1.84	88.4	43.0	111.0	0.29	$A_p = 1\%$	
	1000	2962	7045	3818	1.85	89.4	43.2	111.6	0.29	$A_s = 1\%$	
Metagabbro, Point Sal, California, USA (40 plg(sau), 35 am, 25 cpx)	0	2936								3oD	Ch78
	20	2937	6939	3640	1.91	89.5	38.9	101.9	0.31	satw	
	40	2937	7009	3659	1.92	91.9	39.3	103.3	0.31		
	60	2938	7058	3679	1.92	93.3	39.8	104.5	0.31		
	100	2939	7127	3709	1.92	95.4	40.4	106.3	0.31		
	200	2942	7225	3777	1.91	97.6	42.0	110.2	0.31		
	400	2948	7290	3885	1.88	97.4	44.5	115.8	0.30		
	600	2954	7335	3952	1.86	97.4	46.1	119.5	0.30		
	1000	2966	7365	3976	1.85	98.4	46.9	121.4	0.29		
Gneiss, Karasjok, Norway (40 plg(sau), 33 qu, 16 gar, 5 px, 3 hbl) $\bar{m} = 21.97$	50	2942	5987	3722	1.61	51.1	40.8	96.6	0.19	1D(1S)	Ke81, Ke81a
	100	2944	6267	3876	1.62	56.7	44.2	105.3	0.19	dry	
	200	2951	6431	3974	1.62	59.9	46.6	111.0	0.19		
	400	2963	6568	4032	1.63	63.6	48.2	115.4	0.20	$A_p = 0.84\%$	
	600	2976	6645	4058	1.64	66.1	49.0	117.9	0.20		

(continued)

Table 14 (continued)

Material	P MPa	ϱ kg m⁻³	v_p m s⁻¹	v_s m s⁻¹	v_p/v_s	K GPa	G GPa	E GPa	σ	Notes	Ref.
Amphibolite, S. Alta, Norway (45 hbl, 43 plg, 5 tit, 3 mi) $\bar{m}=22.63$	50	2969	6287	3462	1.82	69.9	35.6	91.3	0.28	1D(1S)	Ke81, Ke81a
	100	2972	6622	3712	1.78	75.7	41.0	104.1	0.27	dry	
	200	2976	6870	3884	1.77	80.6	44.9	113.6	0.27		
	400	2986	6990	3956	1.77	83.6	46.7	118.2	0.26		
	600	2999	7041	3978	1.77	85.4	47.5	120.1	0.27		
Microcline Granulite, Tupper Lake, New York, USA (40.4 am, 27.7 plg, 24.5 scapolite, 2.4 mu, 3.1 sph) $\bar{m}=22.5$	0	2984	6017	3326	1.81	64.0	33.0	84.5	0.28	$A_p=8\%$ 3̄oD(3S)	Ma74
	50	2986	6787	3688	1.84	83.4	40.6	104.8	0.29	dry	
	100	2988	6859	3738	1.84	84.9	41.8	107.6	0.29		
	200	2991	6954	3792	1.83	87.3	43.0	110.8	0.29		
	400	2998	7074	3850	1.84	90.8	44.4	114.6	0.29		
	600	3005	7137	3878	1.84	92.8	45.2	116.6	0.29		
	800	3011	7188	3898	1.84	94.6	45.8	118.2	0.29	$A_p=1.4\%$	
	1000	3017	7221	3897	1.85	96.2	45.8	118.6	0.30	$A_s=3.4\%$	
Metagabbro, Trinity Complex, California, USA (40 plg(sau), 40 am, 13 chl, 7 sp)	0	3038	6819	3710	1.84	85.6	41.8	107.9	0.29	3̄oD	Ch78
	20	3039	6839	3709	1.84	86.4	41.8	108.0	0.29	satw	
	40	3039	6858	3729	1.84	86.6	42.3	109.1	0.29		
	60	3040	6877	3749	1.84	86.9	42.7	110.2	0.29		
	100	3042	6925	3787	1.83	87.8	43.7	112.4	0.29		
	200	3045	6979	3824	1.83	89.2	44.6	114.7	0.29		
	400	3052	7024	3841	1.83	90.7	45.1	116.1	0.29		
	600	3059	7094	3856	1.84	93.7	45.7	117.9	0.29		
	1000	3072									
Gabbroic Granulite, Santa Clara, New York, USA (54 plg, 13.5 cpx, 13.5 am, 9.0 opx, 4.9 gar, 4.1 op) $\bar{m}=21.9$	0	3041	6714	3653	1.84	83.0	40.6	104.7	0.29	3̄oD(3S)	Ma74
	50	3043	6978	3900	1.79	86.5	46.3	117.8	0.27	dry	
	100	3045	7027	3924	1.79	87.8	46.9	119.4	0.27		
	200	3048	7064	3941	1.79	89.0	47.3	120.6	0.27		
	400	3055	7110	3969	1.79	90.3	48.1	122.6	0.27		
	600	3062	7146	3992	1.79	91.3	48.8	124.3	0.27		
	800	3068	7176	4013	1.79	92.1	49.4	125.7	0.27	$A_p=1.8\%$	
	1000	3075	7201	4032	1.79	92.8	50.0	127.1	0.27	$A_s=2.7\%$	

(continued)

Gebrande

Table 14 (continued)

Material	P MPa	ϱ kg m⁻³	v_p m s⁻¹	v_s m s⁻¹	v_p/v_s	K GPa	G GPa	E GPa	σ	Notes	Ref.
Serpentinized Peridotite, Mikabu, Japan	0	3060	6407	3187	2.01	84.2	31.1	83.0	0.34	3oD(3S)	Ka65,
(54 ol, 40 serp, 3 plg(sau), 1 cpx)	200	3067	6582	3297	2.00	88.4	33.3	88.9	0.33	dry ?	Ka70
	400	3074	6693	3362	1.99	91.4	34.7	92.5	0.33		
	600	3081	6778	3419	1.98	93.5	36.0	95.7	0.33	$A_p=0.6\%$	
	800	3087	6820	3456	1.97	94.4	36.9	97.9	0.33	$A_s=2.7\%$	
$\bar{m}=20.8$	1000	3094	6875	3494	1.97	95.9	37.8	100.1	0.33		
Plagioclase Granulite, Willsboro,	0	3067	5275	3367	1.57	39.0	34.8	80.4	0.16	3oD(3S)	Ma74
New York, USA	50	3070	6595	3767	1.75	75.4	43.6	109.6	0.26	dry	
(56.3 plg, 19.0 cpx, 14.7 opx, 6.3 op,	100	3072	6716	3797	1.77	79.5	44.3	112.1	0.27		
2.9 gar)	200	3076	6847	3837	1.78	83.8	45.3	115.1	0.27		
	400	3083	6955	3880	1.79	87.2	46.4	118.3	0.27		
	600	3090	7026	3908	1.80	89.6	47.2	120.4	0.28		
	800	3097	7076	3930	1.80	91.3	47.8	122.2	0.28	$A_p=2.0\%$	
$\bar{m}=22.3$	1000	3103	7116	3941	1.81	92.9	48.2	123.3	0.28	$A_s=4.0\%$	
Amphibolite, Karasjok, Norway	50	3070	5408	3454	1.57	41.0	36.6	84.6	0.16	1D(1S)	Ke81,
(72 hbl, 24 plg, 3 qu, 1 gar)	100	3072	5799	3625	1.60	49.5	40.4	95.2	0.18	dry	Ke81a
	200	3081	6130	3771	1.63	57.4	43.8	104.8	0.20		
	400	3095	6359	3865	1.65	63.5	46.2	111.6	0.21		
$\bar{m}=22.88$	600	3106	6471	3910	1.66	66.7	47.5	115.1	0.21	$A_p=11\%$	
Amphibolite, origin unknown	50	3071	6794	4048	1.68	74.7	50.3	123.3	0.23	1D(1S)	Ke81a
	100	3073	6883	4091	1.68	77.0	51.4	126.2	0.23	dry	
	200	3083	6959	4123	1.69	79.4	52.4	128.9	0.23		
	400	3094	7037	4147	1.70	82.3	53.2	131.3	0.23		
	600	3107	7078	4161	1.70	83.9	53.8	133.0	0.24	$A_p=4.8\%$	
Amphibolite, Bayrischer Wald,	50	3078	6456	3870	1.67	66.8	46.1	112.4	0.22	1D(1S)	Ke81,
Germany	100	3076	6607	3930	1.68	70.9	47.5	116.5	0.23	dry	Ke81a
(72 hbl, 12 plg, 10 ep)	200	6081	6719	3979	1.69	146.2	96.3	236.8	0.23		
	400	3094	6825	4018	1.70	77.5	50.0	123.4	0.24		
$\bar{m}=20.26$	600	3109	6886	4042	1.70	79.7	50.8	125.7	0.24	$A_p=6.0\%$	
Sillimanite-Gneiss, Karasjok, Norway	50	3078	6658	3947	1.69	72.5	48.0	117.9	0.23	1D(1S)	Ke81,
(27 gar, 25 or, 24 qu, 14 sil, 6 plg)	100	3078	6845	4034	1.70	77.4	50.1	123.6	0.23	dry	Ke81a
$\bar{m}=21.94$	200	3083	6987	4098	1.71	81.5	51.8	128.2	0.24		
	400	3095	7096	4134	1.72	85.3	52.9	131.5	0.24		
(continued)	600	3107	7155	4149	1.73	87.7	53.5	133.4	0.25	$A_p=3.3\%$	

Table 14 (continued)

Material	P MPa	ϱ kg m⁻³	v_p m s⁻¹	v_s m s⁻¹	v_p/v_s	K GPa	G GPa	E GPa	σ	Notes	Ref.
Granulite, Valle d'Ossola, Italy	0	3085	6179	3479	1.78	68.0	37.4	94.8	0.27	$\bar{3}$oD(3S)	Ch75
(60 plg, 26 px, 9 gar, 4 op)	40	3087	6677	3708	1.80	81.1	42.5	108.5	0.28	dry	
	100	3089	7049	3840	1.84	92.9	45.6	117.6	0.29		
	200	3093	7282	3940	1.85	100.2	48.1	124.4	0.29		
	400	3099	7384	3973	1.86	103.9	49.0	127.1	0.30		
	600	3105	7436	3989	1.86	106.0	49.5	128.5	0.30	$A_p = 2.4\%$	
	800	3111	7458	4022	1.85	106.2	50.4	130.6	0.30	$A_s = 2.5\%$	
	1000	3117									
Serpentinized Peridotite, Burro	0	3140	6700	3700	1.81	83.6	43.0	110.1	0.28	$\bar{3}$oD(3S)	Ch66
Mountain, California, USA	10	3140	6999	3879	1.80	90.9	47.3	120.9	0.28	dry	
(62.1 ol/Fo₉₃, 22.4 serp, 14.8 au/En₉₂)	50	3142	7087	3939	1.80	92.9	48.8	124.5	0.28		
	100	3144	7155	3967	1.80	95.1	49.5	126.6	0.28		
	200	3147	7220	3984	1.81	97.6	50.1	128.3	0.28		
	400	3153	7285	4002	1.82	100.2	50.6	129.9	0.28		
	600	3160	7340	4009	1.83	102.7	50.9	131.0	0.29	$A_p = 1.7\%$	
	800	3166	7395	4026	1.84	104.9	51.4	132.6	0.29	$A_s = 4.5\%$	
	1000	3172									
Granulite, Elizabethtown, New York,	0	3244	5397	3327	1.62	46.6	35.9	85.7	0.19	$\bar{3}$oD(3S)	Ma74
USA	50	3246	6860	3836	1.79	89.1	47.8	121.6	0.27	dry	
(47 plg/An₂₀, 23.3 cpx, 12.4 op, 11.1 gar,	100	3248	7001	3903	1.79	93.2	49.5	126.1	0.28		
6.6 opx)	200	3252	7083	3981	1.78	94.4	51.5	130.8	0.27		
$\bar{m} = 22.4$	400	3259	7175	4057	1.77	96.3	53.6	135.7	0.27		
	600	3265	7228	4092	1.77	97.7	54.7	138.2	0.26		
	800	3272	7263	4112	1.77	98.8	55.3	139.9	0.26	$A_p = 1.4\%$	
	1000	3279	7292	4125	1.77	100.0	55.8	141.1	0.27	$A_s = 4.0\%$	
Eclogite, Rödhaugen, Norway	0	3270	6490	3800	1.71	74.8	47.2	117.0	0.24	$\bar{3}$oD(3S)	Ma74
(47 gar, 37 cpx, 10 opx, 4 ol, 1 mu)	50	3272	7515	4151	1.81	109.6	56.4	144.4	0.28	dry	
$\bar{m} = 21.1$	100	3273	7690	4227	1.82	115.6	58.5	150.1	0.28		
	200	3276	7842	4301	1.82	120.7	60.6	155.7	0.29		
	400	3281	8003	4374	1.83	126.4	62.8	161.6	0.29		
	600	3286	8066	4411	1.83	128.5	63.9	164.5	0.29		
	800	3292	8095	4432	1.83	129.5	64.7	166.3	0.29	$A_p = 2.8\%$	
	1000	3297	8142	4444	1.83	131.7	65.1	167.7	0.29	$A_s = 2.0\%$	

(continued)

Gebrande

Table 14 (continued)

Material	P MPa	ϱ kg m⁻³	v_p m s⁻¹	v_s m s⁻¹	v_p/v_s	K GPa	G GPa	E GPa	σ	Notes	Ref.
Eclogite, Valley Ford, California, USA (40 cpx, 32.8 ep, zo, rutile, 9.4 sph, 14.2 chl, 2.2 gar) $\bar{m}=22.0$	0	3364	7508	4095	1.83	114.4	56.4	145.3	0.29	$\bar{3}$oD(3S)	Ma74
	50	3365	7693	4154	1.85	121.7	58.1	150.3	0.29	dry	
	100	3367	7774	4190	1.86	124.7	59.1	153.1	0.30		
	200	3369	7877	4236	1.86	128.4	60.5	156.8	0.30		
	400	3375	7958	4285	1.86	131.1	62.0	160.6	0.30		
	600	3380	8028	4321	1.86	133.7	63.1	163.6	0.30		
	800	3385	8065	4330	1.86	135.6	63.5	164.7	0.30	$A_p=2.7\%$	
	1000	3390	8092	4342	1.86	136.8	63.9	165.9	0.30	$A_s=1.4\%$	
Eclogite, Saualpe, Austria (45 px, 29 gar, 10 zo, 9 sym, 4 hbl, 2 qu) $\bar{m}=22.37$	50	3406	7232	4328	1.67	93.1	63.8	155.8	0.22	1D(1S)	Ke81, Ke81a
	100	3410	7543	4474	1.69	103.0	68.3	167.7	0.23	dry	
	200	3420	7749	4570	1.70	110.1	71.4	176.2	0.23		
	400	3432	7894	4636	1.70	115.5	73.8	182.5	0.24		
	600	3442	7976	4670	1.71	118.9	75.1	186.0	0.24	$A_p=0.83\%$	
Eclogite, Sonoma, California, USA (34.2 gar, 29 cpx, 24.2 chl, 6.2 mu, 1.7 ep) $\bar{m}=21.8$	0	3422	5823	3243	1.80	68.0	36.0	91.8	0.28	$\bar{3}$oD(3S)	Ma74
	50	3424	7337	3971	1.85	112.3	54.0	139.6	0.29	dry	
	100	3425	7629	4102	1.86	122.5	57.6	149.5	0.30		
	200	3428	7846	4216	1.86	129.8	60.9	158.1	0.30		
	400	3433	7995	4293	1.86	135.1	63.3	164.2	0.30		
	600	3438	8062	4324	1.86	137.7	64.3	166.9	0.30		
	800	3443	8109	4346	1.87	139.7	65.0	168.9	0.30	$A_p=1.9\%$	
	1000	3448	8146	4360	1.87	141.4	65.5	170.3	0.30	$A_s=2.5\%$	
Eclogite, Healdsburg, California, USA (78 cpx, 18 gar, 3 chl, 1 sph) $\bar{m}=22.0$	0	3422	7600	4287	1.77	113.8	62.9	159.3	0.27	$\bar{3}$oD(3S)	Ma74
	50	3423	7789	4361	1.79	120.9	65.1	165.6	0.27	dry	
	100	3425	7864	4407	1.78	123.1	66.5	169.1	0.27		
	200	3428	7946	4444	1.79	126.2	67.7	172.3	0.27		
	400	3433	8033	4499	1.79	128.9	69.5	176.7	0.27		
	600	3438	8090	4527	1.79	131.1	70.5	179.3	0.27		
	800	3444	8113	4541	1.79	132.0	71.0	180.7	0.27	$A_p=1.9\%$	
	1000	3449	8168	4551	1.80	134.9	71.4	182.1	0.28	$A_s=2.5\%$	

(continued)

Gebrande

Table 14 (continued)

Material	P MPa	ϱ kg m^{-3}	v_p m s^{-1}	v_s m s^{-1}	$v_\mathrm{p}/v_\mathrm{s}$	K GPa	G GPa	E GPa	σ	Notes	Ref.
Eclogite, Kimberley, S. Africa (62 gar, 28.8 cpx, 8.2 plg, 1.0 op) $\bar m=22.0$	0	3424	7037	4045	1.74	94.9	56.0	140.4	0.25	$\bar{3}$oD(3S)	Ma74
	50	3426	7531	4160	1.81	115.3	59.3	151.8	0.28	dry	
	100	3427	7633	4193	1.82	119.3	60.3	154.7	0.28		
	200	3430	7706	4222	1.83	122.2	61.1	157.2	0.29		
	400	3435	7786	4254	1.83	125.4	62.2	160.0	0.29		
	600	3441	7836	4277	1.83	127.4	62.9	162.1	0.29		
	800	3446	7872	4289	1.84	129.0	63.4	163.4	0.29	$A_\mathrm{p}=2\,\%$	
	1000	3452	7900	4299	1.84	130.4	63.8	164.6	0.29	$A_\mathrm{s}=3.3\,\%$	
Eclogite, Russian River, California, USA (74.8 cpx, 23 gar, 0.8 qu, 0.8 sph) $\bar m=22.1$	0	3442	6650	3842	1.73	84.5	50.8	127.0	0.25	$\bar{3}$oD(3S)	Ma74
	50	3444	7685	4289	1.79	118.9	63.4	161.4	0.27	dry	
	100	3445	7813	4383	1.78	122.1	66.2	168.2	0.27		
	200	3448	7942	4471	1.78	125.6	68.9	174.8	0.27		
	400	3453	8063	4542	1.78	129.5	71.2	180.6	0.27		
	600	3459	8124	4584	1.77	131.4	72.7	184.1	0.27		
	800	3464	8177	4614	1.77	133.3	73.7	186.8	0.27	$A_\mathrm{p}=0.7\,\%$	
	1000	3469	8219	4637	1.77	134.9	74.6	188.9	0.27	$A_\mathrm{s}=1.3\,\%$	
Eclogite, Healdsburg, California, USA (72 px, 24 gar) $\bar m=22.2$	0	3442	7310	4260	1.72	100.6	62.5	155.3	0.24	$\bar{3}$oD(6S)	Bi60,
	100	3445	7688	4429	1.74	113.5	67.6	169.1	0.25	dry	Bi61,
	200	3448	7805	4477	1.74	117.9	69.1	173.5	0.26		Si64a
	400	3454	7881	4525	1.74	120.2	70.7	177.4	0.25		
	600	3460	7926	4542	1.75	122.2	71.4	179.2	0.26	$A_\mathrm{p}=1.0\,\%$	
	1000	3471	7988	4567	1.75	124.9	72.4	182.0	0.26	$A_\mathrm{s}=2.4\,\%$	
Eclogite, Fichtelgebirge, Germany (44 px, 35 gar, 7 qu, 6 hbl, 4 sym) $\bar m=23.03$	50	3467	7140	4333	1.65	90.0	65.1	157.3	0.21	1D(1S)	Ke81,
	100	3466	7424	4456	1.67	99.3	68.8	167.7	0.22	dry	Ke81a
	200	3473	7647	4548	1.68	107.3	71.8	176.2	0.23		
	400	3483	7807	4608	1.69	113.7	74.0	182.3	0.23		
	600	3495	7887	4638	1.70	117.2	75.2	185.8	0.24	$A_\mathrm{p}=0.33\,\%$	
Eclogite, Higashiakaishi, Japan (53 gar, 42 px, 5 hbl) $\bar m=22.2$	0	3510	6987	4040	1.73	95.0	57.3	143.1	0.25	$\bar{3}$oD(3S)	Ka65,
	200	3516	8092	4243	1.91	145.9	63.3	165.9	0.31	dry(?)	Ka70
	400	3521	8299	4356	1.91	153.4	66.8	175.0	0.31		
	600	3525	8391	4427	1.90	156.1	69.1	180.6	0.31		
	800	3530	8438	4482	1.88	156.8	70.9	184.8	0.30	$A_\mathrm{p}=1.9\,\%$	
	1000	3534	8477	4510	1.88	158.1	71.9	187.3	0.30	$A_\mathrm{s}=4.4\,\%$	

(continued)

Table 14 (continued)

Material	P MPa	ϱ kg m^{-3}	v_p m s^{-1}	v_s m s^{-1}	v_p/v_s	K GPa	G GPa	E GPa	σ	Notes	Ref.
Eclogite, Sunmore, Norway	0	3539	7167	4203	1.71	98.4	62.5	154.8	0.24	$\bar{3}$oD(3S)	Ma74
(46 gar, 37.4 cpx, 13 bi, 2.8 chl)	50	3541	7911	4504	1.76	125.8	71.8	181.0	0.26	dry	
$\bar{m}=22.0$	100	3542	7984	4591	1.74	126.2	74.7	187.1	0.25		
	200	3545	8056	4621	1.74	129.1	75.7	190.0	0.26		
	400	3550	8126	4684	1.74	130.6	77.9	194.9	0.25		
	600	3556	8168	4717	1.73	131.7	79.1	197.8	0.25		
	800	3561	8200	4737	1.73	132.9	79.9	199.7	0.25	$A_p=2.4\%$	
	1000	3566	8229	4748	1.73	134.3	80.4	201.1	0.25	$A_s=1.0\%$	
Eclogite, Grytinvaag, Norway	0	3585	7577	4302	1.76	117.4	66.3	167.5	0.26	$\bar{3}$oD(3S)	Ma74
(50.9 gar, 24.1 opx, 23.4 cpx, 1.6 rutile)	50	3586	7917	4448	1.78	130.2	70.9	180.1	0.27	dry	
$\bar{m}=22.4$	100	3588	8052	4501	1.79	135.7	72.7	185.0	0.27		
	200	3590	8162	4553	1.79	139.9	74.4	189.6	0.27		
	400	3595	8278	4602	1.80	144.8	76.1	194.4	0.28		
	600	3600	8348	4641	1.80	147.5	77.5	197.9	0.28		
	800	3605	8392	4665	1.80	149.3	78.5	200.3	0.28	$A_p=1.2\%$	
	1000	3610	8427	4686	1.80	150.7	79.3	202.3	0.28	$A_s=1.4\%$	
Pyroxenite Gneiss, Willsboro,	0	3714	6688	3717	1.80	97.7	51.3	131.0	0.28	$\bar{3}$oD(3S)	Ma74
New York, USA	50	3716	7006	3823	1.83	110.0	54.3	139.9	0.29	dry	
(44.6 opx, 19.5 gar, 19.2 op, 8.8 plg,	100	3717	7095	3855	1.84	113.5	55.2	142.6	0.29		
2.9 cpx, 2.2 ap)	200	3721	7163	3883	1.85	116.1	56.1	145.0	0.29		
$\bar{m}=24.6$	400	3727	7225	3909	1.85	118.6	57.0	147.3	0.29		
	600	3733	7265	3929	1.85	120.2	57.6	149.1	0.29		
	800	3740	7293	3942	1.85	121.4	58.1	150.4	0.29	$A_p=2.2\%$	
	1000	3746	7318	3950	1.85	122.7	58.4	151.3	0.29	$A_s=2.5\%$	

3.1.3.1.5 Monomineralic rocks and miscellaneous materials — Monomineralische Gesteine und Verschiedenes

Table 15. Density, elastic wave velocities and elastic constants as function of pressure for some mineral aggregates and other materials at room temperature; densities and velocities are corrected for compression; abbreviations are explained in section 3.1.1.1.

Material	P MPa	ϱ kg m^{-3}	v_p m s^{-1}	v_s m s^{-1}	v_p/v_s	K GPa	G GPa	E GPa	σ	Notes	Ref.
Fused quartz, delay line quality, Syncor, Colorado, USA $\bar{m}=20.03$	0	2200	5969	3758	1.59	37.0	31.1	72.8	0.17		Si65a
	40	2202	5959	3750	1.59	36.9	31.0	72.6	0.17		
	70	2204	5952	3743	1.59	36.9	30.9	72.4	0.17		
	100	2206	5945	3737	1.59	36.9	30.8	72.3	0.17		
	200	2212	5909	3717	1.59	36.5	30.6	71.7	0.17		
	300	2218	5874	3697	1.59	36.1	30.3	71.1	0.17		
	500	2231	5766	3648	1.58	34.6	29.7	69.3	0.17		
	1000	2265	5531	3528	1.57	31.7	28.2	65.2	0.16		
Aluminium, alloy 2011-T3	0	2828	6331	3095	2.05	77.2	27.1	72.8	0.34		Si65a, Br65
	70	2831	6353	3112	2.04	77.7	27.4	73.6	0.34		
	100	2832	6362	3117	2.04	77.9	27.5	73.9	0.34		
	200	2835	6391	3131	2.04	78.7	27.8	74.6	0.34		
	300	2839	6414	3143	2.04	79.4	28.0	75.3	0.34		
	500	2846	6438	3168	2.03	79.9	28.6	76.6	0.34		
	1000	2864	6485	3222	2.01	80.8	29.7	79.5	0.34		
Monticellite, Crestmore, California, USA (83 monticellite, 17 ca) $\bar{m}=22.7$	0	2973	6990	3800	1.84	88.0	42.9	110.8	0.29	1D(2S) dry	Si64, Si64a
	50	2975	7149	3829	1.87	93.9	43.6	113.3	0.30		
	100	2976	7161	3879	1.85	92.9	44.8	115.7	0.29		
	200	2979	7215	3917	1.84	94.1	45.7	118.0	0.29		
	400	2986	7260	3954	1.84	95.1	46.7	120.4	0.29		
	600	2992	7305	3972	1.84	96.7	47.2	121.8	0.29		
	1000	3004	7414	4006	1.85	100.9	48.2	124.8	0.29		
Idocrasite (Vesuvianite), Crestmore, California, USA (60 idocrase, 36 ca, 4 diopside) $\bar{m}=22.8$	0	3144	5520	2950	1.87	59.3	27.4	71.1	0.30	$\bar{2}$oD(2S) dry	Si64, Si64a
	50	3147	6098	3569	1.71	63.6	40.1	99.4	0.24		
	100	3149	6537	3793	1.72	74.1	45.3	112.9	0.25		
	200	3153	6944	3981	1.74	85.4	50.0	125.4	0.26		
	400	3160	7258	4163	1.74	93.4	54.8	137.4	0.26	$A_p=0.8\%$	
	600	3166	7382	4235	1.74	96.9	56.8	142.5	0.26	$A_s=0.2\%$	
	1000	3179	7512	4309	1.74	100.7	59.0	148.1	0.26		

(continued)

Gebrande

Table 15 (continued)

Material	P MPa	ϱ kg m⁻³	v_p m s⁻¹	v_s m s⁻¹	v_p/v_s	K GPa	G GPa	E GPa	σ	Notes	Ref.
Sillimanite, Williamstown, Australia	0	3187	9400	4930	1.91	178.3	77.5	203.0	0.31	ōoD(1S)	Si64,
(99 sil, Al₂SiO₅)	50	3188	9509	5040	1.89	180.3	81.0	211.3	0.31	dry	Si64a
$\bar{m} = 20.25$	100	3189	9548	5059	1.89	181.9	81.6	213.0	0.31		
	200	3191	9596	5078	1.89	184.1	82.3	214.8	0.31		
	400	3194	9643	5106	1.89	186.0	83.3	217.4	0.31	$A_p = 4.7\%$	
	600	3197	9670	5124	1.89	187.0	84.0	219.1	0.31	$A_s = 1.9\%$	
	1000	3204	9713	5141	1.89	189.4	84.7	221.1	0.31		
Jadeite, Japan	0	3192	7600	4650	1.63	92.3	69.0	165.8	0.20	ōoD(6S)	Ei60,
$\bar{m} = 20.4$	100	3195	8207	4719	1.74	120.4	71.1	178.3	0.25	dry	Si65a
	200	3198	8215	4747	1.73	119.7	72.1	180.1	0.25		
	400	3203	8221	4775	1.72	119.1	73.0	181.9	0.25	$A_p = 1.3\%$	
	600	3208	8226	4782	1.72	119.3	73.4	182.6	0.25	$A_s = 0.6\%$	
	1000	3219	8257	4806	1.72	120.3	74.4	185.0	0.24		
Steel, annealed	0	7808	5876	3215	1.83	162.0	80.7	207.6	0.29		Si65a,
	70	7811	5921	3220	1.84	165.9	81.0	209.0	0.29		Er65
	100	7813	5927	3222	1.84	166.3	81.1	209.3	0.29		
	200	7817	5942	3227	1.84	167.5	81.4	210.2	0.29		
	300	7822	5957	3232	1.84	168.6	81.7	211.0	0.29		
	500	7831	5984	3238	1.85	170.9	82.1	212.3	0.29		
	1000	7854	6007	3256	1.85	172.4	83.3	215.2	0.29		

3.1.3.1.6 Influence of porosity and pore content — Einfluß von Porosität und Porenfüllung

As already mentioned above, the strong increase of seismic velocities up to pressures of about 0.1 to 0.2 GPa must be attributed to the decrease of porosity, especially crack-porosity (Fig. 24). If a fluid or a fluid-gas mixture fills the pore space, two additional influencing factors have to be considered:

1. the degree of saturation of the pore space with pore fluid,
2. the pressure of the pore fluid (P_{pore}).

A typical example for the difference of P-wave-velocities in water-saturated samples with pore pressure $P_{pore}=0$ und $P_{pore}=P_{ext}$ has already been shown in Fig. 23. The difference between dry and water-saturated samples, both at $P_{pore}=0$, is demonstrated in Fig. 26. Water saturation increases the P-wave velocities at small pressures and all the more the greater the crack porosity of the rock is [Nu69]. The difference decreases with increasing pressure and may be neglected above about 0.2 GPa. Fig. 27 shows that – in contrast to P-wave velocities – S-wave velocities are barely influenced by the presence of pore fluids with zero pore pressure. If, however, the pore pressure equals the confining pressure, microcracks and grain boundaries are kept open and P- as well as S-wave velocities remain much smaller with increasing pressure than values measured in dry rocks. The pressure derivatives ($\mathrm{d}v_p/\mathrm{d}P$ and $\mathrm{d}v_s/\mathrm{d}P$) are constant in the case $P_{pore}=P_{ext}$ and are identical with the corresponding derivatives in dry rocks at high pressures.

Wie bereits erwähnt, muß die starke Zunahme der seismischen Geschwindigkeiten bis zu Drucken von etwa 0.1 bis 0.2 GPa auf die Abnahme der Porosität und insbesondere der Rißporosität zurückgeführt werden (Fig. 24). Bei Anwesenheit einer flüssigen oder teils gasförmigen und teils flüssigen Porenfüllung sind als zusätzliche Einflußfaktoren zu berücksichtigen:
1. der Grad der Sättigung des Porenraumes mit Porenflüssigkeit,
2. der Druck der Porenflüssigkeit (P_{pore}).

In Fig. 23 ist bereits ein typisches Beispiel für den Unterschied der P-Wellen-Geschwindigkeiten in wassergesättigten Proben bei $P_{pore}=0$ und $P_{pore}=P_{ext}$ gezeigt worden. Fig. 26 verdeutlicht den Unterschied zwischen trockenen und wassergesättigten Proben bei $P_{pore}=0$. Wassersättigung erhöht die P-Wellen-Geschwindigkeit bei kleinen Drucken, und zwar um so mehr je größer die Riß-Porosität der Gesteine ist [Nu69]. Der Unterschied wird mit zunehmendem Druck kleiner und ist oberhalb von etwa 0.2 GPa zu vernachlässigen. Die Fig. 27 zeigt, daß – im Gegensatz zu den P-Wellen-Geschwindigkeiten – die S-Wellen-Geschwindigkeiten durch die Porenflüssigkeit kaum beeinflußt werden, sofern der Porendruck Null ist. Ist der Porendruck aber gleich dem Umschließungsdruck, so bleiben Mikrorisse und Korngrenzen geöffnet und sowohl P- wie auch S-Wellen-Geschwindigkeiten bleiben bei zunehmendem Druck deutlich unter den Werten, die im trockenen Gestein gemessen werden. Die Druckkoeffizienten ($\mathrm{d}v_p/\mathrm{d}P$ und $\mathrm{d}v_s/\mathrm{d}P$) sind im Fall $P_{pore}=P_{ext}$ konstant und identisch mit den entsprechenden Koeffizienten des trockenen Gesteins bei hohen Drucken.

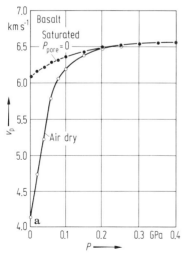
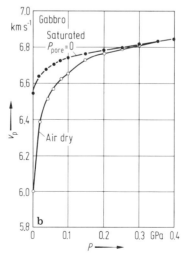

Fig. 26. Compressional wave velocities for air-dried and water-saturated rocks ($P_{pore}=0$) as functions of confining pressure (a) for basalt, (b) for gabbro [Ch75a].

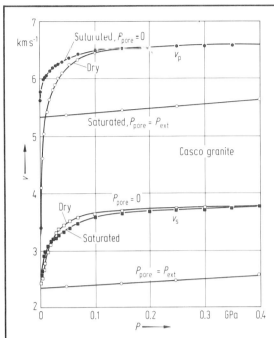

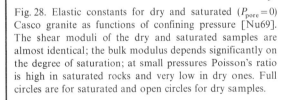

Fig. 27. Velocities of elastic waves in dry and water-saturated Casco granite [Nu69]. The compressional wave velocity depends significantly on the degree of saturation at low pressures but shear velocity is almost independent; both velocities are strongly influenced by pore pressure. Total porosity of Casco granite is only 0.7%, crack porosity is 0.45%.

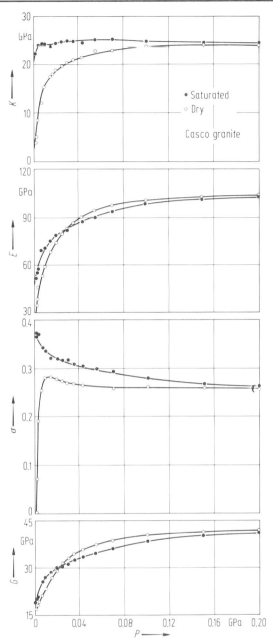

Fig. 28. Elastic constants for dry and saturated ($P_{pore}=0$) Casco granite as functions of confining pressure [Nu69]. The shear moduli of the dry and saturated samples are almost identical; the bulk modulus depends significantly on the degree of saturation; at small pressures Poisson's ratio is high in saturated rocks and very low in dry ones. Full circles are for saturated and open circles for dry samples.

The pressure dependence of elastic constants of Casco granite in the dry and water-saturated state ($P_{pore}=0$) is shown in Fig. 28. The bulk modulus in the water-saturated rock is nearly pressure-independent and is identical to the bulk modulus of the dry rock at high pressures. The shear modulus is only slightly influenced by water saturation. Poisson's ratio of dry rocks with crack-porosity is very small at small pressures (and may even be negative according to [Nu69]), but is abnormally high in saturated rocks.

Die Druckabhängigkeit der Elastizitäts-Konstanten von Casco-Granit im trockenen und wassergesättigten Zustand ($P_{pore}=0$) ist in Fig. 28 dargestellt. Der Kompressionsmodul im wassergesättigten Gestein ist nahezu druckunabhängig und identisch mit dem Kompressionsmodul des trockenen Gesteins bei hohen Drucken. Der Scherungsmodul wird durch die Wassersättigung nur unwesentlich beeinflußt. Die Poisson-Zahl ist bei kleinen Drucken im trockenen Gestein mit Rißporosität sehr klein (in Einzelfällen nach [Nu69] sogar negativ) und im wassergesättigten Gestein anomal groß.

So far velocity measurements under *controlled pore-fluid pressure* were carried out mainly for sediments. Fig. 29 shows pertinent P-wave velocities for two sandstones according to [Hi56]. For constant differential pressure

$$\Delta P = P_{\text{ext}} - P_{\text{pore}} \tag{23}$$

velocities are almost constant. That is why measurements under controlled pore pressure are usually plotted against differential pressure as e.g. in Figs. 30···33, taken from [Do77, Jo80]. The water-saturated samples from [Jo80] were measured at pore pressure $P_{\text{pore}} = 0.465 \cdot P_{\text{ext}}$; this ratio of pore pressure to confining pressure approximately equals the ratio of hydrostatic to lithostatic pressure *in situ*. The differences in S-wave velocities for the sandstone samples can largely be explained by the greater bulk densities of water-saturated samples.

The velocity hysteresis for Bedford limestone ($\phi = 11.9\,\%$) is caused by pore collapse, beginning at about 100 MPa during the loading cycle [Jo80]. In that way additional microcracks which reduce the velocities after unloading are produced. The effect is observed in both dry and water-saturated rocks at about the same differential pressure; it was also described in [Nu69].

Geschwindigkeitsmessungen bei *kontrolliertem Druck der Porenflüssigkeit* sind bisher vor allem an Sedimenten durchgeführt worden. Die Fig. 29 gibt entsprechende Messungen der P-Wellengeschwindigkeiten für zwei Sandsteine nach [Hi56] wieder. Die Geschwindigkeiten sind für konstanten Differenzdruck nahezu konstant. Messungen bei kontrolliertem Porendruck werden daher oft als Funktion des Differenzdrucks dargestellt wie z.B. in Fig. 30···33 nach [Do77, Jo80]. Die wassergesättigten Proben von [Jo80] wurden bei einem Porendruck $P_{\text{pore}} = 0.465\,P_{\text{ext}}$ gemessen; dieses Verhältnis von Poren- zu Umschließungsdruck ist etwa gleich dem Verhältnis von hydrostatischem zu lithostatischem Druck *in situ*. Die unterschiedlichen S-Wellengeschwindigkeiten der Sandsteine für trockene und wassergesättigte Proben können weitgehend durch die größeren Dichten der wassergesättigten Proben erklärt werden.

Die Hysterese der Geschwindigkeiten für Bedford-Kalkstein ($\phi = 11.9\,\%$) ist durch das Zusammenbrechen von Poren bei zunehmendem Differenzdruck, beginnend bei etwa 100 MPa, zu erklären [Jo80]. Dadurch werden zusätzliche Mikrorisse geschaffen, die nach Druckentlastung zu geringeren Geschwindigkeiten führen. Der Effekt ist für trockene und wassergesättigte Proben bei etwa gleichem Differenzdruck zu beobachten; er wurde auch in [Nu69] beschrieben.

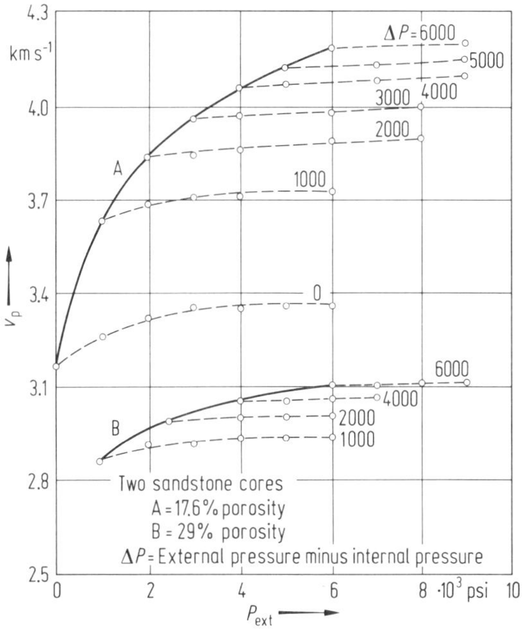

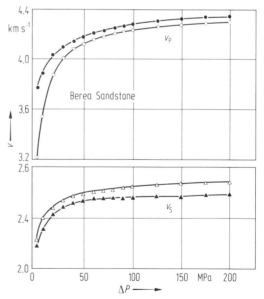

Fig. 29. Compressional wave velocities v_{p} in two sandstone samples as functions of external (confining) pressure [Hi56]. The solid curves are for zero pore pressure; dashed lines are for constant differential pressure. Wave velocities depend predominantly on differential pressure. All pressures are given in PSI-units; 1000 PSI = 6.895 MPa.

Fig. 30. Compressional and shear wave velocities, v_{p}, v_{s}, for dry (open symbols) and water-saturated (solid symbols) Berea sandstone as function of differential pressure $\Delta P = P_{\text{ext}} - P_{\text{pore}}$ [Jo80].

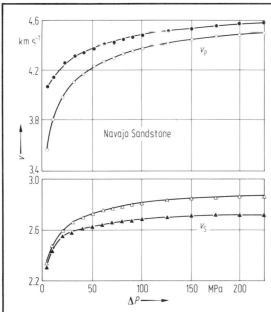

Fig. 31. Same as Fig. 30 for Navajo sandstone [Jo80].

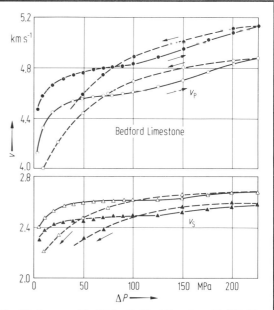

Fig. 32. Same as Fig. 30 for Bedford limestone [Jo80]. The unloading path is denoted by the dashed lines. Hysteresis is due to pore collapse, initiated at about 100 MPa during loading.

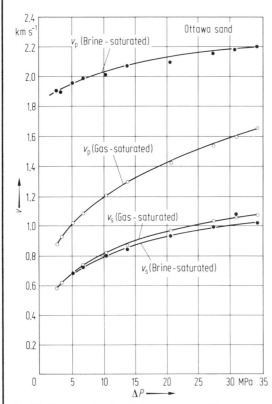

Fig. 33. Compressional and shear wave velocities, v_p, v_s, vs. differential pressure for gas- and brine-saturated unconsolidated sand (Ottawa sand; initial porosity $\phi = 38.3\%$) [Do77].

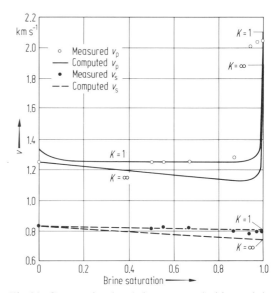

Fig. 34. Compressional and shear wave velocities vs. brine saturation (= volume fraction of brine in the pore space) for Ottawa sand (unconsolidated; initial porosity $\phi = 38.3\%$) at constant differential pressure $\Delta P = 10.34$ MPa [Do77]. Solid and dashed lines are theoretical curves for different coupling factors K between pore fluid and solid frame; for details of calculation, see reference.

At constant differential pressure, velocity variations are possible depending on the nature of the pore content. Fig. 34 shows experimentally and theoretically obtained seismic velocities in unconsolidated sand as function of pore fluid composition; it was a brine-gas mixture in this case. S-wave velocities are almost independent of brine saturation. P-wave velocities are nearly constant up to about 90 % brine content and then increase drastically. This is easy to understand, because the effective bulk modulus of the brine-gas mixture is very small as long as a significant gas content is present. Similar results were obtained in [El75].

Bei konstantem Differenzdruck sind Geschwindigkeitsvariationen in Abhängigkeit von der Art der Porenfüllung möglich. Die Fig. 34 zeigt experimentelle und theoretische Ergebnisse über die seismischen Geschwindigkeiten in einem unverfestigten Sand in Abhängigkeit von der Zusammensetzung der Porenflüssigkeit, die durch ein Sole-Gas-Gemisch gebildet wird. Die S-Wellengeschwindigkeiten sind vom Solegehalt weitgehend unabhängig. Die P-Wellengeschwindigkeiten sind bis zu Solegehalten von etwa 90 % nahezu konstant und nehmen dann drastisch zu. Dies ist verständlich, da der effektive Kompressionsmodul der Porenfüllung sehr klein ist, solange noch ein signifikanter Gasgehalt vorhanden ist. Ähnliche Ergebnisse wurden von [El75] erhalten.

3.1.3.1.7 Velocity-density relations — Geschwindigkeits-Dichte-Beziehungen

Several empirical and semi-empirical relations between seismic velocities and density have been proposed in the literature [Wo59, Ta59, Bi61, Si64, An67, Ch72a, Sh74, Ma74a]. Two of them, the "Woollard-" and the "Nafe-Drake curve", have already been shown in section 3.1.2.3 (Figs. 16 and 17). Birch [Bi61] demonstrated on the basis of data shown in Fig. 35 that compressional wave velocities of rocks and minerals are linearly related to density:

Verschiedene empirische und halb-empirische Beziehungen zwischen seismischen Geschwindigkeiten und Dichte sind in der Literatur vorgeschlagen worden [Wo59, Ta59, Bi61, Si64, An67, Ch72a, Sh74, Ma74a]. Zwei davon, die „Woollard-" und die „Nafe-Drake-Kurve" sind bereits im Abschnitt 3.1.2.3 (Fig. 16 und 17) vorgestellt worden. Birch [Bi61] hat auf der Grundlage der Daten von Fig. 35 gezeigt, daß P-Wellengeschwindigkeiten in Gesteinen und Mineralen linear mit der Dichte gemäß

$$v_p = a(\bar{m}) + b\varrho \qquad (24)$$

where a depends on the mean atomic weight $\bar{m}$ only and b is a constant. Eq. (24) is known as Birch's law. The

verknüpft sind, wobei a nur vom mittleren Atomgewicht $\bar{m}$ abhängt und b eine Konstante ist. Gl. (24)

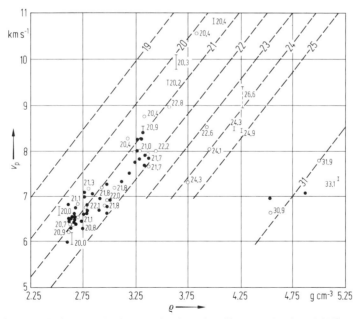

Fig. 35. Compressional wave velocity v_p at 1.0 GPa vs. density ϱ for silicates and oxides [Bi61]. Values calculated from single-crystal elastic constants are indicated as vertical lines; the numbers attached to open circles are mean atomic weights; dashed lines suggest variation for constant mean atomic weight.

Gebrande

constants a and b are usually determined by a least squares fit of data for samples with similar $\bar{m}$. Regression line parameters for different rock groups and from different sources are compiled in Table 16 for compressional and shear wave velocities. It should be emphasized that no universally valid parameters are found and that considerable standard deviations between regression lines and experimental data points are observed. This is true even at high pressures when the disturbing influence of porosity is diminished. At zero pressure the scatter is, however, much larger (see Figs. 13···15).

Velocity-density plots and linear least squares solutions v_p and v_s on ϱ for plutonic and metamorphic rocks from Tables 11 and 14 are shown in Fig. 36 for different pressures. It should be noted that the regression lines v on ϱ are not identical with lines of constant $\bar{m}$, because correlation coefficients between ϱ and $\bar{m}$ are positive for the data of Figs. 36a, b, c. Two-dimensional regression analyses v on ϱ and $\bar{m}$ give the following results for the same sets of plutonic and metamorphic rocks:

wird auch als Birch-Beziehung bezeichnet. Die Größen a und b werden gewöhnlich durch lineare Regression der Daten von Proben mit ähnlichem $\bar{m}$ bestimmt. Regressions-Parameter für verschiedene Gesteinsgruppen und aus verschiedenen Quellen sind in Tab. 16 für P- und S-Wellengeschwindigkeiten zusammengestellt. Es muß betont werden, daß keine allgemein gültigen Parameter gefunden werden und daß beträchtliche Standardabweichungen zwischen den Regressionsgeraden und den experimentellen Daten beobachtet werden. Dies gilt auch bei hohen Drucken, wo der Einfluß der Porosität abnimmt. Bei Normaldruck ist die Streuung allerdings noch viel größer (vgl. Fig. 13···15).

Geschwindigkeits-Dichte-Diagramme und Ausgleichsgeraden für v_p und v_s als Funktionen der Dichte ϱ sind in Fig. 36 für Plutonite und Metamorphite der Tabellen 11 und 14 und für verschiedene Drucke dargestellt. Es ist zu vermerken, daß die Regressionsgeraden für v nach ϱ nicht identisch mit Linien für konstantes $\bar{m}$ sind, da für die Daten der Fig. 36a, b, c eine positive Korrelation zwischen ϱ und $\bar{m}$ besteht. Durch zweidimensionale Regressionsanalyse v nach ϱ und $\bar{m}$ erhält man für die gleichen Plutonite und Metamorphite folgende Ergebnisse:

0.2 GPa:
$$v_p = [-0.30 + 2.41\varrho - 0.18\,(\bar{m} - 21.5) \pm 0.23]\ \mathrm{km\ s^{-1}} \tag{25a}$$
$$v_s = [+0.41 + 1.15\varrho - 0.02\,(\bar{m} - 21.5) \pm 0.14]\ \mathrm{km\ s^{-1}} \tag{25b}$$

0.6 GPa:
$$v_p = [-0.22 + 2.42\varrho - 0.14\,(\bar{m} - 21.5) \pm 0.21]\ \mathrm{km\ s^{-1}} \tag{25c}$$
$$v_s = [+0.43 + 1.17\varrho + 0.002\,(\bar{m} - 21.5) \pm 0.19]\ \mathrm{km\ s^{-1}} \tag{25d}$$

1.0 GPa:
$$v_p = [-0.38 + 2.49\varrho - 0.20\,(\bar{m} - 21.5) \pm 0.22]\ \mathrm{km\ s^{-1}} \tag{25e}$$
$$v_s = [0.0 + 1.30\varrho - 0.09\,(\bar{m} - 21.5) \pm 0.18]\ \mathrm{km\ s^{-1}} \tag{25f}$$

Eqs. (25a···f) show that for plutonic and metamorphic rocks the dependence of seismic velocities on density is much more important than the dependence on mean atomic weight. Shear wave velocities are almost independent on mean atomic weight within the usual range of variation of $\bar{m}$. Comparison of the standard deviations of Eqs. (25a···f) with the results of one-dimensional regression analyses in Table 16 shows that the prediction error of shear wave velocities is not at all, and that of compressional wave velocities is only slightly improved by taking the dependence on $\bar{m}$ into account. The remaining standard error is still larger than the velocity variation due to a variation of $\bar{m}$ by one unit. Thus there is not much hope that small $\bar{m}$-variations in the earth's interior can be derived from seismic velocities and density on the basis of such simple relations as (24).

For some data sets a nonlinear velocity-density relation, as suggested in [Ch75a],

Die Gln. (25a···f) zeigen, daß für Plutonite und Metamorphite die Abhängigkeit der seismischen Geschwindigkeiten von der Dichte sehr viel bedeutender ist als die Abhängigkeit vom mittleren Atomgewicht. Die S-Wellengeschwindigkeiten sind innerhalb des üblichen Variationsbereichs von $\bar{m}$ vom mittleren Atomgewicht nahezu unabhängig. Der Vergleich der Standardabweichungen der Gl. (25a···f) mit den Ergebnissen der eindimensionalen Regressions-Analysen der Tab. 16 zeigt, daß der Approximationsfehler für S-Wellen gar nicht und für P-Wellen nur geringfügig durch die Hinzunahme der Abhängigkeit von $\bar{m}$ verbessert wird. Der verbleibende Standardfehler ist immer noch größer als die Geschwindigkeitsvariation, die durch die Variation von $\bar{m}$ um eine Einheit bedingt wird. Es besteht daher wenig Hoffnung, daß kleine $\bar{m}$-Variationen im Erdinnern auf der Grundlage so einfacher Beziehungen wie (24) aus seismischen Geschwindigkeiten und Dichte abgeleitet werden können.

Für manche Daten scheint eine nicht-lineare Geschwindigkeits-Dichte-Beziehung

$$v = a + b\varrho^c \tag{26}$$

seems to give better approximations; this is demonstrated in Fig. 37 for water-saturated oceanic basalts at 0.05 GPa, from [Ch75a]. In this example the standard

wie sie in [Ch75a] vorgeschlagen wurde, bessere Approximationen zu ermöglichen; dies ist in Fig. 37 für wassergesättigte ozeanische Basalte bei 0.05 GPa

deviations for v_p and v_s are 0.20 and 0.17 km s^{-1} for the linear relation (Eq. 24) and 0.18 and 0.14 km s^{-1}, respectively, for the nonlinear relation (Eq. 26).

gezeigt, nach [Ch75a]. In diesem Beispiel sind die Standardabweichungen für v_p und v_s 0.20 und 0.17 km s^{-1} für die lineare Beziehung (Gl. 24), bzw. 0.18 und 0.14 km s^{-1} für die nicht-lineare Beziehung (Gl. 26).

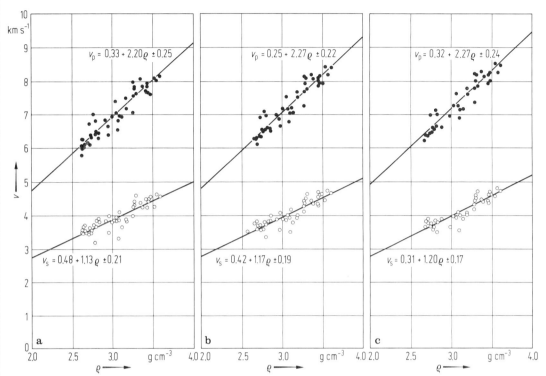

a $v_p = 0.33 + 2.20\varrho \pm 0.25$
     $v_s = 0.48 + 1.13\varrho \pm 0.21$

b $v_p = 0.25 + 2.27\varrho \pm 0.22$
     $v_s = 0.42 + 1.17\varrho \pm 0.19$

c $v_p = 0.32 + 2.27\varrho \pm 0.24$
     $v_s = 0.31 + 1.20\varrho \pm 0.17$

Fig. 36. Velocity-density relations for plutonic and quasi-isotropic metamorphic rocks at (a) 0.2, (b) 0.6, and (c) 1.0 GPa; data are from Table 11 and Table 14. Solid lines are least squares solutions; the indicated errors are standard deviations. Densities and velocities are corrected for length changes due to compression. Further details are given in Table 16.

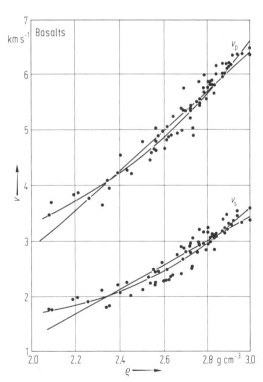

Fig. 37. Velocity-density relations for water-saturated oceanic basalts at 0.05 GPa [Ch75a]. The standard errors for the estimation of v on ϱ are 10 to 20% smaller for the nonlinear least squares solutions (eq. 26) than for the linear solutions (eq. 24).

Table 16. Empirical relations between compressional wave velocities v_p and shear wave velocities v_s and density ϱ at elevated pressures P. s.d.: standard deviation of linear least squares fit with parameters a, b; r: correlation coefficient.

Material	P GPa	No. of data points	$v_p = a + b\varrho$ a km s^{-1}	b km cm^3 g^{-1} s^{-1}	s.d. km s^{-1}	r	Ref.
Granites[1]	1.0	15	−0.30	2.54±0.21	0.08	0.56	Bi61
Diabase, gabbro, eclogite[1]	1.0	18	−1.00	2.67±0.09	0.22	0.92	Bi61
Different rocks, $\bar m = 20\cdots22$, except basic rocks	1.0	45	−1.87	3.05	0.26	0.94	Bi61
as before plus monomineralic aggregates[2]	1.0	46	−2.55	3.31	0.28		Bi61
Plagioclase[1]	1.0	7	−4.94	4.41	0.06		Bi61
Plutonic rocks (mainly)[1]	0.6	33	−1.3	2.8	≈0.5		Ka65
Granulite+eclogite, $\bar m = 21±0.5$	1.0	7+2	−0.63	2.67		0.97	Ma74
Granulite+eclogite+basalt	1.0	7+13+6	−1.85	2.87		0.98	Ma74
Granulites	0.6	10	0.31	2.27		0.95	Ch75
Plutonites+metamorphites, $20.7 \le \bar m \le 23.0$; $\bar{\bar m} = 21.55$	0.2	47	0.33	2.20	0.25	0.94	Tables 11, 14
as before, but $\bar{\bar m} = 21.58$	0.6	45	0.25	2.27	0.22	0.94	Tables 11, 14
as before, but $\bar{\bar m} = 21.49$	1.0	38	0.32	2.27	0.24	0.95	Tables 11, 14
Serpentinite+serpentinized peridotite+peridotite	1.0	9	−4.39	3.83		0.99	Ch66
Basalts with $\phi \le 7\%$ dry, $\bar m = 21.6±0.5$[1]	1.0	8	−1.88	2.90	0.57	0.94	Ma68
Basalts and ultrabasic rocks, $\bar m$ and ϕ as before	1.0	13	−3.10	3.35	0.12	0.98	Ma68
Oceanic basalts, satw	0.6	32	−2.61	3.06	0.17	0.97	Ch73a
	1.0	32	−2.01	2.88	0.17	0.96	Ch73a
$v_s = a + b\varrho$							
Granulite+eclogite, $\bar m = 21±0.5$	1.0	9	−0.66	1.56		0.98	Ma74
Granulite+eclogite	1.0	20	−0.33	1.40		0.94	Ma74
Plutonites+metamorphites, $\bar{\bar m} = 20.6\cdots21.3$[1]	1.0	22	−0.88	1.63			Ch68
Different rocks, $\bar m = 21±0.5$	1.0		−0.24	1.49			Ko69
Plutonites+metamorphites, $20.7 \le \bar m \le 23.0$, $\bar{\bar m} = 21.55$	0.2	47	0.48	1.13	0.21	0.87	Tables 11, 14
as before, but $\bar{\bar m} = 21.58$	0.6	45	0.42	1.17	0.19	0.84	Tables 11, 14
as before, but $\bar{\bar m} = 21.49$	1.0	38	0.31	1.20	0.18	0.90	Tables 11, 14
Oceanic basalts, satw	0.6	32	−1.53	1.66	0.14	0.92	Ch73a
	1.0	32	−1.11	1.52	0.15	0.91	Ch73a

[1] Uncorrected for length and density changes.
[2] Only ϱ is corrected.

Birch [Bi61] pointed out that systematic exceptions from the linear relation (24) arise for calcium-rich plagioclase rocks and Simmons [Si64] developed an improved empirical relation which takes the CaO-content into account:

Birch [Bi61] hat darauf hingewiesen, daß systematische Abweichungen von der einfachen Beziehung (24) für Kalzium-reiche plagioklashaltige Gesteine auftreten, und Simmons [Si64] hat eine verbesserte empirische Beziehung aufgestellt, die den CaO-Gehalt berücksichtigt:

$$v_p = \{-0.98 + 0.7\,(21 - \bar{m}) + 2.76\varrho + 4.60\,[CaO]\}\ \mathrm{km\ s^{-1}} \tag{27}$$

[CaO] is the weight fraction of CaO in the rock. Manghnani et al. [Ma74] found

[CaO] ist der relative Gewichtsanteil von CaO im Gestein. Manghnani et al. [Ma74] fanden

$$v_p = \{-0.53 + 0.7\,(21 - \bar{m}) + 2.58\varrho + 4.60\,[CaO]\}\ \mathrm{km\ s^{-1}} \tag{27a}$$

more appropriate for granulites and eclogites at 1.0 GPa and obtained a similar relation for shear wave velocities for these rocks:

für Granulite und Eklogite bei 1.0 GPa besser geeignet und erhielten für die S-Wellengeschwindigkeiten in diesen Gesteinen:

$$v_s = \{-0.63 + 0.21\,(21 - \bar{m}) + 1.56\varrho + 0.016\,[CaO]\}\ \mathrm{km\ s^{-1}} \tag{27b}$$

The coefficient of [CaO] in (27b) is much smaller than in (27), and the effect of CaO content on v_s is therefore not as significant as that on v_p. The CaO-correction should not be applied calcium-carbonate! The relatively large differences between the coefficients of $\bar{m}$ in Eqs. (25) on the one and Eqs. (27) on the other side are surprising and require further investigations.

Der Koeffizient von [CaO] in (27b) ist sehr viel kleiner als in (27) und daher wirkt sich der CaO Gehalt auf v_s weniger signifikant als auf v_p aus. Auf Kalziumkarbonat sollte die CaO-Korrektur nicht angewandt werden! Die relativ großen Unterschiede der Koeffizienten von $\bar{m}$ in den Gln. (25) einerseits und den Gln. (27) andererseits sind überraschend und bedürfen weiterer Untersuchungen.

Other velocity-density relations make use of the bulk velocity v_φ (called hydrodynamic velocity by others):

In anderen Geschwindigkeits-Dichte-Beziehungen wird von der hydroakustischen Geschwindigkeit v_φ (engl.: bulk velocity, hydrodynamic wave velocity)

$$v_\varphi = \sqrt{\varphi} = \sqrt{\frac{K}{\varrho}} = \sqrt{v_p^2 - \tfrac{4}{3}v_s^2} = \sqrt{\left(\frac{\partial P}{\partial \varrho}\right)_s} \tag{28}$$

v_φ is a purely arithmetic quantity, but has the advantage of beeing related to volume changes only. According to [Wa68, Wa69, Ch73] Birch's law (24) is also valid for the bulk velocity. A least squares fit [Wa69] of 16 rocks and minerals with $\bar{m}$ between 20 and 21 and at a pressure of 1.0 GPa gives:

Gebrauch gemacht. v_φ ist eine rein rechnerische Größe, hat aber den Vorteil, ausschließlich mit Volumenänderungen verknüpft zu sein. Nach [Wa68, Wa69, Ch73] gilt die Birch-Beziehung (Gl. 24) auch für die hydroakustische Geschwindigkeit. Für 16 Gesteine und Minerale mit $\bar{m}$ zwischen 20 und 21 ergibt sich nach [Wa69] bei 1.0 GPa:

$$v_\varphi = \{-2.32 + 2.59\varrho \pm 0.28\}\ \mathrm{km\ s^{-1}} \tag{29a}$$

For 22 rocks and minerals with similar $\bar{m}$ Chung [Ch73] obtained:

Chung [Ch73] erhält für 22 Gesteine und Minerale mit ähnlichem $\bar{m}$:

$$v_\varphi = \{-2.59 + 2.65\varrho \pm 0.14\}\ \mathrm{km\ s^{-1}} \tag{29b}$$

In both equations ϱ should be used in g cm^{-3} and the error indicated is the standard deviation.

Anderson [An67] developed a nonlinear relation

ϱ ist in beiden Gl. in gcm^{-3} einzusetzen und der angegebene Fehler ist jeweils die Standardabweichung.

Anderson [An67] gibt eine als „seismische Zustandsgleichung" bezeichnete nicht-lineare Beziehung

$$\frac{\varrho}{\bar{m}} = A\,\varphi^n \tag{30}$$

which was called "seismic equation of state", and he obtained the numerical form by a least squares fit of 31 rocks and minerals with $\bar{m}$ between 18.6 and 33.08:

an und erhält durch Ausgleichsrechnung für 31 Gesteine und Minerale mit $\bar{m}$ zwischen 18.6 und 33.08 die konkrete Form:

$$\frac{\varrho}{\bar{m}} = 0.048\,\varphi^{0.323} \tag{30a}$$

If φ is inserted in $km^2 s^{-2}$, ϱ is obtained in $g\,cm^{-3}$ The standard deviation ± 0.12 given in [An67] for $\varrho/\bar{m}$ is incorrect and apparently refers to ϱ. For oxides which may be important in the Earth's lower mantle, the version

Wenn φ in $km^2 s^{-2}$ eingesetzt wird, ergibt sich ϱ in gcm^{-3}. Die in [An67] für $\varrho/\bar{m}$ angegebene Standardabweichung ± 0.12 ist falsch und gilt anscheinend für ϱ. Für Oxyde, die im unteren Erdmantel von Bedeutung sein mögen, ist nach [An69] und [Da71]

$$\frac{\varrho}{\bar{m}} = 0.0492\,\varphi^{1/3} \qquad (30b)$$

is recommended [An69, Da71]. Eq. (30a) is illustrated in Fig. 38.

zu empfehlen. Fig. 38 illustriert Gl. (30a).

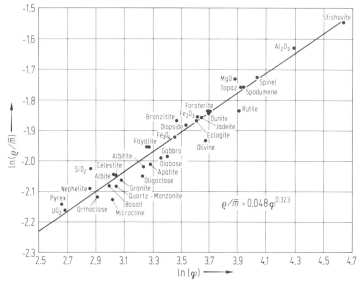

Fig. 38. $\ln(\varrho/\bar{m})$ vs. $\ln(\varphi)$ for selected rocks and minerals; solid line is a least squares fit to this data with parameters as shown [An67].

A more general relation was derived by Shankland [Sh72, Sh74, Sh77]:

Ein allgemeineres Gesetz wird von Shankland [Sh72, Sh74, Sh77] angegeben:

$$v_\varphi = B\,\varrho^\lambda(\bar{m}_0/\bar{m})^{\frac{1}{2} + \lambda(1-c)} \qquad (31)$$

The essential difference to (30) is, that ϱ and $\bar{m}$ appear with different powers in (31). B is a universal constant which was determined as 1.42 (standard deviation about 8.5 %) from bulk velocities in 26 minerals [Sh74]. λ is related to a Grüneisen parameter γ by $\lambda = \gamma - 1/3 \approx 1.25$, $\bar{m}_0 = 20.2$ is a reference mean atomic weight, and $c \approx 1.3 \cdots 1.4$ is defined by the logarithmic derivative of density versus mean atomic weight at constant crystal structure x:

Der wesentliche Unterschied zu (30) besteht darin, daß ϱ und $\bar{m}$ mit verschiedenen Exponenten erscheinen. B ist eine allgemeine Konstante, die aus den hydroakustischen Geschwindigkeiten für 26 Minerale zu 1.42 (Standardabweichung etwa 8.5 %) bestimmt wurde [Sh74]. λ ergibt sich aus einem Grüneisen-Parameter γ gemäß $\lambda = \gamma - 1/3 \approx 1.25$ und $c \approx 1.3 \cdots 1.4$ ist durch die logarithmische Ableitung der Dichte nach dem mittleren Atomgewicht bei gleichbleibender Kristallstruktur x definiert:

$$1 - c = (\partial \ln \varrho / \partial \ln \bar{m})_x$$

The values predicted from (31) for minerals are usually within 5 % of the measured values.

und $\bar{m}_0 = 20.2$. Nach (31) errechnete Werte stimmen für Minerale im allgemeinen innerhalb 5 % mit den Meßwerten überein.

Mao [Ma74a] introduced Poisson's ratio σ as additional parameter into velocity-density systematics and defined a corrected bulk velocity:

$$v_\varphi^* = -1.84 + 7.51\,\sigma + 1.98\,\varrho \tag{32a}$$

v_φ^* is an empirical approximation of v_φ for different minerals with $\bar{m} \approx 20.2$. For arbitrary $\bar{m}$ the quantities v_φ^*, v_φ, ϱ, and $\bar{m}$ are connected by the following relation [Ma74a]:

$$(v_\varphi^* - v_\varphi)(\bar{m} - 20.2) = 0.058 + 2.28\,\varrho/\bar{m} \tag{32b}$$

If v_p and v_s and either ϱ or $\bar{m}$ are known, the fourth quantity can be calculated from (32b) in connection with (32a).

An excellent review of velocity-density systematics and their foundations in the physics of solids is given in [Sh77].

Mao [Ma74a] führte als zusätzlichen Parameter in Geschwindigkeits-Dichte-Beziehungen die Poisson-Zahl σ ein und definierte eine korrigierte hydro-akustische Geschwindigkeit:

v_φ^* ist eine empirische Approximation für v_φ für verschiedene Minerale mit $\bar{m} \approx 20.2$. Für beliebiges $\bar{m}$ sind nach [Ma74a] v_φ, v_φ^*, ϱ und $\bar{m}$ durch folgende Beziehung verknüpft:

Wenn v_p und v_s und entweder ϱ oder $\bar{m}$ gegeben sind, kann die 4. Größe aus (32b) in Verbindung mit (32a) errechnet werden.

Ein vorzüglicher Überblick über Geschwindigkeits-Dichte-Beziehungen und deren Festkörper-physikalische Hintergründe ist in [Sh77] gegeben.

3.1.3.1.8 Comparison between static and dynamic measurements — Vergleich zwischen statischen und dynamischen Messungen

Whereas dynamic measurements at different frequencies usually give consistent values of elastic constants (compare 3.1.1.2), very large differences are sometimes observed between static and dynamic measurements. This is true in particular for *in-situ* measurements (compare e.g. [Ju64, Cl66]), which are not discussed here. However, the effect is also observed under laboratory conditions. In Table 17 dynamically and statically measured compressibilities [Si65a] for different materials and pressures are compared. Static and dynamic measurements were carried out on the same specimens; each value is a mean value for samples taken in three mutually perpendicular directions. The static compressibility was calculated from the measured linear compressibilities of the three samples [Br65], the dynamic compressibility was obtained from P- and S-wave velocities corrected for length changes; in addition the statically measured values were corrected to adiabatic conditions by Eq. (4a). Values for the two metals, fused quartz and limestone ($\phi = 0$!) are in agreement within the experimental errors. For the other rocks static compressibilities are systematically greater than dynamic values; at elevated pressures, however, the differences are within the range of the accuracy. On the other hand, differences up to the threefold are observed at normal pressure. This effect is associated to the crack porosity existing at moderate pressures [Si65a], but details are not yet fully understood.

Während dynamische Messungen bei unterschiedlichen Frequenzen im allgemeinen konsistente Werte für die Elastizitätskonstanten liefern (vgl. 3.1.1.2), werden z.T. große Unterschiede zwischen den statisch und den dynamisch bestimmten Konstanten beobachtet. Dies trifft vor allem für *in-situ*-Messungen zu (vgl. z.B. [Ju64, Cl66]), die hier nicht diskutiert werden sollen. Der Effekt ist aber auch unter Laboratoriumsbedingungen zu beobachten. In Tab. 17 sind nach [Si65a] dynamisch und statisch bestimmte Kompressibilitäten für verschiedene Materialien und Drucke gegenübergestellt. Die statischen und dynamischen Messungen wurden jeweils an denselben Proben durchgeführt; die Meßwerte sind Mittelwerte für jeweils drei senkrecht zueinander entnommene zylindrische Proben. Die statische Kompressibilität wurde aus den linearen Kompressibilitäten der Einzelproben berechnet [Br65], die dynamische Kompressibilität aus längenkorrigierten P- und S-Wellengeschwindigkeiten; die statisch bestimmten Werte wurden darüber hinaus nach Gl. (4a) auf adiabatische Bedingungen korrigiert. Die Werte für die Metalle, Quarzglas und Kalkstein ($\phi = 0$!) stimmen im Rahmen der Meßgenauigkeit überein. Die statischen Kompressibilitäten der Gesteine sind systematisch größer als die dynamischen; der Unterschied ist für höhere Drucke allerdings im Bereich der Meßgenauigkeit. Für Normaldruck werden dagegen Unterschiede bis zu einem Faktor drei beobachtet. Der Effekt wird mit der bei schwachen Drucken vorhandenen Rißporositäten in Verbindung gebracht [Si65a], wird aber im Einzelnen noch nicht voll verstanden.

Table 17. Comparison of static (β_{sta}) and dynamic (β_{dyn}) measurements of compressibility; static (isothermal) measurements were reduced to adiabatic conditions by means of eq. (4a); porosities ϕ are for zero pressure; [Si65a, Br65].

Material	β [(TPa)$^{-1}$]	P [GPa]			
		0.0	0.3	0.5	0.9
Quartzite, Rutland, Vermont, USA $\phi = 0.6\%$	β_{dyn}	23.7	24.1	23.0	21.6
	β_{sta}	76	25.1	24.6	22.4
Granite, Westerly, Rhode Island, USA $\phi = 1.1\%$	β_{dyn}	56	19.5	18.4	17.4
	β_{sta}	82	20.5	19.8	
Granite, Stone Mountain, Georgia, USA $\phi = 0.3\%$	β_{dyn}	71.8	19.2	18.5	17.4
	β_{sta}	156	20.0	19.1	17.6
Diabase, Frederick, Maryland, USA $\phi = 0.1\%$	β_{dyn}	12.9	11.8	11.5	11.1
	β_{sta}	12.4	12.5	12.3	11.7
Limestone, Oak Hall, Pennsylvania, USA $\phi = 0.0\%$	β_{dyn}	14.8	13.1	12.9	12.8
	β_{sta}	13.5	13.2	13.1	
Aluminum, 2011-T3 $\phi = 0.0\%$	β_{dyn}	13.0	12.6	12.5	12.4
	β_{sta}	12.6	12.5	12.4	12.1
Steel, annealed $\phi = 0.0\%$	β_{dyn}	6.18	5.93	5.86	5.80
	β_{sta}	5.7	5.8	5.8	
Fused Quartz	β_{dyn}	27.1	27.8	29.0	31.1
	β_{sta}	9	28.3	29.3	31.3

3.1.3.2 Anisotropic rocks — Anisotrope Gesteine

This subsection will be relatively short in view of the abundance of anisotropic rocks. However, only few rocks have been investigated at elevated pressures with sufficient detail to give reliable results on the type of anisotropy or even to enable the calculation of complete sets of elastic constants (see 3.1.1.3). Measurements of P- and S-wave velocities in three mutually perpendicular directions are useful to check isotropy, but are not sufficient to resolve the character of anisotropy. Such data is included in the Tables 11···15, if the anisotropy seems to be less than about 5 % or are omitted otherwise. This subsection contains results of more detailed studies only. Table 18 gives P- and S-wave velocities in metamorphic and sedimentary rocks with known directions of wave propagation and S-wave polarization relative to rock fabric. The z-axis is chosen orthogonal to foliation or bedding. The first index of v gives the direction of wave propagation and the second index the direction of polarization, thus v_{xx} is a (quasi-)longitudinal and v_{yz} a (quasi-)transverse wave. In the case of transverse isotropy (hexagonal symmetry) the following identities apply:

Dieser Abschnitt ist relativ kurz angesichts der weiten Verbreitung anisotroper Gesteine. Allerdings sind bisher nur wenige Gesteine ausführlich genug untersucht worden, um verläßliche Ergebnisse über den Typ der Anisotropie oder gar die Berechnung vollständiger Sätze elastischer Konstanten zu ermöglichen (vgl. 3.1.1.3). Messungen von P- und S-Wellengeschwindigkeiten in drei zueinander senkrechten Richtungen sind wichtig um die Isotropie zu kontrollieren, aber sie reichen nicht aus, um den Charakter der Anisotropie zu erfassen. Solche Daten sind in die Tab. 11···15 aufgenommen worden, sofern die Anisotropie kleiner als etwa 5 % zu sein scheint, oder wurden andernfalls weggelassen. Der vorliegende Abschnitt enthält nur Ergebnisse detaillierterer Untersuchungen. In Tab. 18 sind P- und S-Wellengeschwindigkeiten zusammengestellt für metamorphe und sedimentäre Gesteine, in denen die Ausbreitungsrichtungen und Polarisationsrichtungen relativ zum Gesteinsgefüge bekannt sind. Die z-Achse ist senkrecht zur Schieferungs- oder Schichtungsebene gewählt. Der erste Index von v gibt die Richtung der Wellenausbreitung und der zweite die Polarisationsrichtung an; daher ist v_{xx} eine (quasi-)longitudinale und v_{yz} eine (quasi-)transversale Welle. Im Falle der Transversal-Isotropie (hexagonale Symmetrie) gelten die folgenden Identitäten:

$$v_{xx} = v_{yy}; \quad v_{xy} = v_{yx}; \quad v_{xz} = v_{yz} = v_{zx} = v_{zy} \tag{33}$$

It can be seen that this is approximately true for some rocks of Table 18, in particular at elevated pressures. Table 19 contains the most extensive sets of velocity determinations published so far [Ch71, Ba72]. The authors found that the anisotropy is clearly related to rock fabric. Both types of anisotropy, hexagonal and orthorhombic symmetry (approximately), were observed for Twin Sisters Dunite.

Fig. 39 shows P-wave anisotropy as function of pressure for different rocks according to [Ke78]. From the pressure dependence it may be concluded that two components of anisotropy exist: a pressure-independent component due to preferred orientation of crystallographic axes and a pressure-sensitive component due to preferred orientation of microcracks ([Ke78]; see also 3.1.4.4.1.3 and Fig. 7 there). The pressure range of strong anisotropy decrease coincides with the pressure range of strong velocity increase (about 0 to 0.2 GPa). Anisotropy and velocities above some 0.2 GPa can be considered identical to the same quantities in crack-free rocks. The decrease of v_p due to cracks is strong for a propagation direction normal to the preferred plane of the cracks and less pronounced for propagation in this plane. It can be concluded from the data of Table 18 that v_s is also distinctly decreased, but relatively less than v_p, for propagation normal to the preferred plane

Wie aus Tab. 18 zu ersehen, trifft dies für einige Gesteine, insbesondere bei höheren Drucken, näherungsweise zu. Die Tab. 19 enthält die bisher umfangreichsten Geschwindigkeitsbestimmungen für einzelne Gesteine [Ch71, Ba72]. Beide Autoren fanden, daß die Anisotropie in klarer Beziehung zum Gesteinsgefüge steht. Im Twin-Sisters-Dunit wurde näherungsweise sowohl hexagonale, wie auch rhombische Symmetrie beobachtet.

Die Fig. 39 zeigt die Anisotropie der P-Wellengeschwindigkeiten für verschiedene Gesteine nach [Ke78]. Aus der Druckabhängigkeit kann gefolgert werden, daß die Anisotropie zwei Anteile besitzt: einen Druck-unabhängigen Anteil durch bevorzugte Orientierung kristallographischer Achsen und einen Druck-empfindlichen Anteil durch bevorzugte Orientierung von Mikrorissen ([Ke78]; siehe auch 3.1.4.4.1.3 und die dortige Fig. 7). Der Druckbereich von 0 bis etwa 0.2 GPa, in dem die Anisotropie stark abnimmt, stimmt mit dem Bereich überein, in dem die Geschwindigkeiten stark zunehmen. Anisotropie und Geschwindigkeiten oberhalb etwa 0.2 GPa können mit den betreffenden Eigenschaften im Riß-freien Gestein gleichgesetzt werden. Die Abnahme von v_p aufgrund von Rissen ist stark für die Ausbreitungsrichtung senkrecht zur bevorzugten Ebene der Risse und weniger ausgeprägt für die Ausbreitung in dieser Ebene. Aus den

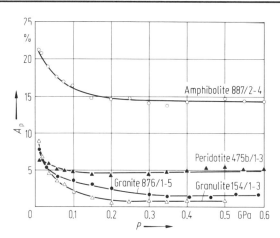

Fig. 39. Anisotropy A_p of P-wave velocity vs. pressure P for dry rocks at room temperature [Ke78]; the pressure dependence correlates well with the crack porosities of the samples: 0.55 % for amphibolite, 0.1 % for peridotite, 0.45 % for granite, and 0.50 % for granulite.

(independent of polarization direction), and is affected to the same degree for propagation in the preferred plane, if the polarization is normal to this plane. For propagation *and* polarization in the preferred plane of cracks the influence on v_s is very small. This is in agreement with the results of theoretical studies, e.g. in [An74], which are illustrated in Fig. 40a, b. This Fig. also shows the influence of fluid filled cracks depending on the bulk modulus of the fluid. These predictions are qualitatively in good agreement with the experimental results shown in Fig. 10 [Wa75]. P-wave velocities strongly increase and P-wave anisotropy decreases when cracks are water-saturated. S-wave velocities and anisotropy are almost independent on water saturation. Anisotropy due to cracks is greater for P-waves than for S-waves for dry cracks, and the reverse is true for water-saturated cracks. For a more comprehensive discussion of these phenomena, see [Pa78].

Daten der Tab. 18 kann geschlossen werden, daß auch v_s – wenngleich weniger stark als v_p – in Richtung senkrecht zur Vorzugsebene der Risse deutlich vermindert ist (unabhängig von der Polarisations-Richtung) und daß v_s im selben Maße bei Ausbreitung in der Vorzugsebene beeinflußt wird, wenn die Polarisations-Richtung senkrecht hierzu steht. Wenn Ausbreitungs- *und* Polarisations-Richtung in der Vorzugsrichtung der Risse liegen, ist der Einfluß auf v_s nur gering. Dies ist in Übereinstimmung mit Ergebnissen theoretischer Untersuchungen, z.B. [An74], die in Fig. 40a, b dargestellt sind. Diese Abbildung zeigt auch den Einfluß Flüssigkeits-gefüllter Risse in Abhängigkeit vom Kompressionsmodul der Flüssigkeit. Diese Prognosen sind qualitativ in guter Übereinstimmung mit den in Fig. 10 [Wa75] gezeigten experimentellen Ergebnissen. Die P-Wellengeschwindigkeiten nehmen stark zu und ihre Anisotropie nimmt ab, wenn die Risse mit Wasser gefüllt werden. Die S-Wellengeschwindigkeiten und deren Anisotropie sind von der Wassersättigung nahezu unabhängig. Die durch trockene Risse erzeugte Anisotropie ist für P-Wellen stärker als für S-Wellen; für wassergefüllte Risse gilt gerade das Gegenteil. Eine ausführlichere Diskussion dieser Phänomene ist in [Pa78] zu finden.

Gebrande

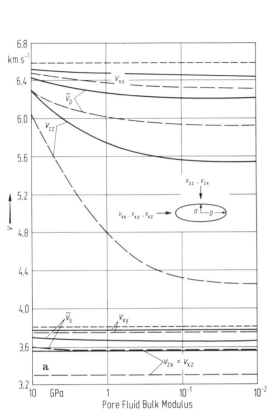

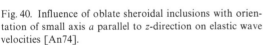

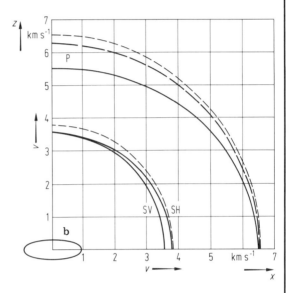

Fig. 40. Influence of oblate sheroidal inclusions with orientation of small axis a parallel to z-direction on elastic wave velocities [An74].

a) Velocities normal ($v_{xx} = v_{yy}$, $v_{xy} = v_{yx} \neq v_{xz} = v_{yz}$) and parallel ($v_{zz}$, $v_{zx} = v_{zy} = v_{xz}$) to the z-direction as functions of pore fluid bulk modulus, calculated for a granite matrix containing dilute inclusions with aspect ratio $\alpha = a/b = 0.05$. Solid curves are for porosity of 1%, and long-dashed curves for porosity of 2%. First and second index of v refer to the directions of wave propagation and polarization, respectively. $\bar{v}_p$ and $\bar{v}_s$ are velocities for a matrix with randomly oriented cracks. Short-dashed lines are velocities in the crack-free granitic matrix.

b) Velocities as functions of angle and fluid properties for porosity of 1% and inclusions with aspect ratio $\alpha = 0.05$. The short-dashed curves give the velocity surface for the isotropic crack-free matrix, the long-dashed curves the surface for liquid-filled cracks ($K_{por} = 10$ GPa), and the solid curves the surfaces for gas-filled cracks ($K_{por} = 10$ MPa). The two shear velocities do not depend on the pore fluid bulk modulus. P stands for the quasi-longitudinal wave, SV for the quasi-transverse wave with polarization in the zx-plane, and SH for the quasi-transverse wave polarized normal to the zx-plane.

Table 18. Elastic wave velocities v in anisotropic metamorphic and sedimentary rocks at elevated pressures P. The first and second indices of v give the directions of wave propagation and polarization respectively. The z-axis is orthogonal to planes of foliation or bedding. Velocities taken from [Ke81a] are corrected for change of dimensions under pressure; velocities taken from [Ch65, Ch66a] are uncorrected.

Material	P MPa	ϱ kg m^{-3}	v_{xx} m s^{-1}	v_{yy}	v_{zz}	v_{xy}	v_{yx}	v_{xz}	v_{yz}	v_{zx}, v_{zy}	Ref.
Metamorphites:											
Garnet schist, Thomaston, Connecticut, USA (35 plg/An$_{22}$, 30 bi, 29 qu, 5 gar) $\varrho_x = 2750$ kg m^{-3} $\varrho_y = 2760$ kg m^{-3} $\varrho_z = 2760$ kg m^{-3}	10		5400	5300	5000	3400	3700	2600	2800	2600	Ch65, Ch66ε
	20		5700	5600	5300	3600	3700	2700	2900	2800	
	60		6300	6200	5610	3710	3810	2920	3080	2940	
	100		6440	6440	5710	3770	3860	2960	3120	3010	
	200		6580	6540	5840	3820	3880	3020	3140	3040	
	400		6700	6640	5959	3850	3910	3060	3200	3070	
	600		6740	6710	6050	3880	3920	3090	3250	3110	
	800		6770	6760	6110	3900	3940	3140	3280	3140	
	1000		6820	6770	6170	3910	3960	3180	3300	3170	
Gneiss 5, Torrington, Connecticut, USA (50 plg/An$_{35}$, 21.5 hbl, 18.3 bi, 8.9 qu, 1.3 zo) $\varrho_x = 2845$ kg m^{-3} $\varrho_y = 2850$ kg m^{-3} $\varrho_z = 2848$ kg m^{-3}	10		5500	5700	5200	3400	3000	3100	2700	3100	Ch65, Ch66a
	20		5800	5700	5400	3500	3200	3200	2800	3200	
	60		6100	5810	5900	3580	3350	3290	3000	3330	
	100		6290	5880	5980	3650	3450	3340	3120	3370	
	200		6350	5990	6090	3680	3510	3360	3200	3410	
	400		6470	6160	6180	3710	3550	3390	3330	3450	
	600		6540	6360	6240	3730	3600	3460	3380	3480	
	800		6580	6490	6290	3760	3620	3520	3450	3490	
	1000		6630	6580	6330	3780	3680	3570	3500	3520	
Kyanite schist 1, Torrington, Connecticut, USA (27 kyanite, 23 plg/An$_{25}$, 23 mu, 14 bi, 8 qu, 2 gar, 3 op) $\varrho_x = 2990$ kg m^{-3} $\varrho_y = 2950$ kg m^{-3} $\varrho_z = 3070$ kg m^{-3}	10		5000	5600	4600	2900	3100	2600	2700	2600	Ch65, Ch66a
	20		5400	6100	5100	3100	3300	2700	2800	2800	
	60		6200	6550	5800	3280	3300	2950	2910	3010	
	100		6530	6780	6120	3320	3350	3070	3000	3080	
	200		6720	7010	6610	3400	3420	3090	3130	3150	
	400		7070	7170	6920	3480	3520	3150	3170	3230	
	600		7270	7290	7030	3510	3550	3210	3260	3300	
	800		7410	7350	7190	3560	3570	3270	3320	3360	
	1000		7500	7480	7260	3600	3590	3320	3380	3400	

(continued)

Table 18 (continued)

Material	P MPa	ϱ kg m⁻³	v_{xx} m s⁻¹	v_{yy}	v_{zz}	v_{xy}	v_{yx}	v_{xz}	v_{yz}	v_{zx}, v_{zy}	Ref.
Gneiss 2, Torrrington, Connecticut, USA (41.4 kfsp, 26 plg/An₂₃, 22 qu, 10.6 mi) $\varrho_x = 2661$ kg m⁻³ $\varrho_y = 2651$ kg m⁻³ $\varrho_z = 2650$ kg m⁻³	10		4500	4500	4600	2800	2800	2400	2500	2800	Ch65, Ch66a
	20		4800	4900	5100	3000	3000	2500	2700	3000	
	60		5500	5700	5700	3310	3290	2780	3060	3240	
	100		5790	5910	5850	3420	3390	2920	3220	3350	
	200		6040	6090	6040	3530	3490	3230	3350	3390	
	400		6220	6220	6090	3570	3530	3300	3380	3420	
	600		6290	6270	6150	3590	3560	3390	3400	3430	
	800		6350	6310	6210	3610	3590	3440	3430	3450	
	1000		6390	6350	6250	3620	3610	3480	3450	3460	
Serpentinite, origin unknown	50	2752	7465	7550	5659	4488	4486			3201	Ke81a
	100	2751	7533	7624	5771	4508	4508			3226	
	200	2754	7581	7691	5885	4518	4528			3248	
	400	2767	7626	7742	6017	4528	4542			3266	
	600	2780	7652	7770	6129	4532	4548			3277	
Staurolite-garnet schist, Litchfield, Connecticut, USA (44 qu, 23 bi, 12 mu, 13 plg/An₂₅) $\varrho_x = 2750$ kg m⁻³ $\varrho_y = 2760$ kg m⁻³ $\varrho_z = 2750$ kg m⁻³	10		5900	5600		3800	3500	2500	2600	2500	Ch65, Ch66a
	20		6400	6000	4800	3900	3700	2600	2800	2600	
	60		6740	6410	5250	3970	3830	2750	2920	2890	
	100		6870	6540	5410	4000	3900	2790	2950	2930	
	200		6970	6660	5540	4040	3950	2830	3000	2980	
	400		7160	6770	5630	4080	3980	2890	3020	3060	
	600		7200	6850	5700	4120	4000	2950	3050	3090	
	800		7230	6890	5780	4140	4030	2990	3080	3160	
	1000		7250	6930	5850	4170	4040	3040	3100	3210	
Gneiss 6, Goshen, Connecticut, USA (39.4 qu, 31.9 plg/An₃₂, 24.9 mi, 2.8 gar) $\varrho_x = 2760$ kg m⁻³ $\varrho_y = 2750$ kg m⁻³ $\varrho_z = 2760$ kg m⁻³	10		5400	5200	4100	3600	3400	2800	2800	2600	Ch65, Ch66a
	20		5700	5500	4400	3700	3500	2900	2900	2800	
	60		6000	5900	4900	3760	3640	2990	3110	2960	
	100		6170	6050	5160	3840	3730	3120	3210	3110	
	200		6340	6240	5430	3920	3850	3220	3330	3220	
	400		6490	6440	5660	3990	3950	3350	3410	3360	
	600		6550	6520	5810	4050	3980	3400	3450	3410	
	800		6600	6570	5900	4080	4010	3440	3480	3460	
	1000		6650	6630	5990	4110	4030	3470	3510	3500	

(continued)

Table 18 (continued)

Material	ϱ kg m⁻³	P MPa	v_{xx} m s⁻¹	v_{yy}	v_{zz}	v_{xy}	v_{yx}	v_{xz}	v_{yz}	v_{zx}, v_{zy}	Ref.
Sediments:											
Slate, Poultney, Virginia, USA		10	6280	6290	4940	3800	3800	2700	2700	2700	Ch65,
$\varrho_x = 2770$ kg m⁻³		20	6310	6310	4970	3900	3900	2700	2800	2700	Ch66a
$\varrho_y = 2750$ kg m⁻³		60	6350	6330	5040	3890	3900	2740	2820	2770	
$\varrho_z = 2770$ kg m⁻³		100	6370	6360	5090	3920	3930	2780	2850	2800	
		200	6410	6400	5150	3930	3950	2850	2870	2830	
		400	6480	6460	5290	3980	4010	2930	2980	2920	
		600	6540	6510	5410	4020	4050	2970	3050	3000	
		800	6590	6570	5500	4050	4080	3030	3090	3050	
		1000	6660	6630	5590	4070	4100	3090	3110	3080	
Slate 949, Rheinisches Schiefergebirge, Germany		500	6272	6473	4915	3892	3900			2679	Ke81a
(85 phy, 8 ab+qu, 5 car, 2 mi)	2797	500									
	2799	1000	6341	6530	5028	3917	3928			2741	
	2808	2000	6409	6600	5186	3945	3962			2846	
	2831	4000	6512	6706	5428	3987	4009			2988	
	2859	6000	6593	6787	5599	4014	4038			3071	

Table 19. Detailed investigations of elastic wave propagation in anisotropic ultrabasic rocks at elevated pressures [Ch71, Ba72]; ϱ = bulk density, v = velocity.

Material	ϱ kg m⁻³	Direction of wave propagation	Direction of wave displacement	v [m s⁻¹] 50	100	200	400	600	800	1000 [MPa]	Ref.
Dunite A, Twin Sisters, Washington, USA	3260	[001]	[001]		8568	8622	8688	8726	8750	8762	Ch71
(83.1 ol, 12.1 serp, 2.7 op, 2.1 en)			[100]		4507	4530	4560	4578	4589	4597	
$\bar{m} = 20.84$			[010]		4506	4531	4565	4587	4595	4598	
strong concentration of olivine,	3224	[100]	[100]		7854	7896	7946	7972	7992	8007	
a axes parallel to [001];			[010]		4352	4381	4414	4430	4440	4447	
girdles of b and c axes ⊥ to [001]			[001]		4572	4599	4635	4660	4671	4677	
	3248	[010]	[010]		7797	7839	7896	7922	7945	7957	
			[001]		4597	4621	4643	4657	4667	4676	
			[100]		4319	4341	4375	4393	4400	4405	
	3296	[110]	[110]		7769	7813	7857	7884	7900	7912	
			[1̄10]		4344	4360	4379	4392	4402	4415	
			[001]		4577	4618	4665	4683	4694	4696	

(continued)

Table 19 (continued)

Material	ϱ kg m⁻³	Direction of wave propagation	Direction of wave displacement	v [m s⁻¹] 50	100	200	400	600	800	1000 [MPa]	Ref.
Dunite A (continued)	3269	[110]	[1̄10]		7852	7905	7965	8001	8026	8046	Ch71
			[110]		4279	4304	4335	4350	4358	4362	
			[001]		4617	4647	4677	4687	4689	4691	
	3286	[011]	[011̄] ∥		8339	8383	8444	8481	8501	8504	
	3256	[101]	[101] ∥		8346	8393	8464	8508	8530	8536	
Dunite B, Twin Sisters, Washington, USA (96.3 ol, 2.6 op, 0.9 en, 0.2 serp) $\bar{m}=20.98$	3300	[100]	[100]		8939	9002	9069	9103	9128	9150	
			[010]		4667	4709	4763	4786	4814	4832	
			[001]		4809	4842	4885	4912	4934	4946	
strong concentrations of olivine, a, b and c axes parallel to [100], [010] and [001], respectively	3322	[010]	[010]		7664	7704	7756	7788	7810	7831	
			[001]		4631	4673	4733	4774	4799	4814	
			[100]		4645	4692	4753	4788	4806	4815	
	3329	[001]	[001]		8062	8119	8182	8222	8250	8272	
			[100]		4921	4931	4948	4962	4971	4980	
			[010]		4634	4663	4704	4727	4740	4747	
	3315	[011]	[011] ∥		7915	7975	8042	8092	8133	8166	
	3327	[101]	[101] ∥		8438	8511	8584	8627	8665	8701	
	3308	[110]	[110] ∥		8450	8515	8576	8613	8650	8686	
Dunite C, Twin Sisters, Washington, USA (94 ol/Fo₉₀, 5 en, 1 chr) petrofabric similar to sample B of [Ch71]	3310	[100]	[100]	8960	9000	9040	9090	9140	9190	9240	Ba72
			[010]	4880	4890	4920	4960	4980	4990	5000	
			[001]	4730	4760	4780	4810	4830	4840	4850	
	3310	[101̄]	[101] ∥	8380	8410	8440	8490	8530	8560	8590	
			[101] ∥	4880	4900	4930	4970	4990	5000	5000	
			[010] ∥	4850	4880	4910	4940	4960	4980	5000	
	3300	[001]	[001]	7850	7900	7950	8000	8050	8090	8140	
			[100]	4830	4850	4870	4910	4930	4940	4950	
			[010]	4490	4510	4550	4580	4590	4600	4600	
	3300	[101̄]	[101] ∥	8000	8050	8080	8140	8180	8220	8270	
			[101] ∥	4710	4760	4800	4830	4860	4880	4900	
			[010] ∥	4560	4590	4620	4660	4680	4700	4710	
	3280	[11̄0]	[110] ∥	8310	8350	8390	8450	8490	8540	8580	
			[110] ∥	4850	4890	4920	4950	4970	4990	5010	
			[001] ∥	4780	4820	4860	4890	4920	4940	4960	

(continued)

Table 19 (continued)

Material	ϱ kg m⁻³	Direction of wave propagation	Direction of wave displacement	v [m s⁻¹] 50	100	200	400	600	800	1000 [MPa]	Ref.
Dunite C (continued)	3280	$[\bar{1}11]$	$\parallel [\bar{1}11]$	7920	7980	8020	8090	8140	8190	8230	Ba72
			$\parallel [111]$	4470	4500	4540	4580	4610	4640	4670	
			$\parallel [1\bar{1}0]$	4650	4700	4750	4770	4790	4800	4810	
	3280	$[\bar{1}11]$	$\parallel [\bar{1}1\bar{1}]$	7690	7730	7780	7850	7910	7950	7980	
			$\parallel [1\bar{1}1]$	4630	4670	4710	4740	4750	4760	4770	
			$\parallel [1\bar{1}0]$	4630	4650	4690	4720	4730	4740	4750	
	3300	$[010]$	$[010]$	7880	7930	8010	8100	8140	8170	8190	
			$[100]$	4750	4770	4800	4860	4890	4900	4910	
			$[001]$	4420	4450	4490	4520	4540	4550	4570	
	3260	$[011]$	$[011]$	7800	7850	7900	7950	7980	8020	8060	
			$[01\bar{1}]$	4430	4450	4470	4500	4530	4560	4580	
			$[100]$	4710	4730	4760	4790	4810	4830	4850	
	3320	$[01\bar{1}]$	$[01\bar{1}]$	7920	7960	8010	8060	8100	8120	8150	
			$[011]$	4480	4500	4520	4540	4570	4600	4630	
			$[100]$	4890	4910	4930	4960	4980	5000	5010	
	3270	$[\bar{1}10]$	$\parallel [\bar{1}10]$	8520	8560	8600	8650	8670	8690	8710	
			$\parallel [1\bar{1}0]$	4880	4920	4940	4960	4980	4990	5010	
			$\parallel [001]$	4800	4820	4850	4890	4910	4930	4940	
	3300	$[\bar{1}11]$	$\parallel [11\bar{1}]$	8400	8430	8460	8530	8600	8650	8700	
			$\parallel [1\bar{1}1]$	4650	4670	4700	4730	4740	4740	4750	
			$\parallel [1\bar{1}0]$	4830	4850	4870	4900	4910	4920	4930	
	3300	$[\bar{1}11]$	$\parallel [11\bar{1}]$	7740	7830	7890	7960	8010	8040	8060	
			$\parallel [1\bar{1}1]$	4670	4700	4760	4800	4820	4830	4840	
			$\parallel [1\bar{1}0]$	4630	4650	4690	4740	4770	4780	4790	
Bronzitite, Stillwater, Montana, USA (93 br/En$_{83}$, 3.5 am, 3 ol, 0.5 ma) petrofabric is more random than that of dunite; weak concentration of crystallographic a axes of bronzite in the plane normal to [010]	3290	$[100]$	$[100]$	7640	7670	7720	7810	7880	7930	7970	
			$[001]$	4450	4480	4500	4520	4550	4570	4590	
			$[010]$	4450	4470	4490	4520	4540	4570	4590	
	3290	$[101]$	$[101]$	7760	7790	7830	7890	7940	7980	8020	
			$[10\bar{1}]$	4500	4520	4540	4570	4590	4610	4630	
			$[010]$	4500	4520	4550	4580	4600	4620	4640	
	3290	$[00\bar{1}]$	$[001]$	7760	7810	7860	7940	8010	8080	8140	
			$[100]$	4530	4560	4580	4600	4610	4620	4630	
			$[010]$	4550	4580	4600	4620	4630	4650	4660	

(continued)

Table 19 (continued)

Material	ϱ kg m^{-3}	Direction of wave propagation	Direction of wave displacement	v [m s^{-1}] 50	100	200	400	600	800	1000 [MPa]	Ref.
Bronzitite (continued)	3290	$[10\bar1]$	$[10\bar1]$	7730	7770	7800	7860	7930	7990	8050	
			$[\bar10\bar1]$	4550	4580	4620	4670	4700	4720	4730	
			$[010]$	4530	4570	4620	4680	4720	4740	4750	
	3260	$[\bar110]$	$[110]$	7620	7660	7690	7750	7810	7850	7890	
			$[\bar110]$	4470	4480	4510	4540	4560	4570	4590	
			$[001]$	4570	4590	4600	4620	4630	4640	4660	
	3220	$[\bar111]$	$[\bar111]$	7450	7490	7540	7630	7710	7760	7820	
			$[1\bar11]$	4290	4310	4330	4350	4370	4380	4390	
			$[110]$	4330	4360	4390	4420	4440	4460	4480	
	3250	$[111]$	$[111]$	7550	7600	7650	7730	7780	7830	7870	
			$[1\bar11]$	4400	4430	4460	4500	4510	4520	4530	
			$[110]$	4390	4430	4450	4470	4490	4500	4510	
	3240	$[010]$	$[010]$	7350	7400	7470	7560	7610	7650	7670	
			$[001]$	4290	4330	4370	4400	4420	4440	4460	
			$[100]$	4240	4270	4330	4390	4420	4440	4460	
	3280	$[011]$	$[011]$	7690	7730	7770	7830	7880	7940	7980	
			$[01\bar1]$	4550	4570	4590	4610	4630	4640	4660	
			$[100]$	4600	4630	4650	4670	4690	4700	4720	
	3210	$[01\bar1]$	$[011]$	7320	7370	7410	7490	7570	7620	7650	
			$[01\bar1]$	4240	4260	4280	4320	4340	4360	4370	
			$[100]$	4340	4370	4400	4430	4450	4470	4480	
	3260	$[\bar110]$	$[110]$	7500	7550	7620	7660	7710	7750	7780	
			$[\bar110]$	4480	4500	4530	4570	4590	4610	4630	
			$[001]$	4470	4500	4520	4560	4580	4600	4620	
	3270	$[111]$	$[\bar111]$	7710	7760	7800	7870	7920	7980	8020	
			$[\bar110]$	4580	4620	4650	4680	4700	4720	4730	
			$[1\bar10]$	4670	4720	4740	4760	4780	4800	4810	
	3240	$[\bar111]$	$[111]$	7480	7510	7550	7620	7680	7730	7770	
			$[1\bar11]$	4340	4370	4390	4430	4460	4480	4490	
			$[110]$	4410	4430	4450	4470	4490	4500	4510	

References for 3.1.1 ··· 3.1.3 — Literatur zu 3.1.1 ··· 3.1.3

A161 Aleksandrov, K.S., Ryzhova, T.V.: Izv. Akad. Nauk SSSR, Ser. Geofiz. **12** (1961) 1799.
A161a Aleksandrov, K.S., Ryzhova, T.V.: Sov. Phys. Cryst. (English Transl.) **6** (1961) 228.
A162 Aleksandrov, K.S., Ryzhova, T.V.: Izv., Acad. Sci. USSR, Geophys. Ser. (English Transl.) **1962**, 129.
A164 Aleksandrov, K.S., Ryzhova, T.V., Belikov, B.P.: Sov. Phys. Cryst. (English Transl.) **9** (1964) 589.
An67 Anderson, D.L.: Geophys. J. R. Astron. Soc. **13** (1967) 9.
An68 Anderson, O.L., Liebermann, R.C., in: Physical Acoustics, Vol. IVB, ed.: W.P. Mason, New York: Academic Press **1968**, p. 329.
An69 Anderson, D.L.: J. Geophys. Res. **74** (1969) 3857.
An69a Anderson, D.L., Sammis, C.G., Phinney, R.A., in: The Application of Modern Physics to the Earth and Planetary Interiors, ed.: S.K. Runcorn, Wiley and Sons **1969**, p. 465.
An74 Anderson, D.L., Minster, B., Cole, D.: J. Geophys. Res. **79** (1974) 4011.
As76 Aslanyan, A.T., Volarovich, M.P., Levykin, A.I., Beguni, A.T., Artunyan, A.V., Skvortsova, L.S.: Izv., Acad. Sci. USSR, Phys. Solid Earth (English Transl.) **12** (1976) 96.
At41 Atanasoff, J.V., Hart, P.J.: Phys. Rev. **59** (1941) 85.
Ba65 Babuška, V.: Geofys. Sb. **223** (1965) 275.
Ba72 Babuška, V.: J. Geophys. Res. **77** (1972) 6955.
Ba72a Baker, D.W., Charter, N.L., in: Flow and Fracture of Rocks, eds.: H.C. Heard et al., Am. Geophys. Union, Washington, **1972**, p. 157.
Ba78 Babuška, V., Fiala, J., Mayson, D.J., Liebermann, R.C.: Stud. Geophys. Geod. **22** (1978) 349.
Be54 Bergmann, L.: Der Ultraschall, Stuttgart: S. Hirzel Verlag, **1954**.
Bi60 Birch, F.: J. Geophys. Res. **65** (1960) 1083.
Bi61 Birch, F.: J. Geophys. Res. **66** (1961) 2199.
Bo77 Bonczar, L.J., Graham, E.K., Wang, H.: J. Geophys. Res. **82** (1977) 2529.
Bo80 Boyce, R.E., in: Initial Reports of the Deep Sea Drilling Project **50**, eds.: Y. Lancelot et al., Washington **1980**.
Br65 Brace, W.F.: J. Geophys. Res. **70** (1965) 391.
Br65a Brace, W.F.: J. Geophys. Res. **70** (1965) 5657.
Br65b Brace, W.F.: J. Geophys. Res. **70** (1965) 5669.
Ch65 Christensen, N.I.: J. Geophys. Res. **70** (1965) 6147.
Ch66 Christensen, N.I.: J. Geophys. Res. **71** (1966) 5921.
Ch66a Christensen, N.I.: J. Geophys. Res. **71** (1966) 3549.
Ch68 Christensen, N.I.: Tectonophysics **6** (1968) 331.
Ch70 Christensen, N.I., Shaw, G.H.: Geophys. J.R. Astron. Soc. **20** (1970) 271.
Ch71 Christensen, N.I., Ramananantoandro, R.: J. Geophys. Res. **76** (1971) 4003.
Ch72 Christensen, N.I.: Geophys. J. R. Astron. Soc. **28** (1972) 425.
Ch72a Chung, D.H.: Science **177** (1972) 261.
Ch73 Chung, D.H.: Earth Planet. Sci. Lett. **18** (1973) 125.
Ch73a Christensen, N.I., Salisbury, M.H.: Earth Planet. Sci. Lett. **19** (1973) 461.
Ch74 Christensen, N.I.: J. Geophys. Res. **79** (1974) 407.
Ch75 Christensen, N.I., Fountain, D.M.: Geol. Soc. Am. Bull. **86** (1975) 227.
Ch75a Christensen, N.I., Salisbury, M.H.: Rev. Geophys. Space Phys. **13** (1975) 57.
Ch78 Christensen, N.I.: Tectonophysics **47** (1978) 131.
Cl66 Clark, G.B.: Deformation moduli of rocks; in: Testing techniques for rock mechanics, Am. Soc. Testing and Materials, Philadelphia **1966**, p. 133.
Cr71 Crosson, R.S., Lin, J.-W.: J. Geophys. Res. **76** (1971) 570.
Da68 Dandekhar, D.P.: J. Appl. Phys. **39** (1968) 3694.
Da68a Dandekhar, D.P., Ruoff, A.L.: J. Appl. Phys. **39** (1968) 6004.
Da71 Davies, G.D., Anderson, D.L.: J. Geophys. Res. **76** (1971) 2617.
Do77 Domenico, S.N.: Geophysics **42** (1977) 1339.
Dr74 Dreyer, W.: Gebirgsmechanik im Salz, Stuttgart: Enke-Verlag **1974**.
El75 Elliott, S.E., Wiley, B.F.: Geophysics **40** (1975) 949.
Fi71 Fielitz, K.: Z. Geophys. **37** (1971) 943.
Fr72 Frisillo, A.L., Barsch, G.R.: J. Geophys. Res. **77** (1972) 6360.
Ga74 Gardner, G.H.F., Gardner, L.W., Gregory, A.R.: Geophysics **39** (1974) 770.

Ge61	Geertsma, J., Smit, D.C.: Geophysics **26** (1961) 169.
Go76	Goto, T., Ohno, I., Sumino, Y.: J. Phys. Earth **24** (1976) 149.
Gr65	Grant, F.S., West, G.F.: Interpretation theory in applied geophysics; New York: McGraw-Hill **1965**.
Gr69	Graham, E.K., Barsch, G.R.: J. Geophys. Res. **74** (1969) 5949.
Ha62	Hashin, Z., Shtrikman, S.: J. Mech. Phys. Solids **10** (1962) 335.
Ha62a	Hashin, Z., Shtrikman, S.: J. Mech. Phys. Solids **10** (1962) 343.
Ha70	Hamilton, E.L.: J. Geophys. Res. **75** (1970) 4423.
Ha71	Hamilton, E.L.: Geophysics **36** (1971) 266.
Ha79	Hamano, Y., in: Initial Reports of the Deep Sea Drilling Project **51, 52, 53,** eds.: T. Donnelly et al., Washington **1979**, p. 1457.
He61	Hearmon, R.F.S.: An Introduction to Applied Anisotropic Elasticity, Oxford: University Press **1961**.
He66	Hearmon, R.F.S.: The elastic constants of non-piezoelectric crystals; in: Landolt-Börnstein, N.S., Vol. III/1, Berlin: Springer-Verlag **1966**, S. 1.
He69	Hearmon, R.F.S.: The elastic constants of non-piezoelectric crystals; in: Landolt-Börnstein, N.S. Vol. III/2, Berlin: Springer-Verlag **1969**, S. 1.
He79	Hearmon, R.F.S., in: Landolt-Börnstein, N.S. Vol. III/11, Berlin: Springer-Verlag **1979**, S. 1.
Hi52	Hill, R.: Proc. Phys. Soc. London, Sect. A **65** (1952) 349.
Hi56	Hicks, G.W., Berry, J.E.: Geophysics **21** (1956) 739.
Hi63	Hill, R.: J. Mech. Phys. Solids **11** (1963) 357.
Hö70	Höchli, U.T.: Solid State Commun. **8** (1970) 1487.
Hu50	Hughes, D.S., Jones, H.J.: Bull. Geol. Soc. Am. **61** (1950) 843.
Hu51	Hughes, D.S., Cross, J.H.: Geophysics **16** (1951) 577.
Hu52	Hughes, D.S., Kelly, J.L.: Geophysics **17** (1952) 739.
Hu56	Hughes, D.S., Maurette, Ch.: Geophysics **21** (1956) 277.
Hu57	Hughes, D.S., Maurette, Ch.: Geophysics **22** (1957) 23.
Hu68	Hurtig, E.: Geol. Geophys. **12** (1968) 3.
Id37	Ide, J.M.: J. Geol. **45** (1937) 689.
Is76	Isaak, D.G., Graham, E.K.: J. Geophys. Res. **81** (1976) 2483.
Jo74	Johnson, L.R., Wenk, H.-R.: Tectonophysics **23** (1974) 79.
Jo80	Johnston, D.H., Toksök, M.N.: J. Geophys. Res. **85** (1980) 925.
Ju64	Judd, W.R., in: State of stress in the earth's crust, ed.: W.R. Judd, New York: Am. Elsevier **1964**.
Ka48	Kammer, E.W., Pardue, T.E., Frissel, H.F.: J. Appl. Phys. **19** (1948) 265.
Ka65	Kanamori, H., Mizutani, H.: Bull Earthquake Res. Inst. **43** (1965) 173.
Ka70	Kanamori, H., Mizutani, H.: Addenda, Bull. Earthquake Res. Inst. **48** (1970) 1009.
Ke75	Kern, H., Fakhimi, M.: Tectonophysics **28** (1975) 277.
Ku78	Kern, H.: Tectonophysics **44** (1978) 185.
Ke79	Kern, H.: Phys. Chem. Minerals **4** (1979) 161.
Ke79a	Kern, H., Richter, A.: Tectonophysics **54** (1979) 231.
Ke81	Kern, H., Richter, A.: J. Geophys. **49** (1981) 47.
Ke81a	Kern, H.: Private Commun.
Kl68	Klima, K., Babuška, V.: Stud. Geophys. Geod. **12** (1968) 377.
Ko69	Kovach, R.L., Russell, R.: Bull. Seismol. Soc. Am. **59** (1969) 1653.
Ko80	Kono, M., Hamano, Y., Morgan, W.J., in: Initial Reports of the Deep Sea Drilling Project **55**, eds.: E.D. Jackson et al., Washington **1980**, p. 715.
Kr67	Kröner, E.: J. Mech. Phys. Solids **15** (1967) 319.
Kr76	Kroenke, L.W., Manghnani, M.H., Rai, C.S., Fryer, P., Ramananantoandro, R., in: The Geophysics of the Pacific Ocean Basins and its Margins, eds.: G.H. Sutton et al., Am. Geophys. Union, Washington, D.C. **1976**.
Ku54	Kubotera, A.: J. Phys. Earth **2** (1954) 33.
Ku69	Kumazawa, M., Anderson, O.L.: J. Geophys. Res. **74** (1969) 5961.
Ku69a	Kumazawa, M.: J. Geophys. Res. **74** (1969) 5973.
Ku69b	Kumazawa, M.: J. Geophys. Res. **74** (1969) 5311.
La78	Lama, R.D., Vutukuri, V.S.: Handbook on Mechanical Properties of Rocks, – Testing Techniques and Results – Vol. II, Trans. Tech. Publications, Clausthal **1978**.
Le79	Levien, L., Weidner, D.J., Prewitt, C.T.: Phys. Chem. Minerals **4** (1979) 105.

Ma68	Manghnani, M.H., Woollard, G.P., in: The Crust and Upper Mantle of the Pacific Area, eds.: L. Knopoff et al., Am. Geophys. Union, Washington **1968**
Ma70	Mao, N.-H., Ito, J., Hays, J.F., Drake, J., Birch, F.: J. Geophys. Res. **75** (1970) 4071.
Ma74	Manghnani, M.H., Ramananantoandro, R.: J. Geophys. Res. **79** (1974) 5427.
Ma74a	Mao, N.H.: J. Geophys. Res. **79** (1974) 5447.
Ma76	Martinez, E.: Dipl.-Thesis, Inst. f. Allg. Angew. Geophys., Univ. München **1976**.
Me66	Meister, R., Peselnick, L.: J. Appl. Phys. **37** (1966) 4121.
Mo60	Molotova, L.V., Vassil'ev, Y.I.: Izv., Acad. Sci. USSR, Geophys. Ser. (English Transl.) **1960**, 731.
Mü79	Müller, H.J., Raab, S., Seipold, U., in: Theoretical and experimental investigations of physical properties of rocks and minerals under extreme p, T-conditions, eds.: H. Stiller et al., Berlin: Akademie-Verlag **1979**, p. 15.
Na57	Nafe, J.E., Drake, C.L.: Geophysics **22** (1957) 523.
Nu69	Nur, A., Simmons, G.: Earth Planet. Sci. Lett. **7** (1969) 183.
Nu69a	Nur, A., Simmons, G.: J. Geophys. Res. **74** (1969) 6667.
Ny57	Nye, J.F.: Physical Properties of Crystals, Oxford: Clarendon Press **1957**.
Oh76	Ohno, I.: J. Phys. Earth **24** (1976) 355.
Pa78	Paterson, M.S.: Experimental Rock Deformation – The Brittle Field, Berlin-Heidelberg-New York: Springer **1978**.
Pe61	Peselnick, L., Outerbridge, W.L.: J. Geophys. Res. **66** (1961) 581.
Pe62	Peselnick, L.: J. Geophys. Res. **67** (1962) 4441.
Pe65	Peselnick, L., Meister, R.: J. Appl. Phys. **36** (1965) 2879.
Pe68	Peselnick, L., Wilson, W.H.: J. Geophys. Res. **73** (1968) 3271.
Pr62	Pros, Z., Vaněk, J., Klima, K.: Stud. Geophys. Geod. **6** (1962) 347.
Pr66	Press, F., in: Handbook of Physical Constants, ed.: S.P. Clark jr., Geol. Soc. Am. Mem. 97, **1966**, p. 195.
Re29	Reuss, A.: Z. Angew. Math. Mech. **9** (1929) 49.
Ry62	Ryzhova, T.V., Aleksandrov, S.K.: Izv., Acad. Sci. USSR, Geophys. Ser. (English Transl.) **1962**, 1125.
Ry64	Ryzhova, T.V.: Izv. Akad. Nauk SSSR, Ser. Geofiz. **7** (1964) 1049.
Ry65	Ryzhova, T.V., Aleksandrov, K.S.: Izv., Acad. Sci. USSR, Phys. Solid Earth (English Transl.) **1965**, 53.
Ry66	Ryzhova, T.V., Aleksandrov, K.S., Korobkova, V.M.: Izv., Acad. Sci. USSR, Phys. Solid Earth (English Transl.) **1966**, 111.
Sch65	Schwerdtner, M.W., Tou, J.C., Hertz, P.B.: Can. J. Earth Sci. **2** (1965) 673.
Sch67	Schaafs, W., in: Landolt-Börnstein, N.S., Vol. II/5, Molekularakustik, Berlin-Heidelberg-New York: Springer **1967**, S. 71.
Sh58	Shumway, G.: Geophysics **23** (1958) 494.
Sh72	Shankland, T.J.: EOS **53** (1972) 1120.
Sh74	Shankland, T.J., Chung, D.H.: Phys. Earth Planet. Interiors **8** (1974) 121.
Sh77	Shankland, T.J.: Geophys. Surveys **3** (1977) 69.
Si64	Simmons, G.: J. Geophys. Res. **69** (1964) 1117.
Si64a	Simmons, G.: J. Geophys. Res. **69** (1964) 1123.
Si65	Simmons, G.: Proc. IEEE **53** (1965) 1337.
Si65a	Simmons, G., Brace, W.F.: J. Geophys. Res. **70** (1965) 5649.
Si71	Simmons, G., Wang, H.: Single Crystal Elastic Constants and Calculated Aggregate Properties: A Handbook, Cambridge, Mass.: M.I.T. Press **1971**.
Sm74	Smagin, A.G., Milstein, B.G.: Kristallografiya **19** (1974) 832.
So67	Soga, N., Anderson, O.L.: J. Am. Ceram. Soc. **50** (1967) 239.
So79	Sondergeld, C.H., Schreiber, E.: Phys. Chem. Minerals **5** (1979) 21.
Sp70	Spetzler, H.: J. Geophys. Res. **75** (1970) 2073.
Sp72	Spetzler, H., Schreiber, E., O'Connell, R.: J. Geophys. Res. **77** (1972) 4938.
St73	Städtler, G.: Dipl.-Thesis, Inst. f. Allg. Angew. Geophys., Univ. München **1973**.
Ta59	Talwani, M., Sutton, G.H., Worzel, J.L.: J. Geophys. Res. **64** (1959) 1545.
Th72	Thomsen, L.: J. Geophys. Res. **77** (1972) 315.
Ti68	Timur, A.: Geophysics **33** (1968) 584.
Tr69	Truell, R., Elbaum, Ch., Chick, B.C.: Ultrasonic Methods in Solid State Physics, New York-London: Academic Press **1969**.

Va78	Vaughan, M.T., Weidner, D.J.: Phys. Chem. Minerals **3** (1978) 133.
Ve60	Verma, R.K.: J. Geophys. Res. **65** (1960) 757.
Vo28	Voigt, W.: Lehrbuch der Kristallphysik, Leipzig: Teubner, **1928**.
Vo75	Volarovich, M.P., Bayuk, Y.I., Valyus, V.P., Galkin, I.N.: Izv., Acad. Sci. USSR, Phys. Solid Earth (English Transl.) **11** (1975) 319.
Wa65	Walsh, J.B.: J. Geophys. Res. **70** (1965) 381.
Wa68	Wang, C.Y.: Nature **218** (1968) 74.
Wa69	Wang, C.Y.: J. Geophys. Res. **74** (1969) 1451.
Wa72	Wang, H., Simmons, G.: J. Geophys. Res. **77** (1972) 4379.
Wa74	Wang, H., Simmons, G.: J. Geophys. Res. **79** (1974) 2607.
Wa75	Wang, C.Y., Lin, W., Wenk, H.-R.: J. Geophys. Res. **80** (1975) 1065.
Wa76	Watt, J.P., Davies, G.F., O'Connell, R.J.: Rev. Geophys. Space Phys. **14** (1976) 541.
We69	Wenk, H.-R., Wenk, E.: Schweiz. Min. Petr. Mitt. **49** (1969) 343.
We75	Weidner, D.J., Swyler, K., Carleton, H.R.: Geophys. Res. Lett. **2** (1975) 189.
We77	Weidner, D.J., Carleton, H.R.: J. Geophys. Res. **82** (1977) 1334.
We78	Weidner, D.J., Wang, H., Ito, J.: Phys. Earth Planet. Interiors **17** (1978) P7.
Wo41	Wood, A.B.: A Textbook of Sound, New York: MacMillan **1964**.
Wo59	Woollard, G.P.: J. Geophys. Res. **64** (1959) 1521.
Wo63	Woeber, A.F., Katz, S., Ahrens, T.J.: Geophysics **28** (1963) 658.
Wy56	Wyllie, M.R.J., Gregory, A.R., Gardner, L.W.: Geophysics **21** (1956) 41.
Wy58	Wyllie, M.R.J., Gregory, A.R., Gardner, G.H.F.: Geophysics **23** (1958) 459.
Za74	Zaslavskiy, B.I., Krupnyy, A.I., Aleksandrov, K.S.: Izv., Acad. Sci. USSR, Phys. Solid Earth (English Transl.) **1974**, 515.
Za74a	Zaslavskiy, B.I., Usol'Tsev, Y.K., Aleksandrov, K.S.: Izv., Acad. Sci. USSR, Phys. Solid Earth (English Transl.) **1974**, 835.
Zu62	Zubov, V.G., Firsova, M.M.: Kristallografiya **7** (1962) 469.

3.1.4 Elastic wave velocities and constants of elasticity of rocks at elevated pressures and temperatures — Geschwindigkeiten elastischer Wellen und Elastizitäts-Konstanten bei erhöhten Drucken und Temperaturen

3.1.4.0 Introduction — Einführung

The physical properties of polycrystalline rock material under the conditions of greater depth are controlled by the interaction of physical and lithologic parameters (Table 1). Pressure and temperature increase with depth; in addition, the mineralogical composition as well as the rock microstructure may change with depth.

Die physikalischen Eigenschaften polykristalliner Gesteine werden unter den Bedingungen größerer Erdtiefe durch das Zusammenwirken von physikalischen und lithologischen Parametern bestimmt (Tab. 1). Druck und Temperatur nehmen mit der Tiefe zu; darüberhinaus ändern sich die Gesteine hinsichtlich ihrer mineralogischen Zusammensetzung und Mikrostruktur mit der Tiefe.

Table 1. Factors controlling the elastic properties of rocks. – Faktoren, die die elastischen Eigenschaften von Gesteinen bestimmen.

Physical parameters	Lithologic parameters	Lithologische Parameter
Temperature (T)	Chemical composition	Chemische Zusammensetzung
Effective pressure (P_{eff})	Mineralogical composition	Mineralogische Zusammensetzung
($P_{eff} = P_c - n \cdot P_f$)	Rock fabric	Gesteinsgefüge
$n \lesssim 1$ (e.g. [Ro73])	distribution and orientation of minerals (shape and lattice orientation)	Verteilung und Orientierung der Minerale (Orientierung nach der Kornform und dem Kristallgitter)
	distribution and orientation of micro-cracks (pore space)	Verteilung und Orientierung von Mikrorissen (Porenraum)
	grain size and pore size	Korngröße und Porengröße
	fluid content	Gehalte an Flüssigkeiten

In dry, low-porous rocks P_{eff} is identical to P_c; in porous fluid-filled rocks, P_f is linked to P_{eff} [Hu59, Ha63] by the equation

In trockenen, porenarmen Gesteinen ist P_{eff} identisch mit P_c, in porösen, mit Flüssigkeit gefüllten Gesteinen ist P_f [Hu59, Ha63] gemäß der Gleichung

$$P_{eff} = P_c - n \cdot P_f \quad (n \lesssim 1)$$

Every change in the pore-fluid pressure under otherwise constant conditions automatically results in a change of the effective pressure.

Compared with the wealth of data about the pressure dependence of ultrasonic wave velocities in rocks (e.g. [Bi60, Bi61, Si64, Ch65, Ch66, Ch74]), the literature about the influence of temperature on wave velocities at constant confining pressure is scarce.

Reference to authors working in this field will be made at appropriate places in the text and in the tables.

mit P_{eff} verbunden. Jede Änderung des Porendruckes führt bei sonst gleichbleibenden Bedingungen automatisch zu einer Änderung des Effektiv-Druckes.

Im Vergleich zu der Vielzahl von Daten über die Druckabhängigkeit der Geschwindigkeit von Ultraschallwellen in Gesteinen (z.B. [Bi60, Bi61, Si64, Ch65, Ch66, Ch74]) gibt es im Schrifttum relativ wenig Arbeiten über den Temperatureinfluß bei konstantem Umschließungsdruck.

Auf die einschlägigen Arbeiten wird an geeigneter Stelle im Text und in den Tabellen Bezug genommen.

3.1.4.1 Symbols and abbreviations—Symbole und Abkürzungen

P_c	[bar]	confining pressure
P_{eff}	[bar]	effective pressure (matrix pressure, grain to grain pressure)
P_f	[bar]	pore fluid pressure
T	[°C]	temperature
$\varrho(*)$	[g cm^{-3}]	bulk density (* at pressure)
v_P	[km s^{-1}]	P-wave velocity
v_S	[km s^{-1}]	S-wave velocity
K	[Mbar]	bulk modulus
G	[Mbar]	shear modulus
λ	[Mbar]	Lamé constant
E	[Mbar]	Young's modulus
σ	[Mbar]	Poisson's ratio
φ		seismic parameter
Φ	[vol-%]	porosity
$\bar{m}(*)$		mean atomic weight (* calculated by the present author)

Under "Notes"

$\bar{3}\perp$	average of data measured in three orthogonal directions in the same sample
$\bar{3}\cdot1\perp$	average of data measured in three orthogonal directions in three different samples
C	parameters, derivatives etc. calculated by the present author
A	% velocity anisotropy $\left(A = \dfrac{v_{max} - v_{min}}{v_{max}} \cdot 100\right)$

3.1.4.2 Experimental techniques — Experimentelle Methoden

In general, the experimental data has been obtained by the method of "pulse transmission" through rock samples of a few centimeters length and using piezoelectric transducers with natural frequencies of about 1···2 MHz. Distinction of two groups of experimental techniques can be made on the basis of the method by which pressure is generated and of the position of the transducers relative to the specimen:

Im allgemeinen wurden die Daten an Gesteinsproben von einigen Zentimetern Größe nach der „Durchschallungsmethode" ermittelt. Die natürlichen Frequenzen der verwendeten piezoelektrischen Wandler sind etwa gleich 1···2 MHz. Nach Art der Druckerzeugung und nach Lage der Wandler zur Probe sind zwei verschiedene Meßverfahren zu unterscheiden:

(1) Measurements on jacketed cylindric samples in internally heated fluid and gas apparatures with the transducers placed directly onto the sealed specimen [Hu51, Hu56, Hu57, Sp76, Kr76, St77, St78, Ra78, Ch79]. Transmitting and receiving transducers are exposed to pressure and temperature during the experiments. Because the Curie temperature limits the peak temperature of the available transducer material, successful measurements are at present possible up to about 450···500 °C. Sealing of the specimens allows the study of the effect of pore pressure on wave velocities.

(2) Experiments on unjacketed specimens in cubic pressure apparatus with heated anvils. A state of near hydrostatic stress is achieved by pressing six pyramidal pistons in the three mutually orthogonal directions onto cube-shaped specimens [Fi71, Ke75, Ke78, Ke79, Ke81, Ke82]. Transducers are placed on the low temperature side of the pistons. Thus, no special requirement of the Curie temperature is needed for the transducer material and measurements are possible up to temperatures of about 750 °C. The geometry of the piston-sample arrangement allows simultaneous measurements in the three orthogonal directions of the sample cubes, thereby enabling us the determination of the directional dependence of wave velocity and the calculation of the change of volume (i.e. change of density) at temperature and pressure from piston displacement.

In general, the experiments are carried out on oven-dried specimens ("dry rocks")[1]. Only few experiments have been done with confining pressure and temperature under controlled fluid pressure conditions ("wet rocks"), see subsection 3.1.4.4.2.

(1) Messung an ummantelten zylindrischen Proben in innengeheizten Flüssigkeits- und Gasapparaturen [Hu51, Hu56, Hu57, Sp76, St77, St78, Ra78, Ch79]. Die piezoelektrischen Wandler sind direkt auf die ummantelte Probe montiert und während des Meßvorganges Drucken und Temperaturen ausgesetzt. Die maximalen Meßtemperaturen sind deshalb durch die Curie-Temperatur des Wandlermaterials festgelegt. Derzeit sind Experimente bis zu Temperaturen von 450···500 °C möglich. Durch Abkapselung der Proben kann auch der Einfluß des Porendruckes auf die Wellengeschwindigkeiten untersucht werden.

(2) Experimente an nicht-ummantelten Würfelproben in dreiaxialen Stempelpressen [Fi71, Ke75, Ke78, Ke79, Ke81, Ke82]. Die Aufheizung der Probe erfolgt über die Stempel. Durch allseitig gleichen Stempelvorschub wird in den Würfelproben ein quasi-hydrostatischer Druck erzeugt. Die Wandler sind jeweils am gekühlten Stempelende angebracht, so daß hinsichtlich der maximalen Meßtemperatur der Curie-Punkt des Wandlermaterials keine Rolle spielt und die Geschwindigkeiten elastischer Wellen bis etwa 750 °C gemessen werden können. Probenform und Stempelanordnung erlauben die gleichzeitige Messung der Wellengeschwindigkeiten in drei orthogonalen Richtungen, so daß auch Aussagen über die Richtungsabhängigkeit der Wellenausbreitung gemacht werden können. Darüberhinaus lassen sich aus dem Stempelvorschub Änderungen des Probenvolumens und der Probendichte ermitteln.

Im allgemeinen wurden die Messungen an im Trockenschrank gelagerten Proben durchgeführt („trockene Gesteine")[1]. In einigen wenigen Experimenten wurden Umschließungsdruck und Temperatur unter kontrollierten Flüssig-Druck-Bedingungen gemessen („nasse Gesteine"), s. Abschnitt 3.1.4.4.2.

3.1.4.3 Sample description — Proben-Beschreibung

A complete listing is not attempted here, as priority in selection is given to recent measurements on well described rocks. Table 2 summarizes the main characteristics of the rock specimens included in this study: Rock type, locality, room temperature densities, mean atomic weights as calculated from the chemical analyses (if available), percentages of major minerals and essential features of the rock microstructure.

The rocks are divided into four groups: plutonic rocks, volcanic rocks, metamorphic rocks, and sedimentary rocks. The plutonic rocks are subdivided according to their position in the main sequence. Within each group (subgroup) the rocks are listed in the order of increasing density.

Es werden hier bevorzugt Messungen an gut beschriebenen Gesteinsproben berücksichtigt, ohne daß Anspruch auf Vollständigkeit besteht. In Tabelle 2 sind die Hauptcharakteristiken der Gesteinsproben zusammengestellt: Gesteinstyp, Herkunft, Dichten bei Raumtemperatur, mittleres Atomgewicht gemäß chemischer Analysen (wenn vorhanden), Volumenanteile der Hauptminerale und wichtigste Gefügemerkmale.

Die Gesteine sind in vier Gruppen eingeteilt: Plutonische Gesteine, vulkanische Gesteine, metamorphe Gesteine und sedimentäre Gesteine. Die plutonischen Gesteine sind entsprechend der genetischen Normalabfolge in Untergruppen unterteilt. Innerhalb jeder Gruppe (Untergruppe) sind die Gesteine nach zunehmender Dichte geordnet.

---

[1]) It should be noted, however, that it is nearly impossible to remove the fluid completely. At least a mono-molecular layer of fluid will probably remain in the thinnest cracks.

[1]) Es ist darauf hinzuweisen, daß die Feuchtigkeit nicht vollständig aus dem Gestein entfernt werden kann. Zumindest bleibt ein monomolekularer Film an Korngrenzen und Rissen erhalten.

Table 2. Description of rocks.

No.	Rock type, locality	ρ g/cm³	m̄	Microstructure	Modal analysis	Ref.
Plutonic rocks						
20	Granite, Wurmberg, Harz, Germany	2.57		Medium grained, inequigranular	33 quartz, 43 alkali feldspar, 22 plagioclase, 2 mica	Fi71
	Gray Granite, Llano County, Texas	2.609		Fine grained, equigranular	major minerals: quartz, microcline, biotite, plagioclase; accessories: zircon, opaques, magnetite, ilmenite	Hu56
	Granite, Westerly, USA	2.646		fine-grained, lath-shaped feldspars	28 quartz, 35 alkali feldspar, 31 plagioclase (An₁₇), 5 mica	Br65
	Granite, Cape Anne, Massachusetts	2.621		Holocrystalline, coarse grained, inequigranular	30 quartz, 40 perthite, 20 plagioclase, (An₁₂), 10 biotite	Ch79
	Pink granite, Llano County, Texas	2.636		Coarse grained, equigranular	major minerals: quartz, microcline, biotite, plagioclase; accessories: zircon, apatite, magnetite, ilmenite	Hu56
876	Granite, source unknown	2.649		Inequigranular, medium to fine grained, interlobate	22 quartz, 33 orthoclase, 31 plagioclase, 13 mica, 1 chlorite	Ke78
5001/1	Anorthosite, Tellnes, Norway	2.692		Coarse grained, inequigranular	98 plagioclase (An₄₀₋₅₀), 1 orthopyroxene, 1 ore	Ke32a
29	Anorthosite, Tellnes, Norway	2.745		Coarse grained, inequigranular	95 plagioclase (An₄₀₋₅₀), 3 orthopyroxene, 2 ore	Ke32a
	Anorthosite, Lake St. John, Quebec	2.971		Holocrystalline, coarse grained	70 plagioclase (An₅₇), 7 hypersthene, 15 ilmenite, 5 hornblende, 2 biotite, 1 spinel	Ch79
	Bytownite-Gabbro, Duluth, Minnesota	2.885		Coarse grained, ophitic	70 plagioclase, 15 olivine, 10 (clinopyroxene, ilmenite)	Hu57
	Gabbro, Mid-Atlantic Ridge	2.901		Holocrystalline, medium grained	50 plagioclase (An₅₀), 35 clinopyroxene, 15 olivine (partly replaced by serpentine, tremolite, chlorite)	Ch79
	Hornblende-Gabbro, source unknown	2.933		Fine grained, hypidiomorphic granular	40 plagioclase (An₅₅), 35 hornblende, 15 pyroxene, 10 biotite	Hu57
13	Gabbro, Radautal, Harz, Germany	2.97		Medium to coarse grained	42 plagioclase, 12 hornblende, 38 clinopyroxene, 5 mica, 1 ore, 2 others	Fi71
	Gabbro, San Marcos, California	2.993		Medium to fine grained, hypidiomorphic granular	55 plagioclase, 35 hornblende, 10 (clinopyroxene, biotite, others)	Hu57
2612c	Gabbro, Papua, New Guinea	3.03	21.5	Subophitic, fine grained, cumulate, not layered	46 plagioclase, 3 orthopyroxene, 50 clinopyroxene	Kr76

(continued)

Table 2 continued

No.	Rock type, locality	ϱ g/cm³	$\bar{m}$	Microstructure	Modal analysis	Ref.
84	Norite, Radautal, Harz, Germany	2.906	21.67	Granulose, fine grained, inequigranular	1 quartz, 1 mica, 57 plagioclase, 8 hornblende, 28 pyroxene, 5 saussurite	Ke81
84a	Norite, Radautal, Harz, Germany	2.988		Granulose, medium grained inequigranular	58 plagioclase, 31 orthopyroxene, 4 clinopyroxene, 4 biotite, 1 ilmenite, 2 accessories	Ke82a
5001/2	Ilmenite-Norite, Tellnes, Norway	3.428		Medium to fine grained, inequigranular, xenomorphic	53 plagioclase, (An$_{40-50}$), 10 orthopyroxene, 1 clinopyroxene, 29 ilmenite, 3 magnetite, 3 biotite, 1 olivine	Ke82a
2611c	Pyroxenite, Papua, New Guinea	3.28	21.5	Anhedral, granular, medium grained, not layered	14 orthopyroxene, 77 clinopyroxene, 8 olivine	Kr76
2602c	Pyroxenite-Harzburgite, Papua, New Guinea	3.30	21.3	Anhedral, granular, fine grained, cumulate, layered	62 orthopyroxene, 35 olivine, 1 chromite, 1 serpentine	Kr76
	Harzburgite, Antalya, Turkey	3.32		fine grained, foliated and lineated	74 olivine, 22 orthopyroxene, 1.5 clinopyroxene, 1.5 spinel, 1 serpentine	Pe78
	Lherzolite, Salt Lake Crater, Hawaii	3.24		foliated and lineated	71 olivine, 22 orthopyroxene, 5 clinopyroxene, 0.7 amphibole, 0.3 others	Pe78
10	Peridotite, Stubachtal, Hohe Tauern, Austria	3.25			92 olivine, 1 clinopyroxene, 3 ore, 4 others	Fi71
475	Peridotite Finero, Italy	3.260	21.54	Inequigranular, coarse grained	15 pyroxene, 80 olivine, 3 serpentine, 2 ore	Ke81
	Peridotite (Spinel lherzolite) nodule (from an alkali basalt), British Columbia	–	22.59*	Medium to fine grained, hot pressed at 1000 °C at 14 kbar, resp. at 1100 °C at 5 kbar	olivine, orthopyroxene, clinopyroxene, spinel	Mu76 Mu78
	Dunite, Twin Sisters Mountain, Washington	3.160		Medium grained olivine mosaic with large clinopyroxene crystals	90 olivine, 10 clinopyroxene	Hu51
	Dunite, Twin Sisters Mountain, Washington	–		Inequigranular, medium grained	98 olivine (Fo$_{90}$), 2 opaques	Ra78
	Dunite, Jackson County, N. Carolina	3.198		Fine to medium grained, peculiar granular	99 olivine, chromite	Hu57
1675	Dunite, Åheim, Norway	3.231	20.78	Granulose, fine grained	6 pyroxene, 92 olivine, 2 chlorite	Ke81
	Dunite, Twin Sisters Mountain, Washington	3.306		Inequigranular, medium grained	99 olivine, 1 opaques	Ch79

(continued)

Kern

Table 2 continued

Volcanic rocks

No.	Rock type, locality	ϱ g/cm³	$\bar{m}$	Microstructure	Modal analysis	Ref.
12/2	Analcite–Labradorite–Trachyte, source unknown	2.712		Holocrystalline, porphyric, phenocrysts of clinopyroxene, plagioclase	Phenocrysts: 60 alkalifeldspar, 15···20 plagioclase, Groundmass: 10···15 pyroxene, 5 analcite	Hu57
	Tholeiite–Basalt, Suduroy Faeroe Islands	2.504		Aphyric, zeolite-filled bubbles	Phenocrysts: 2 plagioclase, 20 zeolite; Groundmass: 25 (plagioclase + pyroxene), 8 phyllosilicates + 9 ore, 36 zeolite	Ke79a Ke82
2305	Tholeiite–Basalt, Eysturoy Faeroe Islands	2.752	22.57	Moderately phyric, random zeolite-filled amygdules	Phenocrysts: 6 plagioclase, 3 zeolite; Groundmass: 70 (plagioclase, pyroxene), 18 ore, 2 phyllosilicates	Ke79a Ke82
2311	Tholeiite–Basalt, Streymoy, Faeroe Islands	2.803	23.36	Coarsely phyric, ophitic	Phenocrysts: 27 plagioclase; Groundmass: 23 plagioclase, 25 pyroxene, 19 phyllosilicates, 6 ilmenite	Ke79a Ke82
2306	Tholeiite–Basalt, Vagar, Faeroe Islands	2.850	22.76	Moderately phyric, zeolite-filled amygdules	Phenocrysts: 2 plagioclase, 12 olivine, 12 zeolite; Groundmass: 74 (plagioclase, pyroxene, zeolite)	Ke79a Ke82
2304	Tholeiite–Basalt, Vidoy (?), Faeroe Islands	2.886	21.91	Sparsely phyric, irregular pores filled with chapasite	Phenocrysts: 40 { plagioclase; Groundmass: 48 (pyroxene + olivine) 8 zeolithe, 4 ilmenite	Ke79a Ke82
2308	Tholeiite–Basalt, Vagar, Faeroe Islands	2.932	22.22	Sparsely phyric, microphenocrysts of plagioclase and clinopyroxene	Phenocrysts: 37 { plagioclase, 5 olivine; Groundmass: 39 { pyroxene, 11 phyllosilicates, 8 ore	Ke79a; Ke82
2303	Tholeiite–Basalt, Suduroy, Faeroe Islands	2.934	21.27	Aphyric	Phenocrysts: 39 plagioclase (rare); Groundmass: 39 pyroxene, 11 phyllosilicates	Ke79a Ke82
2301	Tholeiite–Basalt, Suduroy, Faeroe Islands	2.947	22.60	Aphyric	Phenocrysts: 43 { plagioclase, 33 { pyroxene; Groundmass: 43 { plagioclase, pyroxene, 17 phyllosilicates, 7 ore	Ke79a Ke82
6a/2	Tholeiite–Basalt, Sandoy, Faeroe Islands	2.967		Coarsely phyric, random	Phenocrysts: 43 plagioclase; Groundmass: 21 plagioclase, 21 pyroxene, 11 phyllosilicates, 4 ore	Ke79a Ke82
2302	Tholeiite–Basalt, Suduroy, Faeroe Islands	2.990	22.99	Aphyric, microphenocrysts of plagioclase and clinopyroxene	Phenocrysts: 44 plagioclase; Groundmass: 37 clinopyroxene, 9 chlorite, 10 ilmenite	Ke79a Ke82
	Basalt, Chaffee County (?), Colorado	2.586		Porphyrite, phenocrysts of plagioclase	Phenocrysts: ≈50 plagioclase, olivine; Groundmass: plagioclase, magnetite, ilmenite, calcite	Hu57
	Basalt, East Pacific Rise	2.882		Intergranular, variolitic	Phenocrysts: 60 plagioclase, 20 pyroxene; Groundmass: 20 (composition not determined)	Ch79

(continued)

Kern

Table 2 continued

No.	Rock type, locality	ϱ g/cm³	$\bar{m}$	Microstructure	Modal analysis	Ref.
12	Metabasalt, Koitsøy Island, Stavanger, Norway	2.99		Fine grained, greenschist	chlorite, actinolite, albite, epidote	Bu82
Metamorphic rocks						
	Serpentinite, Mid-Atlantic Ridge	2.509		Massive with mesh structure and bastite pseudomorphs	98 chrysotile + lizardite, 2 opaques	Ch79
987	Serpentinite, source unknown	2.748		Massive with mesh structure	75 serpentinite, 20 olivine, 4 ore, 1 others	Ke82
1389	Quartz–Phyllite, S. Alta, Norway	2.663	–	Gneissose, xenoblastic, medium to fine grained	48 quartz, 21 orthoclase, 5 plagioclase, 23 muscovite, 3 ore	Ke82
	Quartzite, Baraboo, Wisconsin	2.647		Granulose, granoblastic, xenoblastic	95 quartz, 4 muscovite, 1 graphite	Ch79
1419	Quartzite, Koli, Finland	2.662		Fine grained, xenoblastic	100 quartz	Ke79
1452	Quartzite, Bayr. Wald, Germany	2.687	20.37	Inequigranular, medium to fine grained (irregular, polyhedral), xenoblastic	81 quartz, 15 mica, 1 garnet, 2 chloritoid, 1 staurolite	Ke81
938	Carrara Marble	2.729	20.02	Granulose, medium grained, equigranular	99 calcite, 1 quartz	Ke82
154	Granulite Gneiss, Inari, Finland	2.612	20.65	Inequigranular, medium grained, platelike xenoblasts	62 orthoclase, 28 quartz, 6 plagioclase, 2 sillimannite, 2 garnet	Ke78
	Granulite, New Jersey Highlands	2.680		Granulose, hypidioblastic, inequigranular	30 quartz, 50 plagioclase (An$_{30}$), 17 hypersthene, 2 opaques, 1 hornblende	Ch79
	Granulite, Saranac Lake, New York	2.848		Granulose, coarse grained, xenoblastic, inequigranular	35 perthite, 15 myrmekite, 2 plagioclase, 3 quartz, 15 garnet, 13 clinopyroxene, 5 biotite, 4 hornblende, 4 opaques, 3 apatite, 1 zircon	Ch79
	Granulite, Adirondack Mountains, New York	2.911		Gneissose, medium grained, xenoblastic	32 plagioclase (An$_{32}$), 30 perthite, 38 pyroxene	Ch79
	Granulite, Valle d'Ossola, Italy	3.125		Gneissose, medium to coarse grained	60 plagioclase (An$_{40}$), 26 pyroxene, 9 garnet, 4 opaques, 1 microcline	Ch79
1400	Biotite–Orthopyroxene–Plagioclase–Gneiss Karasjok, Norway	2.784	21.54	Gneissose, medium grained, xenoblastic, inequigranular	15 quartz, 3 mica, 66 plagioclase, 13 pyroxene, 2 ore, 1 apatite	Ke81

(continued)

Kern

Table 2 continued

No.	Rock type, locality	ϱ g/cm³	$\bar{m}$	Microstructure	Modal analysis	Ref.
268	Plagioclase–Amphibolite–Gneiss Karasjok, Norway	2.817	21.65	Gneissose, coarse to medium grained, xenoblastic, inequigranular	33 quartz, 40 plagioclase, 24 hornblende, 1 garnet, 1 ore, 1 mica, clinopyroxene	Ke81
160	Biotite–Gneiss, Floitental, Tirol	2.845	–	Gneissose, medium grained, inequigranular	47 quartz, 7 plagioclase, 28 biotite, 18 epidote	Ke82
1398	Pyroxene–Plagioclase–Garnet–Gneiss Karasjok, Norway	2.940	21.97	Gneissose, medium to fine grained, xenoblastic, inequigranular	33 quartz, 40 plagioclase, 3 hornblende, 16 garnet, 5 pyroxene, 2 ore	Ke81
298	Plagioclase–Quartz–Amphibolite–Gneiss Karasjok, Norway	3.070	22.90	Gneissose, fine grained, xenoblastic, inequigranular	27 quartz, 18 plagioclase, 48 hornblende, 4 epidote, 2 ore, 1 titanite	Ke81
1403	Sillimanite–Garnet–Gneiss Karasjok, Norway	3.072	21.94	Gneissose, medium to fine-grained, xenoblastic, inequigranular	24 quartz, 6 plagioclase, 24 orthoclase, 27 garnet, 14 sillimanite, 1 rutile	Ke81
	Amphibolite, Indian Ocean, Central Ridge	2.930		Holocrystalline, medium grained, inequigranular	50 hornblende, 40 plagioclase (An_{30-50}), 5 orthopyroxene, 5 actinolite+chlorite	Ch79
1387	Amphibolite S. Alta, Norway	2.962	22.63	Gneissose, fine-grained, xenoblastic	3 mica, 43 plagioclase, 45 hornblende, 3 ore, 5 titanite	Ke8?
1396	Amphibolite, Karasjok, Norway	3.067	22.88	Gneissose, medium grained, xenoblastic to hypidioblastic, inequigranular	3 quartz, 24 plagioclase, 72 hornblende, 1 garnet	Ke8?
1454	Epidote–Amphibolite, Bayrischer Wald, Germany	3.073	20.26	Gneissose, medium to fine grained, xenoblastic, inequigranular	12 plagioclase, 72 hornblende, 10 epidote, 4 serpentine, 2 titanite	Ke81
886	Eclogite, Saualpe, Austria	3.402	22.37	Granulose, medium grained	2 quartz, 4 hornblende, 29 garnet, 45 pyroxene, 10 zoisite, 9 symplectite, 1 rutile	Ke81
11	Eclogite, Fichtelgebirge, Germany	3.463	23.03	Granulose, medium grained	7 quartz, 1 mica, 6 hornblende, 35 garnet, 44 pyroxene, 4 symplectite, 2 rutile	Ke81
15	Eclogite, Münchberger Gneissmasse, Fichtelgebirge, Germany	3.51		Granulose, medium grained	4 quartz, 47 clinopyroxene, 35 garnet, 7 hornblende, 7 others	Fi71
	Eclogite, Nove' Dvory, Czechoslovakia	3.559		Granulose, medium to fine grained, symplecite intergrowth xenoblastic	49 garnet, 45 omphacite, 5 opaques+rutile	Ch79

(continued)

Kern

Table 2 continued

Sedimentary rocks

No.	Rock type, locality	ϱ g/cm³	Φ vol-%	Microstructure	Modal analysis	Ref.
	Sandstone, Caplen Dome, Galveston County	2.543	5.1			Hu51
L 071-5	Arkosic Sandstone, Diabolo Range, California	2.696	0.7	Pumpellyite-bearing	4 quartz, 40 plagioclase, 10 chlorite, 8 mica, 1 epidote, 17 phyllosilicates, 3 fragments, 9 others	St77, St78
	Sandstone, Berea (Mississippian), Ohio	–	17.1	Fine and very fine grained, moderately well sorted, grains are equant and sub-rounded	major minerals: quartz + fragments, feldspar; accessories: phyllosilicates, carbonate	Ti68, Ti77
	Sandstone, Boise (Tertiary)	–	30.84	Fine to medium grained, poorly sorted, grains are elongated, angular to sub-angular and randomly orientated	major minerals: volcanic fragments, quartz, feldspar, mica; accessories: zeolite	Ti68, Ti77
	Limestone, Solnhofen, Bavaria, Germany	2.656	4–5	Extremely fine grained (5…20 µm)	99 calcite, traces of phyllosilicates, quartz, ore	Hu51, Ke71, Ke82
	Argillaceous limestone, Upton County, Texas	2.739	–	well bedded	70 calcite, 30 phyllosilicates	Hu51
BCS 57	Shale, Central Belt, California	2.646	3.2	Slightly recrystallized	18 quartz, 1 plagioclase, 13 chlorite, 22 mica 2 epidote, 45 others (undifferentiated fine grained mica, chlorite, plagioclase, etc.)	St77, St78
949	Shale, Rheinisches Schiefergebirge, Germany	2.797		Microcrystalline to fine grained, bedding shistosity	85 phyllosilicates, 8 (albite + quartz), 2 mica, 5 carbonate	Ke82a
W 2	Graywacke, Costal Belt, California	2.549	6.2	Not recrystallized	20 quartz, 26 plagioclase, 10 alkali feldspar, 12 chlorite, 1 mica, 2 epidote, 6 carbonate, 3 leucite, 15 fragments, 5 others	St77, St78
W 5	Graywacke, Costal Belt, California	2.631	4.1	Not recrystallized	19 quartz, 17 plagioclase, 12 alkalifeldspar, 7 chlorite, 1 mica, 4 epidote, 1 leucite, 30 fragments, 9 others	St77, St78
SR 70-1	Graywacke, Central Belt, California	2.643	1.0	Slightly recrystallized	15 quartz, 40 plagioclase, 22 chlorite, 3 mica, 1 carbonate, 5 phyllosilicates, 4 fragments, 5 others	St77, St78
IV 5	Graywacke, Diabolo Range, California	2.658	1.2	Recrystallized	26 quartz, 44 plagioclase, 3 chlorite, 8 mica, 1 epidote, 6 pumpellyite, 11 phyllosilicates, 1 fragments, 1 others	St77, St78

(continued)

Table 2 continued

No.	Rock type, locality	ϱ g/cm³	Φ vol-%	Microstructure	Modal analysis	Ref.
SP 740	Graywacke, Central Belt, California	2.658	1.9	Slightly recrystallized	26 quartz, 21 plagioclase, 14 chlorite, 2 mica, 2 epidote, 1 carbonate, 12 phyllosilicates, 19 fragments, 3 others	St77, St78
	Graywacke, Steimkertal, Harz, Germany	2.68	0.5	Very fine grained, slightly recrystallized	major minerals: quartz, feldspar; accessories: chlorite, serpentine, biotite, ore	Bu82
BCSS 55	Graywacke, Central Belt, California	2.714	1.4	Pumpellyite-bearing	19 quartz, 46 plagioclase, 6 chlorite, 1 mica, 2 epidote, 2 carbonate, 3 pumpellyite, 9 fragments, 12 others	St77, St78
21 RGC 60	Metagraywacke, Diabolo Range, California	2.815	0.4	Jadeite-bearing	43 quartz, 17 mica, 7 lawsonite, 25 jadeite, 2 carbonate, 6 others	St77, St78
P3	Metagraywacke, Diabolo Range, California	2.816	1.3	Jadeite-bearing	35 quartz, 1 plagioclase, 6 chlorite, 14 mica, 7 lawsonite, 31 jadeite, 2 carbonate, 3 pumpellyite, 1 others	St77, St78
P5	Metagraywacke, Diabolo Range, California	2.926	1.2	Jadeite-bearing	36 quartz, 1 chlorite, 10 mica, 5 lawsonite, 35 jadeite, 4 carbonate, 3 leucite, 1 fragments, 5 others	St77, St78
	Carnallitite, Salt Mine Asse II, Wolfenbüttel, Germany	1.685		coarse to medium grained, xenoblastic	84 carnallite, 8 kieserite, 7 halite, 1 anhydrite	Ke82a
2574/2	Rock salt, Salt Mine Siegfried-Giesen, Hildesheim, Germany	2.153		coarse grained, inequigranular	100 halite	Ke82a
	Kieseritic rock salt, Salt Mine Herfa, Germany	2.233		medium grained, inequigranular	50 halite, 22 sylvite, 25 kieserite, 3 anhydrite	Ke82a
	Polyhalite–Halite, Salt Mine Herfa, Germany	2.531		medium grained, rhythmic layers of polyhalite and halite, in places xenoblasts of halite	65 polyhalite, 35 halite	Ke82a
2574/1	Anhydrite, Salt Mine Siegfried Giesen, Hildesheim, Germany	2.931		medium grained, in places spherolitic, sedimentary layering indicated by clay minerals	95 anhydrite, 5 clay minerals	Ke82a

3.1.4.4 Velocity-temperature relations at confining pressure —
Geschwindigkeit-Temperatur-Beziehungen bei konstanten Drucken

3.1.4.4.1 "Dry", low porous rocks — „Trockene", porenarme Gesteine

Elastic wave propagation through dry natural rocks is known to be sensitive to the state of microcracking of the polycrystalline aggregates. Determination of the intrinsic effect of the temperature on wave velocity is difficult, because differential thermal expansion of the constituent minerals of rocks may cause grain boundaries to widen and new cracks to open. Measurements of thermal volumetric strain on several metamorphic and plutonic rocks [Ke78] have shown that microfracturing induced by the rapid thermal change of volume of the mineral phases will be increasingly suppressed as pressure is raised until at pressures of some kbars it does not take place at all.

At low P_c non-linear slope and significant hysteresis indicating microfracturing is observed in the velocity-temperature curves. In contrast, near linear slope and reversibility is obtained at high P_c (Fig. 1). The minimum pressure needed to prevent the opening of cracks and other non-linear effects was estimated to be around 1 kbar per 100 °C [Ke78]. Only those physical parameters that are obtained at or above these crack-

In trockenen Gesteinen wird die Ausbreitung elastischer Wellen stark durch den Grad der Mikrorißbildung beeinflußt. Die Bestimmung des reinen Temperatureinflusses ist schwierig, da die unterschiedliche thermische Ausdehnung der beteiligten Gesteinsminerale zu einer Gefügeauflockerung führen kann. Messungen der thermischen Volumenänderung an verschiedenen metamorphen und magmatischen Gesteinen [Ke78] haben gezeigt, daß die durch schnelle thermische Ausdehnung der einzelnen Mineralphasen bedingte Gefügeauflockerung mit steigendem äußeren Druck zunehmend verhindert wird und bei Drucken von einigen kbar überhaupt keine Gefügeauflockerung mehr erfolgt.

Bei niedrigem P_c sind die Geschwindigkeit-Temperatur-Beziehungen infolge der Bildung von Mikrorissen nicht-linear und irreversibel; bei hohem P_c sind sie linear und reversibel (Fig. 1). Der zur Verhinderung einer Gefügeauflockerung und anderer nicht-linearer Effekte erforderliche Mindestdruck ist etwa gleich 1 kbar pro 100 °C Temperaturerhöhung [Ke78]. Daher zeigen nur jene physikalischen Parameter das eigent-

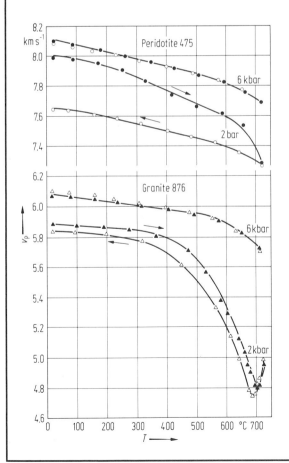

◄ Fig. 1. Compressional wave velocities, v_P, for peridotite and granite as a function of temperature, T, at 2 kbar and 6 kbar confining pressure. Solid symbols indicate measurements with increasing temperature; open symbols indicate measurements with decreasing temperature. The velocity inversion in granite at 2 kbar is caused by the high-low transition in the constituent quartz minerals [Ke81].

Kern

closing pressure conditions seem to be nearly correct indicators for intrinsic properties of the compact aggregates.

It is interesting to note that this slope is also the slope along which there is no change of volume for most rock-forming minerals.

liche Materialverhalten an, die bei oder oberhalb dieses, eine Gefügeauflockerung verhindernden Mindestdruckes ermittelt wurden.

Es ist hervorzuheben, daß obige Druck-Temperatur-Bedingungen mit jenen identisch sind, bei denen in den wichtigsten gesteinsbildenden Mineralen keine Volumenänderung erfolgt.

$$\frac{1}{V}\left(\frac{dV}{dT}\right)_P = \alpha$$

α for typical minerals: $10^{-5}/°C$

α für typische Minerale: $10^{-5}/°C$

$$-\frac{1}{V}\left(\frac{dV}{dP}\right)_T = \frac{1}{K}$$

K for typical minerals: 10^6 bars.

K für typische Minerale: 10^6 bar.

Plots of v_P and v_S vs. temperature for 6 kbar confining pressure for a series of selected rocks (Fig. 2) indicate almost linear behavior from room temperature up to about 500 °C. Beyond this temperature, however, it becomes more and more non-linear, indicating the onset of thermal cracking. Minimum velocities due to the high-low quartz transition (see

Die Geschwindigkeitskurven von v_P und v_S als Funktion der Temperatur sind bei 6 kbar Umschließungsdruck bis etwa 500 °C linear. Bei höheren Temperaturen weichen sie zunehmend von der Linearität ab, was auf eine thermische Rißbildung hinweist. In quarzführenden Gesteinen wird hier ein Geschwindigkeitsminimum als Folge der Hoch-Tief-Umwandlung

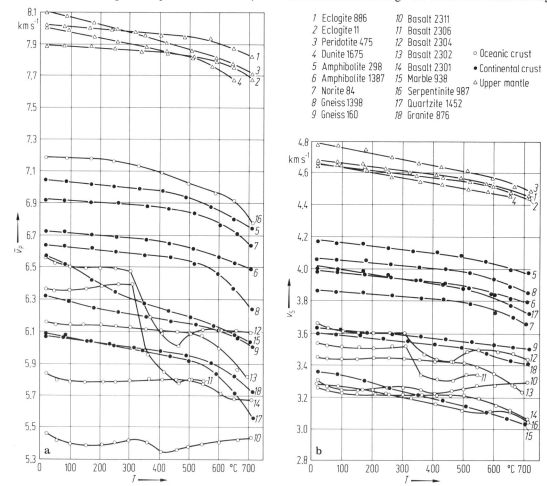

1	Eclogite 886	10	Basalt 2311
2	Eclogite 11	11	Basalt 2306
3	Peridotite 475	12	Basalt 2304
4	Dunite 1675	13	Basalt 2302
5	Amphibolite 298	14	Basalt 2301
6	Amphibolite 1387	15	Marble 938
7	Norite 84	16	Serpentinite 987
8	Gneiss 1398	17	Quartzite 1452
9	Gneiss 160	18	Granite 876

○ Oceanic crust
● Continental crust
△ Upper mantle

Fig. 2. Velocities of (a) compressional waves, $\bar{v}_P$, and (b) shear waves, v_S, as a function of temperature, T, at 6 kbar confining pressure for a series of selected rocks. $\bar{v}_P$ is the mean of the velocities measured in three orthogonal directions of the sample cubes. v_S is the velocity in one direction [Ke82].

3.1.4.4.1.1) are not observed in the velocity curves of quartz-bearing rocks, because the pressure dependence and the shift of the $\alpha - \beta$ quartz transition temperature in polycrystalline aggregates [Ke79] places the transition temperature outside the temperature range investigated (compare Fig. 4).

Compared to most plutonic and metamorphic rocks the slopes of the velocity curves of the basalts are low. According to [Hu50] the presence of glass in volcanic rocks would reduce the slope, because in glass sound velocity increases with temperature. The thermal volumetric strain curves shown in Fig. 3 indicate that the low temperature sensitivity of v_P and v_S in basalts must be attributed to more powerful additional factors. In contrast to most plutonic and metamorphic rocks, the bulk volume of the basalts decreases significantly with temperature at confining pressure, thus indicating a further reduction of porosity. Basalts crystallize at relatively low pressure, and grain boundary cracks and pores associated with cooling will not be closed at these low pressure conditions. Thus, in laboratory experiments, part of the grain boundary cracks and especially more equant pores cannot be closed by confining pressure alone. Under simultaneous action of temperature and high confining pressure, however, cracks and pore spaces are closed more effectively,

des Quarzes (s. 3.1.4.4.1.1) nicht beobachtet, da durch die Druckabhängigkeit der Transformation und des in polykristallinem Material zu beobachtenden „shifts" der Umwandlungstemperatur [Ke79] die Transformation außerhalb des untersuchten Temperaturbereichs erfolgt (vergleiche Fig. 4).

Im Vergleich zu den meisten plutonischen und metamorphen Gesteinen ist die temperaturbedingte Abnahme der Geschwindigkeiten der elastischen Wellen in Basalten im allgemeinen gering. Nach [Hu50] wirken Glasanteile in Basalten einer Geschwindigkeitsabnahme entgegen, da in Gläsern die Geschwindigkeiten mit der Temperatur zunehmen. In Fig. 3 zeigen die Diagramme zur thermischen Volumenausdehnung, daß die im allgemeinen geringe Temperaturabhängigkeit von v_P und v_S in Basalten auf andere, wirksamere Faktoren zurückgeführt werden muß. Im Gegensatz zu den meisten plutonischen und metamorphen Gesteinen ist in Basalten bei hohem Druck eine deutliche Volumenabnahme mit steigender Temperatur zu beobachten, was auf eine weitere Reduktion des Porenvolumens hinweist. Basalte kristallieren bei niedrigen Drucken, und die im Zuge der Abkühlung entstehenden Mikrorisse und Poren können durch den niedrigen Druck nicht vollständig geschlossen werden. In Laborexperimenten ist es daher nicht möglich, besonders die iso-

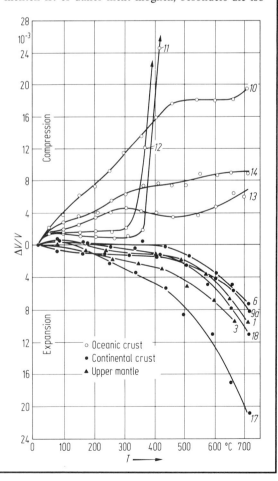

Fig. 3. Thermal volumetric strain ($\Delta V/V$) at 6 kbar confining pressure for selected samples of igneous and metamorphic rocks [Ke82]. For curve numbers, see Fig. 2, except 9a: Gneiss 1403.

thereby giving rise to the observed significantly lower velocity decrease or even increase for the basalts with increasing temperature. For the reasons for the abrupt velocity decrease around 350 °C in basalt No. 2304 and 2306 (Fig. 2), see 3.1.4.4.1.2.

metrischen Poren allein durch hydrostatischen Druck zu schließen. Durch gleichzeitige Einwirkung von Druck und Temperatur wird jedoch eine effektivere Reduzierung des Riß- und Porenvolumens erzielt, und die Wellengeschwindigkeiten nehmen im Vergleich zu den übrigen Gesteinen mit steigender Temperatur weniger stark ab oder sogar geringfügig zu. Für die deutliche Geschwindigkeitsabnahme bei etwa 350 °C in den Basaltproben Nr. 2304 und 2306 (Fig. 2) wird in Abschnitt 3.1.4.4.1.2 eine Begründung gegeben.

3.1.4.4.1.1 Elastic wave velocities across the α−β quartz transition— Geschwindigkeiten elastischer Wellen im Bereich der α−β-Quarz-Transformation

Solid-solid phase transitions have an appreciable affect on elastic wave velocities. In quartz-bearing rocks (granite, gneiss, quartzite) the α−β transition is associated with a pronounced velocity decrease when approaching the transition and with a significant velocity increase after the transition (Fig. 4).

There is a general positive correlation between volume change and velocity decrease with increasing temperature up to the region of the α−β quartz transition. However, the volumetric strain curves [Ke79] reveal no positive correlation in the β-field. The large drop of v_P may be explained by the elastic softening of the structure of the constituent quartz minerals near the α−β transition and by the opening of grain-boundary cracks, which is caused by the very high volumetric thermal expansion of quartz [Sk66] relative to other component minerals. The velocity increase in the β-field may be attributed to an elastic hardening of the quartz structure [Yo62, Yo64]. The significant drop of v_P associated with the α−β transition of the constituent quartz crystals may, possibly, account for low velocities in the earth's crust in regions with anomalous high heat flow.

Fest-fest-Phasenumwandlungen üben einen bedeutenden Einfluß auf die Geschwindigkeit elastischer Wellen aus. In Quarz-führenden Gesteinen (Granit, Gneis, Quarzit) ist die α−β-Umwandlung bei Annäherung an die Umwandlungstemperatur mit einem bedeutenden Geschwindigkeitsabfall verbunden. Nach der Umwandlung steigt die Geschwindigkeit wieder stark an (Fig. 4).

Im allgemeinen besteht mit steigender Temperatur bis zur α−β-Quarz-Umwandlung zwischen Volumenänderung des Gesteins und Geschwindigkeitsabfall eine positive Korrelation. Im β-Feld ist jedoch keine derartige Beziehung erkennbar [Ke79]. Der starke Abfall von v_P im Bereich der α−β-Umwandlung ist durch ein „elastic softening" der Struktur der beteiligten Quarzminerale und durch Mikrorisse erklärbar, die besonders durch die starke thermische Ausdehnung des Quarzes [Sk66] induziert werden. Der Geschwindigkeitsanstieg im β-Feld wird auf ein „elastic hardening" der Quarzstruktur zurückgeführt [Yo62, Yo64]. Der mit der α−β-Umwandlung des Quarzes verbundene starke Geschwindigkeitsabfall von v_P ist möglicherweise Ursache für „low velocity channels" (Nieder-Geschwindigkeits-Zonen) in der Erdkruste, in Bereichen mit anomal hohem Wärmefluß.

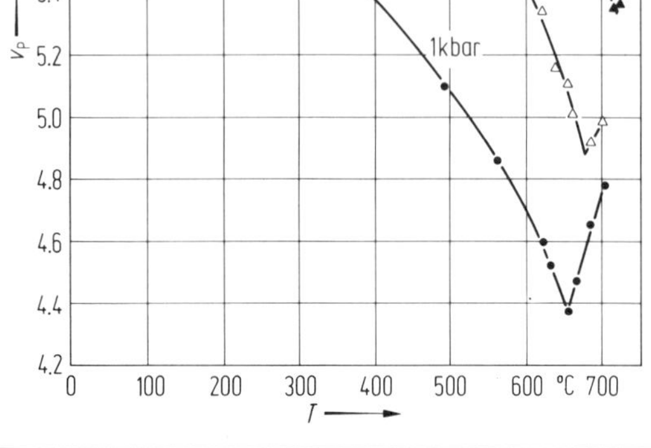

Fig. 4. Compressional wave velocity, v_P, in granite as a function of temperature, T, at different confining pressures up to 6 kbar [Ke78].

Stress heterogeneities as a consequence of different thermal expansion of the constituent minerals cause the $\alpha - \beta$ quartz transition temperature to shift to higher values than might be expected on grounds of single crystal behavior. This internal pressure may exceed more than two times the confining pressure applied to the boundaries of the specimen (Fig. 5). Although such large stress heterogeneities can probably not exist for geological periods of time, they may induce and accelerate reactions in metamorphic rocks and thus define the peak formation conditions of mineral assemblages.

In polykristallinem Material bewirken Spannungs-Inhomogenitäten als Folge unterschiedlicher thermischer Ausdehnung der beteiligten Gesteinsminerale eine Verschiebung der $\alpha - \beta$-Quarz-Umwandlungstemperatur zu höheren Werten als in Einkristallen. Die an Korngrenzen entstehenden Druckkonzentrationen können den von außen auf die Probe wirkenden Druck um mehr als das 2-fache übersteigen (Fig. 5). Derartig hohe Spitzendrucke sind zwar nicht über geologische Zeiten existent, dennoch ist es möglich, daß sie metamorphe Mineralreaktionen initiieren und damit die obere Grenze der Bildungsbedingungen von Mineralvergesellschaftungen festlegen.

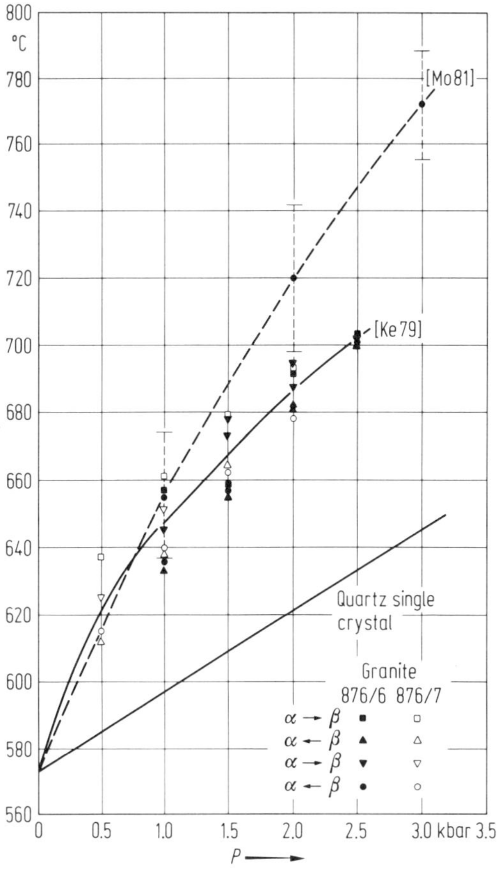

◄

Fig. 5. The quartz $\alpha - \beta$ transition temperature, $T_{\alpha-\beta}$, in granite as a function of confining pressure, P, as determined by v_P measurements [Ke79]. The results of [Mo81] obtained by thermal expansion measurements on Delegate aplite (dashed line) and the pressure dependence of $T_{\alpha-\beta}$ for single crystals of quartz are also included [Ke82].

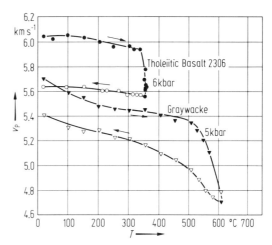

Fig. 6. Velocities of compressional waves, v_P, for jacketed samples of tholeiitic basalts (No 2306) as a function of temperature, T, at 6 kbar confining pressure [Ke79a] and for Graywacke at 5 kbar confining pressure [Bu82]. The solid symbols represent measurements on the way to higher temperatures, the open symbols represent measurements on the way back to room temperature.

3.1.4.4.1.2 The effect of dehydration reactions on wave velocities —
Der Einfluß von Entwässerungsreaktionen auf die Wellengeschwindigkeiten

Release of chemically-bound water with increasing temperature (during prograde metamorphism) may also have a tremendous effect on sound velocities. In some basalts containing zeolites [Ke79a] and in chlorite-bearing metagraywacke and metabasalt [Bu82], a discontinuous velocity decrease of P- and S-wave velocities is observed in the velocity temperature plot (Fig. 6).

Eine mit ansteigender Temperatur erfolgende Freisetzung chemisch gebundenen Wassers (während aufsteigender Metamorphose) vermag die Ultraschall-Geschwindigkeiten ebenfalls entscheidend zu beeinflussen. Die Geschwindigkeit-Temperatur-Kurven einiger zeolithführender Basalte [Ke79a] und chloritführender Metagrauwacken und Metabasalte [Bu82] zeigen eine starke Geschwindigkeitsabnahme der P- und S-Wellen (Fig. 6).

Zeolites as well as chlorites contain 10···20% by weight of structurally bound water. In some zeolites (natrolite, scolecite) and in chlorite, a part or all of this water is released abruptly on heating from room temperature to about 350 °C or about 500 °C, respectively. Dehydration reactions produce solid-fluid systems and as a consequence P_{eff} decreases, thereby giving rise to widening of old cracks and to formation of new cracks.

v_P and v_S exhibit irreversible velocity declines on the velocity-temperature curves and further thermal cycling is of no significant influence. This indicates that the fluid phase itself contributes only weakly to the decrease of wave velocities. A most effective contribution, however, comes from reconstitution of pore geometry and an increase of pore spaces by the internally-created P_f as dehydration proceeds. In general, dehydration reactions produce denser mineral phases and allow an increase of pore spaces and a reduction of bulk volume at the same time (Fig. 3).

It should be noted that not all hydrous mineral phases will dehydrate sharply. Minerals like amphibole, mica, and serpentine (Fig. 2) dehydrate continuously over a range of conditions and not at such a sharp point as some zeolites and chlorites.

Those low pressure phenomena [Br72] offer an explanation for low velocity channels in the oceanic crust alternative to that given recently by [Ja80].

In addition, hydrous to anhydrous phase transitions producing low effective pressure will change the physical strength of rocks [Ra65]. At a low effective pressure, brittle fracture is more dominant than ductile behavior and fault creep more dominant than stick slip [Br72]. Thus, it is possible that some intermediate-focus earthquakes result from brittle or frictional instability related to high pore fluid pressure generated by dehydration reactions. Dehydration reactions are most important along subduction zones.

3.1.4.4.1.3 Velocity anisotropy at elevated pressures and temperatures —
Geschwindigkeits-Anisotropien bei erhöhten Drucken und Temperaturen

Some magmatic rocks (peridotite, dunite) and most metamorphic rocks (amphibolite, serpentinite, mica-gneiss, marble, quartzite), as well as some sediments (e.g. shale) show significant velocity anisotropy of P- and S-waves (see e.g. [Ke75, Ra78, Ke81]).

Highest values of velocity anisotropy are observed at room temperature and at a low pressure of some 100 bar (Fig. 7). As pressure is raised, velocity anisotropy decreases at a smaller and smaller rate until constant

Zeolithe und Chlorite enthalten 10···20 Gew.-% chemisch gebundenen Wassers in ihrer Struktur. In einigen Zeolithen (Natrolith, Scolezit) und in Chlorit wird das Kristallwasser beim Aufheizen spontan ganz oder teilweise bei etwa 350 °C, bzw. bei etwa 500 °C abgegeben. Diese Entwässerungsreaktionen erzeugen Fest-Flüssig-Systeme, die eine Erniedrigung von P_{eff} bewirken, so daß bestehende Risse aufgeweitet und neue entstehen können.

v_P und v_S zeigen im Geschwindigkeit-Temperatur-Diagramm eine irreversible Abnahme, die sich auch bei weiterer cyclischer thermischer Belastung des Gesteins nicht wesentlich ändert. Dies deutet darauf hin, daß die flüssige Phase selbst nur wenig zur Abnahme der Wellen-geschwindigkeiten beiträgt. Vielmehr ist eine Neuordnung der Porengeometrie und eine Zunahme des Poren-volumens infolge eines Anstiegs von P_f mit zunehmender Entwässerung für die Geschwindigkeitsabnahme wesentlich verantwortlich. Im allgemeinen führen Entwässerungsreaktionen zur Bildung dichterer Mineralphasen. Dadurch ist es möglich, daß der Porenraum der Gesteinsprobe trotz gleichzeitiger Abnahme des Brutto-volumens größer wird. (Fig. 3).

Es ist darauf hinzuweisen, daß nicht alle kristall-wasser-haltigen Mineralphasen spontan entwässern. Minerale wie Amphibole, Glimmer und Serpentin (Fig. 2) geben ihr Kristallwasser kontinuierlich über ein größeres Temperatur-Intervall ab und nicht inner-halb eines engen Temperatur-Intervalls wie einige Zeolithe und Chlorite.

Derartige Niederdruck-Phänomene [Br72] sind eine weitere Erklärungsmöglichkeit für „low velocity channels" im Bereich der ozeanischen Kruste, alternativ zu jener, die kürzlich von [Ja80] gegeben wurde.

Darüberhinaus wird beim Übergang von wasser-haltigen zu wasserfreien Mineralphasen die mechanische Festigkeit des Gesteins wesentlich verändert [Ra65]. Bei niedrigem Effektiv-Druck dominiert sprödes Ver-halten über duktiles und Bruchfließen über „stick-slip" [Br72]. Es ist daher möglich, daß auch Erdbeben mit mittleren Herdtiefen aus einer mechanischen Instabili-tät resultieren, die durch hohe Porendrucke als Folge von Entwässerungsreaktionen erzeugt wurde. Dehydra-tions-Reaktionen spielen in Subduktionszonen eine bedeutende Rolle.

Einige magmatische Gesteine (Dunit, Peridotit) und die meisten metamorphen Gesteine (Amphibolit, Ser-pentinit, Glimmer-Gneis, Marmor, Quarzit) und auch einige Sedimentgesteine (z.B. Tonschiefer) zeigen be-deutende Geschwindigkeits-Anisotropien der P- und S-Wellen (s. z.B. [Ke75, Ra78.. Ke81]).

Die Geschwindigkeits-Anisotropien sind bei Raum-temperatur und Drucken von wenigen 100 bar am größten (Fig. 7). Mit steigender Temperatur nehmen sie progressiv ab, bis sie bei etwa 1.5 kbar konstante Werte

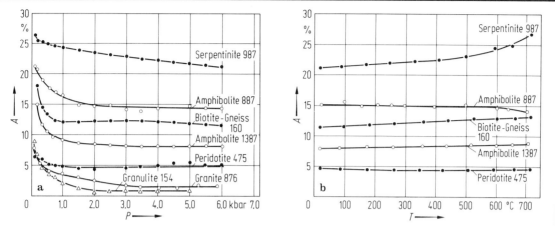

Fig. 7. Velocity anisotropy, A, of compressional wave velocities in possible crustal and mantle rocks as a function of pressure, P, at room temperature (a) and as a function of temperature, T, at 6 kbar confining pressure (b) [Ke78, Ke82].

values are reached at pressures above about 1.5 kbar. In most of the rocks investigated, pronounced velocity anisotropy is preserved, however, even at pressures around 6 kbar.

The directional dependence of elastic wave propagation in rocks is controlled by at least two factors: (1) preferred lattice orientation of major mineral phases and (2) microcracks orientated parallel to grain boundaries. From the slope of the anisotropy-pressure curves shown in Fig. 7a, it may be concluded that the part of velocity anisotropy which can be attributed to orientated cracks is eliminated at higher confining pressure. Observations [Br65, Ch66, Wa74, Ke78] indicate that flat microcracks are generally closed above about 2 kbar. Thus, it may be concluded that the other part of velocity anisotropy unaffected by pressure is caused preferred lattice orientation of predominant minerals. Fig. 8 (dunite) and Fig. 9 (amphibolite) indicate excellent qualitative relationship between fabric anisotropy and velocity anisotropy. Analoguous positive correlations were observed in serpentinite, quartzite, mica-gneiss and marble [Ke74, Ke75, Ke81].

As is evident from the anisotropy-temperature relations at 6 kbar confining pressure (Fig. 7b) this lattice-induced seismic anisotropy is generally unaffected even by temperatures of 700 °C.

Preferred lattice orientation is probably the most important factor in producing anisotropy in dense aggregates under high pressure and temperature, and is the most plausible explanation for seismic anisotropies in the earth's crust and mantle.

erreichen. In den meisten der genannten Gesteine sind aber auch noch oberhalb 6 kbar bedeutende Geschwindigkeits-Anisotropien zu beobachten.

Die Richtungsabhängigkeit der Ausbreitung elastischer Wellen wird durch mindestens zwei Einflußgrößen bestimmt: (1) durch bevorzugte Einregelung der Gesteinsminerale nach dem Kristallgitter und (2) durch Einregelung von Mikrorissen, hauptsächlich entlang Korngrenzen. Aus dem Verlauf der in Fig. 7a dargestellten Anisotropie-Druck-Beziehung wird ersichtlich, daß der durch Risse induzierte Anisotropie-Anteil durch höhere Drucke eliminiert wird. Nach Beobachtungen von [Br65, Ch66, Wa74, Ke78] werden in trockenen Gesteinen flache Mikrorisse im allgemeinen oberhalb von etwa 2 kbar geschlossen. Daraus folgt, daß der druckunabhängige Anisotropie-Anteil durch Vorzugsregelung der Hauptminerale nach dem Kristallgitter hervorgerufen wird. Fig. 8 (Dunit) und Fig. 9 (Amphibolit) lassen eine ausgezeichnete Übereinstimmung von Gefüge- und Geschwindigkeits-Anisotropie erkennen. Analoge Beobachtungen wurden auch in Serpentinit, Quarzit, Glimmer-Gneis und Marmor gemacht [Ke74, Ke75, Ke81].

Aus den Anisotropie-Temperatur-Beziehungen bei 6 kbar Umschließungsdruck (Fig. 7b) wird ersichtlich, daß die Gitter-induzierte elastische Anisotropie im Bereich bis 700 °C durch die Temperatur nicht wesentlich beeinflußt wird.

Unter erhöhten Druck-Temperatur-Bedingungen werden Geschwindigkeits-Anisotropien in erster Linie durch Einregelung der Gesteinsminerale nach den Kristallgittern erzeugt. Seismische Anisotropien in der Erdkruste und im Erdmantel können demzufolge am einfachsten durch Einregelung der Gesteinsminerale nach dem Kristallgitter erklärt werden.

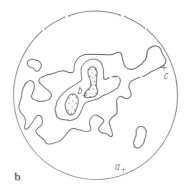

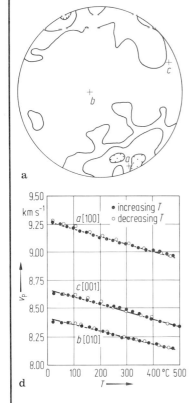

Fig. 8a···d. Lower-hemisphere, equal-area diagrams showing preferred orientations of olivine crystallographic axes in dunites (Twin Sisters Mtn.) a) a [100] axis, b) b [010] axis c) c [001] axis. Density contours 5, 3, 1 % per 1 % area. d) Temperature dependence of v_P at 10 kbar confining pressure, measured parallel to concentrations of a [100], b [010], and c [001], respectively [Ra78].

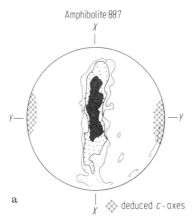

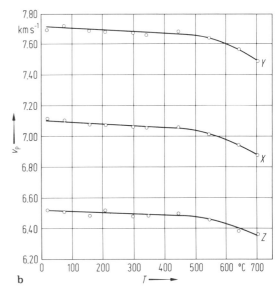

Fig. 9a, b. Relations of fabric anisotropy to anisotropy of elastic wave propagation in an amphibolite rock [Ke82]. a) Pole figure by X-ray goniometer. Lower hemisphere of equal projection. Pole density given in multiples of a uniform distribution. Amphibolite No 887. Hornblende [110]: contours: 2.0–1.6–1.2–1.0 b) Compressional wave velocities for the three orthogonal directions of the sample cube as a function of temperature at 6 kbar confining pressure.

3.1.4.4.1.4 Dynamic elastic parameters and T derivatives at high pressure and temperature —
Dynamische elastische Konstanten und T-Gradienten bei hohen Drucken und Temperaturen

v_P and v_S, ρ (room temperature densities and *in-situ* densities, respectively), along with v_P/v_S, σ, K, G, E, and λ listed in Table 3, are based on the formulae for isotropic bodies as proposed by [Bi60]. The data for

In Tab. 3 sind v_P, v_S, ρ (Dichten bei Raum- bzw. unter *in-situ*-Bedingungen) mit den berechneten Werten von v_P/v_S, σ, K, G, E und λ zusammengestellt. Die elastischen Parameter wurden gemäß der von [Bi60] für

significantly anisotropic rocks must, therefore, be treated with caution.

In general, the values of K, G, E, and φ increase with pressure at room temperature and decrease with temperature at high confining pressure. Fig. 10 illustrates the effect of temperature at 2 kbar on the dynamic elastic parameters in a quartzite. In quartz-bearing rocks the bulk modulus, typically, decreases much more than the shear modulus. Across the high-low quartz transition, the gradual onset of change in slope and the temperature range in which this takes place corresponds well to the feature that appear on the velocity curves (see Fig. 4).

For most minerals and crack-free rocks Poisson's ratios are around 0.2 to 0.3 (see e.g. [Bi66, Ja69]). The Poisson's ratios computed from the velocity data of the quartz-bearing rocks behave very anomalously in this respect: they change to negative values within the transition region. This observation must be attributed to the fact that v_P is very sensitive to the temperature of the high-low quartz transition whereas v_S is nearly unaffected by that temperature, thus leading to very low v_P/v_S ratios. In addition, the presence of cracks at moderate confining pressure gives rise to further reduction of the Poisson's ratio.

Considering the relationship between shear modulus, Young's modulus and Poisson's ratio $\left(G = \dfrac{E}{2(1+\sigma)}\right)$, one might conclude that the Poisson's ratio may become negative, but it has to be greater than -1.

Best-fit solutions of the temperature derivatives of v_P and v_S at confining pressure are given in Tables 4 and 5 respectively.

isotrope Körper aufgestellten Formeln berechnet. Die Daten der stärker anisotropen Gesteine sind daher mit entsprechendem Vorbehalt zu verwenden.

Im allgemeinen steigen K, G, E und φ bei Raumtemperatur mit dem Druck an und fallen mit steigender Temperatur bei konstantem Druck ab. Fig. 10 illustriert den Temperatureinfluß auf die dynamischen elastischen Parameter bei 2 kbar in einem Quarzit. Bezeichnenderweise fällt in quarzführenden Gesteinen der Kompressionsmodul stärker ab als der Schermodul. Im Bereich der Hoch-Tief-Quarz-Umwandlung ist die Kurvenform qualitativ identisch mit der Form der Geschwindigkeitskurven (vgl. Fig. 4).

Für die meisten Minerale und rißfreien Gesteine ist das Poisson-Verhältnis etwa 0.2···0.3 (s. z.B. [Bi66, Ja69]). Die aus den Geschwindigkeitsdaten der quarzführenden Gesteine bestimmten Poisson-Zahlen sind in dieser Hinsicht sehr anomal: sie werden im Inversionsbereich negativ. Dies ist darauf zurückzuführen, daß v_P im Bereich der Hoch-Tief-Umwandlung des Quarzes sehr temperaturabhängig ist, während v_S durch die Temperatur wenig beeinflußt wird, so daß sich niedrige Poisson-Verhältnisse einstellen. Weiterhin führt bei niedrigen Drucken Mikrorißbildung zu einer weiteren Erniedrigung der Poissonschen Zahlen.

Betrachtet man die Beziehungen zwischen Schermodul, E-Modul und Poisson-Verhältnis $\left(G = \dfrac{E}{2(1+\sigma)}\right)$, so ist zu folgern, daß das Poisson-Verhältnis zwar negativ werden kann, nicht aber kleiner als -1.

In den Tabellen 4 bzw. 5 sind die Temperatur-Gradienten für v_P und v_S bei konstanten Drucken aus Regressionsrechnungen (best-fit solutions) aufgeführt.

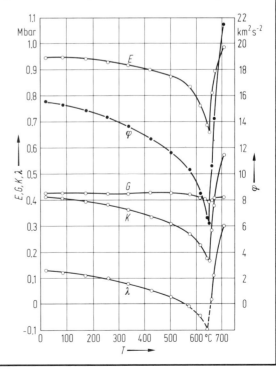

Fig. 10. Values of bulk modulus K, the shear modulus G, Young's modulus E, the Lamé's constant λ, and the seismic parameter φ as a function of temperature, T, at 2 kbar confining pressure for quartzite, calculated from the velocities according to the formulas for isotropic bodies [Ke79].

Table 3. v_P, v_S, ϱ and elastic constants as a function of temperature at confining pressure in rocks.

No.	Rock type, locality	T °C	ϱ g cm⁻³	v_P km s⁻¹	v_S km s⁻¹	v_P/v_S	σ	K Mbar	G Mbar	E Mbar	λ Mbar	Notes	Ref.
Plutonic rocks													
20	Granite, Wurmberg, Harz, Germany	20	2.570	6.100	3.500	1.742	0.254	0.536	0.314	0.789	0.326	4.2 kbar	Fi71
		200	2.570	6.057	3.483	1.739	0.252	0.527	0.311	0.781	0.319		
		300	2.570	6.021	3.472	1.734	0.250	0.518	0.309	0.775	0.312		
		400	2.570	5.972	3.455	1.728	0.248	0.507	0.306	0.766	0.303		
		500	2.570	5.893	3.434	1.716	0.242	0.488	0.303	0.753	0.286		
		600	2.570	5.771	3.402	1.696	0.233	0.459	0.297	0.733	0.261		
	Gray granite, Llano County, Texas	25	2.609	6.300	3.610	1.745	0.255	0.582	0.340	0.853	0.355	6 kbar	Hu55
		100	2.609	6.240	3.600	1.733	0.250	0.565	0.338	0.845	0.339		
		200	2.609	6.180	3.590	1.721	0.245	0.548	0.336	0.837	0.323	C	
	Pink granite, Llano County, Texas	25	2.636	6.540	3.390	1.929	0.316	0.723	0.302	0.797	0.521	6 kbar	Hu56
		100	2.636	6.520	3.360	1.940	0.319	0.723	0.297	0.785	0.525		
		200	2.636	6.460	3.360	1.922	0.314	0.703	0.297	0.782	0.504	C	
876	Granite, Source unknown	20	2.701	5.989	3.397	1.763	0.262	0.553	0.311	0.787	0.345	6 kbar	Ke81
		160	2.695	5.955	3.383	1.760	0.261	0.544	0.308	0.778	0.338		
		315	2.693	5.911	3.346	1.766	0.264	0.538	0.301	0.762	0.337	A = 2.0 %	
		490	2.690	5.872	3.304	1.777	0.268	0.535	0.293	0.744	0.340		
		710	2.667	5.648	3.201	1.764	0.263	0.486	0.273	0.690	0.304		
5001/1	Anorthosite, Tellnes, Norway	20	2.735	6.837	3.454	1.979	0.328	0.843	0.326	0.867	0.625	6 kbar	Ke82a
		93	2.735	6.815	3.444	1.978	0.328	0.837	0.324	0.861	0.621	A = 2.53 %	
		180	2.733	6.806	3.434	1.981	0.329	0.836	0.322	0.856	0.621		
		270	2.735	6.787	3.422	1.983	0.329	0.832	0.320	0.851	0.619	1	
29	Anorthosite Tellnes, Norway	20	2.778	6.926	3.829	1.808	0.279	0.789	0.407	1.042	0.518	4 kbar	Ke82a
		167	2.778	6.872	3.801	1.807	0.279	0.776	0.401	1.027	0.509		
		320	2.775	6.842	3.774	1.812	0.281	0.772	0.395	1.012	0.508	A = 2.99 %	
		483	2.774	6.740	3.663	1.840	0.290	0.763	0.372	0.960	0.515		
		715	2.755	6.533	3.612	1.808	0.279	0.696	0.359	0.920	0.456	1	
	Bytownite–Gabbro, Duluth, Minnesota	25	2.885	6.840	3.540	1.932	0.317	0.867	0.361	0.952	0.626	6 kbar	Hu57
		100	2.885	6.790	3.500	1.940	0.319	0.858	0.353	0.932	0.623	C	
	Hornblende–Gabbro, source unknown	25	2.933	6.890	3.720	1.852	0.294	0.851	0.405	1.050	0.580	6 kbar	Hu57
		100	2.933	6.880	3.710	1.854	0.295	0.850	0.403	1.045	0.580		
		200	2.933	6.840	3.740	1.828	0.286	0.825	0.410	1.055	0.551		
		300	2.933	6.810	3.710	1.835	0.288	0.821	0.403	1.040	0.552	C	

(continued)

Kern

Table 3 continued

No.	Rock type, locality	T °C	ϱ g cm^{-3}	v_P km s^{-1}	v_S km s^{-1}	v_P/v_S	σ	K Mbar	G Mbar	E Mbar	λ Mbar	Notes	Ref.
	Gabbro, Radautal, Harz, Germany	20	2.970	6.800	3.900	1.743	0.254	0.771	0.451	1.133	0.469	4.1 kbar	Fi71
		200	2.970	6.750	3.884	1.740	0.253	0.759	0.448	1.123	0.460		
		300	2.970	6.725	3.865	1.739	0.253	0.751	0.443	1.112	0.455	C	
		400	2.970	6.678	3.845	1.736	0.252	0.739	0.439	1.099	0.446		
		500	2.970	6.616	3.826	1.729	0.248	0.720	0.434	1.085	0.430		
		600	2.970	6.542	3.795	1.723	0.246	0.700	0.427	1.066	0.415		
	Gabbro, San Marcos, California	25	2.993	7.050	3.540	1.991	0.331	0.987	0.375	0.998	0.737	6 kbar	Hu57
		100	2.993	7.020	3.540	1.983	0.329	0.974	0.375	0.997	0.724		
		200	2.993	7.010	3.540	1.980	0.329	0.970	0.375	0.996	0.720	C	
		300	2.993	7.000	3.520	1.988	0.331	0.972	0.370	0.987	0.724		
34	Norite, Radautal, Harz	20	2.930	6.976	3.658	1.907	0.310	0.903	0.392	1.027	0.641	6 kbar	Ke81
		170	2.934	6.958	3.648	1.907	0.310	0.899	0.390	1.023	0.639		
		360	2.935	6.926	3.611	1.918	0.313	0.897	0.382	1.005	0.642	$A=1.6\%$	
		520	2.929	6.870	3.580	1.918	0.313	0.881	0.375	0.986	0.631		
		700	2.918	6.678	3.454	1.933	0.317	0.837	0.348	0.917	0.605		
84a	Norite, Radautal, Harz, Germany	20	3.030	6.669	3.699	1.802	0.277	0.794	0.414	1.059	0.518	6 kbar	Ke82a
		165	3.028	6.631	3.681	1.801	0.277	0.784	0.410	1.048	0.510		
		341	3.026	6.589	3.672	1.794	0.274	0.769	0.408	1.040	0.497		
		504	3.023	6.500	3.603	1.804	0.278	0.753	0.392	1.003	0.492		
		711	2.998	6.248	3.480	1.795	0.275	0.686	0.363	0.925	0.444		
5001/2	Ilmenite–Norite, Tellnes, Norway	20	3.475	7.009	3.750	1.869	0.299	1.055	0.488	1.270	0.729	6 kbar	Ke82a
		180	3.472	6.985	3.756	1.859	0.296	1.040	0.489	1.270	0.714		
		360	3.468	6.914	3.721	1.858	0.296	1.017	0.480	1.244	0.697		
		521	3.462	6.791	3.656	1.857	0.295	0.979	0.462	1.199	0.671		
		712	3.441	6.641	3.592	1.848	0.293	0.925	0.443	1.148	0.629		
10	Peridotite, Stubachtal, Hohe Tauern, Austria	20	3.250	7.800	4.500	1.733	0.250	1.099	0.658	1.646	0.661	4.1 kbar	Fi71
		200	3.250	7.730	4.446	1.738	0.252	1.085	0.642	1.609	0.657	C	
		300	3.250	7.691	4.410	1.743	0.255	1.079	0.632	1.586	0.658		
		400	3.250	7.644	4.370	1.749	0.257	1.071	0.620	1.560	0.657		
		500	3.250	7.597	4.338	1.751	0.258	1.060	0.611	1.538	0.652		
		600	3.250	7.543	4.298	1.755	0.259	1.048	0.600	1.512	0.648		

(continued)

Kern

Table 3 continued

No.	Rock type, locality	T °C	ϱ g cm⁻³	v_P km s⁻¹	v_S km s⁻¹	v_P/v_S	σ	K Mbar	G Mbar	E Mbar	λ Mbar	Notes	Ref.
475	Peridotite,	20	3.290	8.010	4.575	1.750	0.257	1.192	0.688	1.732	0.733	6 kbar	Ke81
475b	Finero, Italy	80	3.288	7.955	4.543	1.751	0.258	1.175	0.678	1.707	0.723		
		350	3.282	7.842	4.448	1.763	0.262	1.152	0.649	1.640	0.719		
		510	3.272	7.733	4.404	1.755	0.259	1.110	0.634	1.599	0.687		
		715	3.259	7.474	4.280	1.746	0.256	1.024	0.596	1.499	0.626	$A=4.7\%$	
	Dunite, Twin Sisters	24	3.160	8.977	4.529	1.982	0.329	1.682	0.648	1.723	1.250	5.1 kbar	Hu51
	Mountain, Washington	100	3.160	8.987	4.559	1.971	0.326	1.676	0.656	1.742	1.238	C	
		200	3.160	8.895	4.514	1.970	0.326	1.641	0.643	1.708	1.212		
		300	3.160	8.792	4.459	1.971	0.326	1.604	0.628	1.667	1.186		
	Dunite, Jackson County,	25	3.198	7.910	4.170	1.896	0.307	1.259	0.556	1.454	0.888	5 kbar	Hu57
	N. Carolina	100	3.198	7.860	4.120	1.907	0.310	1.251	0.542	1.422	0.890	C	
		200	3.198	7.480	3.830	1.953	0.322	1.163	0.469	1.240	0.851		
1675	Dunite, Åheim,	20	3.278	8.227	4.462	1.843	0.291	1.348	0.652	1.685	0.913	6 kbar	Ke81
	Norway	180	3.280	8.163	4.408	1.851	0.294	1.335	0.637	1.649	0.910		
		355	3.271	8.072	4.337	1.861	0.297	1.310	0.615	1.596	0.900	$A=4.8\%$	
		510	3.261	8.022	4.292	1.869	0.299	1.297	0.600	1.561	0.897		
		710	3.271	7.925	4.291	1.846	0.292	1.251	0.602	1.557	0.849		

Volcanic rocks

No.	Rock type, locality	T °C	ϱ g cm⁻³	v_P km s⁻¹	v_S km s⁻¹	v_P/v_S	σ	K Mbar	G Mbar	E Mbar	λ Mbar	Notes	Ref.
12/2	Tholeiite–Basalt,	20	2.615	5.300	2.902	1.826	0.286	0.441	0.220	0.566	0.294	6 kbar	Ke79a
	Suduroy, Faeroe	100	2.615	5.168	2.818	1.833	0.288	0.422	0.208	0.535	0.283		
	Islands	200	2.615	5.235	2.786	1.879	0.302	0.446	0.203	0.529	0.311		
		315	2.615	5.529	2.964	1.865	0.298	0.493	0.230	0.597	0.340		
		410	2.615	5.369	2.994	1.793	0.274	0.441	0.234	0.597	0.285		
2305	Tholeiite–Basalt,	20	2.834	5.415	3.043	1.779	0.269	0.481	0.262	0.666	0.306	6 kbar	Ke79a
	Eysturoy, Faeroe	100	2.834	5.387	3.020	1.783	0.271	0.478	0.258	0.657	0.305		
	Islands	205	2.834	5.396	3.036	1.777	0.268	0.477	0.261	0.663	0.303		
		315	2.834	5.461	3.079	1.773	0.267	0.487	0.269	0.681	0.308		
		500	2.834	5.554	3.165	1.754	0.260	0.496	0.284	0.715	0.306		
		700	2.834	5.632	3.137	1.795	0.275	0.527	0.279	0.711	0.341		

(continued)

Kern

Table 3 continued

No.	Rock type, locality	T °C	ϱ g cm^{-3}	v_P km s^{-1}	v_S km s^{-1}	v_P/v_S	σ	K Mbar	G Mbar	E Mbar	λ Mbar	Notes	Ref.
2311	Tholeiite–Basalt, Streymoy, Faeroe Islands	20	2.990	5.457	3.267	1.670	0.221	0.465	0.319	0.779	0.252	6 kbar	Ke79a
		100	2.990	5.396	3.199	1.686	0.229	0.463	0.306	0.752	0.259		
		200	2.990	5.399	3.203	1.685	0.228	0.463	0.307	0.754	0.258		
		300	2.990	5.412	3.215	1.683	0.227	0.464	0.309	0.759	0.258		
		495	2.990	5.369	3.197	1.679	0.225	0.454	0.306	0.749	0.251		
		705	2.990	5.440	3.239	1.679	0.225	0.467	0.314	0.769	0.257		
2306	Tholeiite–Basalt, Vagar, Faeroe Islands	20	2.898	6.354	3.540	1.794	0.275	0.686	0.363	0.926	0.444	6 kbar	Ke79a
		100	2.898	6.351	3.518	1.805	0.279	0.680	0.353	0.904	0.445		
		205	2.898	6.369	3.506	1.816	0.283	0.690	0.351	0.900	0.456		
		310	2.898	6.371	3.521	1.809	0.280	0.687	0.354	0.906	0.451		
		415	2.898	5.817	3.318	1.753	0.259	0.547	0.314	0.791	0.337		
		510	2.898	5.775	3.347	1.725	0.247	0.526	0.320	0.798	0.313		
2304	Tholeiite–Basalt, Vidoy, Faeroe Islands	20	2.902	6.551	3.658	1.790	0.273	0.728	0.388	0.989	0.469	6 kbar	Ke79a
		100	2.902	6.499	3.611	1.799	0.277	0.721	0.378	0.966	0.469		
		205	2.902	6.485	3.604	1.799	0.277	0.718	0.377	0.962	0.467		
		305	2.902	6.459	3.604	1.792	0.274	0.708	0.377	0.960	0.457		
		510	2.902	6.075	3.487	1.742	0.254	0.601	0.353	0.885	0.365		
		705	2.902	6.082	3.438	1.769	0.265	0.616	0.343	0.868	0.387		
2308	Tholeiite–Basalt, Vagar, Faeroe Islands	20	3.006	6.192	3.475	1.781	0.270	0.669	0.363	0.922	0.427	6 kbar	Ke79a
		100	3.006	6.164	3.446	1.788	0.273	0.666	0.357	0.909	0.428		
		200	3.006	6.178	3.444	1.793	0.275	0.672	0.357	0.909	0.434		
		300	3.006	6.197	3.463	1.789	0.273	0.674	0.360	0.918	0.433		
		495	3.006	6.224	3.486	1.785	0.271	0.677	0.365	0.929	0.434		
		700	3.006	6.016	3.269	1.840	0.291	0.660	0.321	0.829	0.445		
2303	Tholeiite–Basalt, Suduroy, Faeroe Islands	20	2.981	6.107	3.475	1.757	0.261	0.632	0.360	0.908	0.392	6 kbar	Ke79a
		100	2.981	6.130	3.486	1.758	0.261	0.637	0.362	0.914	0.396		
		200	2.981	6.125	3.477	1.761	0.262	0.628	0.360	0.910	0.398		
		305	2.981	6.081	3.462	1.756	0.260	0.626	0.357	0.901	0.388		
		510	2.981	5.962	3.373	1.767	0.265	0.607	0.339	0.858	0.381		
		710	2.981	5.906	3.327	1.775	0.268	0.600	0.330	0.837	0.380		

(continued)

Kern

Table 3 continued

No.	Rock type, locality	T °C	ϱ g cm^{-3}	v_P km s^{-1}	v_S km s^{-1}	v_P/v_S	σ	K Mbar	G Mbar	E Mbar	λ Mbar	Notes	Ref.
2301	Tholeiite–Basalt, Suduroy, Faeroe Islands	20	3.023	5.854	3.257	1.797	0.276	0.608	0.321	0.818	0.395	6 kbar	Ke79a
		100	3.023	5.804	3.220	1.802	0.278	0.600	0.313	0.801	0.391		
		210	3.023	5.795	3.212	1.804	0.278	0.599	0.312	0.797	0.391		
		300	3.023	5.798	3.216	1.802	0.278	0.599	0.313	0.799	0.391		
		500	3.023	5.810	3.215	1.807	0.279	0.604	0.312	0.799	0.396		
		710	3.023	5.674	3.055	1.857	0.296	0.597	0.282	0.731	0.409		
6a/2	Tholeiite–Basalt, Sandoy, Faeroe Islands	20	3.022	6.148	3.480	1.766	0.264	0.654	0.366	0.925	0.410	6 kbar	Ke79a
		100	3.022	6.094	3.460	1.761	0.262	0.640	0.362	0.913	0.399		
		200	3.022	6.082	3.459	1.758	0.261	0.638	0.362	0.912	0.395		
		305	3.022	6.092	3.469	1.756	0.260	0.637	0.364	0.916	0.394		
		505	3.022	6.115	3.469	1.762	0.263	0.645	0.364	0.918	0.403		
		710	3.022	6.262	3.491	1.793	0.275	0.694	0.368	0.939	0.448		
2302	Tholeiite–Basalt, Suduroy, Faeroe Islands	20	3.049	6.141	3.449	1.780	0.270	0.666	0.363	0.921	0.424	6 kbar	Ke79a
		100	3.049	6.120	3.436	1.781	0.270	0.662	0.360	0.914	0.422		
		205	3.049	6.127	3.444	1.779	0.269	0.662	0.362	0.918	0.421		
		305	3.049	6.103	3.436	1.776	0.268	0.656	0.360	0.913	0.416		
		500	3.049	6.067	3.412	1.778	0.269	0.649	0.355	0.901	0.412		
		695	3.049	5.785	3.225	1.793	0.275	0.598	0.317	0.808	0.386		
12	Metabasalt, Koitsøy Island, Stavanger, Norway	22	2.990	6.460	3.670	1.760	0.261	0.710	0.402	1.016	0.442	5 kbar	B₁82
		157	2.990	6.400	3.610	1.772	0.266	0.705	0.389	0.987	0.445		
		357	2.990	6.290	3.560	1.766	0.264	0.677	0.378	0.958	0.425	C	
		450	2.990	6.250	3.530	1.770	0.265	0.671	0.372	0.943	0.422		
		537	2.990	6.210	3.530	1.759	0.261	0.656	0.372	0.939	0.407		
	Metamorphic rocks												
1419	Quartzite, Koli, Finland	20	2.719	6.038	3.879	1.556	0.148	0.445	0.409	0.939	0.173	4 kbar	Ke82a
		82	2.719	6.033	3.891	1.550	0.143	0.447	0.411	0.941	0.166		
		169	2.719	6.005	3.894	1.542	0.137	0.430	0.412	0.937	0.155		
		261	2.719	5.951	3.857	1.542	0.137	0.423	0.404	0.920	0.153		
		346	2.719	5.913	3.855	1.533	0.130	0.411	0.404	0.913	0.142		
		435	2.719	5.865	3.854	1.521	0.120	0.396	0.403	0.904	0.127		
		511	2.719	5.808	3.844	1.510	0.110	0.381	0.401	0.892	0.113		
		549	2.719	5.701	3.834	1.486	0.087	0.350	0.399	0.869	0.084		

(continued)

Kern

Table 3 continued

No.	Rock type, locality	T °C	ϱ g cm⁻³	v_P km s⁻¹	v_S km s⁻¹	v_P/v_S	σ	K Mbar	G Mbar	E Mbar	λ Mbar	Notes	Ref.
1419	Quartzite (continued)	612	2.719	5.616	3.832	1.465	0.064	0.325	0.399	0.849	0.059		
		642	2.719	5.552	3.830	1.449	0.046	0.306	0.398	0.834	0.040		
		667	2.719	5.473	3.838	1.426	0.016	0.280	0.400	0.814	0.013		
		680	2.719	5.387	3.830	1.406	−0.011	0.257	0.398	0.788	−0.008		
		710	2.719	5.196	3.824	1.358	−0.090	0.203	0.397	0.722	−0.061		
		716	2.719	5.264	3.815	1.379	−0.053	0.225	0.395	0.749	−0.038		
		725	2.719	5.696	3.841	1.482	0.083	0.347	0.401	0.868	0.079		
		731	2.719	5.989	3.857	1.552	0.145	0.435	0.404	0.926	0.166		
1452	Quartzite, Bayrischer Wald, Germany	20	2.735	5.883	3.800	1.548	0.142	0.419	0.394	0.902	0.156	6 kbar	Ke81
		170	2.733	5.849	3.796	1.540	0.136	0.409	0.393	0.894	0.147		
		340	2.726	5.763	3.723	1.547	0.141	0.401	0.377	0.862	0.149	$A = 6.0\%$	
		500	2.711	5.680	3.674	1.545	0.140	0.386	0.365	0.834	0.142		
		710	2.680	5.337	3.520	1.516	0.115	0.320	0.332	0.740	0.099		
1400	Biotite–Orthopyroxene–Plagioclase–Gneiss, Karasjok, Norway	20	2.816	6.578	3.574	1.840	0.290	0.738	0.359	0.928	0.499	6 kbar	Ke81
		170	2.817	6.532	3.556	1.836	0.289	0.726	0.356	0.918	0.489		
		360	2.815	6.442	3.489	1.846	0.292	0.711	0.342	0.885	0.482	$A = 2.4\%$	
		515	2.810	6.340	3.436	1.845	0.292	0.687	0.331	0.857	0.465		
		710	2.801	6.103	3.362	1.815	0.282	0.621	0.316	0.811	0.410		
268	Plagioclase–Amphibolite–Gneiss, Karasjok, Norway	20	2.857	6.499	3.769	1.724	0.246	0.665	0.405	1.011	0.395	6 kbar	Ke81
		170	2.860	6.481	3.769	1.719	0.244	0.659	0.406	1.011	0.388		
		335	2.860	6.423	3.726	1.723	0.246	0.650	0.397	0.989	0.385	$A = 1.5\%$	
		505	2.854	6.371	3.679	1.731	0.249	0.643	0.386	0.965	0.385		
		710	2.832	6.136	3.572	1.717	0.243	0.584	0.361	0.898	0.343		
1398	Pyroxene–Plagioclase–Garnet–Gneiss, Karasjok, Norway	20	2.976	6.644	3.858	1.722	0.245	0.723	0.442	1.103	0.427	6 kbar	Ke81
		180	2.980	6.625	3.856	1.718	0.243	0.717	0.443	1.102	0.421		
		350	2.978	6.571	3.810	1.724	0.246	0.709	0.432	1.077	0.421	$A = 0.8\%$	
		520	2.967	6.520	3.777	1.726	0.247	0.696	0.423	1.056	0.414		
		710	2.946	6.256	3.649	1.714	0.242	0.629	0.392	0.974	0.368		
298	Plagioclase–Quartz–Amphibolite–Gneiss, Karasjok, Norway	20	3.107	7.077	3.961	1.786	0.271	0.906	0.487	1.240	0.581	6 kbar	Ke81
		170	3.109	7.065	3.958	1.784	0.271	0.902	0.487	1.238	0.577		
		330	3.107	7.019	3.910	1.795	0.275	0.897	0.475	1.211	0.580	$A = 4.8\%$	
		490	3.102	6.970	3.882	1.795	0.275	0.883	0.467	1.192	0.572		
		710	3.082	6.793	3.777	1.798	0.276	0.835	0.439	1.122	0.542		

(continued)

Kern

Table 3 continued

No.	Rock type, locality	T °C	ϱ g cm^{-3}	v_P km s^{-1}	v_S km s^{-1}	v_P/v_S	σ	K Mbar	G Mbar	E Mbar	λ Mbar	Notes	Ref.
1403	Sillimanite–Garnet-Gneiss, Karasjok, Norway	20	3.107	7.155	3.949	1.811	0.280	0.944	0.484	1.241	0.621	6 kbar	Ke81
		175	3.108	7.121	3.938	1.808	0.279	0.933	0.481	1.233	0.612		
		340	3.106	7.054	3.888	1.814	0.281	0.919	0.469	1.203	0.606	$A=3.3\%$	
		505	3.099	6.957	3.844	1.809	0.280	0.889	0.457	1.172	0.584		
		710	3.081	6.741	3.798	1.774	0.267	0.807	0.444	1.126	0.511		
1387	Amphibolite, S. Alta, Norway	20	2.999	7.041	3.778	1.863	0.297	0.916	0.428	1.111	0.630	6 kbar	Ke81
		175	2.999	7.026	3.761	1.868	0.299	0.914	0.424	1.102	0.632		
		360	3.001	6.991	3.724	1.877	0.301	0.911	0.416	1.083	0.634	$A=8.0\%$	
		515	2.995	6.935	3.673	1.888	0.305	0.901	0.404	1.054	0.632		
		710	2.976	6.809	3.586	1.898	0.308	0.869	0.382	1.001	0.614		
1396	Amphibolite, Karasjok, Norway	20	3.106	6.471	3.711	1.743	0.254	0.730	0.427	1.073	0.445	6 kbar	Ke81
		170	3.111	6.455	3.709	1.740	0.253	0.725	0.427	1.072	0.440		
		335	3.114	6.406	3.666	1.747	0.256	0.719	0.418	1.051	0.440	$A=11\%$	
		510	3.109	6.355	3.646	1.743	0.254	0.704	0.413	1.037	0.429		
		710	3.087	6.270	3.580	1.751	0.258	0.686	0.395	0.995	0.422		
1454	Epidote–Amphibolite, Bayr. Wald, Germany	20	3.109	6.885	3.842	1.792	0.273	0.861	0.458	1.169	0.555	6 kbar	Ke81
		170	3.107	6.853	3.822	1.793	0.274	0.854	0.453	1.156	0.551		
		335	3.109	6.781	3.759	1.803	0.278	0.843	0.439	1.123	0.550	$A=6.0\%$	
		490	3.102	6.712	3.718	1.805	0.278	0.825	0.428	1.096	0.539		
		710	3.084	6.515	3.614	1.802	0.277	0.771	0.402	1.029	0.503		
886	Eclogite, Saualpe, Austria	20	3.442	7.975	4.470	1.784	0.270	1.272	0.687	1.748	0.813	6 kbar	Ke31
		170	3.443	7.954	4.457	1.784	0.271	1.266	0.683	1.738	0.810		
		320	3.437	7.893	4.395	1.795	0.275	1.256	0.663	1.693	0.813	$A=0.8\%$	
		485	3.435	7.863	4.365	1.801	0.277	1.251	0.654	1.671	0.814		
		710	3.409	7.730	4.256	1.816	0.282	1.213	0.617	1.583	0.801		
11	Eclogite, Fichtelgebirge, Germany	20	3.495	7.886	4.438	1.776	0.268	1.255	0.688	1.746	0.796	6 kbar	Ke81
		175	3.499	7.880	4.436	1.776	0.268	1.254	0.688	1.746	0.795		
		340	3.496	7.822	4.372	1.789	0.272	1.247	0.668	1.701	0.802	$A=0.3\%$	
		515	3.488	7.783	4.346	1.790	0.273	1.234	0.658	1.677	0.795		
		710	3.463	7.622	4.242	1.796	0.275	1.180	0.623	1.589	0.765		

(continued)

Kern

Table 3 continued

No.	Rock type, locality	T °C	ϱ g cm^{-3}	v_P km s^{-1}	v_S km s^{-1}	v_P/v_S	σ	K Mbar	G Mbar	E Mbar	λ Mbar	Notes	Ref.
15	Eclogite, Münchberger Gneismasse, Fichtelgebirge, Germany	20	3.510	7.900	4.600	1.717	0.243	1.200	0.742	1.847	0.705	4.1 kbar	Fi71
		200	3.510	7.853	4.572	1.717	0.243	1.186	0.733	1.824	0.697		
		300	3.510	7.813	4.549	1.717	0.243	1.174	0.726	1.806	0.689	C	
		400	3.510	7.766	4.522	1.717	0.243	1.159	0.717	1.785	0.681		
		500	3.510	7.703	4.485	1.717	0.243	1.141	0.706	1.756	0.670		
		600	3.510	7.639	4.448	1.717	0.243	1.122	0.694	1.727	0.659		
987	Serpentinite, source unknown	20	2.780	6.128	3.278	1.869	0.299	0.645	0.298	0.776	0.446	6 kbar	Ke82a
		172	2.775	6.075	3.234	1.878	0.302	0.637	0.290	0.755	0.443		
		331	2.774	5.990	3.181	1.883	0.303	0.621	0.280	0.731	0.433	$v_\perp$	
		498	2.770	5.851	3.104	1.884	0.304	0.592	0.266	0.696	0.414		
		713	2.791	5.437	3.013	1.804	0.278	0.487	0.253	0.647	0.318		
		20	2.780	7.651	4.532	1.688	0.229	0.866	0.570	1.404	0.485		
		172	2.775	7.659	4.506	1.699	0.235	0.876	0.563	1.392	0.500	$v_\parallel$	
		331	2.774	7.646	4.483	1.705	0.238	0.878	0.557	1.380	0.506		
		498	2.770	7.536	4.393	1.715	0.242	0.860	0.534	1.328	0.503		
		713	2.791	7.362	4.306	1.709	0.240	0.822	0.517	1.283	0.477		
		20	2.780	7.770	4.548	1.708	0.239	0.911	0.575	1.425	0.528		
		172	2.775	7.767	4.513	1.721	0.245	0.920	0.565	1.407	0.543	$v_\parallel$	
		331	2.774	7.713	4.462	1.728	0.248	0.913	0.552	1.379	0.545		
		498	2.770	7.611	4.384	1.736	0.251	0.894	0.532	1.332	0.539		
		713	2.791	7.427	4.291	1.730	0.249	0.854	0.513	1.284	0.511		

Sedimentary rocks

No.	Rock type, locality	T °C	ϱ g cm^{-3}	v_P km s^{-1}	v_S km s^{-1}	v_P/v_S	σ	K Mbar	G Mbar	E Mbar	λ Mbar	Notes	Ref.
	Sandstone, Caplen Dome, Galveston County	27	2.543	5.363	2.971	1.805	0.278	0.432	0.224	0.574	0.282	5 kbar	Hu51
		100	2.543	5.345	2.973	1.797	0.276	0.426	0.224	0.573	0.276		
		200	2.543	5.312	2.975	1.785	0.271	0.417	0.225	0.572	0.267	C	
	Limestone, Solnhofen, Bavaria, Germany	25	2.656	6.134	3.040	2.017	0.337	0.672	0.245	0.656	0.508	5 kbar	Hu51
		100	2.656	6.037	3.004	2.009	0.335	0.648	0.239	0.640	0.488		
		200	2.656	5.928	2.974	1.993	0.331	0.620	0.234	0.625	0.463	C	
	Argillaceous limestone, Upton County, Texas	26	2.731	6.368	3.291	1.934	0.317	0.713	0.295	0.779	0.515	5 kbar	Hu51
		100	2.731	6.293	3.282	1.917	0.313	0.689	0.294	0.772	0.493		
		200	2.731	6.207	3.253	1.908	0.310	0.666	0.288	0.757	0.474	C	

(continued)

Kern

Table 3 continued

No.	Rock type, locality	T °C	ϱ g cm⁻³	v_P km s⁻¹	v_S km s⁻¹	v_P/v_S	σ	K Mbar	G Mbar	E Mbar	λ Mbar	Notes	Ref.
	Graywacke, Steimkertal, Harz, Germany	22	2.680	5.700	3.540	1.610	0.186	0.422	0.335	0.796	0.199	5 kbar	Bu82
		156	2.680	5.550	3.470	1.599	0.179	0.395	0.322	0.760	0.180		
		256	2.680	5.460	3.460	1.578	0.164	0.371	0.320	0.747	0.157	C	
		356	2.680	5.460	3.440	1.587	0.170	0.376	0.317	0.742	0.164		
		456	2.680	5.370	3.390	1.584	0.168	0.362	0.307	0.719	0.156		
		568	2.680	5.110	3.220	1.586	0.170	0.329	0.277	0.650	0.144		
949	Shale, Rheinisches Schiefergebirge, Germany	20	2.859	5.598	3.071	1.822	0.284	0.536	0.269	0.692	0.356	6 kbar	Ke82
		188	2.849	5.460	2.877	1.897	0.307	0.534	0.235	0.616	0.377	$v_\perp$	
		358	2.844	5.323	2.759	1.929	0.316	0.517	0.216	0.569	0.372		
		530	2.835	5.130	2.584	1.985	0.330	0.493	0.189	0.503	0.367		
		680	2.770	4.235	2.370	1.786	0.272	0.289	0.155	0.395	0.185		
		20	2.859	6.592	4.014	1.642	0.205	0.628	0.460	1.110	0.321	$v_\parallel$	
		188	2.849	6.475	3.930	1.647	0.208	0.607	0.440	1.063	0.314		
		358	2.844	6.365	3.860	1.648	0.209	0.587	0.423	1.024	0.304		
		530	2.835	6.224	3.771	1.650	0.209	0.560	0.403	0.975	0.291		
		680	2.770	5.858	3.644	1.607	0.184	0.460	0.367	0.871	0.214		
		20	2.859	6.788	4.038	1.681	0.226	0.695	0.466	1.143	0.384	$v_\parallel$	
		188	2.849	6.676	3.952	1.689	0.230	0.676	0.444	1.094	0.379		
		358	2.844	6.563	3.878	1.692	0.231	0.654	0.427	1.053	0.369		
		530	2.835	6.425	3.790	1.695	0.233	0.627	0.407	1.004	0.355		
		680	2.770	6.059	3.640	1.664	0.217	0.527	0.367	0.893	0.282		
	Carnallitite, Salt Mine Asse II, Wolfenbüttel, Germany	19	1.691	4.005	2.203	1.817	0.283	0.161	0.082	0.210	0.107	0.4 kbar	Ke82a
		63	1.691	4.005	2.145	1.867	0.298	0.167	0.077	0.202	0.115		
		80	1.691	3.983	2.121	1.877	0.302	0.166	0.076	0.198	0.116		
		101	1.691	3.953	2.083	1.897	0.307	0.166	0.073	0.191	0.117		
		121	1.691	3.904	2.047	1.907	0.310	0.163	0.070	0.185	0.116		
		141	1.691	3.851	1.998	1.927	0.315	0.160	0.067	0.177	0.115		
		150	1.691	3.824	1.966	1.945	0.320	0.160	0.065	0.172	0.116		

(continued)

Kern

Table 3 continued

No.	Rock type, locality	T °C	ϱ g cm⁻³	v_P km s⁻¹	v_S km s⁻¹	v_P/v_S	σ	K Mbar	G Mbar	E Mbar	λ Mbar	Notes	Ref.
2574/2	Rock salt, Salt Mine Siegfried Giesen, Hildesheim, Germany	20	2.174	4.732	2.617	1.808	0.279	0.288	0.148	0.381	0.189	2 kbar	Ke82a
		48	2.172	4.700	2.598	1.809	0.280	0.284	0.146	0.375	0.186		
		100	2.164	4.637	2.572	1.802	0.277	0.274	0.143	0.365	0.178		
		153	2.154	4.572	2.537	1.802	0.277	0.265	0.138	0.354	0.172		
		204	2.143	4.506	2.499	1.803	0.277	0.256	0.133	0.342	0.167		
		254	2.132	4.451	2.471	1.801	0.277	0.248	0.130	0.332	0.162		
		300	2.123	4.398	2.439	1.803	0.277	0.242	0.126	0.322	0.158		
		359	2.111	4.326	2.401	1.801	0.277	0.232	0.121	0.310	0.151		
		410	2.102	4.263	2.366	1.801	0.277	0.225	0.117	0.300	0.146		
	Kieseritic rock salt, Salt Mine Herfa, Germany	20	2.256	4.860	2.767	1.756	0.260	0.302	0.172	0.435	0.187	2 kbar	Ke82a
		47	2.257	4.849	2.760	1.756	0.260	0.301	0.171	0.433	0.186		
		97	2.253	4.833	2.733	1.768	0.264	0.301	0.168	0.425	0.189		
		149	2.249	4.809	2.701	1.780	0.269	0.301	0.164	0.416	0.191		
		198	2.243	4.769	2.668	1.787	0.272	0.297	0.159	0.406	0.190		
		251	2.236	4.735	2.630	1.800	0.276	0.295	0.154	0.394	0.191		
		302	2.228	4.690	2.597	1.805	0.278	0.289	0.150	0.384	0.189		
		354	2.220	4.613	2.552	1.807	0.279	0.284	0.147	0.376	0.186		
	Polyhalite – Halite, Salt Mine Herfa, Germany	18	2.534	4.979	2.864	1.738	0.252	0.351	0.207	0.520	0.212	0.4 kbar	Ke82a
		61	2.533	4.935	2.832	1.742	0.254	0.346	0.203	0.509	0.210		
		80	2.531	4.915	2.815	1.746	0.255	0.344	0.200	0.503	0.210		
		101	2.529	4.893	2.800	1.747	0.256	0.341	0.198	0.498	0.208		
		122	2.527	4.870	2.783	1.749	0.257	0.338	0.195	0.492	0.207		
		140	2.525	4.858	2.776	1.750	0.257	0.336	0.194	0.489	0.206		
		161	2.524	4.847	2.768	1.751	0.258	0.335	0.193	0.486	0.206		
2574/1	Anhydrite, Salt Mine Siegfried Giesen, Hildesheim, Germany	20	2.938	6.054	3.309	1.829	0.286	0.647	0.321	0.828	0.433	2 kbar	Ke82a
		49	2.939	6.041	3.299	1.831	0.287	0.646	0.319	0.823	0.432		
		100	2.936	6.017	3.270	1.840	0.290	0.644	0.313	0.810	0.435		
		152	2.935	5.997	3.256	1.841	0.290	0.640	0.311	0.803	0.433		
		201	2.935	5.968	3.259	1.831	0.287	0.629	0.311	0.802	0.421		
		249	2.935	5.939	3.259	1.822	0.284	0.619	0.311	0.800	0.411		
		301	2.934	5.908	3.247	1.819	0.283	0.611	0.309	0.794	0.405		
		354	2.932	5.880	3.229	1.820	0.284	0.606	0.305	0.785	0.402		
		405	2.932	5.846	3.209	1.821	0.284	0.599	0.301	0.775	0.398		

Table 4. T derivatives of v_p in rocks.

No., Rock type	$-\,dv_p/dT$ $\dfrac{10^{-4}\,\text{km}}{\text{s}\,°C}$	T °C	P_c kbar	Ref.	Notes
Plutonic rocks					
Granite, Westerly, USA	1.5···3	100···400	2.5	Sp76	
20 Granite, Wurmberg, Harz, Germany	4.2	20···500	4.2	Fi71	C
Gray granite, Llano County, Texas	7.1	20···300	6	Hu56	C
Granite, Cape Ann, Massachusetts	3.9	25···300	2	Ch79	
Pink granite, Llano County, Texas	6.2	20···300	6	Hu56	C
876 Granite, source unknown	2.7	20···500	6	Ke81	
2628 Granite, Caucasus, USSR	15.8	20···200	4	Ba74	C
2025 Granite, Caucasus, USSR	16.8	20···200	4	Ba74	C
5001/1 Anorthosite, Tellnes, Norway	3.4	20···500	6	Ke82a	
29 Anorthosite, Tellnes, Norway	2.4	20···400	4	Ke82a	
Anorthosite, Lake St. John, Quebec	4.1	25···300	2	Ch79	
Gabbro, Mid-Atlantic Ridge	5.7	25···300	2	Ch79	
Hornblende-Gabbro, source unkown	3.1	25···300	6	Hu57	C
13 Gabbro, Radautal, Harz, Germany	3.8	20···500	4.1	Fi71	C
Gabbro, San Marcos, California	1.7	25···300	6	Hu57	C
2612c Gabbro, Papua, New Guinea	15.0	150···250	3	Kr76	
2208 Gabbro (amphibolized), Voronezh anteclise, USSR	19.8	20···200	3	Ba74	C
84 Norite, Radautal, Harz, Germany	1.7	20···500	6	Ke81	
84a Norite, Radautal, Harz, Germany	3.2	20···500	6	Ke82a	
5001/2 Ilmenite-Norite, Tellnes, Norway	4.3	20···500	6	Ke82a	$\bar{3}\perp$
2611c Pyroxenite, Papua, New Guinea	14.8	150···250	3	Kr76	
2602c Pyroxenite-Harzburgite, Papua, New Guinea	13.9	150···250	3	Kr76	
Harzburgite, Antalya, Turkey	6.3	25···275	8	Pe78	$\bar{3}\cdot1\perp$ $A=8.2\%$
Lherzolite, Salt Lake Crater, Hawaii	6.9	25···275	8	Pe78	$\bar{2}\cdot1\perp$
10 Peridotite, Stubachtal, Hohe Tauern, Austria	4.2	20···500	4.1	Fi71	C
475 Peridotite, Finero, Italy	4.9	20···500	6	Ke82	$\bar{3}\perp$
Dunite, Twin Sisters Mountain, Washington	7.1	25···300	5.17	Hu51	C
Dunite, Twin Sisters Mountain, Washington	6.1	20···500	10	Ra78	$\bar{3}\cdot1\perp$
Dunite, Jackson County, N. Carolina	25.3	25···200	5	Hu57	C

(continued)

Table 4 continued

No., Rock type		$-dv_P/dT$ $\frac{10^{-4}\,km}{s\,°C}$	T °C	P_c kbar	Ref.	Notes
1675	Dunite, Åheim, Norway	4.1	20···500	6	Ke81	$A=9.15\%$
	Dunite, Twin Sisters Mountain, Washington	5.6	25···300	2	Ch79	
Volcanic rocks						
	Analcite-Labradorite-Trachyre, source unknown	1.3	25···300	6	Hu57	C
2305	Tholeiite-Basalt, Eysturoy, Faeroe Islands	−3.7	20···500	6	Ke82a	$\bar{3}\perp$
2311	Tholeiite-Basalt, Streymoy, Faeroe Islands	1.2	20···400	6	Ke82	$\bar{3}\perp$
2306	Tholeiite-Basalt, Vagar, Faeroe Islands	−1.4	20···310	6	Ke82	$\bar{3}\perp$
2304	Tholeiite-Basalt, Vidoy(?), Faeroe Islands	2.5	20···305	6	Ke82	$\bar{3}\perp$
2308	Tholeiite-Basalt, Vagar, Faeroe Islands	−1.2	20···495	6	Ke82a	$\bar{3}\perp$
2303	Tholeiite-Basalt, Suduroy, Faeroe Islands	2.9	20···510	6	Ke82a	$\bar{3}\perp$
2301	Tholeiite-Basalt, Suduroy, Faeroe Islands	0.1	20···500	6	Ke82	$\bar{3}\perp$
6a/2	Tholeiite-Basalt Sandoy, Faeroe Islands	−0.1	20···500	6	Ke82a	$\bar{3}\perp$
2302	Tholeiite-Basalt, Suduroy, Faeroe Islands	1.3	20···500	6	Ke82	$\bar{3}\perp$
	Basalt, Chaffee County(?), Colorado	−0.5	25···300	6	Hu57	C
	Basalt, East Pacific Rise	3.9	20···300	2	Ch79	
12	Metabasalt, Koitsøy Island, Stavanger, Norway	5.1	22···500	5	Bu82	C
	Diabase, Kola, Peninsula	20.2	20···300	4	Ba74	C
Metamorphic rocks						
	Serpentinite, Mid-Atlantic Ridge	6.8	25···300	2	Ch79	
987	Serpentinite, source unknown	3.7	20···500	6	Ke82a	$\bar{3}\perp$
2606	Serpentinite, Voronezh anteclise, USSR	24.3	25···500	3	Ba74	C
1389	Quartz-Phyllite, S. Alta, Norway	5.2	20···500	6	Ke82	$\bar{3}\perp$
	Quartzite, Baraboo, Wisconsin	5.4	25···300	2	Ch79	

(continued)

Kern

Table 4 continued

No., Rock type		$-\mathrm{d}v_\mathrm{p}/\mathrm{d}T$ $\dfrac{10^{-4}\,\mathrm{km}}{\mathrm{s}\,°\mathrm{C}}$	T °C	P_c kbar	Ref.	Notes
1419	Quartzite, Koli, Finland	4.8	20···500	4	Ke82a	
1452	Quartzite, Bayrischer Wald, Germany	4.0	20···500	6	Ke81	$\bar{3}\perp$
938	Marble, Carrara, Italy	8.2	20···500	6	Ke82	$\bar{3}\perp$
154	Granulite Gneiss, Inari, Finland	3.2	20···500	6	Ke81	$\bar{3}\perp$
	Granulite, New Jersey Highlands	4.9	25···300	2	Ch79	
	Granulite, Saranac Lake, New York	5.1	25···300	2	Ch79	
	Granulite, Adirondack Mountains, New York	6.0	25···300	2	Ch79	
	Granulite, Valle d'Ossola, Italy	5.2	25···300	2	Ch79	
1400	Biotite-Orthopyroxene-Plagioclase-Gneiss, Karasjok, Norway	4.8	20···500	6	Ke81	$\bar{3}\perp$
268	Plagioclase-Amphibolite-Gneiss, Karasjok, Norway	2.5	20···500	6	Ke81	$\bar{3}\perp$
160	Biotite-Gneiss, Floitental, Tirol	3.6	20···500	6	Ke82	$\bar{3}\perp$
1398	Pyroxene-Plagioclase-Garnet-Gneiss, Karasjok, Norway	2.5	20···500	6	Ke81	$\bar{3}\perp$
298	Plagioclase-Quartz-Amphibolite-Gneiss, Karasjok, Norway	2.0	20···500	6	Ke81	$\bar{3}\perp$
1403	Sillimanite-Garnet-Gneiss, Karasjok, Norway	3.6	20···500	6	Ke81	$\bar{3}\perp$
	Amphibolite, Indian Ocean, Central Ridge	5.5	25···300	2	Ch79	
1387	Amphibolite, S. Alta, Norway	2.4	20···500	6	Ke81	$\bar{3}\perp$
1396	Amphibolite, Karasjok, Norway	0.7	20···500	6	Ke81	$\bar{3}\perp$
1454	Epidote-Amphibolite, Bayrischer Wald, Germany	3.0	20···500	6	Ke81	$\bar{3}\perp$
886	Eclogite, Saualpe, Austria	1.9	25···500	6	Ke81	$\bar{3}\perp$
11	Eclogite, Fichtelgebirge, Germany	1.6	25···500	6	Ke81	$\bar{3}\perp$
15	Eclogite, Münchberger Gneismasse, Fichtelgebirge, Germany	4.1	20···500	4.1	Fi71	C
	Eclogite, Nove' Dvory, Czechoslowakia	5.3	25···300	2	Ch79	

(continued)

Kern

Table 4 continued

No., Rock type		$-\mathrm{d}v_\mathrm{P}/\mathrm{d}T$ $\dfrac{10^{-4}\,\mathrm{km}}{\mathrm{s}\,°\mathrm{C}}$	T °C	P_c kbar	Ref.	Notes
Sedimentary rocks						
	Sandstone, Caplen Dome, Galveston County	2.9	27···200	5	Hu51	*C*
L071-5	Arkosic Sandstone, Diabolo Range, California	11.1	20···300	8	St78	
	Limestone, Solnhofen, Bavaria, Germany	11.7	25···200	5	Hu51	*C*
	Argillaceous limestone, Upton County, Texas	9.2	26···200	5	Hu51	*C*
2892	Limestone, Caucasus, USSR	18.1	20···200	4	Ba74	
33	Limestone, White Russia	13.6	20···150	0.3	Ba74	
BCS57	Shale, Central Belt, California	10.5	20···300	8	St78	
	Shale, Rheinisches Schiefer- gebirge, Germany	7.7	20···500	6	Ke82a	
W2	Graywacke, Costal Belt, California	10.5	20···300	8	St78	
W5	Graywacke, Costal Belt, California	11.1	20···300	8	St78	
SR70-1	Graywacke, Central Belt, California	9.5	20···300	8	St78	
IV5	Graywacke, Diabolo Range, California	6.8	20···300	8	St78	
SP740	Graywacke, Central Belt, California	7.5	20···300	8	St78	
	Graywacke, Steimkertal, Harz, Germany	6.4	22···500	5	Bu82	*C*
BCSS55	Graywacke, Central Belt, California	8.4	20···300	8	St78	
21RGC60	Metagraywacke, Diabolo Range, California	5.5···8.3	20···300	8	St78	
P3	Metagraywacke, Diabolo Range, California	6.3	20···300	8	St78	
P5	Metagraywacke, Diabolo Range, California	6.7	20···300	8	St78	
	Carnallitite, Salt Mine Asse II, Wolfenbüttel, Germany	15.1	19···151	0.4	Ke82a	
2574/2	Rock salt, Salt Mine Siegfried Giesen, Hildesheim, Germany	12.0	20···410	2	Ke82a	
	Kieseritic rock salt, Salt Mine Herfa, Germany	7.0	20···355	2	Ke82a	
	Polyhalite – Halite, Salt Mine Herfa, Germany	9.6	19···151	0.4	Ke82a	
2574/1	Anhydrite, Salt Mine Siegfried Giesen, Hildesheim, Germany	5.4	20···405	2	Ke82a	

Table 5. T derivatives of v_S in rocks.

No.	Rock type	$-dv_S/dT$ $\dfrac{10^{-4}\,km}{s\,°C}$	T °C	P_c kbar	Ref.	Notes
Plutonic rocks						
	Granite, Westerly, USA	1···5	100···400	2.5	Sp76	
20	Granite, Wurmberg, Harz, Germany	1.3	20···500	4.2	Fi71	C
	Gray granite, Llano County, Texas	1.1	25···200	6	Hu56	C
	Pink granite, Llano County, Texas	1.6	25···200	6	Hu56	C
876	Granite, source unknown	2.1	20···500	6	Ke81	
5001	Anorthosite, Tellnes, Norway	1.3	20···500	6	Ke82a	
29	Anorthosite, Tellnes, Norway	1.8···1.6	20···400	4	Ke82a	$\bar{2}\perp$
	Hornblende-Gabbro, source unknown	0.0	25···300	6	Hu57	C
13	Gabbro, Radautal, Harz, Germany	1.6	20···500	4.1	Fi71	C
	Gabbro, San Marcos, California	0.7	25···300	6	Hu57	C
84	Norite, Radautal, Harz, Germany	1.5	20···500	6	Ke81	
84a	Norite, Radautal, Harz, Germany	1.7	20···500	6	Ke82a	$\bar{3}\perp$
5001/2	Ilmenite-Norite, Tellnes, Norway	1.9	20···500	6	Ke82a	$\bar{3}\perp$
10	Peridotite, Stubachtal, Hohe Tauern, Austria	3.4	20···500	4.1	Fi71	C
475	Peridotite, Finero, Italy	3.9	20···500	6	Ke81	
	Dunite, Twin Sisters Mountain, Washington	2.9	25···300	5.17	Hu51	C
	Dunite, Jackson County, N. Carolina	19.9	25···200	5	Hu57	C
1675	Dunite, Åheim, Norway	3.5	20···500	6	Ke81	
Volcanic rocks						
2305	Tholeiite-Basalt, Eysturoy, Faeroe Islands	−3.5	20···500	6	Ke82a	
2311	Tholeiite-Basalt, Streymoy, Faeroe Islands	1.0	20···400	6	Ke82	
2304	Tholeiite-Basalt, Vidoy(?), Faeroe Islands	1.6	20···305	6	Ke82	
2308	Tholeiite-Basalt, Vagar, Faeroe Islands	−1.5	20···495	6	Ke82a	
2303	Tholeiite-Basalt, Suduroy, Faeroe Islands	2.2	20···510	6	Ke82a	
2301	Tholeiite-Basalt, Suduroy, Faeroe Islands	0.6	20···500	6	Ke82	
6a/2	Tholeiite-Basalt, Sandoy, Faeroe Islands	0.1	20···505	6	Ke82a	

(continued)

Table 5 continued

No. Rock type		$-\mathrm{d}v_\mathrm{S}/\mathrm{d}T$ $\dfrac{10^{-4}\,\mathrm{km}}{\mathrm{s}\,°\mathrm{C}}$	T °C	P_c kbar	Ref.	Notes
2302	Tholeiite-Basalt, Suduroy, Faeroe Islands	0.6	20···500	6	Ke82a	
	Basalt, Chaffee County(?), Colorado	0.8	25···300	6	Hu57	C
12	Metabasalt, Koitsøy Island, Stavanger, Norway	2.9	22···500	5	Bu82	C
Metamorphic rocks						
987	Serpentinite source unknown	3.1	20···500	6	Ke82a	3⊥
1389	Quartz-Phyllite, S. Alta, Norway	1.6	20···500	6	Ke82	
1419	Quartzite, Koli, Finland	0.9	20···500	4	Ke82a	
1452	Quartzite, Bayrischer Wald, Germany	2.9	20···500	6	Ke81	
938	Marble, Carrara, Italy	4.5	20···500	6	Ke82	
1400	Biotite-Orthopyroxene-Plagioclase-Gneiss, Karasjok, Norway	2.9	20···500	6	Ke81	
268	Plagioclase-Amphibolite-Gneiss, Karasjok, Norway	2.0	20···500	6	Ke81	
160	Biotite-Gneiss, Floitental, Tirol	1.7	20···500	6	Ke82	
1398	Pyroxene-Plagioclase-Garnet-Gneiss, Karasjok, Norway	1.7	20···500	6	Ke81	
298	Plagioclase-Quartz-Amphibolite-Gneiss, Karasjok, Norway	1.8	20···500	6	Ke81	
1403	Sillimannite-Garnet-Gneiss, Karasjok, Norway	2.2	20···500	6	Ke81	
1387	Amphibolite, S. Alta, Norway	2.1	20···500	6	Ke81	
1396	Amphibolite, Karasjok, Norway	1.4	20···500	6	Ke81	
1454	Epidote-Amphibolite, Bayrischer Wald, Germany	2.7	20···500	6	Ke81	
886	Eclogite, Saualpe, Austria	2.5	20···500	6	Ke81	
11	Eclogite, Fichtelgebirge, Germany	2.1	20···500	6	Ke81	
15	Eclogite, Münchberger Gneismasse, Fichtelgebirge, Germany	2.4	20···500	4.1	Fi71	C
Sedimentary rocks						
	Sandstone, Caplen Dome, Galveston County	−0.2	27···200	5	Hu51	C
	Limestone, Solnhofen, Bavaria, Germany	3.7	25···200	5	Hu51	C
	Argillaceous limestone, Upton County, Texas	2.2	26···200	5	Hu51	C

(continued)

Table 5 continued

No. Rock type		$-\mathrm{d}v_S/\mathrm{d}T$ $\dfrac{10^{-4}\,\mathrm{km}}{\mathrm{s}\,^\circ\mathrm{C}}$	T °C	P_c kbar	Ref.	Notes
949	Shale, Rheinisches Schiefer-gebirge, Germany	6.1	20···500	6	Ke82a	
	Graywacke, Steimkertal, Harz, Germany	3.1	20···500	5	Bu82	C
	Carnallitite, Salt Mine Asse II, Wolfenbüttel, Germany	17.8	19···151	0.4	Ke82a	
2574/2	Rock salt, Salt Mine Siegfried Giesen, Hildesheim, Germany	6.4	20···410	2	Ke82a	
	Kieseritic rock salt, Salt Mine Herfa, Germany	6.5	20···355	2	Ke82a	
	Polyhalite, Salt Mine Herfa, Germany	6.9	19···151	0.4	Ke82a	
2574/1	Anhydrite, Salt Mine Siegfried Giesen, Hildesheim, Germany	2.2	20···405	2	Ke82a	

3.1.4.4.2 Elastic wave velocities as a function of temperature in "wet" rocks — Geschwindigkeiten elastischer Wellen als Funktion der Temperatur in „nassen" Gesteinen

There are only few laboratory measurements of elastic wave velocities on rocks with water at elevated temperature [Sp76, Ti77, It79].

In *low-porous* granite (Fig. 11) v_P and K decrease continuously with increasing temperature at low pore-fluid pressure, whereas v_S and G decrease only slightly. The low v_P and nearly unchanged v_S cause a decrease in Poisson's ratio. At high P_f (low P_{eff}) G and v_S decrease with increasing temperature, while v_P is unaffected and σ increases [Sp76].

The temperature dependence of wave velocities in brine-saturated *high-porous* rocks (sandstone, carbonates) at confining pressure and controlled pore pressure is summarized in Table 6 [Ti77].

Bisher gibt es nur wenige Labor-Messungen der Geschwindigkeiten elastischer Wellen bei erhöhten Temperaturen [Sp76, Ti77, It79].

Im *porenarmen* Granit (Fig. 11) nehmen bei niedrigem Porendruck v_P und K mit steigender Temperatur kontinuierlich ab, während v_S und G nur geringfügig abfallen. Die niedrigen Werte für v_P und die nahezu gleichbleibenden Werte für v_S führen zu einer Verringerung des Poisson-Verhältnisses. Bei hohem P_f (niedriger P_{eff}) nehmen G und v_S mit der Temperatur ab, während v_P sich wenig ändert und σ ansteigt [Sp76].

In Tab. 6 ist die Temperaturabhängigkeit der Wellengeschwindigkeiten in laugengesättigten, *porenreichen* Gesteinen (Sandstein, Karbonate) bei hydrostatischen Drucken und kontrolliertem Porendruck zusammengefaßt [Ti77].

Table 6. Temperature dependence of P- and S-wave velocities at confining pressure and controlled pore-fluid pressure [Ti77].

Sample	$\varPhi$ %	P_c bar	P_f bar	T °C	v_P km s^{-1}	v_S km s^{-1}	$\mathrm{d}v_P/\mathrm{d}T$ 10^{-4} km s^{-1} °C^{-1}	$-\mathrm{d}v_S/\mathrm{d}T$
Berea Sandstone	17.11	1380	600	15.6	4.397	2.601	1.25	0.03
		345	150	21.1	4.254	2.510	1.59	1.18
		138	60	17.2	4.129	2.339	1.13	1.25
Boise Sandstone	30.84	690	345	18.3	3.350	1.806	2.57	2.45
Carbonate-1	2.92	915	398	26.1	5.961	3.301	1.39	0.51
Carbonate-2	3.63	916	398	22.2	5.840	3.414	1.67	0.35
Carbonate-3	1.30	917	399	23.9	6.261	3.237	1.81	0.72
Carbonate-4	6.40	920	400	26.1	5.936	3.128	1.29	0.18
Carbonate-5	9.92	921	400	17.2	5.609	2.925	2.01	0.55
Carbonate-6	8.69	953	414	23.9	5.589	2.963	2.03	1.57
Carbonate-7	8.47	953	414	23.9	5.491	2.918	2.46	0.99

Within the range $20\cdots200\,°C$ v_P and v_S decrease linearly (Fig. 12); the maximum velocity decrease correlates with the highest porosity (Boise sandstone). The general velocity decrease is attributed to the effects of temperature on both the moduli of the rock's matrix minerals and the pore fluid [Ti77].

A low P_{eff} (high P_f) causes the wave velocities to decrease more rapidly, possibly as a consequence of formation of new cracks. Velocity measurements in saturated rocks during water-steam transition [It79] are shown in Fig. 13. At 145 °C and 4 bars there is a sharp minimum of v_P. This pressure is found to be very close to the water-vapour transition pressure at 145 °C [Ke69].

v_P in steam-filled rock is lower than in water-filled rock, the reverse is true for v_S. At 198 °C both v_S and v_P decrease from steam-saturated (low P_f) to water-saturated (high P_f) rock (Fig. 13b). σ and v_P/v_S increase from steam saturation to water saturation, both at 145 °C and at 198 °C (Fig. 14).

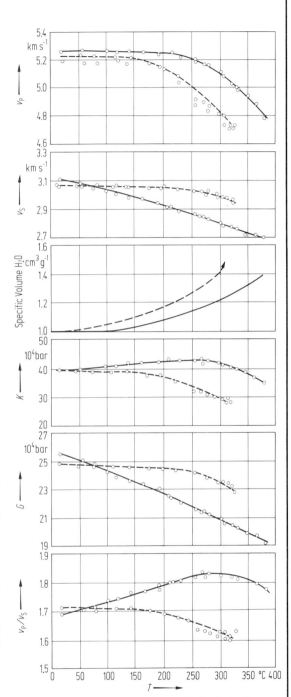

Fig. 11. Velocities, v_P, v_S, and moduli K, G, in Westerly granite with controlled pore pressures and low confining pressure, as a function of temperature [Sp76]. Dashed lines: $P_c = 1$ kbar; $P_f = 100$ bar. Solid lines: $P_c = 1$ kbar; $P_f = 100\cdots1000$ bar.

Innerhalb des Temperatur-Intervalls $20\cdots200\,°C$ nehmen v_P und v_S linear ab (Fig. 12); die maximale Geschwindigkeitsabnahme korreliert mit der höchsten Porosität (Boise Sandstein). Die generelle Geschwindigkeitsabnahme ist auf die Temperaturabhängigkeit der Moduli sowohl der Gesteinsmatrix-Minerale als auch der Porenflüssigkeit zurückzuführen [Ti77].

Niedriger P_{eff} (hoher P_f) bewirkt eine stärkere Geschwindigkeitsabnahme, möglicherweise als Folge zusätzlicher Rißbildung. Fig. 13 zeigt die P-Wellengeschwindigkeiten in gesättigtem Gestein beim Übergang von Wasser in Dampf [It79]. Bei 145 °C durchläuft v_P bei 4 bar ein scharfes Minimum. Dieser Druck entspricht etwa dem Transformationsdruck von Wasser bei 145 °C [Ke69].

In dampfgesättigtem Gestein ist v_P niedriger als in wassergesättigtem Gestein; v_S verhält sich gerade umgekehrt. Bei 198 °C nehmen sowohl v_S als auch v_P von der Dampfsättigung (niedriger P_f) zur Wassersättigung (hoher P_f) ab (Fig. 13b). σ und v_P/v_S steigen sowohl bei 145 °C als auch bei 198 °C beim Übergang vom Bereich der Dampfsättigung zum Bereich der Wassersättigung an (Fig. 14).

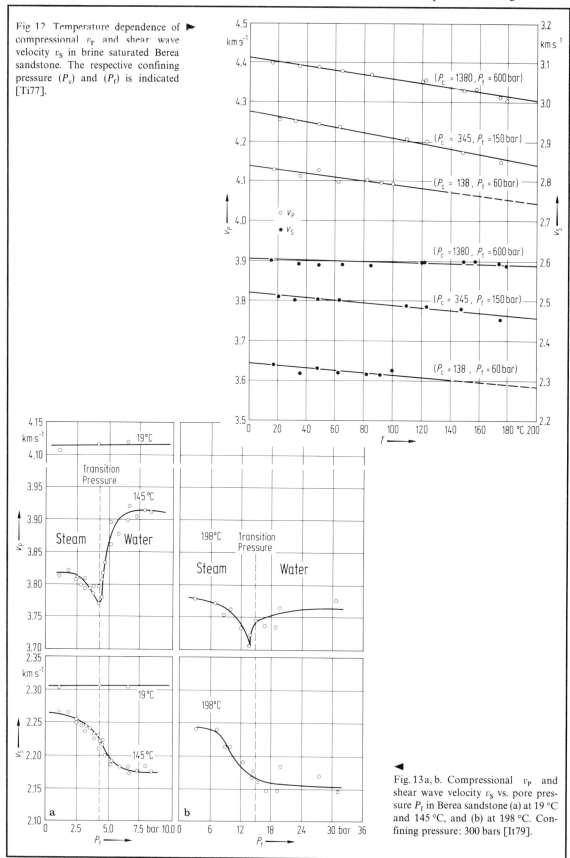

Fig 12 Temperature dependence of ▶ compressional v_p and shear wave velocity v_s in brine saturated Berea sandstone. The respective confining pressure (P_c) and (P_f) is indicated [Ti77].

◀ Fig. 13a, b. Compressional v_p and shear wave velocity v_s vs. pore pressure P_f in Berea sandstone (a) at 19 °C and 145 °C, and (b) at 198 °C. Confining pressure: 300 bars [It79].

Kern

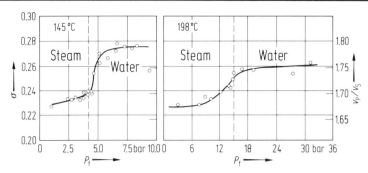

Fig. 14. Poisson's (σ) and v_P/v_S ratios vs. pore pressure at 145 °C and 198 °C in Berea sandstone. Confining pressure: 300 bars [It79].

Higher-than-normal P_f may be produced by a variety of natural processes such as hydrocarbon enrichment, artesian conditions, hydrous to anhydrous phase changes (see also 3.1.4.4.1.2), osmotic conditions etc. In natural subsurface pore fluids which are approximated by the H_2O system, a 1 °C increase in temperature produces at least a 10 bar increase in P_f provided the pore volume remains constant [Kn77]. Difference in thermal expansion of pore fluid and enclosing minerals may drastically decrease P_{eff}. An only moderate increase in temperature (100 °C) produces a large increase in P_f (>1 kbar).

Ein erhöhter P_f kann durch mehrere natürliche Prozesse erzeugt werden: Anreicherung von Kohlenwasserstoffen, artesische Bedingungen, Entwässerungsreaktionen (vergleiche 3.1.4.4.1.2), osmotische Drucke etc. In natürlichen oberflächennahen Porenflüssigkeiten bringt ein Temperaturanstieg von 1 °C bei gleichbleibendem Porenvolumen eine Druckerhöhung von mindestens 10 bar mit sich [Kn77]. Unterschiedliche thermische Volumenexpansion von Porenlösung und Matrixmineralen kann zu einem drastischen Anstieg von P_f führen. Ein nur geringer Temperaturanstieg von 100 °C vermag P_f um mehr als 1 kbar zu erhöhen.

3.1.4.4.3 The effect of partial melting on elastic wave velocities — Der Einfluß partieller Schmelzbildung auf die Geschwindigkeiten elastischer Wellen

Studies of seismic wave velocities in solid-melt mixtures of silicates are scarce. Either they are theoretical or have been based on experimental measurements on simple, non-silicate systems [Sp68, St75].

The only acoustic experiments on partially melted silicates have been reported by [Be74] for granite and by [Mu76] and [Mu78] for peridotite. The results for granite are somewhat puzzling. Granite is not an ideal material to study the effect of partial melting on P- and S-wave velocities. Large changes in velocity, particularly in v_P may be expected in the vicinity of the quartz $\alpha - \beta$ transition [Fi71, Ke78, Ke79]. The solidus of the system granite-H_2O is 670 °C at 3 kbar. Any effects due to partial melting will therefore be superimposed on them and may be difficult to separate.

In dry peridotite, v_P decreases approximately linearly with temperature (Fig. 15). As melting begins, the rate of velocity decrease accelerates and at temperatures 20 °C higher it drops sharply. Melting is pressure dependent. High pressure shifts the begin of melting to higher temperatures. Velocities normalized to the

Die bisher durchgeführten Untersuchungen der seismischen Geschwindigkeiten in Fest-Flüssig-Gemischen von Silikaten sind nicht sehr zahlreich. Meist sind die Untersuchungen theoretischer Natur oder aber sie erfolgten an einfachen, nicht-silikatischen Systemen [Sp68, St75].

Schmelzuntersuchungen an natürlichen Silikatgesteinen mit Ultraschall sind bisher nur von Granit [Be74] und Peridotit [Mu78] bekannt. Die Ergebnisse an Granit sind etwas widersprüchlich. Granit ist zur Untersuchung des Einflusses der Schmelzbildung auf v_P und v_S nicht sehr geeignet. Im Bereich der $\alpha - \beta$-Quarz-Transformation nimmt besonders v_P stark ab [Fi71, Ke78, Ke79]. Im System Granit-H_2O ist die Solidustemperatur bei einem Druck von 3 kbar etwa gleich 676 °C. Jegliche Effekte, die durch die Bildung partieller Schmelzen bedingt sind, werden daher von dem Einfluß der Hoch-Tief-Quarz-Umwandlung überlagert. Es ist daher schwer, diese Einflüsse voneinander zu unterscheiden.

In trockenem Peridotit nimmt v_P quasi-linear mit der Temperatur ab (Fig. 15). Bei Schmelzbeginn wird die Geschwindigkeitsabnahme stärker und nach 20 °C Temperatursteigerung sehr steil. Die Schmelzbildung hängt von den Druckbedingungen ab. Hohe Drucke verschieben den Schmelzbeginn zu höheren Tempera-

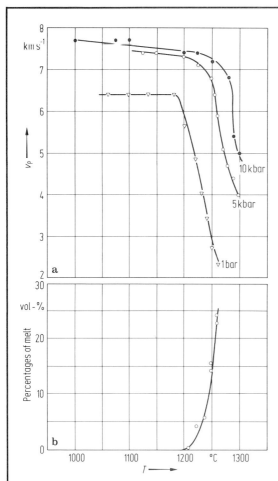

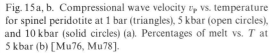

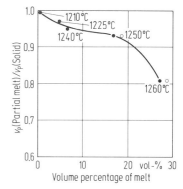

Fig. 15a, b. Compressional wave velocity v_P vs. temperature for spinel peridotite at 1 bar (triangles), 5 kbar (open circles), and 10 kbar (solid circles) (a). Percentages of melt vs. T at 5 kbar (b) [Mu76, Mu78].

Fig. 16. Normalized compressional wave velocities in partial melts vs. volume percentage of melt for spinel peridotite at 5 kbar [Mu78].

velocity of the unmelted sample at its solidus temperature (1200 °C at 5 kbar) indicate a relatively small rate of decrease with melting to about 18 % melt, then the rate decreases rapidly (Fig. 16).

Since the suggestion by [Pr59] it has been common to explain the seismic velocity and attenuation characteristics of the low-velocity zone (LVZ) as being caused by partial melting. Melt volume percentages assumed for the LVZ on seismological and petrological grounds are usually in the range of several percent depending strongly on the textural distribution of the melt. According to [Bi69], the LVZ could be explained by 6 vol.-% of spherical droplets of basaltic melt in aggregates of olivine. Estimates indicating less than 1 % melt correspond to an assumption of complete wetting of all grains with melt [An70, St75].

turen. Die auf die Solidus-Bedingungen (1200 °C, 5 kbar) bezogenen Kompressionswellen-Geschwindigkeiten zeigen bis zu einem Schmelzvolumen von 18 % eine verhältnismäßig geringe Abnahme. Bei höheren Schmelzanteilen nimmt die Geschwindigkeit dann drastisch ab (Fig. 16).

Seit den Überlegungen von [Pr59] ist es üblich, die Geschwindigkeitsabnahme und die Dämpfung elastischer Wellen in der „low-velocity zone" (LVZ) durch partielle Schmelzbildung zu erklären. Die auf Grund seismologischer und petrologischer Überlegungen für die LVZ angenommenen Schmelzvolumina sind entsprechend der texturellen Verteilung der Schmelze etwa gleich einigen Prozenten. Nach [Bi69] würden 6 vol.-% sphärischer Tropfen einer Basaltschmelze in Olivin-Aggregaten die LVZ hervorrufen können. Bei einer vollständigen Benetzung aller Körner durch Schmelze würden weniger als 1 % Schmelze ausreichen [An70, St75].

Direct melting of natural rocks in the laboratory [Me73, Ar78] indicates that the melt distribution is controlled by already existing fractures and the location of low melting point phases. According to [Ar78a], however, annealing for laboratory times appears to be unable to redistribute the melt into a geometry controlled by surface energies. Therefore, the experimental results obtained in dry peridotite allow only an approximate statement in respect to melt quantities.

Schmelzexperimente an natürlichen Gesteinen im Labor [Me73, Ar78] weisen darauf hin, daß die Verteilung der Schmelze im Gefüge durch vorhandene Risse und die Lage der niedrig schmelzenden Mineralphasen bestimmt wird. Nach [Ar78a] werden innerhalb der üblichen Experimentierzeiten die Schmelzanteile nicht in eine durch Oberflächenenenergien kontrollierte Konfiguration gebracht. Die an trockenen Peridotiten erzielten experimentellen Ergebnisse liefern daher im Hinblick auf die zur Erklärung der LVZ erforderlichen Schmelzvolumina nur grobe Anhaltspunkte.

3.1.4.5 References for 3.1.4 — Literatur zu 3.1.4

An70	Anderson, D.L., Spetzler, H.: Phys. Earth Planet. Interior 4 (1970) 62-64.
Ar78	Arzi, A.A.: Tectonophysics 44 (1978) 173-184.
Ar78a	Arzi, A.A.: J. Petrol. 19 (1978) 153-169.
Ba74	Bayuk, Ye.I., Tedeyev, R.V.: Izv. Acad. Sci. (USSR) Earth Physics 8 (1974) 63-70.
Be74	Benzing, W.M., III.: PhD thesis, University of Chicago, 1974.
Bi60	Birch, F.: J. Geophys. Res. 65 (1960) 1083-1102.
Bi61	Birch, F.: J. Geophys. Res. 66 (1961) 2199-2224.
Bi66	Birch, F., in: Handbook of Phys. Constants (S.P. Clarke, jr., ed). Geol. Soc. Am. Mem. 97 (1966) 97-173.
Bi69	Birch, F.: Geophys. Monogr. Am. Geophys. Union 13 (1969) 18-36.
Br65	Brace, W.F.: J. Geophys. Res. 70 (1965) 391-398.
Br72	Brace, W.F.: Geophys. Monogr. Am. Geophys. Union, Washington, D.C. (1972) 265-273.
Bu82	Burkhardt, H., Keller, F., Sommer, J.: Written communication. Stuttgart: E. Schweizerbart, 1982
Ch65	Christensen, N.I.: J. Geophys. Res. 70 (1965) 6147-6164.
Ch66	Christensen, N.I.: J. Geophys. Res. 71 (1966) 3549-3556.
Ch74	Christensen, N.I.: J. Geophys. Res. 79 (1974) 407-412.
Ch79	Christensen, N.I.: J. Geophys. Res. 84 (1979) 6849-6857.
Fi71	Fielitz, K.: Geophys. 37 (1971) 943-956; (written communication 1981).
Go77	Goetze, C., in: High pressure research, applications in geophysics (M.H. Manghnani, S. Akimoto, eds.), New York: Academic Press 1977, p. 3-23.
Ha63	Handin, J., Hager, R.V., Friedman, M., Feather, J.: Am. Soc. Petrol. Geol. Bull. 47 (1963) 717-755.
Hu50	Hughes, D.S., Jones, H.J.: Geol. Soc. Am. Bull. 61 (1950) 843-856.
Hu51	Hughes, D.S., Cross, J.J.: Geophys. 16 (1951) 577-593.
Hu56	Hughes, D.S., Maurette, C.: Geophys. 21 (1956) 277-284.
Hu57	Hughes, D.S., Maurette, C.: Geophys. 22 (1957) 23-31.
Hu59	Hubbert, M., Rubey, W.W.: Geol. Soc. Am. Bull. 70 (1959) 115-166.
It79	Ito, H., De Vilbiss, J., Nur, A.: J. Geol. Res. 84 (1979) 4731-4735.
Ja69	Jaeger, J.C.: Elasticity, fracture, flow. London: Methuen & Co., Science Paperbacks 1969, pp. 268.
Ja80	Jacobi, W.R., Girardin, N.: J. Geophys. 47 (1980) 271-277.
Ke69	Keenan, J.H., Keyes, F.G., Hill, P.G., Moore, J.G.: Steam tables, New York: John Wiley, 1969
Ke71	Kern, H.: Contrib. Mineral. Petrol. 31 (1971) 39-66.
Ke74	Kern, H.: Contrib. Mineral. Petrol. 43 (1974) 47-54.
Ke75	Kern, H., Fakhimi, M.: Tectonophys. 28 (1975) 227-244.
Ke78	Kern, H.: Tectonophys. 44 (1978) 185-203.
Ke79	Kern, H.: Phys. Chem. Minerals 4 (1979) 161-167.
Ke79a	Kern, H., Richter, A.: Tectonophys. 54 (1979) 231-252.
Ke81	Kern, H., Richter, A.: J. Geophys. 49 (1981) 47-56.
Ke82	Kern, H.: Stuttgart: to be published by E. Schweizerbart, 1982.
Ke82a	Kern, H.: unpublished data 1982.
Kn77	Knapp, R.B., Knight, J.E.: J. Geophys. Res. 82 (1977) 2515-2522.
Kr76	Kroenke, L.W., Manghnani, M.H., Rai, C.S., Fryer, P., Ramananantoandro, R., in: The Geophysics of the Pacific Basin and its margin (G.H. Sutton, M.H. Manghnani, R. Moberly, eds.), Am. Geophys. Union Monogr. 19 (1976) 407-421.

Me73	Mehnert, K. R., Büsch, W., Schneider, G.: N. Jb. Mineral. Monatsh. **4** (1973) 165–183.
Mo81	Molen van der, I.: Tectonophys **73** (1981) 323–342.
Mu76	Murase, T., Kushiro, I., Fujii, N.: Carnegie Inst. Wash. Year Book **76** (1976/77) 414–416.
Mu78	Murase, T., Kushiro, I.: Carnegie Inst. Wash. Year Book **78** (1978/79) 559–562.
Nu70	Nur, A. M., Simmons, G.: Int. J. Rock Mech. Min. Sci. **7** (1970) 307–314.
Pe74	Peselnick, L., Nicolas, A., Stevenson, P. R.: J. Geophys. Res. **79** (1974) 1175–1182.
Pe78	Peselnick, L., Nicolas, A.: J. Geophys. Res. **83** (1978) 1227–1235.
Pr59	Press, F.: J. Geophys. Res. **64** (1959) 565–568.
Ra65	Raleigh, C. B., Paterson, M.: J. Geophys. Res. **70** (1965) 3965–3985.
Ra78	Ramananantoandro, R., Manghnani, M. H.: Tectonophys. **47** (1978) 73–84.
Ro73	Robin, P. F.: J. Geophys. Res. **78** (1973) 2434.
Si64	Simmons, G.: J. Geophys. Res. **69** (1964) 1123.
Sk66	Skinner, B. J.: Geol. Soc. Am. Mem. **97** (1966) 75–96.
Sp68	Spetzler, H., Anderson, D. L.: J. Geophys. Res. **73** (1968) 6051–6060.
Sp76	Spencer, J. W., Jr., Nur, A. M.: J. Geophys. Res. **81** (1976) 899–904.
St75	Stocker, R. L., Gordon, R. B.: J. Geophys. Res. **80** (1975) 4828–4836.
St77	Stewart, R., Peselnick, L.: J. Geophys. Res. **82** (1977) 2027–2039.
St78	Stewart, R., Peselnick, L.: J. Geophys. Res. **83** (1978) 831–839.
Ti68	Timur, A.: Geophys. **33** (1968) 584–596.
Ti77	Timur, A.: Geophys. **42** (1977) 950–956.
Wa74	Wang, Ch. Y.: J. Geophys. Res. **79** (1974) 771–772.
Yo62	Young, R. A.: Air Force Office Sci. Final Rept. 2569, 156 pp., Washington D.C., **1962**.
Yo64	Young, R. A.: Proc. 12th Natl. Conf. on Clay and Clay Minerals, Pergamon, **1964** p. 83.

3.2 Fracture and flow of rocks and minerals — Bruch und Inelastizität von Gesteinen und Mineralen

3.2.1 Strength and deformability — Festigkeit und Verformbarkeit

3.2.1.1 Introduction — Einleitung

3.2.1.1.1 General remarks — Allgemeine Bemerkungen

Most of the practical problems in rock mechanics such as tunnelling, dam constructions, or slope stability, but also earthquake mechanisms, tectonic faulting, and large-scale plate movements involve rock deformations associated with disintegration or structural collapse. Although field geology can provide enormous understanding of the kinematics of such deformation and fracture processes, it is difficult to derive final conclusions on the fracture processes or the dynamics of rock deformation from such observations. The physics of rock fracture can only be studied through controlled *in-situ* or laboratory experiments which "realistically simulate the natural environmental conditions" [Han66], the state of stress, the temperature, the material, and the time duration of tectonic processes.

During the past decade both *in-situ* and laboratory data on strength and fracture deformability of rocks and minerals have multiplied. This section summarizes most of the data availably in geoscience literature until 1979 and includes most of the data presented already in 1966 by Handin [Han66]. In order to allow comparison, the presentation of data is very much the same as done by Handin. In addition to the laboratory stress-strain data, compressive strength, and tensile strength data this section includes also data on fracture toughness and surface energy with respect to the new concepts used in rock fracture mechanics. Further, considering the large scatter in experimental data for a specific rock type such as granite it was decided to present the original data as reported in the literature, thus indicating the specific location or origin of the rock, the experimental parameters and the investigator.

Die meisten praktischen Probleme in der Felsmechanik wie die Konstruktion von Untertagebauten, Staudämmen oder von Böschungen beinhalten Gesteinsdeformationen, die mit Bruchvorgängen verbunden sind. Obwohl eine Gelände-geologische Betrachtung viel zum Verständnis der Kinematik solcher Verformungs- und Bruchvorgänge beitragen kann, ist es meist sehr schwierig, aus solchen Geländebeobachtungen auf den Verformungsmechanismus oder auf den Bruchprozeß zu schließen. Die Physik des Bruchvorgangs in geologischen Materialien kann nur in kontrollierten Experimenten untersucht werden, bei denen die in der Natur vorgegebenen Bedingungen möglichst realistisch simuliert werden [Han66]. Die Parameter sind insbesondere das Spannungsfeld, die Temperatur und die Verformungsgeschwindigkeit eines tektonischen Prozesses.

Während des letzten Jahrzehnts wurden eine Vielzahl von derartigen kontrollierten Experimenten *in situ* und im Laboratorium an Gesteinsproben und Einkristallen durchgeführt. Im folgenden sind die meisten Daten dieser Untersuchungen zusammengestellt (bis 1979), wobei Daten einer ähnlichen älteren Zusammenstellung [Han66] berücksichtigt wurden. Um einen Vergleich mit früheren Darstellungen zu ermöglichen, ist die Form der Zusammenstellung ähnlich der von Handin [Han66]. Zusätzlich zu den Angaben über die Spannungs-Verformungs-Beziehungen und die Festigkeitswerte enthält dieser Abschnitt neuere Daten über die Bruchzähigkeit und die spezifische Bruchflächenenergie von Gesteinen. Wegen der großen Streuung der Einzelwerte wurde im wesentlichen auf die Bildung von Mittelwerten verzichtet. Stattdessen sind die meisten Angaben die Originaldaten, wie sie in der Literatur von verschiedenen Forschern für das jeweils untersuchte Gestein berichtet wurden.

3.2.1.1.2 Definitions — Definitionen

a) Compressive stresses are regarded as positive thoughout the three sections. S_1, S_2 and S_3 are the maximum, intermediate, and minimum principal stresses, respectively.

b) The term *triaxial* is used for compression tests on cylindrical specimens jacketed by thin, impermeable rubber, polyurethan, copper, or aluminium tubings to prevent the confining pressure medium (fluids or gas) to penetrate into the rock, which are loaded at constant confining pressure $p_m = S_2 = S_3$ with increasing axial

a) Druckspannungen haben hier ein positives Vorzeichen. S_1, S_2 und S_3 sind die größte, die mittlere und die kleinste Hauptnormalspannung.

b) Als *Triaxialversuche* werden hier Druckversuche an zylindrischen Probekörpern bei konstantem Manteldruck verstanden ($p_m = S_2 = S_3$). Die Proben sind dabei mit Gummi- oder Metallfolien ummantelt. Mit *Differenzspannung* σ_d bei Triaxialversuchen ist die Differenz $S_1 - p_m$ bezeichnet. Tatsächliche Triaxial-

stress S_1. The *differential stress* σ_d is the difference $S_1 - p_m$. Tests under general triaxial stress $(S_2 \neq S_3)$ have recently been conducted by Mogi [Mog79]. Some results are given in Figs. 5···10.

c) *Uniaxial unconfined tests* are compression or tension tests on cylindrical or cubical specimens with $p_m = 0$. Both, in triaxial and most uniaxial unconfined tests the ratio of length to diameter of the specimen is 2 to 3, large enough to minimize the effect of end constraint.

d) Depending on the experimental parameters confining pressure, temperature, and strain rate the material may deform *brittle* or *ductile*. The term brittle here is used for deformation behaviour, which is characterized by a considerable stress decrease or sudden stress drop in the post-failure range of deformation. In general, brittle behaviour is characterized by a defined peak in the stress-strain curve. Ductile deformation, in contrast, is characterized by strength increase or constant strength with increasing inelastic deformation. The various types of brittle, ductile, and transitional deformation are indicated in Fig. 1. (It should be mentioned that other authors prefer a definition of brittle or ductile in terms of the nature of failure as viewed macroscopically or microscopically rather than in terms of the observed stress-strain curve.)

e) *Strength* is qualitatively defined as the resistance of the material to inelastic deformation − fracture or ductile flow. Fracture is used in the sense to imply increasing loss of material cohesion, loss of load-bearing capacity at controlled failure, seperation of the specimen into two or more parts, or release of stored elastic strain energy. The term flow is used to indicate inelastic deformation without loss of material cohesion or deformation with strength increase (strain-hardening). With respect to compressive or tensile stress conditions we differentiate between compressive and tensile strength.

f) *Ultimate strength* is the maximum differential stress a material can withstand under given experimental conditions. In soft loading systems at low confining pressure brittle materials at ultimate strength will explosively disintegrate or seperate into parts leading to significant sudden strength reduction. In *stiff* loading systems or *fast-reacting* servo-systems deformation may be controlled beyond the deformation at ultimate strength (case *c* or *d* in Fig. 1). In this so-called post-failure region of the stress-strain curve progressive microfracturing occurs resulting in a continuous decrease of strength. Thus, the term *breaking strength* may only be adequate for violent fracture in soft systems at ultimate strength (case *a* or *b* in Fig. 1) or for the case of fracture instability in the post-failure region indicating the occurence of a macroscopic shear fracture (case *c* in Fig. 1). Stiff and servo-controlled loading systems have been introduced since about

versuche $(S_1 \neq S_2 \neq S_3)$ wurden neuerdings von Mogi [Mog79] durchgeführt. Einige der Ergebnisse sind in Fig. 5···10 angegeben.

c) *Einachsige Versuche* sind Druck- oder Zugversuche an zylindrischen oder würfelförmigen Proben ohne Manteldruck. Sowohl bei triaxialen wie auch bei einachsigen Druckversuchen ist das Verhältnis von Probenlänge zu Probendurchmesser meist 2.

d) In Abhängigkeit von den experimentellen Parametern Manteldruck, Temperatur und Verformungsgeschwindigkeit kann sich ein Material *spröde* oder *duktil* (zäh) verhalten. Die Bezeichnung spröde wird hier für das Verformungsverhalten verwendet, bei dem die Festigkeit des Materials nach Eintritt des Bruchs drastisch abnimmt. Demgegenüber wird hier unter duktil ein Verformungsverhalten verstanden, bei dem die Festigkeit während der Verformung angenähert konstant bleibt oder sogar ansteigt. Verschiedene Spannungs-Verformungskurven für die spröde und die duktile Verformung sind in Fig. 1 zusammengestellt. (Es sollte erwähnt werden, daß andere Autoren die Bezeichnungen spröde oder duktil nicht aufgrund des Verlaufs der Spannungs-Verformungskurven verwenden, sondern damit den Bruchprozeß beschreiben, wie er sich makroskopisch oder mikroskopisch darstellt.)

e) Mit dem Begriff *Festigkeit* bezeichnet man qualitativ das Widerstandsvermögen eines Materials gegen inelastische Verformung, sei es Sprödbruchverformung oder duktiles Fließen. Die Bruchverformung ist gekennzeichnet durch Kohäsionsverlust, Festigkeitsabnahme oder die Aufspaltung des Körpers in zwei oder mehr Teile, wobei elastische Verformungsenergie freigesetzt wird. Der Begriff Fließen wird für inelastische Verformung verwendet, bei der der Verbund des Materials erhalten bleibt und die Festigkeit des Materials eventuell sogar zunimmt (Verformungsverfestigung). Bezüglich der Wirkung von Druck- oder Zugspannungen unterscheidet man zwischen Druck- und Zugfestigkeit.

f) Das Spannungsmaximum einer vollständigen Spannungs-Verformungs-Kurve bezeichnet man als *Bruchfestigkeit*. In konventionellen Belastungssystemen erfolgt der Bruch spröder Materialien bei kleinen Manteldrücken explosionsartig, sobald der Wert der Bruchfestigkeit erreicht wird. In *steifen* oder schnellen *elektronisch-geregelten* Belastungssystemen kann der Verformungsprozeß auch im sogenannten Post-Failure-Bereich der Spannungs-Verformungs-Kurve kontrolliert werden (Fig. 1c oder d). Die Verformung des Materials ereignet sich als progressiver Bruch durch Entstehung und Ausbreitung von Mikrorissen, wobei die Festigkeit des Materials kontinuierlich abnimmt. Daher ist der Begriff *Bruchfestigkeit* eigentlich nur bei Verformungsprozessen sinnvoll, bei denen sich der Bruch bei Maximallast plötzlich ereignet (Fig. 1a oder b). Anderenfalls ist das Maximum der Spannungs-Verformungs-Kurve lediglich ein momentanes Stadium

1968 and are currently used by increasingly more investigators.

g) *Yield strength* is the differential stress at the onset of inelastic permanent deformation, ideally marked by a sudden break in the stress-strain curve (case *e* or *g* in Fig. 1). Below this stress level the deformation is essentially elastic. However, for most rocks the yield strength is difficult to define (case *b*, *c*, or *d* in Fig. 1).

während des fortschreitenden Bruchgeschehens. Steife und servo-geregelte Belastungssysteme werden seit 1968 bei Laborversuchen verwendet und finden neuerdings verbreitete Anwendung.

g) Als *Nachgebespannung* bezeichnet man die Differenzspannung, bei der die inelastische Verformung einsetzt. Im Idealfall weist dabei die Spannungs-Verformungs-Kurve einen deutlichen Knick auf (Fig. 1*e* oder *g*). Unterhalb der Nachgebespannung ist die Verformung des Materials im wesentlichen elastisch und reversibel. Allerdings ist es im allgemeinen bei Gesteinen schwierig, den Beginn der inelastischen Verformung exakt festzustellen (Fig. 1*b*, *c* oder *d*).

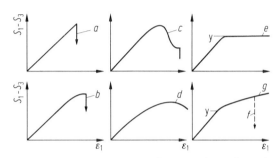

Fig. 1. Representative stress-strain curves for rocks under constant strain rate $\dot{\varepsilon}_1$ as given in Table 1. $S_1 - S_3$ differential stress, ε_1 axial (longitudinal) strain.

Type *a*: very brittle, strain essentially elastic prior to sudden failure;

type *b*: brittle, small inelastic strain prior to sudden rupture;

type *c*: controlled failure, continuous decrease of strength in the post-failure region, significant peak, sudden rupture may occur after extensive deformation;

type *d*: moderately ductile, without well defined yield strength, strength decreases in the post-failure region;

type *e*: ductile, with well defined yield strength;

type *f* and *g*: ductile with poorly defined yield strength and continuous strength increase (work hardening) during inelastic deformation (rupture may occur after large axial inelastic deformation).

h) *Fatigue strength* is the strength of a material under cyclic loading. Only little research has been conducted in this field of rock mechanics.

i) *Tensile strength* characterizes the critical macroscopic condition for the generation of extension fractures in intact rock which involves separation across an internal surface without any relative displacements parallel to this surface. Since extension fractures may occur even when all macroscopic far field stresses are compressive (e.g. cleavage fractures under unconfined compressive stress), the term "tension fracture" is properly only used if the least principal effective stress is tensile.

h) Die Festigkeit eines Materials bei Wechselbelastung bezeichnet man häufig als *Ermüdungsfestigkeit*. Bei Gesteinen liegen hierzu nur wenige Untersuchungen vor.

i) Die *Zugfestigkeit* kennzeichnet die kritische Spannung, bei der makroskopische Dehnungs- oder Trennbrüche auftreten. Da Dehnungsbrüche auch bei reiner Druckbeanspruchung eines Materials entstehen können, ist die Bezeichnung Zugbruch oder Zugfestigkeit dann verwendet, wenn die kleinste Hauptspannung eine Zugspannung ist ($S_3 < 0$).

Tensile strength data are obtained by different experimental tests (Fig. 2) and may vary considerably:

direct tensile strength: obtained from direct pull tests (dp) on cylindrical or dog bone specimens

Brazilian tensile strength: obtained from diametral compression of circular rock discs (br)

ring tensile strength: obtained from diametral compression of hollow discs (r)

bending tensile strength: obtained from three point bending tests on rock beams (b) or from cleavage tests (cl)

hydraulic fracturing strength: obtained from fluid injection tests into drill holes in laboratory specimens or *in-situ* intact rock masses. (hf).

It is typical for rock that its tensile strength is much smaller than its unconfined compressive ultimate strength.

Werte der Zugfestigkeit werden mittels verschiedener Versuchsanordnungen bestimmt (Fig. 2). Je nach der verwendeten Meßanordnung unterscheiden sich die ermittelten Zahlenwerte erheblich:

Zugfestigkeit bei direkter Zugbeanspruchung zylindrischer Proben (dp)

Zugfestigkeit beim Brasilianischen Versuch (br), wobei Kreisscheiben diametral beansprucht werden

Zugfestigkeit von Ring-förmigen Proben bei diametraler Beanspruchung (r)

Biegezugfestigkeit von prismatischen Probekörpern bei Dreipunktbeanspruchung (b) oder Spaltversuchen (cl)

Zugfestigkeit bei der Injektion von Flüssigkeit in Bohrungen eines Probekörpers (hf)

Näherungsweise kann bei Gesteinen angenommen werden, daß die Zugfestigkeit etwa zehnmal kleiner ist als die einachsige Druckfestigkeit.

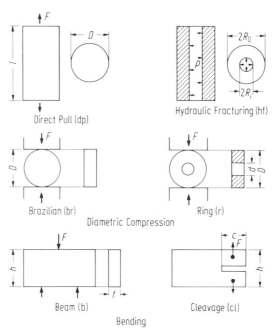

Fig. 2. Various loading configurations for the experimental determination of tensile strength σ_t of rocks.

j) *Fracture surface energy* is the energy required to create unit area of new crack surface. Experimental data are obtained from the Gilman double cantilever beam test (cleavage tests), uniaxial compression tests, direct pull tests on unnotched or notched specimens, three-point bend tests on notched beams, or from indentation tests.

j) Die *spezifische Bruchflächenenergie* ist die Arbeit, die aufgewendet werden muß, um eine Flächeneinheit neuer Bruch- oder Rißflächen zu erzeugen. Meist versteht man darunter die Erzeugung von Trennbruchflächen. Zu ihrer Bestimmung werden verschiedene Methoden angewendet: Spaltversuche an gekerbten Proben, Zugversuche, einachsige Druckversuche an Proben mit einem vorgegebenen Riß, Biegeversuche an gekerbten Proben oder Eindrückversuche.

k) *Fracture toughness:* In the concept of fracture mechanics the stability of existing cracks of given length is described by critical stress intensity factors which are also known as fracture toughness. Presently, experimental fracture toughness data of rocks are only available for the pure crack opening mode, K_{IC}, without sliding or tearing parallel to the crack surface. For experimental determination of K_{IC} of rocks currently the double torsion test, the three-point loading test and hydraulic fracturing tests are used. Research in rock fracture mechanics is relatively new, thus only few data on K_{IC}-values are available.

k) *Bruchzähigkeit*: In der heute in den Werkstoffwissenschaften verwendeten Bruchmechanik wird die Instabilität eines Risses gegebener Länge durch Angabe des kritischen Spannungsintensitätsfaktors oder der Bruchzähigkeit beschrieben. Z.Zt. liegen für Gesteine nur Zahlenwerte für die Rißöffnungsart I vor, K_{IC}. Zur experimentellen Bestimmung wurden bisher an Gesteinen folgende Methoden verwendet: Torsionsversuche an gekerbten Platten, Biegeversuche an gekerbten Prismen und Hydraulic-Fracturing-Versuche. Bisher liegen für Gesteine nur wenige Daten vor.

3.2.1.1.3 List of symbols — Liste der verwendeten Symbole

G_{IC} [J m^{-2}]	critical strain energy release rate	Bruchenergie-Freisetzungsrate
K_I [MN m$^{-3/2}$]	stress intensity factor	Spannungsintensitätsfaktor
K_{IC} [MN m$^{-3/2}$]	fracture toughness	Bruchzähigkeit
P, N, U	orientation of maximum principal stress S_1 is parallel, normal, or unknown with respect to anisotropy planes	kennzeichnet Orientierung der größten Druckspannung S_1 gegenüber Anisotropieflächen; P parallel, N senkrecht, U Orientierung unbekannt
S_1, S_2, S_3 [bar, Pa, MPa]	maximum, intermediate, and minimum principal stress	größte, mittlere und kleinste Hauptnormalspannung
T [°C]	temperature	Temperatur
k	frac gradient with respect to confining pressure p_m	Frac-Gradient bezüglich des Manteldrucks
l [cm]	specimen length	Probenlänge
Δl	specimen shortening under compression	Probenverkürzung bei Druckbeanspruchung
n [%]	porosity	Porosität
p_c [bar, MPa]	hydraulic fracturing strength at external pressure	hydraulische Zugfestigkeit bei äußerer Beanspruchung der Proben
p_{c0}	hydraulic fracturing tensile strength at $p_m = 0$	hydraulische Zugfestigkeit von unbeanspruchten Proben
p_m [MPa]	confining pressure in triaxial tests $(p_m = S_2 = S_3)$	Manteldruck bei Triaxialversuchen
$S_1 - p_m = \sigma_d$	differential stress in triaxial tests	Differenzspannung in Triaxialversuchen
γ [J m^{-2}]	specific fracture surface energy	spezifische Bruchflächenenergie
ε_1 [%]	$\varepsilon_1 = \Delta l/l$ relative specimen shortening, axial strain	axiale Probenstauchung, relative axiale Probenverkürzung, axiale Probenverformung
ε_1^0	strain at σ_1^0 (mostly at failure)	Verformung bei σ_1^0 (meist beim Bruch)
$\dot{\varepsilon}_1$ [s^{-1}]	axial strain rate	axiale Verformungsgeschwindigkeit
ϱ [g cm^{-3}]	bulk density	Gesteinsdichte
σ [MPa]	normal stress	Normalspannung
σ_c [MPa]	uniaxial compressive strength	einachsige Druckfestigkeit
σ_d [MPa]	differential stress $S_1 - p_m$	Differenzspannung
σ_f [MPa]	fatigue strength	Festigkeit beim Ermüdungsversuch
σ_m [MPa]	peak strength in triaxial compression tests	maximale Druckfestigkeit bei Triaxialversuchen
σ_t [MPa]	tensile strength	Zugfestigkeit
σ_u [MPa]	ultimate strength at macroscopic rupture	Druckfestigkeit beim makroskopischen Bruch
σ_y [MPa]	yield strength	Spannung bei Beginn der inelastischen Verformung, Nachgebespannung
σ_1^0 [MPa]	axial stress at a strain of ε_1^0	axiale Spannung bei ε_1^0
τ [MPa, GPa]	shear stress	Scherspannung

3.2.1.1.4 Units, conversions — Einheiten, Umrechnung

SI-units are used throughout this chapter.

Pressure, stress, strength — Druck, Spannung, Festigkeit

$$1\,\text{Pa} = 1\,\text{Nm}^{-2} = 10^{-5}\,\text{bar} = 10\,\text{dyn cm}^{-2}$$
$$1\,\text{MPa} = 10^{6}\,\text{Pa} = 10\,\text{bar}$$
$$1\,\text{GPa} = 10^{9}\,\text{Pa} = 10^{4}\,\text{bar} = 10\,\text{kbar}$$
$$1\,\text{bar} = 0.1\,\text{MPa} = 10^{6}\,\text{dyn cm}^{-2}$$
$$1\,\text{kbar} = 100\,\text{MPa} = 0.1\,\text{GPa}$$
$$1\,\text{kp mm}^{-2} = 9.81\,\text{N mm}^{-2}$$
$$1\,\text{kp cm}^{-2} = 9.81\,\text{N cm}^{-2} = 0.0981\,\text{MPa} = 0.981\,\text{bar}$$
$$1\,\text{atm} = 1.013\,\text{bar} = 0.0981\,\text{MPa}$$
$$1\,\text{psi} = 6.9\,\text{kPa} = 0.069\,\text{bar}$$

Strain — Verformung

$$1\,\mu\text{m/m} = 10^{-6} = 1\ \text{microstrain, or in }\%$$

Strain rate — Verformungsgeschwindigkeit

$$1\,\mu\text{m/m per second} = 10^{-6}\,\text{s}^{-1}$$

Fracture surface energy — Bruchflächenenergie

$$1\,\text{Jm}^{-2} = 1\,\text{N m}^{-1} = 1\,\text{Ws m}^{-2} = 10^{3}\,\text{erg cm}^{-2}$$

Fracture toughness — Bruchzähigkeit

$$1\,\text{MN m}^{-3/2} = 1\,\text{MPa m}^{1/2} = 10^{3}\,\text{N mm}^{-2}\,\sqrt{\text{mm}}$$

3.2.1.2 Data — Daten

Table 1. Stress-strain relation of rocks and minerals from triaxial compression tests at constant confining pressures and constant temperatures.

Data refer to uniaxial and triaxial compression tests ($p_m = S_2 = S_3$) on cylindrical rock samples of the order of 1 to 5 cm in diameter and 3 to 10 cm in length. Samples for triaxial tests were jacketed in rubber, plastic, or thin copper or aluminum tubes.

For comparison with previous presentations, the set-up of this table is analogous to that by Handin [Han66].

Column 1 gives the rock type and the origin of the sample, together with some data on mineral composition, porosity n and axial strain rate $\dot\varepsilon_1$ (most tests were conducted at constant strain rates of the order of 10^{-3} to $10^{-5}\,\text{s}^{-1}$). Column 2 indicates the orientation of the axial stress S_1 with respect to anisotropy planes such as bedding, foliation, schistosity, or cleavage planes (P $= S_1$ parallel, N $= S_1$ normal to anisotropy plane, U = unknown orientation). Temperatures (T in [°C], R = room temperature) and confining pressures (p_m in [MPa]) are given in columns 3 and 4. Column 5 describes the type of stress-strain curve obtained as classified in Fig. 1.

Columns 6 to 9, $S_1 - p_m$, present values of the acting differential stress at 1, 2, 5, and 10 % axial (longitudinal) strain ($\varepsilon_1 = \Delta l / l$ relative specimen shortening, where l is the original length of the specimen). Columns 10 and 11, σ_1 and ε_1, indicate particular points on the stress-strain curve, where σ_1 gives the differential stress and ε_1 the corresponding strain coordinate. A letter behind the stress value in column 10 indicates ultimate strength (u), maximum strength (m), or yield strength (y). An asterisk with the stress value signifies that the test was terminated before rupture and the stress-strain curve is still rising (type g curve, see Fig. 1, p. 143).

All stress data are given in units of MPa, strain data are given in %. The stress and strain data given allow to construct the approximate form of the stress-strain curve in most cases.

Table 1

Material	Orient-ation	T °C	p_m MPa	Type	$S_1 - p_m$ [MPa] during axial deformation ε_1				σ_1 MPa	ε_1 %	Ref.
					1%	2%	5%	10%			
alabaster, pure, fine-grained, Italy, $\dot\varepsilon_1 = 3 \cdot 10^{-4}$ s^{-1}	U	25	200	e	150	222	240	238	243	3	Hea66
	U	25	500	f	180	270	290	310*	310*	10	
	U	100	200	e	130	192	218	220	223	9	
	U	100	500	e	150	235	276	270	276	5	
	U	125	200	d	85	115	118	87	124	3	
	U	125	500	d	148	205	205	160	224	3	
	U	177	200	e	18	18	15	10	18	0.5	
	U	177	500	e	22	21	20	20	25	0.2	
	U	200	500	e	18	17	13	10	19	0.2	
amphibolite, Hudson Highland complex, New York, USA, medium grained, granoblastic	N	150	100	b	103	574			654	2.9	Bor66
	P	150	100	a	103	261			659	3.2	
	N	500	500	c	472	829	1040	840	1138	3.3	
	P	500	500	c	550	1123			1350	3.6	
andesite, "Shinkomatsu", Kanagawa Pref., Japan, pyroxene andesite, n=16%	U	25	13	e	73	80*			64y	0.2	Mog65
	U	25	28	e	85	105*			75y	0.2	
	U	25	40	e	80	86*			70y	0.2	
	U	25	100	f	55	90*			30y	0.2	
andesite, "Shirochaba", Kanagawa Pref., Japan, pyroxene andesite, n=5%	U	25	0.1	b	190				140u	1.6	Yam70
	U	25	28	c	240				250u		Mog65
	U	25	80	d		380			380m	2	
	U	25	130	d	300	530*			600*	3	
andesite, Idaho, USA, hypersthene andesite	U	25	0.1	a					131u		Han66
anhydrite, Alberta, Canada, 30% gypsum	N	25	0.1	b	103	108	86	78	41u	0.5	Bre57
	N	25	17	d	107	141	142	129	108m	2	
	N	25	35	d	140	189	203		145m	3	
	N	25	69	e				208*	208*	10	
anhydrite, Osterode, Harz, FRG, $\dot\varepsilon_1 = 5 \cdot 10^{-4}$ s^{-1}	‖[010]	25	50	d	190	220	300		300m	5	Mül74
	‖[010]	25	100	e	190	290	370	400	400	10	
	‖[010]	25	300	f	205	350	470	565*	565*	10	
	‖[010]	25	500	f	210	370	515	650*	650*	10	
	‖[010]	150	50	c	110	200	265		265m	5	
	‖[010]	150	100	d	110	200	310	350	350m	10	
	‖[010]	150	300	f	110	200	410	540	730*	20	

(continued)　　　　　　　　　　　　　　　　(continued)

Table 1 (continued)

Material	Orientation	T °C	p_m MPa	Type	$S_1 - p_m$ [MPa] during axial deformation ε_1				σ_1 MPa	ε_1 %	Ref.
					1%	2%	5%	10%			
anhydrite (continued)	∥ [010]	300	100	f	50	110	150	370	500*	20	Mil74
	∥ [010]	300	300	f	100	200	330	470	640*	20	
anhydrite, Südheim/Northeim, FRG, Zechstein	U	25	0	b					78		Dre62
anhydrite, Blaine, Oklahoma, USA	N	25	0.1	b	103				128u	1.2	Han58
	N	25	50	d	122	224			276m	4	
	N	25	100	e	178	294	384		384m	5	
	N	25	200	e	242	356	471	517	517m	10	
	N	110	105	d	147	294	368		371u	6.3	Han66
	N	150	100	f	170	240	322	311*	311*	10	Han58
	N	150	200	f	200	280	380	490*	490*	10	
	N	300	100	f	157	227	316	385*	417*	10	
	N	300	200	g	117	210	380	480	623*	24	
anhydrite, Hockley, Texas, USA	U	25	310	f	216	300	460		554	9.3	Han66
anorthosite, Adirondacks, USA 90···95% plagioclase, 5···10% mafics	U	400	440	g	440	900	1280		1450*	8	Sei69
	U	400	1120	g	620	1250	2200		2200*	5	
	U	400	1480	g	600	1200	2080		2080*	5	
	U	600	440	g	440	750	1100		1200*	8	
	U	600	1120	g	550	950	1350		1550*	8	
	U	600	1480	g	550	1030	1490		1720*	8	
	U	800	440	g		380	500		560*	8	
	U	800	1120	g	170	270	450		600*	8	
	U	800	1480	g	230	400	650		830*	8	
	U	1000	1120	g		80	260		420*	8	
	U	1000	1480	g		110	300		390*	8	
anorthosite, Marcy, New York, USA, precambrian, 89% plagioclase, coarse-grained	U	150	100	a	204	236			594u	2.6	Bor66
	U	500	500	d	292	464	729	904	940m	13	
basalt, Mendig, Eifel, FRG, porous	N	R	0.1	a					100···140m	0.3	Rum77
	N	R	19	a					210u	0.9	
	N	R	38	a	234				234u	1.0	
	N	R	58	a	240				250u	1.2	
	N	R	92	b	250				278u	1.4	
	N	R	140	e	260	260			260	1.5	

(continued)

Table 1 (continued)

Material	Orientation	T °C	p_m MPa	Type	$S_1 - p_m$ [MPa] during axial deformation ε_1				σ_1 MPa	ε_1 %	Ref.
					1%	2%	5%	10%			
basalt, Blairsden, California, USA olivine basalt	U	R	505	f	588	965	1530		1540u	6.6	Gri60
	U	300	500	d	588	960	1380	1030	1380m	5	
	U	500	505	d	617	860	1020	1030	1030m	10	
	U	700	507	e	270	437	521	531	531	11	
	U	800	507	e	185	216	240	263	263	16	
basalt, "Knippa", Texas, USA	U	R	0.1	a	262				262u	1.0	Bre57
	U	R	69	a	462				462u	1.1	
	U	R	103	a	551				551u	1.7	
biotite, single crystal, Mt. Isa, Queensland, Australia, $\dot{\varepsilon}_1 = 10^{-4}\,s^{-1}$	‖ [010]	400	300	c	1110	900	700	620	1110m	1	Eth73
	‖ [010]	600	300	b	500	620	600	500	640m	2.3	
	‖ [010]	700	300	c	245	255	205	300	340m	1.3	
biotite, single crystal calcite, $CaCO_3$, single crystal (see special section)	‖ c-axis	500	507	d	30	80	50	40	80m	2	Han58
chalcopyrite, $CuFeS_2$, poly-crystalline ore, Timagami, Ontario, Canada, rather pure	U	R	0.1	a	58				58u	1.0	Lan58
	U	R	50	b	210	370			382u	2.6	
	U	R	100	d	380	415	470		≈500	≈8	
	U	R	200	d	300	400	500	580	≈680	≈15	
	U	R	400	g	380	450	600	760	≈1030*	≈25	
chalcopyrite, $CuFeS_2$, polycryst. ore, Mitterberg, Austria, rather pure	U	R	0.1	a	50	450			80	1.3	Lan58
	U	R	50	b	250	550	600		493	2.1	
	U	R	100	d		530	630	720	600m	5	
	U	R	200	d		530	700	820	789m	15	
	U	R	400	g		520	740	870	1050*	23	
	U	R	500	g					1150*	25	
chalcopyrite, $CuFeS_2$, polycryst., Murgul, Turkey, 75% $CuFeS_2$, 11% pyrite	U	R	100	d		380	385	550	454	2.3	Eth73
	U	R	200	d		400	550	780	610	7	
	U	R	500	g		430	600		950	20	
chalcopyrite, $CuFeS_2$, polycryst., Ergani-Maden, Turkey, 61% $CuFeS_2$, 28% pyrite	U	R	100	b		615			615	2	Eth73
	U	R	200	d		760	≈830		869	4	
	U	R	400	g		820	1080	1150	1191*	15	

(continued)

Table 1 (continued)

Material	Orient-ation	T °C	p_m MPa	Type	$S_1 - p_m$ [MPa] during axial deformation ε_1				σ_1 MPa	ε_1 %	Ref.
					1%	2%	5%	10%			
chalcopyrite, $CuFeS_2$, Mt. Isa, Australia, 80% $CuFeS_2$, 2% pyrite, 13% gangue, $\dot{\varepsilon}_1 = 3 \cdot 10^{-5}\ s^{-1}$	U	R	0.1	a					85	0.3	La68
	U	R	51	b	370				400	1.5	
	U	R	103	d	460	480	520		520 m	10	
	U	R	292	g	480	520	600	700	710	>10	
	U	200	0.1	b					85 u	0.8	
	U	200	57	e	230	260	280	290	290	10	
	U	200	141	g	230	260	305	382	382*	10	
	U	200	267	g	230	260	325	405	405*	10	
	U	400	0.1	d	58	70	40		70 m	2	
	U	400	50	g	75	95	120	135*	135*	10	
	U	400	340	g	98	123	160	185*	185*	10	
chalk, France, 95% calcite, $n=40\%$	U	R	10	e		15	18	20	24*	20	Day70
	U	R	30	d		10	22	40	75*	25	
	U	R	50	d		17	35	66	120*	30	
	U	R	90	c		30	65	105	187*	32	
chalk, Annona, Caddo Parish, Louisiana, USA (from TIPCO-Noel-D-4-wellbore, depth 472 m), 82% calcite, 6.4% quartz	N	R	0.1	b					27 u	3	Wee75
	N	R	10	b					48 u	3	
	N	R	50	g		68	97*		97*	5	
	N	R	100	e, g		72	100*		100*	5	
	N	R	200	e, g		105	132*		132*	5	
	N	R	300	e, g		110	141*		141*	5	
	N	R	500	e		110	131*		131*	5	
chlorite, Wawa, Ontario, Canada, 10…30 calcite	U	R	0.1	b					122 u		Coa66
clay, air-dried	P	R	0.1	b					1 u	0.9	Kie51
	P	R	7	e	1.5	2.5	2.5	2.5	2.5 u	10	
	P	R	13	f	9.5	19	29	36	36 u	14	
	P	R	29	f	13.5	22	37	43	46 u	18	
	P	R	54	f	19.5	37	54	64	69 u	20	
	P	R	70	g	29.5	45	69	85*	88*	14	
	P	R	92	g	29.5	39	59	79*	93*	13	
claystone, "Red Beds", Wyoming, USA	N	R	0.1	a					235 u	0.9	Bre57
	N	R	69	b	345	430			432 u	2.2	
	N	R	103	d	345	414	341	338	438 m	3.5	

(continued)

Table 1 (continued)

Material	Orientation	T °C	p_m MPa	Type	$S_1 - p_m$ [MPa] during axial deformation ε_1				σ_1 MPa	ε_1 %	Ref.
					1%	2%	5%	10%			
coal, O.F.S. Colleries, S. Africa	U	R	0.1	b					32···63 u		Den65
coal, Witbank Colleries, S. Africa	U	R	0.1	b			7	17*	17*	10	Don75
coal, New Largo Colleries, S. Africa	U	R	0.1	b					15···25 u		Bie68
coal, Barnsley Hards, GB, homogeneous, bituminous, volatile content 36% well-marked bedding planes, distinct cleat planes	N	R	0.1						44 u	1.1	Pom71
	N	R	10						85 u		
	N	R	20						100 u		
	45°	R	0.1						19 u		
	45°	R	10						50 u		
	45°	R	20						58 u		
	30°	R	0.1						9 u		
	30°	R	10						31 u		
	30°	R	20						37 u		
	P	R	0.1						18 u		
	P	R	10						58 u		
	P	R	20						70 u		
coal, Kemmerer Coal Comp., Big Pit area, Adaville 1 seam, Wyoming, USA	N	R	0.1	b	25	40			43 u	2.4	Hea76
	P	R	0.1	b					24 u		
	N	R	50	b					37 u		
	P	R	50	b					44 u		
	N	R	100	b					63 u		
	P	R	100	b					54 u		
	N	R	200	b					72 u		
	P	R	200	b					67 u		
	N	R	400	b					90 u		
	P	R	400	b					95 u		
	N	R	500	b					110 u		
	P	R	500	b					102 u		
	N	R	600	b					133 u		
	P	R	600	b					134 u		
	N	R	700	b					137 u		
	P	R	700	b					131 u		

(continued)

Table 1 (continued)

Material	Orientation	T °C	p_m MPa	Type	$S_1 - p_m$ [MPa] during axial deformation ε_1 1%	2%	5%	10%	σ_1 MPa	ε_1 %	Ref.
cryolite, Greenland	⊥ c-axis	R	500	d	118	384	363	347	384m	2	Han66
diabase, Frederick, Maryland, USA fine-grained, 49% anorthite, 46% pyroxene, n=0.1%	U	R	0.1	a					487u	0.55	Bra64
	U	R	49	a					806u	1.65	
	U	R	160	a	1000				1310u	1.37	
	U	R	318	a	1000	1900			2065u	2.2	
diabase, Tishomingo, Johnston County, Oklahoma, USA, 63% plagioclase, 18% pyroxene, medium-grained	U	150	100	a	181	266	334	494	510u	1.9	Ber66
	U	500	500	f	139				546u	18	
diorite, "Orikaba", Iwate Pref., Japan; fine-grained, 67% feldspar, 10% hornblende, 8% pyroxene, 7% biotite, 5% quartz	U	R	18	a	350				350u	1	Mog65
	U	R	50	a	400				470u	1.3	
	U	R	100	a	480				600u	1.5	
	U	R	200	b	470	840			850u	2.1	
	U	R	250	b	500	860			890u	2.3	
diorite, Swandyke diorite gneiss, Colorado, USA	U	R	0.1	a					63…104u		Han66
diorite, Salem, Essex County, Massachusetts, USA; coarse grained, granular	U	R	150	b	200	537	580		613u	2.6	Ber66
	U	R	500	g	121	216		710*	710*	10	
diorite, Orikaba, Japan	U	25	18	a	345				345u	1	Mog65
	U	25	50	b	400				465u	1.3	
	U	25	100	b	460				600u	1.5	
	U	25	150	b	460	780			780u	2	
	U	25	200	c	470	850			850m	2	
	U	25	250	c	500	855			900m	2.3	
dolomite, Blair, W-Virginia, USA; fine-grained; 85% dolomite, 6% calcite; n=0.1%	U	25	0.1	a	411	549			157u	0.9	Han58
	U	25	100	b	508	688			570u	2.4	
	U	25	200	b					694u	2.4	
	U	25	0.1	a					507u	0.5	Bra64
	U	25	46	a					826u	0.9	
	U	25	94	a					1017u	1.3	
	U	25	157						1125u	1.3	
	U	25	273						1373u	1.3	
(continued)	U	25	349						1471u	3	

Table 1 (continued)

Material	Orientation	T °C	p_m MPa	Type	$S_1 - p_m$ [MPa] during axial deformation ε_1				σ_1 MPa	ε_1 %	Ref.
					1%	2%	5%	10%			
dolomite (continued)	U	25	0.1	a	294				343u	1.2	Rob55
	U	25	49	a	300				600u	1.8	
	U	25	98	a	400				830u	1.9	
	U	25	275	b	450	930			990u	2.6	
	U	25	2450	b	300	600	1520		1810u	7	
	U	R	0.1	a					522		Han67
	U	R	50	a					499		
	U	R	100						950		
	U	R	300						1253		
	U	R	450						1339		
	U	100	350						1167		
	U	100	400						1256		
	U	200	300						563		
	U	200	400						1207		
	U	400	100						437		
	U	400	200						593		
	U	400	300						915		
dolomite, Dunham, Williamstown, Massachusetts, USA	U	R	0.1	a					230	0.4	Bra64
dolomite, Hasmark, Montana, USA	N	R	0.1	a	199				128u	0.7	Han58
	N	R	50	a	262	375	400		338u	4.5	Han66
	N	R	101	f	360	450	530		400u	7.3	Han58
	N	R	202	f	456	570	710	535	544u	13	
	N	R	505	f	169	294			765u	6.5	Han66
	N	150	100	f	398	476	530	514	397u	4.1	Han58
	N	150	200	d	273	371	402		530m	5	
	N	300	100	d	150	343	443		402m	6	
	N	300	200	d	417	520	647		443m	5.5	
	N	300	505	f	373	434	595		716u	9.6	
	N	400	500	g	226	292			685*	7.7	Han66
	P	R	101	f	343	422	520		307u	4.4	Han58
	P	R	202	f	520	662	706		529u	8.9	
(continued)	P	R	505	f					734u	8	Han55

Table 1 (continued)

Material	Orient-ation	T °C	p_m MPa	Type	$S_1 - p_m$ [MPa] during axial deformation ε_1				σ_1 MPa	ε_1 %	Ref.
					1%	2%	5%	10%			
dolomite (continued)	P	300	505	g	392	585	657		726*	9	Han55
	P	400	500	f	415	493	626	834	941 u	19.4	Hen6
	P	500	500	f	157	307	576	758	829 u	17	
dunite, Spruce Pine, USA, 95% olivine	U	25	42	c	220	290	220		290m	2	Bye68
	U	25	76	c	290	370	330		390m	2.5	
	U	25	142	c	290	430	460		560m	3.5	
	U	25	220	e	290	430	690	690*	690*	10	
	U	25	408	g	290	550	1060	1070*	1060*	10	
	U	25	510	g	290	550	1190	1220*	1220*	10	
dunite, Dun Mountain, New Zealand, 80% olivine, $\dot{\varepsilon}_1 = 3\cdots7\cdot10^{-4}\,\text{s}^{-1}$	∥ [100]	25	500	a		2290			2290m	2	Gri60
	∥ [100]	25	500	e					2040m	1.5	
	∥ [100]	300	500	e		1400	1410	1410	1350y	1.5	
	∥ [100]	500	500	g		900	1060	1100*	900y	2.5	
	∥ [100]	800	500	e		550	754	754	700y	3	
eclogite, Stammbach, Bavaria, FRG, medium grained, 30% garnet, 50% pyroxene, 8% hornblende, 9% quartz	N	25	0.1	a	124	230 (204⋯236)			230u	2	Rum67
gabbro, Nahant, USA, 40% pyroxene, 15% olivine, 20 serpentine 10% anorthite, 10% mica	U	25	0.1	a	280				280u		Bye63
	U	25	42	a					480u	1.6	
	U	25	84	a					600u	1.7	
	U	25	153	c		600			700m	2.5	
	U	25	335	e		600	805		870m	4	
	U	25	402	e		600	950		960m	4	
	U	25	517	g		600	1050	1080	1030y	4	
gabbro, Elizabethtown, New York, USA	U	150	100	b	213	372	652		529u	2.2	Bcr66
	U	500	505	d	248	400		809	817m	12	
galena, PbS, single crystal, Broken Hill, New South Wales, Australia, $\dot{\varepsilon}_1 = 5\cdot10^{-4}\,\text{s}^{-1}$	∥ [001]	25	0.1	a	210	280	340		290u	1.9	Lya65
	∥ [001]	25	250	g	270	330	410		340*	5	
	∥ [001]	25	500	g					410*	5	
galena, polycryst. ore, Mt. Isa, Australia, 80% PbS, $\dot{\varepsilon}_1 = 5\cdot10^{-4}\,\text{s}^{-1}$	U	25	50	g	70	105	180	250*	250*	10	Lya65
	U	25	250	g	80	140	235	300*	300*	10	
	U	25	500	g	120	180	285	355*	355*	10	

(continued)

Table 1 (continued)

Material	Orientation	T °C	p_m MPa	Type	$S_1 - p_m$ [MPa] during axial deformation ε_1				σ_1 MPa	ε_1 %	Ref.
					1%	2%	5%	10%			
galena, polycryst. ore, Bulman, Australia, nearly pure PbS	U	25	500	g			185	270*	270*	10	Lya66
galena, polycryst. ore, SE Missouri, USA, coarse grained	U	25	200	g	50	72	120	173*	173*	10	Sal74
	U	200	200	g	45	70	93	119*	119*	10	
	U	400	200	e	19	22	25	30*	30*	10	
glass, Pyrex	U	25	0.1						1105u		Han67
	U	25	100						1934u		
	U	25	200						2412u		
	U	25	400						2652u		
gneiss, biotite gneiss, Fordham, New York, USA	N	500	500	d	490	665	763	746	765m	6	Bor66
	P	500	500	c	163	414	825	735	880m	3.1	
	45°	500	500	d	295	395	480	545	565m	14	
gneiss, biotite gneiss, St. Lawrence, New York, USA	N	150	100	a	115	330			590u	2.5	Bor66
	N	500	500	c	325	645	1130	970	1130m	4.8	
	P	150	100	a	190	270			703u	3.4	
	P	500	500	c	325	700	1050		1075m	4.2	
gneiss, granitic, "Diana", New York, USA	N	150	100	b	390	615			630u	2.9	Bor66
	N	500	500	c	515	900	1100	870	1150m	4	
	P	150	100	b	130	510			610u	2.8	
	P	500	500	c	660	1190	1260		1290m	5.1	
granite, Fichtelgebirge, FRG, coarse-grained	N	25	0.1	c	560				122m	0.2	Rum78c
	N	25	100	c	650				655m	1.5	
	N	25	200	c	650				915m	1.5	
	N	25	300	d	650	1060			1175m	2.8	
granite, Katashirakawa, Japan, coarse-grained	U	25	0.1	a	620				158u	0.2	Mat61
	U	25	102	b	700				666u	1.1	
	U	25	260		700				1040u	1.8	
	U	25	440	b					1215u	2.1	
granite, granodiorite, "Charcoal Grey", St. Cloud, Minnesota, USA	U	25	0.1	c					200…270	0.4	Thi75, Waw68
	U	150	100	a	147	367	925		822u	3.7	Bor66
	U	500	500	d	323	610	925	925	925m	10	

(continued)

Table 1 (continued)

Material	Orientation	T °C	p_m MPa	Type	$S_1 - p_m$ [MPa] during axial deformation ε_1 1%	2%	5%	10%	σ_1 MPa	ε_1 %	Ref.
granite, Westerly, Rhode Island, USA, fine-grained, 36% microcline, 31% anorthite, 5% mica, US rock mechanics standard	U	25	0.1	a					165…320	0.4	Bye69
	U	25	50	a					690u		
	U	25	100	b					850m		Bye69, Hea74a
	U	25	200	c					1135m		
	U	25	300	c					1330m		
	U	25	500	c					1680m		Bye69
	U	25	800	c					2070m		
	U	25	1000	c					2290m		
	P	25	500	c	725	1470	1980		2080m	4	Gri60
	P	300	500	e	610	1170	1620		1660m	4	
	P	500	500	d	600	900	1125	995	1130m	3.5	
	P	800	500	d	250	475	615	605	620m	3.5	
granite, Barre, Vermont, USA, medium to fine, 27% anorthite, 26% quartz, 25% orthoclase, 12% mica	U	25	0.1	a					160…220	0.6	Rob55
	U	25	50	b	314				470u	1.6	
	U	25	100	b	441				608u	1.7	
granite, Aare, Gotthardttunnel, Switzerland, 28% quartz, 59% feldspat, 6% hornblende, coarse-grained	U	25	0.1	a					170u		Kov74
	U	25	10	a					290u		
	U	25	20	a					366u		
	U	25	40	a					475u		
	U	25	60	a					565u		
granite, Tumut, New South Wales Australia	U	25	800	e	650	960	1240		1250m	4	Pat64
granodiorite, Climax Stock, Nevada Test Site, USA	U	25	0.1	a					215	0.34	Sch73
	U	25	50	a					320	0.43	
	U	25	100	a					620		
	U	25	200						980		
	U	25	400						1440		
	U	25	600						1760		
graphite, electro-graphite, grad EY 9, $n = 24\%$, $\dot{\varepsilon}_1 = 4 \cdot 10^{-4}\ \mathrm{s}^{-1}$	U	25	100	g			92	125*	150*	20	Pat72
	U	25	400	g			145	200*	272*	20	
	U	25	800	g			180	260*	330*	20	

(continued)

Table 1 (continued)

Material	Orientation	T °C	p_m MPa	Type	$S_1 - p_m$ [MPa] during axial deformation ε_1				σ_1 MPa	ε_1 %	Ref.
					1%	2%	5%	10%			
halite, artificial	∥ a-axis	25	0.1	f	10	13	20	26	27u	10.8	Han58
	∥ a-axis	25	100	g	11	15	22	30*	58*	30	
	∥ a-axis	25	200	g	14	16	23	32*	63*	30	
	∥ a-axis	150	100	g	8	10	14	18*	26*	24	
	∥ a-axis	150	200	g	9	11	14	19*	33*	28	
	∥ a-axis	300	200	e	8	9	9	9	10*	31	
halite, polycryst., artificial	U	25	200	g			41	47*	47*	10	Hea72
	U	100	200	g			24	27*	27*	10	
	U	200	200	g			12	14*	14*	10	
	U	400	200	e			2.6	3.5	3.5*	10	
halite, Grand Saline, Texas, USA	∥ a-axis	25	100	g	18	19	19	29	61*	30	Han66
	∥ a-axis	300	100	d	18	18	17	16	20m	38	
	∥ a-axis	500	100		18	19	18	17	19m	28	
limestone, "Wombeyan marble", New South Wales, Australia, coarse-grained	N	25	0.1	a	92				72u	0.4	Pat58
	N	25	10	b					103u	1.7	
	N	25	50	e	160	160	160	170	170	20	
	N	25	100	g	188	207	243	285	360*	20	
	N	25	800	g	220	300	400	500	660*	20	
limestone, Wells Station, Australia	N	25	20	a	200				216u	1.6	Pat64
	N	25	34	b	239				240u	3.9	Pat58
	N	25	80	g	291				320*	5	
	N	25	100	g	300				350*	5	
limestone, Alberta, Canada, D-1 formation	U	25	0.1	a	138				160u	0.3	Bre57
	U	25	35	b	390	240			240u	2	
	U	25	52	b	515	510			510u	2	
	U	25	103	c		530	435		600m	3	
limestone, Solnhofen, FRG, Jurassic, lithographic, 99% calcite, micritic, rock mechanics standard for calcitic rocks	U	25	0.1	a	294				294u	1	average
	U	25	50	b	340				370u	1.6	Rob55
	U	25	100	e	400	410	420	400	420m	5	Rob55
	U	25	200	g	350	410	480	660*	510*	8	Hea60
	U	25	400	g	340	410	500	760*	660*	10	Rob55
	U	25	500	g	470	560	650	760*	760*	10	Hea60
	U	25	800	g		370	465	615*	850*	20	Pat64

(continued)

Table 1 (continued)

Material	Orient-ation	T °C	p_m MPa	Type	$S_1 - p_m$ [MPa] during axial deformation ε_1				σ_1 MPa	ε_1 %	Ref.
					1%	2%	5%	10%			
limestone (continued)	U	150	50	b	410				435	1.5	Hea66)
	U	150	100	e	400	450	450	450	455	2	
	U	150	250	g	440	470	510	564*	580	11	
	U	150	500	g	360	470	550	630*	650	12	
	U	300	50	e	340	400			400	2	
	U	300	300	g	350	415	500	550*	550*	10	
	U	300	500	g	260	400	465	515*	515*	10	
	U	500	20	g		318	322*		322*	5	
	U	500	101	g		290	313	330*	330*	10	
	U	500	202	e		290	310	310	310	10	
	U	500	304	g		272	300	315*	315*	10	
	U	800	507	g	330	390	450		470*	7.4	Han66
limestone, Weißenburg, Jurassic, FRG, "Treuchtlinger Marmor"	N	25	0.1	a	155				155 u	1	Rum75
	N	25	0.1	a					50...115	1.6	Rum65
limestone, Alabama, USA	U	25	0.1	a					53 u	0.2	Han66
	U	25	51	g	74	111	262	320	415 u	28	
	U	25	102	g	96	129	195	261	313 u	30	
	U	25	203	g	96	143	255	337	465*	27	
	U	25	505	g	230	291	427	463	538*	16	
	U	300	505	g	162	177	229	287	343*	20	
limestone, Leadville, Colorado, USA, "Yule marble"	N	25	0.1	a					39 u	0.4	Han66
	N	25	50	b					200 u		
	N	25	101	f	107	169	274	346	383 u	23	
	N	25	202	g	173	237	326	431	565*	27	
	N	25	505	g	211	319	389	461	540*	20	
	N	25	1010	g	240	309	387	466	466*	10	
	N	150	101	g	107	138	188	252	252*	10	
	N	150	202	g	98	146	223	326	326*	10	
	N	150	1010	g	133	177	250	334	334*	10	
	N	300	101	g	53	78	125	177	177*	10	
	N	300	202	g	81	106	173	244	244*	10	
	N	300	505	g	123	152	211	275	275*	10	
	N	400	304	g	102	140	188	218	218*	10	

(continued)

Table 1 (continued)

Material	Orient-ation	T °C	p_m MPa	Type	$S_1 - p_m$ [MPa] during axial deformation ε_1				σ_1 MPa	ε_1 %	Ref.
					1%	2%	5%	10%			
limestone, Bedford, Indiana, USA, "Indiana limestone", 99% calcite, fine-grained, $n=15\%$	U	25	0.1						28···62		Hea74
	U	25	50						123		
	U	25	100						171		
	U	25	200						258		
	U	25	300						330		
	U	25	500						430		
	U	25	700						523		
limestone, Knoxville, Tennessee, USA, "Tennessee Marble", 98% calcite, medium grained	U	25	0.1	c					100···200u	0.25	Rum70
	U	25	21	d					200u	0.6	
	U	25	35	d	245	220			245u	1	
	U	25	55	g	290	300			300*	2	
limestone, Beldens, Vermont, USA, "Danby marble"	P	25	0.1	a					45u	0.5	Rob55
	P	25	39	g	137	147	157	186*	186*	10	
	P	25	88	f	177	206			265u	4.2	
	P	25	245	g	226	265	340		453*	8	
	P	25	392	g	245	265	343	432*	432*	10	
limestone, Wolfcamp, Texas, USA	N	25	0.1	a	82				82···110u	1	Han58
	N	25	50	b	197	254			273u	4.4	
	N	25	101	d	345	381	391	388	391m	5	
	N	25	202	g	412	461	516	550	587*	30	
	N	150	50	d	164	218	237	210	239m	5	
	N	150	101	g	157	217	287	327*	327*	10	
	N	150	202	g	297	331	397	472	608*	22	
	N	300	101	g	131	196	263	313	380*	27	
	N	300	202	g	262	295	357	408	460*	27	
magnesite, $MgCO_3$, single crystal	‖ c-axis	25	500	g	420	520	600		660*	8	Hig59
	‖ c-axis	300	500	e	580	580			580	2	
	‖ c-axis	500	500	a	390				425u	1.2	
magnesite, natural, CSSR	U	25	0.1	a					128···160		Her66
magnesium oxide, MgO, polycryst., grain size 0.3···0.5 mm, $\dot{\varepsilon}_1 = 10^{-3}$ s^{-1}	U	25	0.1	a	450				360u	0.1	Pat70
	U	25	100	b	700	470			470u	2	
	U	25	200	b	800	850	920		735u	1.5	
	U	25	500	g	1150	1200	1350	1050*	1050*	10	
(continued)	U	25	1000	g				1600*	1600*	10	

Table 1 (continued)

Material	Orient-ation	T °C	p_m MPa	Type	$S_1 - p_m$ [MPa] during axial deformation ε_1				σ_1 MPa	ε_1 %	Ref.
					1%	2%	5%	10%			
magnesium oxide (continued)	U	300	500	g		1010	1220	1260	1260*	10	Pa70
	U	500	500	g		760	960	1060	1060*	10	
	U	750	500	g		450	600	680	680*	10	
magnetite, ore, Kiruna, Sweden, 95% magnetite, grain size 1.3 mm	U	25	0.1	a	390				90u	0.8	Mül72
	U	25	50	b	480	630			505m	1.8	
	U	25	100	b	550	840			660m	2.2	
	U	25	200	d		950			980m	4	
	U	25	400	g			1220	1400	1460*	15	
	U	300	200	b, d			650	740	750m	11	
marble, Greece, medium grained	N	25	0.1	c	100	80			60m	0.3	Run75
	N	25	10	d	155	160			100m	1	
	N	25	30	e	180				130y	0.5	
	N	25	60	g	180	230*			150y	0.6	
	N	25	88	g	250	280			250y	0.6	
marble, Carrara, Italy, pure calcitic, isotropic, grain size <0.2 mm	U	25	0.1	a			315*		85…134		average Kar11
	U	25	20	b	190	180			200u	4	
	U	25	50	d	245	250	240		250m	2	
	U	25	83	g	275	300	315*		315*	5	
	U	25	162	g	320	360	420	470*	470*	10	
	U	25	320	g	350	400	500*		500*	5	
marble, Oota, "Mito marble", Japan, grain size 0.5…2 mm	U	25	0.1	a	120				80u		Mog65
	U	25	15	b					93u		
	U	25	25	e	230				120m	1	
	U	25	100	g	275	250*			150y	0.3	
	U	25	200	g		300*			210y	0.4	
migmatite, St. Lawrence County, New York, USA, 43% plagioclase, 30% quartz, 19% microcline, 6% biotite	N	150	100	a	125	200	710		725u	2.9	Bor56
	N	500	500	c	250	460	830	860	1000m	8	
	P	150	100	a	230	700	600		750u	2.2	
	P	500	500	g	420	550		670*	540y	1.5	
monzonite, San Juan County, Colorado, USA, 34% plagioclase, 29% orthoclase, 17% quartz, 5% biotite	U	150	100	c	125	200	640		640m	5	Bor66
	U	500	500	c	370	620	960	870	960m	5	

(continued)

Table 1 (continued)

Material	Orientation	T °C	p_m MPa	Type	$S_1 - p_m$ [MPa] during axial deformation ε_1 1%	2%	5%	10%	σ_1 MPa	ε_1 %	Ref.
norite, Witwatersrand, S. Africa	U	25	0.1	c	65	130	350		260m	3.5	Cro71
	U	25	10	c	65	130	400		350m	5	
	U	25	20	c	65	130	425		410m	6.5	
	U	25	50	c	65	150		600	613m	9	
peridotite, Mt. Burnette, Alaska, pure olivine dunite	U	500	500	e	1050	1210	1200	1165	1210m	2	Gri60
	U	800	500	g	330	480	620		780*	10	
peridotite, Snowy Mts., Australia, from peridotite nodule of alkali basalt, $\dot{\varepsilon}_{1800} = 5 \cdot 10^{-5}$, $\dot{\varepsilon}_{1000} = 2 \cdot 10^{-4}$ s^{-1}	U	800	500	g	672	850			915*	3	Ral68
	U	1000	500	g	360	480			565*	3	
peridotite, Haruyama, Japan, "Nabeishi peridotite", 25% olivine, 74% hornblende	U	25	18	a					310u	0.5	Mog65
	U	25	50	b					440u	0.9	
	U	25	80	b	530				540m	1.1	
	U	25	150	b	610				760m	1.7	
	U	25	250	b	600	950			994m	2.2	
peridotite, Dun Mt., New Zealand 80% olivine	‖[100]	25	500	a	1900	2200			2290u	2.2	Gri60
	‖[100]	300	500	e	1200	1400	1460	1540	1540	10	
	‖[100]	500	500	g	540	870	1040	1080	1080*	10	
	‖[100]	800	500	e	420	600	780	800	800	10	
peridotite, Lowell, Vermont, USA, 72% serpentine, 19% augite, 7% olivine, serpentinized	N	150	100	a	150	450			510u	2.4	Bor66
	N	500	500	c	250	400	350	320	400m	2	
	P	150	100	a	150				445u	1.8	
	P	500	500	d	220	340	365	310	365m	4	
phyllite, Nelligen, New South Wales, Australia, fine-grained, chlorite-sericite, $\dot{\varepsilon}_1 = 4 \cdots 8 \cdot 10^{-4}$ s^{-1}	P	25	100	d		275	320	300	330m	4	Pat66
	P	25	250	d		480	480	430	470y	1.7	
	P	25	500	d	660	660	630	590	660y	1.3	
	P	25	1000	c	910	860	850	805	910m	1	
	45°	25	300	g		130	160	185*	130y	2.1	
	45°	25	500	g		185	200	220*	185y	2	
	45°	25	700	g		210	260	330*	200y	1.7	
plagioclase andesine, 30% anorthite, Kragerö, Norway, single crystal	45°	600	1000	g	360		470*		410y	2.7	Bor70
	45°	800	1000	g	220		275*		210y	1.5	

(continued)

Table 1 (continued)

Material	Orientation	T °C	p_m MPa	Type	$S_1 - p_m$ [MPa] during axial deformation ε_1				σ_1 MPa	ε_1 %	Ref.
					1%	2%	5%	10%			
plagioclase (continued)											
labradorite, 55% anorthite, Labrador, Canada, single crystal	45°	600	800	g	280	375*			350y	1.7	Bor70
	45°	800	800	g	105	160	220*		160y	2	
bytownite, 77% anorthite, Crystal Bay, Minnesota, USA, polycryst.	45°	800	1000	g		110	250	390*	250y	5	
	annealed	800	1000	e		480	810*		810y	5	
albite, 2% anorthite, Greenwood, Maine, USA, polycryst.	45°	800	1000	e	190	650	600		600y	1.8	
	annealed	800	1000	g		330	750	990*	750y	5	
anorthite, 95% anorthite, Grass Valley, California, polycryst.	45°	800	1000	e		240	340		240y	2	
pyrite, FeS$_2$, polycryst., Silver-mines, Eire, 92% pyrite, n=7%	U	25	0.1	a	120				120u	1	Atk75
	U	25	70	b	420				420u	1	
	U	25	290	d	660	740	880	810	880m	5	
	U	200	0.1	a	110				110u	1	
	U	200	140	c	470	620	460	390	620m	2	
	U	200	265	g	470	540	680	730*	470y	1	
	U	400	0.1	a					80u	0.5	
	U	400	50	c	310	250			340m	0.3	
	U	400	200	d	610	680	580		690m	2.8	
	U	400	288	g	500	580	720	780*	580y	2	
pyrite, FeS$_2$, polycryst. ore, Skovares, Norway	U	25	100	a	500	900			760u	1.5	Lan68
	U	25	200	a	500	1200			930u	2.1	
	U	25	300	b	600	1200			1210u	2.2	
	U	25	400	b	700	1300			1450u	3	
	U	25	500	c	1100	1500			1650u	3	
pyrite, FeS$_2$, polycryst. ore, Rio Tinto, Spain	U	25	100	a	740				740u	1.2	Lan68
	U	25	200	b	740	815			850u	2	
	U	25	300	d	740	975			975m	2	
	U	25	400	d	950	1230			1240m	2.5	
	U	25	500	d	950	1420			1450m	3	
pyrite, FeS$_2$, Bingham, Utah, USA	U	25	0.1	b	147				147u	0.9	Rob55
	U	25	50	a					500u	0.6	
	U	25	220	a					785u	0.5	

(continued)

Table 1 (continued)

Material	Orientation	T °C	p_m MPa	Type	$S_1 - p_m$ [MPa] during axial deformation ε_1 1%	2%	5%	10%	σ_1 MPa	ε_1 %	Ref.
pyroxenite, Mt. Boardman, Coast Range, California, USA	U	25	507	e	1220	1650	1670	1740	1500y	1.5	Gri60
	U	300	507	g	640	960	1160	1250*	850y	1.5	
	U	500	507	g	440	740	840	910*	700y	1.5	
	U	800	507	e	330	500	670	670	600y	2.2	
pyroxenite, Webster, N. Carolina, USA, 96% enstatite, 4% hornblende	N	150	100	b	160	415			530u	3.1	Jun76
	N	500	500	e	300	570	610	640	570y	2	
pyrrhotite, polycryst. ore, Sudbury, Canada, 80% $Fe_{1-x}S$, grain size 1.5 mm, $\dot{\varepsilon}_1 = 3 \cdot 10^{-4}$ s^{-1}	U	20	150	e	450	490	550	560*	450y	1	Atk75a
	U	200	150	e	250	300	330	330	250y	1	
	U	300	150	g	75	110	130	145*	75y	1	
	U	400	150	g	50	60	75	80*	50y	1	
quartzite, Dover flint nodules, from Dover chalk, England, $\dot{\varepsilon}_1 = 10^{-5}$ s^{-1}	U	400	1500	g		1210	1800*		2200*	8	Gre70
	U	450	1500	g		250	700	1200	1500*	12	
	U	700	1500	g		210	450	550	620*	15	
	U	800	1500	g		150	200	270	400*	20	
quartzite, Sioux, Minnesota, USA	U	25	0.1	a	215	255			360u	2.5	Han58
	U	25	100	a	230	280	930		1079u	5.2	
	U	25	200		650	1300*			1300*	2	
quartzite, Eureka, Nevada, USA	U	500	500	a	560	1160			1690u	3.7	Gri60
quartzite, Canon Creek, USA, $\dot{\varepsilon}_1 = 8 \cdot 10^{-6}$	U	700	600	g			1980*		1980*	5	Lal71
	U	800	600	g			1600*		1600*	5	
	U	900	600	d			400	505	505*	10	
	U	1000	600	e			200	200	200	10	
quartzite, Cheshire, Rutland, Vermont, USA	U	25	0.1	a					461u	0.6	Bra64
	U	25	61	a					1185u	1.3	
	U	25	162	a					2165u	2.7	
	U	25	308	a					2845u	3.4	
quartzite, Sunray Cullen well, Oklahoma, USA, Simpson ortho-quartzite, silica cemented, grain size 0.2 mm, $\dot{\varepsilon}_1 = 10^{-5} \ldots 10^{-6}$ s^{-1}	U	500	800	c		950	1900	2100	2170m	8	Hea68
	U	600	800	c		810	1450	1600	1610m	9	
	U	700	800	c		600	1280	1350	1350y	6	
	U	800	800	d		620	880*		810y	2.5	
	U	900	800	e	220	380	510		460y	3	
	U	1000	800	e		360	450		250y	1	

(continued)

Table 1 (continued)

Material	Orient-ation	T °C	p_m MPa	Type	$S_1 - p_m$ [MPa] during axial deformation ε_1				σ_1 MPa	ε_1 %	Ref.
					1%	2%	5%	10%			
rhyolite, Tishomingo, Oklahoma, USA	U	150	100	a	143	515	1030		800u	2.6	Han66
	U	500	500	g	422	844	1030	927	1050	30	Waa69
rubber, Dunlop Rubber Comp., 65 natural.+styrene/butadiene rubber (used as jackets in rock testing)	U	20	500	g			6	12	12*	10	
	U	20	600	g	25	35	40	50*	30y	1.5	
	U	20	800	g		78	80	95*	75y	1.5	
	U	20	1000	g		125	130	140*	125y	2	
salt, NaCl, artificial, at 200 MPa, 650 °C, polycryst., $n<5\%$, grain size 0.1···0.3 mm	U	20	25	g	30.5	39.5	54	70	95*	20	Ker43
	U	100	25	g	25	32	43	55	70*	20	
	U	200	25	g	18	26	31	38	48*	20	
salt, NaCl, artificial, 140 MPa, 120 °C, polycryst.; 0.5···1.0 mm; $\varrho=2.139$ gcm^{-3}, $n=1\%$, $\dot{\varepsilon}_1=10^{-4}$ s^{-1}	U	20	0.1	b	65				61u	3	Hea75
	U	20	0.5	b	65	77			62.3u	3	
	U	20	1.5	b	65				77.4u	3	
	U	20	10	f	65		83.6		81y	3	
	U	20	20	f	65	75	83.6		75.4y	3	
	U	20	50	g	65	75	82	89*	78y	3	
	U	20	100	g	65	75	82	84*	79y	3	
	U	20	200	g	65	74	82	91*	78y	3	
	U	20	400	g	65		88*		86y	3	
salt, NaCl, artificial	U	25	0.1	g	29				38*	1.9	Sch37
	U	25	10	g	31	39			43*	2.4	
	U	25	30	g	33	42			48*	2.6	
	U	25	39	g	36	48			55*	2.7	
	U	25	54	g	34	45			55*	3.2	
salt, rock salt, Blaine, Oklahoma, USA	N	25	13	e	30	61	76	78	78	18	Han66
	N	102	102	g	25	61	54	71*	78*	14	
	N	105	106	e	33	69	120		120	8	
salt, rock salt, Carlsbad, New Mexico, USA, 95% NaCl, 0···15% anhydrite, medium to coarse grained, $\hat{S}_1=12$ MPa s^{-1}	U	25	0.1	g	19	25	35		14···31u	4···5	Waw76
	U	25	3.5	g	12	14	19	22*	36*	5.4	
	U	150	3.5	g					23*	12	
salt, rock salt, "Hockley", Texas, USA	N	25	0.1	b	5	10	26		26	7.7	Han53, Han59
	N	25	2.5	f	6	13	22	38	38*	10	
	N	25	10	f	17	37	54	68	68*	10	
	N	25	53	g	31	48	60	74	74*	10	
(continued)											

Table 1 (continued)

Material	Orientation	T °C	p_m MPa	Type	$S_1 - p_m$ [MPa] during axial deformation ε_1				σ_1 MPa	ε_1 %	Ref.
					1%	2%	5%	10%			
salt (continued)	N	25	118	g	39	54	75	84	84*	10	Han53
	N	25	206	g	39	54	74	93	93*	10	Han59
	N	25	285	g	54	62	84	104*	118*	15	
	N	25	515	g	48	54	64	76	76*	10	
sand, St. Peter formation, Illinois, USA, 99% quartz grains, well-rounded, 250···300 μm	U	25	50	g	6	11	28	65	97*	20	Bor68
	U	25	100	g	20	36	72	122	200*	20	
	U	25	200	g	35	69	158	285	490*	20	
compacted at 500 MPa, 500 °C, pore pressure 100 MPa	U	500	500	d			62	87	1120 m	18	Gri60
sandstone, Gosford, New South Wales, Australia, fine grained (0.2 mm), weakly cemented in clay matrix, n=13%	U	25	0.1	a					37···50u		Edm72
	U	25	100	e	170	225	205	200	200y	1.5	
	U	25	200	g	130	200	290	310*	250y	3.5	
	U	25	400	g	160	260	460	600*	460y	5	
	U	25	600	g	200	330	670	860*	888m	15	
sandstone, Alberta, Canada, 70% sand, 30% carbonate and silica cement	U	25	0.1	a	43				59u	1.3	Bre57
	U	25	35	d	146	152	151	105	160m	3	
	U	25	69	d	180	220	200	185	220	2	
sandstone, Buntsandstein, Ettlingen, SW-Germany, FRG	N	25	0.1	c	100				55···60m	0.3	Gow77
	N	25	10	c	130				105m	1.1	
	N	25	20	c	130				135m	1.2	
	N	25	50	d	130	190	160		190m	2	
	N	25	100	g	130	240	250*		240y	2	
	N	25	200	g	130	280	460*		300y	3	
sandstone, Muttenberg, FRG	U	25	0.1	a	170	195	221		68u	0.6	Kar11
	U	25	28	b	190	250			200u	3.1	
	U	25	55	c	190	270	310*		253m	2	
	U	25	150	g	190	340	400*		324	7.1	
	U	25	240						417	7.4	
sandstone, Ruhrsandstone, Witten, Imberg quarry, FRG, medium grained, quartzitic	N	25	0.1	a					140±10u	0.5	Rum75
	N	25	10	a					213u	0.8	
	N	25	50	c	310	160			320m	1.4	
	N	25	100	c	270	520			225m	2.5	

(continued)

Table 1 (continued)

Material	Orientation	T °C	p_m MPa	Type	$S_1 - p_m$ [MPa] during axial deformation ε_1 1%	2%	5%	10%	σ_1 MPa	ε_1 %	Ref.
sandstone, Pico, California, USA	N	25	0.1	g	4	6	13		28*	6.2	Han66
	N	25	100	g	40	80	132	193	193*	10	
	N	25	200	g	40	80	197	346	346*	10	
	N	300	200	g	83	122	245	395	395*	10	
sandstone, Repetto, California, USA	45°	25	49	a	93	221			294 u	2.4	Han66
	45°	25	103	a	95	270			559 u	4	
	45°	25	101	a	51	106			390 u	3.3	
sandstone, Kayenta formation, Colorado, USA, poorly consolidated, medium to coarse grained, 90% quartz, (700) prepressed to 700 MPa	U	25	0.1	a					32 u		Dub74
	U	25	50	a					173 u		
	U	25	100	a					210 u		
	U	25	(700)	g			246		246*	5	
	U	25	200	c			250		250 m	5	
	U	25	(700)	g			387		387*	5	
	U	25	400	g			394		394*	5	
	U	25	(700)	g			630		630*	5	
	U	25	500	g			830		830*	5	
	U	25	700	g			1035		1035*	5	
sandstone, Weber formation, Rangely Oil Field, Colorado, USA, fine-grained, $n = 5\%$, from 6260 feet depth	U	25	0.1	a					114 u		Ha.75
sandstone, graywacke sandstone from the Cretaceous Mesa Verde Formation at 1968 m depth of the CER Geonuclear hole RB-E-01, Rio Blanco County, Colorado, USA, $n = 7\%$, grain size 100…300 μm, $\varrho = 2.49$ gcm^{-3}	U	25	0.1	a					74 u	0.5	Den75
	U	25	40	a					190 m	1.5	
	U	25	100	b					214 m	2.6	
	U	25	300	c					237 m	5.7	
	U	25	600	g					272 m	4.5	
sandstone, same hole at 1782 m depth, 50% water saturated, Paleocene, Fort Union Formation	U	25	0.1	a					81 u		Sch72a
	U	25	100	c					472 m		
	U	25	200	c					677 m		
	U	25	300	g			856*		856*	5	
	U	25	400	g			1008*		1008*	5	
	U	25	500	g			1200*		1200*	5	
	U	25	600	g			1389*		1389*	5	

(continued)

Table 1 (continued)

Material	Orientation	T °C	p_m MPa	Type	$S_1 - p_m$ [MPa] during axial deformation ε_1 1%	2%	5%	10%	σ_1 MPa	ε_1 %	Ref.
sandstone same hole at 1964 m depth, 50% water saturated, Cretaceous, Mesa Verde Formation	U	25	0.1	a					76u		Sch72a
	U	25	100	c					381m		
	U	25	200	d					544m		
	U	25	300	e					706		
	U	25	400	g			937*		937*	5	
	U	25	500	g			1115*		1115*	5	
	U	25	600	g			1187*		1187*	5	
sandstone, graywacke sandstone from the Cretaceous Mesa Verde Formation at 1942 m depth of the Equity S. Salphur Creek No. 4 borehole, Rio Blanco County, Colorado, USA; $n=18\%$, $\varrho=2.29$ gcm^{-3}, dry	U	25	0.1	a					56u		Sch72a
	U	25	100	c					186m		
	U	25	200	d					275m		
	U	25	300	g			358*		358*	5	
	U	25	400	g			465*		465*	5	
	U	25	500	g			525*		525*	5	
	U	25	600	g			670*		670*	5	
	U	25	700	g			777*		777*	5	
sandstone, "Weeks Island S", Louisiana, USA	N	25	0.1	a	10				10u	1	Han58
	N	25	50	b	73	108	138		138u	6.3	
	N	25	100	g	102	138	165	169	169*	10	
	N	25	200	g	116	150	206	285	285*	10	
sandstone, Supai, Nevada, USA	N	300	200	e	120	180	200		204u	9.2	Han66
	N	300	500	g	160	240	420	690	690*	10	
sandstone, Berea, Ohio, USA	U	24	0.1	a					46···100u	2	Han66
	N	24	50	e	94	137	141	153	153*	10	
	N	24	100	e	122	214	220	224	224*	10	
	N	24	200	g	166	238	300	352	325*	10	
	N	24	410	g	123	202	360	506	506*	10	
	N	300	200	g	94	160	254	330	330*	10	
	N	300	310	g	153	235	357	500	500*	10	
	N	300	412	g	97	177	410	640	640*	10	
sandstone, Bartlesville, Oklahoma, USA	N	24	0.1	a	41				41u	0.6	Han66
	N	24	100	e	220	226	226	226	229*	20	
	N	24	200	g	230	270	350	430	430*	10	
sandstone, Rush Spring, Oklahoma, USA	N	24	0	a	440				190u	0.7	Bre57
	N	24	70	b	520				560u	1.8	
	N	24	100	b					620u	1.8	

(continued)

Table 1 (continued)

Material	Orientation	T °C	p_m MPa	Type	$S_1 - p_m$ [MPa] during axial deformation ε_1 1%	2%	5%	10%	σ_1 MPa	ε_1 %	Ref.
sandstone, Barns, Texas, USA	N	25	0.1	a	75				40u	0.6	Hän58
	N	25	50	b		130			170u	2.8	
	N	25	100	g	83	175	230	240	240*	10	
	N	25	200	g	125	250	360	400	400*	10	
	N	150	100	g	83	175	225	230	230*	10	
	N	150	200	g	100	200	305	380	380*	10	
	N	300	100	e	80	155	220	220	220*	10	
	N	300	200	g	100	200	290	340	340*	10	
	P	25	50	a	66	120			147u	3	
	P	25	100	e	66	140	270	270	270*	10	
	P	25	200	g	85	180	340	370	370*	10	
sandstone, Oil Creek, Texas, USA	N	25	0.1	a	265				96u	0.5	Han58
	N	25	100	a		420			690u	3.3	
	N	25	200	a	113	280			1080u	3.8	
	N	150	100	a	260				740u	2.1	
	N	150	200	a	118	300			990u	2.8	
	N	300	100	a	255				690u	2.1	
	N	300	200	a	108	300			910u	2.4	
sandstone, Tensleep, Wyoming, USA	N	150	100	b	130	317			466u	3.9	Han65
sandstone, fine-grained graywacke	U	25	0.1	a					155u		Sch70,
sandstone from the Cretaceous Lance Formation at 3120m depth of the Wagon Wheel No. 1 hole; Sublette County, Wyoming, USA; $\varrho=2.46\,\mathrm{gcm^{-3}}$, $n=8.5\%$, dry, $\dot{\varepsilon}_1=10^{-4}\,\mathrm{s^{-1}}$	U	25	100	b					533u		Sch73
	U	25	200	c					735u		
	U	25	300	d					925m		
	U	25	400	g			1187*		1187*	5	
	U	25	500	g			1420*		1420*	5	
	U	25	600	g			1735*		1735*	5	
	U	25	700	g			1884		1884*	5	
sandstone, Nugget sandstone, Parleys Canyon, Utah, USA, 99% quartz, $n=2\%$	U	25	0.1 $\dot{\varepsilon}_1=10^{-3}\,\mathrm{s^{-1}}$	a	260				260u	1	Waw75
	U		0.1 $\dot{\varepsilon}_1=10^{-5}\,\mathrm{s^{-1}}$	a					230u	0.8	
schist, Keystone, S. Dakota, USA, mica schist, fine, porphyroblastic, 54% quartz, 40% biotite, 5% mica	N	150	100	d	130	190	200		225m	3	Bor66
	P	150	100	e	100	145	170		170u	8	
	N	500	500	d	200	400	550	600	490y	3	
	P	500	500	d		190	430	490	360y	4	

(continued)

Table 1 (continued)

Material	Orient-ation	T °C	p_m MPa	Type	$S_1 - p_m$ [MPa] during axial deformation ε_1				σ_1 MPa	ε_1 %	Ref.
					1%	2%	5%	10%			
schist, hornblende, New York, USA	30°	25	100	a	85	300			635u	3.5	Han66
	45°	25	100	a	25	86			623u	4	
	60°	25	100	d	25	80	690		680u	7.5	
serpentinite, Wurlitz-Woja, NE Bavaria, FRG	30°	25	0.1	c	300				127m	0.5	Rum78c
	30°	25	50	e					300y	1.2	
	30°	25	140	g	300	450			490*	3	
	30°	25	410	g	300	600	1100		1100*	5	
serpentinite, Cabramurra, New South Wales, Australia, antigorite-chrysotile serpentinite, 5···10% olivine, 10% magnetite	U	25	0.1	a	510				390u	1.4	Ral65
	U	25	100	a	500				600u	2	
	U	25	200	c	500	820			820m	4	
	U	25	500	d	600	980	1100		1110m	3	
	U	25	350	d	500	1050			1100m	1.3	
	U	350	350	e	320	750	720		650y	1.3	
	U	600	350	c		300			390m	0.5	
	U	650	350	a					120u	0.5	
serpentinite, Tumut Pond, New South Wales, Australia, antigorite-chrysotile serpentinite	U	25	20	a					390u		Ral65
	U	25	100						605u		
	U	25	200						780u		
	U	25	500	b	505	1020			1400m	4	
	U	250	500	d	505	900	950		1010m	3.2	
	U	500	500	d	405	750	840		870m	3.5	
	U	600	500	d	340	600	680		710m	3.2	
	U	650	500	d	270	260	240		255y	0.8	
	U	700	500	a					125u	0.6	
shale, Green River, Colorado, USA	N	25	0.1	a	42	58	90	84	64u	2.6	Han58
	N	25	26	d	50	77	135	153	90m	5	
	N	25	50	g	67	96	160	193	153*	10	
	N	25	100	g	94	120	213	265	193*	10	
	N	25	206	g	110	150			265*	10	
	P	25	0.1	a					<100		
	P	25	50	e	10	18	36	44	44	18	
	P	25	100	e	34	50	57		58u	10	
	P	25	200	g	50	115	180	195	195*	10	

(continued)

Table 1 (continued)

Material	Orient-ation	T °C	p_m MPa	Type	$S_1 - p_m$ [MPa] during axial deformation ε_1				σ_1 MPa	ε_1 %	Ref.
					1%	2%	5%	10%			
shale, Muddy, Colorado, USA	N	25	0.1	a	60				39u	0.6	Han58
	N	25	50	a		110			137u	2.5	
	N	25	100	c	80	155	225	223	243u	10	
	N	25	200	g	80	155	345	380	380*	5.8	
	N	150	100	e	80	142	200		201	10	
	N	150	200	f	80	157	328	337	337u	6.2	
	N	300	100	f	65	100	145		147u	5	
	N	300	200	d	80	155	190	177	190	5	
shale, "5900 foot sands", Texas, USA	N	25	0.1	a					74u	0.6	Han58
	N	25	100	b	100	150	230		245u	7.5	
	N	25	200	e	205	240	290		290u	9	
shale, Crockett, Texas, USA	N	110	70	d	34	46	48	44	49m	3	Han66
shale, Paradox, Utah, USA, wet	N	100	69	b	135	183			184u	3.3	Han66
	N	100	69	b	205	235			250u	2.8	Han66
shale, Hucknall, England	U	25	14	b	66				98u	1.7	Hob70
siltstone, Ormonde, England	U	25	14	a	95				111u	1.2	Hob70
siltstone, Bilsthorpe, England, silty mudstone	U	25	14	b	91				121u	1.7	Hob70
siltstone, Repetto, California, USA	45°	25	0.1	a	28				28u	1	Han66
	45°	25	50	c	64	101	105	91	130u	≈4	
	45°	25	100	e	64	107	169	171	171m	10	
	45°	25	200	g	64	114	230	254	254*	10	
	45°	150	100	c	45	80	133	109	133m	5	
	45°	150	200	g	64	106	187	212	212*	10	
	45°	300	100	b	45	66	86		86m	5	
	45°	300	200	d	58	86	125	139	140m	≈10	
siltstone, Permian "red beds", Texas, USA	N	25	0.1	a	34				48u	1.4	Han65
	N	25	50	c	111	145	147	141	147m	5	
	N	25	100	c	121	175	226	217	226m	5	
	N	150	50	b	76	120	137		142u	7.1	
	N	150	100	e	82	120	165	171	171	10	
slate, Mystic River, Sommerville, Massachusetts, USA	N	25	0.1	a					314u	0.7	Rob55
	N	25	30	a	520				520u	1	

(continued)

Table 1 (continued)

Material	Orientation	T °C	p_m MPa	Type	$S_1 - p_m$ [MPa] during axial deformation ε_1 1%	2%	5%	10%	σ_1 MPa	ε_1 %	Ref.
slate, Mettawee formation, Vermont, USA, Cambrian	N	25	0.1	a	83				98u	1.2	Han66
	N	25	101	a	214				296u	1.7	
	N	25	202	a	270	422			451u	2.4	
	N	150	101	b	151	188	233		240u	7.4	
	N	150	202	c	103	219	437		437m	5	
	N	300	101	a	189	257			275u	2.7	
	N	300	202	a	81	180			432u	4	
slate, Mettawee formation, Granville, New York, USA, Cambrian	N	150	100	a	200	340	580		500u	2.7	Bor66
	N	500	500	d	115	230		630	640m	6	
	P	500	500	d	175	340	475	550	580m	16	
	45°	500	500	g	100	180	220	260*	180u	2	
slate, Martinsburg, Pennsylvania, USA, Ordovician, well-developed slaty cleavage	N	25	0.1	a					190u	1.2	Don64
	30°	25	0.1	a					20u	1.8	
	P	25	0.1	a					110u	2	
	15°	25	20	c	110	170	225		130m		
	15°	25	40	c	125	270			190m		
	15°	25	80	c	150	270	370		270m		
	15°	25	200	e	130			390	390m	3	
sphalerite, ZnS, single crystal, Santander, Spain, $\dot{\varepsilon}_1 = 5\cdot10^{-5}\,s^{-1}$ $\dot{\varepsilon}_1 = 5\cdot10^{-4}\,s^{-1}$ $\dot{\varepsilon}_1 = 5\cdot10^{-5}\,s^{-1}$ $\dot{\varepsilon}_1 = 5\cdot10^{-4}\,s^{-1}$	∥ [110]	150	300	g			230	260*	110y	0.4	Sie77
	∥ [110]	300	300	g			125	175*	70y	0.3	
	∥ [110]	450	300	g			85	118*	65y	0.7	
	∥ [110]	25	500	g			200	350*	100y	0.5	
	∥ [100]	25	300	g	120		210	290	290*	10	
	∥ [100]	300	300	g	80		125	140	140*	10	
	∥ [100]	450	300	g	50		90	118	118*	10	
	∥ [100]	25	500	g	100		210	260*	240y	1.5	
sphalerite, ZnS, single crystal, Trepča, Jugoslavia, $\dot{\varepsilon}_1 = 5\cdot10^{-5}\,s^{-1}$	∥ [110]	25	300	g	250		335	410*	270y	1.2	Sie77
	∥ [100]	25	300	g	200		270	355*	210y	1.2	
sphalerite, ZnS, Central Tennessee-Knox, USA	U	25	50	f	150	305	350		352u	6	Cla73
	U	25	100	e	150	305	375	410	410	12	
	U	25	200	g	150	315	405	510	510*	10	
	U	200	100	g		210	250	300*	180y	1	
	U	500	100	g	130	160	175	210	130y	1	

(continued)

Table 1 (continued)

Material	Orient- ation	T °C	p_m MPa	Type	$S_1 - p_m$ [MPa] during axial deformation ε_1 1%	2%	5%	10%	σ_1 MPa	ε_1 %	Ref.
sphalerite, ZnS, Nikolaus-Phoenix- Mine, Markelsbach, Siegkreis, FRG, $\dot{\varepsilon}_1 = 3 \cdot 10^{-4}\,s^{-1}$	‖ [110]	25	0.1	c					50m	0.9	Say70
	‖ [110]	25	50	c		445			445m	2	
	‖ [110]	25	100	c		440	490		500m	4	
	‖ [110]	25	200	d		440	580	650	650m	10	
	‖ [110]	25	500	g		450	670	820*	950*	20	
syenite, Victor, Colorado, USA, medium to coarse grained, 57% altered feldspar, 8% biotite, 13% perthite	U	150	100	a	100	330	430	360	530u	2.4	Bor66
	U	500	500	d	320	435			435m	2	
talk, Three Springs, W-Australia, fine-grained, isotropic, $n = 3.2\%$	U	25	200	e	120	125	115	102*	120y	1	Edin72
	U	25	400	e	120	155	170	170	170*	10	
	U	25	800	g	120	180	205	220*	180y	2	
trachyte, "Mizuho", Tomioka, Japan, plagioclase holocryst. groundmass	U	25	15	a	149				149u	1	Mog65
	U	25	60	d	195	200			223m	1.2	
	U	25	120	e	200	250			260*	3.5	
	U	25	200	g	210	270			300*	4	
tuff, "Aoishi", Izu-Nagaoka, Japan, 64% chlorite, 30% plagioclase, $n = 17\%$	U	25	0.1	a	65	50			35u	0.8	Mog65
	U	25	6.5	c	93	75			79m	0.5	
	U	25	18	d	90	90			58y	0.6	
	U	25	30	e	85	100*			70y	0.4	
	U	25	50	g	58	70*			50y	<0.1	
	U	25	120	g	56	78*			10y	<0.1	
	U	25	200	g					18y		
tuff, "Tatsuyama", Hoden, Japan, pumice tuff, $n = 10\%$	U	25	28	c	210	60			210m	1	Mog65
	U	25	100	e	220	375			375m	2	
	U	25	200	g	220	350			440*	3	
tuff, Mt. Helen, Mercury, Nevada, USA, fine-grained, $n = 35 \cdots 40\%$, 2% H_2O saturated, $\varrho = 1.5\,gcm^{-3}$	U	25	0.1	c					41u	<2	Hea73a
	U	25	100	c					83u	<2	
	U	25	200	e					150*	5	
	U	25	400	e, g					358*	5	
	U	25	700	e, g					490*	5	
tuff, Tunnel U 12e/UG 3, area 12, Nevada Test Site, USA, $n = 18\%$, saturated	U	25	0	a					11	0.35	Dub73
	U	25	50	b					14		
	U	25	100	b					19		
	U	25	200	d					22	5	
	U	25	300	g					24*	>10	

Table 2. Stress-strain relation and strength of quartz single crystals as a function of confining pressure p_m, strain rate $\dot{\varepsilon}_1$, and temperature T.

Table 2a. Strength of quartz single crystals at 3 GPa confining pressure at 24 °C at $\dot{\varepsilon}_1 = 3 \cdot 10^{-4}\,\mathrm{s}^{-1}$ (from [Chr74]).

Specimen orientation with respect to S_1	Strength $S_{1m} - p_m$ GPa	Stresses on the shear fault at rupture	
		normal stress σ GPa	shear stress τ GPa
‖ c	4.65(10)	4.46	2.25
‖ r	4.26(16)	4.32	2.07
‖ z	4.21(15)	4.30	2.04
⊥ m	4.28(18)	3.35	1.53

Table 2b. Stress-strain relation and strength of quartz single crystals at 0.8 GPa confining pressure at various temperatures and strain rates (from [Hea68]).

Temperature T °C	Strain rate $\dot{\varepsilon}_1$ sec^{-1}	Differential stress $S_1 - p_m$ GPa	Strain ε_1 %	Remarks
300	$6.7 \cdot 10^{-6}$	0.40	1	synthetic
		0.79	2	quartz
		1.71	5	8200 ppm H/Si
		2.09 m[1])	7.2	
500	$6.9 \cdot 10^{-6}$	0.20	1	synthetic
		0.22 m[1])	1.4	quartz
		0.21	2	8200 ppm H/Si
		0.18	5	
600	$7.1 \cdot 10^{-6}$	0.10 y[2])	0.6	synthetic
		0.11	1	quartz
		0.12 m[1])	2	8200 ppm H/Si
		0.09	5	
700	$7.4 \cdot 10^{-7}$	0.37	1	synthetic
		0.72	2	quartz
		1.66 m[1])	4.6	8200 ppm H/Si
		1.59	5	
800	$6 \cdot 10^{-4}$	2.1	5	from large optical quality natural single crystal
		2.8 m[1])	8	
800	$5 \cdot 10^{-5}$	2.1	5	average from 2 samples
		2.6 m[1])	7	
800	$7 \cdot 10^{-6}$	1.61 m[1])	4.2	average from 3 samples
		1.6	5	
800	$1 \cdot 10^{-6}$	1.21 m[1])	3.4	
910	$5 \cdot 10^{-4}$	0.85	2	
		1.60	4	
		1.88	5	
		2.10	6.7	
910	$7 \cdot 10^{-5}$	0.74	2	
		1.29	4	
		1.48	5.1	
910	$7 \cdot 10^{-6}$	0.58	2	
		1.08 m[1])	5	
		0.84	5	
910	$7 \cdot 10^{-7}$	0.54	2	
		0.69 m[1])	4	
		0.48	4	
1020	$5 \cdot 10^{-4}$	0.46 y[2])	2	
		0.64	4	
		0.77	10	
1020	$2 \cdot 10^{-5}$	0.18	1	
		0.20 y[2])	1.3	
		0.25	2	
		0.26	6	

[1]) m = maximum differential stress.
[2]) y = yield strength.

Table 3. Unconfined uniaxial compressive strength σ_c of rocks at room temperature.

Most values are determined from uniaxial compression tests on cylindrical rock specimens with length to diameter ratio of about 2; some data were obtained from compression tests on cubes. (n) = normal to foliation.

Strength values in parentheses are obtained from wet samples.

* Indicates ductile deformation.

Rock	σ_c MPa	Average MPa	Ref.
sediments			
anhydrite, average	40···128	82	
Alberta, Canada	41		Bre57a
Blaine, Oklahoma, USA	128		Han58
Nordheim, FRG	74		Mül74
Südheim, FRG	78		Dre62
chalk, average	5···30	22	
coal, average	4···60	28	
dolomite, average		175	
Hasmark, Montana, USA	128		Han58
Luning, Nevada, USA	60		Han58
Webatuk, New York, USA	148		Bra64
Blair, Virginia, USA	507		Bra64
	343	377	Rob55
	280		Hea73
Clear Fox, Texas, USA	233		Han58
Dunham, Williamstown, USA	230		Bra64
Fusselman, Texas, USA	145		Han58
Glorietta, N. Mexico, USA	81		Han58
halite			
artifical, ‖ a-axis	270		Han58
Grand Salina, Texas, USA	61		Han66
Goderich, Ontario, Canada	156		Coa66
hematite			
Bell Island, New Foundland	194		Coa66
limestone, average		104	
Wombeyan, New South Wales, Australia	72		Pat58
Perns quarry, Sao Paulo, Brazil	77		Wie68
Perns quarry, Sao Paolo, Brazil	85(48)		Rui66

(continued)

Table 3 (continued)

Rock	σ_c MPa	Average MPa	Ref.
limestone (continued)			
Apiai, Sao Paolo, Brazil	83···98		Rui66
Sorocaba, Sao Paulo, Brazil	73(62)		Rui66
D-1 formation, Alberta, Canada	159		Bre57a
Gagnon, Quebec, Canada	134		Coa66
Ottawa, Ontario, Canada	278		Coa66
Upper Teesdale, Durham, GB	42		Pri60
Solnhofen, FRG	275 347	294	Rob55 Hea60
Eichstätt, FRG	110···195	163	Wei75
Treuchtlinger Marmor, FRG	50···115 155		Rum65 Rum75
Alabama, USA	53		Han66
Yule Marble, Leadville, Colorado, USA	39		Han66
Marianna, Florida, USA	43		Han58
Bedford, Indiana, USA	28···62	52	Kha71, Han66, Hea74
Carthage Marble, Missouri, USA	52···73		Bre57a
New Scotland, New York, USA	128		Rob55
Becraft, New York, USA	98		Rob55
Tennessee Marble, Knoxville, USA	106···214	140	Coo65, Kut71, Thi75
Austin Chalk, Texas, USA	14		Bre57a
Fusselman, Texas, USA	38		Han58
Lueders, Texas, USA	59		Har72
Wolfcamp, Texas, USA	82···111	95	Han58
Cordova Cream, Texas, USA	14···17	15	Har72
Belders, Danby Marble, Vermont, USA	45		Rob55
Mankato Stone Co., USA	90···100		Har72
Virginia, USA	331		Bre57a
magnesite, $MgCO_3$, natural, CSSR	128···159	145	Her66

(continued)

Table 3 (continued)

Rock	σ_c MPa	Average MPa	Ref.
salt, NaCl			
artificial, polycryst.	61		Hea75
	38*		Han66
rock, Salt Mine Borth, FRG	20···40		Dre61
rock, average German location	21···52		Dre61, Dre62
rock, Slamic Mine, Roumania	11···27		Fod66
rock, Carlsbad, New Mexico, USA	14···31		Waw76
rock, "Hockley", Texas, USA	26		Han66
rock salt, average	10···50	27	
sandstone, average		95	
Ecca Ser., Karoo, S. Africa	38···57		Col65
Gosford, New South Wales, Australia	37, 50		Wie68a, Jae69
Botucato Cuesta, Sao Paolo, Brazil	81(73)		Rui66
Alberta, Canada	59		Bre57a
Darley Dale I, Derbyshire, GB	41		Pri60, Hoe69
Darley Dale II, Derbyshire, GB	72···81		Hoe69
Mersey Tunnel, Liverpool, GB	32		Kni68
Kirkaldy, Scotland	19		Kni68
Pennant, S-Wales, GB	158···197		Hoe69, Fra70
Ettlingen, SW Germany, FRG	85		Hak75
Muttenberg, FRG	68		Kar11
Heilbronn, FRG	45		Dre62
Ruhr, Witten, Imberg Quarry, FRG	140		Rum75
Ruhr, Witten, Rauen Quarry, FRG	80		Wei75
Ruhr, Zeche Walsum, FRG	125		Dre65
Ruhr, Zeche Wester- holt, FRG	115···122		Dre65
Mev, Hungary	110		Hai75
Buchberg, Switzerland	58		Kov74
Kayenta formation, Colorado, USA	32		Dub74
Weber, Rangely, Colorado, USA	120		Hai75
Rio Blanco Gas Well, Colorado, USA	74		Den75
St. Peter, Illinois, USA	39		Rol74

(continued)

Table 3 (continued)

Rock	σ_c MPa	Average MPa	Ref.
sandstone (continued)			
Weeks Island, Louisiana, USA	100		Han58
Berea, Ohio, USA	46···110	80	Har72, Han66, Her66
Bartlesville, Oklahoma, USA	41		Han58
Rush Springs, Oklahoma, USA	187		Bra60
Montrose, Pennsyl- vania, USA	150		Har72
Tennessee, USA	122···136		Har72
Barns, Texas, USA	40		Han58
Oil Creek, Texas, USA	96		Han58
Parleys Canyon, Utah, USA	280		Waw75
Pinedale Wagon Wheel Well, Wyoming, USA	155		Sch73
shale, average		59	
Hucknall Colleries	59	59	Hob70
Green River, Colorado, USA	64		Han66
Muddy, Colorado, USA	39		Han58
"5900 ft. sands", Texas, USA	74		Han58
siltstone, average		65	
Chislet, Kent, GB	90		Pri58
Snowdown, Kent, GB	102		Pri58
Ormonde, GB	56	56	Hob70
Repetto, California, USA	28		Han66
Permian Red Beds, Texas, USA	48		Han66
tuff			
"Aoishi", Izu- Nagaoka, Japan	35		Yam70
"Tatsuyama", Hoden, Japan	250		Mog65
Mt. Helen, Nevada, USA	41		Hea73a
Nevada Test Site, Area 12, USA	11		Dub73
igneous rocks			
andesite			
Shirochoba, Japan	140		Yam70
Shinkomatsu, Japan	126···156	141	Yam70
Salt Lake(?), Idaho, USA	108···158	133	Han66

(continued)

Table 3 (continued)

Rock	σ_c MPa	Average MPa	Ref.
basalt, average		186	
origin unknown	255		Ric70
origin unknown	170		Bie73
Jupia dam, Sao Paolo, Brazil	102···106		Rui66
Barra Bonita dam, Sao Paolo, Brazil	191···223		Rui66
Jurumurim dam, Sao Paolo, Brazil	154		Rui66
Mussa quarry, Piraju, Sao Paolo, Brazil	169		Rui66
Edinburgh Castle rock, olivine basalt	135		Kni68
Mendig, Eifel, FRG	100···140		Rum77
unknown origin, USA	359		Waw68
Martin Co., Pennsylvania, USA	150		Har72
"Knippa", Texas, USA	262		Bra60
Yakuno-Kyoto, Japan	150		Mat60
diabase, average	140···487	253	
Cascata quarry, Sao Paolo, Brazil	138		Rui66
Campinas, Sao Paolo, Brazil	158		Rui66
Elliot Lake, Ontario, Canada	218		Coa66
Frederick, Maryland, USA	487		Bra64
diorite	60···219	152	
unknown origin	219		Ric70
Swandyke diorite gneiss, Colorado, USA	63···104		Han66
dolerite			
Northumberland, GB	313		Fra70
Upper Teesdale, Durham, GB	131		Kni68
Clydesdale Coll., S. Africa	390		Den65
Karoo system, S. Africa	331		Wie68a
dunite			
dunitic rocks from Japan		200	Mat61
eclogite			
Weißenstein, Bavaria, FRG	204···236	230	Rum67
gabbro	220···280	254	
Buena, USA	280		Rol74
Nahaut, USA	280		Bye68

Table 3 (continued)

Rock	σ_c MPa	Average MPa	Ref.
granite, average	100···320	187	
Taourirt, Hoggar nuclear test site, Algeria	212		Sch72
Cantareina quarry Sao Paolo, Brazil	104···119		Rui66
Valinhos quarry Sao Paolo, Brazil	98···114		Rui66
Grenville, Quebec, Canada	172		Coa66
St. Lawrence, New Foundland, Canada	276		Coa66
Blackingstone quarry, Devon, GB	197		Fra70
Carnmenellis, Cornwall, GB	108···124	116	Bat77
Kuru, Finland	290		Hak75
Flossenbürg, Bavaria, FRG	126···158	142	Rum67
Fichtelgeb., Bavaria, FRG	112···132	122	Rum78c
Malsburg, SW Germany, FRG	150···157		Kut74a
Cap de Long, France	100···180		Pet70
Senones, Vosges, France	155···183		Hou73
La Forge, Vosges, France	252		Pet70
St. Germain, Cote d'Or, France	115···129		Pet70
Okazakishi, Japan	155		Mat56
Ajimura, Japan	270		Mat56
Shichikucho, Japan	177		Mat56
Katashirakawa, Japan	158		Mat61
Katakanaki, Japan	213		Mat56
Alto Lindoso dam, Portugal	76···128		Rod66
Alvarenga dam, Portugal	102···129		Rod66
Vilarinho dam, Portugal	99···137		Rod66
Aare, Gotthard-Tunnel, Switzerland	170		Kov74
North Chelmford, Massachussetts, USA	182		Pen72
Charcoal Black, St. Cloud, Minnesota, USA	129···173		Sin75, Coo65
Charcoal Gray, St. Cloud, Minnesota, USA	220···277		Thi75, Waw68, Kut71

(continued)   (continued)

Table 3 (continued)

Rock	σ_c MPa	Average MPa	Ref.
granite (continued)			
Missouri, USA	162		Rol74
Di Rienzi, USA	157···177		Har72
French Creek	276		Har72
Kitledge, USA	171		
Westerly, Rhode Island, USA	165···283		Hea74a
Barre, Vermont, USA	167···214		Rob55, Har72
granodiorite			
Climax Stock, Nevada Test Site, USA	214		Sch73
norite, average	244···345	297	
Witwatersrand, S. Africa	275		Cro71
Stone quarry, Transvaal, S. Africa	314···345		Bie75, Bie68
Marikana, S. Africa	244···264		Bie75
peridotite			
Thetford Mines, Quebec, Canada	197		Coa66
Haruyama, Fuku-shima, Japan	250		Mog65
syenite			
Kiruna, Sweden	214···274	244	Han66a
trachyte			
Bowral, Australia	150		Jae59
metamorphic rocks			
gneiss, average	100···155	135	
Enclides da Cunha dam, Sao Paolo, Brazil	132		Rui66
Graminha dam, Sao Paolo, Brazil	156		Rui66
Sao Jose dos Campos, Sao Paolo, Brazil	113		Rui66
Jaguare quarry, Sao Paolo, Brazil	137		Rui66
marble, average	40···150	77	
Greece, unknown location	60		Rum75
Carrara, Italy	85···134		Kov74, Kar11, Jae69, Hou73
Oota, "Mito", Japan	80		Mog65
Gualba, Barcelona, Spain	30···53	38	Mon74
Green Mountain, USA	61		Har72
West Rutland, Vermont, USA	51···57	54	Bra64

Table 3 (continued)

Rock	σ_c MPa	Average MPa	Ref.
phyllite			
quartz-phyllite, Bayerland mine, Bavaria, FRG	20···22		Rum65
quartzite, average	200···460	303	
Witwatersrand, S. Africa	217···317	267	Den65
Kimberley Elsbg., Witwatersrand, S. Africa	284		Bie68
Jarugua Hill, Sao Paolo, Brazil	256		Rui66
Lake Elliot, Ontario, Canada	260		Coa66
Nilsia, Finland	300		Hak75
Cherbourg, France	320···380	350	Pet70
Orsay, Essone, France	220···280	250	Pet70
Jasper, USA	342		Rol74
Sioux, Minnesota, USA	359···389		Han58, Cai74
Cheshire, Rutland, Vermont, USA	461		Bra64
schist, average	10···130	62	
sericite schist, Calaveras fault, California, USA	15(n)		Han66
biotite-sillimanite schist, Idaho springs, Colorado, USA	8···35(n)		Han66
biotite schist, Idaho springs, Colorado, USA	53···84(n)		Han66
biotite chlorite schist, Idaho springs, Colorado, USA	36··· 118(n)		Han66
Nuttlar, Sauerland, FRG	63···130	95	Wei75
serpentinite			
Wurlitz-Woja, Bavaria, FRG	127		Rum78c
Cabramurra, New South Wales, Australia	307		Ral65
shale			
Hucknall Colleries, GB	54···64	59	Hob70
Green River, Colorado, USA	64(n)		Han58
Muddy, Colorado, USA	39(n)		Han58
"5000 ft. sands", Texas, USA	74(n)		Han58

(continued)

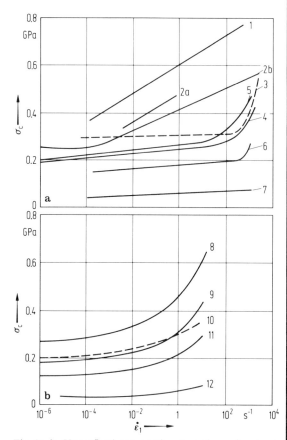

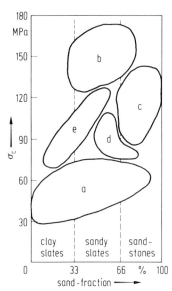

Fig. 3. Uniaxial compressive strength σ_c of slate-sandy carboniferous rocks of the Ruhr-district, FRG, as a function of cement type and sand content. Data refer to uniaxial strength of rock cubes [Dre69a].
a: clay cement, b: carbonatic cement, c: silica cement, d: clay-silica cement, e: clay-carbonatic cement.

Fig. 4a, b. Unconfined compressive strength σ_c of rocks at room temperature as a function of applied strain rate $\dot{\varepsilon}_1$
1:   Dresser basalt [Lin74];
2a: Westerly granite, Rhode Island, USA [Log71];
2b: Westerly granite [Gre72];
3:   Solnhofen limestone, FRG [Gre72];
4:   basalt [Kum68];
5:   granite [Kum68];
6:   porphyric tonalite [Per70];
7:   volcanic tuff [Gre72];
8:   Ishikoshi andesite, Japan [Kob70];
9:   Inada granite, Japan [Kob70];
10: Izumi sandstone, Japan [Kob70];
11: Akiyoshi marble, Japan [Kob70];
12: weak sandstones [Kob70].

Table 4. Maximum triaxial differential compressive strength $(S_1 - p_m)_{max}$ of rocks as a function of confining pressure p_m at room temperature.

Experiments were carried out on cylindrical rock specimens of 1···6 cm diameter and a length to diameter ratio of about 2. Values in parentheses indicate maximum strength during ductile deformation behaviour without development of a single macroscopic shear fracture; otherwise data represent peak strength of the observed stress-strain relation.

Rock	p_m [MPa] 0	10	20	50	100	150	200	250	300	350	400	500	600	800	1000	Ref.
anhydrite, Alberta, Canada	41	85	118	182												Bra60
anhydrite, Blaine, Oklahoma, USA	128	170	206	280	380	420	515									Han66
basalt, Mendig, Eifel, FRG	120		207	250	280	250										Rum77
basalt, Knippa, Texas, USA	262	292	322	410	540											Bra60
chalk, Annona, Louisiana, USA	28	48		(85)	(86)		(124)		(134)		(150)	(139)				Wel75
claystone, Wyoming, USA	235			420	438											Bra60
coal, Kemmerer, Wyoming, USA ⊥	37			37	63		72		84		90	110	133			Hea76
coal =	24			44	54		67		75		95	102	134			Hea76
diorite, Orikaba, Japan			362	471	606	770	853	894								Mog65
dolomite, Blair, Virginia, USA	343				834		950	980								Rob55
dolomite	157				570		694									Han58
dolomite, Clear Fox, Texas, USA	233				593		680	733								Han58
dolomite, Fusselman, Texas, USA	145				584		714									Han58
dolomite, Glorietta, New Mexico, USA	81				389		569									Han58
dolomite, Hasmark, Montana, USA	122			338	389		550	615	660	695	730	765				Han58, Han66
dolomite, Luning, Nevada, USA	60				487		623				471					Han66
dunitic rocks	200			693	960	1166	1342	1497	1637		1887	2107	2306			Mat61
dunite, Spruce Pine, USA				325	450	570	675		(870)		(1050)	(1250)				Bye68
gabbro, Nahant, USA	280				620		735		830							Bye68
granite, Fichtelgebirge, FRG	122			520	655	785	915	1045	1175		930	(1055)				Rum78c
granite, Barre, Vermont, USA	167		280	475	610											Rob55

(continued)

Table 4 (continued)

Rock	p_m [MPa]															Ref.
	0	10	20	50	100	150	200	250	300	350	400	500	600	800	1000	
granite, Westerly, Rhode Island, USA	230			690	880	1026	1150	1258	1355	1490	1530	1683	1822	2068	2285	Bye67, Bye65
granite, Westerly, Rhode Island, USA	165		435		808		1126		1319		1553	1697	1849			Hea74a
granite, Katashirakawa, Japan	158			480	662	776	890	995	1088	1254						Met61
granite, Orikaba, Japan			400	530	741	850	945									Mog65
granite, Hoggar, Algeria	212			720	905	1100	1250	1350	1460	1525	1590	1660	1755	1850	2040	Sch72
granodiorite, Nevada Test Site, USA	210			320	620		980		1280		1440	1700	1950	2480		Sch72
limestone, Solnhofen, FRG	275				395		(600)		(600)		(800)					Rob55
	347				475	490	490	(593)	(633)		390	(735)				Hea60
							255				320		392	(588)	(941)	Gr.36
limestone, Alabama, USA	53			415	310		(465)					(539)				Han65
limestone, Bedford, Indiana, USA saturated	62			(123)	(171)		(258)		(330)		(363)	(430)	(493)			Hea74
dry	39			(87)	(95)		(95)		(113)		(108)	(117)	(138)			Hea74
limestone, Rutland White, Vermont, USA	39			206	(208)				(427)							Rob55
limestone, Danby, Vermont, USA	45			205	280	360	410	(455)	(48)	(510)	(535)					Rob55
limestone, Marianna, Florida, USA	43			115	252		431									Han53
limestone, New Scotland, New York, USA	128			319					(495)							Rob55
limestone, Knoxville, Texas, USA	124	159	183	(243)												Rum70, Waw68
limestone, Wolfcamp, Texas, USA	82	155	190	273	319											Han53
	111			200	391	490	560									Han53
limestone, "Yule marble", Colorado, USA	39				315		(587)		(550)			(470)			(550)	Han65
limestone, Wombeyan marble, Australia	72	103	120	170	(360)		(530)									Pat58
marble, Carrara, Italy	120		155	230	350		(480)	(495)	(500)							Kar11, Kov74

(continued)

Rummel

Table 4 (continued)

Rock	p_m [MPa]															Ref.
	0	10	20	50	100	150	200	250	300	350	400	500	600	800	1000	
marble, Greece	60	90	120	(190)												Rum75
marble, "Mito", Oota, Japan			105	(160)	(196)		(243)									Mog65
norite, Witwatersrand, S. Africa	275	343	411	614												Cro71
peridotite, "Nabe-ishi", Japan			325	444	607	759	838	994								Mog65
pyrite, Utah, USA	147		315	500	650		780									Rob55
pyroxenite, California, USA													1730			Gri60
quartzite, Sioux, Minnesota, USA	359				1075		1290									Han58
salt, artificial NaCl, USA	61	(81)	(75)	(78)	(79)		(78)		(82)		(86)					Hea75
salt, artificial NaCl, 140 MPa/120 °C at 3% strain																
salt, rock salt, Hockley, Texas, USA	26	77		82	100		112		119							Han53, Han59
sandstone, Ruhr, FRG	140	223	268	370	525											Rum75
sandstone, Mutenberg, FRG	68		180	245	(320)		(380)									Kar11
sandstone, Barns, Texas, USA	40			157	255		407									Han58
sandstone, Bartlesville, Oklahoma, USA	41				228		472									Han58
sandstone, Berea, Ohio, USA	46			159	248		425		575		640					Han66
sandstone, Kayenta, Colorado, USA	32			173	210		253		(300)		(394)	(832)				Dub74
					(246)		(387)		(557)		(629)	repressurized	(885)			Dub74
sandstone, Oil Creek, Texas, USA	96				685		1075									Han58
sandstone, Pinedale, Wyoming, USA	155			533		735		893		(1187)	(1420)	(1735)				Sch73, Sch70
sandstone, Rangely, Colorado, USA	10	152	211	328	460	561	646									Bye75
	60	250	328	484	660	795	909									Bye75

(continued)

Rummel

Table 4 (continued)

Rock	p_m [MPa]															Ref.
	0	10	20	50	100	150	200	250	300	350	400	500	600	800	1000	
sandstone, Rio Blanco, Colorado, USA	74			195	214	222	(230)		(237)				(272)			Den75
sandstone, Gosford, New South Wales, Australia	37–50				(225)		(350)				(630)		(900)			Edm72
serpentinite, Wurlitz, Oberfranken, FRG	127			320	420	520	620	720	820	920	1020					Rum78c
serpentinite, Cabramurra, Australia	307				600		820			1040		(1110)				Ral68
serpentinite, Tumut Pond, Australia			390	470	605		780			1120		(1330)				Ral68
shale, Green River, Colorado, USA ⊥	64			161	210		292									Han58
‖	10			43	58		204									Han58
shale, Muddy, Colorado, USA ⊥	39			137	243		390									Han58
silstone, Red Beds, Texas, USA ⊥	48			147	226											Han66
slate, Mettawee, Vermont, USA ⊥	98				296		450									Han58
slate, Martinsburg, Pennsylvania, USA 30°	50		80	130	210											Don75
45°	58		79	128	200											Don75
tuff, Nevada Test Site, USA	25			38	51	44	(42)	(50)	(58)	(67)	(49)	(87)	(111)			Hea71
tuff, Mt. Helen, Nevada, USA dry	41			(85)	(83)	(143)	(149)	(194)	(277)		(358)	(393)	(444)			Hea73a
50% H_2O	25			44	(80)				(135)			(182)				Hea73a
100% H_2O	22			24	27				(38)			(47)				Hea73a

Table 5. Triaxial compressive strength formulae of standard rocks as a function of confining pressure at room temperature.

Column A, uniaxial strength at $p_m = 0$

Column B, triaxial strength $(S_1 - p_m)_{max}$ at $S_3 = p_m = $ constant

Column C, stress range covered by experiments; tests were performed on cylindrical specimens with diameters of 1 to 6 cm and a length to diameter ratio of about 2.

All strength values as well as $p_m = S_3$ in MPa.

Rock	A	B	C	Ref.
dunitic rocks	200	$200(0.22\,S_3 + 1)^{1/2}$	$0\cdots500$	Mat61
		$50 + 92\,S_3^{1/2}$	$100\cdots500$	
dunite, Spruce Pine, USA		$300 + 1.9\,S_3$	$200\cdots500$	Bye68
granite, Weißenstadt, Oberfranken, FRG	122	$395 + 2.6\,S_3$	$50\cdots300$	Rum78c
granite, Westerly, Rhode Island, USA	230	$230 + 65\,S_3^{1/2}$	$0\cdots1000$	Bye67, Bye69
granite, Kitashirakawa, Kyoto-fu, Japan	150	$150(0.17\,S_3 + 1)^{1/2}$	$0\cdots400$	Mat60, Mat61
		$64\,S_3^{1/2}$	$100\cdots400$	
		$395 + 2.3\,S_3$	$100\cdots400$	
granite, Aare, Switzerland	170	$270 + 5.11\,S_3$	$6\cdots60$	Kov74
limestone, Knoxville, USA, "Tennessee marble"	124	$137 + 2.12\,S_3$	$10\cdots50$	Rum70
limestone, Tardos, Hungaria	85	$119 + 10.78\,S_3$	$2\cdots23$	Bod70
limestone, Süttö, Hungaria	70	$60 + 1.58\,S_3$	$0\cdots23$	Bod70
marble, Carrara, Italy	102	$101 + 2.33\,S_3$	$3\cdots60$	Kov74
			$0\cdots150$	Kar11
norite, Witwatersrand, S. Africa	275	$275 + 6.78\,S_3$	$0\cdots55$	Cro71
sandstone, Witten, Ruhr, FRG	140	$205 + 3.2\,S_3$	$30\cdots100$	Rum75
		$85 + 44\,S_3^{1/2}$		
sandstone, Rangely, Colorado, USA	120	$10 + 45\,S_3^{1/2}$	$0\cdots200$	Bye75
		$60 + 60\,S_3^{1/2}$		
sandstone, Gosford, New South Wales, Australia	50	$85 + 1.35\,S_3$	$50\cdots500$	Edm72
sandstone, Mev, Hungaria	110	$110 + 4.4\,S_3$	$0\cdots23$	Bod70
serpentinite, Wurlitz, Oberfranken, FRG	230	$220 + 2.0\,S_3$	$50\cdots410$	Rum78c
serpentinites, Cabramurra and Tumut Pond, New South Wales, Australia		$330 + 2.4\,S_3$	$20\cdots200$	Ral65

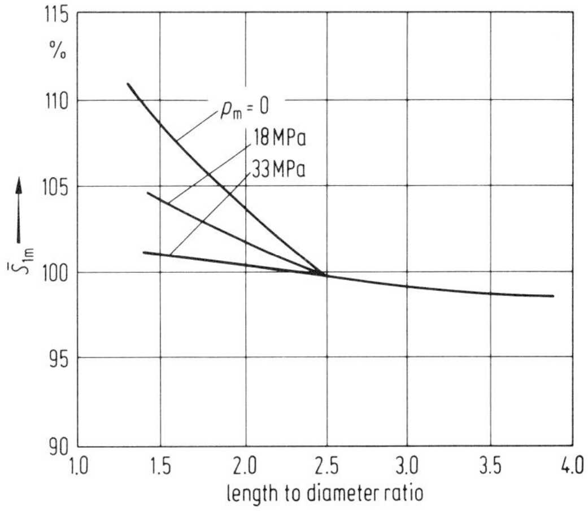

Fig. 5. Relative apparent compressive strength $\bar{S}_{1m}$ of Dunham dolomite, Williamstown, Massachusetts, USA, under various confining pressures p_m as a function of length to diameter ratio of cylindrical specimens. 100 % strength is assumed at length to diameter ratio of 2.5 (after [Mog79]).

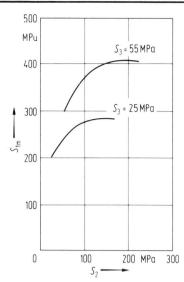

Fig. 6. Influence of the intermediate principal stress on compressive strength. Maximum compressive strength S_{1m} of triaxially loaded $(S_2 \neq S_3)$ specimens of Yamaguchi marble, Japan, as a function of the intermediate principal stress S_2 and of the minimum principal stress S_3 (after [Mog79]).

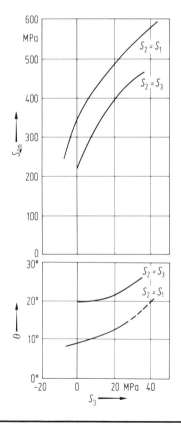

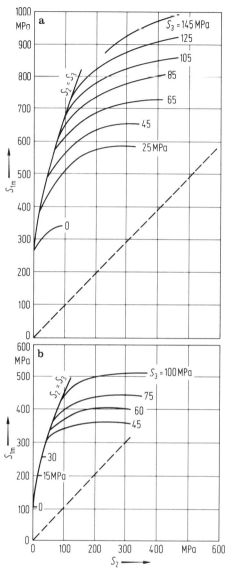

Fig. 8a, b. Maximum compressive strength S_{1m} of Dunham dolomite (a) and Mizuho trachyte (b) under general triaxial loading as a function of intermediate and minimum principal stresses S_2 and S_3 (after [Mog79]).

◀

Fig. 7. Maximum compressive strength S_{1m} and fracture angle θ of Dunham dolomite under general $(S_2 \neq S_3)$ triaxial loading as a function of intermediate and minimum principal stresses S_2 and S_3. θ: angle between S_1 and the shear fracture plane (after [Mog79]).

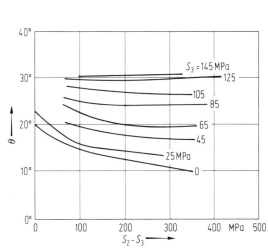

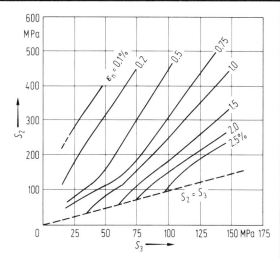

Fig. 9. Fracture angle θ of triaxially loaded specimens of Dunham dolomite as a function of intermediate and minimum principal stresses S_2 and S_3 (after [Mog79]).

Fig. 10. Non-elastic axial strain (permanent strain) ε_n of triaxially loaded Dunham dolomite just before fracture as a function of the minimum principal stress S_3 and the intermediate principal stress S_2 (after [Mog79]).

Table 6. Tensile strength σ_t of rocks at room temperature. (For methods, see Fig. 2.)

dp   direct pull test on cylindrical or dog-bone specimens
br   Brazilian test by diametral compression of circular rock discs
r   ring test by diametral compression of hollow rock discs
b   bending test by three-point loading of rock beams
cl   cleavage test on notched beams
hf   hydraulic fracturing test by fluid injection into drillholes in thick walled cylinders in laboratory tests or fluid injection into *in situ* boreholes

Rock type	Location and rock description	σ_t [MPa]		Method	Ref.
		Range	Mean		
sediments					
chalk	Yorkshire, England; Cenomanian, $n = 29\%$	1.0···3.4	2.6	br	Bel77
	Norfolk, England; Turonian, $n = 33\%$	0.9···2.8	2.1	br	Bel77
	Kent, England; Senonian, $n = 47\%$	0.4···0.6	0.5	br	Bel77
conglomerate	Calumet and Hecla mine, Michigan, USA; copper bearing	3.8		dp	Kre74
limestone	England, (?), calcite, saccharoidal	10···15		r	Pri66
	Eichstätt, Bavarian Jura, FRG; "Treuchtlinger Marmor", 99% calcite, micritic, homogeneous, $\varrho = 2.55$ gcm^{-3}	10···21		hf	Jun76
	Solnhofen, Bavarian Jura, FRG; 99% calcite, micritic, very homogeneous	21		dp	Bra64
	Bedford, Indiana, USA; fine-grained, 99% calcite, $n = 14\%$, "Indiana limestone", $\varrho = 2.29$ gcm^{-3}	2.5···4.5	3.5	dp	Har70
		4.6···5.8	5.2	br	Nar70
		2.5···3.1	2.8	br	Hea74
		8.8···10.8	9.8	r	Har70
		5.4···6.2	5.8	hf	Har70
(continued)	Salem, Massachusetts, USA; notched specimen		4.6	dp	Kre74

Table 6 (continued)

Rock type	Location and rock description	σ_t [MPa]		Method	Ref.
		Range	Mean		
limestone (continued)	Knoxville, Tennessee, USA; "Tennessee Marble", 98% calcite, 2% clay cement	7.3···11.4		dp	Waw68
		5.9···6.2		dp	Thi75
		3.8···6.5		dp	Kim69
		8.4···10.2	9.3	br	Hai78
	Lueders, Texas, USA	5.0···5.8	5.4	dp	Har70
		5.4···6.4	5.9	br	Har70
		7.9···9.5		r	Har70
		7.0···8.6	7.8	hf	Har70
	Texas, USA; "Cordova Cream"		⊥1.7	dp	Har72
			‖ 2.7	dp	Har72
	Mankato Stone Company, USA; dolomitic limestone		15	dp	Har72
	Holston, USA		9	dp	Kre74
sandstone	Gosford, New South Wales, Australia; fine-grained, quartz and feldspar, weakly cemented in clay matrix		3.6	dp	Jae66
			3.7	br	Jae66
			8.3	r	Jae66
			7.9	b	Jae66
	Witten, Ruhr-district, "Ruhr-Sandstein", FRG, fine to medium grained, homogeneous, $n = 5\%$	13···18	17.6	hf	Jun76
			24.3	hf	Zob77
	Izumi, Japan		11.6	br	Yam70
		26···29.5		r	Yam70
	Rangely Oil Well, Colorado, USA; from 2000 m depth, fine-grained, homogeneous, $n = 5\%$	9.2···11.2		br	Hai75
		11.2···15.2	13	hf	Hai75
	Berea, USA	1.1···4.7		dp	Kre74, Rol74, Har72, Har70
		2.8···3.4		br	Har70
		4.9···5.9		r	Har70
		3.8···4.0		hf	Har70
	Montrose, USA; "Pennsylvania Bluestone", 50% quartz, 5% feldspar, 45% fragments		⊥4.8	dp	Har72
			‖ 8.7	dp	Har72
			⊥6.9	br	Har72
			‖ 9.3	br	Har72
	Crab Orchard, USA; fine-grained 84% quartz, $n = 5.5\%$	6.2···7.6		dp	Har70
		10.9···12.3		br	Har70
		13.5···15.7		r	Har70
		11.3···12.5		hf	Har70
	Kayenta formation, Mesa County, Colorado, USA; 90% quartz, poorly consolidated, medium to coarse grained, $n = 24\%$	1.4···1.6	1.5	br	Dub74
	CER Geonuclear RB-E-01 hole, Rio Blanco County, Colorado, USA; from 1782 and 1968 m depth, $n = 7\%$, fine-grained graywacke, 50% water saturated	3.3···4.1	3.7	br	Sch72a
		3.7···4.6	3.9	br	Sch72a
(continued)	Equity S. Sulphur Creek hole, Rio Blanco County, Colorado, USA; from 1942 m depth, fine-grained graywacke, $n = 18\%$, dry	3.5		br	Sch72a

Table 6 (continued)

Rock type	Location and rock description	σ_t [MPa]		Method	Ref.
		Range	Mean		
sandstone (continued)	Tennessee, USA; "Tennessee sandstone"		6.9	dp	Har72
		5.9···6.2		dp	Kim69
tuff	Aoishi, Japan	3.6···4.8		br	Yam70
	Diamond Dust Site, Nevada Test Site		0.9	br	Ste70
	Nevada, USA; $n=9\%$, sample from		0.85	br	Ste70
	8, 42, 109, and 112 m depth		0.55	br	Ste70
		0.5		br	Ste70
	Mt. Helen, Nevada, USA; fine-grained,	2.7···4.3 (dry)		br	Hea73a
	$n=37\%$, $\varrho \approx 1.46$ gcm^{-3}	2.5···2.8 (25% H_2O)		br	Hea73a
		1.3···1.7 (75% H_2O)		br	Hea73a
		1.3···2.0 (100% H_2O)		br	Hea73a
igneous rocks					
andesite	Shinkomatsu, Japan	5.7···7.0		br	Yam70
basalt	Dresser, USA; notched specimen		21.7	dp	Kre74
diabase	Frederick, Maryland, USA; dog-bone specimen	40		dp	Bra64
dolerite	England	60···70		r	Pri66
eclogite	Stammbach, Bavaria, FRG		15	hf	Rum78
		1.5···5.0		hf (in situ)	Rum78
gabbro	Sweden; medium grained, very dense		26.4	hf	Zob77
	Buena, USA		16.7	dp	Rol74
	Ely, Minnesota, USA		17	dp	Sch70a
			16	br	Sch70a
		25···32		hf	Sch70a
granite	Carnmenellis, Cornwall, England medium grained	1.9···2.1	2.0	br	Bat77
	Weißenstadt, Fichtelgebirge, FRG medium grained		16.7	hf	Rum78
	Falkenberg, Bavaria, FRG coarse grained	13.3···19.8	15.1	hf	Rum78
		2.4···5.2		hf (in situ)	Rum78
	Inada, Japan	9.2···11.6		br	Yam70
	Chelmsford, Massachussetts, USA fine-grained quartz monzonite	6.4···8.3		br	Pen72
	St. Cloud, Minnesota, USA;	1.1···13.6		dp	Thi75
	"Charcoal Gray", medium-grained		7.0	dp	Kre74
	Granodiorite	12.5···16.5		br	Har73
		25···31		r	Har73
		11···23		b	Har73
	St. Cloud, Minnesota, USA;		11.1	br	Sch70a
	"Coldspring Pink		24.2	hf	Sch70a
	Missouri, USA	9.9		?	Rol74
	Kitledge, USA	11.2		?	Rol74
	Westerly, Rhode Island, USA;		8.4	dp	Kre74
	homogeneous, fine-grained	21		dp	Bra64
		11···12	11.5	br	Hea74a

(continued)

Table 6 (continued)

Rock type	Location and rock description	σ_t [MPa]		Method	Ref.
		Range	Mean		
granite (continued)	Barre, Vermont, USA; granodiorite with 1···5 mm grain size		9.4	dp	Har70
			7.7	dp	Kre74
			11.5	br	Har70
			18.5	r	Har70
			10.6	hf	Har70
norite	norite, South Africa		18.8	?	Bie68a
syenite	porphyry, Kiruna, Sweden	11···18		br	Han66a
spessarite			17.2	dp	Rie76
trachyte	Bowral, Australia		13.7	dp	Jae66
			12.0	br	Jae66
			25.2	b	Jae66
			24.0	r	Jae66
metamorphic rocks					
gneiss	Urach, Geothermal Drillhole, FRG; from 2300 m depth, paragneiss		18	hf	Rum78
marble	Carrara, Italy		6.9	dp	Jae66
			8.6	br	Jae66
			11.6	b	Jae66
			17.3	r	Jae66
	Gualba, Barcelona, Spain	2.6···3.9		dp	Mon74
		4.1···5.9		br	Mon74
	Rutland, Vermont, USA	5.4		dp	Bra64
quartzite	Witwatersrand, South Africa; Silfontein Mine; 69% quartz, medium grained, recryst. sandstone	22.8		dp	Bie68
	Cheshire, USA	28		dp	Bra64
	Jasper, USA	25.6		dp	Rol74
	Sioux, USA		9.2	dp	Kre74
			18.4	?	Cai74
shale	White Pine Mine, Michigan, USA		33.8	hf	Sch70a

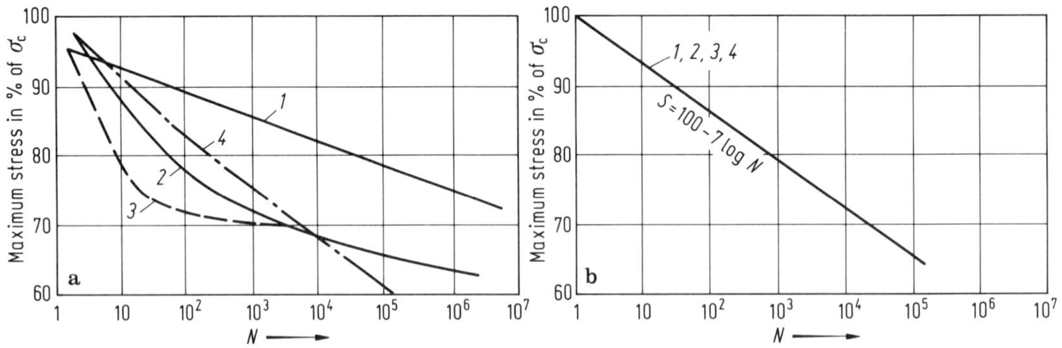

Fig. 11 a, b. Fatigue strength of rocks under cyclic loading [Hai78].
a) cyclic compression (σ_c uniaxial compressive strength, frequency 1 Hz);
b) cyclic tension (σ_t uniaxial tensile strength, frequency 1 Hz).
1 Tennessee "marble"; 2 Indiana limestone; 3 Westerly granite; 4 Berea sandstone; N = Number of cycles.

Table 7. Hydraulic fracturing strength p_c of rock as a function of confining pressure.

p_c is obtained from laboratory hydraulic fracturing tests where fluid is injected into axial boreholes of cylindrical rock specimens compressed by a confining pressure $S_2 = S_3 = p_m$.

p_c defines the critical fluid pressure for unstable tensile fracture propagation. Experimental results may be characterized by

$$p_c = p_{co} + k\,p_m$$

where p_{co} is the hydraulic fracturing tensile strength of externally unloaded specimens and k is the frac gradient with respect to p_m. $\dot{p}$ is the rate of pressurization.

Rock	Location	p_{co} MPa	k	$\dot{p}$ MPa s^{-1}	p_m-range MPa	Ref.
granite	Epprechtstein, Fichtel-gebirge, FRG	13.4	1.15	0.175	0···30	Rum78a
	Falkenberg, NE Bavaria, FRG	15.4	1.06	0.21	0···50	Rum78a
gneiss	Urach Research Drill-Hole, depth 2130 m, FRG	18.0	1.11	0.21	0···60	Rum78a
marble	NE Greece, trade name in FRG "Alexander"	12.8	1.16	0.175	0···30	Rum78a
sandstone	Ruhr, Witten, FRG	17.6	1.14	0.175	0···30	Rum78a
	Ruhr, borehole Werne 6, depth 1276 m, FRG	18.8	1.19	0.21	0···60	Rum78b
	Ruhr, borehole Werne 6, depth 1314 m, FRG	19.9	1.13	0.21	0···60	Rum78b
	Ruhr, borehole Werne 6, depth 1320 m, FRG	20.9	1.18	0.21	0···60	Rum78b
	Ruhr, borehole Werne 7, depth 1024 m, FRG	25.9	1.26	0.21	0···60	Rum78b
conglomerate	Ruhr, borehole Werne 7, depth 1130 m, FRG	23.4	1.19	0.21	0···60	Rum78b
shale	Ruhr, borehole Werne 7, depth 981 m, FRG	24.7	1.20	0.21	0···60	Rum78b

Table 8. Specific fracture surface energy γ of rocks and minerals at room temperature and atmospheric pressure.

Specific surface energy values with asterisk are calculated from thermodynamics, values in parentheses are calculated from K_{IC} data.

cl = cleavage type test; dt = double torsion; nb = three- or four-point bend test on notched specimens; dp = direct pull; ndp = direct pull on notched specimen; ind = indentation test; ts = thermal shock; hf = hydraulic fracturing.

Rock/mineral	Location/name	γ J m^{-2}	Method	Ref.
aragonite	(010)	0.25*		Bra62
barite	(001)	0.48*		Bra62
basalt	"Dresser", USA	50	ndp	Kre74
calcite	(10$\bar{1}$0)	0.38*	d	Gil60
	(10$\bar{1}$1), 77 °C	0.23	cl	Gil60
	(10$\bar{1}$1)	0.347	cl	San68
conglomerate	Calumet and Hecla mine, Michigan, USA	50	dp	Kre74
feldspar	orthoclase, (001)	0.20*		Bra62
		7.77	cl	Bra62
fluorite	CaF$_2$, (111)	0.54*		Gil60
	CaF$_2$, (111)	0.450	cl	Gil60

(continued)

Table 8 (continued)

Rock/mineral	Location/name	γ $J\,m^{-2}$	Method	Ref.
firebrick	Chamotte	30	nb	Nak65
	high alumina content	49	nb	Nak65
	silica-rich	3.0	nb	Nak65
	alkaline	4.2	nb	Nak65
gabbro	Black gabbro	(41)	dt	Atk79a
galena	PbS, (100)	0.625*		Gil60
	(110)	0.440*		Gil60
glass	soda lime glass, thermal shock		ts	Kin71
	soda lime glass	3.8		Wie69
	soda lime glass	4.33	ind	Swa78
	plate glass	3.4···5.5	nb	Nak65
gneiss	Urach, FRG, Geothermal drillhole at 3000 m depth	17···22	nb	Rum79
granite	Epprechtstein, Fichtelgebirge, FRG	29	nb	Win79
	Falkenberg, Bavaria, FRG	7	nb	Rum79
	Carnmenellis, Cornwall, England	121	?	Bat77
	Chelmsford, USA	50	nb	For68
	Barre, Vermont, USA	60	dp	Kre74
		9···38	nb	Sum71
	Charcoal Gray, Minnesota, USA	16	cl	Waw68
	Granodiorite, St. Cloud Gray, Minnesota, USA	102	dp	Kre74
	Orthoclase Pink granite	(22)	dt	Atk79b
	Westerly, Rhode Island, USA	(28)	dt	Atk79b
		30	ind	Swa76
		(7.5)	hf	Zob78
		(58)	nb	Schm78
		139	ndp	Kre74
halite	NaCl, (100) plane	0.33	cl	Gil59
	(100) plane	0.31*		Gil59
	(110) plane	0.345*		Gil59
hematite	Fe_2O_3, (0001) plane	0.98*		Bra62
limestone	Solnhofen, FRG	5.1	nb	Win79
		(6.3···8.4)	dt	Atk79b
	Treuchtlinger „Marmor", FRG	8.0	nb	Win79
		16	nc	Rum75
	Lueders, Texas, USA	9···12	nb	Fri72
		19	cl	Per63
	Austin, USA	8	cl	Per63
	Carthage, USA	39	cl	Per63
	Holston, USA	23	ndp	Kre74
	Tennessee, "Marble", USA	10	cl	Waw68
	Salem, Indiana, USA	16···22	nb	Fri72
		(28)	nb	Sch76
		(14···28)	nb	Sch75
		42	cl	Per63
	Danby	50		For68
	Falerans micrite	(6.7)	nb	Hen77
marble	Carrara, Italy	(3.5)	dt	Atk79b
	Ekeberg, Sweden	20	nb	Ouc80
muscovite	(001) plane	0.375	cl	Obr30

(continued)

Rummel

Table 8 (continued)

Rock/mineral	Location/name	γ $J\,m^{-2}$	Method	Ref.
olivine	forsterite (010)	0.37*		Bra62
	88 % forsterite, 12 % fayalite (010)	8.63*		Swa78
		0.98		Swa78
	88 % forsterite, 12 % fayalite (001)	12.06*		Swa78
		1.26		Swa78
orthoclase	single crystal	7.8	ind	Atk79b
periclase	MgO, (100) plane	1.310*		Gil59
	(100) plane	1.2	cl	Gil60
	(110) plane	2.330		Gil59
pyrite	FeS, (100)	1.55*		Bra62
quartz, single crystal	$(10\bar{1}1)$	0.76*		Bra62
		0.41	cl	Bra62
	$(\bar{1}011)$	0.45*		Bra62
		0.50	cl	Bra62
	$(10\bar{1}0)$	1.03	cl	Bra62
quartz, synthetic	a plane, $\perp z$	(3.49)	dt	Atk79b
	a plane, $\perp r$	(4.83)	dt	Atk79b
quartzite	Chilhowee, USA	50	nb	Fri72
	Sioux, Minnesota, USA	89	ndp	Kre74
	Arkansas novaculite	(12)	dt	Atk79b
rutile	TiO_2	1.42*		Bra62
sandstone	Ruhr sandstone, Witten, FRG	27.7	nb	Win79
	carbonian sandstone from drill-hole	30.8	nb	Rum80
	Werne 7, FRG	31.8	nb	Rum80
	Werne 6	36.8	nb	Rum80
		47.3	nb	Rum80
		44.9	nb	Rum80
	Balderhaar Z1, BH7	7.2	nb	Rum80
	Tennessee sandstone, USA	38	nb	Fri72
		88	cl	Per66
		197	cl	Per66
	Berea, USA	(57)	hf	Zob78
		20	ndp	Kre74
	Coconino, USA	25	nb	Fri72
		127	cl	Per63
	Milsap, USA	114	cl	Per63
	Colorado, USA	96	cl	Per63
	Woodbine	9.5	cl	Per63
	Torpedo	61	cl	Per63
	Boise	44	cl	Per63
sphalerite	ZnS, (110)	0.36		Gil59

Table 9 Fracture toughness K_{IC} and critical strain energy release rate G_{IC} for rocks and minerals.

G_{IC} values are found by the relation $G_{IC} = K_{IC}(1-v^2)/E$ or from measured specific fracture surface energy $\gamma = G_C/2$. Calculated data are presented in parentheses and should be treated with caution.

HF = hydraulic-fracturing test on thick-walled cylindrical specimens; 3 PB = three-point bend test of beams or cores; CL = double cantilever beam test; VI = Vickers indentation test; DT = double torsion test (v = Poisson's ratio, E = Young's modulus).

Material	Location	K_{IC} MNm$^{-3/2}$	G_{IC} J m^{-2}	Method	Ref.
aluminum		25	8000		Ouc80
aluminum	alloy 2124-T851	31.1			Ouc80
calcite	(0$\bar{1}$10)-plane, 77 °C	(0.16)	(0.46)	CL	Gil60
	(0$\bar{1}$10)-plane		0.5	CL	Ouc80
	(0$\bar{1}$10)-plane	(0.20)	(0.70)	CL	San68
fluorite	(111)-plane	(0.30)	(0.90)	CL	Gil60
feldspar	orthoclase (001)-plane	(1.30)	(15.54)	CL	Bra62
gabbro	Black Gabbro	2.884	(82)	DT	Atk79a
	Academy Black, USA	0.87		HF	Zob78
glass	silica glass	0.753		3 PB	Wie74
	soda-lime glass	(0.75)	(7.60)		Wie69
		(0.79)	(8.66)	VI	Swa78
gneiss	Urach, FRG, Geothermal drill-	1.62···1.74		3 PB	Rum79
	hole at 3000 m depth	1.46	29.4	3 PB	Rum80
granite	Epprechtstein, Fichtelgebirge, FRG	1.60	58.4	3 PB	Win79
	Falkenberg, Bavaria, FRG	0.78···0.86	14	3 PB	Rum80
	Carnmenellis, Cornwall, England	2.158		DT	Bat77
	Chelmsford, Massachusetts, USA	0.57···0.64	(6···12)	3 PB	Pen72
	Westerly, Rhode Island, USA	0.90	(15)	HF	Zob78
		1.74	(56)	DT	Atk79b
		(1.79)	(60)	VI	Swa76
		2.5	(117)	3 PB	Schm78
		(3.9)	(278)		Kre74
	Sierra White, USA	0.79		HF	Zob78
	Orthoclase Pink granite	1.66	(44)	DT	Atk79b
	Charcoal, Minnesota, USA		(32)	CL	Waw68
	Barre, USA	(0.75···1.5)	(19···77)	3 PB	Sum71
	Bohus 1, Sweden	1.4		3 PB	Ouc80
	Bohus 2, Sweden	1.32		3 PB	Ouc80
	Stripa-granite	1.8		3 PB	Ouc80
	Finnsjöng-granodiorite	2.5		3 PB	Ouc80
	Syenite-porphyry	2.3		3 PB	Ouc80
halite	NaCl (100)-plane	(0.18)	(0.66)	CL	Gil59
ice	−13 °C	0.116	1.5	3 PB	Goo78
	−16 °C	> 0.30	> 10		Goo78
	−20 °C	0.24	7		Goo78
	−20 °C	0.21	5	VI	Goo78
	−38 °C	0.070	0.5	VI	Goo78
	−12° to −4 °C	0.123···0.149	1.7···2.5		Lui78
	−12° to −4 °C	0.177···0.222	3.6···5.8	CL	Lui78
	−45° to −4 °C	0.140···0.100	3···1	CL	Lui78
	−16.7 °C	0.053···0.060	0.3···0.4	thermal shock	Gold63

(continued)

Rummel

Table 9 (continued)

Material	Location	K_{IC} MNm$^{-3/2}$	G_{IC} J m^{-2}	Method	Ref.
limestone	Solnhofen, FRG	0.78	10.2	3 PB	Win79
	Solnhofen, FRG	0.87···1.006	12.5···16.7	DT	Atk79b
	Treuchtlinger Marmor, FRG	0.99	16	3 PB	Win79
		(1.39)	(31.8)	CL	Rum75
	Lueders, Texas, USA	0.37···0.58	(6.6···16)	HF	Zob78
		(0.61···0.71)	(18···24)	3 PB	Fri72
		(0.89)	(38)	CL	Per63
	Austin, Texas, USA	(0.44)	(16)	CL	Per63
	Carthage, USA	(1.65)	(77)	CL	Per63
	Holston, USA		(46)		Kre74
	"Tennessee Marble", USA		(20)	CL	Waw68
	Falerous micrite, France	1.01	13.5	3 PB	Hen76
marble	Carrara, Italy	0.644	(6.9)	DT	Atk79
	St. Pons marble, France	0.70···1.39	10···42	3 PB	Hen76
	Ekebergs, Sweden	1.5	40	3 PB	Ouc80
muscovite	(001)-plane	(0.21)	(0.75)	CL	Obr30
olivine	88 % forsterite, 12 % fayalite				
	(010)-plane	(0.59)	(1.96)		Swa78
	(001)-plane	(0.73)	(2.52)		Swa78
periclase	MgO (100)-plane	(0.79)	(2.4)	CL	Gil60
plexiglas	PMMA	1.5			Ouc80
quartz	synthetic, a-plane ⊥z	0.852	(6.98)	DT	Atk79b
	⊥r	1.002	(9.65)	DT	Atk79b
quartzite	Arkansas novaculite	1.335	(24)	DT	Atk79
sandstone	Ruhr sandstone, Witten, FRG	1.43	55.4	3 PB	Win79
	Ruhr sandstone,				
	drillhole Werne 7	0.92···1.24		3 PB	Rum80
	Werne 6	0.88···1.14		3 PB	Rum80
	drillhole Balderhaar Z1, Bh6	0.47		3 PB	Rum80
	Tennessee, USA	(1.43)	(76)	3 PB	Fri72
		0.454	(7.22)	DT	Atk79b
	Berea, USA	0.28	113	HF	Zob78
		(0.17)	(40)		Kre74
	Coconino, USA	(0.65)	(50)	3 PB	Fri72
		(1.46)	(254)	CL	Per63
	Milsap, USA	(1.55)	(228)	CL	Per63
	Colorado, USA	(2.53)	(192)	CL	Per63
	Woodbine, USA	(0.06)	(19)	CL	Per63
	Torpedo, USA	(1.0)	(122)	CL	Per63
	Boise, USA	(0.67)	(88)	CL	Per63
	coarse-grained porous	0.47···1.42		HF	Cli76
shale	USA	1.29		HF	Cli76
	$n = 0.2 \%$	0.95		HF	Cli76
	$n = 3.2 \%$	0.61		HF	Cli76
	$n = 6.7 \%$	0.87		HF	Cli76
	oil shale, Colorado, USA typ A				
	normal to bedding	1.0···1.1		3 PB	Sch75
	subparallel to bedding	0.92···0.95		3 PB	Sch75
(continued)	parallel to bedding	0.75		3 PB	Sch75

Table 9 (continued)

Material	Location	K_{IC} MNm$^{-3/2}$	G_{IC} J m^{-2}	Method	Ref.
shale (continued)	oil shale, Colorado, USA, typ B				
	normal to bedding	0.67		3 PB	Sch75
	subparallel to bedding	0.62		3 PB	Sch75
	parallel to bedding	0.35···0.45		3 PB	Sch75
	Rome Basin Grey shale	1.2···1.34			Abo78
steel	construction steel	50	10000		Ouc80

For Fig. 12, see next page.

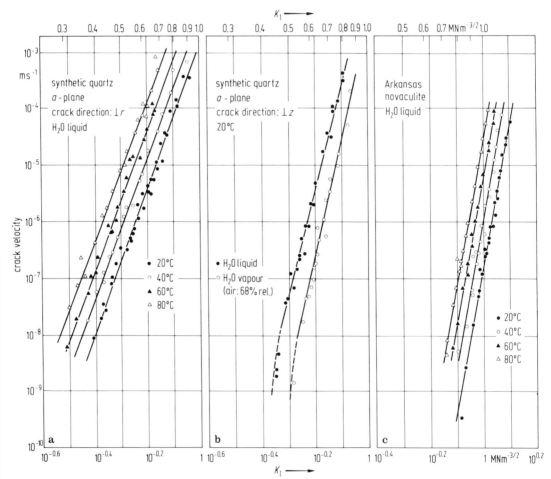

Fig. 13a···c. Crack growth in quartz single crystals and in Arkansas novaculite (fine grained quartz rock) in presence of moist air and water at temperatures up to 80 °C as a function of stress intensity factor $K_1 < K_{IC}$ (data are taken from [Atk78]).

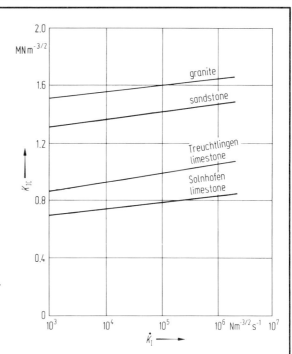

Fig. 12. Fracture toughness K_{IC} of rocks as a function of stress intensity rate $\dot{K}_I$ (after [Win79], Fichtelgebirge granite, Ruhr sandstone, Treuchtlinger "marble", Solnhofen limestone).

3.2.2 Friction — Reibung

3.2.2.1 Introduction — Einleitung

3.2.2.1.1 General remarks — Allgemeine Bemerkungen

Although friction may be considered as an integral part of fracture, friction data here are presented separately. Friction is assumed to play a dominant role in geosciences and in applied rock mechanics. Its effects may arise on all scales, on minute Griffith crack surfaces in otherwise intact rock, along grain boundaries, on existing joint surfaces in rock masses or on fault and fracture planes, and lithospheric plate boundaries. Frictional effects may cause dissipative attenuation of seismic waves and determine the stability of rock slopes as well as of large fault structures where instabilities occur as earthquakes.

Because of increasing engineering application needs and research in earthquake source physics, much experimental work has been done on friction of minerals and rocks during the last two decades. Much of this work is summarized in detailed review articles [Jae66, Bar76, Bye78] which should be used as introduction to the subject.

Obwohl der Bruch im allgemeinen Reibungsprozesse bereits beinhaltet, wird in dieser Darstellung die Reibung getrennt behandelt. Reibungsprozesse spielen in vielen geowissenschaftlichen Problemen sowie in der angewandten Felsmechanik eine bedeutende Rolle. Sie sind sowohl im mikroskopischen Bereich an den Oberflächen von Griffith-Rissen und an Korngrenzen im Gestein, als auch an Trennklüften, Verwerfungs- und Bruchstrukturen der obersten Kruste und an den Grenzen von Lithosphärenplatten wirksam. Reibung wird als eine der möglichen Ursachen für die Dämpfung seismischer Wellen angesehen und bestimmt die Stabilität von Böschungen oder Großverwerfungsstrukturen, wo Instabilität in der Form von Erdbeben auftritt.

In den vergangenen zwei Jahrzehnten wurden aufgrund der Anforderungen der Ingenieurwissenschaften und bei der Erforschung der Physik von Erdbeben zahlreiche experimentelle Arbeiten zur Untersuchung der Reibungsprozesse durchgeführt. Ein großer Teil dieser Arbeiten ist in detaillierten Überblicksartikeln [Jae66, Bar76, Bye78] dargestellt. Sie sollten als Einführung in das Gebiet herangezogen werden.

3.2.2.1.2 Definitions — Definitionen

Internal Friction

Shear failure of intact rock may be described by the Coulomb-criterium $|\tau| = \tau_0^* + \mu^* \sigma$ (Fig. 14), where τ and σ are the shear stress and the normal stress acting across the fracture plane, and τ_0^* and μ^* are constants. In analogy with ordinary frictional sliding, μ^* is called the *coefficient of internal friction*, and τ_0^* is called *cohesion or inherent shear strength*. Experimental data for μ^* and τ_0^* usually are derived from triaxial compression tests on cylindrical rock specimens (see Tables 1⋯4) with

$$\sigma = 0.5\,(S_1 + S_3) + 0.5\,(S_1 - S_3) \cos 2\beta \quad \text{and/und}$$
$$|\tau| = 0.5\,(S_1 - S_3) \sin 2\beta,$$

where S_1 is the axial stress at shear fracture initiation, S_3 is the confining pressure and β is the angle between S_1 and the normal of the fracture plane. Instead of μ^* one often uses the *angle of internal friction*, ϕ^*, where

$$\phi^* = \tan^{-1} \mu^*, \quad \text{or/oder} \quad \mu^* = \tan \phi^*.$$

If μ^* and τ_0^* are determined from direct shear tests (see Fig. 15), τ is the shear stress at shear fracture initiation and σ is the acting normal stress. The linear Coulomb criterium is a good approximation for many rocks at low and intermediate stresses.

Statische innere Reibung

Die Scherbruchbedingung von intakten Gesteinen kann häufig durch das Coulomb'sche Scherbruchkriterium $|\tau| = \tau_0^* + \mu^* \sigma$ angegeben werden (Fig. 14), wobei τ und σ die Scher- und Normalspannung auf der Scherbruchfläche sind. In Analogie zur Reibung auf Scherflächen werden die Konstanten μ^* als *Koeffizient der inneren Reibung* und τ_0^* als *Bruchflächenkohäsion* bezeichnet. Die experimentellen Daten für μ^* und τ_0^* werden meist aus den Ergebnissen triaxialer Druckversuche an zylindrischen Gesteinsproben abgeleitet (vergleiche Tabelle 1⋯4), wobei

S_1 ist die axiale Spannung bei Scherbruchbeginn und S_3 ist der Manteldruck. β ist der Winkel zwischen S_1 und der Bruchflächennormalen. Anstelle von μ^* wird häufig auch der *Winkel der inneren Reibung*, ϕ^*, angegeben, wobei

Falls μ^*, ϕ^* und τ_0^* aus direkten Scherversuchen (vergleiche Fig. 15) gewonnen werden, ist für τ der Wert der Scherspannung bei Scherbruchbeginn und für σ die wirksame Normalspannung einzusetzen. Das lineare Coulombkriterium stellt für die meisten Gesteine bei niedrigen und mittleren Normalspannungen eine gute Näherung zur Beschreibung des Scherbruchs dar.

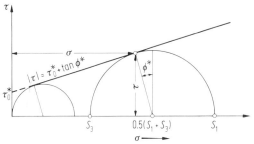

Fig. 14. Shear stress (τ) vs. normal stress (σ) diagram for the presentation of the Coulomb criterion, $|\tau| = \tau_0^* + \sigma \tan \phi^*$ of shear fracture of intact rock. S_1, S_3 principal stresses. Note, the criterion of fracture does not involve the intermediate principal stress S_2.

Fig. 15. Experiments used to determine shear strength parameters of rocks. N = normal force, F = shear force.
a: direct shear used in low normal stress experiment on large rock samples;
b: symmetric shear used for intermediate normal stress experiments;
c: triaxial test used for high temperature and high normal stress experiments;
d: double torsion used for large shear displacement tests;
e: test situation *in situ* at low normal stresses.

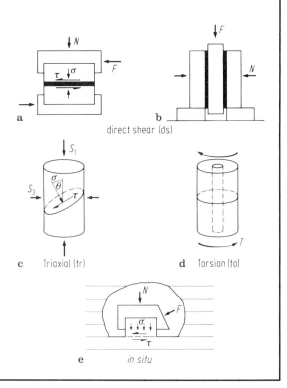

Frictional Sliding

Data for characterizing the frictional behaviour of shear surfaces (polished or saw-cut rock surfaces, fracture surfaces, natural joint surfaces) are determined by various testing configurations (Fig. 15). The coefficient of sliding friction, μ, is then defined as $\mu = \tau/\sigma$, where τ und σ are the shear and the normal stress on the shear plane. Data for μ presented here, in general, refer to initial frictional sliding. Since frictional behaviour changes during shear the coefficient of sliding friction may reach a residual value after extensive sliding. This value is noted as μ_r.

As may be seen from Fig. 16a, the peak shear stress for initial sliding friction of rock surfaces and thus μ may vary significantly at low normal stresses depending mainly on the roughness of the shear surface, but also on the rock type and water content on the shear surface. At high normal stresses (Fig. 16b) friction is nearly independent of the rock type, the normal stress and the roughness of the shear surface. At normal stresses above 200 MPa the coefficient of sliding friction of most rocks is about 0.6 ($\phi = 31°$). The variation of μ with normal stress or with confining pressure is illustrated in Fig. 17 and Fig. 18 for ground surfaces in Westerly granite and for fracture surfaces in peridotite.

Reibung auf Scherflächen

Die Werte zur Beschreibung der Reibung auf Scherflächen (künstliche polierte oder geschliffene Gesteinsoberflächen, Bruchflächen, Kluftflächen) werden mittels verschiedener Versuchsmethoden bestimmt (Fig. 15). Dabei ist der Koeffizient der Gleitreibung μ als $\mu = \tau/\sigma$ definiert, wobei τ und σ die wirksame Scher- und Normalspannung sind. Im allgemeinen beziehen sich die hier angegebenen Werte für μ auf den Beginn des Scherprozesses. Da die Reibung sich während des Scherprozesses im allgemeinen ändert, erreicht der Reibungskoeffizient erst bei großen Scherwegen einen konstanten Wert, der als *Restreibung* μ_r bezeichnet wird.

Wie aus Fig. 16a zu sehen ist, streut die Spitzenscherspannung und damit der Reibungskoeffizient von Gesteinen bei niedrigen Normalspannungen signifikant. Dies ist im wesentlichen durch die Beschaffenheit (Oberflächenrauhigkeit) der Scherfläche, aber auch vom Gestein oder durch das eventuelle Vorhandensein von Wasser auf der Scherfläche bedingt. Dagegen ist die Reibung bei hohen Normalspannungen (Fig. 16b) nahezu unabhängig von der Oberflächenrauhigkeit der Scherflächen, dem Gestein und der wirksamen Normalspannung. Bei Normalspannungen von $\sigma > 200$ MPa ist der Reibungskoeffizient der meisten Gesteine angenähert 0.6 ($\phi = 31°$). Die Variation von μ als Funktion der Normalspannung oder des Manteldrucks ist in Fig. 17 und Fig. 18 für geschliffene Scherflächen in Granit und für Scherbruchflächen in Peridot verdeutlicht.

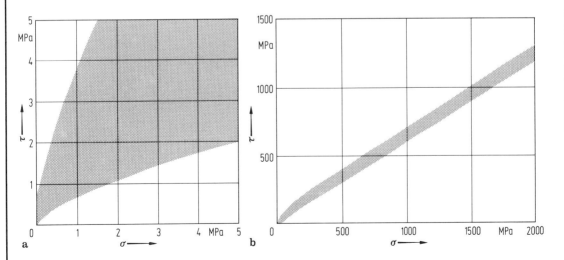

Fig. 16a, b. a) Peak shear stresses, τ, for sliding friction on joint surfaces at low normal stresses, σ, may vary significantly depending on rock type, surface roughness, water content and effective normal stresses.
b) Peak shear stresses, τ, at high normal stresses, σ, are almost independent on rock type and surface roughness, the coefficient of sliding friction, μ, is about 0.6 independent of the effective normal stress [after Bye78].

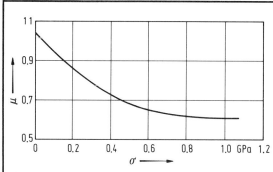

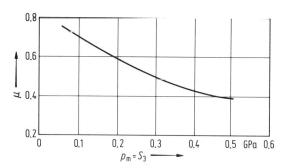

Fig. 17. Coefficient of sliding friction, μ, as a function of normal stress σ for ground surfaces in Westerly granite. Data for low stresses ($\sigma < 0.2$ GPa) are from torsion tests [Chr74], data for high normal stresses follow the equation $\mu = 0.6 + 0.05/\sigma$, where $0.2 < \sigma < 2$ GPa [Bye67].

Fig. 18. Coefficient of sliding friction, μ, of shear fractures in peridotite as a function of confining pressure $p_m = S_3$ (after [Ral65]).

3.2.2.1.3 List of symbols — Liste der verwendeten Symbole

S_1, S_3 [MPa]	maximum and minimum principal stresses; in triaxial tests ($S_2 = S_3$) S_3 corresponds to the confining pressure p_m	größte und kleinste Hauptnormalspannungen; in Triaxialversuchen ($S_2 = S_3$) entspricht S_3 dem Manteldruck p_m
β [deg]	angle between S_1 and the normal of the shear plane	Winkel zwischen der Richtung von S_1 und der Scherflächennormale
σ [MPa]	normal stress acting across the shear plane	Normalspannung auf der Scherfläche
τ [MPa]	shear stress acting along the shear plane	Scherspannung auf der Scherfläche
τ_0^* [MPa]	inherent shear strength (often called "cohesion" in analogy to its notation in soil mechanics)	Bruchflächenkohäsion
μ	coefficient of sliding friction	Koeffizient der Gleitreibung
μ^*	coefficient of internal friction	Koeffizient der inneren Reibung
μ_r	coefficient of residual friction	Koeffizient der Restreibung
ϕ [deg]	friction angle	Reibungswinkel
ϕ^* [deg]	angle of internal friction	Winkel der inneren Reibung

3.2.2.2 Data — Daten

Table 10. Coefficient of internal friction, μ^*, angle of internal friction, ϕ^*, and inherent shear strength τ_0^* of rocks.
tr: triaxial test; ds: double shear test, direct shear; to: rotation shear test
$\sigma =$ normal stress across fracture plane; $p_m =$ confining pressure in triaxial tests.

Rock	Location	μ^*	ϕ^* deg	τ_0^* MPa	Test	Range of σ or p_m MPa	Ref.
chalk	Kent, England	0.31	17	2.6	tr	$0.7 < p_m$ < 10.5	Bel77
limestone	Lueders, USA	0.53	28	15	tr	$p_m < 100$	Han69
	Solnhofen, FRG	0.53	28	105	tr	$p_m < 100$	Han69
	Grauboforsen, Sweden	1.2	50	30	ds	$\sigma < 600$	Lun66
	average	1.19	50	17···22	tr	σ low	Hoe69
	range	0.75···1.6	37···58	3.4···34	tr	σ low	Hoe69
dolomite	Blair, USA	1.0	45	45	tr	$p_m < 100$	Han69
sandstone	Ruhr, Witten, FRG	0.88	41	37	tr	$\sigma < 100$	Rum75
		0.82	39		tr	$100 < \sigma < 200$	
(continued)		0.75	37		tr	$200 < \sigma < 300$	

Table 10 (continued)

Rock	Location	μ^*	ϕ^* deg	τ_0^* M Pa	Test	Range of σ or p_m M Pa	Ref.
sandstone (continued)	Gosford, New South Wales, Australia	0.5	26.5				Jae69
	Buchberg, Switzerland	1.33···1.66	53···59	6.5···10.4	tr	$p_m < 3$	Kov74
		0.81···1.23	39···51	11.5···17.2	tr	$3 < p_m < 6$	
		0.67···0.78	34···38	21···28	tr	$6 < p_m < 60$	
	Tennessee, USA	0.84	40	50	tr	$p_m < 100$	Han69
concrete (Beton)	BH 300, compressive strength $S_{1m} = 42.3$ MPa	0.78	38(7)	8.5(18)	tr	$p_m < 3$	Kov74
		0.87	41(1)	7.9(4)	tr	$3 < p_m < 6$	
		0.70	35	12.4(8)	tr	$6 < p_m < 60$	
	BH 200, compressive strength $S_{1m} = 32.5$ MPa	0.84	40(3)	6.5(6)	tr	$p_m < 3$	Kov74
		0.70	35(2)	8.1(8)	tr	$3 < p_m < 6$	
		0.75	37(1)	8.4(6)	tr	$6 < p_m < 60$	
granite	Epprechtstein, Kristall-granite, Fichtel-gebirge, FRG	1.15	49	15	tr	$p_m < 70$	Rum78c
	Falkenberg, NE Bavaria, FRG	0.67	34	105	tr	$100 < p_m < 300$	Rum78c
	Aare, Gotthardt, Switzerland	2.05	64(2)	19(5)	tr	$p_m < 3$	Kov74
		1.80	61(4)	24(8)	tr	$3 < p_m < 6$	
		1.11	48(1)	46(4)	tr	$6 < p_m < 60$	
	Westerly, Rhode Island, USA	1.03	46	70	tr	$80 < p_m < 150$	Bra64
	Carnmenellis, Cornwall, England	1.28	52	24	tr	$p_m < 7$	Bat77
	Bohuslän, Sweden	2.0	63	60	ds	$\sigma < 600$	Lun66[1]
	Bredseleforsen, Sweden	2.0	63	40	ds	$\sigma < 600$	Lun66[1]
	average	1.43	55	25		σ low	Bie73
diorite	average	1.33···1.43	53···55	14		σ low	Bie73
diabase	Frederick, Maryland, USA	0.82	59	39	tr	$200 < \sigma < 600$	Bra64
dunite	Addic, N. Carolina, USA	0.5	26.5		to	$\sigma < 300$	Rie64
magnetite	ore, Grängesberg, Sweden	1.8	61	30	ds	$\sigma < 600$	Lun66[1]
monzonite	quartz-monzonite	1.48	56	26	(?)	$\sigma < 40$	Hoe69
norite	S. Africa	1.17	50			σ low	Bie73
basalt	average	1.15	49	31		σ low	Bie73
trachyte	Bowral, Australia	1.0	45		tr	σ low	Jae66
marble	Carrara, Italy	0.7	35		tr		Jae66
		1.43	55(3)	13(2)	tr	$p_m < 3$	Kov74
		0.62	32(6)	27(4)	tr	$3 < p_m < 6$	
		0.62	32(1)	28(1)	tr	$6 < p_m < 60$	
quartzite	Elliot Lake, Canada	1.23	51		tr		Bar71
	S. Africa	2.05	64	40	tr	σ low	Bie73
	Witwatersrand, S. Africa	1.0	45		tr	σ low	Hoe65
	Cheshire, Vermont, USA	1.17	49	75	tr	$300 < \sigma < 600$	Bra64
		0.85	40	300	tr	$600 < \sigma < 1000$	
	Gautojarre	2.0	63	60	ds	$\sigma < 600$	Lun66[1]

(continued)

[1]) For comparison with other values see [Lun66].

Table 10 (continued)

Rock	Location	μ^*	ϕ^* deg	τ_0^* MPa	Test	Range of σ or p_m MPa	Ref.
gneiss	Urach Geothermal Research drillhole, FRG,						
	core 37, 2938 m	0.81		20	tr	$15 < \sigma < 120$	Rum79
		0.53		48	tr	$100 < \sigma < 260$	
	core 41, 3257 m	0.78		61	tr	$80 < \sigma < 180$	
		0.55		105	tr	$180 < \sigma < 340$	
	pegmatite gneiss, Valdemarsvik, Sweden	2.5	68	50	ds	$\sigma < 600$	Lun66[1]
	gneiss-granite, Valdemarsvik, Sweden	2.5	68	60	ds	$\sigma < 600$	Lun66[1]
	mica-gneiss, Videlvorsen, Sweden	1.2	50	50	ds	$\sigma < 600$	Lun66[1]
serpentinite	Tumut Pond and Cabramurra, New South Wales, Australia	0.65	33	90	tr	$p_m < 200$	Ral65
	Wurlitz, NE Bavaria, FRG	0.57	30	57	tr	$200 < \sigma < 900$	Rum78c
schist (Schiefer)	average	0.49$\cdots$2.7	26$\cdots$70	9		σ low	Bie73
shale (Schiefer)	average	1$\cdots$2	45$\cdots$64	8		σ low	Bie73
slate (Schiefer)	average	1$\cdots$1.7	45$\cdots$60	0.7		σ low	Bie73
	grey slate, Granbovorsen, Sweden	1.8	61	30	ds	$\sigma < 600$	Lun66[1]
	black slate, Gautojarre	1.0	45	60	ds	$\sigma < 600$	Lun66[1]

[1] For comparison with other values see [Lun66].

Table 11. Coefficient of sliding friction, μ, and coefficient of residual friction, μ_r, as a function of friction surface type, testing method and normal stress for rocks.

ds = direct shear; tr = triaxial test; to = torsion shear; g = ground surface; rs = rough sawn; sb = sand-blasted; p = polished; sf = shear fracture; tf = tensile fracture; j = joint; bpl = bedding plane; clpl = cleavage plane

Rock	Origin	Type of shear plane	Test	μ	μ_r	σ (normal stress) range MPa	Ref.
amphibolite	unknown	rs	ds	0.62		0.1$\cdots$4	Bar76
andesite	Manazuru, Japan	g	ds	0.56	0.87[1]	$\sigma < 100$	Ohn75
	S. Africa	sf	tr	0.53$\cdots$0.58			Bie73
basalt	unknown	rs	ds	0.70$\cdots$0.78		0.1$\cdots$8.5	Bar76
		rs, wet	ds	0.60$\cdots$0.73		0.1$\cdots$7.9	
chalk	unknown	rs	ds	0.58		$\sigma < 0.4$	Bar76
clay	Opalinuston, Jurassic, Heiningen, Baden-Württemberg, FRG	rs	ds	0.37		0.3$\cdots$2.0	Sch77

(continued)

[1] At 1.8 mm shear displacement.

Table 11 (continued)

Rock	Origin	Type of shear plane	Test	μ	μ_r	σ (normal stress) range MPa	Ref.
conglomerate	unknown	rs	ds	0.70		0.3···3.4	Bar76
diabase	Rächolshausen, Rhön, FRG	tf	ds	0.70		$\sigma < 2.5$	Wei77
diorite	unknown	dry		0.47			Lac63
		wet		0.36			
dolomite	Blair, W. Virginia, USA	g	tr	0.55···0.40		50···330	Han69
	Knox, USA	rs	tr	0.60···0.47		50···380	Han69
	Beckmantown, USA	tf	ds	1.8···0.8		0.8···1.8	Mau65
	Oneota, USA	g	ds	0.5···0.62		0.1···7	Cou70
		rs	ds	0.61···0.77		0.1···7	
	Dunham, USA	g	ds	0.46	0.31	$\sigma < 100$	Ohn75
gabbro	unknown	p	ds	0.18		0.8···5	Hos68
		g	ds	0.32		0.8···5	
		r	ds	0.66		0.8···5	
gabbro	San Marcos, California, USA	sf	tr	0.6[7]		400···700	Ste74
		sf	tr	0.52[8]		400···700	
		sf	tr	0.45[9]		300···650	
gneiss	granitic gneiss	rs	tr	0.59		20···100	Jae59
		sf	tr	0.71		20···100	
		sf, wet	tr	0.61		20···100	
	schistose gneiss, USA	rs	tr	0.49		$\sigma < 140$	Dek65
		rs	tr	0.29		140···260	
		sf, wet	tr	0.58		30···140	Dek67
	Contra dam, Verzasca, Switzerland	j	ds	0.90		$\sigma < 20$	Lom66
		bpl	ds	0.65		$\sigma < 20$	
		fault	ds	0.78		$\sigma < 20$	
	schistose gneiss	rs	ds	0.49···0.55		0.1···8	Cou70
		rs, wet	ds	0.42···0.49		0.1···7.9	
	Urach Geothermal drillhole, core 37, 2938 m	sf	tr	0.75		30···100	Rum79
		sf	tr	0.55		100···250	
	core 41, 3257 m	sf	tr	0.70		$\sigma < 250$	
granite	Epprechtstein, Fichtelgebirge, FRG	p	tr	0.46		$\sigma < 300$	Rum78c
		p	tr	0.32		300···600	
		sf	tr	0.67		150···600	
	Malsburg, Black Forest, FRG	p	ds	0.48		0.3···0.6	Kut74a
		rs	ds	0.51		0.3···0.6	
		g	ds	0.57		0.3···0.6	
		tf	ds	0.79		0.3···0.6	
		tf	ds	0.60		$\sigma < 4$	Ren71
	Malsburg, FRG	tf	ds		0.65[2]	$\sigma < 1.6$	Sch76
	Aare, Gotthardtunnel, Switzerland	sf	tr	1.07		15···200	Kov74
	Australia	p, 50 μ	ds	0.42		0.8···5	Cou70,
		p, 200 μ	ds	0.50		0.8···5	Hos68
(continued)		p, 450 μ	ds	0.64		0.8···5	

[2] Residual friction after about 10 cm shear displacement.
[7] $T < 400\,°C$, $\tau_0 = 70\,\mathrm{MPa}$.
[8] $T = 600\,°C$, $\tau_0 = 52\,\mathrm{MPa}$.
[9] $T = 700\,°C$.

Table 11 (continued)

Rock	Origin	Type of shear plane	Test	μ	μ_r	σ (normal stress) range MPa	Ref.
granite (continued)	Inada, Japan	g	ds	0.59	0.76[1])	$\sigma < 70$	Ohn75
	Westerly, Rhode Island, USA	tf	tr	1.03		$\sigma < 20$	Waw74
		tf	tr	0.90		20⋯40	Waw74
		sf	tr	0.60[6])		200⋯1000	Bye 67
		p	tr	0.60[6])		200⋯1000	
		p	to	0.9⋯0.6		$\sigma < 200$	Chr74
		p, 200 μ	ds	0.50		0.2⋯2	Bye66
		p, 800 μ	ds	0.60		0.2⋯2	
		p(?)	to	0.8		500	Abe73
		p(?)	to	0.15		6000	
	Georgia, USA	tf	ds	0.75⋯1.3		14⋯70	Mau65
	Grand Coulee, USA	g	ds	0.73		0.1⋯7	Cou70,
		rs	ds	0.66		0.1⋯7	Hos68
limestone	Solnhofen, Jura, Bavaria, FRG	p	tr	0.62		$\sigma < 300$	Han69
		p	tr	0.55⋯0.67		80⋯450	
		sf	tr	0.70⋯0.72		50⋯200	Don75
		sf	tr	0.91		?	Bar71
		g	ds	0.71⋯0.89		0.1⋯7	Cou70,
		sb	ds	0.2⋯0.78		0.1⋯7	Hos68
		g	ds	0.46	0.42[1])	$\sigma < 100$	Ohn75
	Solnhofen, FRG	tf	ds	0.55	0.51[2])	$\sigma < 6.0$	Sch76
	Eichstätt, Jura, FRG	tf	ds	0.73		$\sigma < 2$	Wei75
	Weißenburg, Jura, FRG (Treuchtlinger Marmor)	tf	ds	0.84		$\sigma < 2.5$	Wei77
	Boningen, Born Mt., Switzerland	j	ds	1.04		σ low	Loc70
		bpl	in situ	1.23		σ low	
	Veytaux, Switzerland	j	ds	0.6		σ low	Loc70
	Crown Point, USA	sf	tr	0.50[3])		180⋯310	Ols74
				0.44[4])		180⋯310	
				0.47[5])		140⋯240	
	Lueders, Texas, USA	g	tr	0.60		$\sigma < 200$	Han69
	Crown Point, New York, USA	sf	tr	0.71 (0.68⋯0.74)		50⋯100	Don75
	Indiana, USA	tf	ds	0.7⋯1.4		7⋯28	Mau65
	Chico, USA	tf	ds	1.2⋯2		7⋯28	Mau65
	Bedford, Indiana, USA	g	ds	0.5⋯0.79		0.1⋯7	Cou70
		sb	ds	0.75⋯0.83		0.1⋯7	
marble	Carrara, Italy	p	ds	0.41		$\sigma < 5$	
		tf	ds	0.84		$\sigma < 2.5$	Wei75
		sf	tr	0.62		2⋯100	Jae59
		sf	tr	0.93⋯1.42		$\sigma < 5$	Kov74
		sf	tr	0.84		15⋯100	
	Greece	p	tr	0.36⋯0.5		$\sigma < 150$	Alh75
		sf	tr	0.53⋯0.68		$\sigma < 150$	
	Yamaguchi	g	ds	0.31	0.35[1])	$\sigma < 100$	Ohn75
	Carthage, Montana, USA	sf	ds	1.1⋯1.8		13⋯55	Mau65
	Wombeyan, Australia	rs	ds	0.75		1⋯5	Hos68
(continued)		sf	tr	0.62		$\sigma < 100$	Jae59

[1]) At 1.8 mm shear displacement.
[2]) Residual friction after about 10 cm shear displacement.
[3]) At 10^{-3} cm s^{-1}, 25 °C.
[4]) At 10^{-3} cm s^{-1}, 100 °C.
[5]) At 10^{-3} cm s^{-1}, 300 °C.
[6]) $T < 600$ °C, $\tau_0 = 80$ MPa.

Table 11 (continued)

Rock	Origin	Type of shear plane	Test	μ	μ_r	σ (normal stress) range MPa	Ref.
monzonite	Orikaba, Japan	g	ds	0.29	0.68[1])	$\sigma < 100$	Ohn75
peridotite	see Fig. 18						
porphyry	unknown	sf	tr	0.86		20···100	Jae59
	Atalaga, Spain	j	ds	0.59		0.1···3.5	Kut74
quartzite	Contra dam, Verzasca, Switzerland	j	ds	0.87		$\sigma < 20$	Lom66
	Emmelshausen, FRG	sf	tr	0.75···1.11		$\sigma < 1$	Wol70
	S. Africa	g	tr	0.48			Wie68
		sf	tr	0.67			
	Elliot Lake, Canada	sf	tr	0.94		σ low	Bar71
sandstone	Ruhr, FRG	sf	tr	0.67···1.07		$\sigma < 10$	Wol70
	Ruhr, Witten, FRG	rs	tr	0.60		30···200	Alh75
		p	tr	0.54		30···200	
		sf	tr	0.80		30···200	
		tf	ds	0.51		$\sigma < 2$	Wei75
	Ettlingen, Black Forest, FRG	p	ds	0.62		0.3···0.6	Kut74a
		tf	ds	0.63		0.3···0.6	
	Buchberg, Switzerland	sf	tr	0.70		15···100	Kov74
	Burgdorf, Switzerland	bpl	ds	0.67		σ low	Loc70
	Darley Dale, Derbyshire, England	tf	ds	0.80		$\sigma < 5$	Kut74
	Gosford, New South Wales, Australia	g	ds	0.51		0.8···5	Hos68
		g	ds, wet	0.61		0.8···5	
		sf	tr	0.52			Jae59
	Rangely, Colorado, USA (Weber)	rs	tr	0.85		$\sigma < 200$	Bye75
		sf	tr	0.85		$\sigma < 200$	
		rs	tr	0.60		200···350	
		sf	tr	0.60		200···350	
	Buntsandstone, Trifels-Schichten, Landau, FRG	tf	ds	1.15	0.73[2])	6.0	
	Portage, New York, USA	sf	tr	0.57···0.63 (0.61)		50···200	Don75
	Berea, USA	tf	ds	1.25···1.6		6···18	Mau65
		g	ds	0.47···0.64		0.1···7	Cou70
	Navajo, USA	g	ds	0.5···0.65		0.1···7	Cou70
	Tennessee, USA	g	tr	0.70		$\sigma < 300$	Han69
	Rush Springs, USA	tf	ds	0.9···1.5		13···100(?)	Mau65
serpentinite	Wurlitz, NE-Bavaria, FRG	p	tr	0.43		$\sigma < 150$	Rum78c
		p	tr	0.34		150···600	
		sf	tr	0.78		$\sigma < 200$	
		sf	tr	0.72		200···800	
	Fidalgo Island, Washington, USA	sf	tr	0.4···0.75		100···500	Ral65
	Cabramurra, New South Wales, Australia	sf	tr	0.7···0.9		100···500	Ral65
	Tumut Pond, New South Wales, Australia	sf	tr	0.6···0.9		100···500	Ral65
(continued)	unknown	g	ds	0.37	0.53[1])	$\sigma < 100$	Ohn75

[1]) At 1.8 mm shear displacement.   [2]) Residual friction after about 10 cm shear displacement.

Table 11 (continued)

Rock	Origin	Type of shear plane	Test	μ	μ_r	σ (normal stress) range MPa	Ref.
slate	Nuttlar, Sauerland, FRG	tf	ds	0.42		$\sigma < 2$	Wei75
	Delabole, Cornwall, U.K.	clpl	ds	0.42		$\sigma < 3$	Kut74
	unknown	g	ds	0.52	0.74[1]	$\sigma < 100$	Ohn75
	Martinsburg, Pennsylvania, USA	sf	tr	0.49···0.54		40···150	Don75
	S. Africa	sf	tr	0.45···0.67			Bie73
trachyte	Mizuho, Japan	g	ds	0.56	0.78[1]	$\sigma < 100$	Ohn75
tuff	unknown	g	ds	0.52	0.74[1]	$\sigma < 100$	Ohn75

[1]) At 1.8 mm shear displacement.

Table 12. Coefficient of sliding friction, μ, and friction angle ϕ of minerals as a function of friction surface type and normal stress.

Mineral	Friction surface		μ	ϕ deg	Ref.
Al_2O_3			0.40	22	Bow64
augite	rough,	dry	0.29	16	Fry53
		wet	0.45	24	
biotite, Canada	cleavage,	dry	0.31	17	Hor62
		wet	0.13	7.4	
calcite, Kansas, USA	polished,	dry	0.14	8	Hor62
		wet	0.68	34	
chlorite, Vermont, USA	polished,	dry	0.53	28	Hor62
		wet	0.22	12	
diamond	polished, dry, σ low		0.15	8.5	Bow64
		σ high	0.05	3	
	chem. clean, polished				
		σ low	0.35	19	Sea58
		σ high	0.50	26.5	
glass			0.7	35	Bow64
graphite	dry		0.15···0.20	8···11	Sea58, Bow64
	chem. clean surface, dry		0.55	29	Sea58
hornblende	rough/polished, dry		0.48	26	Fry53
		wet	0.61	31	
ice			0.50	26.5	Bow64
MgO	cleavage,	dry	0.20	11	Bow64a
microcline	polished,	dry	0.12	7	Hor62
		wet	0.77	38	
	rough,	dry	0.40	22	Fry53
		wet	0.62	32	
muscovite	cleavage,	dry	0.41···0.45	23	Hor62
		wet	0.22···0.26	13	
NaCl	cleavage,	dry	0.7···0.8	35···38	Bow64, Kin54
		wet	0.15	8.5	

(continued)

Table 12 (continued)

Mineral	Friction surface	μ	ϕ deg	Ref.
olivine	$\sigma < 3$ GPa, dry	0.35	19	Rie64
PbS		0.6	31	Bow64
phlogopite	cleavage, dry	0.29···0.31	17	Hor62
	wet	0.15···0.16	9	
pyroxene	dry	0.51	27	Lac63
	wet	0.48	25.6	
pyrophyllite	dry	0.16	9	Tsc48
	wet	0.12	7	
quartz	polished, dry	0.11···0.14	6···8	Hor62
	polished, wet	0.42···0.51	23···27	Hor62
	rough, dry	0.30···0.51	17···27	Bro66
	rough, wet	0.5	27	
	chem. clean surface, polished, dry or wet	0.9	42	
sulfur		0.5	27	Bow64
sapphire	polished, dry	0.06	3.4	Rie64a
serpentine	polished, dry	0.62···0.76	32···37	Hor62
	wet	0.29···0.48	16···26	
steatite, N-Caroline, USA	polished, dry	0.38	21	Hor62
	wet	0.23	13	
talc	polished, dry	0.36	20	Hor62
	wet	0.16	9	
	rough/polished, dry	0.12	7	Fry53
	rough/polished, wet	0.18	10	

3.2.3 Rheology — Rheologie

Deformation mechanisms – creep diagrams – constitutive equations – rheological constants.

Verformungsmechanismen – Verformungsdiagramme – Kriechgesetze – rheologische Konstanten

3.2.3.1 Introduction — Einleitung

3.2.3.1.1 General remarks — Allgemeine Bemerkungen

At or near the Earth's surface most rocks and minerals under short-time loading can be regarded as brittle elastic solids, to which the theory of elasticity can be applied before brittle rupture occurs. However, under geologic conditions (higher temperature, high lithostatic pressure, geologic times) rocks may flow and deform irrecoverably without permanent loss of internal cohesion. Such time-dependent deformation is often described by idealized *rheological models* which are based on three fundamental elements, the *elastic* (Hooke) body, the *viscous* (Newton) fluid and the *plastic* (St. Venant) body. The very simplest model for describing the mechanical behaviour of rocks would seem to be a series and parallel combination of such elements as suggested in Fig. 19.

Although such idealized mathematical models are conceptually useful as a first approximation, the physics

Gesteine an oder nahe der Erdoberfläche können bei kurzzeitiger Belastung im allgemeinen als spröde, elastische Körper angesehen werden. Bei Erreichen einer kritischen Belastung erfolgt ein Sprödbruch. Unter geologischen Bedingungen (erhöhte Temperatur, hoher lithostatischer Druck, Dauer geologischer Prozesse) können Gesteine jedoch Fließeigenschaften aufweisen, wobei sich das Gestein permanent verformt ohne an Kohäsion zu verlieren. Eine derartige zeitabhängige Verformung wird häufig mit Hilfe mathematisch idealisierter *rheologischer Modellkörper* angenähert beschrieben. Sie basieren auf den drei rheologischen Grundelementen, dem *elastischen* oder Hooke'schen Körper, der *viskosen* oder der Newton'schen Flüssigkeit und dem *plastischen* oder St. Venant Körper. Eine gute Näherungsdarstellung des Verformungsverhaltens vieler Gesteine wird bereits durch einen einfachen Modellkörper erreicht, wie er in Fig. 19 angedeutet ist.

Obwohl das Konzept idealistischer rheologischer Körper sicherlich häufig für die quantitative Behand-

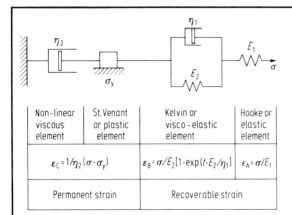

Non-linear viscous element	St.Venant or plastic element	Kelvin or visco-elastic element	Hooke or elastic element
$\varepsilon_C = 1/\eta_2(\sigma-\sigma_y)$		$\varepsilon_B = \sigma/E_2[1-\exp(t\cdot E_2/\eta_1)]$	$\varepsilon_A = \sigma/E_1$
Permanent strain		Recoverable strain	

◄

Fig. 19. Example of an idealized rheological model for rocks, consisting of an elastic (Hooke) element, a visco-elastic (Kelvin) and a non-linear Bingham element. The upper part of the figure shows the mechanical analogies, consisting of two perfect springs, two frictionless dashpots and a friction element. E_1, E_2 = elastic moduli; η_1, η_2 = coefficients of viscosity; σ_y = yield strength of the St. Venant element; $\varepsilon_A, \varepsilon_B, \varepsilon_C$ = strain components of the three elements, σ = stress.

of inelastic behaviour must take into account the real material behaviour, which consists of various flow processes such as cataclasis, intra-crystalline glide, diffusion, dislocation movement, grain growth, or re-crystallization. This requires detailed laboratory studies of the dynamics of rock deformation under controlled conditions, in which lithospheric parameters are simulated realistically.

lung in erster Näherung von Nutzen ist, wird dabei die Physik inelastischer Prozesse nicht berücksichtigt. Die Untersuchung der wirksamen Verformungsprozesse wie Kataklase, intra-kristallines Gleiten, Diffusion, Versetzungsbewegungen, Kornwachstum oder Rekristallisation verlangt detaillierte Laborversuche bei währenden Bedingungen.

3.2.3.1.2 Laboratory testing methods — Untersuchungsmethoden

Laboratory measurements of inelastic behaviour of rocks and minerals are generally conducted as triaxial compression tests (S_1, $S_2 = S_3 = p_m$) on cylindrical rock specimens which are encased in impermeable jackets and then subjected to confining pressure p_m and temperature T (sometimes triaxial extension tests are used with $S_1 = S_2 = p_m$ and S_3 acting axially). Confining pressures equivalent to lithostatic pressures acting at depth to about 100 km, as well as temperatures to that of melting can be attained today in laboratory rock deformation testing. In most cases confining pressure and temperature are maintained constant.

Specimen deformation is usually performed at constant axial strain (*constant strain-rate test*, $\dot{\varepsilon}_1$ = const.) or at constant axial stress S_1 (*creep test*, S_1 = const). In the first case, the differential axial stress ($\sigma = S_1 - p_m$) sustained by the rock is plotted against the axial strain (ε_1) as shown in Fig. 20 (a). In the creep test, the axial strain at various constant axial stress levels is observed as a function of time (Fig. 20 (b)). The duration of creep tests is obviously very short as compared to times of geological processes; therefore, extrapolation of measured data beyond the laboratory time scale can only be made under careful assumptions.

Laboruntersuchungen der Inelastizität von Gesteinen und Mineralen werden im allgemeinen als triaxiale Druckversuche (S_1, $S_2 = S_3 = p_m$) an zylindrischen Probenkörpern durchgeführt. Die Proben werden ummantelt und bei konstantem Manteldruck und konstanter Temperatur axial verformt. Seltener werden triaxiale Dehnungsversuche angewendet, bei denen $p_m = S_1 = S_2$ = konstant ist und S_3 axial wirkt). Manteldrucke von ca. 3 GPa, was einem lithostatischen Druck in einer Tiefe von 100 km entspricht, und Temperaturen bis zum Schmelzpunkt von Gesteinen können heute bei Laboruntersuchungen ohne besondere Schwierigkeiten erreicht werden.

Im allgemeinen werden die Proben entweder bei *konstanter axialer* Verformungsrate ($\dot{\varepsilon}_1$ = konstant) oder bei konstanter axialler Spannung (S_1 = konstant oder S_3 = konstant) axial verformt. Letztere Methode wird als *Kriechversuch* bezeichnet. Im Fall konstanter axialer Verformungsgeschwindigkeit wird die wirksame axiale Differenzspannung $\sigma_1 = S_1 - p_m$, in Abhängigkeit von der axialen Verformung ε_1 aufgetragen (vergleiche Fig. 20 (a)). Beim Kriechversuch wird die axiale Verformung ε_1 als Funktion der Zeit t (Fig. 20 (b)) aufgezeichnet (im folgenden wird $\varepsilon_1 = \varepsilon$ und $\sigma_1 = \sigma$ bezeichnet). Die Dauer von Kriechversuchen bei Laboruntersuchungen ist bedingtermaßen kurz gegenüber der Dauer tektonischer Verformungsprozesse, bei denen die Verformungsgeschwindigkeit $\varepsilon \approx 10^{-14}\,\text{s}^{-1}$ beträgt. Es sei deshalb angemerkt, daß Extrapolationen von Ergebnissen von Laborversuchen für geologische Zeitmaßstäbe nur unter bestimmten Bedingungen möglich sind (gleicher Verformungsmechanismus, keine Materialveränderung).

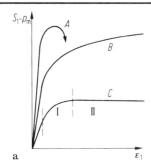
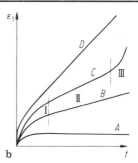

a

b

Fig. 20a, b. a) Representative constant strain-rate curves. Curve A: at high strain-rate or at low temperature; Curve C: at low strain-rate or at high temperature; Curve B: at intermediate strain-rate or temperature (after [Car78]).

b) Representative creep curves. Curve A: at low temperature and low differential stress; Curve D: at high temperature or high differential stress; Curves B, C: intermediate stages; I, II, and III: primary or transient creep, secondary or steady-state creep, accelerating or tertiary creep, respectively (after [Rum67]).
$S_1 - p_m$ = differential stress; ε_1 = axial strain; t = time.

Neglecting the instantaneous, mainly elastic strain, typical creep tests generally yield three major stages: *primary* or *transient* creep (I in Fig. 20), *secondary* or *steady-state* creep (II), and eventually *tertiary* creep (III) which occurs just prior to fracture. Any of these three stages may dominate the deformation process, depending on the material or the testing conditions (curves A to D).

Most studies on rocks have been conducted in triaxial constant strain-rate tests, where the axial stress is allowed to vary as a function of axial strain. The strain-rates imposed vary between 10^{-1} to 10^{-7} s^{-1}. In direct creep tests sometimes strain-rates of 10^{-9} s^{-1} are achieved. Recently, *stress-relaxation tests* are becoming more popular, in which the stress and the stress-rate vary with time by loading the rock specimen to some fixed value of stress or strain, and then allowing the stress to decay.

With respect to testing conditions the results of creep tests are generally divided into the following categories:

a) **Low temperature creep** ($T < 0.3\ T_m$, T_m melting temperature)
occurs in halite, calcite, gypsum, and layered silicates, although high differential stresses at high confining pressures are required to induce significant deformation. Low temperature creep in silicate rocks is associated with brittle microfracturing and dilatancy [Bye78]. Transient creep ε_t at low temperature often is approximated by logarithmic creep equations of the form $\varepsilon_t = \alpha \log (1 + \beta t)$, where α and β are constants.

Läßt man die beim Aufbringen der Belastung sofort auftretende Verformung (im wesentlichen elastischer Verformungsanteil) unberücksichtigt, weisen typische Kriechversuche im allgemeinen drei verschiedene Verformungsstadien auf: *primäres Kriechen* (Übergangskriechen, I in Fig. 20), *sekundäres* oder *stationäres Kriechen* (II) und eventuell *tertiäres Kriechen* (Bruchfließen, III in Fig. 20). Jedes dieser Stadien kann den Verformungsprozeß je nach Material oder Versuchsbedingungen auszeichnen (Kurventypen $A - D$ in Fig. 20).

Die meisten der bisher vorliegenden Ergebnisse wurden bei Triaxialversuchen mit konstanter axialer Verformungsgeschwindigkeit gewonnen. Die verwendeten Verformungsgeschwindigkeiten variieren zwischen $10^{-1} < \dot{\varepsilon} < 10^{-7}$ s^{-1}. Bei einigen Kriechversuchen wurden Verformungsraten bis zu $\dot{\varepsilon} = 10^{-9}$ s^{-1} erreicht. Neuerdings werden auch *Spannungs-Relaxationsversuche* angewendet, bei denen die axiale Spannung oder Verformung bis zu einem bestimmten Wert erhöht wird und anschließend der Spannungsabfall beobachtet wird.

Bezüglich der Versuchsparameter Druck und Temperatur unterscheidet man folgende Kriechversuche (im folgenden soll der Ausdruck Kriechen auch Versuche bei konstanter Verformungsgeschwindigkeit umfassen):

a) **Kriechversuche bei gemäßigten Temperaturen** ($T < 0.3\ T_m$, T_m Schmelztemperatur)
Bei niedriger Temperatur beobachtet man Kriechen vorwiegend in Nicht-Silikaten (z.B. Steinsalz, Kalzit, Gips) und Schichtsilikaten, obwohl zur Erzeugung signifikanter Verformungsbeträge auch hier bereits hohe Differenzspannungen (Manteldrucke) notwendig sind. In Silikaten sind Kriechverformungen bei niedrigen Temperaturen immer mit Mikrorißbildung und Dilatanz verbunden [Bye78]. Das hierbei vorwiegend auftretende Übergangskriechen ε_t kann meist durch eine logarithmische Zeitfunktion der Form $\varepsilon_t = \alpha \log (1 + \beta t)$ annähernd dargestellt werden (α, β Konstanten).

b) **High temperature creep**

with dominating steady-state creep is observed in laboratory tests at temperatures $T > 0.5\ T_m$. High temperature transient creep may be approximated by two principal types of equations:

$$\varepsilon_t = \varepsilon_{t\infty}[1 - \exp(-t/t_R)],$$

where $\varepsilon_{t\infty}$ is the total transient creep strain and t_R is a relaxation time coefficient, and

$$\varepsilon_t = \beta_0\,\sigma^n\,t^m\,\exp\,(-Q/RT),$$

where β_0 is a constant, σ is the differential stress (in MPa), n is the stress exponent, m is the time exponent, Q the activation energy (in kJ mol^{-1}), R the gas constant, and T the absolute temperature. Some data on high pressure transient creep of rocks are given in Table 13 and Fig. 22, which are taken from [Han80].

Present data on *steady-state-creep* of rocks and minerals (as well as for metals and ceramics) may be best described by a relation of the form

$$\dot\varepsilon = A\,\exp(-Q_c/RT)\,\sigma^n,$$

where $\dot\varepsilon$ is the axial strain rate, σ the differential stress, n the stress exponent, A is a nearly temperature-insensitive material constant. Q_c, which is the activation energy for creep in the steady-state region is nearly identical to that for self-diffusion of the least-mobile atomic species of the rock or mineral. Some values of the parameters of A, Q_c and n are listed in Table 14.

b) **Kriechen bei hohen Temperaturen** ($T > 0.5\ T_m$)

Zur mathematischen Darstellung des hierbei auftretenden Übergangskriechens werden zwei verschiedene Beziehungen verwendet:

und

Hierin sind $\varepsilon_{t\infty}$ der Gesamtbetrag des Übergangskriechens, t_R eine Relaxationszeit, σ die Differenzspannung, Q die Aktivierungsenergie, R die Gaskonstante, T die absolute Temperatur, n der Spannungs- und m der Zeit-Exponent und β_0 eine Konstante. Einige Zahlenwerte für Gesteine sind in Tab. 13 und Fig. 22 dargestellt (aus [Han80]).

Die *stationäre* Kriechverformung von Gesteinen und Mineralen (auch Metalle und keramische Werkstoffe) kann durch folgende Beziehung allgemein beschrieben werden:

Die Konstante A ist dabei temperaturunabhängige Materialkonstante. Die Aktivierungsenergie Q_c des stationären Kriechens ist angenähert gleich der Aktivierungsenergie der langsamsten Atomart des Festkörpers. Einige Werte von A, Q_c und n sind in Tab. 14 angegeben.

3.2.3.1.3 Deformation mechanisms — Verformungsmechanismen

Considering deformation mechanisms acting during rock or mineral deformation the following mechanisms are generally differentiated:

a) Steady-state deformation processes
1. Dislocation glide
2. Dislocation creep
3. Difffusional creep by diffusive flow of single ions or vacancies
   i) Nabarro-Herring creep (lattice diffusion)

   ii) Coble creep (grain boundary creep)
4. Mass transport by diffusion through a fluid phase

5. Viscous flow of a liquid
b) Non-steady deformation processes
1. Anelastic or recoverable creep
2. Transient or primary creep with work-hardening effects
3. Twinning or kinking
4. Superplastic flow

Die Untersuchung der bei der zeitabhängigen Verformung allgemein beteiligten Prozesse hat folgende formungsmechanismen aufgezeigt:

a) Stationäre Verformungsprozesse
1. Versetzungsgleiten
2. Versetzungskriechen
3. Diffusionskriechen (Diffusion von Einzelionen oder Gitterfehlstellen)
   i) Nabarro-Herring-Kriechen (Diffusion im Kristallgitter)
   ii) Coble-Kriechen (Diffusion auf Korngrenzen)
4. Massentransport durch Diffusion in einer flüssigen Phase
5. viskoses Fließen einer Flüssigkeit
b) Nicht-stationäre Verformungsprozesse
1. Anelastizität oder reversibles Kriechen
2. Übergangs- oder logarithmisches Kriechen mit Verfestigung
3. Zwillingsbildung, Kink-Bildung
4. Superplastizität

c) Other mechanisms which influence creep but do not themselves produce deformations
   1. Recovery and polygonization
   2. Recrystallization
   3. Grain growth
   4. Dissolution or precipitation of phases
   5. Phase changes.

c) Andere Mechanismen, die Kriechprozesse beeinflussen, aber nicht selbst Deformationen hervorrufen.
   1. Polygonisierung
   2. Rekristallisation
   3. Kornwachstum
   4. Ausfällung und Lösung von bestimmten Stoffen
   5. Phasenumwandlungen

3.2.3.1.4 Creep diagrams — Verformungsdiagramme

In order to simply depict the rheological behaviour of polycrystalline solids creep diagrams in the stress-temperature plane are used (deformation maps [Ash72]) These diagrams indicate the fields in which different mechanisms of plastic deformation are dominant. In general, the diagrams relate to a given microstructural state characteristic of a solid and also show strain rate contours superimposed (Fig. 21). Thus, knowing two of the three experimental variables stress, temperature, and strain-rate, the other can simply be taken from the diagram.

Zur Darstellung der Rheologie polykristalliner Festkörper werden Verformungsdiagramme in der Spannungs-Temperatur-Ebene verwendet (Verformungskarten nach [Ash72]). Aus diesen Diagrammen sind die Bereiche ersichtlich, in denen verschiedene Verformungsmechanismen dominieren. Im allgemeinen beziehen sich derartige Darstellungen auf ein Material bestimmter Mikrostruktur (Korngröße) und zeigen außerdem Kurven gleicher Verformungsgeschwindigkeit (Fig. 21). Kennt man daher zwei der drei experimentellen Parameter Spannung, Temperatur und Verformungsgeschwindigkeit, so kann die andere Größe aus den Diagrammen ermittelt werden.

◄ Fig. 21. Schematic creep diagram showing various types of deformation mechanisms as a function of normalized shear stress τ/μ (μ = shear modulus) and homologous temperature T/T_m (T_m = melting temperature in [K]). $\dot{\varepsilon}$ = strain rate, τ_c = critical shear stress for dislocation motion ($\tau_c < 10$ MPa). [$\tau = \frac{1}{2}(S_1 - p_m)$]. In practice, creep diagrams are given in terms of differential compressive stress σ ($\sigma = S_1 - p_m$, p_m = confining pressure in symmetric triaxial tests) and axial compressive strain rate $\dot{\varepsilon}$ ($\dot{\varepsilon}_1$, $\dot{\varepsilon}_2$ = two different strain rates).

As mentioned generally in Section 3.2.3.1.2 the data are derived from triaxial compression or extension experiments on cylindrical test samples. The stress value given in such diagrams is the acting differential stress $\sigma = S_1 - p_m$ where S_1 is the axial stress (i.e. at 10% deformation) and p_m is the confining pressure.

Anelastic and *recoverable creep* occurs at stresses lower than the critical shear stress for large-scale dislocation motion in the solid. Low temperature creep proceeds above this critical shear stress by *multiplication* and *glide* of *slip dislocations* or by *twin gliding* and *kinking* combined with work hardening because of limited dislocation mobility (*transient creep*). At higher temperatures screw dislocations may cross-slip, and ion and vacancy diffusion allow edge segments to climb around barriers to slip, leading to *steady-state dislocation glide* and *creep*. At very high temperatures and low

Wie bereits in Abschnitt 3.2.3.1.2 allgemein erwähnt werden die rheologischen Daten in Triaxialversuchen an zylindrischen Proben ermittelt. Die Spannungswerte in den Diagrammen stellen meist die Differenzspannungen $\sigma = S_1 - p_m$ dar, wobei S_1 die axiale Spannung (im allgemeinen bei 10% Verformung) und p_m der Manteldruck sind.

Anelastizität beobachtet man bei Scherspannungen, die kleiner als die kritische Scherspannung zur Aktivierung von Versetzungen sind. Niedertemperaturkriechen erfolgt oberhalb dieser kritischen Scherspannung in der Form von *Vervielfachung* und *Gleiten* von *Stufenversetzungen* oder durch *Zwillingsgleitung* oder *Kink-Bildung*, wobei im allgemeinen aufgrund von begrenzter Versetzungs-Mobilität Verfestigung eintritt (*Übergangskriechen*). Bei erhöhten Temperaturen können sich Schraubenversetzungen sich kreuzen und die Diffusion von Ionen und Fehlstellen ermöglichen ein

stresses, creep may occur by bulk diffusion (*Nabarro Herring creep*) or by grain-boundary diffusion (*Coble creep*).

Other forms of presenting creep data are to plot axial creep-strain ε vs. time t, axial strain rate $\dot{\varepsilon}$ vs. differential stress σ, strain rate vs. inverse temperature ($1/T$, T in K), $\log \sigma$ vs. $\log \dot{\varepsilon}$, or σ vs. $\log \dot{\varepsilon}$. In the following, data are presented in various forms mainly following the original investigator.

Klettern von Versetzungen über Hindernisse, was zu *stationären Versetzungsgleiten* und *Versetzungskriechen* führt. Bei sehr hohen Temperaturen und niedrigen Scherspannungen erfolgt die Verformung durch Diffusion von Fehlstellen im Kristallgitter (*Nabarro-Herring-Kriechen*) oder durch Diffusion auf den Korngrenzen (*Coble-Kriechen*).

In anderen Darstellungen von Kriechdaten werden die axiale Kriechverformung ε als Funktion der Zeit t, die axiale Kriechgeschwindigkeit $\dot{\varepsilon}$ als Funktion der Differenzspannung $\sigma = S_1 - p_m$, oder die axiale Kriechverformungsgeschwindigkeit $\dot{\varepsilon}$ als Funktion des Kehrwerts der absoluten Temperatur ($1/T$, T in K) wiedergegeben. Als logarithmische Darstellungen sind üblich: $\log \sigma$ vs. $\log \dot{\varepsilon}$ und σ vs. $\log \dot{\varepsilon}$. Wird statt σ die Scherspannung τ in der Darstellung verwendet, so bedeutet $\tau = \frac{1}{2}(S_1 - p_m)$. Im folgenden sind verschiedene Darstellungen gewählt, meist in Anlehnung an die Originalarbeit.

3.2.3.1.5 Constitutive equations — Kriechgesetze

Temperature dependence

Steady-state creep rates in crystalline materials above $0.5\, T_m$ (T_m melting temperature) are related to temperature and stress by a relation of the form

Temperaturabhängigkeit

Die Verformungsgeschwindigkeit $\dot{\varepsilon}$ kristalliner Festkörper während der stationären Verformungsphase ($\dot{\varepsilon} =$ konstant) bei Temperaturen $T > 0.5\, T_m$ (T_m Schmelztemperatur) ist von der wirksamen Spannung σ und der Temperatur T abhängig:

$$\dot{\varepsilon} = A\, \mathrm{f}(\sigma) \exp(-Q_c/RT). \tag{1}$$

In this relation A is a constant, $\mathrm{f}(\sigma)$ is a stress function, Q_c is the activation energy of creep, R ist the gas constant and T is the absolute temperature. Evidently, the temperature dependence of the creep rate is given by the relation

Dabei ist A eine Konstante, $\mathrm{f}(\sigma)$ eine Spannungsfunktion, Q_c die Aktivierungsenergie des Kriechens, R die Gaskonstante und T die absolute Temperatur. Offensichtlich wird die reine Temperaturabhängigkeit durch folgenden Ausdruck beschrieben:

$$\dot{\varepsilon} = C \exp(-Q_c/RT) \tag{2}$$

where C is a constant for a given constant stress value. At temperatures $T > 0.5\, T_m$, Q_c is nearly identical to the activation energy of self-diffusion, Q_D, of the slowest-moving atomic species. That is, the creep rate is proportional to the diffusion coefficient

Hierin ist C eine Konstante bei konstanter Spannung. Bei $T > 0.5\, T_m$ ist Q_c identisch mit der Aktivierungsenergie der Selbstdiffusion, Q_D, der langsamsten Atomart im Festkörper. Daher ist die Verformungsgeschwindigkeit proportional zum Diffusionskoeffizienten:

$$D = D_0 \exp(-Q_D/RT) \tag{3}$$

where D_0 is a constant. D can also be expressed by the empirical relation

D_0 ist eine Konstante. D kann auch durch folgende empirische Beziehung ausgedrückt werden:

$$D = D_0 \exp(-g\, T_m/T) \tag{4}$$

where g is a dimensionless constant and T_m is the melting temperature at hydrostatic pressure.

Hierin ist g eine dimensionslose Konstante und T_m die Schmelztemperatur bei hydrostatischem Druck.

Stress dependence

The effect of a hydrostatic pressure P on the creep rate is given by

Spannungsabhängigkeit

Die Wirkung eines hydrostatischen Drucks auf die Verformungsgeschwindigkeit ist durch folgende Beziehung beschrieben:

$$\dot{\varepsilon} = A\, \mathrm{f}(\sigma) \exp[-(Q_c + PV_c)/RT] \tag{5}$$

where V_c is the activation volume for creep.

Considering the stress function $\mathrm{f}(\sigma)$ we may differentiate between the following relations

V_c ist dabei das Aktivierungsvolumen.

Bei Betrachtung der Spannungsabhängigkeit können wir folgende Relationen unterscheiden:

$f(\sigma) = C_1\,\sigma$ (Nabarro-Herring creep at low differential stresses σ — Nabarro-Herring-Kriechen bei sehr kleinen Differenzspannungen σ)   (6)

$f(\sigma) = C_2\,\sigma^n$ (power law creep at intermediate stresses — Potenzgesetz bei mittlerer Belastung)   (7)

$f(\sigma) = C_3 \exp(B_1\,\sigma)$ (8)

or

$f(\sigma) = C_4 \sin h(B_2\,\sigma)^m$ (8a)

(exponential creep at high stresses —   Exponentialgesetz bei hohen Spannungen).

With respect to the dominating deformation mechanism during steady state creep various creep equations are used. In general, the equations are presented in terms of the experimental variables axial differential compressive stress σ and the axial compressive strain rate $\dot\varepsilon$. If shear stress τ is used, $\tau = \frac{1}{2}(S_1 - p_m)$.

Im Hinblick auf die verschiedenen möglichen Verformungsmechanismen werden in der Literatur verschiedene Kriechgesetze angegeben. Im allgemeinen werden die Kriechgesetze als Funktionen der experimentellen Parameter Differenzspannung σ und axiale Verformungsgeschwindigkeit $\dot\varepsilon$ dargestellt. Bei Verwendung der Scherspannung τ gilt $\tau = \frac{1}{2}(S_1 - p_m)$.

a) Dislocation glide   a) Versetzungsgleiten

$$\dot\varepsilon = K_1 \exp(-Q/RT)\exp(B_1\,\sigma) \tag{9}$$

(after [Atk76, Atk77])   (nach [Atk76, Atk77])

$$\dot\varepsilon = K_2\,\sigma^3 = \alpha D_v(\sigma/\mu)^2(\sigma V/RT) \tag{9a}$$

(after [Wee70])   (nach [Wee70])

K_1, K_2, B_1, α empirical constants α, K_1, K_2, B_1 empirische Konstanten
α depending on obstacles for gliding motion α abhängig von der Versetzungsmobilität
μ shear modulus μ Schermodul
V activation volume V Aktivierungsvolumen
D_v effective volume diffusion constant. D_v Diffusionskoeffizient

b) Dislocation creep   b) Versetzungskriechen

$$\dot\varepsilon = (K_3 D_v \mu b/kT)(\sigma/\mu)^n \tag{10}$$

(after [Muk69, Wee70])   (nach [Muk 69, Wee 70])

K_3, n empirical constants K_3, n empirische Konstanten
D_v effective volume diffusion coefficient D_v Diffusionskoeffizient
b Burger's vector b Burger-Vector

or/oder

$$\dot\varepsilon = K_4 D_v(\sigma/\mu)^m(\sigma V/RT) = \beta\,\sigma^n \tag{10a}$$

(after [Muk69])   (nach [Muk 69])

K_4, m, n, β empirical constants K_4, n, m, β empirische Konstanten
$n = m+1$ $n = m+1$

c) Diffusion creep   c) Diffusionskriechen

$$\dot\varepsilon = \frac{K_5 D_v}{d^2}\left(\frac{\sigma V}{RT}\right)\left(1 + \frac{\pi\delta}{d}\frac{D_B}{D_v}\right) \tag{11}$$

K_5 empirical constant K_5 empirische Konstante
D_v volume diffusion coefficient D_v Diffusionskoeffizient
D_B grain-boundary diffusion coefficient D_B Korngrenzendiffusionskoeffizient
d grain diameter d Korndurchmesser
δ grain boundary width (thickness of grain boundary diffusion path) δ Korngrenzendicke

Eq. (11) will yield to Nabarro-Herring creep for $(\pi\delta/d)(D_B/D_v) \ll 1$

Gleichung (11) ergibt Nabarro-Herring-Kriechen für $(\pi\delta/d)(D_B/D_v) \ll 1$

$$\dot\varepsilon = \frac{K_5 D_v}{d^2}\frac{\sigma V}{RT}, \tag{12}$$

and subgrain creep for coarse-grained materials containing subgrains within the grains

zu Kriechen in der Kristallitstruktur bei grobkristallinen Materialien

$$\dot\varepsilon = \frac{K_5 D_v}{d_0^2}\left(\frac{\sigma}{\mu}\right)^2\frac{\sigma V}{RT}, \tag{13}$$

$d - d_0 \, (\mu/\sigma)$

$d_0 =$ constant

and Coble creep for $(\pi\delta/d)(D_B/D_v) \gg 1$ with mass transport along grain boundaries by diffusion

$d = d_0 \, (\mu/\sigma)$

$d_0 =$ Konstante

und zu Coble-Kriechen für $(\pi\delta/d)(D_B/D_v) \gg 1$ mit Diffusion entlang der Korngrenzen

$$\dot{\varepsilon} = \left(\frac{K_5 \delta D_B}{d^3}\right) \cdot \pi \cdot \left(\frac{\sigma V}{RT}\right). \tag{14}$$

3.2.3.1.6 List of symbols — Liste der verwendeten Symbole

$A, B_1, B_2,$ $C, C_1 \cdots C_4$	constants in constitutive equations	Konstanten in den Kriechgesetzen
D [cm^2 s^{-1}]	diffusion coefficient	Diffusionskoeffizient
D_0, D_{OB}, D_{OV}	constants	Konstanten
D_B	diffusion coefficient for grain boundaries	Korngrenzendiffusionskoeffizient
D_v	volumetric diffusion coefficient	Volumendiffusionskoeffizient
$K_1 \cdots K_5$	constants	Konstanten
Q [kJ mol^{-1}, k cal mol^{-1}]	activation energy	Aktivierungsenergie
Q_c	activation energy for creep	Aktivierungsenergie des Kriechprozesses
Q_D	activation energy of selfdiffusion	Aktivierungsenergie der Selbstdiffusion
Q_{VD}, Q_{BD}	activation energies of selfdiffusion, volumetric or on grain boundaries	Aktivierungsenergie der Selbstdiffusion, volumetrisch oder auf Korngrenzen
P [Pa, bar]	hydrostatic pressure ($S_1 = S_2 = S_3$)	hydrostatischer Druck ($S_1 = S_2 = S_3$)
R	gas constant	Gaskonstante
S_1, S_2, S_3 [Pa, bar]	principal stresses	Hauptnormalspannungen
T [°C, K]	temperature	Temperatur
T_m [°C, K]	melting temperature	Schmelztemperatur
V [cm^3 mol^{-1}]	molar volume	Molvolumen
V_a [cm]	activation volume	Aktivierungsvolumen
a, b, c [10^{-8} cm]	lattice parameters	Gitterparameter
b [cm]	Burger's vector	Burger-Vektor
d [cm]	grain diameter	Korndurchmesser
g	constant	Konstante
l [cm]	grain size	Korngröße
m	exponent	Exponent
n	exponent	Exponent
p_m [Pa, bar]	confining pressure	Manteldruck
t	time	Zeit
t_R	relaxation time coefficient	Relaxationszeit
α, β	constants	Konstanten
δ [cm]	thickness of grain-boundary	Korngrenzendicke
ε	axial (creep)-strain	axiale Verformung
$\dot{\varepsilon}$ [s^{-1}]	axial strain rate	axiale Verformungsgeschwindigkeit
σ [Pa]	differential stress $S_1 - p_m$	Differenzspannung $S_1 - p_m$
τ [Pa]	shear stress $(S_1 - p_m)/2$	Scherspannung $(S_1 - p_m)/2$
μ [Pa]	shear modulus	Schermodul
μ_0 [Pa]	shear modulus at room temperature	Schermodul bei Normaltemperatur
Ω [cm^3]	molecular volume	Molekularvolumen
Ω_A [cm^3]	atomic volume	Atomvolumen

Conversions:

Umrechnung:

$1 \, \text{kbar} = 100 \, \text{MPa}$

$1 \, \text{kcal mol}^{-1} = 4.19 \, \text{kJ mol}^{-1}$

3.2.3.1.7 Review literature on creep — Literaturangaben zur Rheologie

Reviews on creep of rocks and minerals are presented in [Wee70, Wee70a, Sto73, Car76, Atk76a,

Überblicksdarstellungen zur Rheologie von Gesteinen und Mineralen sind in folgenden Beiträgen ge-

Bye78a, Car78, Lan79, Han80]. General reviews on plasticity of solids are given by [Fre58, Nab67, See58]. The constitution equations presented above are taken from [Muc69, Wee70, Raj71, Sto73, Atk75b, Atk76, Car76, Car78, Han80]. Equations for fluid phase transport and viscous flow are given in [Sto73].

geben: [Wee70, Wee70a, Sto73, Car76, Atk76a, Bye78a, Car78, Lan79, Han80]. Allgemeine Darstellungen über die Plastizität von Festkörpern sind gegeben durch [Fre58, Nab67, See58]. Die oben aufgeführten Kriechgesetze sind folgenden Artikeln entnommen: [Muc69, Wee70, Raj71, Sto73, Atk75b, Atk76, Car76, Car78, Han80]. Die Beziehungen über den Transport flüssiger Phasen und viskoses Fließen sind in [Sto73] angegeben.

3.2.3.2 Data — Daten

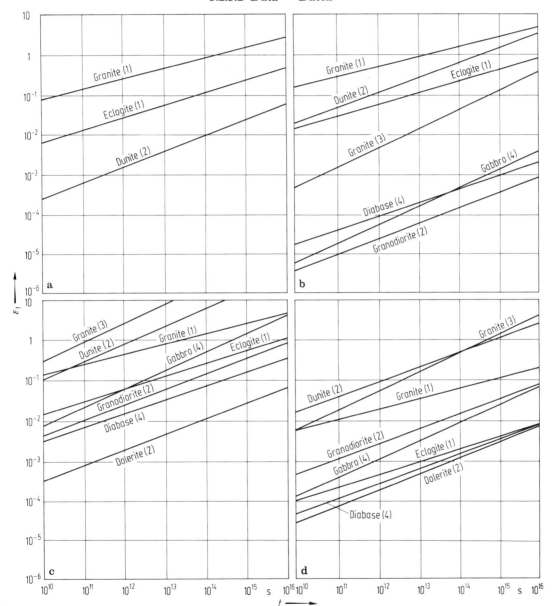

Fig. 22a···d. Transient creep strain ε_t of crystalline rocks as a function of time t. a) $T = 200\,°C$, $\sigma = 100\,MPa$; b) $T = 400\,°C$, $\sigma = 100\,MPa$; c) $T = 600\,°C$, $\sigma = 100\,MPa$; d) $T = 600\,°C$, $\sigma = 10\,MPa$. Extrapolation of experimental data (Table 13) by the creep equation $\varepsilon_t = \varepsilon_{t\infty}[1 - \exp(-t/t_R)]$ (see text, Sect. 3.2.3.1.2) (after [Car78]); (1) [Rum67, 69]; (2) [Mur73], (3) [Goe71], (4) [Goe72].

Table 13. Parameters of transient creep of rocks according to creep equation $\varepsilon_t = \beta_0\,\sigma^n\,t^m\exp(-Q/RT)$ for small axial strains (after [Han80]). p_m – confining pressure, σ = differential stress $S_1 - p_m$, β_0 = constant, n: stress exponent, m: time exponent, Q: activation energy, ε_t: transient axial creep strain.

Material	T °C	p_m MPa	σ MPa	β_0 $(MPa)^{-1}\,s^{-1}$	n	m	Q $(kJ\,mol^{-1})$	Ref.
granite	20···400	0	87	$2\cdot10^{-6}$	1.35	0.25	6	Rum69, Rum67
granite	711	400	64	3.9	1.7	0.49	159	Goe71
granodiorite	830···1045	0	100	$3\cdot10^2$	1	0.37	176	Mur76, Mur73
dolerite	675···1045	0	13	$9\cdot10^3$	1	0.38	222	Mur76, Mur73
diabase	920···1007	420	90	$6.6\cdot10^{-3}$	1.8	0.35	125	Goe72
gabbro	505···860	420	90	2.8	1.8	0.44	176	Goe72
eclogite	24···400	0	164	$2\cdot10^{-9}$	2.2	0.30	8	Rum69, Rum67
dunite	585···1045	0	13···15	$4\cdot10^{-4}$	1	0.38	54	Mur73, Mur76
dunite	700···1000	0	40				146	Eat68, Mur76
peridotite	1045	420	90		2.0	0.33		Goe72
lherzolite	900	0	10			0.5	59	Mur76
peridotite	780	0	10			0.44	96	Mur76

▶

Fig. 23. Steady-state flow stress of rocks as a function of absolute temperature at a strain-rate $\dot{\varepsilon}=10^{-14}\,s^{-1}$. For extrapolation data of Table 14 together with creep function $\dot{\varepsilon}=A\exp(-Q_c/RT)\,\sigma^n$ is used. $\sigma=S_1-p_m$ differential stress. Solid lines: dry materials; dashed lines: wet materials (after [Car76, Han80]).

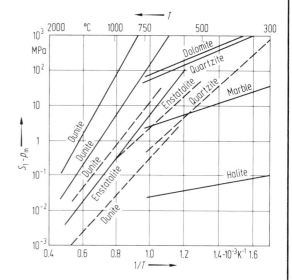

Table 14. Steady-state creep data for different rocks according to steady-state creep function $\dot{\varepsilon}=A\exp(-Q_c/RT)\,\sigma^n$ (see text section 3.2.3.1.2). $\dot{\varepsilon}$: axial strain rate, Q_c: activation energy for creep, R: gas constant, σ: differential stress (from [Han80]).

Material	A $(MPa)^{-1}\,s^{-1}$	Q_c $kJ\,mol^{-1}$	n
halite	$3\cdot10^{-5}$	98	5.5
marble	$6\cdot10^{-12}$	259	8.3
dolomite	$1\cdot10^{-21}$	347	9.1
quartzite (dry)	$6.7\cdot10^2$	268	6.5
quartzite (wet)	$7\cdot10^1$	230	2.6
dunite (dry)	$1.8\cdot10^6$	418	3.0
dunite (dry)	$4.3\cdot10^6$	527	3.0
dunite (dry)	$5.1\cdot10^7$	464	3.3
dunite (wet)	$1.2\cdot10^1$	226	2.1
dunite (wet)	$4.3\cdot10^6$	393	3.0
enstatolite (dry)	$2\cdot10^2$	293	2.4
enstatolite (wet)	$2.3\cdot10^1$	272	2.8

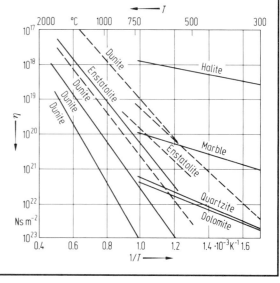

▶

Fig. 24. Calculated dynamic viscosities η ($\eta=\sigma/3\dot{\varepsilon}$) of rocks as a function of temperature at a strain-rate of $\dot{\varepsilon}=10^{-14}\,s^{-1}$. Extrapolation of data from Table 14 using $\dot{\varepsilon}=A\exp(-Q_c/RT)\,\sigma^n$. Solid lines: dry materials; dashed lines: wet materials (after [Car76]).

Table 15. Transient hot creep of rocks.

A. Data on Westerly granite, San Marcos gabbro, Maryland diabase, Mt. Albert peridotite [Goe72].

1. Experimental parameters
   creep test at constant shear stress τ of 45 MPa
   confining pressure: $p_m = 600$ MPa
   internal water pressure: $P = 180$ MPa
   temperature: $T = 500 \cdots 1000$ °C

2. Creep law

   $$\dot{\varepsilon} = A\,\tau^n\,t^m \exp(-Q_c/RT) \quad (t \text{ in s})$$

3. Numerical data

Material	n	m	A	Q_c kJ mol^{-1}
Westerly granite	1.7	0.49	$1.5 \cdot 10^6$	$314 \cdots 335$
San Marcos gabbro	1.8	0.44	$3.5 \cdot 10^4$	$377 \cdots 420$
Maryland diabase	1.8	0.35	$4 \cdot 10^3$	$335 \cdots 377$
Mt. Albert peridotite	2.0	0.33		

C. Data on Westerly granite [Goe71]

1. Experimental parameters
   Creep tests at constant axial stresses S_1 (in the order of $10 \cdots 70$ M Pa),
   confining pressure: $p_m = 400 \cdots 500$ M Pa
   temperature: $T = 430 \cdots 715$ °C
   internal water pressure: $P < 200$ MPa
   melting temperature at those conditions:
   $T_m = 993$ K

2. Creep equation

   $$(\varepsilon^*/\varepsilon_0) = (\sigma/64.3)^{0.66}(t/t_R)^{0.5}$$

   for $10^{-2} < (\varepsilon^*/\varepsilon_0) < 2$ and $12.8 < \sigma < 64.3$ MPa;
   ε^* creep strain; ε_0 elastic strain; t time in [s];
   σ axial stress S_1; $t_R = (2.2 \cdot 10^{-14}) \exp(327/RT)$ in [s].

B. Data from [Mis65]

1. Experimental parameters
   Uniaxial creep tests at constant axial stress without lateral confinement
   temperature range: $T = 500 \cdots 750$ °C
   axial stress: $50 < S_1 < 120$ MPa
   strength σ_m and estimated melting temperature T_m

Rock	σ_m MPa	T_m K
anhydrite	$98 \cdots 110$	1723
granodiorite	$380 \cdots 390$	1573
peridotite	$139 \cdots 155$	1823

2. Creep equations

Rock	T K	S_1 MPa	$\varepsilon = A\,t^m$ (t in s, $t > 60$ s)
anhydrite	500	57	$0.26\,t^{0.26} \cdot 10^{-3}$
	550	57	$0.62\,t^{0.33} \cdot 10^{-3}$
granodiorite	475	167	$11.2\,t^{0.31} \cdot 10^{-6}$
	475	208	$12.4\,t^{0.34} \cdot 10^{-6}$
	520	120	$7.3\,t^{0.36} \cdot 10^{-6}$
	630	120	$13.0\,t^{0.36} \cdot 10^{-6}$
peridotite	700	78	$30.4\,t^{0.27} \cdot 10^{-6}$
	750	82	$37.1\,t^{0.29} \cdot 10^{-6}$
	750	92	$49.4\,t^{0.30} \cdot 10^{-6}$
	750	102	$58.5\,t^{0.29} \cdot 10^{-6}$
	750	112	$63.6\,t^{0.29} \cdot 10^{-6}$
	750	117	$78.0\,t^{0.29} \cdot 10^{-6}$

Table 16. Steady-state creep in granite [Aue79].

A. Experimental parameters
   Fine-grained granite (aplite) from Schauinsland, Black Forest, FRG
   confining pressure: $p_m = 420$ MPa
   temperature: $900 < T < 1100$ °C
   differential stress: $0.4 < \sigma < 10$ MPa

B. Creep equation
   a) $\dot{\varepsilon} \propto \sigma^{1.2} \exp(-335/RT)$
   for $\dot{\varepsilon} < 3 \cdot 10^{-6}$ s^{-1}; $0.4 < \sigma < 1.5$ MPa;
   　$850 < T < 1012$ °C
   b) $\dot{\varepsilon} \propto \sigma^4 \exp(-870/RT)$
   for $\dot{\varepsilon} > 3 \cdot 10^{-6}$ s^{-1}; $1.5 < \sigma < 5$ MPa;
   　$1012 < T < 1070$ °C.

Rummel

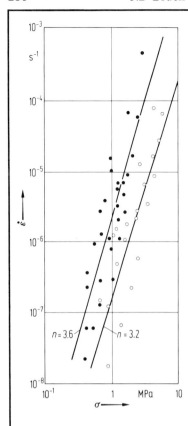

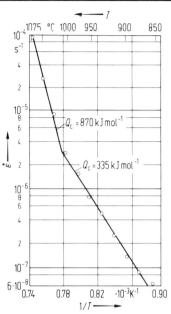

Fig. 25. Steady-state creep rate $\dot{\varepsilon}$ vs. differential stress σ for fine-grained granite at a confining pressure p_m = 420 MPa and at T = 982 °C (open circles) and at T = 1007 °C (full circles). Solid lines show the creep law for stress exponent n = 3.2 and n = 3.6, respectively. Granite is from Schauinsland, Schwarzwald, FRG (after [Aue79]).

Fig. 26. Steady-state creep rate $\dot{\varepsilon}$ vs. absolute temperature T for fine-grained Schauinsland granite (Schwarzwald, FRG) at p_m = 420 MPa and a differential stress σ = 1.1 MPa. Q_c = activation energy (after [Aue79]).

Table 17. Steady-state creep in peridotite [Ber79].

A. Experimental parameters
  Dry lherzolite from Ivrea-Verbano zone at Balmuccia, Italy
  Mineral composition: olivine (96 % forsterite) 58 %, orthopyroxene 27 %, clinopyroxene 11 %, spinel 4 %; grain size 1···2 mm
  Creep tests at $0.1 < \sigma < 30$ MPa
  $T \approx 1300$ °C
  $p_m = 0$

B. Creep equation
  a) $\dot{\varepsilon} = 45\,\sigma^4 \exp(-Q_c/RT)$,
     for $\sigma > 100$ bar (σ in bar!); $Q_c = 125$ kcal mol^{-1}; $1200 < T < 1400$ °C.

  b) $\dot{\varepsilon} = 3 \cdot 10^7\,\sigma^{1.4} \exp(-Q_c/RT)$,
     for $20 < \sigma < 100$ bar; $Q_c = 125$ kcal mol^{-1}; $1200 < T < 1400$ °C.

  c) $\dot{\varepsilon} = 4.2 \cdot 10^6\,\sigma^2 \exp(-Q_c/RT)$,
     for $5 < \sigma < 20$ bar; $Q_c = 125$ kcal mol^{-1}; $1200 < T < 1400$ °C.

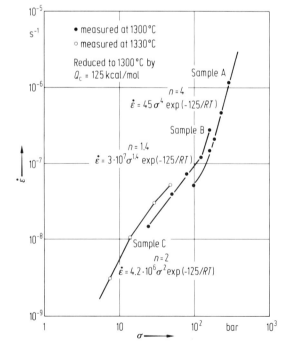

Fig. 27. Steady-state creep rate $\dot{\varepsilon}$ of Ivrea peridotite (3 samples) (Lherzolite from Balmuccia, Italy) as a function of differential stress σ at T = 1300 and 1330 °C (after [Ber79]).

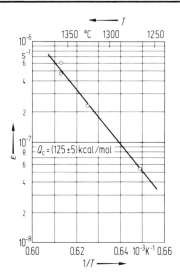

Fig. 28. Steady-state creep rate $\dot\varepsilon$ of Ivrea peridotite as a function of absolute temperature at differential stress $\sigma = 0.48$ MPa (after [Ber79]).

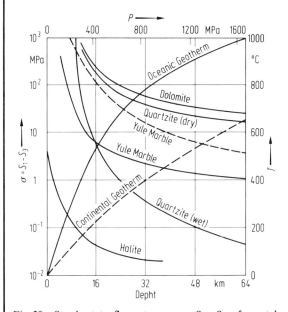

Fig. 29. Steady-state flow stresses $\sigma = S_1 - S_3$ of crustal rocks as a function of depth at a strain-rate of $\dot\varepsilon = 10^{-14}\,\mathrm{s}^{-1}$. Calculated from data of Table 14. Geotherms are pyroxene geotherms given by [Mer75]. $P =$ lithostatic pressure (after [Mer77, Han80]).

Table 18. Melting temperature diffusion coefficients $D(T_\mathrm{m})$ for materials of various crystal types [Sto73].

Material	$D(T_\mathrm{m})$ cm^{2}s^{-1}	mean $D(T_\mathrm{m})$ cm^{2}s^{-1}
body-centered cubic metals Raum-zentrierte kubische Metalle (Na, α-Fe, Li, Nb, Cr, Ta, β-Zr, β-Tl, γ-U)	10^{-8} to $2 \cdot 10^{-7}$	$8 \cdot 10^{-8}$
face-centered cubic and hexagonal close-packed metals Flächen-zentrierte kubische und hexagonal dichtest-gepackte Metalle (Pt, Au, Cu, β-Co, γ-Fe, Ni, Pb, Mg, Zn, Cd, α-Tl, α-Zr, Ag)	10^{-9} to 10^{-8}	$4 \cdot 10^{-9}$
diamond cubic and related structures Diamantstrukturen (Ge, β-Sn, H$_2$O)	$4 \cdot 10^{-12}$ to 10^{-10}	$3 \cdot 10^{-11}$
carbides (diffusivity of metal ion) Karbide (Diffusion des Metallions) (TiC)	$\approx 4 \cdot 10^{-7}$	$\approx 4 \cdot 10^{-7}$
oxides with close-packed oxygen lattice (oxygen diffusion) Oxide mit dichter Sauerstoffpackung (Sauerstoffdiffusion) (Al$_2$O$_3$, BeO, CoO, MgO, NiO)	10^{-12} to $4 \cdot 10^{-9}$	10^{-10}

Table 19. Oxygen diffusion coefficient in oxides at the melting point (see eq. (3) [Sto73]).

Material	Oxygen packing	D_0 cm^{2}s^{-1}	Q_D kJ mol^{-1}	$D(T_\mathrm{m})$ cm^{2}s^{-1}
Al$_2$O$_3$	hep	$1.9 \cdot 10^3$	637 (105)	$9.7 \cdot 10^{-12}$
Al$_2$O$_3$	hep	2	461 (63)	$9.1 \cdot 10^{-11}$
BeO	fcc	$2.9 \cdot 10^{-5}$	285	$1.3 \cdot 10^{-10}$
CoO	fcc	$5.9 \cdot 10^{-6}$	151	$9.5 \cdot 10^{-10}$
CoO	fcc	$5 \cdot 10^1$	398	$4.9 \cdot 10^{-9}$
MgO	fcc	$2.5 \cdot 10^{-6}$	260	$1.1 \cdot 10^{-10}$
MgO	fcc	$4.3 \cdot 10^{-5}$	344	$7.8 \cdot 10^{-11}$
NiO	fcc	$6.2 \cdot 10^{-4}$	243	$1.7 \cdot 10^{-9}$

Table 20. Activation energy Q_D and constant g in eq. (4), p. 210, for ice and some metals (after [Wee70]).

Material	T_m [1]) K	Q_D [2]) kJ mol^{-1}	g [3])
ice	273	58	25.5
K	337	41	14.5
Na	371	44	14
In	430	78.5	22
Li	452	55	15
Sn	504	≈ 106	25
Cd	594	≈ 80	16
Pb	601	106	21
Zn	693	≈ 96	17
Ag	1234	185	18

[1]) T_m at atmospheric pressure.
[2]) Activation energy for diffusion.
[3]) $g = Q_D/RT_m$ at moderate pressures.

Table 21. Anhydrite, $CaSO_4$ [Mül78].

A. Structure insensitive parameters

1. Lattice parameters
   $a = 6.238 \cdot 10^{-8}$ cm
   $b = 6.991 \cdot 10^{-8}$ cm
   $c = 6.996 \cdot 10^{-8}$ cm

2. Molar volume
   $V = 45.94$ cm^3 mol^{-1}

   Molecular volume
   $\Omega = 9.12 \cdot 10^{-23}$ cm^3

B. Experimental parameters
   Fine-grained, pure anhydrite (90 % $CaSO_4$) from Riburg, Switzerland, drillhole No. 51;
   grain-size of anhydrite needles:
   $l = 3 \cdot 10^{-2}$ cm,
   $d = 5 \cdot 10^{-3}$ cm
   confining pressure: $p_m = 150$ MPa
   temperature: $T = 20 \cdots 450$ °C
   strain rates: $10^{-7} < \dot{\varepsilon} < 10^{-4}$ s^{-4}

C. Creep parameters

1. Work-hardening occurs at $T < 300$ °C and $\dot{\varepsilon} > 10^{-6}$ s^{-1} resulting in a strength of $\sigma > 130$ MPa at 10 % deformation.

2. Best fit for steady-state deformation at higher temperature for $\sigma < 330$ MPa.
   $$\dot{\varepsilon} = A \exp(-Q_c/RT) (\sin h(\sigma/\sigma_0))^n$$
   $A = 2.07 \cdot 10^5$ s^{-1}; $Q_c = 152\,(9)$ kJ mol^{-1};
   $\sigma_0 = 80$ MPa; $n = 2$.

3. Creep mechanisms
   a) $56 < \sigma < 330$ MPa
      twinning and dislocation glide dominant
   b) $\sigma < 56$ MPa
      grain boundary sliding

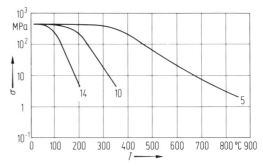

Fig. 30. Deformation map for polycrystalline anhydrite with anhydrite needles of 0.3 mm length and 0.05 mm diameter. σ = differential stress. Contours show lines of constant strain rate $\dot{\varepsilon}$, numbers refer to $-\log \dot{\varepsilon}$, $\dot{\varepsilon}$ in [s^{-1}]. Confining pressure $p_m = 150$ MPa (after [Mül78]).

Table 22. Calcite, $CaCO_3$ [Rut76, Hea63, Car76].

A. Structure insensitive parameters

  1. Lattice parameters
    $a = 4.99 \cdot 10^{-8}$ cm
    $c = 17.06 \cdot 10^{-8}$ cm

  2. Molar volume
    $V = 36.94$ cm^3 mol^{-1}

    Molecular volume
    $\Omega = 6.13 \cdot 10^{-23}$ cm^3

  3. Shear modulus at 293 K
    $\mu_0 = 25$ GPa

  4. Melting temperature
    $T_m = 1613$ K at 104 MPa

B. Experimental parameters

  [Rut76]: Carrara marble, Solnhofen limestone;
        confining pressure: $p_m = 150$ MPa
        temperature: $20 < T < 500$ °C
        strain rates: $10^{-4} < \dot{\varepsilon} < 10^{-8}$ s^{-1}

  [Hea63]: Yule marble;
        confining pressure: $p_m = 500$ MPa
        temperatures: $500 < T < 800$ °C
        strain rates: $10^{-3} < \dot{\varepsilon} < 10^{-7}$ s^{-1}

C. Intrinsic diffusion parameters

	[Rut76]	[Hea63, Car76]
D_{OV} [cm^2 s^{-1}]	$5 \cdot 10^{-2}$	0.41
Q_{VD} [kJ mol^{-1}]	197 at $T < 673$ K	
	250 at $T > 673$ K	257 (12) at $T > 773$ K
D_V [cm^2 s^{-1}]	10^{-11} at $T = 1373$ K	
D_{OB} [cm^2 s^{-1}]	$5 \cdot 10^{-2}$	
Q_{BD} [kJ mol^{-1}]	131 at $T < 673$ K	
	167 at $T > 673$ K	
δ [cm]	10^{-7}	
d [cm]	10^{-2}	$6 \cdot 10^{-2}$
d^* [cm]		$2 \cdot 10^{-3}$
Ω [cm^3]	$6.14 \cdot 10^{-23}$	$5.8 \cdot 10^{-23}$

D. Creep equations

  1. Best fit to experimental data
    $\dot{\varepsilon} = A \sigma^n \exp(-Q_c/RT)$

	Solnhofen limestone [2]	Carrara marble [2]	Yule marble
$\log A$ [1]	-33.5	-24.3	-4
Q_c [kJ mol^{-1}]	188 (25)	239 (50)	230 (40)
n	16.6 (20)	15.8 (20)	8 (3)

  [1]) For σ in MPa
  [2]) An exponential stress function is given by [Rut74].

  2. Extrapolation
    Nabarro-Herring creep:
    $\dot{\varepsilon} = 2 \cdot 10^{-5} D_V \sigma/d^2 T$
    for Yule marble [Car76].

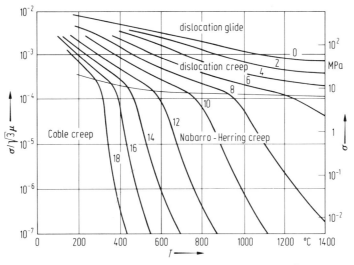

Fig. 31. Deformation mechanism map for calcite with grain size of $d = 0.1$ mm. σ = differential stress. μ = shear modulus. Contours show lines of constant strain rate $\dot{\varepsilon}$. Numbers refer to $-\log \dot{\varepsilon}$, $\dot{\varepsilon}$ in [s^{-1}] (after [Rut76]).

Table 23. Comparison of rheological parameters of calcite rocks (after [Sch79]).

	Solnhofen limestone		Carrara marble		Yule marble	
					1-orientation[2]	T-orientation
grain size [μm]	4		200		300…400	
study by	[Rut74]	[Sch77]	[Rut74]	[Sch79]	[Hea63, Hea72a]	
temperatures [°C]	T<500	600<T<900	T<500	600<T<1050	T<800	
confining pressure [kbar]	1.5	3	1.5	3	5 (extension tests, $S_1=S_2$, S_3)	
flow law in regime 1 ($\dot{\varepsilon}=$)[1], [3]	$10^{-0.12} \cdot \exp\left(-\dfrac{47}{RT}+\dfrac{\sigma}{160}\right)$	consistent with [Rut74]	$10^{5.8} \cdot \exp\left(-\dfrac{62}{RT}+\dfrac{\sigma}{114}\right)$	consistent with [Rut74]	$10^{7.8}\exp\left(-\dfrac{62}{RT}+\dfrac{\sigma}{91}\right)$	$10^{7.0}\exp\left(-\dfrac{57}{RT}+\dfrac{\sigma}{138}\right)$
stress at transition to regime 2 [bars]	1900		1000		1100	1400
flow law in regime 2 ($\dot{\varepsilon}=$)[1], [3]	$10^{-1.33}\exp\left(-\dfrac{71}{RT}\right)\sigma^{4.7}$		$10^{-4.5}\exp\left(-\dfrac{100}{RT}\right)\sigma^{7.6}$		$10^{-12.2}\exp\left(-\dfrac{62}{RT}\right)\sigma^{8.3}$	$10^{-11.3}\exp\left(-\dfrac{61}{RT}\right)\sigma^{7.7}$
stress at transition to regime 3 [bars][3]	1000 (600°) to 400 (900°)		200			
flow law in regime 3 ($\dot{\varepsilon}=$)[1], [3]	$10^{2.7}\exp\left(-\dfrac{51}{RT}\right)\sigma^{1.7}$		$10^{3.9}\exp\left(-\dfrac{102}{RT}\right)\sigma^{4.2}$			

[1] Units of s, bar, kcal mol⁻¹, °C; $\dot{\varepsilon}$=strain rate, R=gas constant, σ=differential stress, T=absolute temperature; the exponential fits for regime 1 in Yule marble are taken from [Rut74]; the flow law in regime 1 refers to the stress at 10% strain in all cases.
[2] The 1-orientation is more favourable to twinning in extension than is the T-orientation.
[3] Regime 1: $\sigma>1000$ bars
   Regime 2: $1000>\sigma>200$ bars
   Regime 3: $\sigma<200$ bars.

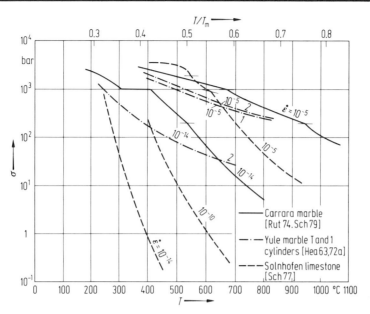

Fig. 32. Plot of differential stress σ as a function of temperature T for calcitic rocks. T_m = melting temperature in [K]. (1) 1-direction, (2) T-direction of Yule marble cylinders (after [Sch79]).

Table 24. Dolomite, CaMg(CO$_3$)$_2$ [Car76, Hea76a].	Table 25. Enstatite.

Table 24. Dolomite, CaMg(CO₃)₂ [Car76, Hea76a].

A. Structure insensitive parameters

  1. Lattice parameters
    $a = 4.808 \cdot 10^{-8}$ cm
    $c = 16.01 \cdot 10^{-8}$ cm

  2. Molar volume
    $V = 64.35$ cm^3 mol^{-1}

    Molecular volume
    $\Omega = 1.07 \cdot 10^{-22}$ cm^3

B. Experimental parameters

  a) Single crystals at 500 MPa and
    $24 < T < 500\,°C$ at $\dot\varepsilon = 10^{-4}$ s^{-1}

  b) Polycrystalline dolomites:
    confining pressure: $p_m = 300, 500, 1500$ MPa
    temperature: $T = 20 \cdots 300, 300 \cdots 800, 1000\,°C$
    strain rates: $\dot\varepsilon = 8 \cdot 10^{-7} \cdots 10^{-4}$ s^{-1}

C. Creep parameters
  Best fit to experimental data on Crevola
  dolomite in steady-state regime
  $\dot\varepsilon = A\,\sigma^n \exp(-Q_c/RT)$
  $Q_c = 349\,(25)$ kJ mol^{-1}
  $A = 1.26 \cdot 10^{-13}$ (σ in MPa)
  $n = 9.1\,(9)$

Table 25. Enstatite.

Orthoenstatite/Clinoenstatite, MgSiO$_3$ [Car76, Kir76, Coe75, Ral71]

A. Structure insensitive parameters

  1. Lattice parameters
    Orthoenstatite
    $a = 8.829 \cdot 10^{-8}$ cm
    $b = 18.22 \cdot 10^{-8}$ cm
    $c = 5.192 \cdot 10^{-8}$ cm

    Clinoenstatite
    $a = 9.618 \cdot 10^{-8}$ cm
    $b = 8.825 \cdot 10^{-8}$ cm
    $c = 5.186 \cdot 10^{-8}$ cm

  2. Molar volume
    $V_{enstatite} = 31.40$ cm^3 mol^{-1}
    $V_{clinoenstatite} = 31.47$ cm^3 mol^{-1}

    Molecular volume
    $\Omega_{enstatite} = 5.214 \cdot 10^{-8}$ cm^3
    $\Omega_{clinoenstatite} = 5.226 \cdot 10^{-8}$ cm^3

  3. Predominant slip system
    (100) [001]

  4. Burger's vector on (100) [001] (enstatite)
    $b = 5.175 \cdot 10^{-8}$ cm

  5. Clinoenstatite to enstatite transformation at
    hydrostatic pressure P
    $T(°C) = 538 + 32.9 \cdot P*)$
    $T(°C) = 639 + 26 \cdot P*)$
    $T(°C) = 566 + 45 \cdot P$ (continued)

    *) Three experimental data sets, P in GPa.

Table 25 (continued)

6. Melting temperature of enstatite

P G Pa	T_m K
0	
1	1943
2	2033
3	2113
5	2243

B. Experimental parameters
Experiments were performed on enstatite single crystals of orthobronzite $(Mg_{0.85}Fe_{0.14}Ca_{0.005})SiO_3$ [Coe75] and on enstatite-bearing rock (enstatolite) [Ral71].
confining pressure: $500 < p_m < 1500$ MPa
temperature: $800 < T < 1250\,°C$
strain rates: $10^{-3} < \dot{\varepsilon} < 10^{-7}\,s^{-1}$
differential stress: $100 < \sigma < 500$ MPa

C. Enstatite to clinoenstatite transformation at non-hydrostatic pressure [Ral71].
The boundary depends on the temperature and on the strain-rate.

D. Creep equations for enstatite
$$\dot{\varepsilon} = A\,\sigma^n \exp(-Q_c/RT)$$

Material	A	n	Q_c kJ mol^{-1}
enstatolite	0.32	2.4	293[1]
dry	$3.2 \cdot 10^{-25}$	8.0	42[2]
enstatite $+ H_2O$	$9 \cdot 10^{-2}$	3	268

[1] For slip.
[2] For transformation.

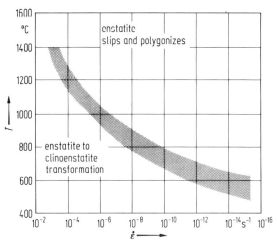

Fig. 33. Temperature vs. strain rate $\dot{\varepsilon}$ plot of boundary between the fields in which enstatite slips and pologonizes and in which it transforms to clino-enstatite during deformation (after [Car76, Ral71]).

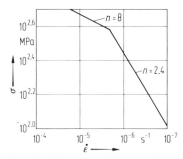

Fig. 34. Differential stress σ vs. strain rate $\dot{\varepsilon}$ plot for dry enstatite at $T = 1100\,°C$ and a confining pressure $p_m = 1$ GPa. $n =$ stress exponent in the power law (after [Ral71]).

Table 26. Galena, PbS [Atk77, Atk76].

A. Structure insensitive parameters
  1. Lattice parameter
     $a = 5.936 \cdot 10^{-8}$ cm
  2. Burger's vector for glide in $\langle 110 \rangle$ type directions
     $b = 4.2 \cdot 10^{-8}$ cm
  3. Molar volume
     $V = 31.5$ cm^3 mol^{-1}
     Molecular volume
     $\Omega = 5.23 \cdot 10^{-23}$ cm3
  4 Shear modulus at 293 °K
     $\mu_0 = 32.6$ GPa
     normalized temperature coefficient of μ at $P = 0.1$ MPa
     $(1/\mu_0)(\partial\mu/\partial T) = 3.34 \cdot 10^{-4}\,K^{-1}$
  5. Melting temperature
     $T_m = 1400$ K

B. Experimental parameters
Artificial, polycrystalline, stoichiometric galena of grain sizes of $10 \cdots 10^3$ μm (average: 35 μm)
temperature: $500 < T < 800\,°C$
strain rate: $10^{-4} < \dot{\varepsilon} < 10^{-9}$

C. Intrinsic diffusion parameters
$D_{OV} = 8.6 \cdot 10^{-5}$ cm^2 s^{-1}
$Q_{VD} = 147$ kJ mol^{-1}
$D_{OB} = 8.6 \cdot 10^{-5}$ cm^2 s^{-1}
$Q_{BD} = 98$ kJ mol^{-1}
$$\delta = \begin{cases} 8.4 \cdot 10^{-8} = 2b \\ 4.2 \cdot 10^{-7} = 10b \end{cases}$$

(continued)

Table 26 (continued)

D. Creep equations

1. Dislocation glide
   $$\dot{\varepsilon} = K \exp(-Q_c/RT) \exp(B\sigma)$$
   $$K = 3.72 \cdot 10^{-3}\ \text{s}^{-1}$$
   $$Q_c = 94.5\ \text{kJ mol}^{-1}$$
   $$B = 1.16 \cdot 10^{-7}\ \text{Pa}^{-1}$$

2. Dislocation creep
   $$\dot{\varepsilon} = (B \mu b D_V / k\,T)\,(\sigma/\mu)^n$$

	200···400 °C	700 °C
B [Pa^{-1}]	$3.5 \cdot 10^{13}$	$4.5 \cdot 10^6$
n	7	5

3. Diffusion creep
   $$\dot{\varepsilon} = 21\,(\sigma \Omega D_V / k\,T d^2) \cdot \left\{ 1 + \frac{\pi \delta}{d}\left(\frac{D_B}{D_V}\right) \right\}$$
   $$10 < d < 10^3\ \mu\text{m}.$$

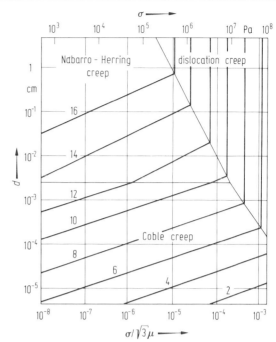

Fig. 36. Deformation mechanism map (stress σ vs. grain size d) for polycrystalline galena (PbS) at 0.5 T_m (427 °C). The contours show lines of constant strain rate $\dot{\varepsilon}$ numbered in terms of $-\log \dot{\varepsilon}$, $\dot{\varepsilon}$ in [s^{-1}] μ: shear modulus (after [Atk77]).

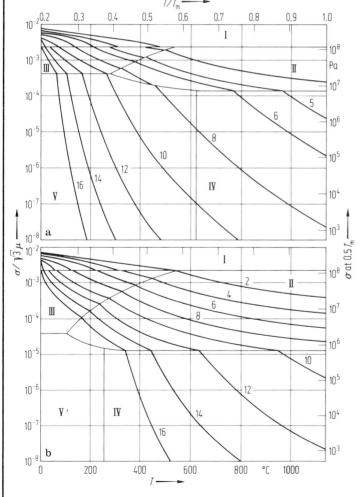

Fig. 35a, b. Deformation mechanism maps for stoichiometric galena with grain size 10 μm (a) and 10^3 μm (b). T_m = melting temperature in [K]. μ = shear modulus. Numbers on contours of constant strain rate refer to $-\log \dot{\varepsilon}$, $\dot{\varepsilon}$ in [s^{-1}] (after [Atk77]).

I, dislocation glide;
II, high-temperature dislocation creep;
III, low temperature dislocation creep;
IV, Nabarro-Herring creep;
V, Coble creep.

Table 27. Halite, NaCl [Car70].

A. Structure insensitive parameters

    1. Lattice parameter
       $a = 5.64 \cdot 10^{-8}$ cm^3

    2. Molar volume
       $V = 27.02$ cm^3 mol^{-1}

       Molecular volume
       $\Omega = 1.8 \cdot 10^{-22}$ cm^3

    3. Melting temperature
       $T_m = 1093$ K at 0.1 MPa
       $T_m = 1273$ K at 1 GPa
       $T_m = 1473$ K at 2 GPa

B. Experimental parameters

Ref.	Material	σ MPa	p_m MPa	T °C	$\dot{\varepsilon}$ s^{-1}
Chr56	natural polycrystalline	≈ 1		600$\cdots$760	$10^{-4}\cdots10^{-6}$
Geg64	artificial single crystal	0.01$\cdots$1		750$\cdots$780	
LeC65	artificial polycrystalline	7$\cdots$14	≤ 100	< 300	transient creep
Tho65	natural polycrystalline		21	33	
Blu67	artificial single crystal	0.4$\cdots$3.6		550$\cdots$800	$10^{-2}\cdots10^{-6}$
Bur68	artificial polycrystalline	0.3$\cdots$14		365$\cdots$740	$10^{-2}\cdots10^{-7}$
Blu69	artificial polycrystalline	2$\cdots$22		250$\cdots$780	$10^{-1}\cdots10^{-6}$
Sch70b	artificial single crystal	0.1$\cdots$20		260$\cdots$780	$10^{-1}\cdots10^{-6}$
	artificial single crystal	1.5$\cdots$40	200	20$\cdots$500	$10^{-1}\cdots10^{-8}$
	artificial polycrystalline	1.5$\cdots$47	200	20$\cdots$400	$10^{-1}\cdots10^{-8}$
	artificial single crystal	0.5$\cdots$1.5		500$\cdots$750	$10^{-3}\cdots2\cdot10^{-8}$ (transient creep)
	natural polycrystalline	2$\cdots$18		27$\cdots$300	$10^{-7}\cdots10^{-10}$

C. Intrinsic diffusion parameters
    $D_{OV} = 10^{-4}$ cm^2 s^{-1} for Na$^+$
    $D_V\ = 2 \cdot 10^{-9}$ cm^2 s^{-1} for Na$^+$ at T_{mO}
    $Q_{VD} =\ 98\ (8)$ kJ mol^{-1} for Na$^+$
    $Q_{VD} = 201\ (25)$ kJ mol^{-1} for Cl$^-$

D. Creep equations

    1. Best fit to experimental data
      a) Dislocation creep (for $0.5 < \sigma < 10$ MPa)
         $\dot{\varepsilon} = A\,\sigma^n \exp(-Q_c/RT)$
         $A = 0.95$ (σ in MPa)
         $Q_c = 98\ (8)$ kJ mol^{-1}
         $n = 5.5\ (4)$

      b) Dislocation glide with strain hardening
        (for $10 < \sigma < 35$ MPa)
         $\dot{\varepsilon} = A \exp(-Q_G/RT) \sin h(B\,\sigma)$
         $A = 3.2 \cdot 10^4$ ($\log A = 4.5$)
         $Q_G = 109\ (10)$ kJ mol^{-1}
         $B = 5.2 \cdot 10^{-1}$ (for $10 < \sigma < 35$ MPa)

    2. Extrapolations
      Nabarro-Herring creep for $\sigma < 0.5$ MPa
      $\dot{\varepsilon} = \alpha\,\Omega\,D_v\,\sigma/d^2\,kT$
      $= 6.5 \cdot 10^{-5}\,(D\,\sigma/T)$ for $d = 1$ cm, $\alpha = 5$

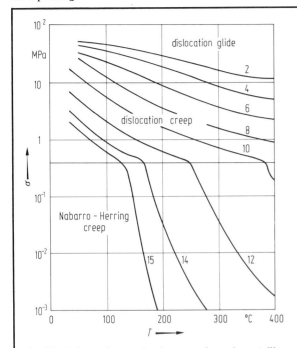

Fig. 37. Deformation mechanism map for polycrystalline halite. $\sigma =$ differential stress. Contours show lines of constant strain rate; numbers refer to $-\log \dot{\varepsilon}$, $\dot{\varepsilon}$ in [s^{-1}] (after [Hea71]).

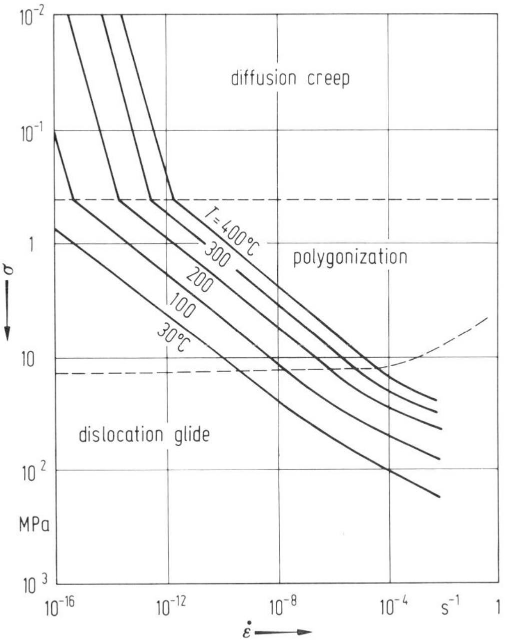

Fig. 38. Deformation map of polycrystalline halite (strain rate $\dot{\varepsilon}$ vs. differential stress σ). Contours show lines of constant temperature T in [°C] (after [Hun79]).

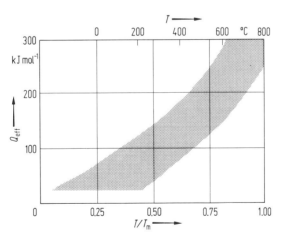

Fig. 39. Effective activation energy Q_{eff} for power law steady-state creep of NaCl as a function of temperature. T_{m} melting temperature of NaCl; T, T_{m} in [°C] (after [Hun79]).

Table 28. Comparison of rheological data of rock salt (after [Lan79]).

Type of rock salt[1]	σ p_m bar	$\dot{\varepsilon}$[2] s^{-1}	T °C	creep law[3]	P: A E: C s^{-1} bar^{-n} s^{-1}	Q_c Q_{power} kcal mol^{-1} kcal mol^{-1}	n B — bar^{-1}	Ref.
np	20··180 —	10^{-7}··10^{-10}	27··30	P	$4.8 \cdot 10^{-11}$	13.7	5.0	Alb78
ap	20··220 —	10^{-1}··10^{-6}	250··780	(P)	—	23··80	4··15	Blu69
as	4··36 —	10^{-2}··10^{-6}	550··800	P	—	58.2	4.04	Blu67
ap	3··138 —	10^{-2}··10^{-7}	365··740	P	$5 \cdot 10^{-3}$	$T<550$ °C: 38 $T>500$ °C: 48	5.5	Bur68
as	15··400 2000	10^{-1}··10^{-8}	25··500	300··500 °C: P 25··400 °C: E	— —	33 20	7 $5.3 \cdot 10^{-2}$	Car70
as	0.1··10 —	—	750··780	P	—	57.6	4.2··3.8	Geg64, Bur68
ap	16··470 2000	10^{-1}··10^{-8}	23··400	100··400 °C: P 23··400 °C: E	$2.5 \cdot 10^{-6}$ $3.2 \cdot 10^{4}$	23.5 26	5.5 $4.9 \cdot 10^{-2}$	Hea72
as	5··15 —	10^{-3}··10^{-8}	500··700	(P)	$3 \cdot 10^{3}$	51	5.3	Hs63, Bur68
np	— 0··207	—	33	(P)	$7.4 \cdot 10^{-11}$	13	5.24	Tho65
np	20··180 —	10^{-7}··10^{-11}	27··300	P	$2.1 \cdot 10^{-11}$	12.9	5.0	Wal79

[1] ap: artificial and polycrystalline; as: artificial single crystal; np: natural polycrystalline.
[2] Strain rate of steady-state creep.
[3] P: power creep law $\dot{\varepsilon} = A \exp(-Q_c/RT) \sigma^n$; E: exponential creep law $\dot{\varepsilon} = C \exp(-Q_c/RT) \sin h(B\sigma)$.

Table 29. Hematite, α-Fe_2O_3 [Atk77].

A. Structure insensitive parameters

  1. Lattice parameters
    $a = 5.033 \cdot 10^{-8}$ cm
    $c = 13.75 \cdot 10^{-8}$ cm

  2. Burger's vector
    $b = 6.41 \cdot 10^{-8}$ cm

  3. Molar volume
    $V = 30.28$ cm^3 mol^{-1}

    Molecular volume
    $\Omega = 5.028 \cdot 10^{-23}$ cm^3

  4. Shear modulus at 293 K
    $\mu_0 = 91$ GPa

    normalized temperature coefficient of shear modulus at
    $P = 0.1$ MPa
    $(1/\mu_0)(\partial\mu/\partial T) = 1.09 \cdot 10^{-4}$ K^{-1}

  5. Melting temperature
    $T_m = 1895$ K

B. Experimental parameters
No experimental creep data are available for both dislocation glide and low-temperature dislocation creep.

C. Intrinsic diffusion parameters

D_{OV} [cm^2 s^{-1}]	Fe^{3+}-diffusion	$8.5 \cdot 10^5$
Q_{VD} [kJ mol^{-1}]	Fe^{3+}-diffusion	444
D_V [cm^2 s^{-1}]	Fe^{3+}-diffusion at 0.5 T_m	$2.94 \cdot 10^{-19}$
D_{OB} [cm^2 s^{-1}]		$8.5 \cdot 10^5$
Q_{BD} [kJ mol^{-1}]		296
D_{BD} [cm^2 s^{-1}]	at 0.5 T_m	$4.19 \cdot 10^{-11}$
δ [cm]		$6.41 \cdot 10^{-7}$
		$= 10\, b$

D. Creep equations

  1. Dislocation creep (at 0.5 T_m)
    $\dot{\varepsilon} = (B\,\mu\,b\,D_V/kT)(\sigma/\mu)^n$
    $B = 0.574$ for $n = 3$
    $n = 3.07$ (29) log B

  2. Coble creep at 0.5 T_m
    $\dot{\varepsilon} = 21\,(\sigma\,\Omega\,D_V/kT d^2)\left\{1 + \dfrac{\pi\,\delta}{d}\left(\dfrac{D_B}{D_V}\right)\right\}$
    10^{-5} cm $< d < 1$ cm

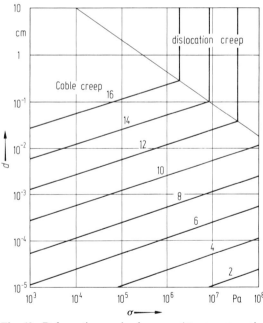

Fig. 40. Deformation mechanism map (stress σ vs. grainsize d) for polycrystalline hematite ($\alpha - Fe_2O_3$) at 0.5 T_m (675 °C). The contours show lines of constant strain rate numbered in terms of $-\log\dot{\varepsilon}$, $\dot{\varepsilon}$ in [s^{-1}] (after [Atk77]).

Table 30. Magnetite Fe_3O_4 [Atk77]

A. Structure insensitive parameters

1. Lattice parameter
   $a = 8.394 \cdot 10^{-8}$ cm

2. Burger's vector for glide $\langle 110 \rangle$-type directions
   $\boldsymbol{b} = 5.94 \cdot 10^{-8}$ cm

3. Molar volume
   $V = 44.53$ cm^3 mol^{-1}

   Molecular volume
   $\Omega = 7.395 \cdot 10^{-23}$ cm^3

4. Shear modulus at 293 K
   $\mu_0 = 98$ GPa
   normalized temperature coefficient of shear
   modulus at $P = 0.1$ MPa
   $(1/\mu_0)(\partial \mu / \partial T) = 1.12 \cdot 10^{-4}$ K^{-1}

5. Melting temperature
   $T_m = 1870$ K

B. Experimental parameters
   No experimental creep data available.

C. Intrinsic dissusion parameters

D_{OV} [cm^2 s^{-1}]	O^{2-}-diffusion		$3.2 \cdot 10^{-10}$
Q_{VD} [kJ mol^{-1}]	O^{2-}-diffusion		71.2
D_V [cm^2 s^{-1}]	O^{2-}-diffusion at 0.5 T_m		$3.4 \cdot 10^{-18}$
D_{OB} [cm^2 s^{-1}]			$3.2 \cdot 10^{-11}$
Q_{BD} [kJ mol^{-1}]			46.45
D_B [cm^2 s^{-1}]	for O^{2-} at 0.5 T_m		$7.2 \cdot 10^{-17}$
δ [cm]			$5.94 \cdot 10^{-7}$

D. Creep equations

1. Dislocation creep
   $\dot{\varepsilon} = (B \mu b D_V / kT)(\sigma/\mu)^n$
   $B = 0.5736$ for $n = 3$; $n = 3.07 + 0.29 \log B$.

2. Dislocation glide
   Critical shear stress for transition from
   dislocation creep to dislocation glide is
   assumed to be equal to 50 MPa.

3. Nabarro-Herring creep
   $\dot{\varepsilon} = 21 (\sigma \Omega D_V / kT d^2)$
   10^{-5} cm $< d < 1$ cm

4. Coble creep
   $\dot{\varepsilon} = 21 (\sigma \Omega D_V / kT d^2)\left[1 + \dfrac{\pi \delta}{d}\left(\dfrac{D_B}{D_V}\right)\right].$

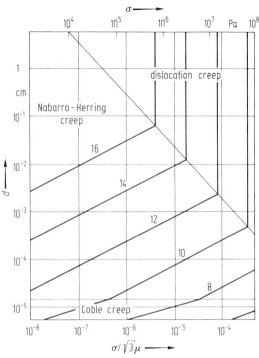

Fig. 41. Deformation mechanism map (stress σ vs. grain-size d) for polycrystalline magnetite (Fe_3O_4) at 0.5 T_m (662 °C). Contours show lines of constant strain rate numbered in terms of $-\log \dot{\varepsilon}$, $\dot{\varepsilon}$ in [s^{-1}] (after [Atk77]). μ = shear modulus.

Table 31. Olivine, $(Mg_xFe_{1-x})_2SiO_4$ [Sto73, Ral70, Koh76, Car76, Atk75a].

A. Structure insensitive parameters

1. Lattice parameters
   forsterite Mg_2SiO_4
   $a = 4.758 \cdot 10^{-8}$ cm
   $b = 10.214 \cdot 10^{-8}$ cm
   $c = 5.984 \cdot 10^{-8}$ cm

   fayalite Fe_2SiO_4
   $a = 4.817 \cdot 10^{-8}$ cm
   $b = 10.477 \cdot 10^{-8}$ cm
   $c = 6.105 \cdot 10^{-8}$ cm

2. Molar volume
   forsterite: $V = 43.79$ cm^3 mol^{-1}
   fayalite: $V = 46.39$ cm^3 mol^{-1}

   Molecular volume
   forsterite: $\Omega = 7.27 \cdot 10^{-23}$ cm^3
   fayalite: $\Omega = 7.70 \cdot 10^{-23}$ cm^3

   Atomic volumes in forsterite Fo_{85}
   $\Omega_{O^{--}} = 1.15 \cdot 10^{-22}$ cm^3
   $\Omega_{Mg^{++}} = 1.56 \cdot 10^{-24}$ cm^3
   $\Omega_{Si^{4+}} = 7.36 \cdot 10^{-26}$ cm^3

(continued)

Table 31 (continued)

3. Burger's vector (average for forsterite Fo_{90})
   $b = 6.89 \cdot 10^{-8}$ cm

4. Shear modulus at 293 K for forsterite Fo_{93}
   single crystal (average)
   $\mu_0 = 79.1$ GPa

   normalized temperature coefficient of μ for
   forsterite for $T > 500$ K at 0.1 MPa
   $(1/\mu_0)(\partial \mu / T) = -1.36 \cdot 10^{-4}$ K^{-1}

   normalized pressure coefficient of μ for
   forsterite Fo_{93} at 293 K
   $(1/\mu_0)(\partial \mu / P) = 0.023$ GPa^{-1}

5. Melting temperature
   forsterite: $T_m = 2173$ K at 0.1 MPa
   fayalite: $T_m = 1478$ K at 0.1 MPa
   Effect of pressure on melting temperature:

P GPa	Mg_2SiO_4 T_m K	$MgSiO_3$ T_m K
0	2173	
1	2223	1943
2	2263	2033
3	2313	2113
5	2413	2243

C. Intrinsic diffusion parameters

1. Lattice diffusion

B. Experimental parameters

Experimental creep data on olivine and olivine-bearing rocks (dunites, eclogites, peridotites, lherzolite) have been rapidly multiplied during the past decade in response to the increasing interest in upper-mantle flow accociated with large-scale plate motions. Investigations of high-temperature, steady-state creep include experiments on natural olivine-rich polycrystalline materials as well as on natural olivine single crystals. Data from polycrystalline samples have the advantage to allow grain boundary diffusion processes, however, the presence of other mineral phases (e.g. pyroxene) lowers the melting temperature to about 1400 °C. On the other hand, experiments on olivine single crystals allow to select favourable glide systems and data from single crystal creep experiments provide a lower limit for a polycrystalline flow law. The experimental investigations cover the following experimental range:

a) Confining pressure during triaxial deformation tests on about 0.5 cm diameter specimens:
   $p_m < 200$ MPa
b) differential stress: $10 < \sigma < 1000$ MPa
c) temperature for high temperature creep tests:
   $700 < T < 1650$ °C
d) strain rates: $10^{-4} < \dot{\varepsilon} < 10^{-7}$ s^{-1}

Ref.	D_0 $cm^2 s^{-1}$	Q $kJ\,mol^{-1}$	V [5] $cm^3\,mol^{-1}$	D_{VD} at T_m $cm^2 s^{-1}$	$-g$
[Sto73]	$1.2 \cdot 10^6$	645[6]	11[8]	$3.6 \cdot 10^{-10}$	36
[Sto73]	$1.4 \cdot 10^5$	608[6]	37[9]	$3.2 \cdot 10^{-10}$	34
[Goe73]	$3 \cdot 10^4$	566[4]		$7.3 \cdot 10^{-10}$	31
[Kir73]	$4 \cdot 10^2$	545[7]	11	$3 \cdot 10^{-11}$	30
[Kir73]	0.3	402[8]	11	$6 \cdot 10^{-11}$	22
[Ral70]	$1.2 \cdot 10^2$	419[1]		$1 \cdot 10^{-8}$	23
[Ral70]	$5 \cdot 10^2$	444[2]		$1 \cdot 10^{-8}$	25
[Ral70]	$1.9 \cdot 10^3$	478[3]		$1 \cdot 10^{-8}$	26

[1] Dry, fine-grained artificial dunite at 1000 °C and 1.5 GPa.
[2] Lherzolite at 1000 °C and 1.5 GPa.
[3] Dry Mt. Burnet dunite at 1000 to 1200 °C.
[4] Annealing of natural deformed peridotites at 1290 to 1450 °C.
[5] Activation volume.
[6] 85 % forsterite.
[7] 90 % forsterite.
[8] Diffusion of O^{--} in olivine.
[9] For SiO_4^{4-} diffusion.

2. Grain boundary diffusion [Sto73]

	(a) [1]	(b) [1]
D_{OB} $[cm^2\,s^{-1}]$	$1.2 \cdot 10^6$	$1.4 \cdot 10^{-5}$
Q_{BD} $[kJ\,mol^{-1}]$	432	406
V_a $[cm^3\,mol^{-1}]$	11	37
δ [cm]	$1.4 \cdot 10^{-7}$	$1.4 \cdot 10^{-7}$

[1] Two alternative sets of data assuming O^{--} or SiO_4^{4-} diffusion.

(continued)

Rummel

Table 31 (continued)

D. Creep equations

1. Empirical creep equations and best fit to experimental data

$$\dot\varepsilon = A\,\sigma^n \exp(-Q_c/RT) \quad (\sigma \text{ in MPa})$$

Material	Temperatures/stress difference σ or confining pressure p_m	Q_c kJ mol^{-1}	A MPa^{-1}s^{-1}	n	Ref.
single crystals	1400···1650 °C 10···200 MPa	524 (63)	$3 \cdot 10^5$	3.0	[Koh76]
dunite, Mt. Burnet, Canada, dry	1100···1350 °C p_m: 500···3000 MPa	465	$1.3 \cdot 10^3$	3.3	[Car76]
	1000 °C p_m: 1500 MPa	419	$1.8 \cdot 10^2$	3.0	[Kir73]
	725···1325 °C	545	$6.8 \cdot 10^2$	3.2	[Pos73]
wet	725···1325 °C	390	$6.8 \cdot 10^2$	3.2	[Pos73]
	1000···1350 °C	226	$8 \cdot 10^{-2}$	2.1	[Car76]
dunite, Åheim, Norway	<1600 °C σ: 1···20 MPa	561	$5.2 \cdot 10^8$	2.0	[Ber78]
peridotite, Ivrea, Italy	<1600 °C σ: 1···20 MPa	524	$7.5 \cdot 10^8$	1.4	[Ber78]
lherzolite	1000···1100 °C p_m: 1500 MPa	444	$1 \cdot 10^{-2}$	5	[Ral70]

2. Extrapolations

a) Dislocation creep [Sto73]

$$\dot\varepsilon = B(D_V\,\mu\,b/k\,T)(\sigma/\mu)^n$$

	1)	2)
B	0.7	$1.2 \cdot 10^4$
μ	3.0	4.2

1), 2) Different sets of values depending on D_V (see above)

b) Nabarro-Herring creep [Ral70]

$$\dot\varepsilon = \frac{\alpha\,\Omega}{l^2}\,\frac{\sigma\,D_V}{k\,T}$$

$$D_V = D_0 \exp(-26.4\,T_m/T)$$

$\alpha = 5$

$\Omega = 1.1 \cdot 10^{-23}$ cm^3

l = grain size (e.g. l = 1 cm)

c) subgrain diffusion creep [Ral70]

$$\dot\varepsilon = \frac{\alpha\,\Omega}{(l_0\,\mu)^2}\,\frac{D_V\,\sigma^3}{k\,T}$$

$l_0\,\mu = 2.4 \cdot 10^{-2}$ MPa cm (l_0 is a constant and related to the subgrain diameter l by the relation $l = l_0\,\mu/\sigma$)

$\Omega = 1.1 \cdot 10^{-23}$ cm^3

$\alpha = 5$

For Fig. 42, see next page.

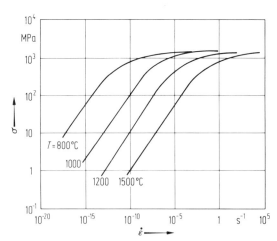

Fig. 43. Deformation diagram for dry olivine with an activation energy $V_a = 524$ kJ mol^{-1} in the temperature interval $800 < T < 1500$ °C (after [Koh76]).

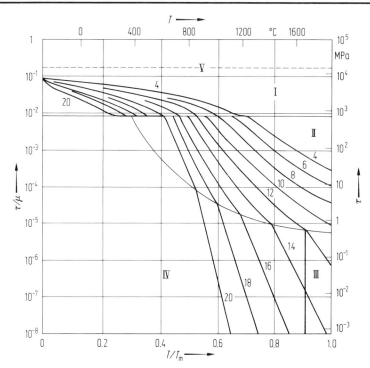

Fig. 42. Deformation mechanism map for polycrystalline olivine $(Mg_{0.85}Fe_{0.15})_2SiO_4$ with a grain size $d = 1$ mm and an activation volume $V_a = 50\ cm^3\ mol^{-1}$ at pressure $P = 1$ GPa. $\tau =$ shear stress, $\mu =$ shear modulus, $T_m =$ melting temperature in [K]. I, dislocation glide; II, dislocation creep; III, Nabarro-Herring creep; IV, Coble creep; V, theoretical strength. Contours show lines of constant strain rate $\dot\varepsilon$, numbers refer to $-\log \dot\varepsilon$, $\dot\varepsilon$ in [s^{-1}] (after [Sto73]).

Table 32. Pyrrhotite, $Fe_{1-x}S$ [Atk75a].

A. Structure insensitive parameter

  1. Lattice parameters
    $a = 3.44 \cdot 10^{-8}$ cm
    $b = 5.7\ (1) \cdot 10^{-8}$ cm

B. Experimental parameters
  Natural polycrystalline pyrrhotite ore with 80 vol% pyrrhotite with a mean grain diameter of 1.53 mm; bulk composition of the pyrrhotic is 46.9 at $-\%$ Fe. confining pressure: $p_m = 150$ MPa temperature: $20 < T < 400\,°C$ strain rates: $10^{-4} < \dot\varepsilon < 10^{-8}\ s^{-1}$

C. Creep equation
  Best fit to experimental data
  $\dot\varepsilon = K \exp(B\sigma)$ (σ differential stress at 10% strain)

T °C	$\log K$ [K in s^{-1}]	B MPa^{-1}
20	-21.90	$7.7\ (64) \cdot 10^{-2}$
200	-11.47	$5.5\ \ (7) \cdot 10^{-2}$
300	-10.94	$11.5\ (13) \cdot 10^{-2}$
400	-13.18	$25.7\ (54) \cdot 10^{-2}$

D. Creep mechanism
  $20 < T < 200\,°C$: dislocation glide plus kinking, accompanied by small component of cataclasis
  $300 < T < 400\,°C$: twinning accompanies dislocation glide and kinking
  $T > 400\,°C$ and $\dot\varepsilon < 10^{-8}\ s^{-1}$: probably dislocation creep

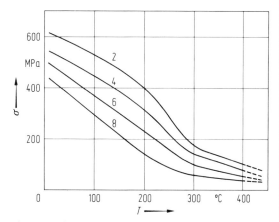

Fig. 44. Deformation map for polycrystalline pyrrhotite at a confining pressure $p_m = 150$ MPa. $\sigma =$ differential stress. Contours show lines of constant strain rate $\dot\varepsilon$, numbers refer to $-\log \dot\varepsilon$, $\dot\varepsilon$ in [s^{-1}]. Grain size $d = 1.53$ mm (after [Atk75a]).

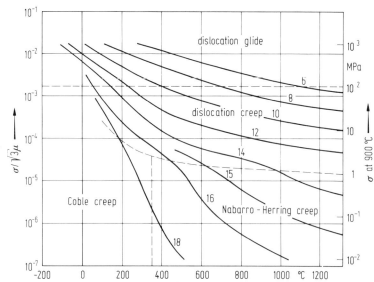

Fig. 45. Deformation mechanism map for quartz with grain size $d = 0.1$ mm. σ = differential stress, μ = shear modulus. Contours show lines of constant strain rate $\dot{\varepsilon}$. Numbers refer to $-\log \dot{\varepsilon}$, $\dot{\varepsilon}$ in [s^{-1}] (after [Rut76]).

Table 33. Quartz, α-SiO$_2$, β-SiO$_2$.

A. Structure insensitive parameters

   1. Lattice parameters
$$a_\alpha = 4.914 \cdot 10^{-8} \text{ cm}$$
$$a_\beta = 5.999 \cdot 10^{-8} \text{ cm}$$
$$c_\alpha = 5.405 \cdot 10^{-8} \text{ cm}$$
$$c_\beta = 5.459 \cdot 10^{-8} \text{ cm}$$

   2. Molar volume
$$V_\alpha = 22.69 \text{ cm}^3 \text{ mol}^{-1}$$
$$V_\beta = 23.72 \text{ cm}^3 \text{ mol}^{-1}$$

   Molecular volume
$$\Omega_\alpha = 3.77 \cdot 10^{-23} \text{ cm}^3$$
$$\Omega_\beta = 3.94 \cdot 10^{-23} \text{ cm}^3$$

   3. Burger's vector
$$b = 5 \cdot 10^{-8} \text{ cm}$$

   4. Shear modulus at 293 K
$$\mu_0 = 42 \text{ GPa (for quartzite)}$$

   normalized temperature coefficient of shear modulus at 1100 K
$$(l/\mu_0)(\partial\mu/\partial T) = 2.17 \cdot 10^{-4} \text{ K}^{-1}$$

   5. α-quartz – β-quartz transition

P MPa	T K
0.1	845
200	899
400	952
600	1001
800	1046
1000	1088

B. Experimental parameters (see [Car76, Rut76, Kir77, Hob72, Hea68, Chr64])
Natural and synthetic quartz single crystals and quartzites at various pressures (up to 3 GPa) and various temperatures (up to 1400 K) and at strain rates from $10^{-4} \cdots 10^{-8} \text{ s}^{-1}$.

C. Intrinsic diffusion parameters

D_{OV} [cm^2 s^{-1}]	for OH$^-$ and H$^-$ diffusion	$4 \cdot 10^{-14}$
Q_{VD} [kJ mol^{-1}]	for OH$^-$ and H$^-$ diffusion	84
D_V [cm^2 s^{-1}]	for OH$^-$ and H$^-$ diffusion (at 900 K)	$1.2 \cdot 10^{-17}$
D_{OB} [cm^2 s^{-1}]		$4 \cdot 10^{-14}$
Q_{BD} [kJ mol^{-1}]		56
δ [cm]		10^{-8}

D. Creep equations

   1. Best fit to experimental data

    a) Hydrolytically-weakened synthetic quartz [Kir77]
$$\dot{\varepsilon} = A \exp(-Q_c/RT)$$
$Q_c = 163\ (9) \text{ kJ mol}^{-1}$ for α-SiO$_2$
$Q_c = 60\ (20) \text{ kJ mol}^{-1}$ for β-SiO$_2$

    b) Dry Simpson sandstone [Car76]
$$\dot{\varepsilon} = A \sigma^n \exp(-Q_c/RT);$$
$Q_c = 268\ (23) \text{ kJ mol}^{-1}$; $n = 6.5\ (8)$;
$A = 6.7 \cdot 10^{-12}$ (σ in MPa)

    c) Canyon Creek quartzite in the presence of H$_2$O [Car76]
$$\dot{\varepsilon} = A \sigma^n \exp(-Q_c/RT)$$
$Q_c = 230\ (29) \text{ kJ mol}^{-1}$; $n = 2.6\ (4)$;
$A = 4.4 \cdot 10^{-2}$ (σ in MPa)

   2. Extrapolations [Rut76]

    a) Dislocation creep
$$\dot{\varepsilon} = (B \mu b D_V/RT)(\sigma/\mu)^n$$
$n = 4$; $B = 1.6 \cdot 10^3$; $n = 3.64$ (for hydrolytically-weakened quartz at 500 °C);
$n = 4.2$ (for synthetic quartz at 900 °C).

    b) Diffusion creep
$$\dot{\varepsilon} = 21 (\sigma \Omega D_V/k\, T d^2) \left\{ 1 + \frac{\pi \delta}{d} \left(\frac{D_B}{D_V} \right) \right\}$$

3.2.4 References for 3.2.1 ··· 3.2.3 — Literatur zu 3.2.1 ··· 3.2.3

Abe73	Abey, A.E., Heard, H.C.: Abstr., Trans. Am. Geoph. Union **54** (1973) 464–465.
Alb78	Albrecht, H., Meister, D., Wallner, M.: Proc. 3. Nat. Tagg. Felsmechanik, Aachen-Essen **(1978)**, 189–207.
Abo78	Abou-Sayed, A.S., Brechtel, C.E., Clifton, R.J.: J. Geophys. Res. **83** (1978) 2851–2862.
Alh75	Alheid, H.J.: Dipl. Arbeit, Geophysik, Univ. Bochum, **1975**.
Ash72	Ashby, M.F.: Acta Met. **20** (1972) 887–897.
Atk74	Atkinson, B.K.: Trans. Inst. Min. Metall. **83** (1974) B19–B28.
Atk75	Atkinson, B.K.: Economic Geol. **70** (1975) 473–487.
Atk75a	Atkinson, B.K.: N. Jb. Min. Mh. **75** (1975) 483–499.
Atk75b	Atkinson, B.K.: Econ. Geol. **71** (1976) 513–525.
Atk76	Atkinson, B.K.: Earth and Panet. Sci. Lett. **29** (1976) 210–218.
Atk76a	Atkinson, B.K.: J. Geol. Sci. **132** (1976) 555–562.
Atk77	Atkinson, B.K.: Geol. Föreningens i. Stockholm Förhandlingar **99** (1977) 186–197.
Atk78	Atkinson, B.K., Rutter, E.R., Sipson, R.H., White, S.H.: Rep. 1978 to USGS, I.C. London, Geol. Dept., Oct. **1978**.
Atk79	Atkinson, B.K.: Int. J. Rock Mech. Min. Sci. **16** (1979) 49–53.
Atk79a	Atkinson, B.K., Rawlings, R.D.: Proc. Inst. Acoustics, London **1979**, unpublished data. (see [Atk79b]).
Atk79b	Atkinson, B.K.: Final Tech. Rep. for FY 1979, US Nat. Earthquake Hazard Reduction Program, Imp. Coll. London, **1979**, 2.1–2.67.
Aue79	Auer, F., Berckhemer, H.: unpublished data **1979**.
Bar71	Barron, K.: Int. J. Rock M. Min. Sci. **8** (1971) 541–551.
Bar76	Barton, N.: Int. J. Rock Mech. Min. Sci. **13** (1976) 255–279.
Bat77	Batchelor, A.: A synopsis of work undertaken on the granite of S.W. England at Camborne School of Mines. 1972–1977.
Bel77	Bell, F.G.: Eng. Geol. **11** (1977) 217–255.
Ber78	Berckhemer, H., Auer, F.: Protokoll. Koll. Bad Honnef, DFG, 15–17, **1978** (unpubl. data).
Ber79	Berckhemer, H., Auer, F., Drisler, J.: Phys. Earth Plan. Int. **20** (1979) 48–59.
Bie68	Bieniawski, Z.T.: South African CSIR, Rep. R-MEG 273, Pretoria, **1968**.
Bie68a	Bieniawski, Z.T.: Int. J. Rock Mech. Min. Sci. **4** (1968) 407–413.
Bie70	Bieniawski, Z.T., Vogler, U.W.: Proc. 2nd Congr. Int. Soc. Rock Mech., 2–12, Belgrad, **1970**.
Bie73	Bieniawski, Z.T.: Tunnelling in Rock, SAICE, 105–123, Pretoria, **1973**.
Bie75	Bieniawski, Z.T.: Eng. Geol. **9** (1975) 1–11.
Blu67	Blum, W., Ilschner, B.: Phys. Status Solidi **20** (1967) 629–642.
Blu69	Blum, W.: Diss., Univ. Erlangen, **1969**.
Bod70	Bodonyi, J.: Proc. 2nd Congr. I.S.R.M., 2–17, Belgrad, **1970**.
Bor66	Borg, I., Handin, J.: Tectonophysics, Spec. Issue **3** (1966) 249–368.
Bor68	Borg, I., Friedman, M., Handin, J., Higgs, D.V.: A Geol. Soc. Am. Mem. **79** (1960) 133–191.
Bor70	Borg, I., Heard, H.C.: Experimental and natural rock deformation, p. 375–403, Berlin: Springer, **1970**.
Bow64	Bowden, F.G., Tabor, D.: The friction and lubrication of solids. Vol. 1/2, Oxford: Clarendon Press, 1950/1964.
Bow64a	Bowden, F.G., Brooks, C.A., Hanwell, A.E.: Nature (London) **203** (1964), No. 4940, (1964) 27.
Bra60	Brace, W.F.: Discussion, Geol. Soc. Amer., Mem. **79** (1960) 9–20.
Bra62	Brace, W.F., Walsh, J.B.: The Am. Miner. **47** (1962) 1111–1122.
Bra64	Brace, W.F., in: State of Stress in the Earth's Crust. Judd (ed.), New York: Elsevier, p. 111–174, **1964**.
Bre57	Bredthauer, R.O.: Trans. Am. Soc. Mech. Eng. **79** (1957) 695–703.
Bre57a	Bredthauer, R.O.: Trans. Am. Soc. Mech. Eng. **79** (1957) 695.
Bro66	Bromwell, L.G.: Thesis, MIT, **1966**.
Bur68	Burke, P.M.: PhD thesis, Stanford Univ. **1968**.
Bye66	Byerlee, J.D.: PhD thesis, MIT, Cambride/USA, **1966**.
Bye67	Byerlee, J.D.: J. Geophys. Res. **72** (1967) 3639–3648.
Bye68	Byerlee, J.D.: J. Geophys. Res. **73** (1968) 4741–4750.
Bye69	Byerlee, J.D.: Reply, J.G. Res. **74** (1969) 5349–5350.

Bye75	Byerlee, J.D.: Int. J. Rock Mech. Min. Sci. **12** (1975) 1 4.
Bye78	Byerlee, J.D.: Pageoph **116** (1978) 615–626.
Bye78a	Byerlee, J.D.: Pageoph **116** (1978) 586–602.
Cai74	Cain, P.J., Peng, S.S., Podnieks, E.R.: Proc. 3rd Congr., Int. Soc. R. Mech., Vol. 2 A, p. 367–372, Denver, **1974**.
Car70	Carter, N.L., Heard, H.C.: Am. J. Sci. **269** (1970) 193–249.
Car76	Carter, N.L.: Rev. Geophys. Space Phys. **14** (1976) 301–360.
Car78	Carter, N.L., Kirby, S.H.: Pageoph **116** (1978) 807–839.
Chr56	Christie, R.W.: Acta Metall. **4** (1956) 441–443.
Chr64	Christie, J.M., Heard, H.C., La Mori, P.N.: Am. J. Sci. **262** (1964) 26–55.
Chr74	Christensen, R.J., Swanson, S.R., Brown, W.S.: Proc. 3rd Congr. Int. Soc. R. Mech., Vol. 2A, p. 221, Denver, **1974**.
Cla73	Clark, B.R., Kelly, W.C.: Econ. Geol. **68** (1973) 332–352.
Cli76	Clifton, R.J., Simonson, E.R., Jones, A.H., Green, S.J.: Experim. Mech. **16** (1976) 233–238.
Coa66	Coates, D.F., Parson, R.C.: Int. J.R.M. Min. Sci. **3** (1966) 181–189.
Coe75	Coe, R.S., Kirby, S.H.: Contr. Mineral. Petrol. **52** (1975) 29–55.
Col65	Colback, P.S.B., Wird, B.L.: 3rd Canad. Symp. Rock Mech., **1965**.
Coo65	Cook, N.G.W.: Int. J. Rock Mech. Min. Sci. **2** (1965) 389–403.
Cou70	Coulson, J.H.: Tech. Rep. MRD-2-70, Miss. River Div., Corps of Eng., Omaha, Nebraska, **1970**.
Cro71	Crouch, S.L.: Eng. Geol. **6** (1971) 19–30.
Day70	Dayre, M., Dessenne, J.L., Wack, B.: Proc. 2nd Congr. ISRM, 2–15, Belgrade, **1970**.
Dek65	Deklotz, E.J., Heck, W.J., Neff, T.L.: Corps of Eng., Lab. Rep. No. 64/126, **1965**.
Dek67	Deklotz, E.J., Brown, J.W.: U.S. Army Corps of Eng., Missouri Riv., Tech. Rep., 1–67, **1967**.
Den65	Denkhaus, H.G.: Int. J. Rock Mech. Min. Sci. **2** (1965) 111–126.
Den75	Dengler, L.A.: UCRL Rep. No. 51 919, Livermore, **1975**.
Don64	Donath, F.A.: in: State of stress in the earth's crust, ed. Judd, p. 281–298, New York: Elsevier, **1964**.
Don75	Donath, F.A., Fruth, L.S., Ollson, W.A.: 14th Symp. Rock Mech., 189–222, Penn. State Univ., **1975**.
Dre61	Dreyer, W., Borchert, H.: Zur Druckfestigkeit von Salzgesteinen, Kali und Steinsalz. Heft 7, p. 234–241, Verl. Glückauf, Essen, **1961**.
Dre62	Dreyer, W., Borchert, H.: Bergbautechnik **12** (1962) 265–272.
Dre69	Dreyer, W.: Z. deutsche geol. Ges. **119** (1969) 148–171, Hannover.
Dre69a	Dreyer, W.: Bergbau-Wissenschaften **16** (1969) 191–195.
Dub73	Duba, A.G., Abey, A.E., Heard, H.C.: UCID-Rep. No. 16377, Livermore, **1973**.
Dub74	Duba, A.G., Abey, A.E., Bonner, B.P., Heard, H.C., Schock, R.N.: UCRL-Rep. No. 51 526, Livermore, **1974**.
Eat68	Eaton, S.F.: PhD thesis, Princeton Univ. **1968**.
Edm72	Edmond, J.M., Paterson, M.S.: Int. J. Rock Mech. Min. Sci. **9** (1972) 161–182.
Eth73	Etheridge, M.A., Hobbs, B.E., Paterson, M.S.: Contr. Mineral and Petrol. **38** (1973) 21–36.
Fod66	Fodor, I., Tökes, T.: Proc. 1st. Congr. ISRM, 1, p. 705–709, Lisbon, **1966**.
For68	Forootan-Rad, P., Moavenzadek, F.: MIT; Dept. Civil Eng., Rep. R. 68–29, **1968**.
Fra70	Franklin, J.A., Hoek, E.: Rock Mech. **2** (1970) 223–228.
Fre58	Freudenthal, A.M., Geiringer, H.: The inelastic continuum. Handbuch d. Physik, Bd. VI, Springer Verlag. Berlin, **1958**.
Fri72	Friedman, M., Handin, J., Alani, G.: Int. J. Rock Mech. Min. Sci. **9** (1972) 757–766.
Fry53	Fry, F.S.: MS thesis, Univ. Illinois, **1953**.
Geg64	Geguzin, Y.E., Rabets, V.L., Chernyshof, A.A.: Sov. Phys. Solid State (English Transl.) **5** (1964) 1387–1392.
Gil59	Gilman, J.J.: Proc. Fracture Conf. at Swampscott, Massachusetts, USA, (**1959**), 193–222.
Gil60	Gilman, J.J.: J. Appl. Phys. **31** (1960) 2208–2218.
Goe71	Goetze, C.: J. Geophys. Res. **76** (1971) 1223–1230.
Goe72	Goetze, C., Brace, W.F.: Tectonophys. **13** (1972) 583–600.
Goe73	Goetze, C., Kohlstedt, D.L.: J. Geophys. Res. **78**/26 (1973) 5961–5971.
Gol63	Gold, L.W.: Can. J. Phys. **14** (1963) 1712–1728.
Goo78	Goodman, D.J., Tabor, D.: J. Glaciology **21** (1978) 651–660.
Gow77	Gowd, T.N., Rummel, F.: Int. J. Rock Mech. Min. Sci. **14** (1977) 203–208.

Gre70	Green, H.W., Griggs, D.T., Christie, J.M.: in: Exp. and Nat. Rock Deformation (ed. Paulitsch), p. 272–335, Springer, **1970**.
Gre72	Green, S.J., Perkins, R.D.: Basis and Appl. Rock Mech., 10th Symp. Rock Mech., p. 35, AIME, N.Y., **1972**.
Gri36	Griggs, D.T.: J. Geol. **44** (1936) 541.
Gri60	Griggs, D.T., Turner, F.J., Heard, H.C.: Geol. Soc. Am. Mem. **79** (1960) 39–104.
Hai75	Haimson, B.C.: Proc. 14. Symp. Rock Mech., Pen. State, p. 689–708, **1975**.
Hai78	Haimson, B.C.: Geotechn. Testing, ASTM STP 654, **1978**, 228–245.
Hak75	Hakaletho, K.O.: 14. Symp. Rock Mech., Pen. State, USA, p. 613–621, **1975**.
Han53	Handin, J.: Trans. Am. Soc. Mech. Eng. **75** (1953) 315.
Han55	Handin, J., Fairbairn, H.W.: Bull. Geol. Soc. Am. **664** (1955) 1257–1273.
Han58	Handin, J., Hager, R.V., Jr.: Bull. Am. Assoc. Petr. Geol. **41** (1957) 1; **42** (1958) 2892.
Han59	Handin, J.: J. Petr. Tech. **11** (1959) 15.
Han66	Handin, J.: Strength and ductility. Handbook of physical constants, ed. Clark. Geol. Soc. Am. Mem. **97** (1966) 223–289.
Han66a	Hansagi, I.: Proc. 1st Cong. Rock Mech. Int. Soc., Lisbon, 1, p. 179–183, **1966**.
Han67	Handin, J., Heard, H.C., Magouirk, J.N.: J. Geophys. Res. **72** (1967) 611–640.
Han69	Handin, J.: J. Geophys. Res. **74** (1969) 5343–5348.
Han80	Handin, J., Carter, N.: Proc. 4th Int. Congr. Rock Mechanics, ISRM, Montreux, Vol. III, **1979** 97–106, publ. 1980.
Har70	Hardy, H.R., Jayaraman, N.I.: Proc. 2nd Congr. Int. Soc. Rock Mech., Vol. 3, 5–12, Belgrad, **1970**.
Har72	Hardy, H.R., Mozumdar, B.K., Kimble, E.J.: Int. Rep. RML-IR/72-18, Penn. State Univ., **1972**.
Har73	Hardy, M.P.: PhD thesis, Univ. Minn. **1973**.
Hea60	Heard, H.C.: Geol. Soc. Amer., Mem. **79** (1960) 193–226.
Hea63	Heard, H.C.: J. Geol. **71** (1963) 162–195.
Hea66	Heard, H.C., Rubey, W.W.: Bull. Geol. Soc. Am. **77** (1966) 741–760.
Hea68	Heard, H.C., Carter, N.: Am. J. Sci. **266** (1968) 1–42.
Hea71	Heard, H.C., Schock, R.N., Stephens, D.R.: UCRL-Report No. 51099, Livermore, **1971**.
Hea72	Heard, H.C.: Geophys. Monogr. Ser., **16** "Griggs Vol." (1972) 191–209.
Hea72a	Heard, H.C., Raleigh, C.B.: Geol. Soc. Am. Bull. **83** (1977) 935–956.
Hea73	Heard, H.C., Duba, A., Abey, A.E., Schock, R.N.: UCRL-51465, Livermore, **1973**.
Hea73a	Heard, H.C., Bonner, B.P., Duba, A.G., Schock, R.N., Stephens, D.R.: UCID-Rep. No. 16261, Livermore, **1973**.
Hea74	Heard, H.C., Abey, A.E., Bonner, B.P.: UCID-Rep. No. 16501, Livermore, **1974**.
Hea74a	Heard, H.C., Abey, A.E., Bonner, B.P., Schock, R.N.: UCRL-Rep. No. 51642, Livermore, **1974**.
Hea75	Heard, H.C., Abey, A.E., Bonner, B.P., Duba, A.: UCRL-Rep. 51743, Livermore, **1975**.
Hea76	Heard, H.C., Bonner, B.P., Constantino, M.S., Schock, R.N., Weed, H.C.: UCRL-Rep. 52063, Livermore, **1976**.
Hea76a	Heard, H.C.: Phil. Trans. Royal Soc. Ser. A **283** (1976) 173.
Hen76	Henry, J.-P., Paquet, J.: Bull. Soc. géol. France **18** (1976) 1573–1582.
Her66	Herel, J.: Proc. 1st. Int. Soc. Rock Mech. 1, 503–507, Lisbon, **1966**.
Hig59	Higgs, Handin, J.: Bull. Geol. Soc. Am. **70** (1959) 245.
Hob70	Hobbs, D.W.: Int. J.R.M. Min. Sci. **7** (1970) 125–148.
Hob72	Hobbs, B.E., McLaren, A.C., Paterson, M.S.: Geophys. Monogr. Ser. **16** (1972) 29–53.
Hoe65	Hoek, E.: Nat. Mech. Eng. Res. Inst., CSIR, Rep. No. MEG 383, Pretoria, **1965**.
Hoe69	Hoek, E., in: Rock Mechanics, p. 117, London: Wiley **1969**.
Hor62	Horne, H.M., Deere, D.U.: Geotechnique **12** (1962) 319–335.
Hos68	Hoskins, E.R., Jaeger, J.C., Rosengren, K.J.: Int. J. Rock Mech. Min. Sci. **5** (1968) 143–154.
Hou73	Houpert, R.: PhD-thesis, Univ. Nancy, France, **1973**.
Hun79	Hunsche, U.: Private communication, **1979**.
Ils63	Ilschner, B., Reppich, B.: Phys. Status Solidi **3** (1963) 2093–2100.
Irw58	Irwin, G.R.: Handb. d. Physik **6** (1958) 551.
Jae59	Jaeger, J.C.: Geot. Dura Appl. **43** (1959) 148–158.
Jae66	Jaeger, J.C., Hoskins, E.R.: Brit. J. Appl. Phys. **17** (1966) 685–692.
Jae69	Jaeger, J.C., Cook, N.G.W.: Fundamentals of rock mechanics, p. 178, London: Methuen, **1969**.
Jun76	Jung, R.: Dipl. Arbeit, Geophys., Univ. Bochum, **1976**.
Kar11	v. Karman, Th.: Zeitschr. Verein Deutsch. Ingenieure **55** (1911) 1749–1757.

Ker73	Kern, H., Braun, G.: Contr. Mineral. and Petrol. **40** (1973) 169–181.
Kha71	Khair, A.W.: Int. Rep. RML-IR/71-19, Penn. State Univ., **1971**.
Kie51	Kienow: Neues Jahrbuch Geolog. Palaento. Monatsbl. **2** (1951) 39–52.
Kim69	Kim, Y.S.: PhD-thesis, Penn. State Univ., **1969**.
Kin54	King, R.F., Tabor, D.: Proc. Roy. Soc. London, Ser. A **223** (1954) 225.
Kin71	King, C.Y., Webble, W.W.: J. Appl. Phys. **42** (1971) 2386–2395.
Kir73	Kirby, S.H., Raleigh, C.B.: Tectonophys. **19** (1973) 165–194.
Kir76	Kirby, S.H., in: Electron Microscopy in Mineralogy, ed. by Wenk et al., Chapter 6.7, p. 465–472, Springer, **1976**.
Kir77	Kirby, S.H.: Geoph. Res. Lett. **4/3** (1977) 97–100.
Kni68	Knill, J.K., Franklin, J.A., Malone, A.W.: J.R.M. Min. Sci. **5** (1968) 87–121.
Kob70	Kobayashi, R.: Rock Mech. in Japan **1** (1970) 56.
Koh76	Kohlstedt, D.L., Goetze, C., Durham, W.B., in: The Physics and Chemistry of Minerals and Rocks, ed. Strens, p. 35–49, London: J. Wiley **1976**.
Kov74	Kovari, K., Tisa, A.: Mitteilg. Inst. Straßenbau, ETH Zürich, No. 26, **1974**.
Kre74	Krech, W.W.: Proc. 3rd Cong., Int. Soc. Denver, **1974**.
Kum68	Kumar, A.: Geophysics **33** (1968) 501.
Kut71	Kutter, H., Fairhurt, C.: Int. J.R.M. Min. Sci. **8** (1971) 181–202.
Kut74	Kutter, H.: Rock Mech. Res. Rep. No. 28, Imp. College, London, **1974**.
Kut74a	Kutter, H.: Festschr. L. Müller, p. 45–55, Rock Mech., Karlsruhe, **1974**.
Lac63	Lacy, W.C.: Trans. Am. Inst. Mining, Metall., Petr. Eng. **226** (1963) 272.
Lal71	Lallement-Ave, H.G., Carter, N.L.: Am. J. Sci. **270** (1971) 218–235.
Lan68	Lang, H.: Diss. TH Aachen, Germany, **1968**.
Lan79	Langer, M.: General Report, Proc. 4th Int. Congr. on Rock Mechanics, ISRM, Montreux **1979**, Vol. III, 29–96.
LeC65	Le Compte, P.: J. Geol. **73** (1965) 469–484.
Lin74	Lindholm, U.S., Yeakley, L.M., Nagy, A.: Int. J. Rock Mech. Min. Sci. **11** (1974) 181.
Loc70	Lochner, H.G., Rieder, U.G.: Proc. 2nd Congr. ISRM, 2, 3-1, Belgrad, **1970**.
Log71	Logan, J.M., Handin, J.: Dyn. Rock Mech., 12th Symp. Rock Mech., AIME, N.Y., p. 167, **1971**.
Lom66	Lombardi, G., Vesco, E.D.: Proc. 1st. Congr. ISRM, 1, p. 571–576, Lisbon, **1966.**
Lui78	Lui, H.W.: Private communication, cited in [Goo78].
Lun66	Lundborg, N.: Proc. 1st. Congr. ISRM, 1, p. 251–254, Lisbon, **1966.**
Lya66	Lyall, K.D., Paterson, M.S.: Acta Metall. **14** (1966) 371–383.
Mat56	Matsushima, S.: Zisin. Ser., II, **8** (1956) 173–183.
Mat60	Matsushima, S.: Dis. Prev. Res. Inst., Bull. No. 36, Kyoto Univ., **1960.**
Mat61	Matsushima, S.: Dis. Prev. Res. Inst., Bull. No. 43, Kyoto Univ., **1961**.
Mau65	Maurer, W.C.: J. Soc. Petr. Eng. **5** (1965) 167.
Mer75	Mercier, J.-C.C., Carter, N.L.: J. Geophys. Res. **80** (1975) 3349–3362.
Mer77	Mercier, J.-C.C., Anderson, D.A., Carter, N.L.: Pageoph **115** (1977) 199–226.
Mis65	Misra, A.K., Murrel, S.A.F.: Geophys. J. **9** (1965) 509–535.
Mog65	Mogi, K.: Bull. Earthqu. Res. Inst., **43** (1965) 349–379.
Mog79	Mogi, K.: General Report, Int. Congr. Mog 79 on Rock Mechanics, ISRM, Montreux **1979**.
Mon74	Montoto, M., Ordaz, J.: Proc. 3rd. Congr., ISRM, Vol. IIa, p. 187–192, Denver, **1974**.
Muk69	Mukherjee, A.K., Bird, J.E., Dorn, J.E.: Am. Soc. Metals Trans. **62** (1969) 155–179.
Mül72	Müller, P., Siemes, H.: N. Jb. Miner. Abh. **117** (1972) 39–60.
Mül74	Müller, P., Siemes, H.: Tectonophys. **23** (1974) 105–127.
Mül78	Müller, W.H., Briegel, U.: Eclogae geol. Helv., **71** (1978) 397–407.
Mur73	Murrell, S.A.F., Chakravarthy, S.: Geophys. J., Roy. Astron. Soc. **34** (1973) 211–250.
Mur76	Murrell, S.A.F.: Tectonophys. **36** (1976) 5–24.
Nab67	Nabarro, F.R.N.: Theory of crystal dislocations. Oxford Univ. Press, New York, **1967**.
Nak65	Nakayama, J.: J. Am. Ceram. Soc. **48** (1965) 583–587.
Obr30	Obreimoff, J.W.: Proc. Royal Soc. London, Ser. A **127** (1930) 290–297.
Ohn75	Ohnaka, M.: J. Phys. Earth **23** (1975) 87–112.
Ols74	Olsson, W.A.: Int. J. Rock Mech. **11** (1974) 267–278.
Ouc74	Ouchterlony, F.: PhD thesis, Imperial College London **(1974)**.
Ouc80	Ouchterlony, F.: Swedish Detonic Research Foundation, Rep. DS 1980: 4 **(1980)**.
Out74	Outwater, J.O., Murphy, M.C., Kumble, R.G., Berry, J.T.: ASTM-STP 559, Am. Soc. Test. Mater. **1974**, 127–138.

Pat58	Paterson, M.S.: Bull. Geol. Soc. Am. **69** (1958) 465–475.
Pat64	Paterson, M.S.: J. Inst. of Eng., Australia **1964**, 23–30.
Pat66	Paterson, M.S., Weiss, L.E.: Geol. Soc. Am. Bull. **77** (1966) 343–373.
Pat70	Paterson, M.S., Weaver, C.W.: J. Am. Ceram. Soc. **53** (1970) 463–471.
Pat72	Paterson, M.S., Edmond, J.M.: Carbon **10** (1972) 29–34.
Pen72	Peng, S., Johnson, A.M.: Int. J.R.M. Min. Sci. **9** (1972) 37–86.
Per63	Perkins, T.K., Bartlett, L.E.: Soc. Petr. Eng. J. **3** (1963) 307.
Per66	Perkins, T.K., Krech, W.W.: Soc. Petr. Eng. J. **6** (1966) 308–314.
Per70	Perkins, R.D., Green, S.J., Friedman, M.: Int. J. Rock Mech. Min. Sci. **7** (1970) 527.
Pet70	Peter, A., Ragot, J.P., Sima, A.: Proc. 2nd Congr. Int. Soc. Rock Mech., Vol. 1, p. 1–30, Belgrad, **1970**.
Pom71	Pomeroy, C.D., Hobbs, D.W., Mahmoud, A.: Int. J.R.M. Min. Sci. **8** (1971) 227–238.
Pos73	Post, R.L., Griggs, D.T.: Science **181** (1973) 1242–1244.
Pri58	Price, N.J.: Mechanical properties of non-metallic brittle materials. 122, Butterworth's, London, **1958**.
Pri60	Price, N.J.: Nat. Coal Board, MRE Rep. No. 2159, England, **1960**.
Pri66	Price, D.G., Knill, J.L.: Proc. 1st. Congr. Int. S. Rock Mech., Vol. 1, p. 439–442, Lisbon, **1966**.
Raj71	Raj, R., Ashby, M.F.: Metall. Trans. **2** (1971) 1113–1127.
Ral65	Raleigh, C.B., Paterson, M.S.: J. Geophys. Res. **70** (1965) 3965–3985.
Ral68	Raleigh, C.B.: J. Geophys. Res. **73** (1968) 5391–5406.
Ral70	Raleigh, C.B., Kirby, S.H.: Min. Soc. Am., Spec. Pap. **3** (1970) 113–121.
Ral71	Raleigh, C.B., Kirby, S.H., Carter, N.L., Ave'Lallement, H.G.: J. Geophys. Res. **76** (1971) 4011–4022.
Ren71	Rengers, N.: Veröffentl. Int. Felsmech., Heft 47, Univ. Karlsruhe, **1971**.
Ric70	Ricketts, T.F., Goldsmith, W.: Int. J.R.M. Min. Sci. **7** (1970) 315–335.
Rie64	Riecker, R.E., Seifert, K.E.: Geol. Soc. Am. Bull. **75** (1964) 571–574.
Rie64a	Riege, C.H., Weber, H.S.: Wear **7** (1964) 67–81.
Rob55	Robertson, E.C.: Bull. Geol. Soc. Am. **66** (1955) 1275–1314.
Rod66	Rodrigues, F.P.: Int. Soc. Rock Mech., Vol. 1, 721–731, Lisbon, **1966.**
Rol74	Rollins, R.R., Clark, G.B., Brown, J.W.: 3rd Congr. Int. Soc. Rock Mech., Vol. II B, p. 1384–1390, Denver, **1974**.
Rui66	Ruiz, M.D.: Proc. 1. Congr. Int. S. Rock Mech., Lisbon, 1, p. 115–119, **1966**.
Rum65	Rummel, F.: Boll. Geot. Theor. Appl. **12** (1965) 165–174.
Rum67	Rummel, F.: Diss., Univ. München, **1967**.
Rum69	Rummel, F.: Z. f. Geophys. **35** (1969) 17–42.
Rum70	Rummel, F., Fairhurst, C.: Rock Mech. **2** (1970) 189–204.
Rum75	Rummel, F.: Ber. Inst. Geophys. 4 Univ. Bochum, **1975**.
Rum77	Rummel, F.: Unpublished data on compression tests on Eifel basalt, **1977**.
Rum78	Rummel, F.: Unpublished data, 1978.
Rum78a	Rummel, F.: Rep. 1, Ruhr Univ., Bochum, **1978**.
Rum78b	Rummel, F.: BMFT-Research Contr. ET-3023 A-1, Rep. 3, **1978**.
Rum78c	Rummel, F., Alheid, H.J., Frohn, C.: Pageoph **116** (1978) 743–764.
Rum79	Rummel, F.: Unpublished data on Falkenberg granite and gneisses from the Urach geothermal deep drill hole, **1979**.
Rum80	Rummel, F., Winter, R.B.: BMFT-DGMK Bericht ET 3023 A, Ruhr-Univ. Bochum, **1980**.
Rut74	Rutter, E.H.: Tectonophys. **22** (1974) 311–334.
Rut76	Rutter, E.H.: Phil. Trans. R. Soc. London Ser. A, **283** (1976) 203–219.
Sal74	Salmon, B.C., Clark, B.R., Kelly, W.C.: Econ. Geol. **69** (1974) 1–16.
San68	Santhanan, A.T., Gupta, Y.P.: Int. J. Rock Mech. Min. Sci. **5** (1968) 253–259.
Say70	Saynisch, H.J.: Festigkeits- und Gefügeuntersuchungen an experimentell und natürlich verformten Zinkblendeerzen. Exp. and Nat. Rock Deform., ed. Paulitsch, 209–252, Springer, **1970**.
Sch37	Schmidt, W.: Z. Angew. Min. **1** (1937) 1.
Sch70	Schock, R.N., Heard, H.C., Stephens, D.R.: UCRL-Rep. No. 50963, Livermore, **1970**.
Sch70a	v. Schoenfeldt, H.: PhD-thesis, Univ. Minn., USA, **1970**.
Sch70b	Schuh, F., Blum, W., Ilschner, B.: Proc. British Ceram. Soc. **15** (1970) 143–156.
Sch72	Schock, R.N., Abey, A.E., Heard, H.C., Louis, H.: UCRL-Rep. 51296, Livermore, **1972**.
Sch72a	Schock, R.N., Heard, H.C., Stephens, R.D.: UCRL report 51260, LLL, Livermore **1972**.

Sch73	Schock, R.N., Heard, H.C., Stephens, R.D.: J. Geophys. Res. **78** (1973) 5922–5941.
Sch75	Schmidt, R.A.: Closed Loop Mag., MTS Corp., Minnpolis, Nov. **1975**
Sch77	Schmidt, R.A., Huddle, C.W.: Int. J. Rock Mech. Min. Sci. **14** (1977) 289–293.
Sch76	Schneider, H.: Rock Mech. **8** (1976) 169–184.
Sch76a	Schmidt, R.A.: Exp. Mech. **16** (1976) 161–167.
Sch77	Schneider, H.: Bull. Int. Assoc. Eng. Geol. **16** (1977) 235–239.
Sch78	Schmidt, R.A., Lutz, T.J.: Nat. Sym. Fracture Mech., Blacksburg, USA **(1978)**, unpublished data.
Sch80	Schmid, S.M., Paterson, M.S., Boland, J.N.: Tectonophys. **65** (1980) 245–280.
Sea58	Seal, M.: Proc. Roy. Soc. London Ser. A **248** (1958) 379–393.
See58	Seeger, A.: Kristallplastizität. Handbuch d. Physik, Bd. VII/2. Springer Verlag, Berlin **1958**
Sei69	Seifert, K.E.: Geol. Soc. Am. Bull. **80** (1969) 2053–2059.
Sie77	Siemes, H., Borges, B.: Pers. communication, **1977**.
Sin75	Singh, M.M., Huck, P.J.: 14. Symp. Rock Mech., Penn. State Univ., p. 35–60, **1975**.
Ste70	Stephens, D.R., Heard, H.C., Schock, R.N.: UCRL report 50858, LLL, Livermore **(1970)**.
Ste74	Stesky, R.M., Brace, W.F., Riley, D.K., Robin, P.: Tectonophys. **23** (1974) 177–203.
Sto73	Stocker, R.L., Ashby, M.F.: Rev. Geophys. Space Phys. **11** (1973) 391–426.
Sum71	Summers, D.A., Corwine, J., Li-king Chen: 12 Sym. Rock. Mech., Univ. Missouri, Rolla, USA, **1971**.
Swa76	Swain, M.V., Lawn, R.B.: Int. J. Rock Mech. Min. Sci. **13** (1976) 311–319.
Swa78	Swain, M.V., Atkinson, B.K.: Pageoph **116** (1978), 866–872.
Thi75	Thill, R.E.: 14. Symp. Rock Mech., Penn. State Univ., p. 649–687, **1975**.
Tho65	Thompson, E.G.: PhD-thesis, Univ. Texas, **1965**.
Tsc48	Tschebotarioff, G.P., Welch, J.D.: Proc. 2nd. Congr. Soil Mech. Foundation Engin., 7, p. 135–138, **1948**.
Wal79	Wallner, M., Caninenberg, C., Gonther, H.: Proc. 4th Int. Congr. Rock Mech., ISRM, Montreux **1979**
Waw68	Wawersik, W.R.: PhD-thesis, Univ. Minn., **1968**.
Waw74	Wawersik, W.R.: Proc. 3rd Congr. ISRM, Vol. IIa, p. 357–363, Denver, **1974**.
Waw75	Wawersik, W.R.: 14. Symp. Rock Mech., Penn. State, USA, p. 85–106, **1975**.
Waw76	Wawersik, W.R., Callender, J.F., Weaver, B., Dropek, R.K.: U.S. Rock Mech. Symp., Salt Lake City, **1976**.
Wea69	Weaver, C.W., Paterson, M.S.: J. Polymer Sci., Pt. A-2, **7** (1969) 387–391.
Wee70	Weertman, J.: Rev. Geophys. Space Phys. **8** (1970) 145–168.
Wee70a	Weertman, J., Weertman, J.R.: Mechanical properties, strongly temperature dependent. Phys. Metall., 2nd ed., ed. R.W. Cahn, Amsterdam: North Holl. Publ. Co., **1970**.
Wee75	Weed, H.C., Heard, H.C.: UCID-Rep. No. 16675, Livermore, **1975**.
Wei75	Weißbach, G.: Dipl. Arbeit, Geol., Univ. Bochum, **1975**.
Wei77	Weißbach, G.: personal communications unpublished, **1977**.
Wie68	Wiebols, G.A., Cook, N.G.W.: I.J. Rock Mech. Min. Sci., **1968**.
Wie68a	Wiebols, G.A., Jaeger, J.C., Cook, N.G.W.: 10th Rock Mech. Symp., Rice Univ., Houston, **1968**.
Wie69	Wiederhorn, S.M.: J. Am. Ceram. Soc. **52** (1969) 99–105.
Wie74	Wiederhorn, S.M., Evans, A.G., Roberts, D.E. in: Fracture Mech. of Ceramics 2, p. 829–847, New York: Plenum Press, **1974**.
Win79	Winter, R.B.: Diplomarbeit, Ruhr-Univ. Bochum, **1979**.
Wol70	Wolters, R.: Rock Mech., Suppl. 1 **(1970)** 3–19.
Yam70	Yamayuchi, U.: Int. J.R.M. Min. Sci. **7** (1970) 209–227.
Zob78	Zoback, M.D.: 19th U.S. Symp. Rock Mech. **1978**.
Zob77	Zoback, M.D., Rummel, F., Jung, R., Raleigh, C.B.: Int. J. Rock Mech. Min. Sci. **14** (1977) 49–58.

4 Thermal properties — Thermische Eigenschaften

4.1 Thermal conductivity and specific heat of minerals and rocks — Wärmeleitfähigkeit und Wärmekapazität der Minerale und Gesteine

(V. Čermák, L. Rybach)

4.2 Thermal conductivity of soil — Wärmeleitfähigkeit des Bodens

(M. Schuch)

4.3 Melting temperature of rocks — Schmelztemperaturen der Gesteine

(R. Schmid)

4.4 Radioactive heat generation in rocks — Radioaktive Wärmeproduktion in Gesteinen

(L. Rybach, V. Čermák)

See Subvolume V/1a, p. 305···371

5 Electrical properties — Elektrische Eigenschaften

5.1 Electrical conductivity (resistivity) of minerals and rocks at ordinary temperatures and pressures — Die elektrische Leitfähigkeit (spezifischer Widerstand) der Minerale und Gesteine bei normalen Temperaturen und Drucken

5.1.0 Introduction — Einleitung

The electrical conductivity σ, or its reciprocal, the resistivity $\varrho = 1/\sigma$ of minerals and rocks is an extremely variable property and depends on a number of factors, e.g. the porosity, the nature and the saturation degree of the pore-electrolyte, temperature, and pressure.

The resistivity increases as the amount of pore-electrolyte decreases. The resistivity of a rock depends on the moisture content, which may, for example, be effected by meteorological conditions. The electrical constants of rocks may be obtained by *laboratory* measurements, using rock specimens, and by "*in situ*" or field measurements in which average resistivities of subsurface materials and outcrops are measured in place. Only in situ measurements give truly representative values. The extremely strong dependence on

Die elektrische Leitfähigkeit σ (Kehrwert des Widerstandes $\varrho = 1/\sigma$) hängt ab von: Wassergehalt der Gesteins-Matrix, Porosität, Sättigungsgrad des Porenelektrolytes, Temperatur und Druck.

Die Leitfähigkeit nimmt ab mit Abnahme des Gehaltes an Elektrolyt im Porenraum. Dementsprechend nimmt bei Austrocknung der Proben die Leitfähigkeit mit abnehmendem Wassergehalt stark ab. Ein und dieselbe Probe kann daher zu verschiedenen Zeiten verschiedene Leitfähigkeit haben, wenn z.B. der Feuchtigkeitsgehalt witterungsbedingten Schwankungen unterworfen ist. Für die Bestimmung der elektrischen Leitfähigkeit gilt daher ganz besonders, daß eine Messung an aus dem natürlichen Verband gelösten Proben *im*

moisture content and some other physical and chemical factors like crystal structure, impurities, and weathering results in the wide variation of resistivity, especially for laboratory measurements.

Most rocks are electrically anisotropic. In the case of microanisotropy, orientation of elongated or flat grains in the rock may result in the resistivity being a function of the direction of current flow.

Macroanisotropy is observed when a section of rocks is made up of layers with different resistivities. The resistivity measured along the bedding direction (the longitudinal resistivity) is less than the resistivity measured across the bedding (the transverse resistivity). The square root of the ratio of these two resistivities is the coefficient of macroanisotropy.

Ores and minerals can be rather roughly divided into two classes, good conductors and bad ones. Minerals having a metallic lustre and their ores, as well as graphite and graphitic shales are good conductors. The resistivity of ore deposists depends upon the content and distribution of conducting minerals.

Most of the rock forming minerals are bad conductors: quartz, feldspar, mica as well as rock salt, sulfur, livid ores, and lignite.

Silicates and oxides predominate in the group of intermediate conductors.

In general, sediments are better conductors than igneous rocks, basic rocks conduct better than acid rocks, shales better than sandstones, and soils better than consolidated rocks.

The resistivity of material is defined as the resistance [in ohms] between opposite faces of a unit cube of the uniform material. The unit of resistivity is the ohm-meter [Ωm]. The corresponding conductivity unit is mho/meter, 1/ohm-meter [$\Omega^{-1}\,m^{-1}$].

In the tables, the resistivity ϱ is given in [Ωm].

These data were obtained from direct current or low frequency alternating current resistivity methods, otherwise the frequency is given in the tables.

Labor nur einen angenäherten Wert der Leitfähigkeit liefert. Benötigt man repräsentative Werte der Leitfähigkeit, so muß man "*in situ*", d.h. an Ort und Stelle, bei möglichst ungestörter Durchfeuchtung und Temperatur messen. Diese starke Abhängigkeit vom Wassergehalt sowie anderen physikalisch-chemischen Zustandsänderungen wie Änderungen in der Kristallstruktur, Verunreinigungen, Zersetzung und Verwitterung bewirken in extremen Fällen Unterschiede der Leitfähigkeit um einige Zehnerpotenzen. Hieraus folgt die starke Streuung der im Labor gemessenen Werte.

Die meisten Gesteine sind elektrisch anisotrop. Im Falle der Mikroanisotropie kann die Orientierung oder Form der Kornstrukturen der Gesteine Unterschiede in der Leitfähigkeit als Funktion der Stromflußrichtung bewirken.

Man beobachtet Makroanisotropie, wenn ein Gesteinspaket aus Schichten verschiedener Leitfähigkeit aufgebaut ist. In solch einem Fall ist die Leitfähigkeit parallel zur Schichtung (Longitudinal-Leitfähigkeit) größer als die Leitfähigkeit senkrecht zur Schichtung (Transversal-Leitfähigkeit). Als Wert der Makroanisotropie bezeichnet man die Quadratwurzel aus dem Verhältnis der Leitfähigkeiten transversal zu longitudinal.

Erze und Minerale können grob in zwei Klassen — gut oder schlecht leitende — eingeteilt werden. Verallgemeinert kann man sagen: Minerale mit metallischem Glanz sowie deren Erze sind gute Leiter, ebenso auch Graphit und Graphitschiefer. Die Leitfähigkeit von Erzlagerstätten hängt jedoch vom Gehalt an leitfähigem Material und dessen Verteilung ab.

Die meisten gesteinsbildenden Minerale sind sehr schlechte Leiter — Nichtleiter. Hierzu gehören unter anderen Quarz, Feldspat, Glimmer sowie Steinsalz, Schwefel, die Fahlerze und Braunkohle.

Zur Gruppe der Halbleiter gehören vorwiegend Silikate und Oxide.

Allgemein leiten Sedimente wegen ihres Gehaltes an Elektrolyt im Poren-Raum besser als magmatische und metamorphe Gesteine, Tongesteine besser als Sandgesteine, quellungsfähige Böden besser als verfestigte Gesteine, basische Gesteine besser als saure.

Statt der elektrischen Leitfähigkeit σ wird oft der Kehrwert, der Widerstand $\varrho = 1/\sigma$ verwendet. Der spezifische Widerstand ϱ des Materials ist gleich dem Widerstand [in Ohm] zwischen entgegengesetzten Seiten eines Würfels einheitlichen Materials. Die Einheit des spezifischen Widerstands ist das Ohm-Meter [Ωm]. Die entsprechende Einheit der Leitfähigkeit σ ist 1/Ohm-Meter [$\Omega^{-1}\,m^{-1}$], im Angelsächsischen mho/meter.

In den Tabellen wird der spezifische Widerstand ϱ in [Ωm] angegeben.

Nahezu alle Tabellenwerte sind mit Gleichstrom- oder niederfrequenten Wechselstrommeßmethoden gewonnen. Einige Tabellenwerte sind mit höherfrequenten Meßmethoden bestimmt worden; die jeweilige Frequenz ist angegeben.

5.1.1 Electrical conductivity of minerals — Elektrische Leitfähigkeit der Minerale

Element, Mineral		Resistivity [Ωm]			Ref.
		low	high	average	
5.1.1.1 Elements — Elemente					
Antimony — Antimon	Sb	$5.0 \cdot 10^{-7}$	$1.0 \cdot 10^{-6}$		Kel66
Arsenic — Arsen	As	$1.5 \cdot 10^{-7}$	$3.3 \cdot 10^{-7}$		Kel66
Bismuth — Wismut	Bi	$1.0 \cdot 10^{-6}$	$1.3 \cdot 10^{-6}$		Kel66
		$1.3 \cdot 10^{-1}$	$1.4 \cdot 10^{-1}$		Par67
Copper — Kupfer	Cu	$1.2 \cdot 10^{-8}$	$3.0 \cdot 10^{-7}$		Kcl66, Mei43, Gri72
		$1.0 \cdot 10^{-3}$	$1.7 \cdot 10^{-3}$		Par67, Haa58
Diamond — Diamant	C	$1.0 \cdot 10^{12}$	$1.0 \cdot 10^{16}$	$1.0 \cdot 10^{12}$	Kel66, Par67, Jak40/50, Haa58
Gold	Au			$2.4 \cdot 10^{-8}$	Kel66
				$2.2 \cdot 10^{-3}$	Par67
Graphite — Graphit	C	$3.6 \cdot 10^{-7}$	$6.0 \cdot 10^{-2}$	$1.0 \cdot 10^{-5}$ $\cdots 1.0 \cdot 10^{-3}$	Kel66, Hei51, Gra65, Par67, Jak40/50, Dob60, Par56
Iron — Eisen	Fe			$1.0 \cdot 10^{-8}$	Kel66
Mercury — Quecksilber	Hg	$9.4 \cdot 10^{-3}$	$9.6 \cdot 10^{-2}$		Par67, Mei43
Nickel	Ni	$1.0 \cdot 10^{-7}$	$1.5 \cdot 10^{-3}$		Jak40/50
Platinum — Platin	Pt			$1.0 \cdot 10^{-8}$	Kel66
				$1.1 \cdot 10^{-2}$	Par67
Silver — Silber	Ag	$1.6 \cdot 10^{-8}$	$2.0 \cdot 10^{-8}$		Kel66, Mei43
				$1.6 \cdot 10^{-3}$	Par67
Sulfur — Schwefel	S	$1.0 \cdot 10^{5}$	$1.0 \cdot 10^{16}$	$1.0 \cdot 10^{15}$	Kel66, Hei51, Par67, Fri60, Jak40/50, Mei43, Haa58
Tellurium — Tellur	Te	$1.1 \cdot 10^{-4}$	$2.5 \cdot 10^{-3}$		Kel66
Tin — Zinn	Sn			$1.3 \cdot 10^{-2}$	Par67

5.1.1.2 Sulfides, antimonides, arsenides, tellurides — Sulfide, Antimonide, Arsenide, Telluride

Element, Mineral		Resistivity [Ωm]			Ref.
Sulfides — Sulfide					
Argentite — Argentit	Ag_2S	$1.5 \cdot 10^{-3}$	20		Kel66, Par67
Bismuthinite — Bismuthinit, Wismutglanz	Bi_2S_3	18	$5.7 \cdot 10^{2}$		Kel66, Par67
Bornite — Bornit, Buntkupferkies	$Fe_2S_3 \cdot Cu_2S$	$1.6 \cdot 10^{-6}$	$6.0 \cdot 10^{-3}$	$3.0 \cdot 10^{-3}$	Kel66, Hei51, Par67, Haa58
		$5.0 \cdot 10^{-2}$	$5.0 \cdot 10^{-1}$		Jak40/50
Chalcocite — Chalkosin, Kupferglanz	Cu_2S	$8.0 \cdot 10^{-5}$	23	$1.0 \cdot 10^{-2}$	Kel66, Par67
Chalcopyrite — Chalcopyrit, Kupferkies	$Fe_2S_3 \cdot Cu_2S$	$1.2 \cdot 10^{-5}$	$2.0 \cdot 10^{-1}$	$1.0 \cdot 10^{-4}$ $\cdots 1.0 \cdot 10^{-3}$	Kel66, Hei51, Gra65, Par67, Jak40/50, Haa58, Par56, Gri72
Cinnabarite — Cinnabarit, Zinnober	HgS	$2.0 \cdot 10^{7}$	$2.0 \cdot 10^{9}$		Hei51, Par67, Fri60

(continued)

Mineral		Resistivity [Ωm]			Ref
		low	high	average	
5.1.1.2 (continued)					
Covellite — Covellin, Kupferindig	CuS	$3.0 \cdot 10^{-7}$	$1.0 \cdot 10^{-3}$	$1.0 \cdot 10^{-6}$ $\cdots 1.0 \cdot 10^{-5}$	Kel66, Hei51, Par67, Jak40/50, Haa58
Galena — Galenit, Bleiglanz	PbS	$1.0 \cdot 10^{-5}$	$1.0 \cdot 10^{-1}$	$1.0 \cdot 10^{-3}$ $\cdots 1.0 \cdot 10^{-1}$	Kel66, Hei51, Gra65, Par67, Dob60, Haa58, Par56, Gri72
Haureite — Haureit	MnS$_2$	10	20		Kel66
Marcasite — Markasit	FeS$_2$	$1.0 \cdot 10^{-3}$	3.5	$1.0 \cdot 10^{-2}$ $\cdots 1.0 \cdot 10^{-1}$	Kel66, Hei51, Par67, Jak40/50
Metacinnabarite — Metacinnabarit	4(HgS)	$2.0 \cdot 10^{-6}$	$1.0 \cdot 10^{-3}$		Kel66
Millerite — Millerit, Haarkies	NiS	$2.0 \cdot 10^{-7}$	$4.0 \cdot 10^{-7}$		Kel66
Molybdenite — Molybdänit, Molybdänglanz	MoS$_2$	$8.0 \cdot 10^{-2}$	$1.6 \cdot 10^{2}$		Kel66, Hei51, Par67, Haa58
Pentlandite — Pentlandit	(Ni,Fe)$_9$S$_8$	$1.0 \cdot 10^{-6}$	$1.1 \cdot 10^{-5}$		Kel66
Pyrite — Pyrit	FeS$_2$	$1.0 \cdot 10^{-5}$	1.5	$1.0 \cdot 10^{-3}$ $\cdots 1.0 \cdot 10^{-1}$	Kel66, Hei51, Par67, Dob60, Haa58, Par56, Gri72
Pyrrhotite — Pyrrhotit	Fe$_7$S$_8$	$2.0 \cdot 10^{-6}$	$1.0 \cdot 10^{-3}$	$1.0 \cdot 10^{-5}$ $\cdots 1.0 \cdot 10^{-4}$	Kel66, Hei51, Gra65, Par67, Haa58, Par56
Sphalerite — Sphalerit, Zinkblende	ZnS	1.0	$1.0 \cdot 10^{5}$		Kel66, Hei51, Par67, Haa58
		$1.8 \cdot 10^{-2}$	$4.0 \cdot 10^{-2}$		Kel66, Gri72
Stannite — Stannin, Zinnkies	Cu$_2$FeSnS$_5$	$1.0 \cdot 10^{-3}$	$6.0 \cdot 10^{3}$		Kel66
Stibnite — Stibnit, Antimonit, Antimonglanz	Sb$_2$S$_3$	$1.0 \cdot 10^{5}$	$1.0 \cdot 10^{12}$		Hei51, Par67
Sulfo-antimonides — Sulfide-Antimonide					
Berthierite — Berthierit	FeSb$_2$S$_4$	$8.3 \cdot 10^{-3}$	2.0		Kel66
Boulangerite — Boulangerit	Pb$_5$Sb$_4$S$_{11}$	$2.0 \cdot 10^{3}$	$4.0 \cdot 10^{5}$	$1.8 \cdot 10^{4}$	Kel66
Cylindrite — Cylindrit	Pb$_3$Sn$_4$Sb$_2$S$_{14}$	2.5	60	13	Kel66
Franckeite — Franckeit	Pb$_5$Sn$_3$Sb$_2$S$_{14}$	1.2	4.0		Kel66
Hauchecornite — Hauchecornit	Ni$_9$(Bi,Sb)$_2$S$_8$	$1.0 \cdot 10^{-6}$	$8.3 \cdot 10^{-5}$		Kel66
Jamesonite — Jamesonit	Pb$_4$FeSb$_6$S$_{14}$	$2.0 \cdot 10^{-2}$	$1.5 \cdot 10^{-1}$		Kel66
Tetrahedrite — Tetrahedrit	Cu$_3$SbS$_3$	$3.0 \cdot 10^{-1}$	$3.0 \cdot 10^{4}$		Kel66
Ullmannite — Ullmannit	NiSbS	$9.0 \cdot 10^{-8}$	$1.2 \cdot 10^{-6}$		Kel66
Sulfo-arsenides — Sulfo-Arsenide					
Arsenopyrite — Arsenopyrit, Arsenkies	FeAsS	$2.0 \cdot 10^{-5}$	$2.0 \cdot 10^{-2}$	$3.0 \cdot 10^{-4}$ $2.0 \cdot 10^{-1}$	Kel66, Gra65, Par67, Par56, Jak40/50, Haa58
Cobaltite — Cobaltin, Kobaltglanz	CoAsS	$1.0 \cdot 10^{-5}$	$1.3 \cdot 10^{-1}$	$1.0 \cdot 10^{-2}$	Kel66, Par67, Par56
Enargite — Enargit	Cu$_3$AsS$_4$	$2.0 \cdot 10^{-4}$	$9.0 \cdot 10^{-1}$	$2.0 \cdot 10^{-2}$	Kel66

(continued)

Mineral		Resistivity [Ωm]			Ref.
		low	high	average	
5.1.1.2 (continued)					
Gersdorffite — Gersdorffit	NiAsS	$1.0 \cdot 10^{-6}$	$1.6 \cdot 10^{-4}$	$1.5 \cdot 10^{-5}$	Kel66
Glaucodote — Glaucodot	(Co, Fe)AsS	$5.0 \cdot 10^{-6}$	$1.0 \cdot 10^{-4}$		Kel66
Tennantite — Tennantit	Cu_3AsS_3	$7.0 \cdot 10^{-4}$	$4.0 \cdot 10^{-1}$	$2.7 \cdot 10^{-3}$	Kel66
Antimonides — Antimonide					
Breithauptite — Breithauptit	NiSb	$3.0 \cdot 10^{-8}$	$5.0 \cdot 10^{-7}$		Kel66
				$1.6 \cdot 10^{-4}$	Kel66
Dyscrasite — Dyskrasit, Antimonsilber	Ag_3Sb	$1.2 \cdot 10^{-7}$	$1.2 \cdot 10^{-6}$		Kel66
Arsenides — Arsenide					
Allemonite — Allemonit	$SbAs_3$	70	$6.0 \cdot 10^4$	$2.0 \cdot 10^3$	Kel66
Löllingite — Löllingit	$FeAs_2$	$2.0 \cdot 10^{-6}$	$1.5 \cdot 10^{-4}$		Kel66, Par56
Niccolite — Niccolit, Nickelin, Rotnickelkies	NiAs	$1.1 \cdot 10^{-7}$	$2.0 \cdot 10^{-5}$		Kel66, Hei51, Par67, Haa58
Skutterudite — Skutterudit	$CoAs_3$	$1.1 \cdot 10^{-6}$	$1.6 \cdot 10^{-4}$		Kel66
Smaltite — Smaltin	$CoAs_2$	$1.1 \cdot 10^{-6}$	$1.2 \cdot 10^{-5}$		Kel66
Tellurides — Telluride					
Altaite — Altait	PbTe	$2.0 \cdot 10^{-5}$	$2.0 \cdot 10^{-4}$		Kel66
Calaverite — Calaverit	$AuTe_2$	$6.0 \cdot 10^{-6}$	$1.2 \cdot 10^{-5}$		Kel66
Coloradoite — Coloradoit	HgTe	$4.0 \cdot 10^{-6}$	$1.0 \cdot 10^{-4}$		Kel66
Hessite — Hessit	Ag_2Te	$4.0 \cdot 10^{-6}$	$1.0 \cdot 10^{-4}$		Kel66
Nagyagite — Nagyagit	$Pb_6Au(S,Te)_{14}$	$2.0 \cdot 10^{-5}$	$8.0 \cdot 10^{-5}$		Kel66
Sylvanite — Sylvanit	$AgAuTe_4$	$4.0 \cdot 10^{-6}$	$2.0 \cdot 10^{-5}$		Kel66

5.1.1.3 Halides — Halogenide

Mineral		Resistivity [Ωm]			Ref.
		low	high	average	
Fluorite — Fluorit, Flußspat	CaF_2			$7.9 \cdot 10^{13}$	Par67
Halite — Halit, Steinsalz	NaCl	30	$5.0 \cdot 10^3$		Hei51
		$1.0 \cdot 10^{10}$	$1.0 \cdot 10^{15}$		Par67, Fri60
Sylvite — Sylvin	KCl	$1.0 \cdot 10^{11}$	$1.0 \cdot 10^{13}$		Par67

5.1.1.4 Oxides, hydroxides — Oxide, Hydroxide

Mineral		Resistivity [Ωm]			Ref.
		low	high	average	
Oxides — Oxide					
Barium priderite	$BaTiO_3$, synthetic			$1.6 \cdot 10^7$	Kel66
Braunite — Braunit	Mn_2O_3	$1.6 \cdot 10^{-1}$	1.2	$4.3 \cdot 10^{-1}$	Kel66, Par67
Bromellite	BeO, synthetic			$2.3 \cdot 10^9$	Kel66
Cassiterite — Cassiterit, Zinnstein	SnO_2	$4.5 \cdot 10^{-4}$	$1.0 \cdot 10^4$	$1.73 \cdot 10^{-1}$	Kel66, Par67
Chromite — Chromit, Chromeisenerz	$FeCr_2O_4$	1.0	$1.7 \cdot 10^6$	$2.0 \cdot 10^4$	Jak40/50, Haa58
Cuprite — Cuprit, Rotkupfererz	Cu_2O	10	$3.0 \cdot 10^2$	$1.0 \cdot 10^{-3}$	Kel66, Jak40/50 Par67
Hematite — Hämatit, Blutstein	Fe_2O_3	$2.1 \cdot 10^{-3}$	$6.0 \cdot 10^2$	$3.5 \cdot 10^{-3}$ $\cdots 1.5 \cdot 10^{-2}$	Kel66, Hei51, Par67, Haa58
		10	$1.0 \cdot 10^7$		Kel66, Hei51, Gra65, Haa58

(continued)

Mineral		Resistivity [Ωm]			Ref.
		low	high	average	
5.1.1.4 (continued)					
Ice I — Eis I	H_2O			$4.7 \cdot 10^5$	Kel66
Ilmenite — Ilmenit, Titaneisen	$FeTiO_3$	$1.0 \cdot 10^{-3}$	50	5.0	Kel66, Hei51, Par67
Magnetite — Magnetit	Fe_3O_4	$1.5 \cdot 10^{-5}$	$7.5 \cdot 10^5$	$1.0 \cdot 10^{-4}$ $\cdots 1.0 \cdot 10^{-3}$	Kel66, Hei51, Gra65, Par67, Dob60, Haa58, Par56
Melaconite — Melaconit	CuO			$6.0 \cdot 10^3$	Kel66
Periclase — Periklas	MgO			$6.2 \cdot 10^{10}$	Kel66
Pyrolusite — Pyrolusit	MnO_2	$1.5 \cdot 10^{-3}$	5.0		Kel66, Hei51, Par67, Haa58
Quartz — Quarz	SiO_2	$3.8 \cdot 10^{10}$	$2.0 \cdot 10^{14}$		Hei51, Par67, Haa58
Rutile — Rutil	TiO_2	29	$9.1 \cdot 10^2$		Kel66
Uraninite — Uraninit	UO_2	1.5	$2.0 \cdot 10^2$		Kel66
Hydroxides — Hydroxide					
Bauxite — Bauxit	$Al_2O_3 \cdot nH_2O$	$2.0 \cdot 10^2$	$6.0 \cdot 10^3$		Kel66
Hollandite — Hollandit	$(Ba, Na, K)Mn_2$ $\cdot Mn_6O_{16}$ $\cdot H_2O$	$2.0 \cdot 10^{-3}$	$1.0 \cdot 10^{-1}$	$2.0 \cdot 10^{-2}$	Kel66
Limonite — Limonit	$Fe_2O_3 \cdot nH_2O$			$1.0 \cdot 10^7$	Par67
Manganite — Manganit	$MnO \cdot OH$	$1.2 \cdot 10^{-2}$	$5.0 \cdot 10^{-1}$	$7.1 \cdot 10^{-2}$	Kel66, Par67
Psilomelane — Psilomelan	$KMnO \cdot MnO_2,$ nH_2O	$4.1 \cdot 10^{-2}$	1.2		Kel66

5.1.1.5 Carbonates, sulfates, wolframates — Carbonate, Sulfate, Wolframate

Mineral		low	high	average	Ref.
Carbonates — Carbonate					
Calcite — Calcit, Kalkspat	$CaCO_3$	$5.0 \cdot 10^{12}$	$1.0 \cdot 10^{14}$		Hei51, Par67, Fri60, Haa58
		$1.0 \cdot 10^4$	$1.0 \cdot 10^5$		Fri60, Jak40/50
Siderite — Siderit, Eisenspat	$FeCO_3$			70	Hei51, Par67, Jak40/50, Dob60, Haa58
Sulfates — Sulfate					
Anhydrite — Anhydrit	$CaSO_4$			$1.0 \cdot 10^9$	Par67
		$1.0 \cdot 10^3$	$1.0 \cdot 10^5$		Jak40/50
Wolframates — Wolframate					
Wolframite — Wolframit	$(Mn, Fe)[WO_4]$	10	$1.0 \cdot 10^5$		Hei51, Haa58

5.1.1.6 Silicates — Silikate

Mineral		low	high	average	Ref.
Hornblende				$1.0 \cdot 10^7$	Hei51, Haa58
Labradorite — Labradorit				$1.0 \cdot 10^5$	Fri60
Muscovite — Muskovit	$KAl_2[OH, F]_2$ $\cdot AlSi_3O_{10}$	$1.0 \cdot 10^{12}$	$1.0 \cdot 10^{14}$		Par67
Phlogopite — Phlogopit	KMg_3 $\cdot [Si_3AlO_{10}]$ $\cdot (F, OH)_2$	$1.0 \cdot 10^{11}$	$1.0 \cdot 10^{12}$		Par67
Serpentine — Serpentin				$2.0 \cdot 10^2$	Hei51

5.1.2 Electrical conductivity of rocks — Elektrische Leitfähigkeit der Gesteine

Rock — Gestein		Resistivity [Ωm]						Ref.
		laboratory			*in situ*			
		low	high	average	low	high	average	

5.1.2.1 Igneous rocks — Magmatische Gesteine

Rock — Gestein		low	high	average	low	high	average	Ref.
Albite — Albit	wet	$2.9 \cdot 10^3$	$4.5 \cdot 10^3$					Par67
	dry			$4.0 \cdot 10^6$				Par67
Amphibolite —					90	$2.0 \cdot 10^2$		Ebe52
Amphibolit	wet	79	$2.0 \cdot 10^2$					Fri60
Andesite — Andesit	wet	$1.0 \cdot 10^2$	$4.5 \cdot 10^4$					Par67, Fri60
	dry			$1.7 \cdot 10^2$				Par67
Anhydrite —	wet	$1.0 \cdot 10^3$	$1.0 \cdot 10^5$					Fri60
Anhydrit	dry			$1.0 \cdot 10^9$				Fri60
Aplite — Aplit	wet			$5.0 \cdot 10^5$				Fri60
Basalt					$7.0 \cdot 10^3$	$1.0 \cdot 10^4$		Ram66
	wet	$1.6 \cdot 10^3$	$2.3 \cdot 10^4$					Par67, Fri60, Jak40/50
	dry	$1.0 \cdot 10^6$	$1.0 \cdot 10^9$					Par67, Fri60
Basalt-tuff —	wet			$1.0 \cdot 10^4$				Fri60
Basalt-Tuff	dry			$2.0 \cdot 10^6$				Fri60
Dacite — Dacit	wet			$2.1 \cdot 10^4$				Par67
Diabase — Diabas	16 Hz						$4.5 \cdot 10^2$	Hei51
							$2.0 \cdot 10^2$	Ebe52
	wet	$2.9 \cdot 10^2$	$4.6 \cdot 10^7$					Hei51, Par67, Fri60, Jak40/50, Dob60, Haa58
	dry	$3.3 \cdot 10^5$	$1.7 \cdot 10^{10}$					Par67, Fri60
Diorite — Diorit							$7.0 \cdot 10^3$	Ebe52
	wet	$6.3 \cdot 10^3$	$2.0 \cdot 10^6$					Hei51, Par67, Fri60, Jak40/50, Dob60
	dry	$1.8 \cdot 10^5$	$5.0 \cdot 10^7$					Par67, Fri60
Felspars — Feldspat	wet			$4.0 \cdot 10^3$				Fri60
Gabbro							$4.9 \cdot 10^2$	Ebe52
	wet	$1.0 \cdot 10^2$	$2.0 \cdot 10^4$					Fri60, Jak40/50
	dry	$1.0 \cdot 10^5$	$1.4 \cdot 10^7$					Hei51, Dob60, Haa58
Granite — Granit	16 Hz						$5.0 \cdot 10^3$	Hei51
					$1.6 \cdot 10^2$	$6.5 \cdot 10^2$		Ebe52
							$4.3 \cdot 10^3$	Ebe52
	wet	$1.6 \cdot 10^2$	$3.6 \cdot 10^6$					Hei51, Par67, Fri60, Jak40/50, Dob60, Haa58
	dry	$1.0 \cdot 10^6$	$3.2 \cdot 10^{16}$					Par67, Fri60

(continued)

Rock — Gestein		Resistivity [Ωm]						Ref.
		laboratory			*in situ*			
		low	high	average	low	high	average	
5.1.2.1 (continued)								
Greenstone —							$1.1 \cdot 10^3$	Hei51
Grünstein	porous						$1.6 \cdot 10^2$	Hei51
	wet	$2.5 \cdot 10^4$	$1.0 \cdot 10^2$					Fri60
	dry	$8.5 \cdot 10^2$	$1.0 \cdot 10^4$					Fri60, Haa58
Keweenawan lavas—10···15 Hz					$1.2 \cdot 10^2$	$5.0 \cdot 10^4$		Hei51
Keweenawanlava								
Lava	foamy				$6.0 \cdot 10^3$	$1.2 \cdot 10^4$		Ram66
	fresh				$3.2 \cdot 10^3$	$6.6 \cdot 10^3$		Ebe52
	weather —						$1.7 \cdot 10^2$	Ebe52
	beaten							
	wet	$1.0 \cdot 10^2$	$1.0 \cdot 10^4$					Fri60
	dry	$1.0 \cdot 10^5$	$1.0 \cdot 10^6$					Fri60
Mica — Glimmer	wet	$2.0 \cdot 10^2$	$1.0 \cdot 10^5$					Fri60
	dry	$1.0 \cdot 10^7$	$2.0 \cdot 10^{15}$					Fri60, Haa58
Nepheline-	wet	$1.0 \cdot 10^4$	$1.0 \cdot 10^5$					Fri60
tephrite —								
Nephelintephrit								
Olivine — Olivin	wet	$1.0 \cdot 10^3$	$6.0 \cdot 10^4$					Par67
Peridotite —	wet			$3.0 \cdot 10^3$				Par67
Peridotit	dry			$6.5 \cdot 10^3$				Par67
Porphyry —	wet	10	$2.0 \cdot 10^5$					Hei51, Par67,
Porphyr								Fri60, Jak40/50,
								Haa58
	dry	$3.3 \cdot 10^3$	$1.0 \cdot 10^7$					Par67, Fri60
P., carbonitized	wet			$2.5 \cdot 10^3$				Par67
	dry			$5.9 \cdot 10^4$				Par67
P.-diabase	wet			$9.6 \cdot 10^2$				Par67
	dry			$1.7 \cdot 10^5$				Par67
P.-diorite	wet			$1.9 \cdot 10^3$				Par67
	dry			$2.8 \cdot 10^4$				Par67
P.-felspars	wet			$4.0 \cdot 10^3$				Par67
P.-granite							$1.0 \cdot 10^3$	Ebe52
	wet	$4.5 \cdot 10^3$	$7.0 \cdot 10^3$					Par67
	dry	$1.0 \cdot 10^5$	$1.3 \cdot 10^6$					Par67
P.-quartz							$3.4 \cdot 10^2$	Hei51, Ebe52
	wet			$9.2 \cdot 10^5$				Par67
Porphyrite —	wet	63	$1.0 \cdot 10^4$					Fri60
Porphyrit	dry			$1.0 \cdot 10^6$				Fri60
Pumice — Bims							$3.0 \cdot 10^3$	Ram66
Quartz — Quarz	200 Hz						$2.0 \cdot 10^4$	Hei51
					$1.0 \cdot 10^5$	$1.3 \cdot 10^5$		Ebe52
	wet	$1.0 \cdot 10^4$	$6.0 \cdot 10^7$					Par67, Fri60
	dry	$1.0 \cdot 10^7$	$1.0 \cdot 10^{16}$					Par67, Fri60
Serpentine —	200 Hz						$2.1 \cdot 10^2$	Hei51
Serpentin							$5.3 \cdot 10^2$	Hei51
	wet	$1.0 \cdot 10^2$	$2.0 \cdot 10^4$					Fri60,
								Jak40/50,
								Dob60, Haa58
	dry			$2.0 \cdot 10^5$				Fri60
								(continued)

Rock — Gestein		Resistivity [Ωm]						
		laboratory			in situ			
		low	high	average	low	high	average	
5.1.2.1 (continued)								
Syenite — Syenit	200 Hz						$2.4 \cdot 10^3$	Hei51
	wet	$1.0 \cdot 10^2$	$1.0 \cdot 10^7$					Fri60, Jak40/50
	dry			$1.0 \cdot 10^9$				Fri60
Trachyte — Trachyt	wet	10	$1.0 \cdot 10^5$					Fri60, Jak40/50
	dry			$1.0 \cdot 10^9$				Fri60

5.1.2.2 Metamorphic rocks — Metamorphe Gesteine

Rock — Gestein		laboratory low	laboratory high	laboratory average	in situ low	in situ high	in situ average	
Gneiss					$5.0 \cdot 10^2$	$9.0 \cdot 10^2$		Ebe52
	wet	$2.0 \cdot 10^2$	$6.8 \cdot 10^4$					Par67, Fri60, Jak40/50
	dry	$3.2 \cdot 10^6$	$1.0 \cdot 10^9$					Par67, Fri60
Biotite gneiss				$4.0 \cdot 10^6$				Hei51
Garnet gneiss				$2.0 \cdot 10^5$				Hei51, Dob60
Hornblende gneiss		$1.0 \cdot 10^6$	$6.0 \cdot 10^6$					Hei51, Dob60
Hornfels	wet	$8.1 \cdot 10^3$	$6.0 \cdot 10^5$					Par67
	dry	$6.0 \cdot 10^6$	$6.0 \cdot 10^7$					Par67
Marble — Marmor	wet	$1.0 \cdot 10^2$	$1.0 \cdot 10^5$					Par67, Fri60, Jak40/50
	dry	$1.0 \cdot 10^7$	$1.0 \cdot 10^{11}$					Par67, Fri60
	wet			$7.1 \cdot 10^9$				Par67
	dry			$1.8 \cdot 10^{18}$				Par67
Phyllite — Phyllit	wet			$4.0 \cdot 10^2$				Fri60
	dry			$1.0 \cdot 10^7$				Fri60
Quarzite — Quarzit					$6.0 \cdot 10^3$	$8.0 \cdot 10^3$		Ebe52
	wet	10	$4.7 \cdot 10^6$					Par67, Fri60, Jak40/50
	dry	$2.0 \cdot 10^8$	$5.0 \cdot 10^9$					Par67, Fri60
Slate — Schieferton	wet	$1.0 \cdot 10^3$	$6.4 \cdot 10^4$					Par67
	dry	$1.6 \cdot 10^5$	$3.6 \cdot 10^7$					Par67
Schist — Schiefer					20	$6.0 \cdot 10^2$		Hei51
							$1.3 \cdot 10^4$	Ebe52
					$1.0 \cdot 10^3$	$5.0 \cdot 10^3$		Beb74
Mica schist					20	$6.0 \cdot 10^2$		Hei51, Ebe52
	wet			$5.0 \cdot 10^2$				Fri60
	dry			$4.0 \cdot 10^8$				Fri60
		5.0	$6.5 \cdot 10^4$					Fri60, Jak40/50, Dob60, Haa58
Skarn				$2.5 \cdot 10^2$				Par67
Tuff	wet	$1.0 \cdot 10^2$	$1.0 \cdot 10^4$					Par67, Fri60
	dry	$1.0 \cdot 10^5$	$1.0 \cdot 10^8$					Par67, Fri60

Rock — Gestein	Resistivity [Ωm]						Ref.
	laboratory			*in situ*			
	low	high	average	low	high	average	
5.1.2.3 Sediments, unconsolidated — Sedimente, locker							
Bog — Moor flat bog				80	$1.3 \cdot 10^2$		Ebe52
high bog						$4.3 \cdot 10^2$	Ebe52
Clay — Ton 16 Hz				1.0	2.0		Hei51
40 MHz						51	Hei51
wet	1.0	$3.2 \cdot 10^3$		5.0	$1.5 \cdot 10^3$	40	Hei51, Jak40/50, Ebe52
dry	$1.0 \cdot 10^3$	$1.0 \cdot 10^5$					Fri60, Haa58
Glacial sediments — Glacial-Sedimente						$9.5 \cdot 10^3$	Jak40/50
wet	8.0	$4.0 \cdot 10^3$					Fri60, Dob60, Haa58
Loam — Lehm				21	$1.6 \cdot 10^2$		Ebe52
	5.0	50					Haa58
Loess — Löß						$1.0 \cdot 10^2$	Ram66
				25	40		Ebe52
Marl — Mergel 16 Hz				3.0	50		Hei51
				12	70		Hei51, Jak40/50, Dob60, Ebe52
	3.0	70					Fri60, Haa58
Molasse sediments — Molasse-Sedimente				20	30		Kem77
Nagelfluh	$1.3 \cdot 10^2$	$2.5 \cdot 10^2$					Fri60
Sand silty				$2.8 \cdot 10^2$	$6.3 \cdot 10^2$		Ebe52
gravelly						$1.3 \cdot 10^3$	Ebe52
valley sand				$3.6 \cdot 10^2$	$1.5 \cdot 10^3$		Ebe52
dune sand				$6.2 \cdot 10^{3\,-}$	$7.7 \cdot 10^3$		Ebe52
river sand	$1.7 \cdot 10^2$	$8.3 \cdot 10^2$					Haa58
Silt moist						23	Hei51
dry				13	20		Hei51
glacial, dry				13	21		Hei51
glacial, wet				$3.9 \cdot 10^2$	$8.4 \cdot 10^2$		Hei51
gravel, wet				$1.2 \cdot 10^2$	$1.4 \cdot 10^2$		Hei51
river sand				$1.7 \cdot 10^2$	$8.3 \cdot 10^2$		Hei51
stream gravel						$3.3 \cdot 10^2$	Hei51
river gravel				$4.8 \cdot 10^2$	$8.9 \cdot 10^2$		Hei51
	$3.3 \cdot 10^2$	$5.0 \cdot 10^3$					Haa58
Soil — Böden Humus				10	$1.0 \cdot 10^2$		Mei43
Humus	10	$1.0 \cdot 10^2$					Fri60, Haa58
clayish	6.3	$6.3 \cdot 10^2$					Fri60, Haa58
sandy	10	$4.0 \cdot 10^4$					Fri60, Haa58
stony	$1.0 \cdot 10^2$	$1.0 \cdot 10^4$					Fri60

Rock — Gestein	Resistivity [Ωm]						Ref.
	laboratory			in situ			
	low	high	average	low	high	average	
5.1.2.4 Sediments, consolidated — Sedimente, fest							
Argillite — Argillit	16 Hz			74	$8.4 \cdot 10^2$		Hei51
Clay — Ton						21	Hei51
	sea clays			30	50		Kem77
Conglomerate — Konglomerate, Nagelfluh				$2.0 \cdot 10^3$	$1.3 \cdot 10^4$		Hei51, Jak40/50
	wet	25	$1.0 \cdot 10^4$				Fri60
	dry	$1.1 \cdot 10^3$	$1.3 \cdot 10^4$				Haa58
Dolomite — Dolomit				$7.0 \cdot 10^2$	$2.5 \cdot 10^3$		Kem77
	wet	$3.5 \cdot 10^2$	$1.0 \cdot 10^4$				Par67, Fri60
	dry			$5.0 \cdot 10^6$			Fri60
Graywacke — Grauwacke				$1.5 \cdot 10^3$	$1.0 \cdot 10^4$		Jak40/50, Ebe52
				$4.0 \cdot 10^2$	$1.2 \cdot 10^3$		Beb74
Limestone — Kalkstein	16 Hz			$1.2 \cdot 10^2$	$3.0 \cdot 10^3$		Hei51, Dob60, Ebe52
				$3.5 \cdot 10^2$	$6.0 \cdot 10^4$		Beb74, Kem77
	wet	$1.3 \cdot 10^3$	$8.4 \cdot 10^6$				Par67, Fri60, Haa58
	dry	$1.0 \cdot 10^4$	$1.0 \cdot 10^9$				Par67, Fri60
Sandstone — Sandstein				35	$1.2 \cdot 10^2$		Hei51, Dob60
	16 Hz			$1.0 \cdot 10^3$	$4.0 \cdot 10^3$		Hei51, Dob60
				30	$1.0 \cdot 10^5$		Jak40/50
				$1.3 \cdot 10^2$	$3.2 \cdot 10^2$		Ebe52
	wet	$1.0 \cdot 10^2$	$1.0 \cdot 10^5$				Par67, Fri60, Haa58
	dry	$3.1 \cdot 10^5$	$1.0 \cdot 10^9$				Par67, Fri60
Shale — Tonschiefer	$10 \cdots 60$ Hz			20	$2.0 \cdot 10^3$		Hei51, Dob60
				8.0	$1.0 \cdot 10^4$		Jak40/50
	wet	$4.0 \cdot 10^{-1}$	$2.0 \cdot 10^3$				Fri60
	dry	$1.0 \cdot 10^6$	$2.0 \cdot 10^6$				Fri60
Slate — Tonschiefer				$3.4 \cdot 10^2$	$1.6 \cdot 10^3$		Hei51, Ebe52
	wet	$2.0 \cdot 10^{-4}$	$2.0 \cdot 10^7$				Fri60
	dry	$1.0 \cdot 10^4$	$1.0 \cdot 10^9$				Fri60
5.1.2.5 Oil — Oel							
Oil — Erdoel				$1.0 \cdot 10^9$	$1.0 \cdot 10^{16}$		Mei43
Oil chalk — Oelkreide		$1.0 \cdot 10^4$	$1.0 \cdot 10^7$				Fri60
Oilsand — Oelsand	electrical log	4.0	70				Hei51, Dob60
	dry	$1.0 \cdot 10^4$	$1.0 \cdot 10^{16}$				Fri60

Rock — Gestein	Resistivity [Ωm]						Ref.
	laboratory			*in situ*			
	low	high	average	low	high	average	
5.1.2.6 Coal — Kohle							
Antracite — Antrazit	1.0	$2.0 \cdot 10^5$					Hei51, Jak40/50, Haa58
Bituminous Coal — bituminöse Kohle	$6.0 \cdot 10^{-1}$	$1.0 \cdot 10^5$					Hei51, Jak40/50
Coal — Steinkohle wet dry	$1.6 \cdot 10^2$	$1.5 \cdot 10^5$	$1.0 \cdot 10^{-2}$ $1.0 \cdot 10^9$				Hei51, Haa58 Fri60 Fri60
Coal seam — Flözkohle	$4.0 \cdot 10^2$	$1.0 \cdot 10^3$					Hei51
Lignite — Lignit, Braunkohle	9.0	$2.0 \cdot 10^2$					Jak40/50
5.1.2.7 Salt — Salz							
Rock salt — Steinsalz				$1.0 \cdot 10^2$	$1.0 \cdot 10^5$		Mei43
	30	$1.0 \cdot 10^5$					Hei51, Jak40/50, Dob60, Haa58
dry	32	$1.0 \cdot 10^6$					Fri60, Haa58
5.1.2.8 Ores — Erze							
Argentite — wet Argentit	$7.1 \cdot 10^{-1}$	79					Fri60
Arsenopyrite — Arsenopyrit, wet Arsenkies 60% FeAsS	$1.0 \cdot 10^{-4}$ $6.3 \cdot 10^{-3}$	$1.0 \cdot 10^{-1}$ $1.2 \cdot 10^{-1}$	$3.9 \cdot 10^{-1}$				Par67, Par56 Fri60 Par67
Bismuthinite — wet Bismuthinit, dry Wismutglanz	$1.5 \cdot 10^{-2}$	$1.0 \cdot 10^{-1}$	18				Fri60 Fri60
Bornite — Bornit, rich Buntkupferkies 40% Cu₂S wet	$6.0 \cdot 10^{-4}$	$2.0 \cdot 10^{-2}$		$1.0 \cdot 10^{-3}$ $3.0 \cdot 10^{-3}$ $7.0 \cdot 10^{-2}$	$5.0 \cdot 10^{-3}$		Ebe52 Par67 Par67 Fri60
Chalcocite — Chalcosin, Kupferglanz	$4.4 \cdot 10^{-5}$	$6.0 \cdot 10^{-1}$					Hei51, Par67, Fri60
Chalcopyrite — Chalcopyrit, Kupferkies	$1.0 \cdot 10^{-4}$	$6.5 \cdot 10^{-1}$		$2.0 \cdot 10^{-3}$	$5.0 \cdot 10^{-1}$		Ebe52 Hei51, Par67, Fri60, Jak40/50, Par56
Chromite — Chromit, wet Chromeisenerz	$1.0 \cdot 10^3$ $1.0 \cdot 10^3$	$1.7 \cdot 10^6$ $2.0 \cdot 10^4$	$1.0 \cdot 10^4$				Hei51, Par67 Fri60

(continued)

Rock — Gestein	Resistivity [Ωm]						Ref.
	laboratory			*in situ*			
	low	high	average	low	high	average	
5.1.2.8 (continued)							
Cinnabarite — Zinnober						$2.0 \cdot 10^9$	Ebe52
Cobalt Iron — Kobalt-Eisen	wet		$2.0 \cdot 10^3$ $5.0 \cdot 10^{-4}$				Fri60 Jak40/50
Copper — Kupfer	$1.5 \cdot 10^{-8}$	$2.0 \cdot 10^{-3}$	$1.0 \cdot 10^3$				Hei51, Fri60, Jak40/50
Galena — Galenit, Bleiglanz	$1.0 \cdot 10^{-5}$	$8.2 \cdot 10^{-1}$		$1.0 \cdot 10^{-7}$	$5.0 \cdot 10^{-2}$		Ebe52 Hei51, Gra65, Par67, Fri60, Jak40/50, Par56
	50···80% PbS	1.0	3.0				Par67
	G.-Sphalerite	$6.0 \cdot 10^{-2}$	$3.0 \cdot 10^2$				Jak40/50, Par56
Graphite — Graphit	16 Hz				10 $1.0 \cdot 10^{-2}$	$1.0 \cdot 10^2$ 2.0		Hei51 Ebe52
	pure	$1.0 \cdot 10^{-4}$ $2.0 \cdot 10^{-3}$	$5.0 \cdot 10^{-3}$ 3.5					Par67 Hei51, Gra65, Par67
	wet	$2.5 \cdot 10^{-4}$	1.9					Fri60
	dry			3.2				Fri60
Hematite — Hämatit	wet	$1.0 \cdot 10^{-2}$	$6.3 \cdot 10^3$					Par67, Fri60, Jak40/50
	dry	$1.0 \cdot 10^2$	$1.0 \cdot 10^6$					Hei51, Fri60, Jak40/50
Ironglance, Specular iron ore — Eisenglanz	wet	$1.0 \cdot 10^{-5}$	$4.0 \cdot 10^3$					Fri60
Magnetite — Magnetit		$1.0 \cdot 10^{-4}$	$1.0 \cdot 10^2$		$1.0 \cdot 10^{-3}$	$1.0 \cdot 10^{-1}$		Ebe52 Hei51, Gra65, Par67, Fri60, Jak40/50, Par56
		$5.5 \cdot 10^3$	$1.0 \cdot 10^6$					Hei51, Par67
Manganic ore — Manganerz	wet			$5.0 \cdot 10^{-3}$				Fri60
Meteoric iron — Meteoreisen	dry dry	$1.0 \cdot 10^{-8}$ $1.0 \cdot 10^4$	$3.0 \cdot 10^{-8}$ $3.0 \cdot 10^4$	10				Fri60 Fri60 Jak40/50
Millerite — Millerit, Haarkies	wet	$3.2 \cdot 10^{-5}$	$7.0 \cdot 10^{-5}$					Fri60
Molybdenite — Molybdänit, Molybdänglanz		$1.0 \cdot 10^{-3}$	$4.0 \cdot 10^3$				$5.0 \cdot 10^{-1}$	Ebe52 Par67, Fri60, Jak40/50
Nickel		$1.0 \cdot 10^{-7}$	$1.5 \cdot 10^{-3}$					Jak40/50

(continued)

Rock — Gestein	Resistivity [Ωm]						Ref.
	laboratory			*in situ*			
	low	high	average	low	high	average	
5.1.2.8 (continued)							
Nickel-Cobalt ore — Nickel-Kobalt-Erz	$5.0 \cdot 10^{-4}$	$6.0 \cdot 10^{-3}$					Fri60, Jak40/50
Pyrite — Pyrit	$1.0 \cdot 10^{-4}$	$1.0 \cdot 10^{2}$					Hei51, Gra65, Fri60, Jak40/50, Par56
95···18% FeS$_2$	$1.0 \cdot 10^{-1}$	$3.0 \cdot 10^{2}$					Par67
Pyrolusite — Pyrolusit			1.6				Par67
Pyrrhotite — Pyrrhotit, Magnetkies	$1.0 \cdot 10^{-1}$	$8.3 \cdot 10^{3}$					Hei51
	$1.0 \cdot 10^{-5}$	$7.0 \cdot 10^{-2}$					Hei51, Par67, Par56
95···41% Fe$_7$S$_8$	$1.4 \cdot 10^{-3}$	$2.2 \cdot 10^{-2}$					Par67
Rubric — Roteisen				$1.0 \cdot 10^{4}$	$1.0 \cdot 10^{7}$		Ebe52
dry	$1.0 \cdot 10^{4}$	$1.0 \cdot 10^{7}$					Fri60
Siderite — wet	$1.0 \cdot 10^{-1}$	$1.0 \cdot 10^{4}$					Fri60
Siderit, Spateisenstein dry	$1.0 \cdot 10^{6}$	$1.0 \cdot 10^{7}$					Fri60
Silver ore — Silbererz wet			25				Fri60
Titanium iron — Titan-Eisen wet	2.2	10					Fri60
Sphalerite — Sphalerit, Zinkblende	1.5	$5.0 \cdot 10^{5}$		$1.0 \cdot 10^{6}$	$1.5 \cdot 10^{8}$		Ebe52 Hei51, Fri60, Jak40/50
wet	1.5	$1.0 \cdot 10^{6}$					Fri60
dry			$7.0 \cdot 10^{8}$				Fri60
30···90% ZnS	$7.5 \cdot 10^{-1}$	$1.3 \cdot 10^{2}$					Par67
Stibnite —	$1.0 \cdot 10^{3}$	$3.5 \cdot 10^{7}$					Par67,
Stibnit, wet	$1.0 \cdot 10^{3}$	$1.0 \cdot 10^{5}$					Jak40/50
Antimonglanz dry		$1.0 \cdot 10^{12}$					Fri60 Fri60
Wolframite —						$1.0 \cdot 10^{5}$	Ebe52
Wolframit	$1.0 \cdot 10^{3}$	$1.0 \cdot 10^{7}$					Par67
wet	10	$1.0 \cdot 10^{5}$					Fri60, Jak40/50
dry		$1.0 \cdot 10^{5}$					Fri60
Zinc ore — wet	$1.0 \cdot 10^{-4}$	$3.5 \cdot 10^{-2}$					Fri60
Zink-Erz dry	$1.0 \cdot 10^{-1}$	35					Fri60

5.1.3 Bibliography — Bibliographie

5.1.3.1 References for 5.1.1 and 5.1.2 — Literatur zu 5.1.1 und 5.1.2

Beb74   Beblo, M.: Diss., Fak. f. Geowissenschaften, Univ. München, **1974**.

Dob60   Dobrin, M.B.: Introduction to geophysical prospecting; Mc Graw-Hill Book Company, Inc., New York, **1960**.

Ebe52   Ebert, L.: Elektrische Eigenschaften von Gesteinen. Landolt-Börnstein. 6. Aufl., Band III, Springer Verlag, **1952**.

Fri60   Fritsch, V.: Elektrische Messungen an räumlich ausgedehnten Leitern; C. Braun Verlag, Karlsruhe, **1960**.

Gra65   Grant, F.S., West, G.F.: Interpretation theory in applied geophysics; Mc Graw-Hill Bock Company, New York, **1965**.

Gri72   Grissemann, Ch.: Untersuchung der komplexen frequenzabhängigen Leitfähigkeit erzhaltiger Gesteine mit Hilfe künstlicher Modelle. In: Geoelektronik, Eds.: Bitterlich, W., Wöbking, H., Springer Verlag, Wien, New York, **1972**.

Haa58   Haalck, H.: Lehrbuch der angewandten Geophysik, Teil II. Verlag Gebrüder Bornträger, Berlin-Nikolassee, **1958**.

Hal43   Hallenbach, F.: Elektrische Eigenschaften der Gesteine. In: Taschenbuch der angewandten Geophysik, Eds.: H. Reich, R. v. Zwerger, Akademische Verlagsgesellschaft, Becker u. Erler Kom.-Ges. Leipzig, **1943**.

Hei51   Heiland, G.A.: Resistivities and dielectric constants of minerals, ores, rocks and formations. Geophysical Exploration, Prentice – Hall Inc., New York, **1951**.

Jak40/50 Jakosky, J.J.: Exploration geophysics. Trija Publishing Co., Los Angeles, **1940**, **1950**.

Kel66   Keller, G.V.: Electrical properties of rocks and minerals. Handbook of physical constants, Clark S.P. (Editor), The Geological Society of America, Inc Memoir 97, **1966**.

Kem77   Kemmerle, K.: Diss., Fak. f. Geowissenschaften, Univ. München, **1977**.

Mei43   Meissner, O.: Praktische Geophysik. Verlag von Theodor Steinkopff, Dresden und Leipzig, **1943**.

Par56   Parasnis, D.S.: The electrical resistivity of some sulphide and oxide minerals and their ores. Geophysical prospecting **4** (1956).

Par67   Parkhomenko, E.J.: Electrical properties of rocks; Plenum Press, New York, **1967**.

Ram66   Rammer, R.: Z. Geophys. **32** (1966) 532–538.

5.1.3.2 Further references — Weitere Literatur

Bitterlich, W., Wöbking, H.: Geoelektronik. Springer Verlag, Wien New York, **1972**

Ere, A.S., Keys, D.A.: Applied geophysics in the search for minerals. Cambridge University Press, **1954**.

Kohlrausch, F.: Praktische Physik.

Kunetz, G.: Principles of direct current resistivity prospecting. Gebrüder Bornträger, Berlin-Nikolassee, **1966**.

Malmqvist, D.: Geophysical surveys in mining, hydrological, and engineering projects. European Association of Exploration Geophysicists, **1958**.

Parasnis, D.S.: Methods in geochemistry and geophysics, **3**, Mining Geophysics. Elsevier Scientific Publishing Company, Amsterdam, London, New York, **1973**.

Reich, H.: Grundlagen der Angewandten Geophysik für Geologen. Akademische Verlagsbuchhandlung, Leipzig, **1960**.

5.2 The dielectric constant ε_r of minerals and rocks —
Die Dielektrizitätskonstante ε_r der Minerale und Gesteine

5.2.0 Introduction — Einleitung

The dielectric constant (relative dielectric constant ε_r) is defined in terms of the ratio: specific capacity of material (ε) to specific capacity of vacuum (ε_0), $\varepsilon_r = \varepsilon/\varepsilon_0$. The values of ε_0 are different in different unit-systems.

As other physical properties the dielectric constant depends on pressure and temperature and very strongly on frequency. The dielectric constant decreases with increasing frequency.

The dielectric constant of an inhomogeneous material depends on the dielectric constant of the components. The values for wet rocks depend on the dielectric constants of the dry minerals, the air within the porous volume, and the electrolyte. The dielectric constant of air is nearly one, that of the electrolyte near the value for water ($\varepsilon_r = 81$). The dielectric constant of dry rocks is very low ($\varepsilon_r \approx 10$), therefore even a small water content in the matrix increases the dielectric constant of the rock. This significant dependence upon the water content should be considered using the values given in the tables.

Die Dielektrizitätskonstante (relative Dielektrizitätskonstante ε_r) ist definiert als das Verhältnis der Kapazität eines Kondensators mit Dielektrikum (ε) zu der eines Kondensators gleicher Gestalt mit Vakuum (ε_0), $\varepsilon_r = \varepsilon/\varepsilon_0$. Es ist zu beachten, daß die Größen ε und ε_0 je nach den verwendeten Maßsystemen unterschiedliche Zahlenwerte annehmen.

Die Größe der Dielektrizitätskonstanten hängt wie alle Materialkonstanten vom Zustand der Substanz, z.B. Druck, Temperatur, und von den Untersuchungsbedingungen ab. Die Frequenz des Spannungsfeldes ist hierbei der wichtigste äußere Faktor. Die Dielektrizitätskonstante nimmt im allgemeinen mit steigender Frequenz ab.

Die Dielektrizitätskonstante eines inhomogenen Körpers wird bestimmt von den Dielektrizitätskonstanten seiner einzelnen Komponenten. Die Dielektrizitätskonstante des feuchten Gesteins hängt ab: von der der trockenen Minerale, der der Luft im Porenraum und der der wässrigen Lösung. Die Dielektrizitätskonstante der Luft ist nahezu 1, die der elektrolytischen Flüssigkeit nahe der des Wassers ($\varepsilon_r = 81$). Die Dielektrizitätskonstante trockener Gesteine ist im allgemeinen klein ($\varepsilon_r \approx 10$), sodaß durch die hohe Dielektrizitätskonstante des Wassers schon geringe Wasseranteile im Gesteinsgefüge die Dielektrizitätskonstante des Gesteins selber entscheidend bestimmen. Diese starke Abhängigkeit der Dielektrizitätskonstanten vom Feuchtigkeitsgehalt sollte beim Gebrauch der Tabellenwerte immer beachtet werden.

5.2.1 Dielectric constant ε_r of minerals — Dielektrizitätskonstante ε_r der Minerale

Mineral		Frequency [Hz]	ε_r	Ref.
a) Elements — Elemente				
Sulfur — Schwefel	S	$4 \cdot 10^8$	3.6	Hei51, Haa58
		$4 \cdot 10^8$	3.9$\cdots$4.7	Hei51
b) Sulfides — Sulfide				
Galena — Galenit (Bleiglanz)	PbS	$>10^6$	17.9	Kel66
Sphalerite — Sphalerit (Zinkblende)	ZnS		8.3	Hei51, Haa58
		$>10^6$	7.9	Kel66
		$>10^6$	12.1	Kel66
		$>10^{12}$	5.6$\cdots$6.2	Kel66
			7.8	Ebe52

(continued)

5.2.1 (continued)

Mineral		Frequency [Hz]	ε_r	Ref.
c) Halides — Halogenide				
Fluorite — Fluorit	CaF_2	$>10^6$	6.26$\cdots$6.79	Kel66
(Flußspat)		$>10^{12}$	2.06	Kel66
Halite — Halit	NaCl	$>10^6$	5.70$\cdots$6.20	Kel66
(Steinsalz)		$10^2\cdots10^7$	5.90	Kel66
		$4\cdot10^8$	5.6	Fri60, Haa58
		$>10^{12}$	2.39	Kel66
Sylvite — Sylvin	KCl	$>10^6$	4.39$\cdots$6.20	Kel66
		$>10^{12}$	2.20	Kel66
d) Oxides, hydroxides — Oxide, Hydroxide				
Oxides — Oxide				
Cassiterite — Cassiterit	SnO_2	$4\cdot10^8$	12.7	Hei51
(Zinnstein)		$>10^6$	23.4$\cdots$24.0	Kel66
		$>10^{12}$	3.98$\cdots$4.36	Kel66
			24.0	Ebe52
Chalcedony — Chalcedon	SiO_2	10^3	29.1	Kel66
		$10^4\cdots10^7$	8.70$\cdots$5.10	Kel66
Chromite — Chromit	$FeCr_2O_4$	$>10^6$	11.0$\cdots$13.2	Kel66
(Chromeisenerz)		$>10^{12}$	3.10$\cdots$3.14	Kel66
Hematite — Hämatit	Fe_2O_3		25.0	Hei51, Fri60
		$>10^{12}$	8.65$\cdots$10.33	Kel66
Ice — Eis	H_2O		3.2	Hei51, Haa58
		$<10^2$	93.0$\cdots$153.0	Fri60
		$3.2\cdot10^2$	86.0	Fri60
		$5.4\cdot10^3$	12.0	Fri60
		$2.6\cdot10^5$	2.0	Fri60
Quartz — Quarz	SiO_2	$4\cdot10^8$	4.3$\cdots$4.6	Hei51, Haa58
		$10^2\cdots10^7$	4.6	Kel66
		$>10^6$	4.09$\cdots$5.0	Kel66
		$>10^{12}$	2.36$\cdots$2.41	Kel66
Rutile — Rutil	TiO_2	$>10^{12}$	6.82$\cdots$8.42	Kel66
Hydroxides — Hydroxide				
Diaspore — Diaspor	AlO(OH)	$>10^6$	7.27$\cdots$8.30	Kel66
		$>10^{12}$	2.90$\cdots$3.05	Kel66
Limonite — Limonit	$Fe_2O_2\cdot n\,H_2O$		10.0$\cdots$11.0	Hei51, Fri60
e) Carbonates, sulfates, phosphates — Carbonate, Sulfate, Phosphate				
Carbonates — Carbonate				
Aragonite — Aragonit	$CaCO_3$	$>10^6$	6.46$\cdots$9.72	Kel66
		$>10^{12}$	2.34$\cdots$2.84	Kel66
Calcite — Calcit	$CaCO_3$	$4\cdot10^8$	8.0$\cdots$8.5	Hei51, Haa58
(Kalkspat)		$10^2\cdots10^7$	7.31$\cdots$7.60	Kel66
		$>10^{12}$	2.21$\cdots$2.75	Kel66
Dolomite — Dolomit	$CaMg(CO_3)_2$	$>10^6$	6.11$\cdots$8.0	Kel66, Haa58
		$>10^{12}$	2.28$\cdots$2.85	Kel66
Siderite — Siderit	$Fe_2(CO_3)_2$	$4\cdot10^8$	6.9$\cdots$7.9	Hei51, Fri60, Haa58
(Eisenspat)			7.4	Ebe52
Sulfates — Sulfate				
Anglesite — Anglesit	$PbSO_4$	$>10^6$	74$\cdots$500	Kel66
		$>10^{12}$	3.52$\cdots$3.59	Kel66
				(continued)

5.2.1 (continued)

Mineral		Frequency [Hz]	ε_r	Ref.
Anhydrite — Anhydrit	$CaSO_4$		$6.0\cdots7.0$	Hei51, Haa58
		$>10^6$	$5.7\cdots6.3$	Kel66
		$>10^{12}$	$2.48\cdots2.61$	Kel66
Barite — Baryt	$BaSO_4$	$4\cdot10^8$	$7.6\cdots12.2$	Hei51, Haa58
(Schwerspat)			$9.0\cdots11.0$	Fri60
		$>10^6$	$6.72\cdots10.0$	Kel66
		$>10^6$	$7.85\cdots12.31$	Kel66
		$>10^{12}$	$2.38\cdots2.71$	Kel66
Celestite — Cölestin	$SrSO_4$	$>10^6$	$7.60\cdots8.26$	Kel66
		$>10^{12}$	$2.64\cdots2.66$	Kel66
Gypsum — Gips	$CaSO_4\cdot nH_2O$	$4\cdot10^8$	$5.0\cdots9.9$	Hei51, Fri60, Haa58
		$>10^6$	$5.4\cdots12.0$	Kel66
		$>10^{12}$	$2.31\cdots2.34$	Kel66
Phosphates — Phosphate				
Apatite — Apatit	$Ca_5(F,Cl)(PO_4)_3$	$>10^6$	$7.40\cdots10.47$	Kel66
		$>10^6$	$6.07\cdots7.43$	Kel66
		$>10^{12}$	$2.69\cdots2.71$	Kel66
		$10^2\cdots10^7$	$12.0\cdots10.5$	Kel66
Vivianite — Vivianit	$Fe_3(PO_4)_2\cdot8H_2O$	$>10^6$	6.07	Kel66
		$>10^{12}$	$2.49\cdots2.67$	Kel66
f) Silicates — Silikate				
Analcim	$NaAlSi_2O_6\cdot H_2O$	$>10^6$	5.88	Kel66
		$>10^{12}$	2.21	Kel66
Augite — Augit	$Ca(Mg,Fe,Al)$	$4\cdot10^8$	$6.9\cdots8.6$	Hei51, Haa58
	$\cdot(Al,Si)_2O_6$	$>10^6$	$6.9\cdots10.27$	Kel66
		$>10^{12}$	$2.92\cdots3.01$	Kel66
Beryl — Beryll	$Be_3Al_2Si_6O_{18}$	$>10^6$	$5.48\cdots7.80$	Kel66
		$>10^6$	$5.67\cdots6.18$	Kel66
		$>10^{12}$	$2.53\cdots2.56$	Kel66
Biotite — Biotit	$K(Mg,Fe)_3AlSi_3$	$10^2\cdots10^7$	$4.81\cdots4.67$	Kel66
	$\cdot O_{10}(OH)_2$	$>10^6$	$6.19\cdots9.30$	Kel66
		$>10^{12}$	$2.50\cdots2.68$	Kel66
Epidote — Epidot	$Ca_2(Al,Fe)_3(SiO_4)_3OH$	$>10^6$	$7.60\cdots15.36$	Kel66
		$>10^{12}$	$3.01\cdots3.17$	Kel66
Lepidomelane — Lepidomelan		$10^2\cdots10^7$	$4.00\cdots3.89$	Kel66
Leucite — Leuzit	$K(AlSi_2O_6)$	$>10^6$	7.13	Kel66
		$>10^{12}$	2.27	Kel66
Microcline — Mikroklin	$K[AlSiO_3O_8]$	$10^2\cdots10^7$	$8.82\cdots5.62$	Kel66
Muscovite — Muskovit	$KAl_2Si_3O_{10}(OH)_2$	$10^2\cdots10^7$	5.40	Kel66
		$>10^6$	$6.19\cdots8.00$	Kel66
			9.00	Hei51, Fri60, Haa58
		$>10^{12}$	$2.46\cdots2.60$	Kel66
Opal	$SiO_2\cdot nH_2O$	$>10^6$	$7.15\cdots7.43$	Kel66
		$>10^6$	4.21	Kel66
		$>10^{12}$	2.10	Kel66
Orthoclase — Orthoklas	$KAlSi_3O_8$	$10^2\cdots10^7$	$3.27\cdots2.89$	Kel66
		$>10^6$	$4.50\cdots5.55$	Kel66
		$>10^{12}$	$2.30\cdots2.34$	Kel66
Perthite — Pertit		$10^2\cdots10^7$	$6.49\cdots5.80$	Kel66
		$10^2\cdots10^7$	$3.65\cdots3.29$	Kel66

(continued)

5.2.1 (continued)

Mineral		Frequency [Hz]	ε_r	Ref.
Phlogopite — Phlogopit	$KMg_2Al_2Si_3O_{10}(OH)_2$	$10^2 \cdots 10^7$	$6.20 \cdots 6.10$	Kel66
		$> 10^{12}$	$2.44 \cdots 2.58$	Kel66
Plagioclase — Plagioklas	$Ab_{99}An_1$ [1])	$> 10^6$	5.45	Kel66
	$\vdots$		$\vdots$	
	Ab_2An_{98}		7.24	
	$Ab_{99}An_1$ [1])	$> 10^{12}$	2.33	Kel66
	$\vdots$		$\vdots$	
	Ab_2An_{98}		2.51	
Sericite — Serizit		$> 10^6$	$19.55 \cdots 25.35$	Kel66
Sillimanite — Sillimanit	Al_2SiO_5	$> 10^{12}$	$2.78 \cdots 2.84$	Kel66
Spodumene — Spodumen	$LiAl[Si_2O_6]$	$10^2 \cdots 10^7$	$10.4 \cdots 8.46$	Kel66
Topaz — Topas	$Al_2SiO_4(F,OH)_2$	$> 10^6$	$6.30 \cdots 7.60$	Kel66
		$> 10^6$	$6.27 \cdots 6.43$	Kel66
		$> 10^{12}$	$2.66 \cdots 2.68$	Kel66
Tourmaline — Turmalin		$> 10^6$	$5.60 \cdots 7.10$	Kel66
		$> 10^{12}$	$2.76 \cdots 2.89$	Kel66
Zircon — Zirkon	$ZrSiO_4$	$> 10^6$	$8.59 \cdots 12.0$	Kel66
		$> 10^{12}$	3.84	Kel66

[1]) Ab = Albite, An = Anorthite

5.2.2 Dielectric constant ε_r of rocks — Dielektrizitätskonstante ε_r der Gesteine

Rock	Water content	Frequency [Hz]	ε_r	Ref.
a) Igneous rocks — Magmatische Gesteine				
Anorthosite — Anorthosit	dry	$10^2 \cdots 10^4$	$167 \cdots 25$	Kel66
	dry	$10^5 \cdots 10^7$	$10.9 \cdots 9.03$	Kel66
Basalt			12	Hei51, Fri60, Haa58
			13	Ebe52
	dry	$5 \cdot 10^5$	15.6	Par67
	dry	$5 \cdot 10^5$	10.3	Par67
		$> 10^6$	$6.54 \cdots 11.9$	Kel66
Dacite — Dacit		$3 \cdot 10^6$	$6.80 \cdots 8.20$	Kel66, Par67
Diabase — Diabas	dry	$10^2 \cdots 10^7$	$23.5 \cdots 8.50$	Kel66
	dry	$10^2 \cdots 10^7$	$13.4 \cdots 7.76$	Kel66
	dry	$5 \cdot 10^5$	11.6	Par67
		$10^4 \cdots 10^7$	$13.0 \cdots 9.0$	Par67
Diorite — Diorit	dry	$10^2 \cdots 10^7$	$17.0 \cdots 8.57$	Kel66
	dry	$10^2 \cdots 10^5$	$7.21 \cdots 6.05$	Kel66
			8.5	Hei51
	dry	$10^5 \cdots 10^7$	$6.3 \cdots 5.9$	Par67
	dry	$10^4 \cdots 10^7$	$11.5 \cdots 8.5$	Par67
			$9.0 \cdots 10.0$	Ebe52
Dunite — Dunit	dry	$10^2 \cdots 10^7$	$10.0 \cdots 7.18$	Kel66
Fayalite — Fayalit		$> 10^6$	$7.45 \cdots 8.59$	Kel66
	dry		8.32	Par67
Gabbro	dry	$10^2 \cdots 10^7$	$15.0 \cdots 8.78$	Kel66
	dry	$10^2 \cdots 10^7$	$15.6 \cdots 7.30$	Kel66
3.2 $\cdots$ 6.4% Sulfide	dry	$> 10^6$	$12.8 \cdots 39.9$	Kel66, Par67
	dry	$10^4 \cdots 10^7$	$10.0 \cdots 8.8$	Par67

(continued)

5.2.2 (continued)

Rock	Water content	Frequency [Hz]	ε_r	Ref.
Granite — Granit	dry	$>10^6$	4.80···18.9	Kel66
		$>10^6$	14.9	Kel66
	dry	$10^2···10^7$	9.63···5.23	Kel66
	dry	$10^2···10^7$	8.47···6.68	Kel66
	dry		8.0	Hei51, Fri60, Haa58
			7.0···14.0	Ebe52
	dry	$5·10^5$	4.74···5.42	Par67
	dry		7.0···9.0	Par67
Biotite granite	dry		4.8	Par67
Hornblende granite	dry		11.1···7.20	Kel66
Labradorite — Labradorit	dry		7.82	Par67
	0.03 %		8.24	Par67
Luyavrite	dry	$10^5···10^7$	11.4···9.7	Par67
Norite — Norit		$>10^6$	61.3	Kel66
Obsidian		$>10^6$	5.80···10.4	Kel66
		$10^2···10^7$	7.31···6.59	Kal66
Olivine — Olivin (Pyroxenit)	dry	$10^5···10^7$	9.5···8.4	Par67
Peridotite — Peridotit		$>10^6$	8.59	Kel66
	dry	$10^5···10^7$	18.8···15.7	Par67
	dry	$5·10^5$	12.1	Par67
70 % olivine	dry		8.6	Par67
Pitchstone — Pechstein		$>10^6$	18.7	Kel66
Porphyry — Porphyr			9.0···10.0	Hei51, Fri60
			12.0	Ebe52
Augite porphyry	dry	$10^5···10^7$	12.6···9.5	Par67
Granite porphyry		$>10^6$	27.1	Kel66
Quartz porphyry			14.1···49.3	Kel66
Quartz — Quarz	dry	$10^2···10^7$	5.62···4.90	Kel66
			4.0···5.0	Fri60
Syenite — Syenit			12.0	Hei51
		$>10^6$	6.93···12.7	Kel66
	dry	$5·10^5$	6.83	Par67
			13.0···14.0	Par67, Fri60, Ebe52
combined	dry		6.93···9.56	Par67
			8.0···9.0	Ebe52
Trachyte — Trachyt			8.0···9.0	Ebe52
Trap — Trapp		$>10^6$	18.9···39.8	Kel66
Tuff		$>10^6$	3.78···3.99	Kel66
	dry		3.8···4.5	Par67
Urtite — Urtit	dry	$10^5···10^7$	8.5···7.3	Par67
feldspatic	dry		11.9	Par67
b) Sediments, unconsolidated — Sedimente, locker				
Bog — Moorboden		$8.5·10^6$	90.0···100.0	Sch67
Chalk — Kreide	24 %	10^6	21.0	Hei51
	26 %	10^6	38.0	Hei51

(continued)

5.2.2 (continued)

Rock	Water content	Frequency [Hz]	ε_r	Ref.
Clay — Ton	dry	$1 \cdot 10^7 \cdots 4 \cdot 10^7$	$26.5 \cdots 19.5$	Hei51
Stone chips	dry	$1 \cdot 10^7 \cdots 4 \cdot 10^7$	$10.0 \cdots 7.0$	Hei51
	wet	$1 \cdot 10^7 \cdots 2 \cdot 10^7$	$32.0 \cdots 29.0$	Hei51
Blue clay	23 %	10^6	29.0	Hei51, Fri60, Haa58, Ebe52
	25 %	10^6	46.0	Hei51, Fri60, Haa58, Ebe52
	27 %	10^6	75.0	Ebe52
Clay and sand	21 %	10^6	42.0	Hei51, Fri60, Ebe52
	26 %	10^6	48.0	Hei51, Fri60, Ebe52
Loam — Lehm, sandy		$1 \cdot 10^6 \cdots 2 \cdot 10^6$	11.0	Hei51
Loam and clay	15 %	10^6	21.0	Hei51
	33 %	10^6	43.0	Hei51
	dry		$25.0 \cdots 34.0$	Fri60
	21 %		25.0	Ebe52
	wet		$29.0 \cdots 32.0$	Haa58
Sand	wet	$8.5 \cdot 10^6$	96.0	Sch67
	dry	$>10^6$	2.93	Kel66
	1.5 %	$>10^6$	5.00	Kel66
	3.0 %	$>10^6$	11.0	Kel66
	4.5 %	$>10^6$	39.1	Kel66
	6.0 %	$>10^6$	105	Kel66
	$6 \cdots 30$ % kerosine	$>10^6$	$2.89 \cdots 3.59$	Kel66
river sand	dry		$2.0 \cdots 3.0$	Hei51, Haa58
Soil — Böden	11 %	10^6	$8.0 \cdots 10.0$	Hei51
topsoil	dry	$1 \cdot 10^6 \cdots 4 \cdot 10^6$	$15.0 \cdots 12.0$	Hei51
	wet	$2 \cdot 10^6 \cdots 4 \cdot 10^6$	23.0	Hei51
subsoil		$2 \cdot 10^6 \cdots 4 \cdot 10^6$	28.0	Hei51
	7.8 %	$>10^6$	3.95	Kel66
	19.9 %	$>10^6$	4.75	Kel66
	26.8 %	$>10^6$	5.23	Kel66
	32.1 %	$>10^6$	7.93	Kel66
	36.8 %	$>10^6$	21.9	Kel66
	41.4 %	$>10^6$	29.4	Kel66
sandy soil	dry	$10^2 \cdots 10^7$	$3.41 \cdots 2.56$	Kel66
loamy soil	dry	$10^2 \cdots 10^7$	$3.06 \cdots 2.43$	Kel66
clayey soil	dry	$10^2 \cdots 10^7$	$4.72 \cdots 2.44$	Kel66
	dry		$2.6 \cdots 2.8$	Fri60, Haa58
	3.6 %		$2.3 \cdots 5.6$	Fri60, Haa58
	$16 \% \cdots 30 \%$		$18.0 \cdots 30.0$	Fri60, Haa58

c) Sediments, consolidated — Sedimente, fest

Rock	Water content	Frequency [Hz]	ε_r	Ref.
Anhydrite — Anhydrit		$>10^6$	6.19	Kel66
			6.3	Par67
Chert — Hornstein		10^2	63.9	Kel66
		10^3	13.0	Kel66
		10^4	6.19	Kel66
		$10^5 \cdots 10^7$	$4.80 \cdots 4.50$	Kel66

(continued)

5.2.2 (continued)

Rock	Water content	Frequency [Hz]	ε_r	Ref.
Dolomite — Dolomit		$4 \cdot 10^8$	6.8···7.0	Hei51
	dry	$10^2 \cdots 10^7$	11.9···7.72	Kel66
	dry	$10^3 \cdots 10^7$	8.6···8.0	Par67
			8.0···9.0	Ebe52
			7.0···7.3	Ebe52
Graywacke — Grauwacke			6.0	Fri60
			9.0···10.0	Ebe52
Kaolinite — Kaolinit	dry	$10^2 \cdots 10^7$	7.65···4.49	Kel66
Limestone — Kalkstein			8.0···12.0	Hei51
		$>10^6$	15.1	Kel66
	dry	$10^2 \cdots 10^7$	10.4···8.56	Kel66
	dry	$10^2 \cdots 10^7$	15.4···9.22	Kel66
	dry		7.3	Par67
	24%		8.0···12.0	Par67, Fri60, Haa58, Ebe52
	27%		21.0	Ebe52
			24.0···34.0	Ebe52
Novaculite — Wetzschiefer		$10^2 \cdots 10^7$	5.93···4.86	Kel66
Sandstone — Sandstein			9.0···11.0	Hei51, Haa58, Ebe52
	dry	$>10^6$	4.69···4.99	Kel66
	1.5%	$>10^6$	7.40	Kel66
	2.8%	$>10^6$	12.1	Kel66
	4.2%	$>10^6$	10.9	Kel66
arkose sandstone	dry	$10^2 \cdots 10^7$	5.94···5.31	Kel66
graywacke sandstone		$10^2 \cdots 10^7$	11.6···5.87	Kel66
quartzite sandstone		$10^2 \cdots 10^7$	5.15···4.72	Kel66
jurassic sandstone	7%	$10^2 \cdots 10^7$	13.0···5.2	Kel66
	dry	$5 \cdot 10^5$	3.96···4.66	Par67
variegated sandstone			9.0···11.0	Par67
slaty sandstone	dry		5.53	Par67
	0.2%		7.17	Par67
			4.0···7.0	Fri60
Shale — Schiefer	10%	$10^2 \cdots 10^6$	45.0···10.0	Par67
	dry	$10^2 \cdots 10^6$	7.0···4.0	Par67
	dry	$10^2 \cdots 10^6$	10.0···9.5	Par67
d) Metamorphic rocks — Metamorphe Gesteine				
Argillite — Schieferton		$10^2 \cdots 10^7$	11.9···7.97	Kel66
Amphibolite — Amphibolit	dry	$10^5 \cdots 10^7$	8.9···7.9	Par67
Gneiss	dry	$10^2 \cdots 10^7$	9.73···8.07	Kel66
			14.0	Hei51, Ebe52
			8.0···15.0	Par67, Fri60
granite gneiss		$5 \cdot 10^2 \cdots 5 \cdot 10^7$	9.0···8.0	Par67
			7.0···8.0	Ebe52
Hornfels			9.0···12.0	Ebe52
Marble — Marmor			6.0	Hei51, Fri60, Haa58
	dry		8.22	Par67
	0.002%		8.37	Par67
	dry	$10^3 \cdots 10^7$	9.0···8.9	Par67
			8.5	Ebe52

(continued)

5.2.2 (continued)

Rock	Water content	Frequency [Hz]	ε_r	Ref.
Phyllite — Phyllit			13,0	Par67, Ebe52
quartz phyllite			8.0···11.0	Ebe52
Quartzite — Quarzit	dry	$5 \cdot 10^5$	4.36	Par67
	dry	$5 \cdot 10^5$	4.85	Par67
			6.6···7.0	Par67
			9.0	Ebe52
Serpentine — Serpentin		$10^2 \cdots 10^7$	10.1···6.42	Kel66
Schist — Schiefer			16.0···17.0	Hei51, Haa58
mica schist			16.0	Hei51, Fri60, Ebe52
quartzite schist			9.0	Fri60
sericite schist			11.0···12.0	Ebe52
hornblende schist	dry	$10^2 \cdots 10^7$	10.3···8.88	Kel66
talc schist	dry	$10^2 \cdots 10^7$	31.5···7.57	Kel66
Slate — Schieferton	dry	$50 \cdots 5 \cdot 10^7$	34.0···7.5	Par67
	dry	$50 \cdots 5 \cdot 10^7$	10.0···9.0	Par67
	dry		6.71	Par67
	0.1 %		7.74	Par67
e) Oil — Oel				
Oil — Erdoel			2.1	Fri60, Haa58
Asphalt			2.7	Haa58
f) Coal — Kohle				
Anthracite — Antrazit			5.6···6.3	Hei51, Haa58
g) Salt — Salz				
Rocksalt — Steinsalz		$4 \cdot 10^8$	5.6	Hei51, Fri60, Haa58

5.2.3 References for 5.2.1 and 5.2.2 — Literatur zu 5.2.1 und 5.2.2

Ebe52   Ebert, L.: Elektrische Eigenschaften von Gesteinen; Landolt-Börnstein, 6. Aufl., Springer Verlag, **1952**.

Fri60   Fritsch, V.: Elektrische Messungen an räumlich ausgedehnten Leitern; Karlsruhe: C. Braun Verlag, **1960**.

Haa58   Haalck, H.: Lehrbuch der angewandten Geophysik, Teil II; Berlin: Verlag Gebrüder Bornträger, **1958**.

Hei51   Heiland, G.A.: Resistivities and dielectric constants of minerals, ores, rocks and formations. Geophysical Exploration, New York: Prentice-Hall Inc., **1951**.

Kel66   Keller, G.V.: Electrical properties of rocks and minerals. Handbook of physical constants, Clark S.P. (Ed.), The Geological Society of America, Inc., Memoir 97, **1966**.

Par67   Parkhomenko, E.J.: Electrical properties of rocks; New York: Plenum Press, **1967**.

Sch67   Schuch, M., Wanke, R.: Z. Geophys. **33** (1967).

5.3 Electrical conductivity of moisture containing rocks — Die elektrolytische Leitfähigkeit der Gesteine

5.3.1 Electrical conductivity of pure salt solutions and natural waters — Elektrische Leitfähigkeit von Lösungen reiner Salze und von natürlichen Wässern

5.3.1.1 Introduction — Einleitung

The electrical conductivity of a rock depends considerably on the electrical conductivity of the moisture present in the rock. Therefore it seems useful to discuss first the electrical conductivities of natural waters in rocks. As conductivity of natural waters may vary greatly according to salinity and chemical composition, we will discuss too – for better comparison – conductivities of pure salt solutions and their dependence on some physical parameters.

Die elektrolytische Leitfähigkeit eines Gesteins hängt wesentlich von der elektrischen Leitfähigkeit der im Gestein enthaltenen Wässer ab. Deshalb wird zunächst die elektrische Leitfähigkeit von natürlichen Wässern angegeben. Ferner wird für Lösungen reiner Salze die Abhängigkeit der elektrischen Leitfähigkeit von verschiedenen physikalischen Parametern dargestellt.

5.3.1.2 Definitions — Definitionen

In an aqueous solution of electrolytes electric charge is carried by ions. The mobility of ions in aqueous solutions is much smaller than the mobility of electrons in a metal. As a consequence, conductivities of salt solutions are much smaller than conductivities of metals. A generalized Ohm's law is true also for electrolytes:

Bei der Lösung von Elektrolyten in Wasser entstehen Ionen. Diese sind Träger einer elektrischen Ladung. Da Ionen in einer wässrigen Lösung sehr viel weniger beweglich sind als Elektronen in Metallen, ist die elektrische Leitfähigkeit von Elektrolyten in der Regel mehrere Zehnerpotenzen kleiner als die der Metalle. Wie für metallische Leiter gilt das verallgemeinerte Ohmsche Gesetz:

$$j = \sigma \cdot E = \frac{1}{\varrho} \cdot E$$

j, E: current density and electric field, respectively, in the spatial conductor
ϱ: resistivity,
σ: conductivity

j, E: Stromdichte bzw. elektrisches Feld im Leiter
ϱ: spezifischer elektrischer Widerstand;
σ: spezifische elektrische Leitfähigkeit

In geophysical literature the MKSA-system is mainly used and the dimension of resistivity there is Ohm $\cdot$ m (Ωm). Correspondingly the dimension of conductivity is $\Omega^{-1} \cdot m^{-1}$.

Der spezifische elektrische Widerstand wird in der geophysikalischen Literatur überwiegend im MKSA-System in der Dimension Ohm $\cdot$ m (Ωm) angegeben. Die Dimension der spezifischen elektrischen Leitfähigkeit ist entsprechend $\Omega^{-1} \cdot m^{-1}$.

In physics literature the dimension of $\Omega \cdot$ cm is also used for resistivity.

In der physikalischen Literatur wird der spezifische elektrische Widerstand auch in der Dimension $\Omega \cdot$ cm angegeben.

Conversion:
value in the tables of ϱ multiplied by 10^2 = value of ϱ in $\Omega \cdot$ cm and, correspondingly,
value in the tables of σ multiplied by 10^{-2} = value of σ in $\Omega^{-1} \cdot$ cm^{-1}

Für die Umrechnung gilt:
Tabellenwert von $\varrho \cdot 10^2$ = Zahlenwert von ϱ in $\Omega \cdot$ cm und entsprechend
Tabellenwert von $\sigma \cdot 10^{-2}$ = Zahlenwert von σ in $\Omega^{-1} \cdot$ cm^{-1}

Conductivity of an aqueous solution depends on several physical parameters such as
a) the concentration c of ions (amount of salt/unit solvent, degree of dissociation α)

Die elektrische Leitfähigkeit von Elektrolyten hängt von mehreren physikalischen Größen ab,
a) von der Konzentration c der Ladungsträger (Menge des gelösten Salzes in der Volumen- oder Gewichtseinheit des Lösungsmittels, Dissoziations-Grad α)

b) the interaction of ions in the solution

b) von der Wechselwirkung zwischen den Ionen im Elektrolyten

c) the charge number of the ions (kind of substance)
d) the mobility u of the ions in the solution

c) von der Ladungszahl der Ionen (Art der Substanz)
d) von der Beweglichkeit u der Ionen in der wässrigen Lösung

e) as well as on density, viscosity, dielectric constant, pressure, and temperature of the aqueous solution

a) The conductivity of an aqueous solution depends largely on the concentration of the charge carriers. The concentration of the charge carriers depends on the amount of salt/unit solvent and on the degree of dissociation. The degree of dissociation α of an electrolyte is defined by the ratio: number of molecules dissociated into ions/total number of molecules in the electrolyte.

The concentration c of an electrolyte may be defined in different ways. The following definitions are used in the tables of this chapter:

Molar concentration c_V (volume molarity): In one litre of *solution* (not of solvent) a molar solution contains 1 mol of dissolved salt.

Molal concentration c_G (weight molarity): In 1000 g of solvent (e.g. pure water) a molal solution contains one mol of dissolved salt.

Demal concentration: in 1000 g solution a demal solution contains 1 mol of dissolved salt.

Normal or equivalent concentration c_{val}: In one litre of solution a normal solution contains one gram equivalent (val) of dissolved salt.

(One gram equivalent (val) of an atom or a group of atoms is the amount of salt (in grams), that reacts with 1.008 g hydrogen or substitutes that amount of salt in a compound, e.g. 35.5 g chlorine, 8 g oxygen etc.. Gram equivalent may also be defined electrochemically: The amount of a certain type of ion (in grams) which corresponds to an amount of charge of 96490 Coulomb (=1 Faraday) is called one gram equivalent.)

For better comparison between conductivities of different electrolytes the ratio between electrical conductivity and the concentration of the electrolyte is often used. The following definitions are used in the tables:

Molar conductivity

$$\sigma_{mol} = \frac{\text{conductivity } \sigma}{\text{molar concentration } c_V} \text{ in } \frac{\Omega^{-1} \cdot m^{-1}}{mol/l}$$
$$= 10 \, \Omega^{-1} \cdot mol^{-1} \cdot cm^2$$

equivalent conductivity

$$\Lambda = \frac{\text{conductivity } \sigma}{\text{equivalent concentration } c_{val}} \text{ in } \frac{\Omega^{-1} \cdot m^{-1}}{val/l}$$
$$= 10 \, \Omega^{-1} \cdot val^{-1} \cdot cm^2$$

b) Most of the salts dissolved in natural waters are highly dissociated. If we increase the concentration of a solution the distance between the ions will decrease. As a consequence, there is an electrostatic Coulomb interaction between the ions at high concentrations. This interaction will reduce the mobility of the ions at high concentrations. As the dielectric constant of the

e) sowie von Dichte, Viskosität, Dielektrizitäts-Konstante, Druck und Temperatur des Elektrolyten

zu a): Die elektrische Leitfähigkeit von Elektrolyten hängt stark von der Konzentration der Ladungsträger ab. Die Konzentration c der Ladungsträger ist gegeben durch die Menge des gelösten Salzes in der Volumen- oder Gewichtseinheit des Lösungsmittels sowie durch den Dissoziations-Grad des Elektrolyten. Der Dissoziations-Grad α eines Elektrolyten ist definiert als der Quotient zwischen den in Ionen zerfallenen Molekülen und den insgesamt im Elektrolyten gelösten Molekülen.

Die Konzentration c von Elektrolyten kann auf verschiedene Weise angegeben werden. In den Tabellen dieses Kapitels werden die folgenden Definitionen benutzt:

Molare Konzentration c_V (Volumen-Molarität): Eine molare Lösung enthält im Liter *Lösung* (nicht Lösungsmittel) 1 Mol des gelösten Salzes.

Molale Konzentration c_G (Gewichts-Molarität): Eine molale Lösung enthält in 1000 g Lösungsmittel (z.B. reinem Wasser) ein Mol des gelösten Salzes.

Demale Konzentration: eine demale Lösung enthält in 1000 g Lösung 1 Mol des gelösten Salzes.

Normale oder Äquivalent-Konzentration c_{val}: Eine normale Lösung enthält im Liter Lösung ein Gramm-Äquivalent (val) des gelösten Salzes.

(Ein Gramm-Äquivalent (val) eines Atoms oder einer Atomgruppe ist diejenige Menge in Gramm, die mit 1.008 g Wasserstoff reagiert oder diese Menge in Verbindungen ersetzt; also z.B. 35.5 g Chlor, 8 g Sauerstoff usw.. Man kann das Gramm-Äquivalent auch elektrochemisch definieren: Ein Gramm-Äquivalent einer Ionensorte ist diejenige Menge in Gramm, die einer Ladungsmenge von 96490 Coulomb (=1 Faraday) entspricht.)

Um die elektrische Leitfähigkeit verschiedener Elektrolyte besser miteinander vergleichen zu können, wird oft der Quotient zwischen der elektrischen Leitfähigkeit und der Konzentration des Elektrolyten angegeben. In den Tabellen werden die folgenden Definitionen benutzt:

Molare Leitfähigkeit

$$\sigma_{mol} = \frac{\text{spezifische Leitfähigkeit } \sigma}{\text{molare Konzentration } c_V} \text{ in } \frac{\Omega^{-1} \cdot m^{-1}}{Mol/l}$$
$$= 10 \, \Omega^{-1} \cdot Mol^{-1} \cdot cm^2$$

Äquivalent-Leitfähigkeit

$$\Lambda = \frac{\text{spezifische Leitfähigkeit } \sigma}{\text{Äquivalent-Konzentration } c_{val}} \text{ in } \frac{\Omega^{-1} \cdot m^{-1}}{val/l}$$
$$= 10 \, \Omega^{-1} \cdot val^{-1} \cdot cm^2$$

zu b): Die im Bodenelektrolyten gelösten Salze sind im allgemeinen stark dissoziiert. Mit zunehmender Konzentration des Elektrolyten nähern sich die Ionen einander immer mehr, so daß sich bei hohen Konzentrationen die Ionen zunehmend gegenseitig elektrostatisch beeinflussen. Mit der Verringerung des Abstandes der Ionen im zeitlichen Mittel nimmt die

solvent decreases with increasing concentration, the interaction between ions will be intensified with increasing concentration. As a result of both effects, conductivity may increase to a maximum value at a certain concentration and then decrease with further increasing concentration.

Coulombsche Wechselwirkung zu. Durch die Wechselwirkung zwischen den Ionen werden die Ionen in ihrer Beweglichkeit behindert. Die Wechselwirkung zwischen den Ionen verstärkt sich bei hoher Konzentration noch dadurch, daß die Dielektrizitäts-Konstante des Lösungsmittels durch den Elektrolyten erniedrigt wird. Die Coulombsche Wechselwirkung kann dazu führen, daß die elektrische Leitfähigkeit mit zunehmender Konzentration ein Maximum durchläuft.

e) Degree of dissociation α, interaction between ions and mobility of the ions will depend – as a function of density, viscosity, and dielectric constant – on pressure and temperature of the electrolyte.

zu e): Der Dissoziations-Grad α, die Wechselwirkung zwischen den Ionen und die Beweglichkeit der Ionen hängen über Dichte, Viskosität und Dielektrizitäts-Konstante des Elektrolyten auch von Druck und Temperatur ab.

The dependence of conductivity σ on the different parameters may be described as follows:

Die Leitfähigkeit σ eines Elektrolyten läßt sich somit in der folgenden Weise beschreiben:

$$\text{current density} \quad j = n_+ \cdot q_+ \cdot v_+ + n_- \cdot q_- \cdot v_- \qquad \text{Stromdichte} \quad j = n_+ \cdot q_+ \cdot v_+ + n_- \cdot q_- \cdot v_-$$

n_+, n_-: number of positive and negative charges, resp., per m^3

v_+, v_-: drift velocity of the ions in m s^{-1}

q_+, q_-: charge of the ions in A $\cdot$ s

The solution is electrically neutral

n_+, n_-: Anzahl der positiven bzw. negativen Ladungen pro m^3

v_+, v_-: Drift-Geschwindigkeit der Ionen in m s^{-1}

q_+, q_-: Ladung der Ionen in A $\cdot$ s

Da die Flüssigkeit insgesamt ungeladen ist, gilt:

$$n_+ \cdot q_+ = n_- \cdot q_-$$

The mobility u of the ions is defined by the ratio: drift velocity v/field strength E

Die Beweglichkeit u der Ionen ist definiert durch den Quotienten Driftgeschwindigkeit v/angelegte Feldstärke E

$$u = \frac{v}{E}$$

So we get

Damit wird

$$j = (n_+ \cdot q_+ \cdot u_+ + n_- \cdot q_- \cdot u_-) \cdot E$$

Using generalized Ohm's law $\sigma = j/E$ we can write

Mit $\sigma = j/E$ läßt sich für die elektrische Leitfähigkeit schreiben:

$$\sigma = \frac{j}{E} = n_+ \cdot q_+ \cdot u_+ + n_- \cdot q_- \cdot u_-$$

If z is the charge number of an ion its charge will be

Ist z die Ladungszahl eines Ions, so ist dessen Ladung

$$q_+ = z_+ \cdot e, \qquad q_- = z_- \cdot e$$

If a molecule dissociates in v_+ positive and v_- negative ions we get

Zerfällt ein Molekül in v_+ positive und v_- negative Ionen, so ist

$$n_+ = v_+ \cdot n, \qquad n_- = v_- \cdot n$$

The product of v and z is often called electrochemical valence w

Das Produkt aus v und z bezeichnet man oft als elektrochemische Wertigkeit w:

$$w = v_+ \cdot z_+ = v_- \cdot z_-$$

For fully dissociated electrolytes, electrical conductivity can be described by

Bei vollständiger Dissoziation eines Elektrolyten kann man damit die elektrische Leitfähigkeit beschreiben durch

$$\sigma = n \cdot e \cdot w (u_+ + u_-)$$

For partially dissociated electrolytes, the right side of the equation has to be multiplied by the degree of dissociation α.

Bei unvollständiger Dissoziation ist die rechte Seite mit dem Dissoziationsgrad α zu multiplizieren.

5.3.1.3 Laboratory measurements of the conductivity of electrolytic solutions — Messung der elektrischen Leitfähigkeit von Elektrolyten im Labor

The conductivity of an electrolytic solution is measured in a cell of electrically well isolating material (ceramics, quartz glass). In this cell the resistance of the electrolyte between 2 parallel electrodes is measured.

Die spezifische elektrische Leitfähigkeit von Elektrolyt-Lösungen wird in Meßzellen aus elektrisch gut isolierendem Material (Keramik, Quarzglas) bestimmt. Dazu wird der elektrische Widerstand des Elektrolyten zwischen 2 parallelen Flächen-Elektroden in der Meßzelle bestimmt.

$$R = C \cdot \varrho \quad \text{or/oder} \quad \sigma = C \cdot \frac{1}{R}, \quad C \approx \frac{l}{F}$$

R: resistance of the electrolyte between the 2 parallel electrodes

F: area of an electrode, l: distance between the 2 electrodes

C: cell constant

R: elektrischer Widerstand des Elektrolyten zwischen den parallelen Flächen-Elektroden

F: Elektrodenfläche, l: Abstand der Elektroden

C: Meßzellen-Konstante

The resistance R is often measured by a Wheatstone bridge. The cell constant C is determined using reference solutions of known conductivity (e.g. KCl-solutions). Platinum is commonly used for the electrodes. The surface area of the electrodes is often increased by Platinum Mohr. This keeps the boundary electrolyte/metal free from polarization effects. To prevent polarization and electrolysis alternating current is used for the measurement of the resistance. The resistivity of a solution can be determined as a function of the measuring frequency (e.g. $10^3 \cdots 10^6$ cps) and the resulting function can be extrapolated to infinite frequencies. This has been done for the values shown in the Tables $3 \cdots 8$.

Zur Messung des Widerstandes R wird oft die Wheatstone'sche Brückenschaltung benutzt. Die Meßzellen-Konstante C wird mit Hilfe von Eichlösungen bekannter elektrischer Leitfähigkeit (meist KCl-Lösungen) bestimmt. Die Elektroden bestehen meist aus Platin. Oft wird die wirksame Oberfläche der Elektroden mit Platin-Mohr vergrößert. Dadurch sollen eventuelle Polarisationen beim Ladungsübergang an der Grenze Elektrolyt/Metall verhindert werden. Um Polarisationseffekte und Elektrolyse zu vermeiden, werden die Messungen mit Wechselstrom durchgeführt. Dabei kann man die elektrische Leitfähigkeit in Abhängigkeit von der Frequenz bestimmen (z.B. zwischen 10^3 und 10^6 Hz) und dann auf die Leitfähigkeit bei unendlich hoher Frequenz extrapolieren. Dies wurde z.B. für die Werte in den Tabellen $3 \cdots 8$ gemacht.

Conductivity measurements of electrolytic solutions at high pressure and temperature are described in the following papers: [Fra56, Fra61, Fra62, Fra70, Hen64, Hol66, Hwa70, Klo73, Qui63, Qui66, Qui68, Qui69, Qui70, Man69, Ren70, Rit68].

Die Messung der Leitfähigkeit von Elektrolyt-Lösungen bei hohem Druck und bei hoher Temperatur ist in den folgenden Arbeiten [Fra56, Fra61, Fra62, Fra70, Hen64, Hol66, Hwa70, Klo73, Qui63, Qui66, Qui68, Qui69, Qui70, Man69, Ren70, Rit68] ausführlich beschrieben.

5.3.1.4 Conductivity of pure salt solutions as a function of the concentration of the solutions — Elektrische Leitfähigkeit von Lösungen reiner Salze in Abhängigkeit von der Konzentration der Lösungen

For Table 1, see next page.

Table 2. Mobility u of ions in aqueous solutions at 25 °C [kel66].

	H^+	OH^-	SO_4^{2-}	Cl^-	K^+	NO_3^-	Na^+	Li^+	HCO_3^-
u [$\cdot 10^{-8}$ m^2 s^{-1} V^{-1}]	36.2	20.5	8.3	7.9	7.6	7.4	5.2	4.0	4.6

The mobility u of ions in an aqueous solution is a function of the radius of the ions and as a consequence of the frictional resistance in the solvent. In addition, mobility depends also on the hydratation of the ions. In a normal aqueous solution, for example, an ion of Na is apposited with 8 molecules of water, and of K with 4, Mg with 14 and Cl with 3 molecules of water, respectively.

Table 1. Resistivity ϱ of pure salt solutions as a function of the concentration c of the solutions (at 20 °C) [Hem59, hcl60, con57, koh56, par67].

c val/l	NaCl c mg/l	NaCl ϱ $\Omega \cdot$ m	KCl c mg/l	KCl ϱ $\Omega \cdot$ m	$\frac{1}{2}$CaCl$_2$ c mg/l	$\frac{1}{2}$CaCl$_2$ ϱ $\Omega \cdot$ m	$\frac{1}{2}$MgCl$_2$ c mg/l	$\frac{1}{2}$MgCl$_2$ ϱ $\Omega \cdot$ m	$\frac{1}{2}$Na$_2$CO$_3$ c mg/l	$\frac{1}{2}$Na$_2$CO$_3$ ϱ $\Omega \cdot$ m
0.0001	5.85	882	7.456	741	5.55	826	4.76	869	5.3	725
0.0002	11.69	440	14.91	372	11.1	415	9.52	437	10.6	380
0.0005	29.23	178	37.28	149	27.7	168	23.8	177	26.5	162
0.001	58.5	89.4	74.56	75.1	55.5	84.9	47.6	89.3	53.0	84.9
0.002	116.9	45.1	149.1	37.8	111.0	43.3	95.2	45.6	106.0	43.8
0.005	292.3	18.4	372.8	15.4	277.0	17.8	238.0	18.8	265.0	18.5
0.01	585	9.34	745.6	7.81	555	9.22	476	9.71	530	9.90
0.02	1169	4.78	1491	3.98	1110	4.79	952	5.04	1060	5.31
0.05	2923	1.99	3728	1.65	2770	2.04	2380	2.15	2650	2.37
0.1	5850	1.03	7456	0.85	5550	1.08	4760	1.14	5300	1.30
0.2	11690	0.54	14910	0.44	11100	0.57	9520	0.61	10600	0.72
0.5	29230	0.23	37280	0.18	27700	0.25	23800	0.27	26500	0.35
1.0	58440	0.13	74560	0.10	55500		47610	0.16	53000	0.21

c val/l	$\frac{1}{2}$Na$_2$SO$_4$ c mg/l	$\frac{1}{2}$Na$_2$SO$_4$ ϱ $\Omega \cdot$ m	$\frac{1}{2}$MgSO$_4$ c mg/l	$\frac{1}{2}$MgSO$_4$ ϱ $\Omega \cdot$ m	$\frac{1}{2}$CaSO$_4$ c mg/l	$\frac{1}{2}$CaSO$_4$ ϱ $\Omega \cdot$ m	KNO$_3$ c mg/l	KNO$_3$ ϱ $\Omega \cdot$ m	NaNO$_3$ c mg/l	NaNO$_3$ ϱ $\Omega \cdot$ m
0.0001	7.10	864	6.02	869	6.81	831	10.1	762	8.5	915
0.0002	14.20	435	12.04	442	13.61	420	20.2	382	17.0	459
0.0005	35.5	176	30.10	183	34.03	174	50.5	154	42.5	184
0.001	71.0	89.5	60.2	95.6	68.1	91.5	101	77.4	85.0	93.0
0.002	142.0	45.5	120.4	50.7	136.1	49.2	202	39.1	170.0	47.1
0.005	355.0	18.9	301	22.0	340.3	22.2	505	15.9	425.0	19.1
0.01	710	9.85	602	12.5	681	12.4	1010	8.09	850	9.76
0.02	1420	5.19	1204	7.04	1360		2020	4.16	1700	5.00
0.05	3550	2.28	3010	3.35	3400		5050	1.74	4250	2.09
0.1	7100	1.22	6020	1.92	6810		10100	0.91	8500	1.10
0.2	14200	0.67	12040	1.10	13610		20200	0.48	17000	0.58
0.5	35500	0.32	30100	0.55	34030		50500	0.21	42500	0.26
1.0	71000		60200	0.33	68100		101000		85000	

Nomenclature of the salts in Table 1   molecular weight

Sodium chloride — Natriumchlorid	NaCl	58.44
Potassium chloride — Kaliumchlorid	KCl	74.55
Calcium chloride — Calciumchlorid	CaCl$_2$	111.0
Magnesium chloride — Magnesiumchlorid	MgCl$_2$	95.23
Sodium carbonate — Natriumkarbonat	Na$_2$CO$_3$	106.0
Sodium sulfate — Natriumsulfat	Na$_2$SO$_4$	142.05
Magnesium sulfate — Magnesiumsulfat	MgSO$_4$	120.37
Calcium sulfate — Calciumsulfat	CaSO$_4$	136.15
Potassium nitrate — Kaliumnitrat	KNO$_3$	101.1
Sodium nitrate — Natriumnitrat	NaNO$_3$	85.0
Sodium bicarbonate — Natrium-Bicarbonat	Na(HCO$_3$)	84.01
Calcium bicarbonate — Calcium-Bicarbonat	Ca(HCO$_3$)$_2$	162.1

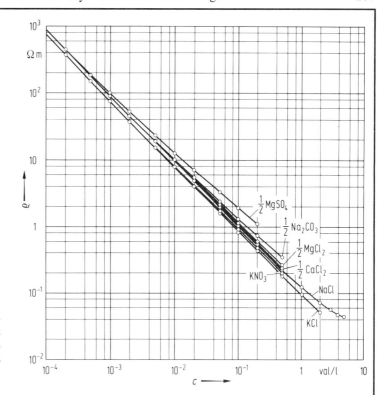

Fig. 1. Resistivity ϱ of some pure salt solutions as a function of concentration $c(T = 20\,°C)$ [Hem59, hel60, con57, koh56, par67].

5.3.1.5 Conductivity σ of pure salt solutions as a function of temperature T and pressure p — Elektrische Leitfähigkeit σ von Lösungen reiner Salze in Abhängigkeit von Temperatur T und Druck p

Viscosity, density, and dielectric constant of water (as a solvent for salts) decrease with increasing temperature, see Fig. 2. Viscosity decreases exponentially with increasing temperature. As a consequence, the mobility u of ions increases. This causes a strong increase of conductivity beginning at low temperatures (see Fig. 3). Increase of conductivity as a function of decrease of viscosity diminishes clearly at higher temperatures ($T > 200\,°C$). Density of water decreases continuously with increasing temperature (at constant pressure). This causes a reduction of conductivity as the number of ions/litre solution decreases. Decrease of the dielectric constant of water with increasing temperature causes an increasing apposition of positive and negative ions and as a consequence a decrease of conductivity. For temperatures below $300\cdots400\,°C$ (see Fig. 3) the temperature dependence of conductivity is mainly determined by viscosity: Conductivity increases considerably with increasing temperature. For temperatures above $300\cdots400\,°C$ the effect of decreasing density and dielectric constant predominates: Conductivity then decreases with increasing temperature.

Dichte, Dielektrizitäts-Konstante und Viskosität des Wassers (als Lösungsmittel für Salze) nehmen mit zunehmender Temperatur ab. Die Viskosität nimmt mit steigender Temperatur exponentiell ab. Dies ist in Fig. 2 dargestellt. Als Folge nimmt die Beweglichkeit u der Ionen und damit die elektrische Leitfähigkeit eines Elektrolyten von niedrigen Temperaturen an stark zu. Dies ist in Fig. 3 zu sehen. Die Zunahme der Leitfähigkeit wegen Abnahme der Viskosität verringert sich bei höheren Temperaturen ($T > 200\,°C$) deutlich. Die Dichte des Wassers nimmt (bei konstantem Druck) mit zunehmender Temperatur stetig ab. Dadurch nimmt die Anzahl der Ionen pro Volumeneinheit und als Folge davon auch die elektrische Leitfähigkeit ab. Die Abnahme der Dielektrizitäts-Konstante des Wassers mit zunehmender Temperatur bewirkt eine zunehmende Anlagerung entgegengesetzt geladener Ionen und damit ebenfalls eine Abnahme der elektrischen Leitfähigkeit. Bei Temperaturen unter $\approx 300\cdots400\,°C$ (siehe dazu Fig. 3) überwiegt der Einfluß der Viskosität auf die Temperaturabhängigkeit der elektrischen Leitfähigkeit: Die elektrische Leitfähigkeit nimmt mit zunehmender Temperatur deutlich zu. Bei Temperaturen über $300\cdots400\,°C$ überwiegt der Einfluß von abnehmender Dichte und Dielektrizitäts-Konstante: Die elektrische Leitfähigkeit nimmt dann mit zunehmender Temperatur wieder ab.

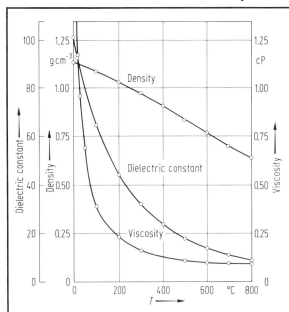

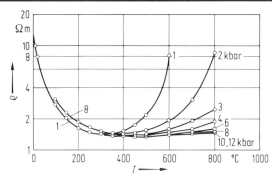

Fig. 3. Resistivity, ϱ, of a 0.01 demal KCl solution (0.745263 g KCl/1000 g solution) as a function of temperature and pressure [Qui70].

◄

Fig. 2. Density, dielectric constant, and viscosity of water as a function of temperature (at a constant pressure of 4000 bars) [Qui68].

In the Tables 3···6 and 8 the equivalent conductivity Λ is presented for the equivalent concentration $c_{val}(p, T)$ of the solutions under the given conditions of pressure and temperature. The equivalent concentration $c_{val}(p_0, T_0)$ under normal conditions is multiplied there with the relative density d/d_0.

Equivalent conductivity

In den Tabellen 3···6 sowie 8 ist die Äquivalent-Leitfähigkeit Λ für die Konzentration der Lösungen $c_{val}(p, T)$ unter den jeweiligen Druck- und Temperaturbedingungen angegeben, wobei die Äquivalent-Konzentration unter Normalbedingungen $c_{val}(p_0, T_0)$ mit der relativen Dichte d/d_0 multipliziert ist.

Äquivalent-Leitfähigkeit

$$\Lambda = \frac{\sigma}{c_{val}(p_0, T_0) \cdot d/d_0} \quad \text{in} \quad \frac{\Omega^{-1} \cdot m^{-1}}{val/l}$$

$\left(= \dfrac{\Omega^{-1} \cdot m^{-1}}{mol/l} = 10 \cdot \Omega^{-1} \cdot mol^{-1} \cdot cm^2 \text{ for the salts NaCl} \right.$
and KCl in the tables with the valences $1-1$)

$\left(= \dfrac{\Omega^{-1} \cdot m^{-1}}{Mol/l} = 10 \cdot \Omega^{-1} \cdot Mol^{-1} \cdot cm^2 \text{ für die in den} \right.$
Tabellen genannten 1—1 wertigen Salze NaCl und KCl)

Where
σ: conductivity in $\Omega^{-1} \cdot m^{-1}$
$c_{val}(p_0, T_0)$: equivalent concentration under normal conditions (e.g. 1 bar, 25 °C)
d_0: density of the electrolyte under normal conditions
d: density of the electrolyte under the pressure and temperature conditions at that time.

Dabei ist:
σ: spezifische elektrische Leitfähigkeit in $\Omega^{-1} \cdot m^{-1}$
$c_{val}(p_0, T_0)$: Äquivalent-Konzentration unter Normalbedingungen (z.B. 1 bar, 25 °C)
d_0: Dichte des Elektrolyten unter Normalbedingungen
d: Dichte des Elektrolyten unter den jeweiligen Druck- und Temperaturbedingungen.

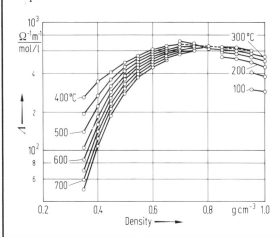

◄

Fig. 4. Equivalent conductivity, Λ, of a 0.1 molal NaCl solution as a function of density and temperature [Qui68].

Table 3. Equivalent conductivity Λ of a 0.001 molal NaCl solution as a function of temperature T and density d [Qui68].

T °C	$\Lambda \left[\dfrac{\Omega^{-1}\,m^{-1}}{mol/l} = 10\,\Omega^{-1}\,mol^{-1}\,cm^2 \right]$														
d g cm^{-3}	0.30	0.35	0.40	0.45	0.50	0.55	0.60	0.65	0.70	0.75	0.80	0.85	0.90	0.95	1.00
100														(357)	340
150														490	450
200												(685)	650	595	550
250												760	715	660	612
300									(960)	920	865	810	758	705	660
350									985	940	885	835	785	730	
400	420	800	1000	1080	1140	1110	1090	1055	1000	950	895	845	795	740	
450	340	650	880	1020	1100	1100	1085	1055	1005	950	905	850	800		
500	280	540	800	980	1070	1090	1080	1050	1000	950	905	840			
550	230	460	730	940	1040	1070	1065	1035	995	945	900				
600	190	400	660	900	1010	1050	1050	1020	985	940	890				
650	155	350	610	860	980	1030	1025	1000	970	930					
700	130	310	560	820	940	1000	1000	975	950	920					
750	110	280	510	780	910	970	970	940							
800	100	250	470	750	880	930	940	910							

Table 4. Equivalent conductivity Λ of a 0.01 molal NaCl solution as a function of temperature T and density d [Qui68].

T °C	$\Lambda \left[\dfrac{\Omega^{-1}\,m^{-1}}{mol/l} = 10\,\Omega^{-1}\,mol^{-1}\,cm^2 \right]$														
d g cm^{-3}	0.30	0.35	0.40	0.45	0.50	0.55	0.60	0.65	0.70	0.75	0.80	0.85	0.90	0.95	1.00
100														(342)	324
150														465	430
200												(640)	606	564	518
250												720	680	630	580
300									(880)	860	820	775	730	680	630
350									900	880	840	805	755	708	
400	230	430	610	710	820	870	915	920	905	885	855	815	770	718	
450	180	340	490	645	770	850	890	910	905	880	855	815	770		
500	140	270	425	590	730	830	870	895	895	870	850	810			
550	110	215	370	545	690	800	845	880	880	860	845				
600	85	180	330	505	650	770	825	860	865	850	830				
650	75	155	295	460	610	730	800	845	845	840					
700	65	140	270	420	570	700	775	825	830	820					
750	60	130	245	385	540	660	750	805							
800	55	120	225	350	500	620	725	780							

Table 5. Equivalent conductivity Λ of a 0.1 molal NaCl solution as a function of temperature T and density d [Qui68].

T °C $\Lambda \left[\dfrac{\Omega^{-1}\,m^{-1}}{mol/l} = 10\,\Omega^{-1}\,mol^{-1}\,cm^2 \right]$

T °C \ d (g cm^{-3})	0.30	0.40	0.45	0.50	0.55	0.60	0.65	0.70	0.75	0.80	0.85	0.90	0.95	1.00
100													(300)	290
150													405	380
200											(535)	520	490	450
250											580	570	535	500
300								(695)	675	635	615	600	575	535
350								680	675	645	635	620	590	
400	260	350	410	490	540	610	650	670	670	655	645	630	600	
450	190	270	360	450	520	590	630	655	660	655	645			
500	140	220	320	410	490	565	610	640	650	655	640			
550	105	185	290	380	470	540	585	625	640	655				
600	85	160	260	360	450	520	565	605	625	645				
650	70	140	235	330	420	495	545	585	610					
700	60	125	215	305	400	470	525	570	600					
750	50	110	195	285	380	450	510							
800	50	100	180	265	350	430	490							

Table 6. Resistivity ϱ and equivalent conductivity Λ $\left[\dfrac{\Omega^{-1}\,m^{-1}}{mol/l} = 10\,\Omega^{-1}\,mol^{-1}\,cm^2 \right]$ of 1 molar NaCl solution as a function of temperature T and pressure p [Klo73].

p bar	ϱ $\Omega\cdot m$	d/d_0	Λ $\dfrac{\Omega^{-1}\cdot m^{-1}}{mol/l}$	p bar	ϱ $\Omega\cdot m$	d/d_0	Λ $\dfrac{\Omega^{-1}\cdot m^{-1}}{mol/l}$
$T=26\,°C$				$T=300\,°C$			
1	0.114	1.000	87.4	500	0.027	0.800	459
500	0.111	1.018	88.8	1000	0.026	0.839	457
1000	0.108	1.036	88.9	1500	0.026	0.868	451
1500	0.107	1.051	88.8	2000	0.025	0.896	443
2000	0.106	1.063	88.6	2500	0.025	0.919	438
2500	0.106	1.074	87.7	3000	0.025	0.942	432
3000	0.105	1.084	87.6				
				$T=400\,°C$			
$T=100\,°C$				1000	0.031	0.729	450
250	0.053	0.958	197	1500	0.028	0.775	457
500	0.053	0.969	196	2000	0.027	0.814	459
1000	0.052	0.990	193	2500	0.026	0.844	460
1500	0.052	1.007	190	3000	0.025	0.875	454
2000	0.052	1.024	188				
2500	0.052	1.040	186	$T=500\,°C$			
3000	0.052	1.056	183	1000	0.058	0.584	297
				1500	0.041	0.668	366
$T=200\,°C$				2000	0.035	0.713	406
500	0.032	0.896	353	2500	0.031	0.754	434
1000	0.031	0.922	346	3000	0.029	0.790	441
1500	0.031	0.947	339				
2000	0.031	0.972	331	$T=600\,°C$			
2500	0.031	0.988	326	1500	0.091	0.558	197
3000	0.031	1.006	321	2000	0.063	0.623	254
				2500	0.049	0.662	307
				3000	0.043	0.703	331

Table 7. Resistivity ϱ of a 0.01 demal KCl solution (0.745263 g KCl/1000 g solution) as a function of temperature T and pressure p [Qui70].

p [bar]	ϱ [$\Omega \cdot$ m] T [°C] 100	150	200	250	300	350	400	450	500	600	700	800
1000	2.72	2.01	1.67	1.51	1.45	1.46	1.54	1.73	2.17	8.00		
2000	2.75	2.04	1.69	1.52	1.43	1.41	1.43	1.47	1.55	1.94	2.98	8.62
3000	2.79	2.07	1.71	1.54	1.44	1.40	1.39	1.41	1.45	1.60	1.89	2.47
4000	2.85	2.08	1,74	1.56	1.46	1.40	1.38	1.38	1.40	1.49	1.64	1.90
6000	2.96	2.15	1.80	1.62	1.50	1.43	1.38	1.36	1.37	1.41	1.49	1.59
8000	3.07	2.24	1.87	1.67	1.55	1.47	1.42	1.40	1.39	1.41	1.44	1.51
10000					1.60	1.50	1.45	1.41	1.41	1.41	1.44	1.48
12000					1.66	1.55	1.48	1.44	1.42	1.42	1.45	1.49

Table 8. Resistivity ϱ and equivalent conductivity Λ of a 1 molar KCl solution as a function of temperature T and pressure p [Klo73, Hwa70].

p bar	ϱ $\Omega \cdot$ m	d/d_0	Λ $\Omega^{-1} \cdot$ m^{-1} mol/l $=10\,\Omega^{-1}$ mol^{-1} cm^2 [Klo73]:	[Hwa70]:
$T=24$ °C				
1	0.091	1.000	110	112
200	0.090	1.008	110	112
500	0.088	1.018	111	
1000	0.086	1.035	112	114
1500	0.085	1.050	113	
2000	0.083	1.064	112	106
2500	0.083	1.077	112	
3000	0.083	1.089	111	103
$T=100$ °C				
250(200)	0.038	0.956	271	265
500	0.038	0.968	272	
1000	0.038	0.990	270	255
1500	0.037	1.004	268	
2000	0.037	1.018	266	234
2500	0.037	1.031	263	
3000	0.037	1.044	260	227
$T=200$ °C				
250(200)	0.024	0.882	476	467
500	0.024	0.896	472	
1000	0.024	0.922	462	438
1500	0.023	0.940	457	
2000	0.023	0.957	450	400
2500	0.023	0.972	443	
3000	0.023	0.987	436	387

p bar	ϱ $\Omega \cdot$ m	d/d_0	Λ $\Omega^{-1} \cdot$ m^{-1} mol/l $=10\,\Omega^{-1}$ mol^{-1} cm^2 [Klo73]:	[Hwa70]:
$T=300$ °C				
250(200)	0.023	0.770	576	559
500	0.022	0.792	582	
1000	0.021	0.836	574	556
1500	0.020	0.864	569	
2000	0.020	0.889	562	495
2500	0.020	0.910	557	
3000	0.020	0.928	550	472
$T=400$ °C				
1000	0.024	0.752	571	535
1500	0.022	0.771	592	
2000	0.021	0.808	598	519
2500	0.020	0.838	600	
3000	0.019	0.862	601	505
$T=500$ °C				
1000	0.036	0.581	473	463
1500	0.030	0.660	512	
2000	0.025	0.720	547	518
2500	0.024	0.760	560	
3000	0.023	0.792	546	528
$T=600$ °C				
1500	0.053	0.576	326	
2000	0.037	0.632	426	441
2500	0.031	0.680	481	
3000	0.027	0.720	510	495

5.3.1.6 Conductivity of natural waters — Elektrische Leitfähigkeit natürlicher Wässer

The resistivities of natural waters vary within a wide range. For comparison, the resistivity of pure water is presented first:

Pure distilled water has a high resistivity of about $2 \cdot 10^5 \, \Omega \cdot m$ at 20 °C. Minute amounts of dissolved compounds markedly decrease the resistivity of pure water. The resistivity of normal distilled water, therefore, is only $2 \cdot 10^4 \cdots 2 \cdot 10^3 \, \Omega \cdot m$.

Atmospheric water, as vapour, rain, or snow, dissolves gases (e.g. CO_2) and other chemical compounds present in the air. The resistivity of atmospheric water therefore is significantly lower than that of pure water. It is about $3 \cdot 10^2 \cdots 2 \cdot 10^3 \, \Omega \cdot m$.

While seeping and flowing through soil and rocks the subterranean waters dissolve ions from the rocks. The following ions are main constituents of natural waters:

Cations
Group A: Na^+, K^+, Ca^{2+}, Mg^{2+}, H^+
Group B: NH_4^+, Al^{3+}, Fe^{2+}, Fe^{3+}, etc.

Anions
Group A: HCO_3^-, CO_3^{2-}, Cl^-, SO_4^{2-}
Group B: NO_2^-, NO_3^-, SO_3^{2-}, OH^-, F^-, SiO_3^{2-}, etc.

Ions of group A are always present in natural waters and sometimes in considerable amounts. They determine the physical properties and the geochemical type of the water. Ions of group B are less common constituents.

On the basis of their chemical composition the natural waters are often subdivided into three major categories: chloride-waters, hydrogencarbonate-waters and sulfate-waters.

The ion concentration of natural waters depends on several different factors such as:

the amount of gases and chemical compounds dissolved in the atmospheric water (the dissolution of calcareous rocks, for example, clearly depends on the amount of CO_2 present in the atmospheric water.)

the chemical composition of the rocks, through which the waters flow,

the duration of the contact between water and rock

the distribution of water in the rocks as a function of porosity, permeability, joints, fissures, etc.

the primary salt concentration in the subsoil

the temperatures to which the subterranean waters had been heated

Der spezifische Widerstand natürlicher Wässer variiert in einem weiten Bereich. Als Extremwert sei zunächst der spezifische Widerstand von reinem Wasser angegeben:

Reinstes destilliertes Wasser hat einen hohen spezifischen Widerstand von $\approx 2 \cdot 10^5 \, \Omega \cdot m$ bei 20 °C. Bereits geringe Mengen gelöster Stoffe erhöhen die elektrische Leitfähigkeit des reinen Wassers deutlich. So hat gewöhnliches destilliertes Wasser nur einen spezifischen Widerstand von etwa $2 \cdot 10^4 \cdots 2 \cdot 10^3 \, \Omega \cdot m$.

Niederschlagswasser nimmt als Wasserdampf, Regen oder Schnee Gase (z.B. CO_2) und andere in der Luft vorhandene chemische Verbindungen auf. Der spezifische Widerstand von Niederschlagswasser ist deshalb deutlich geringer als der von reinem Wasser nämlich etwa $3 \cdot 10^2 \cdots 2 \cdot 10^3 \, \Omega \cdot m$.

Beim Einsickern des Niederschlagswassers in den Untergrund und beim Durchfließen von Böden und Gesteinen werden im Wasser Ionen aus dem Nebengestein gelöst. In den natürlichen Wässern kommen im wesentlichen die folgenden Ionen vor:

Kationen
Gruppe A: Na^+, K^+, Ca^{2+}, Mg^{2+}, H^+
Gruppe B: NH_4^+, Al^{3+}, Fe^{2+}, Fe^{3+} usw.

Anionen
Gruppe A: HCO_3^-, CO_3^{2-}, Cl^-, SO_4^{2-}
Gruppe B: NO_2^-, NO_3^-, SO_3^{2-}, OH^-, F^-, SiO_3^{2-} usw.

Gruppe A enthält Ionen, die regelmäßig und oft in beträchtlicher Menge in natürlichen Wässern vorkommen und die die physikalischen Eigenschaften und den geochemischen Typ des Wassers festlegen. Gruppe B enthält die weniger häufig vorkommenden Ionen.

Bezüglich ihrer chemischen Zusammensetzung unterteilt man die natürlichen Wässer oft nach dem überwiegenden Anion in Chlorid-Wässer, Hydrogenkarbonat-Wässer und Sulfat-Wässer.

Der Gehalt der natürlichen Wässer an Ionen hängt von vielen verschiedenen Faktoren ab so u.a.

vom Gehalt des Niederschlagswassers an gelösten Gasen und anderen chemischen Verbindungen. (So wird z.B. die Auflösung von Kalkgesteinen durch den CO_2-Gehalt des Niederschlagswassers gefördert.)

von der chemischen Zusammensetzung der Gesteine, durch die das Wasser fließt,

von der Dauer des Kontaktes zwischen Wasser und Gestein

von der Art und Verteilung der Hohlräume im Untergrund (Porenraum, Permeabilität, Klüfte, Spalten)

von der primären Salzkonzentration im Untergrund

von den Temperaturen, denen die Wässer ausgesetzt waren

the degree of water exchange with water from the earth's surface.

Resistivities of natural waters vary between about $300\,\Omega\cdot m$ and $1\,\Omega\cdot m$. Subterranean waters in deep sedimentary basins and in geothermal areas may have resistivities of less than $0.1\,\Omega\cdot m$ and can therefore occasionally be better conducting than sea-water. The resistivity of sea-water with an average salt concentration of about $35\%_{oo}$ is about $0.2\,\Omega\cdot m$.

von der Intensität des Wasseraustausches an der Erdoberfläche usw.

Der spezifische Widerstand der natürlichen Wässer variiert etwa zwischen $300\,\Omega\cdot m$ und $1\,\Omega\cdot m$. Wässer in tieferen Teilen von Sedimentbecken und in geothermisch aktiven Gebieten können auch spezifische Widerstände von weniger als $0.1\,\Omega\cdot m$ haben. Die Wässer sind damit zum Teil besser leitfähig als Meerwasser. Der spezifische Widerstand von Meerwasser beträgt bei einem mittleren Salzgehalt von $35\%_{oo}$ etwa $0.2\,\Omega\cdot m$.

Table 9. Resistivity ϱ of sea-water as a function of the salt concentration S and the temperature T.

T [°C]	ϱ [Ω m]							
	S [$\%_{oo}$] 5	10	15	20	25	30	35	40
0	2.14	1.08	0.75	0.57	0.47	0.40	0.34	0.31
15	1.38	0.73	0.50	0.39	0.32	0.27	0.23	0.21
18	1.28	0.68	0.47	0.36	0.29	0.25	0.22	0.19
25	1.11	0.58	0.40	0.31	0.25	0.22	0.19	0.17

Table 10. Resistivity ϱ of waters in various rock matrices (at 20 °C) [Che55, Cla66, kel66].

Origin of water samples	Number of determinations	Average salt concentration p.p.m.	ϱ [Ωm] average	ϱ [Ωm] range
Magmatic rocks				
in Europe	314	650	7.6	3.0···40
in South Africa	175	648	11.0	0.50···80
Metamorphic rocks				
in South Africa	88	640	7.6	0.86···80
in Australia (Precambrian)	31		3.6	1.5···8.6
Pleistocene to recent sedimentary rocks				
in Europe	610	1350	3.9	1.0···27
in Australia	323		3.2	0.38···80
Tertiary sedimentary rocks				
in Europe	993	3827	1.4	0.70···3.5
in Australia (miocene and oligocene)	240		3.2	1.35···10
Mesozoic sedimentary rocks				
in Europe	105	2084	2.5	0.31···47
Paleozoic sedimentary rocks				
in South Africa	161	588	0.93	0.29···7.1
Waters in oil fields				
Chloride waters	967		0.16	0.05···0.95
Sulfate waters	256		1.2	0.43···5.0
Bicarbonate waters	630		0.98	0.24···10

5.3.1.7 Calculation of the conductivity of natural waters from chemical analysis —
Berechnung der elektrischen Leitfähigkeit natürlicher Wässer aus chemischen Analysen

In [Log61] several methods are explained to determine the electrical conductivity of natural waters from chemical analysis. One of these methods, which has been deduced empirically from the analysis of many water probes, is presented here:

B: total concentration of dissolved natural compounds in mval/l.

σ: electrical conductivity in µS $\cdot$ cm^{-1} or $10^{-6} \, \Omega^{-1} \cdot$ cm^{-1}

If $B < 1$ then $\sigma = 100 \, B$

If B between 1 and 3, then $\sigma = 12.27 + 86.38 \, B + 0.835 \, B^2$

If B between 3 and 10, then $\sigma = B(95.5 - 5.54 \cdot \log B)$

If $B > 10$ and HCO$_3^-$ is the main anion, then $\sigma = 90.0 \, B$

If $B > 10$ and Cl$^-$ is the main anion, then $\sigma = 123 \cdot B^{0.9388}$

If $B > 10$ and SO$_4^{2-}$ is the main anion, then $\sigma = 101 \cdot B^{0.9489}$

These empirical relations should be used only for $B < 1000$ mval/l.

In some cases the total concentration (in mval/l) of the anion sum differs from that of the cation sum. Then the average of both sums should be taken for B.

For further information, see also [mat73, pir63, sag58, Dun51, Hem70, Hem59, Moo66].

In [Log61] werden mehrere Methoden referiert, die elektrische Leitfähigkeit von Wässern aus chemischen Analysen zu bestimmen. Eine Methode, die empirisch aus der Analyse vieler natürlicher Wässer abgeleitet worden ist, sei hier wiedergegeben:

B sei die Gesamtkonzentration an gelösten Stoffen in mval/l.

Die elektrische Leitfähigkeit σ wird in µS $\cdot$ cm^{-1} oder $10^{-6} \, \Omega^{-1} \cdot$ cm^{-1} erhalten.

Wenn $B < 1$, dann $\sigma = 100 \, B$

Wenn B zwischen 1 und 3, dann $\sigma = 12.27 + 86.38 \, B + 0.835 \, B^2$

Wenn B zwischen 3 und 10, dann $\sigma = B(95.5 - 5.54 \cdot \log B)$

Wenn $B > 10$ und HCO$_3^-$ vorherrschend, dann $\sigma = 90.0 \, B$

Wenn $B > 10$ und Cl$^-$ vorherrschend, dann $\sigma = 123 \cdot B^{0.9388}$

Wenn $B > 10$ und SO$_4^{2-}$ vorherrschend, dann $\sigma = 101 \cdot B^{0.9489}$

Diese empirisch gefundenen Beziehungen gelten nur für $B < 1000$ mval/l.

Da die Gesamtkonzentration in mval/l der Anionensumme manchmal etwas von der der Kationensumme abweicht, wird der Wert B als Mittel der beiden Summen genommen.

Das Thema wird auch behandelt in [mat73, pir63, sag58, Dun51, Hem70, Hem59, Moo66].

5.3.1.8 References for 5.3.1.1···5.3.1.7 — Literatur zu 5.3.1.1···5.3.1.7

a Reviews

ada76	Adam, A.: Geoelectric and geothermal studies (East-Central Europe, Soviet Asia), KAPG Geophysical Monograph, Akadémiai Kiadó, Budapest, **1976**.
bar52	Bartels, J., Ten Bruggencate, P. (Eds.): Landolt-Börnstein, 6. Aufl., Band III, Astronomie und Geophysik, Berlin-Göttingen-Heidelberg: Springer **1952**.
cam63	Camp, T.R.: Water and its impurities, New York: Reinhold Publishing Co. **1963**.
cla66	Clarke jr., S.P. (Ed.): Handbook of Physical Constants, Rev. Edition, GSA Memoir 97, 587 pp., **1966**.
con57	Conway, B.E.: Elektrochemische Tabellen, Frankfurt: Govi-Verlag **1957**.
dav67	Davis, S.N., de Wiest, R.J.M.: Hydrogeology, 463 pp., New York: Wiley **1967**.
ede72	Eder, F.X.: Moderne Meßmethoden der Physik, Teil 3, Elektrophysik, Berlin: VEB Deutscher Verlag der Wissenschaften **1972**.
hel60	Hellwege, K.-H., Hellwege, A.M., Schäfer, K., Lax, E. (Eds.): Landolt-Börnstein, 6. Aufl., Band II, Teil 7, Eigenschaften der Materie in ihren Aggregatzuständen, Elektrische Eigenschaften II (Elektrochemische Systeme), Berlin-Göttingen-Heidelberg: Springer **1960**.
kel66	Keller, G.V., Frischknecht, F.C.: Electrical methods in geophysical prospecting, London-Paris-Braunschweig: Pergamon Press **1966**.
koh56	Kohlrausch, F.: Praktische Physik, Band 2, Stuttgart: B.G. Teubner Verlagsges. **1956**.
mat73	Mattheß, G.: Lehrbuch der Hydrogeologie, Band 2, Die Beschaffenheit des Grundwassers, Berlin-Stuttgart: Gebr. Borntraeger, 322 S., **1973**.
mol70	Moelwyn-Hughes, E.A.v.: Physikalische Chemie, Stuttgart: Georg Thieme Verlag **1970**.
par67	Parkhomenko, E.I.: Electrical properties of rocks, 314 pp., New York: Plenum Press **1967**.
pir63	Pirson, S.J.: Handbook of Well Log Analysis, Prentice-Hall, Englewood Cliffs, New Jersey, 326 pp., **1963**.

sag58	Sage, J.F.: Water Analysis. In: Subsurface geology in Petroleum Exploration, Colorado School of Mines, Golden, Colorado, pp. 251–264, **1958**.
sou69	Souci, S.W., Quentin, K.E. (Eds.): Handbuch der Lebensmittelchemie, Band VIII, Teil 1, Wasser und Luft, Berlin-Heidelberg-New York: Springer **1969**.
toz59	Tozer, D.C.: Physics and chemistry of the earth, VIII, The electrical properties of the earth's interior, pp. 414–436, New York: Pergamon Press **1959**.
wea76	Weast, R.C. (Ed.): Handbook of Chemistry and Physics, 57th edition, 1976–1977, Cleveland, Ohio: CRC Press **1976**.

b Special references

Car75	Carlé, W.: Die Mineral- und Thermalwässer von Mitteleuropa. Geologie, Chemismus, Genese. Stuttgart: Wissenschaftliche Verlagsgesellschaft **1975**.
Che55	Chebotarev, I.I.: Geochim. Cosmochim. Acta **8** (1955); part 1, pp. 22–48; part 2, pp. 137–170; part 3, pp. 198–212.
Dun51	Dunlap, H.F., Hawthorne, R.R.: Trans. AIME **192** (1951) 373.
Fra56	Franck, E.U.: Z. Phys. Chem. (Frankfurt/Main) **8** (1956) 92.
Fra61	Franck, E.U.: Angew. Chem. **73** (1961) 309.
Fra62	Franck, E.U., Savolainen, J.E., Marshall, W.L.: Rev. Sci. Instr. **33** (1962) 115
Fra70	Franck, E.U.: Pure Appl. Chem. **24** (1970) 13.
Hem59	Hem, J.D.: Study and interpretation of the chemical analyses of natural waters. U.S. Geolog. Surv. Water Supply Papers, 1473, GPO, Washington D.C., 269 pp., **1959**.
Hem70	Hem, J.D.: Study and interpretation of the chemical characteristics of natural water. U.S. Geolog. Surv. Water Supply Papers, 1473, **1970**.
Hen64	Hensel, F., Franck, E.U.: Z. Naturforsch. **19** (1964) 127.
Hol66	Holzapfel, W., Franck, E.U.: Ber. Bunsenges. Physik. Chem. **70** (1966) 1105.
Hwa70	Hwang, J.U., Lüdemann, H.D., Hartmann, D.: High Temperatures-High Pressures **2** (1970) 651.
Klo73	Klostermeier, W.: Die elektrische Leitfähigkeit konzentrierter wässriger Alkalichloridlösungen bei hohen Drucken und Temperaturen. Institut für phys. Chem. u. Elektrochem., Univ. Karlsruhe **1973**.
Log61	Logan, J.: J. Geophys. Res. **66** (1961) 2479.
Man69	Mangold, K., Franck, E.U.: Ber. Bunsenges. Physik. Chem. **73** (1969) 21.
Moo66	Moore, E.J., Szasz, S.E., Whitney, B.F.: Determining formation water resistivity from chemical analysis. J. Petrol. Technol., March, 1966.
Qui63	Quist, A.S., Franck, E.U., Jolley, H.R., Marshall, W.L.: J. Phys. Chem. **67** (1963) 2453.
Qui66	Quist, A.S., Marshall, W.L.: J. Phys. Chem. **70** (1966) 3714.
Qui68	Quist, A.S., Marshall, W.L.: J. Phys. Chem. **72** (1968) 684.
Qui69	Quist, A.S., Marshall, W.L.: J. Phys. Chem. **73** (1969) 978.
Qui70	Quist, A.S., Marshall, W.L., Franck, E.U., v. Osten, W.: J. Phys. Chem. **74** (1970) 2241.
Ren70	Renkert, H., Franck, E.U.: Ber. Bunsenges. Physik. Chem. **74** (1970) 40.
Rit68	Ritzert, G., Franck, E.U.: Ber. Bunsenges. Physik. Chem. **72** (1968) 798.

5.3.2 Electrical conductivity of rocks containing electrolytes — Elektrische Leitfähigkeit von Gesteinen aufgrund von Elektrolyten im Porenraum

5.3.2.1 Introduction — Einleitung

5.3.2.1.1 General remarks — Allgemeine Bemerkungen

In rocks containing electrolytes in their pore space, the main contribution to conductivity is generally made by these electrolytes and any conductivity of the rock matrix (i.e. the solid framework including deposits of fine solid material) is comparatively negligible. Non-negligible matrix conductivity at normal temperatures must be expected only in vein type ores or very rich disseminated ores. Only at rather high temperatures semiconductance of minerals must perhaps be taken into account.

Liquids containing ions and polar groups can contribute to conductivity in two ways:

1. by their intrinsic electrolytic conductivity; here the pore space determines the geometry of the conductive path.

2. by electrochemical interaction with the solid material at their interface; here the internal surface more or less determines the geometry of the conductive path.

The conductivity of water-saturated porous rocks can be described by the following equation:

In Gesteinen, die Elektrolyte in ihrem Porenraum enthalten, liefern im allgemeinen diese Elektrolyte selbst den Hauptbeitrag zur Leitfähigkeit und jegliche Leitfähigkeit der Gesteinsmatrix (d.h. des festen Korngerüstes einschließlich feinen, festen Materials) ist im Vergleich dazu vernachlässigbar. Nicht vernachlässigbare Matrix-Leitfähigkeit ist bei normalen Temperaturen lediglich in Erzadern oder in fein verteilt erzreichem Material zu erwarten. Nur bei ziemlich hohen Temperaturen muß evtl. Halbleitung von Mineralen berücksichtigt werden.

Flüssigkeiten, die Ionen und polare Gruppen enthalten, können auf zweierlei Art zur Leitung beitragen:

1. durch ihre eigene elektrolytische Leitfähigkeit; in diesem Fall bestimmt der Porenraum die Geometrie der leitenden Bahn.

2. durch elektrochemische Wechselwirkung mit dem Matrixmaterial an der Grenzfläche; in diesem Fall bestimmt mehr oder weniger die innere Oberfläche die Geometrie der leitenden Bahn.

Die Leitfähigkeit eines wassergesättigten, porösen Gesteins kann durch folgende Gleichung beschrieben werden:

$$\kappa_0 = \frac{1}{F}\kappa_w + \kappa_{q0} \tag{1}$$

Here κ_0 is the rock (bulk) conductivity, κ_w the electrolytic water conductivity, κ_{q0} a conductivity due to a boundary layer at the internal surface and F a geometrical constant. κ_{q0} can be considered practically independent of κ_w, at least in a very good approximation. If the second term of Eq. (1) is negligible compared with the first term – e.g. due to high salinity of the pore water – this equation is reduced to

wobei κ_0 die Gesteinsleitfähigkeit (Gesteinskonduktivität), κ_w die Leitfähigkeit (Konduktivität) des Elektrolyts, κ_{q0} eine weitere Leitfähigkeit, herrührend von einer Grenzschicht an der inneren Oberfläche (Grenzflächenleitfähigkeit, Grenzflächenkonduktivität), und F eine geometrische Konstante sind. κ_{q0} kann als praktisch unabhängig von κ_w angesehen werden, zumindest in einer sehr guten Näherung. Ist der zweite Term von Gleichung (1) gegenüber dem ersten vernachlässigbar – z.B. im Fall hoher Salinität der Porenflüssigkeit – so reduziert sich diese Gleichung auf

$$\varrho_0 = F\varrho_w \tag{1a}$$

where $\varrho_0 = 1/\kappa_0$ is the resistivity of the rock and $\varrho_w = 1/\kappa_w$ the resistivity of water.

For a not completely water-saturated rock, Eq. (1) is changed to

wobei $\varrho_0 = 1/\kappa_0$ der spezifische Widerstand (Resistivität) des Gesteins und $\varrho_w = 1/\kappa_w$ der spezifische Widerstand (Resistivität) des Wassers sind.

Für ein nicht 100%ig wassergesättigtes Gestein gehen Gl. (1) in

$$\kappa_t = \frac{1}{F\,I(\Sigma_w)}\kappa_w + \kappa_q(\Sigma_w) \tag{2}$$

and Eq. (1a) (under the due conditions) to

und Gl. (1a) (unter den o.a. Bedingungen) in

$$\varrho_t = F\,I(\Sigma_w)\varrho_w \tag{2a}$$

Here κ_t is the true rock conductivity in the under-

über. Hierbei ist κ_t die ‚wahre' Gesteinsleitfähigkeit im

saturated state. I is a geometrical factor which depends on the electrolyte saturation Σ_w given in fractions of the pore volume. The term κ_q is also dependent on Σ_w.

untersättigten Zustand. I, wiederum ein geometrischer Faktor, ist abhängig von der Elektrolytsättigung, Σ_w, die als Anteil des Porenvolumens angegeben wird. κ_q ist ebenfalls abhängig von Σ_w.

5.3.2.1.2 Definitions — Definitionen

The geometrical factor F in Eqs. (1) and (1a) is called formation factor or formation resistivity factor. The first name is more appropriate because it not only applies to electrical resistivity but to all resistive properties of rocks related to phenomena of transport or propagation in the pore space that are proportional to the ratio of cross-section to length of the available ducts. An example is diffusion in the pore space; d'Arcy flow, however, does not belong to this category since here the square of the cross-section enters the geometrical law! (cf. [Scho66]). Thus, the formation factor is a pure rock material constant describing macroscopically the geometrical structure of the pore space.

Der geometrische Faktor F aus den Gleichungen (1) und (1a) ist der sogenannte Formationsfaktor oder Formationswiderstandsfaktor. Die erste Bezeichnung ist die zutreffendere, da sie nicht nur für den elektrischen Widerstand gilt, sondern auch für alle Widerstandseigenschaften des Gesteins gegenüber anderen Phänomenen des Transports oder der Ausbreitung im Porenraum, die proportional zum Verhältnis Querschnittsfläche/Länge der zur Verfügung stehenden Bahnen sind (z.B. für Diffusionsvorgänge im Porenraum; dagegen fällt der d'Arcy-Fluß nicht unter diese Kategorie, da dort das Quadrat der Querschnittsfläche in das geometrische Gesetz eingeht (vgl. [Scho66]). Somit ist der Formationsfaktor eine reine gesteinsspezifische Materialkonstante, welche makroskopisch die geometrische Struktur des Porenraumes beschreibt.

Eq. (1a) is often called "Sundberg Equation" [Sun32, 80]. It is often used for the definition of F. However, care has to be taken in doing so, since this is only allowed in cases where the term κ_{q0} in Eq. (1) is negligible. Generally, one can write

Gl. (1a) wird manchmal auch „Sundberg-Gleichung" genannt [Sun32, 80]. Sie wird oft als Definitionsgleichung für F herangezogen. Damit sollte man jedoch vorsichtig sein, da dies nur in solchen Fällen statthaft ist, in denen κ_{q0} in Gl. (1) vernachlässigbar ist. Ganz allgemein läßt sich natürlich immer schreiben

$$\varrho_0 = F_a \varrho_w \tag{1b}$$

where the "apparent formation factor" F_a is not a rock material constant any longer but depends, among others, on the fluid resistivity ϱ_w itself. The relation between F_a and F can best be seen by dividing Eq. (1) by κ_w:

wobei der sogenannte „scheinbare Formationsfaktor" F_a dann nicht mehr eine gesteinsspezifische Materialkonstante ist, sondern u.a. vom spezifischen Widerstand der Flüssigkeit, ϱ_w, selbst abhängt. Den Zusammenhang zwischen F_a und F kann man am besten erkennen, wenn man Gl. (1) durch κ_w dividiert:

$$\frac{1}{F_a} = \frac{1}{F} + \frac{\kappa_{q0}}{\kappa_w}. \tag{3}$$

For even clearer discerning between F and F_a, F is often named 'true formation factor' and sometimes symbolized by F_t. (For the attention of the reader: in a part of the literature the symbol F^* is used for the true formation factor and the apparent formation factor is then simply symbolized by F!)

Um noch deutlicher den Unterschied zwischen F und F_a zu betonen, wird F manchmal auch „wahrer Formationsfaktor" genannt und durch das Symbol F_t dargestellt. (Die Aufmerksamkeit des Lesers sei an dieser Stelle darauf gelenkt, daß teilweise in der Literatur auch das Symbol F^* für den wahren Formationsfaktor verwendet wird und der scheinbare Formationsfaktor dann einfach nur mit F gekennzeichnet wird!)

The second term κ_{q0} in Eq. (1) is called surface conductivity or interface conductivity. It encompasses the additional conductivity caused by electrochemical fluid/solid interactions at the internal surface. The true nature of the effects generating interface conductivity is not yet fully known (cf. [Wit50, Wax68, Pfa72, Hil56, Rie78]). Another open question is whether phase boundaries between an oil and a water phase can cause interface conductivity, too. - Interface conductivity is observed in every rock material, pure

Der zweite Term in Gl. (1), κ_{q0}, ist die sogenannte Oberflächen- oder Grenzflächenleitfähigkeit. Er stellt eine zusätzliche Leitfähigkeit dar, welche durch elektrochemische Flüssigkeit/Matrix-Wechselwirkungen an der inneren Oberfläche hervorgerufen wird. Die wahre Natur der Effekte, die diese Grenzflächenleitfähigkeit erzeugen, ist bisher noch nicht vollständig bekannt (vgl. [Wit50, Wax68, Pfa72, Hil56, Rie78]). Eine weitere unbeantwortete Frage ist, ob Phasengrenzen zwischen einer Öl- und einer Wasserphase ebenfalls Anlaß für

quartz, glass, ceramics etc. and increases with specific surface. Clay-mineral rich rocks especially show great values.

Interface conductivity is observed to be rather constant and independent of the pore water conductivity for a given rock and electrolyte. Some authors predict a decrease towards very high and very low pore water salinities [Hil56]. Although there is some theoretical support for this, it is still very much open to discussion because of problems in experimentally proving it. At the high-salinity end, the effect itself is negligible, so no change can be observed either. If extremely pure water is used for saturating the rock, an effect in the said direction can sometimes be observed; however these results are not unanimously accepted because of the great influence of experimental error in this case.

With decreasing water saturation the conductive path of the pore space is reduced further. This geometrical effect is macroscopically described by the "resistivity" or "saturation index" I in Eq. (2) or (2a). It is a function of the water saturation and can thus be used as a quantitative indicator of water saturation. For the full range of the water saturation

die Entstehung einer Grenzflächenleitfähigkeit geben. – Die Grenzflächenleitfähigkeit läßt sich in jeglichem Gesteinsmaterial, reinem Quarz, Glas, keramischen Stoffen etc. beobachten; sie wächst mit der spezifischen inneren Oberfläche. An Tonmineralen reiche Gesteine weisen besonders große Werte auf.

Beobachtungen zeigen, daß für ein gegebenes Gestein und für einen gegebenen Elektrolyten die Grenzflächenleitfähigkeit nahezu unabhängig von der Leitfähigkeit des Porenwassers ist. Manche Autoren sprechen von einer Abnahme bei sehr hohen und bei sehr niedrigen Salinitäten des Porenwassers [Hil56]. Obwohl sich dies in gewisser Weise theoretisch untermauern läßt, muß die Diskussion hierüber noch offen bleiben wegen Problemen beim experimentellen Nachweis. Im Bereich hoher Salinitäten ist der Effekt selbst vernachlässigbar, so daß auch keine Änderungen dieses Effektes beobachtet werden können. Dagegen scheint ein solcher Effekt gelegentlich feststellbar, wenn man das Gestein mit extrem reinem Wasser sättigt; jedoch sind diese Beobachtungen noch nicht einstimmig akzeptiert, wegen der großen experimentellen Fehlermöglichkeiten im letzten angesprochenen Fall.

Mit abnehmender Wassersättigung wird die leitende Bahn im Porenraum weiter reduziert. Dieser geometrische Effekt wird makroskopisch durch den „Resistivitäts-" oder „Sättigungsindex" I aus Gl. (2) oder (2a) beschrieben. I ist eine Funktion der Wassersättigung und kann somit als ein quantitativer Indikator für die Wassersättigung herangezogen werden. Für den gesamten Bereich der Wassersättigung

$$0 \leq \Sigma_w \leq 1 \tag{4}$$

we obtain the following range of the saturation index I:

erhalten wir den folgenden Bereich für den Sättigungsindex I:

$$\infty \geq I \geq 1. \tag{5}$$

The interface conductivity κ_q decreases with water saturation, too, but to a lesser degree than the first term of Eq. (2). For the full range of the water saturation

Die Grenzflächenleitfähigkeit κ_q nimmt ebenfalls mit der Wassersättigung ab, aber in einem geringeren Maße als der erste Term aus Gl. (2). Über den gesamten Bereich der Wassersättigung

$$0 \leq \Sigma_w \leq 1 \tag{6}$$

we obtain the following range of the interface conductivity κ_q

erhalten wir folgenden Bereich für die Grenzflächenleitfähigkeit κ_q:

$$0 \leq \kappa_q \leq \kappa_{q0}. \tag{7}$$

5.3.2.1.3 Petrophysical interrelations — Petrophysikalische Zusammenhänge

The formation factor F is tightly connected to porosity ϕ and varies inversely with it. Since porosity not only gives the ratio of conductive fluid volume to total volume but also the ratio of conductive cross-section to total cross-section (cf. 2.1), a theoretical formula can be set up:

Der Formationsfaktor F ist eng verknüpft mit der Porosität ϕ und variiert gegenläufig zur letzteren. Da die Porosität nicht nur das Verhältnis leitfähiges Flüssigkeitsvolumen/totales Volumen sondern auch das Verhältnis leitfähige Querschnittsfläche/totale Querschnittsfläche (vgl. 2.1) angibt, kann eine theoretische Formel folgender Art gefunden werden:

$$F = X/\phi \tag{8}$$

where the proportionality factor X, called (electrical) tortuosity, is another geometrical factor representing all the deviations of the true current path from a straight path with constant current density (cf. [Scho66]). This macroscopical quantity, however, is a purely mathematical quantity and there is no way of independently measuring it. Eq. (8) is credited to M.R.J. Wyllie [Wyl57].

Together with the changes in porosity ϕ during the geological development of rocks, all the other structural properties jointly represented by the tortuosity X undergo changes as well. Thus strict anticorrelation of X and ϕ and hence F and ϕ can be observed in natural rocks:

$$F = \phi^{-m} \tag{9}$$

with

$$1.3 \leq m \leq 2.5. \tag{10}$$

m increases with packing, compaction and cementation of rocks. Eq. (9) is known by the name 'Archie's first equation' [Arc42]. m as well as X are purely mathematical quantities which cannot be measured directly. Only an estimate by an experienced eye is possible from the rock sample.

Eq. (9) has been modified under various names to

$$F = a\,\phi^{-b} \tag{11}$$

in order to obtain a better fit to empirical data [Tun66, 67, Car68, Por71].

For the relation of the saturation index I to water saturation Σ_w, an equation similar to Eq. (9) has also been found empirically by Archie ('Archie's second equation'):

$$I = \Sigma^{-n} \tag{12}$$

with about

$$1.4 \leq n \leq 2.2. \tag{13}$$

In practice $n = 2$ is usually assumed, but with doubtful justification.

The interface conductivity κ_{q0} correlates well with the specific pore space surface S_{por},

$$\kappa_{q0} \propto S_{por}/F \tag{14}$$

and indirectly with other quantities themselves connected with the internal surface e.g. permeability, adsorptivity, electrochemical potential, cation exchange capacity etc. Beside empirical confirmation, there is theoretical justification for such relations. However, research is still going on in this field at a fast pace and no final results can be given as yet.

wobei der Proportionalitätsfaktor X, die sogenannte (elektrische) Tortuosität, als ein weiterer geometrischer Faktor alle Abweichungen der tatsächlichen leitenden Bahn von einer geradlinigen mit konstanter Stromdichte beinhaltet (vgl. [Scho66]). Diese makroskopische Größe ist jedoch eine rein mathematische, und es gibt keine Möglichkeit, sie für sich allein zu bestimmen. Gl. (8) geht wahrscheinlich auf M.R.J. Wyllie [Wyl57] zurück.

Zusammen mit den Änderungen in der Porosität ϕ während der geologischen Entwicklung der Gesteine unterliegen auch alle die anderen Struktureigenschaften, die gemeinsam durch die Tortuosität X repräsentiert werden, ebenfalls Änderungen. Daher läßt sich eine enge, gegenläufige Korrelation von X und ϕ und somit auch von F und ϕ in natürlichen Gesteinen beobachten:

mit

m wächst mit zunehmender Packungsdichte, Kompaktion und Zementation der Gesteine. Gl. (9) ist unter dem Namen „erste Gleichung von Archie" bekannt [Arc42]. m wie X sind rein mathematische Größen, die nicht direkt gemessen werden können. Lediglich eine Abschätzung an Hand von Gesteinsproben nach Augenschein und Erfahrung ist möglich.

Gl. (9) ist unter verschiedenen Namen modifiziert worden zu

um eine bessere Anpassung an die empirischen Daten zu bekommen [Tun66, 67, Car68, Por71].

Eine zu Gl. (9) ähnliche Beziehung zwischen dem Sättigungsindex I und der Wassersättigung Σ_w ist ebenfalls von Archie empirisch formuliert worden („zweite Gleichung von Archie"):

mit etwa

In der Praxis wird im allgemeinen $n = 2$ angenommen, was aber nur wenig gerechtfertigt ist.

Die Grenzflächenleitfähigkeit κ_{q0} korreliert gut mit der spezifischen Oberfläche des Porenraumes, S_{por},

und, durch indirekte Einwirkung, mit weiteren Größen, die selbst mit der inneren Oberfläche zusammenhängen, wie etwa der Permeabilität, der Adsorptivität, dem elektrochemischen Potential, der Kationen-Austausch-Kapazität etc. Neben empirischen Hinweisen besteht eine theoretische Rechtfertigung für die Annahme solcher Beziehungen. Jedoch schreitet die Forschung auf diesem Gebiet schnell voran und endgültige Ergebnisse liegen zur Zeit noch nicht vor.

For the function $\kappa_q(\Sigma_w)$, a power law seems to be applicable too:

Für die Abhängigkeit $\kappa_q(\Sigma_w)$ scheint auch ein Potenzgesetz der Form

$$\kappa_q/\kappa_{q0} = \Sigma_w^v \qquad (15)$$

with about

mit etwa

$$0.4 \leq v \leq 1.2. \qquad (16)$$

zuzutreffen.

The temperature dependence of κ_w in Eqs. (1) and (2) is that of the free electrolyte, while κ_{q0} might have a slightly different temperature coefficient [Ker77]. Whether a non-negligible variation of F_t with temperature exists is still open to question, although there seems to be some indication for this [San73, Bra73].

Die Temperaturabhängigkeit von κ_w in den Gleichungen (1) und (2) ist die des freien Elektrolyten, während κ_{q0} einen geringfügig abweichenden Temperaturkoeffizienten haben dürfte [Ker77]. Weiterhin ist unbeantwortet, ob eine nicht vernachlässigbare Änderung von F_t mit der Temperatur vorkommt oder nicht, obwohl einige Anzeichen dafür sprechen [San73, Bra73].

5.3.2.1.4 Measurement — Messungen

The determination of rock resistivity (ϱ_0, ϱ_t) implies a resistance measurement via plate electrodes using a formatized rock sample (e.g. cylindrical or prismatic) and a calculation of the resistivity from the resistance and known geometry. In view of the accuracy of the measurement, it is recommended to use sample shapes with a high ratio of length to cross-section in order to make contact resistances more negligible.

Die Bestimmung des spezifischen Gesteinswiderstandes (ϱ_0, ϱ_t) verlangt eine Widerstandsmessung über Plattenelektroden an einer formatisierten Gesteinsprobe (z.B. zylindrisch oder prismatisch) und eine Berechnung dieses spezifischen Widerstandes (Resistivität) über den Widerstand und die bekannte Geometrie. Hinsichtlich der Genauigkeit der Messung empfiehlt sich, Probenformen mit einem hohen Verhältnis Länge/ Querschnittsfläche zu benutzen, um die Kontaktwiderstände vernachlässigbarer zu machen.

Contact resistance can be reduced further by placing filter paper, fine tissue or the like, wetted with salt water similar to the pore water, between the plate electrodes and sample faces. However, this is recommended for measurements at full water saturation only, since at partial saturation the wet tissue might disturb the state of saturation.

Der Kontaktwiderstand kann im übrigen heruntergesetzt werden, indem Filterpapier, feines Gewebe oder ähnliches, das mit dem gleichen Salzwasser wie dem im Porenraum befindlichen getränkt ist, zwischen Plattenelektroden und Probenoberflächen eingefügt wird. Dies wird jedoch nur für Messungen bei 100%-iger Wassersättigung empfohlen, da eine partielle Sättigung durch das nasse Gewebe in ihrem Zustand verfälscht werden kann.

Fully saturated samples should be kept in the electrolyte prior to the measurement. On taking them from the liquid, they should be stripped of water on the outside by cloth or leather, but with due care not to remove any water from the pore space.

Vollständig gesättigte Proben sollten bis zur Messung in die elektrolytische Flüssigkeit getaucht bleiben. Nach dem Herausnehmen sollte man das an der Außenfläche haftende Wasser mit einem Tuch oder Leder entfernen, aber mit großer Vorsicht, um kein Wasser dem Porenraum zu entziehen.

Sufficient time should be allowed for equilibration with the surrounding liquid after saturating the sample. An individual beaker is recommended for each sample. The water left in the beaker after removal of the sample for the measurement can then be used for measuring the pore water resistivity ϱ_w in an electrolytic measuring cell.

Genügend Zeit sollte nach Sättigung der Probe verstreichen, damit sich ein Gleichgewicht mit der umgebenden Flüssigkeit einstellen kann. Für jede Probe empfiehlt sich ein eigenes Probenglas. Das im Probenglas nach Entnahme der Probe zurückbleibende Wasser kann für die Messung des spezifischen Widerstandes des Porenwassers, ϱ_w, in einer Elektrolyt-Meßzelle dienen.

The resistance measurement can be made by means of a bridge circuit or by a current and voltage measurement. In the latter case, the voltage drop can be picked

Die Widerstandsmessung läßt sich mit Hilfe einer Brückenschaltung oder über eine Strom- und Spannungsmessung durchführen. Im letzteren Fall kann

up at the current-feeding electrodes themselves ('two-point method') or at separate potential pick-up electrodes ('four-point method'). The four point method provides a better safeguard against errors by contact resistance. It is especially recommended for partially water saturated samples, while for fully saturated ones, the two-point method is generally sufficient.

For determining F and κ_{q0}, a series of about 12 measurements of κ_0 at different salinities (6···200000 ppm) of the pore water is necessary. κ_0 then must be plotted against κ_w. A linear plot is not recommended because of the wide range. Better is a double logarithmic plot (Fig. 1). Here a given master curve (hyperbola) can be shifted parallel to match the data points, then κ_{q0} and F can be read at the intersections of the asymptotes with the y- and x-axis respectively [Rin74].

Another practical plot has been designed by de Witte [Wit50] (Fig. 2). Here ϱ_0 vs. F_a is plotted and a straight line section is obtained within the first quadrant into which the whole range

der Spannungsabfall direkt an den Stromzuführungs-elektroden („Zwei-Punkt-Methode") oder an besonderen Potentialelektroden („Vier-Punkt-Methode") abgegriffen werden. Die Vier-Punkt-Methode stellt einen besseren Schutz gegenüber den durch Kontaktwiderstände hervorgerufenen Fehlern dar. Insbesondere für partiell wassergesättigte Proben sei zu dieser Methode geraten, während bei vollständig gesättigten Proben im allgemeinen die Zwei-Punkt-Methode ausreicht.

Um F und κ_{q0} zu bestimmen, sind eine Reihe von etwa 12 κ_0-Messungen bei verschiedenen Salinitäten des Porenwassers (6···200000 ppm) notwendig. Hiernach muß dann κ_0 gegen κ_w aufgetragen werden. Ein linearer Plot empfiehlt sich aufgrund des großen Bereiches nicht. Wesentlich besser ist ein doppellogarithmischer Plot (Fig. 1). In diesem Fall kann eine Modellkurve (Hyperbel) durch Parallelverschieben in die Meßpunkte eingepaßt werden; dann lassen sich κ_{q0} und F an den Schnittpunkten der Asymptoten mit der y- bzw. x-Achse ablesen [Rin74].

Ein anderer praktischer Plot wurde von deWitte entworfen [Wit50] (Fig. 2). Hierbei wird ϱ_0 gegen F_a aufgetragen, und man erhält eine Strecke innerhalb des ersten Quadranten, auf der der gesamte Bereich

$$0 \leqq \kappa_w \leqq \infty$$

is compressed. The end points of that line on the x- and y-axis give F and ϱ_{q0}. Similar ways of plotting can be applied to the evaluation of I and κ_q.

komprimiert ist. Die Endpunkte dieser Strecke auf der x- und y-Achse liefern F und ϱ_{q0}. Für die Bestimmung von I und κ_q können analoge graphische Methoden angewendet werden.

5.3.2.1.5 List of symbols — Symbolliste

F		formation factor — Formationsfaktor
F_a		apparent formation factor — scheinbarer Formationsfaktor
I		saturation index — Sättigungsindex
K	[md]	permeability — Permeabilität
S_{por}	[m^{-1}]	specific pore space surface — spezifische Oberfläche des Porenraumes
U	[mV]	electrochemical selfpotential — elektrochemisches Eigenpotential
X		electrical tortuosity — elektrische Tortuosität
κ_0	[S m^{-1}]	rock (bulk) conductivity at full water saturation — Leitfähigkeit des Gesteins bei vollständiger Wassersättigung
κ_q	[S m^{-1}]	interface conductivity — Grenzflächenleitfähigkeit
κ_{q0}	[S m^{-1}]	interface conductivity at full water saturation — Grenzflächenleitfähigkeit bei vollständiger Wassersättigung
κ_t	[S m^{-1}]	true rock conductivity — wahre Leitfähigkeit des Gesteins
κ_w	[S m^{-1}]	electrolytic water conductivity — elektrolytische Leitfähigkeit des Wassers
$\varrho_0 = 1/\kappa_0$		rock resistivity at full water saturation — Gesteinsresistivität bei vollständiger Wassersättigung
$\varrho_w = 1/\kappa_w$		water resistivity — Wasserresistivität
Σ_w		electrolyte saturation — Elektrolyt-Sättigung
ϕ	[%]	porosity — Porosität

5.3.2.2 Tables and figures

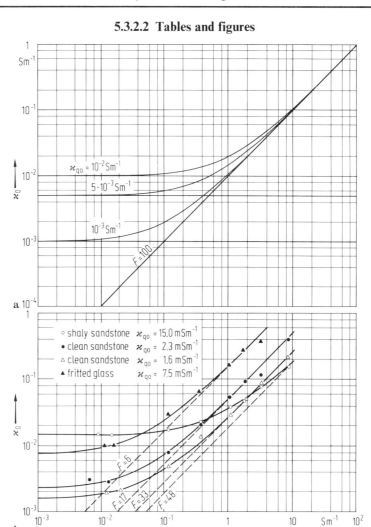

Fig. 1a. Schematic example of a double logarithmic plot of κ_0 vs. κ_w with curves for $F = 100$ and $\kappa_{q0} = 10^{-2}$, $5 \cdot 10^{-3}$, 10^{-3} Sm^{-1}. The curves are obtained by shifting a pre-calculated master curve horizontally and vertically to fit the data points. κ_{q0} then can be read from the intersection of the horizontal asymptote with the ordinate and F is read from the 45° asymptote by the ratio of its abscissa intersection and the ordinate value at the origin.

Fig. 1b. Example of a diagram according to Fig. 1a for actual data [Rin74].

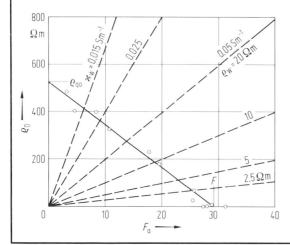

Fig. 2. Schematic example of a linear plot of ϱ_0 vs. F_a after deWitte [Wit50]. At the ordinate intersection of the straight regression line ϱ_{q0} is given and at the abscissa intersection, the true formation factor F. Lines of constant water resistivity (dashed) have been added to the diagram.

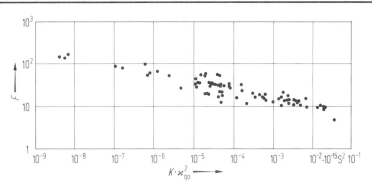

Fig. 3. Plot of the true formation factor F vs. the product of permeability and interface conductivity κ_{q0} squared. Regression leads to the equation $F^5 K \kappa_{q0}^2 = \text{const} \approx 2 \cdot 10^{-12}$ S^2. – The scale unit on the abscissa is chosen for easy permeability extraction, since $10^{-15} S^2 = 1\,\text{md} \cdot 1\,S^2 m^{-2}$ [Rie79].

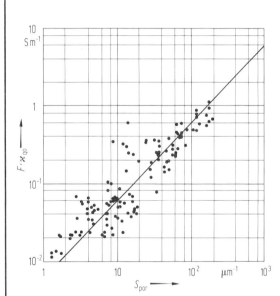

Fig. 4. Plot of the product of the true formation factor F and the interface conductivity κ_{q0} vs. the pore-volume-specific surface S_{por} (cf. 2.2.). The regression line forced into a linear law is given by $S_{por}/F\kappa_{q0} = 1.67 \cdot 10^{-8} S^{-1}$ [Rie79].

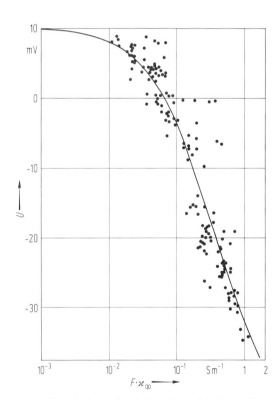

Fig. 5. Plot of electrochemical selfpotential U vs. $F\kappa_{q0}$. Dots measured, and curve calculated for two participating electrolytes with $\kappa_{w1} = 0.95\,\text{Sm}^{-1}$, $\kappa_{w2} = 0.14\,\text{Sm}^{-1}$. The curve converges toward the diffusion potential at the upper end and to the membrane potential at the lower end [Rie79].

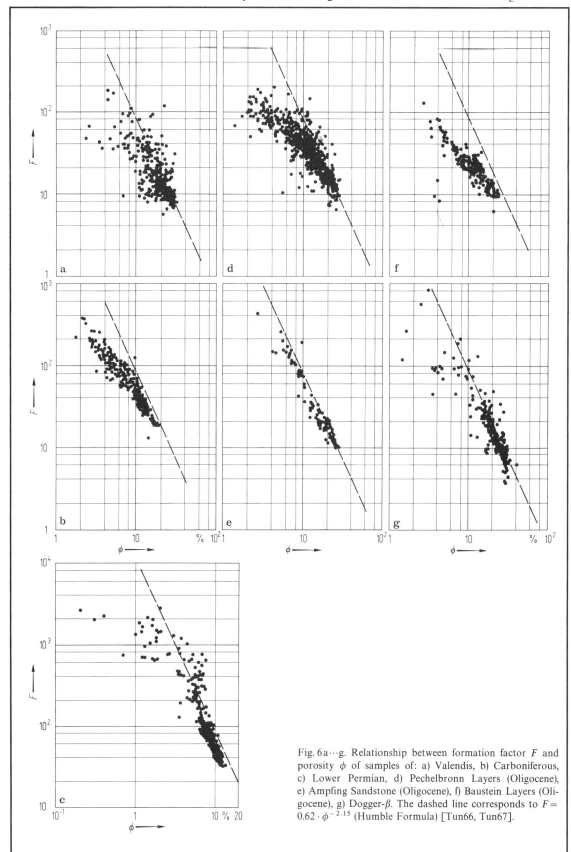

Fig. 6 a⋯g. Relationship between formation factor F and porosity ϕ of samples of: a) Valendis, b) Carboniferous, c) Lower Permian, d) Pechelbronn Layers (Oligocene), e) Ampfing Sandstone (Oligocene), f) Baustein Layers (Oligocene), g) Dogger-β. The dashed line corresponds to $F = 0.62 \cdot \phi^{-2.15}$ (Humble Formula) [Tun66, Tun67].

Table 1. Tabulation of formation factor-porosity relations from Lower Pliocene sandstone, California [Por71].

Well	Formations			
	A	B	C	D
1		$F = 3.7 \cdot \phi^{-0.86}$	$F = 2.7 \cdot \phi^{-1.14}$	$F = 3.5 \cdot \phi^{-1.06}$
2		$F = 3.6 \cdot \phi^{-0.80}$	$F = 3.25 \cdot \phi^{-0.91}$	
3	$F = 3.2 \cdot \phi^{-0.97}$	$F = 3.25 \cdot \phi^{-0.96}$		
4	$F = 3.4 \cdot \phi^{-0.81}$	$F = 3.4 \cdot \phi^{-0.79}$	$F = 3.25 \cdot \phi^{-0.95}$	$F = 1 \cdot \phi^{-1.8}$
5	$F = 1.7 \cdot \phi^{-1.35}$	$F = 3.0 \cdot \phi^{-0.95}$	$F = 3.01 \cdot \phi^{-1.06}$	
6	$F = 3.0 \cdot \phi^{-1.13}$	$F = 3.35 \cdot \phi^{-0.91}$	$F = 3.5 \cdot \phi^{-0.98}$	
7			$F = 3.0 \cdot \phi^{-0.91}$	$F = 1 \cdot \phi^{-1.85}$
8	$F = 4.0 \cdot \phi^{-0.73}$	$F = 3.55 \cdot \phi^{-0.76}$	$F = 2.1 \cdot \phi^{-1.07}$	$F = 3.0 \cdot \phi^{-1.0}$
9	$F = 3.0 \cdot \phi^{-0.73}$	$F = 2.5 \cdot \phi^{-0.72}$	$F = 2.45 \cdot \phi^{-0.89}$	$F = 3.0 \cdot \phi^{-0.81}$
10	$F = 4.0 \cdot \phi^{-0.57}$	$F = 3.4 \cdot \phi^{-0.71}$	$F = 3.35 \cdot \phi^{-0.88}$	
11		$F = 2.2 \cdot \phi^{-1.0}$	$F = 2.7 \cdot \phi^{-1.0}$	$F = 3.4 \cdot \phi^{-0.84}$

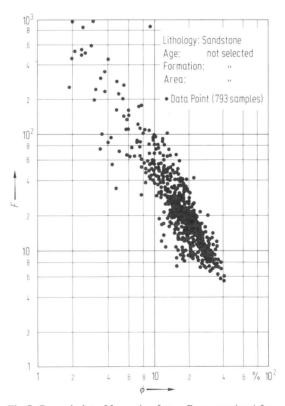

Fig. 7. General plot of formation factor F vs. porosity ϕ for 793 sandstone samples [Car68].

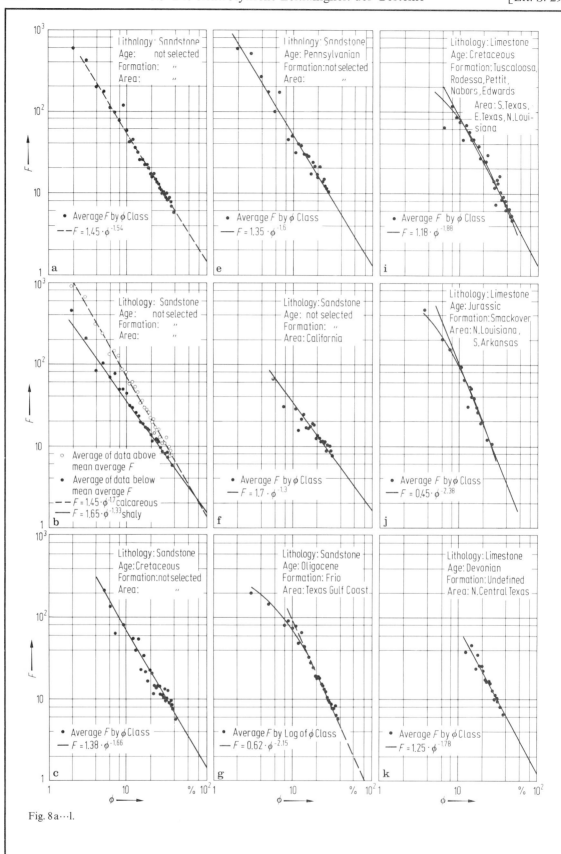

Fig. 8 a···l.

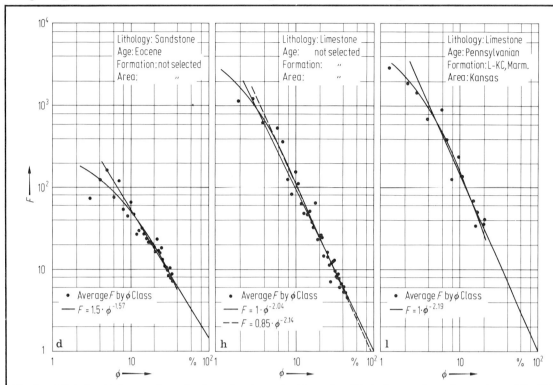

Fig. 8 a···l. Relationship between formation factor F and porosity ϕ for various lithologies, ages, formations and areas [Car68].

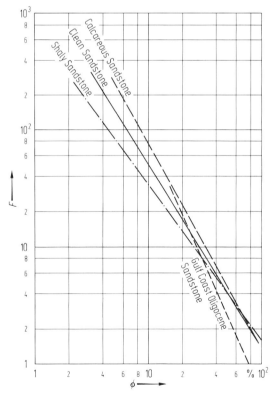

Fig. 9. Relationship between formation factor F and porosity ϕ. Regression lines for clean, calcareous, and shaly sandstones [Car68].

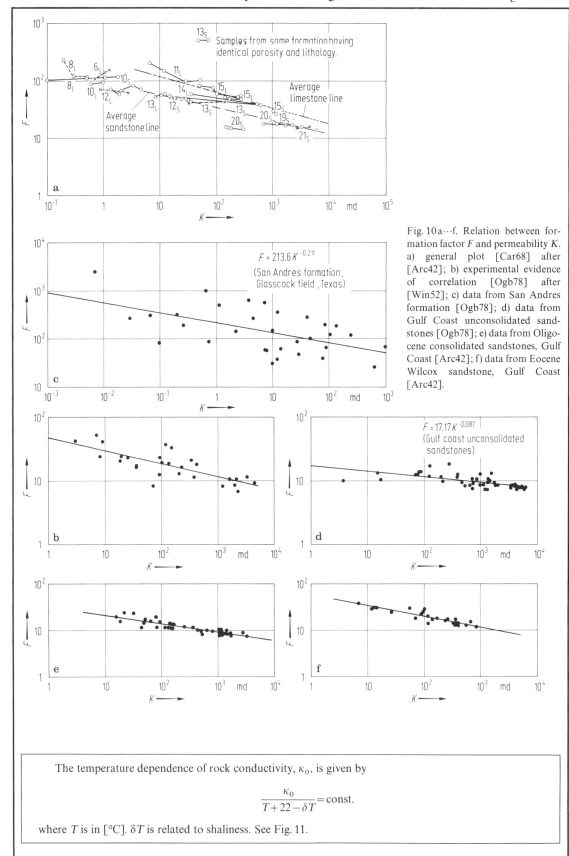

Fig. 10a⋯f. Relation between formation factor F and permeability K. a) general plot [Car68] after [Arc42]; b) experimental evidence of correlation [Ogb78] after [Win52]; c) data from San Andres formation [Ogb78]; d) data from Gulf Coast unconsolidated sandstones [Ogb78]; e) data from Oligocene consolidated sandstones, Gulf Coast [Arc42]; f) data from Eocene Wilcox sandstone, Gulf Coast [Arc42].

The temperature dependence of rock conductivity, κ_0, is given by

$$\frac{\kappa_0}{T + 22 - \delta T} = \text{const.}$$

where T is in [°C]. δT is related to shaliness. See Fig. 11.

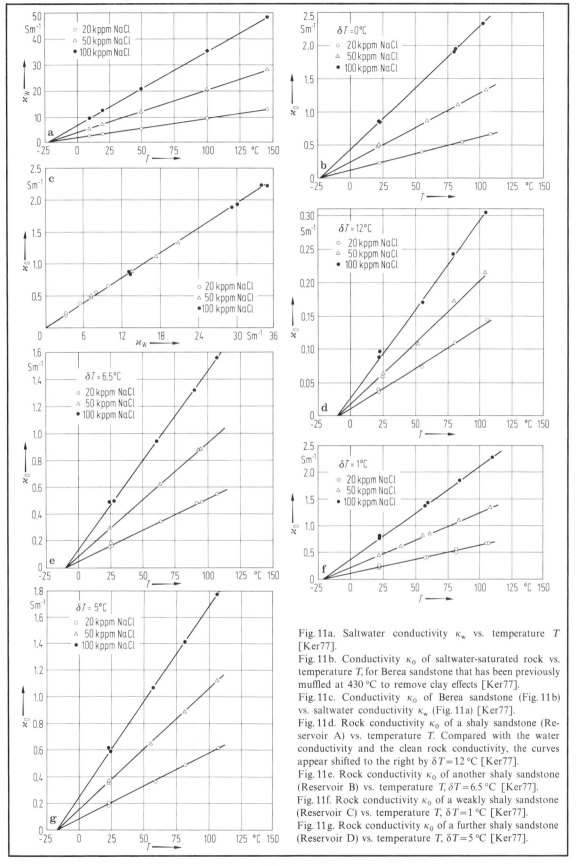

Fig. 11a. Saltwater conductivity κ_{w} vs. temperature T [Ker77].

Fig. 11b. Conductivity κ_0 of saltwater-saturated rock vs. temperature T, for Berea sandstone that has been previously muffled at 430 °C to remove clay effects [Ker77].

Fig. 11c. Conductivity κ_0 of Berea sandstone (Fig. 11b) vs. saltwater conductivity κ_{w} (Fig. 11a) [Ker77].

Fig. 11d. Rock conductivity κ_0 of a shaly sandstone (Reservoir A) vs. temperature T. Compared with the water conductivity and the clean rock conductivity, the curves appear shifted to the right by $\delta T = 12$ °C [Ker77].

Fig. 11e. Rock conductivity κ_0 of another shaly sandstone (Reservoir B) vs. temperature T, $\delta T = 6.5$ °C [Ker77].

Fig. 11f. Rock conductivity κ_0 of a weakly shaly sandstone (Reservoir C) vs. temperature T, $\delta T = 1$ °C [Ker77].

Fig. 11g. Rock conductivity κ_0 of a further shaly sandstone (Reservoir D) vs. temperature T, $\delta T = 5$ °C [Ker77].

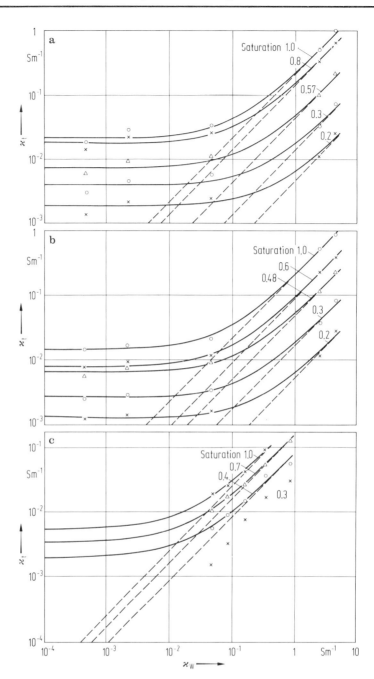

Fig. 12a···c. Rock conductivity κ_t vs. water conductivity κ_w at different water saturations, for loose sands of different grain size: a) 375 µm, b) 750 µm, c) 1500 µm. [Rin73] redrawn after [Rao63].

5.3.2.3 References for 5.3.2

Arc42	Archie, G. E.: Trans. Amer. Inst. Min., Metall., Petrol. Engrs. **146** (1942) 54.
Bra73	Brannan, G.O., Von Gonten, W.D: Ann. Logg. Symp. Trans. **14** (1973) U.
Car68	Carothers, J.E.: Log Analyst **9** (1958) 5, 13.
Hil56	Hill, H.J., Milburn, J.D.: Trans. Amer. Instn. Min., Metall., Petrol. Engrs. **207** (1956) 65.
Ker77	Kern, J.W., Hoyer, W.A., Spann, M.M.: Ann. Logg. Symp. Trans. **18** (1977) U.
Ogb78	Ogbe, D., Bassiouni, Z.: Log Analyst **19** (1978) 5, 21.
Pfa72	Pfannkuch, H.-O., in: Fundamentals of Transport Phenomena in Porous Media. – Amsterdam-London-New York: Elsevier, **1972**
Por71	Porter, C.R., Carothers, J.E.: Log Analyst **12** (1971) 1, 16.
Rao63	Rao, V.B., Sarma, V.V.J.: Geophys. **27** (1962) 470–479; **28** (1963) 310–313.
Rie79	Riepe, L., Rink, M., Schopper, J.R.: Europ. Logg. Symp. Trans. **6** (1979) AA.
Rin73	Rink, M., Schopper, J.R.: Proc. RILEM-IUPAC Internat. Symp. "Pore Structure and Properties of Materials", Praha **1973**.
Rin74	Rink, M., Schopper, J.R.: Ann. Log. Symp. Trans. **15** (1974), I.J; Europ. Logg. Symp. Trans. **3** (1974) M.
San73	Sanyal, S.K., Marsden jr. S.S., Ramey jr., H.J.: Log Analyst **14** (1973) 2, 10–24.
Scho66	Schopper, J.R.: Geophys. Prospect. **14** (1966) 3, 301.
Sun32	Sundberg, K.: Trans. Amer. Soc. Min. Engrs. **97** (1932).
Sun80	dto., Reprint. – Log Analyst 21 (1980) 3, 19.
Tun66	Tunn, W.: Erdöl-Erdgas-Z. **82** (1966).
Tun67	Tunn, W.: Log Analyst **8** (1967) 1, 35.
Wax68	Waxman, M.W., Smits, L.J.M.: Soc. Petrol. Engrs. J. **8** (1968) 107–122.
Wax74	Waxman, M.A., Thomas, E.C.: Trans. Amer. Inst. Min., Metall., Petrol., Engrs. **257** (1974) 213.
Win52	Winsauer, W.O., Shearin jr., H.M., Masson, P.H., Williams, M.: Bull. Amer. Ass. Petrol. Geols. **36** (1952) 253–277.
Wit50	de Witte, L.: Oil Gas J. **49** (1950) 120–132.
Wyl57	Wyllie, M.R.J.: The Fundamentals of Electric Log Interpretation. – New York: Academic Press, **1957**.

5.4 Electrical conductivity of minerals and rocks at high temperatures and pressures — Elektrische Leitfähigkeit von Mineralen und Gesteinen bei hohen Temperaturen und Drucken

5.4.1 Introduction — Einleitung

5.4.1.1 Conduction mechanisms — Leitungsmechanismen

A large number of measurements have confirmed the validity of the following relation for the electrical conductivity of minerals and rocks in the earth's crust and upper mantle as a function of temperature:

Eine große Anzahl von Messungen der elektrischen Leitfähigkeit als Funktion der Temperatur an Mineralen und Gesteinen, die typisch sind für Erdkruste und oberer Mantel, hat folgenden Zusammenhang ergeben

$$\sigma(T) = \sum_i \sigma_{0,i} \exp\left(-E_i/kT\right) = \sum_i \sigma_i(T) \tag{1}$$

where

T [K]
k [eV K^{-1}] = 0.86166 · 10^{-4} eV K^{-1}
E [eV]
σ_0 [S m^{-1}]
i

Es bedeuten

absolute temperature — absolute Temperatur
Boltzmann-constant — Boltzmann-Konstante
activation energy — Aktivierungs-Energie
pre-exponential factor — präexponentieller Faktor
numeration of different conduction mechanisms — Numerierung verschiedener Leitungsmechanismen

The validity of such a relation between conductivity and temperature is confirmed graphically – independent of any physical interpretation – supposed that each of the $\sigma_i(T)$ is dominating in a certain range of temperatures ΔT^i between T_{min}^i and T_{max}^i. In this range, equation

Die Gültigkeit dieser Beziehung zwischen Leitfähigkeit und Temperatur ist rein graphisch – unabhängig von der Interpretation der anderen Größen – vorausgesetzt, daß jeder der $\sigma_i(T)$ in einem bestimmten Temperaturbereich ΔT^i zwischen T_{max}^i und T_{min}^i vorherrscht. In

(1) will simplify to one term only, which will give

diesem Bereich vereinfacht sich dann Gleichung (1) zu einem einzigen Glied und ergibt

$$\ln[\sigma(T)/\sigma_{0,\,i}] = \ln[\sigma_i(T)/\sigma_{0,\,i}] = -E_i/kT \qquad (2)$$

By plotting the ln (or log, see below) of the relative conductivity vs. the inverse temperature, the measured values will follow a straight line with the slope $(-E_i/k)$. The physical interpretation of this relation is to be made in terms of semiconduction.

Thus, over the whole range of $1/T$, the measured values of $\ln\sigma(T)$ will follow a curve consisting of linear sections in the various temperature ranges ΔT^i. This enables the graphical determination of a set of parameters $\ln\sigma_{0,i}$ and E_i, which offer a possibility to discuss the nature of the various processes dominating in the respective temperature ranges.

However, closer investigations on natural minerals and rocks have demonstrated that it is very difficult to distinguish between the different conduction mechanisms unambiguously. In most cases, no particular measurements were made to identify the conduction mechanisms, e.g. by determining the polarity of the charge carrier. On the other hand, as a matter of fact, the original data of $\ln\sigma(T)$ do follow straight lines over $1/T$ in a sequence of temperature ranges ΔT^i, as described above, and values of E_i and $\ln\sigma_{0,i}$ may be formally determined for these ranges. Therefore, the graphically determined "activation"-energies may not be simply interpreted in terms of semiconduction. The increase of the electrical conductivity of a rock sample between its solidus and liquidus temperatures can more plausibly be explained by an increasing amount of the well-conducting liquid matter than by an activation energy of e.g. 5 eV.

Thus, instead of giving the conductivity values themselves as a function of temperature (which would result in an explosion of data), we have decided to represent the values of E_i and $\log\sigma_{0,i}$ for the various "linearizable" temperature intervals $\Delta T^i = T^i_{min} \cdots T^i_{max}$ in the Tables (where i is omitted).

In such an interval, the conductivity $\sigma(T)$ in $[\mathrm{S\,m^{-1}}]$ can be calculated for any temperature T between T_{min} and T_{max} with equation (2), which, for practical purposes, is written (index i omitted)

Trägt man also den ln (oder log, siehe weiter unten) der relativen Leitfähigkeit gegen den Kehrwert der absoluten Temperatur auf, so folgen die Werte einer Geraden mit der Steigung $(-E_i/k)$. Die physikalische Interpretation dieser Beziehung führt auf die Theorie der Halbleitung.

Somit werden die gemeinsamen Werte von $\ln\sigma(T)$ einer Kurve folgen, die aus linearen Abschnitten für die jeweiligen Temperaturbereiche ΔT^i zusammengesetzt ist. Auf diese Weise kann ein Satz von Parametern $\ln\sigma_{0,i}$ und E_i bestimmt werden, die eine Diskussion über die Natur der verschiedenen in den jeweiligen Temperaturbereichen vorherrschenden Prozesse ermöglichen könnten.

Die nähere Untersuchung an natürlichen Mineralen und Gesteinen hat jedoch ergeben, daß es sehr schwierig ist, die verschiedenen Leitungsmechanismen zu identifizieren. In der Mehrzahl der Arbeiten wurden hierzu keine speziellen Messungen, z.B. über die Polarität des Ladungsträgers, unternommen. Genauere Betrachtung der Ergebnisse zeigt zwar, daß viele Folgen von Meßwerten durch eine Anzahl kurzer Geraden „linearisiert" werden können. Aber die hierbei formal bestimmten „Aktivierungs"-Energien dürfen nicht einfach in die Theorie elektrischer Halbleitung einbezogen werden. Die Zunahme der elektrischen Leitfähigkeit eines Gesteins zwischen dessen Solidus- und Liquidus-Temperatur wird besser durch die Zunahme des gut leitfähigen Schmelzanteils als durch eine Aktivierungsenergie von z.B. 5 eV erklärt.

Daher sind in den folgenden Tabellen statt der Werte der Leitfähigkeit selbst als Funktion der Temperatur (was zu einer Flut von Zahlen führen würde), die Werte von E_i und $\log\sigma_{0,i}$ für die verschiedenen „linearisierbaren" Temperaturbereiche $\Delta T^i = T^i_{max} - T^i_{min}$ angegeben, (der Index i ist dabei weggelassen).

In einem solchen Bereich kann für eine Temperatur T zwischen T_{min} und T_{max} die Leitfähigkeit σ in $[\mathrm{S\,m^{-1}}]$ nach der Gleichung (2) berechnet werden, die für den praktischen Gebrauch unter Weglassung von i so geschrieben wird:

$$\log\sigma(T) = \log\sigma_0 - 5040 \cdot E/(T + 273\,^\circ\mathrm{C}) \qquad (3)$$

where E in [eV], T in [°C], and log means Briggsian (base: 10) logarithms.

The values for $\log\sigma_0$ and E are listed in the tables.

The conductivity values are also presented in figures in order to give an immediate impression of the order of magnitude and the variation of the conductivity values for the various minerals and rocks. For nomenclature used in the tables, see 5.4.1.5

mit E in [eV] and T in [°C], und log der dekadische Logarithmus ist.

Die Werte von $\log\sigma_0$ und E sind aus den Tabellen zu entnehmen.

Außer diesen Tabellen werden Figuren gegeben mit dem Zweck, einen guten Überblick über die elektrischen Leitfähigkeiten und deren Streubreite von interessierenden Mineralen und Gesteinen zu erhalten. Zur Nomenklatur in den Tabellen, siehe 5.4.1.5

5.4.1.2 Units — Einheiten

Conductivity

Laboratory data are more frequently published as conductivity $\sigma\,[\mathrm{S\,m^{-1}}=10^{-2}\,\mathrm{S\,cm^{-1}}]$, while field data (e.g. by magnetotelluric methods) are published as resistivity: $\sigma^{-1}=\varrho\,[\mathrm{S^{-1}\,m}=\Omega\,\mathrm{m}=10^{2}\,\Omega\,\mathrm{cm}]$.

Leitfähigkeit

Labordaten werden überwiegend als elektrische Leitfähigkeiten $\sigma\,[\mathrm{S\,m^{-1}}=10^{-2}\,\mathrm{S\,cm^{-1}}]$ veröffentlicht, während Feld-Meßdaten (z.B. mit der Methode der Magnetotellurik) als spezifischer Widerstand $\sigma^{-1}=\varrho\,[\mathrm{S^{-1}\,m}=\Omega\,\mathrm{m}=10^{2}\,\Omega\,\mathrm{cm}]$ angegeben werden.

Pressure

In the past few years, most authors went from the familiar unit kbar on to the SI unit $1\,\mathrm{Pa}=10^{-8}\,\mathrm{kbar}$. But in the official units (DIN-Norm) which correspond to the SI system, the use of kbars is permitted. This unit seems to be much more appropriate for geophysical measurements than the unit Pa, which is suitable for ultra-high-vacuum technique, normal atmospheric pressure being one hundred thousand Pa!

Druck

In den vergangenen 5 Jahren wurden in den meisten Publikationen statt der üblichen Einheit kbar die SI-Einheit $1\,\mathrm{Pa}=10^{-8}\,\mathrm{kbar}$ verwendet. In den geltenden gesetzlichen Einheiten, die dem SI-System entsprechen, ist die Verwendung der Einheit kbar erlaubt. Diese Einheit ist besonders auf die geophysikalischen Größenordnungen zugeschnitten: Man erhält sowohl bei Labormessungen als auch bei den in der Erde herrschenden Drucken „vernünftige" Werte zwischen 0 und 1000 kbar. Die Einheit Pa ist dagegen auf die Höchst-Vakuum-Technik zugeschnitten. Der normale Luftdruck ist einhunderttausend Pa!

Energy

The activation energy E is generally given in eV.

Energie

Die Aktivierungsenergie E wird einheitlich in eV angegeben.

5.4.1.3 Selection of the data — Auswahl des Daten-Materials

The great number of published data needs a critical selection. The only criterion for such a selection should be whether a published value represents a real property of the material (or not). The discussions during approximately the last 5 years about electrical conductivity of olivines have disclosed that many published data represent rather the properties of the measuring equipment or the experimental conditions, etc. Therefore, it would not be reasonable to consider all published data for all minerals, to obtain a representative value for the conductivity. Nearly all results presented here have been published during the past 10 years. Previous data compilations (see references) have not been incorporated. Generally, no mean values from different measurements but individual values have been presented as far as possible. The results not considered here would give a much larger compilation than the tables presented here. An individual justification of the selection would go far beyond the scope of this book.

Die große Menge des veröffentlichten Meßmaterials erfordert eine kritische Auswahl. Einziges Kriterium für die Auswahl sollte die Antwort auf die Frage sein, ob ein veröffentlichter Wert eine reelle Eigenschaft des Materials angibt. Die Diskussion in den vergangenen etwa 5 Jahren über die elektrische Leitfähigkeit von Olivinen hat gelehrt, daß viele veröffentlichte Daten die Eigenschaften der Meßanordnung und der experimentellen Bedingungen wiedergaben. Es ist deshalb unsinnig, eine vollständige Aufstellung aller Minerale in Erdkruste und Erdmantel mit ihren Werten für die elektrische Leitfähigkeit bei hohen Temperaturen und Drücken unter Zuhilfenahme aller veröffentlichten Daten zu suchen. Das hier wiedergegebene Material wurde etwa in den vergangenen 10 Jahren veröffentlicht. Ältere Tabellenwerke werden hier nicht verwendet. Es werden überwiegend individuelle Meßkurven, also keine Mittelwerte, angegeben. Die komplementäre Liste der ausgesonderten Daten ist um ein Vielfaches größer als die der hier gegebenen Daten. Eine Begründung der Auswahl würde den hier gebotenen Rahmen sprengen.

Some particular criteria will be given in the remarks for the individual groups of materials:
Olivines and pyroxenes
Basalts
Ultramafics
Other rocks and minerals

Spezielle Kriterien werden noch individuell bei den einzelnen Gruppen genannt:
Olivine und Pyroxene
Basalte
Ultrabasische Gesteine
Andere Gesteine und Minerale

5.4.1.4 Electrical conductivity under high pressure — Elektrische Leitfähigkeit bei hohen Drucken

Many laboratory measurements have been done to find a correlation between electrical conductivity and pressure. However, many results by different authors contradict each other and cannot be defined as material property. Existing information on the pressure is included in the tables. For further information, see the comments in the texts for the individual groups of substances.

Viele Laborversuche hatten die Druckabhängigkeit der elektrischen Leitfähigkeit zum Ziel. Viele Ergebnisse verschiedener Autoren widersprechen sich, so daß noch keine Ergebnisse als gesicherte Materialeigenschaft zu erkennen sind. Doch wurden Angaben über den Druck jeweils mit den folgenden Tabellen angegeben. Weitere Angaben sind den Einleitungen für die einzelnen Substanzengruppen zu entnehmen.

5.4.1.5 Arrangement of the tables and abbreviations — Anordnung der Tabellen, Abkürzungen

The tables are subdivided into columns:

Column 1: Description of sample, experimental conditions, etc.:

Generally: 1 = yes; 0 = no

Upper line: 5 groups of letters meaning:

i: K1   Control of oxygen partial pressure, fugacity f [bar]

       K0   No control of oxygen partial pressure, or no information on it

ii: NM   Natural monocrystalline mineral

       NP   Natural polycrystalline mineral

       SP   Synthetic polycrystalline material

       00   No information at all about the state of the sample

iii: GP0 Gas-pressure measuring chamber, pressure transfer by gas, no high pressure applied

       GP1 Gas-pressure measuring chamber, under high pressure

       SP0 Solid-pressure measuring chamber, no high pressure

       SP1 Solid-pressure measuring chamber, under high pressure

iv: T0   Time dependence of conductivity was not object of experiment

       T1   Time dependence was object of experiment

v: A1   Chemical or petrographical analysis is presented in the original publication

       A0   No analysis has been given

Lower line: Name and number from the original publication are repeated to facilitate the identification of a sample in the original publication.

(continued)

Die Tabellen sind in Spalten unterteilt:

Spalte 1: Beschreibung der Probe, experimentelle Bedingungen usw.

1 = ja; 0 = nein

Obere Zeile: 5 Buchstabengruppen mit folgender Bedeutung:

i: K1   Kontrolle des Sauerstoff-Partialdruckes, Fugazität f [bar]

       K0   Keine Kontrolle des Sauerstoff-Partialdruckes oder keine Angaben darüber

ii: NM   Natürliches, monokristallines Mineral

       NP   Natürliches, polykristallines, (gemahlenes) Material

       SP   Synthetisches, polykristallines Mineral

       00   keine Angaben

iii: GP0 Gasraum-Meßzelle, Gas als druckübertragendes Medium, kein hoher Druck (Gas-Pressure-nein)

       GP1 Gasraum-Meßzelle, Gas als Druckübertragungsmittel, unter Druck gemessen

       SP0 Fest-(Solid)-Körper als druckübertragendes Mittel, ohne Druck gemessen (Solid-Pressure nein)

       SP1 Festkörper-Presse, mit hohen Drucken gemessen

iv: T0   Die Zeit(Time)-Abhängigkeit war nicht Ziel der Untersuchung

       T1   Es wurde die Zeitabhängigkeit untersucht

v: A1   Es werden Analysen der Probe gegeben (naß-chemisch oder Mikrosonden)

       A0   Es werden keine Analysen gegeben

Untere Zeile: Es wird der Name und die Nummer aus der Originalarbeit wiederholt, um die Identifizierung der Probe bei einem Rückgriff auf das Original zu erleichtern.

v (con-
tinued):   If there is an 1 (yes) for one of the 5
groups, some more information follows:

K1: The buffering medium is given in
parentheses, sometimes also the values of the oxygen partial pressure; e.g.
(MgO) or (log $f = -12$)

GP1: The pressure is given in kbars

A1: Only the direction of crystal orientation is indicated, in brackets, not
the whole analysis

Column 2 and 3: Temperature interval, for which the
results given in

Column 4 and 5 are valid.

Column 6: References

Column 7: The numbers are identical with the numbers
in the figures.

Wurde in der oberen Zeile eine 1 (ja) angemerkt, so folgt hier eine knappe weitere
Angabe:

K1: In Klammern wird das puffernde Medium genannt, manchmal auch die
Werte des Sauerstoff-Partialdruckes;
z.B. (MgO) oder (log $f = -12$)

GP1: Der Druck wird in kbar angegeben

A1: Die Tabellen der Originalarbeit werden zwar nicht wiederholt, aber die
Kristall-Richtung wird gegebenenfalls
aufgeführt, in eckigen Klammern

Spalte 2 und 3: Temperatur-Intervall, für das die in den
folgenden

Spalten 4 und 5 angegebenen Werte gelten

Spalte 6: Literatur

Spalte 7: Die Nummern sind identisch mit den Kur-
vennummern der zugehörigen Abbildungen.

5.4.1.6 Further references — Weitere Literatur

Literature on results not quoted here is represented
at the end of the reference list in form of a few literature
reviews.

Außer den hier zitierten Quellen werden am Ende
des Literaturverzeichnisses auch einige Arbeiten zitiert,
die weitere Ergebnisse zusammenfassen.

5.4.2 Olivines and pyroxenes — Olivine und Pyroxene

5.4.2.1 Criteria for selection — Auswahl-Kriterien

Since the publication of the stability ranges for
olivine by [Nit74], the conductivity of olivine became
a reproducible quantity [Dub73]. No. 1···6 represent
some of these conductivities which indicate the lower
limit of the conductivity for single-crystal olivine. Higher values of conductivity for olivine cannot be ruled
out, but published results may have been influenced
by particular experimental conditions, e.g. unsatis-
factory electrical isolation of the sample in the measuring chamber, high fugacity, e.g. of pyrophyllite, can
increase the conductivity of the sample. No. 7···19
represent higher conductivities, though they have been
measured under controlled thermodynamic conditions
and were reproducible. In the upper group σ has been
measured in gas-pressure chambers, in the lower group
in solid-pressure chambers.

Seit der Angabe der Stabilitätsbereiche für Olivine
durch [Nit74] wurden reproduzierbare Leitfähigkeiten
an Olivinen gemessen [Dub73]. Nr. 1···6 geben diese
Leitfähigkeiten wieder, die die untere Grenze für die
Leitfähigkeit von Olivin-Einkristallen angeben. Höhere
Werte der Leitfähigkeit sind oft kritisch zu beurteilen,
da ungenügende elektrische Isolation der Probe gegenüber der Umgebung, eine hohe f_{H_2O}, z.B. von Pyro-
phyllit, die Leitfähigkeit erhöhen können. Nr. 7···19
geben höhere Werte für die Leitfähigkeit wieder, die
allerdings unter kontrollierten thermodynamischen
Bedingungen und reproduzierbar gemessen wurden. In
der oberen Gruppe wurde σ mit Gasraum-Meßzellen
gemessen, in der unteren Gruppe mit Festkörper-
pressen.

5.4.2.2 Electrical conductivity under high pressure — Elektrische Leitfähigkeit
bei hohen Drucken

Only a slight, if at all, dependence of the conductivity on high pressure has been obtained. According
to [Sch70], summarizing his own results and results
from other authors, the activation energy changed
from $-2.5 \cdot 10^{-6}$ eV bar^{-1} up to $9 \cdot 10^{-6}$ eV bar^{-1}.
[Voi78] obtained an obvious change with pressure
(No. 8···19, pyroxene), while [Sch77] (No. 1 and 2,
pyroxene) obtained a very small, if any at all, change of
conductivity.

Es hat sich eine sehr geringe Änderung der Leitfähigkeit durch hohe Drucke ergeben. Nach [Sch70]
ergibt sich nach eigenen Messungen und den Messungen
anderer Autoren eine Änderung der Aktivierungsenergie von etwa $-2.5 \cdot 10^{-6}$ eV bar^{-1} bis knapp
$9 \cdot 10^{-6}$ eV bar^{-1}. Für Pyroxene erhält [Voi78] deutliche Änderungen der Leitfähigkeit mit dem Druck
(siehe Pyroxene, Nr. 8···19), während [Sch77] (siehe
Pyroxene, Nr. 1 und 2) kaum eine Änderung der Leitfähigkeit beobachten.

Table 1. Olivine and Pyroxene.

Description		T_{min} °C	T_{max} °C	E eV	$\log \sigma_0$ σ_0 in S m^{-1}	Ref.	No.[*]
Olivine							
K0; NM; GP0; T0; A0		700	1100	1.66	2.33	Hug53	1
Red Sea; Fo 90; (in air); [010]		1100	1400	3.01	7.16		
K0; NM; GP1; T0; A1		400	1100	1.21	1.38	Dub74	2
Red Sea; Fo 91; (in argon)							
K1; NM; GP0; T0; A1		800	950	1.07	−0.33	Dub73	3
San Carlos, (CO$_2$/H$_2$), Fo 92		950	1350	1.82	2.81		
K0; NM; GP1; T0; A1		560	1120	0.98	0.13	Dub74	4
Red Sea; Fo 91; 2.5, 5, 8 kbar,		1270	1440	2.33	4.73		
(argon)							
K1; NM; GP0; T0; A1		890	1390	1.51	1.82	Dub74	5
Red Sea; Fo 91; (H$_2$/CO$_2$)		1400	1500	2.74	5.42		
		1510	1540	7.89	20.01		
		1550	1660	3.23	7.07		
		870	1320	1.34	1.24		
		1380	1450	2.36	4.41		
		1450	1490	5.46	13.48		
		1490	1640	3.26	7.21		
		1490	1660	3.13	6.82		
K0; NP; SP1; T0; A1	20 kbar	680	1156	1.14	1.327	Sch77	6
Mt. Leura	50 kbar	680	1156	1.03	0.94		
K1; SP; SP1; T0; A1						Cem80	7
Fo 100 (MgO);	10 kbar	522	970	0.984	0.692		
		970	1075	2.461	6.67		
Fo 90 (Fe/FeO);	10 kbar	593	939	0.777	0.4378	Cem80	8
Fo 90 (Fa/Q/M)[1];	10 kbar	550	950	0.622	0.2667	Cem80	9
Fo 80 (Fe/FeO);	10 kbar	527	939	0.683	1.14	Cem80	10
Fo 80 (Fa/Q/M)[1];	10 kbar	602	887	0.582	1.212	Cem80	11
Fo 60 (Fa/Q/M)[1];	10 kbar	574	1058	0.479	1.53	Cem80	12
Fo 0 (Q/Fe)[1];	10 kbar	301	818	0.523	2.08	Cem80	13
Fo 0 (Q/M)[1];	10 kbar	363	944	0.383	2.12	Cem80	14
K1; NP; SP1; T0; A1		300	600	0.39	−0.40	Sch71	15
Brazilian olivine[2]); (Ni/NiO)	20 kbar	600	1300	1.09	3.85		
Brazilian olivine[2]), (Ni/NiO)	30 kbar	300	600	0.37	−0.45	Sch71	16
		600	1300	1.02	3.56		
K1; NP; SP1; T0; A1		350	650	0.52	0.90	Sch71	17
Dreiser Weiher[3]); (Ni/NiO);	20 kbar	700	1200	0.70	1.74		
	35 kbar	350	650	0.42	0	Sch71	18
		700	1200	0.96	3.08		
	50 kbar	350	650	0.37	−0.33	Sch71	19
		700	1200	0.83	2.4		

(continued)

[1]) Fa = fayalite, Q = quartz, M = magnetite, Fs = ferrosilite.
[2]) Improved chemical analysis [Sch74]: Fe$_2$O$_3$ content = 0.5 wt % of Brazilian olivine sample.
[3]) Improved chemical analysis [Sch74]: Fe$_2$O$_3$ content = 0.4 % of Dreiser Weiher sample.
[*]) Numbers refer to the curves in Figs. 1 and 2.

Table 1 (continued)

Description		T_{min} °C	T_{max} °C	E eV	$\log \sigma_0$ σ_0 in S m^{-1}	Ref.	No.*)
Pyroxene							
K0; NP; SP1; T0; A0	20 kbar	600	1156	1.01	0.865	Sch77	1
Bamle, En88;	50 kbar	600	1156	1.04	1.065	Sch77	2
K0; NP; SP1; T0; A0	10 kbar	800	1046	1.3	1.44	Hea75	3
K1; NM; GP1; T0; A0 [010]	10 kbar	763	1394	1.14	0.53	Hea75	4
K1; NM; GP1; T0; A0 [010]	5 kbar	727	1340	1.064	0.426	Hea75	5
K1; N0; 00; T1; A1 (in vacuo)		25	900	0.62	0.445	Ohl74	6
K0; NP; SP1; T0; A0						Dvo73a	7
Bamle [001]	24 kbar	80	200	0.60	2.91		a
[001]	56 kbar	80	200	0.52	3.25		b
[100]	24 kbar	80	200	0.47	−0.33		c
[100]	56 kbar	80	200	0.56	0.80		d
K1; SP; SP1; T0; A1						Voi78	8
$MgSiO_3$ (Mg_2SiO_4);	10 kbar	550	1000	1.24	2.29		
$MgSiO_3$ (Mg_2SiO_4);	20 kbar	520	980	1.49	3.50	Voi78	9
$MgSiO_3$ (SiO_2);	10 kbar	550	1000	1.13	0.49	Voi78	10
$MgSiO_3$ (SiO_2);	20 kbar	550	1000	1.59	3.50	Voi78	11
En 90 Fs 10 (SiO_2, CCO)	10 kbar	550	1000	1.08		Voi78	12
(SiO_2, CCO)	20 kbar	550	1000	1.40	3.25	Voi78	13
En 80 Fs 20 (SiO_2, CCO)	10 kbar	550	1000	0.83		Voi78	14
(SiO_2, CCO)	20 kbar	550	1000	1.18	3.05	Voi78	15
En 50 Fs 50 (SiO_2, CCO)	10 kbar	550	1000	0.74		Voi78	16
(SiO_2, CCO)	20 kbar	550	1000	0.80	2.40	Voi78	17
$FeSiO_3(Fs, Q, M)^1$);	20 kbar	550	1000	0.37	2.32	Voi78	18
$FeSiO_3(Fs, Q, Fe)^1$);	20 kbar	550	1000	0.42	2.31	Voi78	19
K1; NM; GP0; T1; A1 No. 1269, [100]; $(Co_2/H_2: \log f = -7)^4$)		917	1009	1.26	0.67	Dub73a	20
No. 1269, [100]; $(\log f = -8)^4$)		863	1026	1.30	0.84	Dub73a	21
No. 1269, [100]; $(\log f = -10)^4$)		758	826	1.15	0.17	Dub73a	22
		890	1009	1.36	1.11		
No. 1269, [100]; $(\log f = -12)^4$)		747	890	1.19	0.36	Dub73a	23
		917	1009	1.54	1.92		
No. 1269, [100]; $(\log f = -14)^4$)		737	826	1.25	0.64	Dub73a	24
		838	962	1.60	2.2		
No. 1269, [100]; $(\log f = -14.5)^4$)		962	1043	1.64	2.43	Dub73a	25
K1; NM; GP0; T1; A1		747	863	1.13	0.01	Dub73a	26
No. 1269, 1; [010]; $(\log f = -8)^4$)		917	1009	1.42	1.23		
No. 1269, 2; [010]; $(\log f = -8)^4$)		814	890	1.15	0.31	Dub73a	27
		932	1009	1.38	1.28		
K1; NM; GP0; T1; A1 No. 1269; [001]; $(\log f = -8)^4$)		814	1009	1.18	0.57	Dub73a	28
No. 1269; [001]; $(\log f = -12)^4$)		769	851	1.28	3.2	Dub73a	29
		876	1009	1.76	1.11		
No. 1269; [001]; $(\log f = -14)^4$)		863	1009	1.87	3.72	Dub73a	30
No. 1269; [001]; (air)		747	863	1.15	0.73	Dub73a	31
		863	1009	1.76	3.39		

(continued)

---

^{4}) Fugacity f at 1200 °C, f in bar.

*) Numbers refer to the curves in Fig. 3.

Haak

Table 1 (continued)

Description	T_{min} °C	T_{max} °C	E eV	$\log \sigma_0$ σ_0 in S m^{-1}	Ref.	No.*)
K1; NM; GP0; T1; A1	802	1009	1.60	1.14	Dub73a	32
No. 173 gem; $(\log f = -8)^4$)	1097	1366	1.21	-0.32		
K1; NM; GP0; T1; A1	826	932	1.72	1.81	Dub73a	33
No. 173 gem; $(\log f = -10)^4$)	932	1009	1.86	2.38		
	1097	1340	1.29	0.10		
No. 173 gem; $(\log f = -12)^4$)	791	917	1.79	2.31	Dub73a	34
	917	1009	1.82	2.43		
	1097	1366	1.66	1.36		
K1; NM; GP0; T1; A1	904	1009	1.80	2.31	Dub73a	35
No. 26; [010]; $(\log f = -8)^4$)						
No. 26; [010]; $(\log f = -10)^4$)	904	1009	1.71	2.07	Dub73a	36
No. 26; [010]; $(\log f = -12)^4$)	876	1009	1.75	2.36	Dub73a	37
No. 26; [010]; $(\log f = -14)^4$)	863	1009	1.66	2.13	Dub73a	38
No. 26; [010]; $(\log f : CO_2)^4$)	917	1009	1.50	1.21	Dub73a	39
No. 26; [010]; (air)	838	1009	1.62	2.12	Dub73a	40

^{4}) Fugacity f at 1200 °C, f in bar.   *) Numbers refer to the curves in Fig. 3.

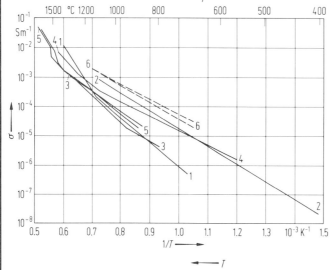

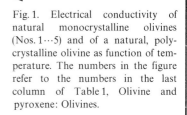

Fig. 1. Electrical conductivity of natural monocrystalline olivines (Nos. 1···5) and of a natural, polycrystalline olivine as function of temperature. The numbers in the figure refer to the numbers in the last column of Table 1, Olivine and pyroxene: Olivines.

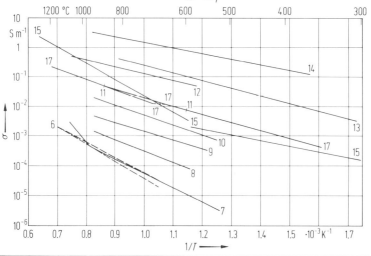

Fig. 2. Electrical conductivity of natural (Nos. 6, 15, 17) and synthetic, polycrystalline olivines as function of temperature. The numbers in the figure refer to the numbers in the last column of Table 1, Olivine and pyroxene: Olivines.

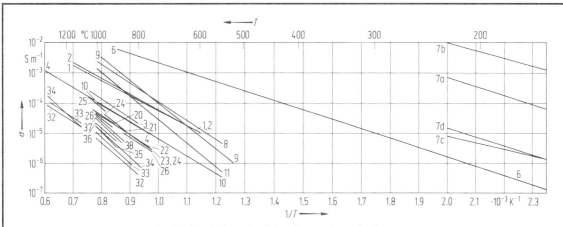

Fig. 3. Electrical conductivity of natural, synthetic, mono- and polycrystalline pyroxenes as function of temperature. The numbers in the figure refer to the numbers of the last column of Table 1, Olivine and pyroxene: Pyroxenes.

5.4.3 Basalts — Basalte

5.4.3.1 Criteria for selection — Auswahl-Kriterien

The correct experimental conditions (see below) are not yet so well known for basalts as they are for olivines. Therefore, no objective criteria for selection could be used. All data which have been cited more frequently or which were obtained from more or less well-described samples have been collected. However, there are more data on conductivity than presented here, which generally are within the range of the data presented here.

Bei den Basalten sind die korrekten Versuchsbedingungen noch nicht so gut bekannt wie bei den Olivinen. Deshalb können noch keine objektiven Kriterien für die Auswahl angegeben werden. Es wurden praktisch alle Ergebnisse zusammengefaßt, die in der Literatur häufiger zitiert wurden. Weitere, hier nicht aufgenommene Ergebnisse liegen generell innerhalb des Streubereichs.

5.4.3.2 Experimental conditions — Versuchsbedingungen

a) *Time:* [Pre72] and [Dub75] did prove a strong variation of the conductivity with time near the solidus temperature. A characteristic length of time is about one week.

b) *Oxygen partial pressure:* [Dub75] did prove a dependence of the conductivity on oxygen partial pressure at sub-solidus temperatures. [Waf75] could show that the conductivity was not (or nearly not) dependent on the oxygen partial pressure at liquidus temperatures.

c) *Water content:* It is supposed that the electrical conductivity will increase notably even with low water content. The presence of water decreases the melting point of basalts, but there are no quantitative data yet.

d) *Effect of high pressure:* Many high-pressure results exist, but they do not agree: It is even unknown whether the conductivity increases or decreases (see [Sha77]). Therefore it seems too early to give some published results in this table.

a) *Zeit:* [Pre72] and [Dub75] haben die Zeitabhängigkeit der Leitfähigkeit in der Nähe der Solidus-Temperatur nachgewiesen. Eine Woche etwa stellt einen charakteristischen Zeitraum dar.

b) *Sauerstoff-Partialdruck:* [Dub75] konnte bei Subsolidus-Temperaturen eine Abhängigkeit der Leitfähigkeit vom Sauerstoff-Partialdruck nachweisen. Dagegen ist die Leitfähigkeit im Liquidus-Bereich fast unabhängig vom Sauerstoff-Partialdruck [Waf75].

c) *Wassergehalt:* Man vermutet, daß die elektrische Leitfähigkeit bei nur wenigen Prozenten H_2O im Hoch-Temperatur-Bereich sehr stark steigt. Quantitative Angaben können aber noch nicht gegeben werden.

d) *Einfluß hoher Drucke:* Es liegen zahlreiche Meßreihen bei hohen Drucken vor. Trotzdem kann man noch nicht einmal erkennen, ob sich die Leitfähigkeit erniedrigt oder erhöht, siehe [Sha77]. Es scheint noch zu früh zu sein, diese Ergebnisse bereits als reelle Eigenschaften von Basalten hier aufzunehmen.

Table 2. Basalts.

Description		T_{min} °C	T_{max} °C	E eV	$\log \sigma_0$ σ_0 in S m^{-1}	Réf.	No.[a]
K0; NP; SP1; T0; A1		753	800	0.44	0.92	Khi70	1
olivine tholeiite;	28 kbar	800	900	0.37	0.6		
see also Nos. 9 and 12		900	1000	0.47	0.99		
		1000	1100	0.51	1.18		
		1100	1200	0.83	2.33		
		1200	1300	3.53	11.6		
		1300	1350	6.45	20.9		
		1350	1400	3.3	11.2		
K0; SP; GP0; T1; A1						Pre72	2
slow measurement		660	1050	1.06	2.04		
		1135	1254	7.4	25		
rapid measurement		660	1050	0.72	2.06		
		1166	1254	4.22	14.5		
K1; NP; GP0; T0; A1		1340	1500	0.815	3.8	Rai78	3
nephelinite (H_2/CO_2)							
K1; NP; GP0; T0; A1		1340	1500	0.957	4.22	Rai78	4
basanite (H_2/CO_2)							
K0; NP; SP0; T0; A0		200	400	0.38	−1.0	Bon72	5
tholeiitic rift basalts		400	800	0.59	0.91		
		800	1000	1.03	2.98		
		1000	1200	2.59	9.17		
K0; NP; SP0; T0; A0		200	400	0.56	0.37	Bon72	6
Indian-ocean island basalts		400	800	0.72	1.51		
		800	1000	1.03	2.98		
		1000	1200	2.59	9.17		
K0; NP; SP0; T0; A0		200	400	0.48	−0.92	Bon72	7
porous alkali basalts		400	800	0.95	2.59		
		800	1000	0.94	2.57		
		1000	1200	1.04	2.95		
K0; NP; SP0; T0; A0		200	400	0.51	−1.00	Bon72	8
alkali basalt No. 5/22		400	800	0.80	1.22		
		800	1000	1.16	2.90		
K0; NP; SP1; T0; A1		800	955	0.78	0.94	Khi70	9
Al-tholeiite;	28 kbar	955	1230	1.19	2.64		
		1230	1395	7.02	22.2		
		1395	1435	3.31	10.95		
		1435	1515	1.19	4.71		
K1; NP; GP; T0; A1		1340	1500	1.17	4.47	Rai78	10
tholeiite (H_2/CO_2)							
K1; NP; GP0; T0; A1		1410	1580	1.024	3.87	Waf75	11
alkali-olivine basalt BCR-2		1326	1410	1.301	4.7		
(H_2/CO_2)		1260	1326	1.41	5.03		
K0; NP; SP1; T0; A1		750	970	0.41	0.01	Khi70	12
quartz-tholeiite;	28 kbar	970	1220	1.32	3.72		
		1220	1355	2.35	7.17		
		1355	1422	8.33	25.73		
		1422	1466	0.33	1.93		
K1; NP; GP0; T0; A1		1235	1580	0.69	2.82	Waf75	13
latite, V31 (H_2/CO_2)							
K1; NP; GP0; T0; A1		1340	1500	1.063	4.25	Rai78	14
alkali-olivine basalt (H_2/CO_2)							

(continued)

*) Numbers refer to the curves in Fig. 4.

Haak

Table 2 (continued)

Description	T_{min} °C	T_{max} °C	E eV	$\log \sigma_0$ σ_0 in S m^{-1}	Ref.	No.*)
K1; NP; GP0; T0; A1	1225	1283	1.59	5.74	Waf75	15
tholeiite basalt PG16 (H$_2$/CO$_2$)	1283	1340	1.76	6.31		
	1340	1415	1.44	5.31		
	1415	1526	1.08	4.23		
K1; NP; GP0; T0; A1	1266	1318	1.87	6.72	Waf75	16
tholeiite basalt 70-15 (H$_2$/CO$_2$)	1318	1380	1.67	6.095		
	1380	1473	1.22	4.72		
K1; NP; GP0; T0; A1	1340	1500	0.96	4.02	Rai78	17
olivine-tholeiite, (H$_2$/CO$_2$)						
K1; NP; GP0; T0; A1	1340	1500	1.03	4.19	Rai78	18
hawaiite, (H$_2$/CO$_2$)						
K1; NP; GP0; T0; A1	1340	1500	1.13	4.47	Rai78	19
olivine-tholeiite, (H$_2$/CO$_2$)						
K1; NP; GP0; T0; A1	1340	1500	0.425	2.26	Rai78	20
trachyte (H$_2$/CO$_2$)						
K1; NP; GP0; T0; A1	1340	1500	0.82	3.32	Rai78	21
mugearite, (H$_2$/CO$_2$)						
K1; NP; GP0; T0; A1	1220	1580	0.84	3.22	Waf75	22
andesite HA, (H$_2$/CO$_2$)						
K1; N0; 00; T1; A0	838	1060	1.03	2.45	Dub76	23
(QFM)^5)						
K1; N0; 00; T1; A0	727	927	0.69	1.59	Dub76	24
(CO$_2$)	927	1060	0.99	2.83		
K0; NP; SP0; T0; A0	400	800	1.03	2.03	Bon72	25
olivine basalt 1483 (Seychelles)	800	1000	0.57	−0.14		
	1000	1200	3.3	10.68		

^{5}) Q = quartz; F = fayalite; M = magnetite.   *) Numbers refer to the curves in Fig. 4.

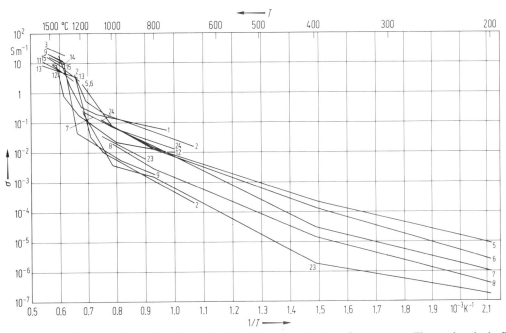

Fig. 4. Electrical conductivity of natural and synthetic (No. 2) basalts as function of temperature. The numbers in the figure refer to the numbers in the last column of Table 2, Basalt.

Haak

5.4.4 Ultramafic rocks — Ultrabasische Gesteine
5.4.4.1 Criteria for selection — Auswahl-Kriterien

Olivine and basalt can be defined as the final products of a differentiation process of an original material which has formed the earth's upper mantle. Ultramafic rocks, however, are less differentiated than basalt or olivine, and should be more similar to the original material. All results have been collected that have been obtained for material which is not basalt or olivine. Eclogite has been included, since it seems to have a conductivity-temperature curve comparable to the one for ultramafics.

Während Basalte und Olivine als Endprodukte eines ursprünglichen Materials, aus dem sich die tieferen Schichten (oberer Mantel) aufbauen, aufgefaßt werden können, sind die ultrabasischen Gesteine nur teilweise differenzierte Produkte. Sie sind dem ursprünglichen Material ähnlicher als Basalte und Olivine. Es wurden praktisch alle Materialien hier aufgenommen, die eindeutig weder Basalt noch Olivin-Einkristalle sind. Eklogit wurde aufgenommen, weil es eine den ultrabasischen Gesteinen vergleichbare Leitfähigkeit-Temperatur-Kurve hat.

5.4.4.2 Experimental conditions — Versuchsbedingungen

They are still rather unknown. Probably they are not too different from the experimental conditions for olivines and basalts. [Rai78a] (see Table 3) have measured the conductivity under controlled oxygen partial pressure. It is obvious that their results do not differ from results which have been obtained without control. The effect of time is unknown. A serious problem in measuring conductivities is the serpentinization of the samples. The change of conductivity with increasing pressure has been investigated but – like for basalts – there are no conclusive results.

Hierüber ist kaum etwas bekannt. [Rai78a] (siehe Tabelle 3) haben bei kontrolliertem Sauerstoff-Partialdruck gemessen. Ihre Ergebnisse unterscheiden sich offenbar nicht von den Ergebnissen mit unkontrolliertem Sauerstoff-Partialdruck. Der Einfluß der Zeit ist noch nicht untersucht worden. Ein ernstes Problem ist die Serpentinisierung der Proben. Die Änderung der Leitfähigkeit mit steigenden Drucken ist zwar untersucht worden, aber noch genauso unsicher wie bei den Basalten.

Table 3. Ultramafics.

Description		T_{min} °C	T_{max} °C	E eV	$\log \sigma_0$ σ_0 in S m^{-1}	Ref.	No. *)
K0; NP; SP0; T0; A0		250	310	0.34	−2.78	Dvo73	4
dunite		310	515	1.23	4.86		
		585	700	2.64	12.06		
K0; NP; SP1; T0; A0		250	350	0.59	−0.29	Dvo73	34
dunite No. 4,	20 kbar	350	505	1.20	4.64		
		605	700	2.54	11.62		
K0; NP; SP0; T0; A0		700	1200	2.0	4.7	Bon73	17
dunite							
K0; NP; SP0; T0; A0		850	1200	2.78	8.85	Bon73	19
pyrope-spinel-dunite							
K1; NP; GP0; T0; A1		930	1100	1.79	4.69	Rai78a	28
eclogite 2E (H$_2$/CO$_2$)		1100	1300	4.6	15.0		
		1300	1395	4.4	14.5		
		1395	1525	1.49	5.6		
K0; NP; SP0; T0; A1		550	900	0.65	−0.35	Las76	18
eclogite							
K0; NP; SP0; T0; A1		310	900	0.8	0.9	Las76	9
eclogite							
K0; NP; SP0; T0; A1		200	900	0.74	1.05	Las76	5
eclogite, 60% amphibole-plagioclase							
20% diopside-plagioclase							
K0; NP; SP0; T0; A1		380	900	0.71	0.69	Las76	6
eclogite, 10% amphibole-plagioclase							(continued)

*) Numbers refer to the curves in Fig. 5.

Table 3 (continued)

Description		T_{min} °C	T_{max} °C	E eV	$\log \sigma_0$ σ_0 in S m^{-1}	Ref.	No.*)
K0; NP; SP0; T0; A1		360	600	0.62	−0.2	Las76	7
eclogite, 5 % diopside-plagioclase		650	900	0.72	1.0		
K0; NP; SP0; T0; A1						Las76	1
eclogite, 20 % diopside-plagioclase		350	700	0.48	0.9		
K0; NP; SP0; T0; A1		360	900	0.65	0.45	Las76	2
eclogite, 20 % diopside-plagioclase							
K0; NP; SP0; T0; A1		350	900	0.89	0.45	Las76	16
eclogite, 5 % diopside-plagioclase							
K1; NP; GP0; T0; A1		747	876	0.48	−0.595	Rai78a	26
garnet-lherzolite 2C (H$_2$/CO$_2$)		876	1076	1.04	1.82		
		1078	1206	6.08	20.6		
		1206	1442	3.39	11.5		
		1442	1579	0.46	2.86		
K1; NP; GP0; T0; A1		917	1060	0.95	1.01	Rai78a	24
spinel-lherzolite 2D (H$_2$/CO$_2$)		1060	1260	2.7	7.64		
		1260	1460	5.87	18.1		
		1460	1580	2.15	7.24		
K0; NP; SP0; T0; A0		250	414	0.64	−1.37	Dvo73	12
olivinite		414	538	1.38	3.96		
K0; NP; SP1; T0; A0		250	406	0.69	−0.57	Dvo73	36
olivinite No. 12;	20 kbar	406	520	1.16	2.94		
		520	700	1.75	6.66		
K0; NP; SP0; T0; A0		300	538	0.83	1.04	Dvo73	10
olivinite		538	1033	2.33	10.33		
K0; NP; SP1; T0; A0		300	538	0.87	1.35	Dvo73	8
olivinite No. 10;	20 kbar	538	600	2.35	10.47		
K0; NP; SP0; T0; A1		550	1200	2.3	6.8	Bon73	13
olivinite + peridotite							
K0; NP; SP0; T0; A0		250	347	0.44	−1.68	Dvo73	3
peridotite		347	508	1.3	5.26		
K0; NP; SP1; T0; A0		250	371	0.61	−0.33	Dvo73	33
peridotite No. 3;	20 kbar	371	503	1.39	5.77		
K0; NP; SP0; T0; A0		250	319	0.06	−6.75	Dvo73	11
peridotite		319	580	1.13	2.33		
K0; NP; SP1; T0; A0		250	318	0.24	−5.01	Dvo73	35
peridotite;	20 kbar	318	573	1.13	2.52		
K0; NP; SP0; T0; A0		810	1050	4.0	12	Par66	14
peridotite							
K1; NP; GP0; T0; A1		957	1115	1.77	4.26	Rai78a	27
garnet peridotite 2B (H$_2$/CO$_2$)		1115	1442	4.59	14.5		
		1442	1580	1.85	6.43		
K1; NP; GP0; T0; A1		932	1115	1.98	5.5	Rai78a	25
garnet-peridotite 2A (H$_2$/CO$_2$)		1115	1395	4.22	13.6		
		1395	1545	2.58	8.65		
K0; NP; SP0; T0; A0		550	1050	2.0	4.95	Bon73	15
plagioclase-peridotite							
K0; NP; SP0; T0; A0		850	1200	3.6	11.9	Bon73	20
pyrope peridotite with ecstatite							
K1; NP; GP0; T0; A1		470	1030	0.5	−0.596	Rai78a	22
garnet websterite 2F (H$_2$/CO$_2$)		1030	1080	3.89	12.53		
		1080	1225	6.66	22.8		
		1225	1347	2.66	9.4		
		1347	1535	0.87	3.82		

*) Numbers refer to the curves in Fig. 5.

Haak

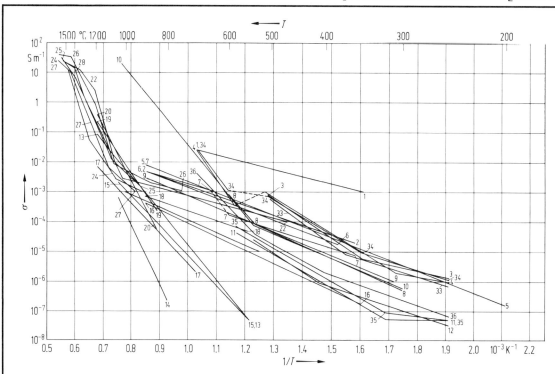

Fig. 5. Electrical conductivity of ultramafic rocks and eclogites as function of temperature. The numbers in the figure refer to the numbers of the last column of Table 3, Ultramafics.

5.4.5 Other minerals — Andere Minerale

5.4.5.1 Criteria for selection — Auswahl-Kriterien

A few minerals and rocks remained which did not fit into the above scheme. They are included for their general importance and the supreme quality of the measuring method. Very important is the time dependence of the conductivity of albite [Piw74], since it implies a time dependence for all feldspar-containing rocks. [End78] measured extremely carefully the conductivity of alumina.

Bei der Einteilung der Meßergebnisse in die oben genannten 4 Gruppen blieben noch die Meßergebnisse einiger Minerale übrig, die nicht in dieses Schema passen. Die hohe Qualität dieser Meßwerte ließ es nicht zu, die Werte einfach zu vernachlässigen. Wichtig ist die von [Piw74] gemessene Zeitabhängigkeit der Leitfähigkeit von Albit. [End78] veröffentlichte sehr sorgfältige Untersuchungen der elektrischen Leitfähigkeit an Al_2O_3: Hierbei verwendete er sowohl hochreine Materialien, wie auch kommerzielle Materialien, die in den Meßzellen zur Bestimmung der Leitfähigkeit als Isolatoren eingesetzt werden.

Table 4. Other minerals.

Description		T_{min} °C	T_{max} °C	E eV	$\log \sigma_0$ σ_0 in S m^{-1}	Ref.	No.*)
K0; NP; SP1; T0; A1		600	1075	0.99	1.7	Khi74	1
albite (An3 Ab97);	2.8 kbar	1075	1136	3.77	12.1		
		1136	1200	13.8	13.8		
		1200	1245	5.34	19		
		560	1245	0.39	2.6		

		T °C	$\log \sigma(t_1)$ S m^{-1}	$t_2 - t_1$ h	$\log \sigma(t_2)$ S m^{-1}		
K1; NM; GP0; T1; A1		1080	−3.7	526	−2.7	Piw74	2
amelia albite, (CO$_2$/CO: f_{O_2}=1.5 bar);		1100	−2.45	1199	−0.25		
T_m=1118 °C		1111	−0.2	1486	+0.6		

Description		T_{min} °C	T_{max} °C	E eV	$\log \sigma_0$ σ_0 in S m^{-1}	Ref.	No.*)
K0; SP; GP0; T0; A1		464	838	1.91	1.98	End78	5
alumina, commercial,	(argon)	838	1420	2.72	5.68		
(Company A)	(air)	375	977	2.08	3.22	End78	6
		977	1417	3.28	7.76		
K0; SP; GP0; T0; A1		352	570	1.18	−1.4	End78	7
alumina, commercial,	(argon)	570	1394	2.11	4.19		
(Company B)	(air)	368	570	1.67	1.3		
		570	1394	2.13	5.07		
K0; SP; GP0; T0; A1		462	1043	1.83	0.66	End78	8
alumina, highly pure:		1043	1512	2.55	3.3		
impurity <10 ppm,	(argon)						
K0; SM; GP0; T0; A1	(air)	677	1273	3.02	5.375	End78	9
synthetic sapphire	(argon)	773	1273	3.02	5.375		
(company C)							
K0; SM; GP0; T0; A1		380	640	1.2	−4.3	End78	10
synthetic sapphire	(argon)	640	1500	3.16	6.6		
(company D)	(air)	400	721	1.05	−4.1	End78	11
		721	1500	3.16	6.6		
K1; SP; GP0; T0; A1		200	1100	1.00	−1.17	Sha69	3
forsterite (Fe/FeO)		1150	1400	3.83	8.53		
K1; SM; GP1; T0; A1		727	1060	1.15	−0.93	Dub72	12
forsterite; (argon)	2.5 kbar	1126	1198	7.99	23.97		
	5.0 kbar	717	1097	1.422	0.19	Dub72	13
	7.5 kbar	779	1116	1.701	1.29	Dub72	14
00; 00; 00; 00; 00		780	1043	0.89	−3.30	Mit59	4
MgO periclase		1043	1393	3.35	6.125		
K1; NP; GP0; T1; A1		527	1009	0.83	0.24	Dub73a	16
lunar pyroxenite, (H$_2$/CO$_2$; $10^{-14.5}$)^4)							
K0; 00; GP0; T0; A0		227	727	1.2	4.8	Gup78	15
quartz							

^{4}) Fugacity f at 1200 °C, f in bar.
*) Numbers refer to the curves in Fig. 6.

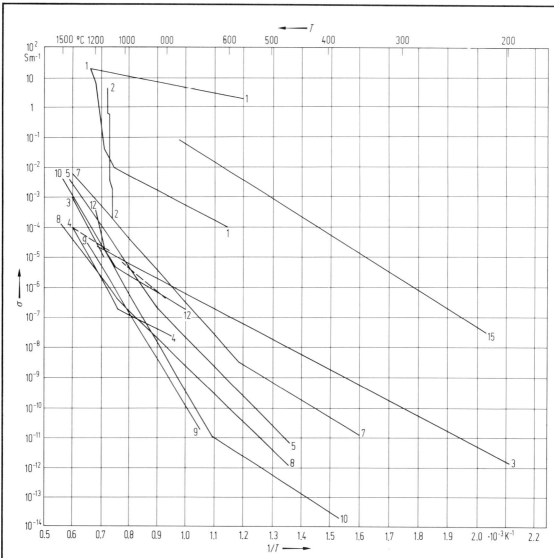

Fig. 6. Electrical conductivity of some selected natural and synthetic minerals as function of temperature. The numbers in the figure refer to the numbers of the last column of Table 4, Other minerals.

5.4.6 Bibliography — Bibliographie

5.4.6.1 References to 5.4.2···5.4.5 — Literatur zu 5.4.2···5.4.5

Bon72	Bondarenko, A.T., Galdin, N.Y.: Izv., Earth Physics **5** (1972) 28.
Bon73	Bondarenko, A.T., Fel'dman, I.S.: Izv., Earth Physics **5** (1973) 63.
Cem80	Cemič, L., Will, G., Hinze, E.: Phys. Chem. Minerals **6** (1980) 95.
Dub72	Duba, A.: J. Geophys. Res. **77** (1972) 2483.
Dub73	Duba, A., Nicholls, I.A.: Earth Planet. Sci. Lett. **18** (1973) 279.
Dub73a	Duba, A., Boland, J.N., Ringwood, A.E.: J. Geol. **81** (1973) 727.
Dub74	Duba, A., Heard, H.C., Schock, R.N.: J. Geophys. Res. **79** (1974) 1667.
Dub75	Duba, A., Ho, P., Piwinskii, A.: Abstract, EOS 56, 1075 (1975).
Dub76	Duba, A.: Acta Geodaet., Geophys. Montanist., Acad. Sci. Hung. Tomus **11** (1976) 485.
Dvo73	Dvořak, Z.: Geophys. **38** (1973) 14.
Dvo73a	Dvořak, Z., Schloessin, H.H.: Geophys. **38** (1973) 25.
End78	Endl, H.: Diss. Fachbereich 17, Technische Universität Berlin **1978**.
Gup78	Gupta, M.L., Sharma, S.R.: Phys. Earth Planet. Int. **17** (1978) 14.
Hea75	Heard, H.C., Duba, A., Piwinskii, A.J., Schock, R.N.: 6th Lunar Sci. Conf. **1975**, p. 355.
Hug53	Hughes, H.: Ph. D. Thesis, Univ. Cambridge, England **1953**.
Khi70	Khitarov, N.I., Slutsky, A.B., Pugin, V.A.: Phys. Earth Planet. Int. **3** (1970) 334.
Khi74	Khitarov, N.I., Slutsky, A.B.: Geokhimiya; cited in [Piw74].
Las76	Laštovičková, M., Parchomenko, E.I.: Pageoph. **114** (1976) 451.
Mit59	Mitoff, S.P.: J. Chem. Phys. **31** (1959) 1261.
Nit74	Nitsan, U.: J. Geophys. Res. **79** (1974) 706.
Olh74	Olhoeft, G.R., Frisillo, A.L., Strangway, D.W., Sharpe, H.: The Moon **9** (1974) 79.
Par66	Parkhomenko, E.I., Berezutskaya, A.A., Urazayev, B.M.: Tr. Inst. Fiz. Zemli, Akad. Nauk SSSR **37** (1966) 204.
Piw74	Piwinskii, A.J., Duba, A.: Geophys. Res. Lett. **1** (1974) 209.
Pre72	Presnall, D.C., Simmons, L.C., Porath, H.: J. Geophys. Res. **77** (1972) 5665.
Rai78	Rai, C.S., Manghnani, M.H., in: H.J.B. Dick, Ed., Magma Genesis. Oregon Dep. Geol. Miner. Ind., Bull 96, p. 296, **1978**.
Rai78a	Rai, C.S., Manghnani, M.H.: Phys. Earth Planet. Int. **17** (1978) 6.
Sch70	Schober, M.: Diss. Naturwiss. Fakultät Ludwig-Maximilians-Universität München, p. 56, **1970**.
Sch71	Schober, M.: Geophys. J. **37** (1971) 283.
Sch74	Schult, A., in: Approaches to taphrogenesis, H. Illies, K. Fuchs, Eds., Stuttgart: Schweizerbarth **1974**, p. 376.
Sch77	Schock, R.N., Duba, A.G., Heard, H.C., Stromberg, H.D., in: M.H. Manghnani, Ed., High Pressure Research, Academic Press **1977**.
Sha69	Shankland, T.J., in: S.K. Runcorn, Ed., The Application of Modern Physics to the Earth and Planetary Interiors. New York: Wiley Interscience, p. 175, **1969**.
Voi78	Voigt, R.: Diss. Rhein. Friedrich-Wilhelms-Universität Bonn **1978**.
Waf75	Waff, H.S., Weill, D.F.: Earth Planet. Sci. Lett. **28** (1975) 254.

5.4.6.2 Tables and review literature — Tabellenwerke und zusammenfassende Literatur

Cla66	Clark, S.P. Jr. (Ed.): Handbook of Physical Constants. Geol. Soc. America, New York, **1966**.
Dub76	Duba, A.: Are Laboratory Electrical Conductivity Data Relevant to the Earth?, Acta Geodaet., Geophys. et Montanist. Acad. Sci. Hung. Tomus **11** (1976) 485–495.
Sha75	Shankland, T.J.: Electrical Conduction in Rocks and Minerals: Parameters for Interpretation. Phys. Earth Planet. Int. **10** (1975) 209–219.
Vol76	Volarovich, M.P., Parkhomenko, F.I.: Electrical Properties of Rocks at High Temperatures and Pressures. In A. Ádám (Ed.), Geoelectric and Geothermal Studies. KAPG Geophysical Monograph. Akadémiai Kiadó, Budapest, p. 321–362, **1976**.

6 Magnetic properties — Magnetische Eigenschaften

6.1 Magnetic properties of natural minerals — Magnetische Eigenschaften natürlicher Minerale

This chapter gives a description of naturally occurring minerals and their synthetic equivalents. Emphasis is laid on those minerals that control the magnetic properties of the major rock types. The data are given in cgs-units, as in nearly all of the original papers. For conversion to SI-units, see Table 6.1.1.2.

In diesem Kapitel werden natürliche Minerale und ihre synthetischen Äquivalente beschrieben. Besonders berücksichtigt werden diejenigen Minerale, die die magnetischen Eigenschaften der wesentlichsten Gesteinstypen bestimmen. Die Werte sind, wie in fast allen Originalarbeiten, in cgs-Einheiten gegeben. Zur Umrechnung in SI-Einheiten siehe Tab. 6.1.1.2.

6.1.1 General introduction — Allgemeine Einleitung

The minerals can be classified with respect to their magnetic properties in the following way:
  diamagnetic minerals
  paramagnetic minerals
  ferromagnetic minerals
  ferrimagnetic minerals
  antiferromagnetic minerals
Minerals consisting of ions without an intrinsic magnetic moment are diamagnetic. Minerals containing ions with an intrinsic magnetic moment are either ferro-, ferri- or antiferro-magnetic (if there exists exchange interaction between these ions) or paramagnetic (if there exists no such interaction).

Necessary for the existence of an intrinsic magnetic moment of a certain ion is the presence of unpaired electrons in the electron shell. Ions with completely filled electron shells have no magnetic moment and are diamagnetic therefore.

The transition elements, the lanthanides and actinides constitute ions with unpaired electrons. The magnetic moment of these ions depends on the number of unpaired electrons and can be characterized by an effective magnetic moment p_{eff} (measured in Bohr magnetons, $\mu_B = 0.9273 \cdot 10^{-20}$ erg/Gauss).

Die Minerale können nach ihren magnetischen Eigenschaften wie folgt klassifiziert werden:
  diamagnetische Minerale
  paramagnetische Minerale
  ferromagnetische Minerale
  ferrimagnetische Minerale
  antiferromagnetische Minerale
Minerale, die aus Ionen ohne permanentes magnetisches Moment bestehen, sind diamagnetisch. Minerale, die Ionen mit permanentem magnetischem Moment enthalten, sind entweder ferro-, ferri- oder antiferromagnetisch (wenn zwischen diesen Ionen eine Austauschwechselwirkung besteht) oder paramagnetisch (wenn keine Wechselwirkung besteht).

Voraussetzung für die Existenz eines permanenten magnetischen Moments in einem bestimmten Ion ist das Vorhandensein von ungepaarten Elektronen in einer Elektronenschale. Ionen mit abgeschlossenen Elektronenschalen haben kein magnetisches Moment und sind diamagnetisch.

Die Ionen der Übergangselemente, der Lanthaniden und der Aktiniden besitzen ungepaarte Elektronen. Das magnetische Moment dieser Ionen hängt von der Anzahl der ungepaarten Elektronen ab und kann durch ein effektives magnetisches Moment p_{eff} dargestellt werden (gemessen in Bohrschen Magnetonen, $\mu_B = 0.9273 \cdot 10^{-20}$ erg/Gauss).

6.1.1.1 List of symbols — Symbolliste

a, b, c	[Å]	lattice parameters — Gitterkonstanten
C_m	[K cm^3 mol^{-1}]	Curie constant per mole — Curiekonstante pro Mol
H	[Oe]	magnetic field strength — Magnetfeldstärke
H_C	[Oe]	coercive force — Koerzitivkraft
H_{RC}	[Oe]	remanence coercivity — Koerzitivkraft der Remanenz
K	[erg cm^{-3}]	anisotropy constant — Anisotropiekonstante
L		Avogadro's number — Loschmidtsche Zahl
(M)	[g mol^{-1}]	molecular weight — Molgewicht
M	[Gauss]	volume magnetization — Magnetisierung pro Volumeneinheit
	(=[emu cm^{-3}])	
M_R	[Gauss]	remanent volume magnetization — remanente Magnetisierung
	(=[emu cm^{-3}])	pro Volumeneinheit
M_S	[Gauss]	saturation volume magnetization — Sättigungsmagnetisierung
	(=[emu cm^{-3}])	pro Volumeneinheit

List of symbols (continued)

M_{TRM} [Gauss]   thermoremanent magnetization — thermoremanente Magnetisierung
          $(=[emu\ cm^{-3}])$ pro Volumeneinheit
n   number of atoms or ions — Anzahl der Atome oder Ionen
p [bar]   pressure — Druck
P [Gauss cm³]   magnetic moment — magnetisches Moment
          $(=[emu])$
p_α $[\mu_B]$ magnetic moment per atom or ion α — magnetisches Moment pro Atom
                oder Ion α
p_{eff} $[\mu_B]$ effective magnetic moment — effektives magnetisches Moment
p_m $[\mu_B]$ magnetic moment per molecule — magnetisches Moment pro Molekel
R $[erg\ mol^{-1}\ K^{-1}]$ gas constant — Gaskonstante
T [°C, K]   temperature — Temperatur
T_C [°C, K]   Curie temperature — Curietemperatur
T_M [°C, K]   Morin temperature (low temperature magnetic transition) —
                Morintemperatur (magnetischer Übergang bei tiefen Temperaturen)
T_N [°C, K]   Néel temperature — Néeltemperatur
T_V [°C, K]   Verwey temperature (magnetic ordering) — Verweytemperatur
                (magnetische Ordnung)
z oxidation parameter — Oxidationsparameter
Θ_a [°C, K]   asymptotic Curie temperature — asymptotische Curietemperatur
Θ_p [°C, K]   paramagnetic Curie temperature — paramagnetische Curietemperatur
λ magnetostriction constant — Magnetostriktionskonstante
μ_B Bohr magneton — Bohrsches Magneton
ϱ $[g\ cm^{-3}]$ density — Dichte
ϱ_X $[g\ cm^{-3}]$ X-ray density — Röntgendichte
σ $[Gauss\ cm^3\ g^{-1}]$ specific magnetization (mass magnetization) — spezifische Magnetisierung
          $(=[emu\ g^{-1}])$
σ_S $[Gauss\ cm^3\ g^{-1}]$ specific saturation magnetization — spezifische Sättigungsmagnetisierung
          $(=[emu\ g^{-1}])$
σ_{RS} $[Gauss\ cm^3\ g^{-1}]$ specific saturation remanent magnetization — spezifische Sättigungsremanenz
          $(=[emu\ g^{-1}])$
σ_{TRM} $[Gauss\ cm^3\ g^{-1}]$ specific thermoremanent magnetization — spezifische thermoremanente
          $(=[emu\ g^{-1}])$ Magnetisierung
χ $[cm^3]\ (=[emu])$ magnetic susceptibility — magnetische Suszeptibilität
χ_g $[cm^3\ g^{-1}]$ mass susceptibility (specific susceptibility) — Suszeptibilität pro Gramm
          $(=[emu\ g^{-1}])$
χ_{gi} $[cm^3\ g^{-1}]$ initial mass susceptibility — Anfangssuszeptibilität pro Gramm
          $(=[emu\ g^{-1}])$
χ_m $[cm^3\ mol^{-1}]$ molar susceptibility — molare Suszeptibilität
          $(=[emu\ mol^{-1}])$
          $[cm^3(gram-$ gram-ion susceptibility — Gramm-Ion-Suszeptibilität
          $ion)^{-1}]$
χ_v $[cm^3\ cm^{-3}=1]$ volume susceptibility — Volumensuszeptibilität
          $(=[Gauss\ Oe^{-1}])$

6.1.1.2 Conversion table of magnetic quantities — Umrechnungstabelle für magnetische Größen

Quantity	Symbol	cgs-unit	Multiplying factor used to convert		SI-unit
			one cgs-unit to SI-units	one SI-unit to cgs-units	
magnetic field	H	Oe	$10^3/4\pi = 7.96 \cdot 10$	$4\pi \cdot 10^{-3}$ $= 1.257 \cdot 10^{-2}$	$A\,m^{-1}$
magnetic induction	B	Gauss	10^{-4}	10^4	T
magnetic moment	P	Gauss cm^3	10^{-3}	10^3	$A\,m^2$
magnetization (=magnetic moment per unit volume)	M	Gauss	10^3	10^{-3}	$A\,m^{-1}$
specific magnetization (=magnetic moment per unit mass)	σ	Gauss cm^3 g^{-1}	1	1	$A\,m^2\,kg^{-1}$
susceptibility	χ_v	cm^3 cm^{-3}	$4\pi = 12.57$	$1/4\pi = 7.96 \cdot 10^{-2}$	$m^3\,m^{-3}$
specific susceptibility	χ_g	cm^3 g^{-1}	$4\pi \cdot 10^{-3} = 12.57 \cdot 10^{-3}$	$10^3/4\pi = 7.96 \cdot 10$	$m^3\,kg^{-1}$
molar susceptibility	χ_m	cm^3 mol^{-1}	$4\pi \cdot 10^{-6} = 12.57 \cdot 10^{-6}$	$10^6/4\pi = 7.96 \cdot 10^4$	$m^3\,mol^{-1}$
permeability	$\mu = \mu_0 \cdot \mu_{rel}$	dimension- less	$4\pi \cdot 10^{-7} = 1.257 \cdot 10^{-6}$	$10^7/4\pi = 7.96 \cdot 10^5$	$H\,m^{-1}$
permeability of vacuum	μ_0	$\mu_0 \equiv 1$	$4\pi \cdot 10^{-7} = 1.257 \cdot 10^{-6}$	$10^7/4\pi = 7.96 \cdot 10^5$	$H\,m^{-1}$
anisotropy constant	K	erg cm^{-3}	10^{-1}	10	$J\,m^{-3}$
demagnetizing factor	N	dimension- less	$1/4\pi = 7.96 \cdot 10^{-2}$	$4\pi = 12.57$	dimension- less

T = tesla (= Wb m^{-2} = V s m^{-2}), H = henry (= V s A^{-1}), J = joule (= kg m^2 s^{-2}),
Oe = oersted (= cm$^{-1/2}$ g$^{1/2}$ s^{-1} = (erg cm^{-3})$^{1/2}$), Gauss (= cm$^{-1/2}$ g$^{1/2}$ s^{-1} = (erg cm^{-3})$^{1/2}$).

Equations of magnetism

cgs-system	SI-system
$B = H + 4\pi M$	$B = \mu_0(H + M)$
$M = \chi_v H$	$M = \chi_v H$

6.1.2 Diamagnetism — Diamagnetismus

6.1.2.1 Introductory remarks — Einleitende Bemerkungen

The number of naturally occurring diamagnetic minerals is much smaller than should be expected from their idealized chemical formula. This is due to the presence of a certain amount of non-stoichiometric Fe- or Mn-ions that frequently occur in natural crystals of a basically diamagnetic mineral. Such content of non-stoichiometric Fe or Mn varies normally from mineral to mineral and is dependent on the genetic conditions. Correspondingly, one and the same mineral is found to be paramagnetic in the one case and diamagnetic in the other. A number of minerals,

Die Anzahl der in der Natur vorkommenden diamagnetischen Minerale ist viel kleiner als man es nach ihrer idealisierten chemischen Formel erwarten könnte. Dies ist auf die Gegenwart einer gewissen Menge von nicht-stöchiometrischen Fe- oder Mn-Ionen zurückzuführen, die häufig in natürlichen Kristallen eines an sich diamagnetischen Minerals vorkommen. Ein solcher Gehalt an nicht-stöchiometrischen Fe oder Mn ändert sich normalerweise von Mineral zu Mineral und hängt von den genetischen Bedingungen ab. Dementsprechend erweist sich ein und dasselbe Mineral in einem

therefore, is listed in both, the table for paramagnetic minerals, and the table for diamagnetic minerals.

Fall als paramagnetisch, in einem anderen als diamagnetisch. Deshalb sind einige Minerale sowohl in der Tabelle für paramagnetische Minerale, als auch in der Tabelle für diamagnetische Minerale aufgeführt.

6.1.2.2 Susceptibility — Suszeptibilität

In contrast to the paramagnetic susceptibility the diamagnetic susceptibility is independent of temperature.

The diamagnetic susceptibility of a mineral can be calculated to a first order approximation by adding up the susceptibilities of the individual ions of the compound. The susceptibilities of the diamagnetic ions constituting the most important rock-forming minerals are listed in Table 1.

Im Gegensatz zu der paramagnetischen Suszeptibilität ist die diamagnetische Suszeptibilität von der Temperatur unabhängig.

Die diamagnetische Suszeptibilität eines Minerals kann in erster Näherung berechnet werden, indem man die Suszeptibilitäten der einzelnen Ionen der Verbindung addiert. Die Suszeptibilitäten der diamagnetischen Ionen, die in den wichtigsten gesteinsbildenden Mineralen enthalten sind, sind in Tab. 1 zusammengestellt.

6.1.2.3 Tables — Tabellen

Table 1. Susceptibilities of the diamagnetic ions constituting the most important rock-forming minerals [Kle36].

Ion	χ_g 10^{-7} cm^3 g^{-1}	χ_m 10^{-7} cm^3 (gram-ion)$^{-1}$
O^{2-}	-7.00	-112
F^{1-}	-3.79	-72
Cl^{1-}	-6.46	-229
S^{2-}	-10.17	-326
CO_3^{2-}	-4.68	-281
SO_4^{2-}	-4.06	-390
Na^{1+}	-1.61	-37
Mg^{2+}	-1.19	-29
Al^{3+}	-0.85	-23
Si^{4+}	-0.68	-19
K^{1+}	-3.35	-131
Ca^{2+}	-2.59	-104

Table 2. Susceptibilities of diamagnetic minerals.

Mineral	Chemical composition	χ_g 10^{-6} cm^3 g^{-1}	χ_v 10^{-6} cm^3 cm^{-3}	Remarks	Ref.
Silicates					
forsterite	Mg_2SiO_4	-0.31		synthetic material	Hoy72
orthoclase	$KAlSi_3O_8$	-0.53	-1.09	$\parallel a$-axis	Fin10
		-0.47	-0.98	$\parallel b$-axis	
		-0.39	-0.81	$\parallel c$-axis	
zircon	$ZrSiO_4$	-0.170	-0.784	$\parallel a$-axis	Voi07
		$+0.732$	$+3.37$	$\parallel c$-axis	
Oxides					
quartz	SiO_2	-0.461	-1.22	$\parallel a$-axis	Voi07
		-0.466	-1.23	$\parallel c$-axis	
opal	amorphous SiO_2	-0.487	-1.023		Voi07

(continued)

Table 2 (continued)

Mineral	Chemical composition	χ_g 10^{-6} cm^3 g^{-1}	χ_v 10^{-6} cm^3 cm^{-3}	Remarks	Ref.
Sulfur compounds					
gypsum	$CaSO_4 \cdot 2H_2O$	−1.0			Pov64
			+40···263	ferromagnetic impurities?	Cra02
galena	PbS	−0.350	−2.63		Voi07
sphalerite	ZnS	−0.261	−1.04		Voi07
celestite	$SrSO_4$	−0.342	−1.35	∥ a-axis	Voi07
		−0.314	−1.24	∥ b-axis	
		−0.359	−1.42	∥ c-axis	
anhydrite	$CaSO_4$	−1.6			Pov64
Carbonates					
aragonite	$CaCO_3$	−0.392	−1.15	∥ a-axis	Voi07
		−0.387	−1.13	∥ b-axis	
		−0.444	−1.30	∥ c-axis	
calcite	$CaCO_3$	−0.363	−0.987	∥ a-axis	Voi07
		−0.405	−1.101	∥ c-axis	
Halogenides					
halite	NaCl	−0.376	−0.816		Voi07
		−1.8			Pov64
blue john	CaF_2	−0.627	−2.00		Voi07
Phosphates					
apatite	$Ca_5[(F, Cl)(PO_4)_3]$	−0.264	−0.845	∥ a-axis	Voi07
		−0.264	−0.845	∥ c-axis	
Elements					
diamond	C	−0.49			Pov64
graphite	C	−6.2			Pov64

6.1.3 Paramagnetism — Paramagnetismus

6.1.3.1 Introductory remark — Einleitende Bemerkung

The most important rock-forming minerals with paramagnetic properties are olivine, orthopyroxene, clinopyroxene, garnet, amphibole, and biotite. It is essentially the content of Fe^{2+}, Fe^{3+} and (less important) of Mn^{2+} that causes the paramagnetism of these minerals (Fig. 2).

Die wichtigsten gesteinsbildenden Minerale mit paramagnetischen Eigenschaften sind Olivin, Ortho-pyroxen, Klinopyroxen, Granat, Amphibol und Biotit. Der Paramagnetismus dieser Minerale wird im wesentlichen durch den Gehalt an Fe^{2+}, Fe^{3+} und (weniger wichtig) an Mn^{2+} bestimmt (Fig. 2).

6.1.3.2 Susceptibility — Suszeptibilität

When no effective interaction between the magnetic ions of a mineral is assumed, the specific magnetic susceptibility (mass-susceptibility in [cm^3 g^{-1}]), χ_g, of the mineral is given according to Langevin's theory, as

Wenn keine effektive Wechselwirkung zwischen den magnetischen Ionen eines Minerals angenommen wird, ist die spezifische magnetische Suszeptibilität (Massensuszeptibilität in [cm^3 g^{-1}]), χ_g, des Minerals nach der Langevinschen Theorie gegeben durch

$$\chi_g = \frac{(L\mu_B)^2}{3RT}(\alpha n_\alpha^2 + \beta n_\beta^2 + \gamma n_\gamma^2 + \cdots) = \frac{L^2}{3RT}(\alpha p_\alpha^2 + \beta p_\beta^2 + \gamma p_\gamma^2 + \cdots) \qquad (1)$$

with $p_\alpha = n_\alpha \cdot \mu_B$, $p_\beta = n_B \cdot \mu_B$, $p_\gamma = n_\gamma \cdot \mu_B$, ...

(Curie law in cgs-units) [Che58].

(Curie-Gesetz im cgs-System) [Che58].

$\alpha, \beta, \gamma, \ldots$ [mol g^{-1}] are the amounts in mol ($\hat{=}$ gram-ion) of Fe^{2+}, Fe^{3+}, Mn^{2+}, ... per gram of paramagnetic mineral.

$n_\alpha, n_\beta, n_\gamma, \ldots$ are the numerical values of the magnetic moment of the individual magnetic ions expressed in units of μ_B (see Table 3).

$\alpha, \beta, \gamma, \ldots$ [mol g^{-1}] geben die Anzahl der Gramm-Ionen Fe^{2+}, Fe^{3+}, Mn^{2+}, ... je Gramm des paramagnetischen Minerals an.

$n_\alpha, n_\beta, n_\gamma, \ldots$ sind die Beträge der magnetischen Momente der einzelnen magnetischen Ionen, in der Einheit μ_B (s. Tab. 3).

$$L \cdot \mu_B = 5589 \text{ emu mol}^{-1} = 5589 \text{ erg Gauss}^{-1} \text{ mol}^{-1}$$

Volume-susceptibility in [cm^{3}cm^{-3}=1] of a mineral:

Volumensuszeptibilität in [cm^3 cm^{-1}=1] eines Minerals:

$$\chi_v = \chi_g \cdot \varrho,$$

ϱ = density of the mineral in [g cm^{-3}].
Mole-susceptibility in [cm^3 mol^{-1}] of a mineral:

ϱ = Dichte des Minerals in [g cm^{-3}].
Molsuszeptibilität in [cm^3 mol^{-1}] eines Minerals:

$$\chi_m = \chi_g \cdot (M) \tag{2}$$

(M) = the molecular weight in [g mol^{-1}] of the mineral.

(M) = Molekulargewicht des Minerals in [g mol^{-1}].

The agreement of measured and calculated values of susceptibility (using formulae (1) or (2)) is satisfactory for most paramagnetic minerals at room temperature (Figs. 3 and 4).

At sufficiently low temperatures ($T < 0$ °C) frequently deviation from paramagnetic behaviour due to antiferromagnetic interaction is observed [Hoy72, Elf65] (Fig. 6). In these cases the paramagnetic susceptibility is more accurately calculated according to the Curie-Weiss law (Fig. 5):

Die Übereinstimmung zwischen gemessenen und berechneten Werten der Suszeptibilität (wobei Formel (1) oder (2) verwendet werden) ist für die meisten paramagnetischen Minerale bei Zimmertemperatur zufriedenstellend (Fig. 3 und 4).

Bei genügend niedrigen Temperaturen ($T < 0$ °C) wird öfters eine Abweichung vom paramagnetischen Verhalten beobachtet, die auf antiferromagnetische Wechselwirkung zurückzuführen ist [Hoy72, Elf65] (Fig. 6). In diesen Fällen wird die paramagnetische Suszeptibilität genauer berechnet nach dem Curie-Weiss-Gesetz (Fig. 5):

$$\chi_g = \frac{(L\mu_B)^2}{3R(T-\Theta_a)}(\alpha n_\alpha^2 + \beta n_\beta^2 + \gamma n_\gamma^2 + \cdots) = \frac{L^2}{3R(T-\Theta_a)}(\alpha p_\alpha^2 + \beta p_\beta^2 + \gamma p_\gamma^2 + \cdots) \tag{3}$$

where Θ_a denotes the asymptotic Curie temperature.

wobei Θ_a die asymptotischen Curie-Temperatur bezeichnet.

Paramagnetic minerals that have been obtained by separation from a host rock frequently contain minute amounts of ferrimagnetic impurities. These impurities cause deviation from paramagnetic behaviour in low magnetic fields [Nag57, Aki58, Che58] as shown in Fig. 1. In these cases the paramagnetic susceptibility has to be determined from the slope of the $\sigma(H)$-curve in strong magnetic fields after saturation of the ferrimagnetic impurity. Susceptibility values of Table 5, obtained by this method, are marked by *).

Paramagnetische Minerale, die durch Abtrennung aus einem Gesteinsverband gewonnen wurden, enthalten häufig winzige Beiträge von ferrimagnetischen Verunreinigungen. Diese Verunreinigungen verursachen eine Abweichung vom paramagnetischen Verhalten in niedrigen magnetischen Feldern [Nag57, Aki58, Che58] (s. Fig. 1). In diesen Fällen muß die paramagnetische Suszeptibilität aus der Steigung der $\sigma(H)$-Kurve in starken magnetischen Feldern nach Sättigung der ferrimagnetischen Verunreinigungen bestimmt werden. Suszeptibilitätswerte in Tab. 5, die nach dieser Methode gewonnen wurden, sind mit *) gekennzeichnet.

6.1.3.3 Figures and tables — Abbildungen und Tabellen

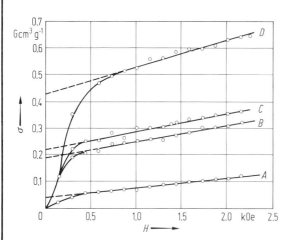

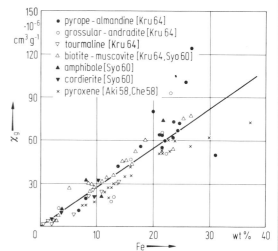

Fig. 1. Specific magnetization, σ, of olivines at room temperature vs. magnetic field, H. Curve A: Olivine fraction extracted from andesite pumice, Hatizyo-Zima Island (Japan). Curves B, C, D: Olivine fractions from dunite, Transvaal (South-Africa).

Ferrimagnetic impurities cause non-linearity of the magnetization curves in magnetic fields below $H = 1000$ Oe. Slopes of the curves B, C, D are nearly equal in high magnetic fields, i.e. nearly equal paramagnetic susceptibility of the Transvaal olivines is observed, whereas the samples differ in the amount of ferrimagnetic impurity [Nag57].

Fig. 2. Correlation between paramagnetic specific susceptibility, χ_g, and total iron content of a wide range of minerals, measured at room temperature. The straight line describes the empirical relationship

$$\chi_g = 2.77 \ (\text{wt\% total Fe}) \cdot 10^{-6} \ \text{cm}^3 \ \text{g}^{-1}$$

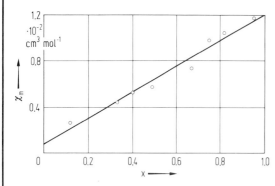

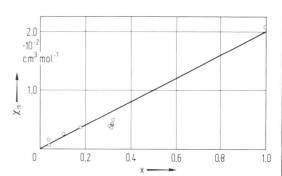

Fig. 3. Variation of paramagnetic molar susceptibility, χ_m, of orthopyroxene $x \, \text{FeSiO}_3 \cdot (1-x) \, \text{MgSiO}_3$ with iron content (composition parameter x) measured at room temperature. Natural crystals. The data can be described by the Curie-law (2) assuming a magnetic moment of Fe^{2+}: $p_\alpha^{Fe^{2+}} = 5.14 \, \mu_B$. The straight line represents the empirical relationship

$$\chi_m = 1.12 \cdot x \cdot 10^{-2} \ \text{cm}^3 \ \text{mol}^{-1} \ [\text{Aki58}].$$

Fig. 4. Variation of paramagnetic molar susceptibility, χ_m, of olivine $x \, \text{Fe}_2\text{SiO}_4 \cdot (1-x) \, \text{Mg}_2\text{SiO}_4$ with iron content (composition parameter x) measured at room temperature. Natural crystals. The data can be described by the Curie-law (2) assuming a magnetic moment of Fe^{2+}: $p_\alpha^{Fe^{2+}} = 5.1 \, \mu_B$. The straight line represents the empirical relationship

$$\chi_m = 2 \cdot x \cdot 10^{-2} \ \text{cm}^3 \ \text{mol}^{-1} \ [\text{Nag57}].$$

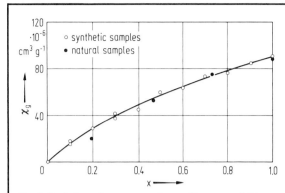

Fig. 5. Variation of paramagnetic specific susceptibility, χ_g, of olivine x $Fe_2SiO_4 \cdot (1-x)\, Mg_2SiO_4$ with iron content (composition parameter x) measured at room temperature. In contrast to natural olivine (Fig. 4) the relationship becomes non-linear for low x-values. The data can be described by the Curie-Weiss law (3) assuming a magnetic moment of Fe^{2+}: $p_\alpha^{Fe^{2+}} = 5.2\ \mu_B$ [Hoy72].

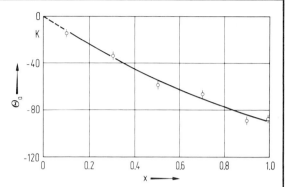

Fig. 6. Variation of asymptotic Curie temperature, Θ_a, of olivine x $Fe_2SiO_4 \cdot (1-x)\, Mg_2SiO_4$ with iron content (composition parameter x). Synthetic crystals. The deviation from a linear dependence is due to variation of exchange energy with changing interatomic distances [Hoy72].

Table 3. Calculated (for pure spin magnetism and maximum multiplicity) and experimental effective magnetic moments, p_{eff}, of the ions that cause the magnetic properties of rock-forming minerals [Pov64].

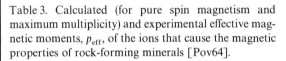

p_{eff} (calculated) $= \sqrt{N(N+2)}\ \mu_B$.

Ions	Number of 3d-electrons		p_{eff} (calculated)	p_{eff} (experimental)
	Total	Unpaired N	μ_B	μ_B
Sc^{3+}	0	0	0.00	0.00
Ti^{4+}				0.00
V^{5+}				0.00
Ti^{3+}	1	1	1.73	1.77···1.79
V^{4+}				
Ti^{2+}	2	2	2.83	2.76···2.85
V^{3+}				
V^{2+}	3	3	3.87	3.81···3.86
Cr^{3+}				3.68···3.86
Mn^{4+}				4.00
Cr^{2+}	4	4	4.90	4.80
Mn^{3+}				5.00
Mn^{2+}	5	5	5.92	5.2···5.96
Fe^{3+}				5.4···6.00
Fe^{2+}	6	4	4.90	5.0···5.5
Co^{3+}				(2.50)
Co^{2+}	7	3	3.87	4.4···5.2
Ni^{2+}	8	2	2.83	2.9···3.4
Cu^{2+}	9	1	1.73	1.8···2.2
Cu^{1+}	10	0	0.00	0.00
Zn^{2+}				0.00

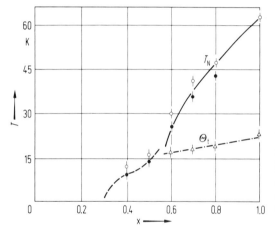

Fig. 7. Variation of Néel temperature, T_N, and spin structure transition temperature, Θ_t, of olivine x $Fe_2SiO_4 \cdot (1-x)$ Mg_2SiO_4 with iron content (composition parameter x). Synthetic crystals. Open circles: measured T_N-values, full circles: measured T_N-values corrected for the variation of exchange energy [Hoy72].

Table 4. Specific susceptibilities, χ_g, of 'pure' paramagnetic garnets at room temperature [Fli59].

Mineral	Chemical composition	χ_g $10^{-6}\ cm^3 g^{-1}$
pyrope	$Mg_3Al_2[SiO_4]_3$	0
almandine	$Fe_3Al_2[SiO_4]_3$	68
spessartine	$Mn_3Al_2[SiO_4]_3$	130
grossular	$Ca_3Al_2[SiO_4]_3$	0
andradite	$Ca_3Fe_2[SiO_4]_3$	49

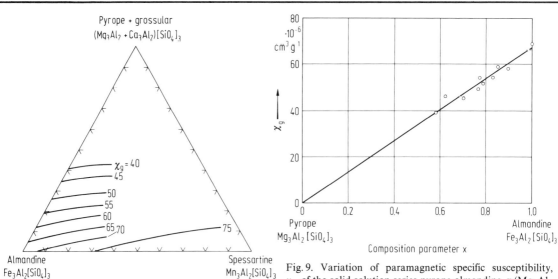

Fig. 8. Paramagnetic specific susceptibilities, χ_g, of natural garnets at room temperature (in 10^{-6} cm^3 g^{-1}) represented in the ternary diagram pyrope + grossular (Mg$_3$Al$_2$ + Ca$_3$Al$_2$)(SiO$_4$)$_3$-almandine (Fe$_3$Al$_2$)(SiO$_4$)$_3$-spessartine (Mn$_3$Al$_2$)(SiO$_4$)$_3$ [Fro60, Syo60, Elf65].

Fig. 9. Variation of paramagnetic specific susceptibility, χ_g, of the solid solution series pyrope-almandine, x (Mg$_3$Al$_2$ [SiO$_4$]$_3$) · (1−x)(Fe$_3$Al$_2$[SiO$_4$]$_3$) with iron content (composition parameter x) measured at room temperature. Natural crystals. The straight line represents the empirical relationship

$$\chi_g = 68 \cdot x \cdot 10^{-6}\ \text{cm}^3\,\text{g}^{-1}\ [\text{Fro60}].$$

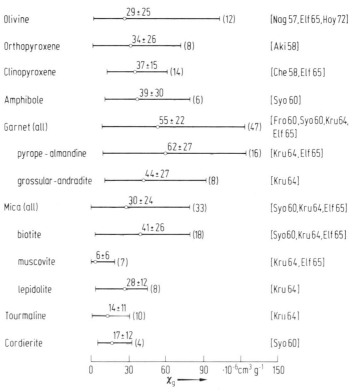

Fig. 10. Range of specific susceptibilities of the most important paramagnetic rock-forming minerals. Room temperature data. Natural minerals. Circles represent mean values with standard deviations, figures in parentheses are numbers of analyses.

Table 5. Paramagnetic susceptibilities of rock-forming minerals at room temperature.

Mineral	Chemical composition	χ_g 10^{-6} cm^3 g^{-1}	χ_v 10^{-6} cm^3 cm^{-3}	Remarks	Ref.
Silicates					
olivine	$(Mg, Fe)_2SiO_4$	3···103 *)		natural samples	Nag57, Elf65, Hoy72
fayalite	Fe_2SiO_4	90		synthetic samples	Hoy72
orthopyroxene	$(Mg, Fe)SiO_3$	3···73 *)		natural samples	Aki58
enstatite	$MgSiO_3$	3 *)		contains 3 % FeO	Aki58
ferrosilite	$FeSiO_3$	73 *)		contains 3.7 % MgO	Aki58
clinopyroxene	$X, Y[Z_2O_6]$, $X = Ca^{2+}, Na^+, K^+,$ $\quad Mn^{2+}$; $Y = Mg^{2+}, Fe^{2+}, Fe^{3+},$ $\quad Al^{3+}, Ti^{3+}, Mn^{3+}$; Z mostly Si, $\quad$ occasionally $\quad Al, Fe^{3+}, V, P$	15···62 *)			Che58, Elf65
diopside	$CaMg[Si_2O_6]$	32 *)		contains 3.8 % Fe_2O_3	Elf65
hedenbergite	$CaFe[Si_2O_6]$	61			Che58
augite	$CaMg[Si_2O_6]$ with Al and Fe	13···26			Fin10
amphibole	$X_2Y_5[(OH,F)Z_4O_{11}]_2$, $X = Ca, Na, K$; $Y = Mg, Fe, Al,$ $\quad Mn, Ti^{3+}$; $Z = Si, Al$	13···80 *)			Fin10, Syo60
actinolite	$Ca_2(Mg, Fe)_5$ $\quad \cdot [OH, Si_4O_{11}]_2$	13		contains 5 % FeO	Syo60
hornblende	Al and Fe bearing amphiboles	18···33 *)			Fin10, Syo60
arfvedsonite	$Na_3Fe_4^{2+}Al$ $\quad \cdot [OH, Si_4O_{11}]_2$	80 *)			Syo60
riebeckite	$Na_2Fe_4^{3+}$ $\quad \cdot [OH, Si_4O_{11}]_2$	75 *)			Syo60
garnet	$X_3^{2+}Y_2^{3+}[SiO_4]_3$, $X^{2+} = Ca, Mg, Fe, Mn,$ $Y^{3+} = Cr, Ti, Fe, Al$	11···124 *)		see also Table 4	Fro60, Syo60, Kru64, Elf65
pyrope	$Mg_3Al_2[SiO_4]_3$	11		contains 3.25 % Fe_2O_3	Kru60
almandine	$Fe_3Al_2[SiO_4]_3$	50···124 *)		CaO + MgO < 10 %	Fro60, Syo60, Kru64, Elf65
spessartine	$Mn_3Al_2[SiO_4]_3$	130			Fro60
andradite	$Ca_3Fe_2[SiO_4]_3$	49···93			Fro60, Kru64
mica	$XY_2[(OH, F)_2Z'Z''_3$ $\quad \cdot O_{10}]$, $X = K, Na, Rb, Cs$; $Y = Al, Fe, Cr, Mn,$ $\quad V, Ti, Mg, Li$; Z' mainly Al^{3+}, also $\quad Si^{4+}$ and Fe^{3+}; Z'' exclusively Si	1···80 *)			Syo60, Kru64, Elf65
muscovite	$KAl_2[(OH, F)_2$ $\quad \cdot AlSi_3O_{10}]$	1···20			Nil38, Kru64, Elf65

(continued)

Table 5 (continued)

Mineral	Chemical composition	χ_g 10^{-6} cm^3 g^{-1}	χ_v 10^{-6} cm^3 cm^{-3}	Remarks	Ref.
biotite	K(Mg, Fe)$_3$[(OH)$_2$ · (Al, Fe)Si$_3$O$_{10}$]	23···80 *)			Nil38, Syo60, Kru64, Elf65
phlogopite	KMg$_3$[(F, OH)$_2$ · AlSi$_3$O$_{10}$]	5···8		contains 4% Fe$_2$O$_3$	Kru60
lepidolite	Li-mica	4···46			Kru60
talc	(Mg, Fe)$_3$(OH)$_2$Si$_4$O$_{10}$	5 *)			Cra02
pyrophyllite	Al$_2$[(OH)$_2$Si$_4$O$_{10}$]	1···2			Kru60
serpentine	Mg$_6$[(OH)$_8$Si$_4$O$_{10}$]	4.7	12.2		Stu18
tourmaline	XY$_3$Z$_6$[(OH)$_4$(BO$_3$)$_3$ · Si$_6$O$_{18}$],	1···39			Wil20, Gre30, Kru64
	X = Na, Ca;	1.118	3.47	‖ a-axis	Voi07
	Y = Mg, Li, Al, Fe^{2+}, Mn;	0.748	2.32	‖ c-axis	
	Z = Al, Fe^{3+}, Ti, Cr				
cordierite	Mg$_2$Al$_3$[AlSi$_5$O$_{18}$]	6···33 *)		contains up to 13% FeO	Syo60
beryl	Al$_2$Be$_3$[Si$_6$O$_{18}$]	0.826	2.23	‖ a-axis	Voi07
		0.386	1.04	‖ c-axis	
epidote	Ca$_2$(Fe, Al)Al$_2$[O, OH, SiO$_4$, Si$_2$O$_7$]	23.8	80.0	‖ a-axis	Fin10
		24.1	80.9	‖ b-axis	
		23.9	80.2	‖ c-axis	
zircon	Zr[SiO$_4$]	−0.25···+6.6			Lew66
		−0.170	−0.784	‖ a-axis	Voi07
		+0.732	+3.37	‖ c-axis crystals are brownish-red	
sphene	CaTi[O, SiO$_4$]	6			Koe98
orthite (allanite)	Ca(Ce, Th) · (Fe^{3+}, Mg, Fe^{2+}) · Al$_2$[O, OH, SiO$_4$, Si$_2$O$_7$]	23···94		frequency maximum between (23···34) · 10^{-6} cm^3 g^{-1}	Fli59
staurolite	2FeO · AlOOH · 4Al$_2$ · [O, SiO$_4$]	17···34		frequency maximum at 22 · 10^{-6} cm^3 g^{-1}	Fli59
thorite	ThSiO$_4$	0···13		frequency maximum at 8 · 10^{-6} cm^3 g^{-1}	Fli59
rhodonite	Mn$_4$(Ca, Fe)[Si$_5$O$_{15}$]	131	457		Stu18
Oxides and hydroxides					
rutile	TiO$_2$	1.96	8.33	‖ a-axis	Voi07
		2.09	8.89	‖ c-axis	
cassiterite	SnO$_2$	13	88	according to [Han66] cassiterite is weakly ferromagnetic due to Fe-content of between 0.05 and 0.6%. T_C = 525 °C	Stu18
pyrolusite	β-MnO$_2$	25···32			Stu18, Pov64
psilomelane	(Ba, H$_2$O)$_2$Mn$_5$O$_{10}$	55	268		Stu18
corundum	Al$_2$O$_3$		143···660	contains presumably ferrimagnetic impurities	Cra02

(continued)

Table 5 (continued)

Mineral	Chemical composition	χ_g 10^{-6} cm^3 g^{-1}	χ_v 10^{-6} cm^3 cm^{-3}	Remarks	Ref.
ilmenite	FeTiO$_3$	146	640		Gre30
columbite	(Fe, Mn)(Nb, Ta)$_2$O$_6$	17···94		frequency maximum between (34···53) · 10^{-6} cm^3 g^{-1}	Fli59
gahnite	ZnAl$_2$O$_4$	13···24			Fli59
limonite		220	59		Stu18
goethite	α-FeOOH	27···220		according to [Hed68] antiferromagnetic at room temperature with weak	Gre30
		21···25		ferrimagnetism superimposed. T_N = 120 °C	Pov64
chromite	(Fe, Mg)Cr$_2$O$_4$	54	245		Stu18
		50···125			Pov64
tenorite	CuO	8.1	38.1		Elf65
pseudobrookite	Fe$_2$TiO$_5$	43		synthetic material	Aki57
ferropseudo-brookite	FeTi$_2$O$_5$	53		synthetic material	Aki57
spinel	MgAl$_2$O$_4$	0.62			Pov64
hausmannite	MnMn$_2$O$_4$	60			Pov64

Sulfur compounds

Mineral	Chemical composition	χ_g 10^{-6} cm^3 g^{-1}	χ_v 10^{-6} cm^3 cm^{-3}	Remarks	Ref.
pyrite	FeS$_2$	0.666	3.36		Voi07
		−0.1···+1			Stu18, Gau43
marcasite	FeS$_2$	1.0···4.0			Stu18, Pov64
sphalerite	ZnS	−0.3···+40		contains up to 20% Fe	Voi07, Stu18, Gau43
		16.3	35	mean value from [Stu18]	
		0.4···7.6			Pov64
chalcopyrite	CuFeS$_2$	6···8		according to [Ter62] chalcopyrite is antiferromagnetic with T_N = 550 °C	Gau43
		7.6	32.2		Stu18
		850			Ter62
bornite	Cu$_5$FeS$_4$	9···14			Gau43, Pov64
tennantite — tetrahedrite	Cu$_3$AsS$_{3.25}$ — Cu$_3$SbS$_{3.25}$	3···13			Gau43
covellite	CuS	3.6···5.2			Gau43
arsenopyrite	FeAsS	0.5···2.3			Gau43
		3.3···8.0			Pov64
		39	237	high value due to ferrimagnetic impurities?	Stu18
enargite	Cu$_3$AsS$_4$	0···3			Gau43
galena	PbS	−0.1···+0.1			Gau43
gypsum	CaSO$_4$ · 2H$_2$O		40···263	high values due to ferrimagnetic impurities?	Cra02
		−1.0			Pov64

(continued)

Table 5 (continued)

Mineral	Chemical composition	χ_g 10^{-6} cm^3 g^{-1}	χ_v 10^{-6} cm^3 cm^{-3}	Remarks	Ref.
Carbonates					
siderite	FeCO$_3$	56···64			Pov64
		94···211			Fli59
		87	331		Stu18
magnesite	MgCO$_3$	1.5	4.5		Stu18
dolomite	CaMg[CO$_3$]$_2$	0.3	0.9		Stu18
rhodochrosite	MnCO$_3$	108	378		Stu18
ankerite	CaFe[CO$_3$]$_2$	8	24		Stu18
malachite	Cu$_2$[CO$_3$(OH)$_2$]	8···19			Pov64
		17.3	69.2		Elf65
Tungsten compounds					
wolframite	(Fe, Mn)WO$_4$	33···53			Stu18, Gre30, Fli59
ferberite	FeWO$_4$	42		exact composition: (Fe$_{0.89}$Mn$_{0.11}$)WO$_4$	Elf65
huebnerite	MnWO$_4$	43		exact composition: (Fe$_{0.03}$Mn$_{0.97}$)WO$_4$	Elf65

*) Susceptibility values obtained from magnetization curves at high magnetic fields after saturation of eventual ferrimagnetic impurities (see also Fig. 1).

6.1.4 Ferromagnetism, ferrimagnetism, antiferromagnetism — Ferromagnetismus, Ferrimagnetismus, Antiferromagnetismus

6.1.4.1 General introduction — Allgemeine Einleitung

Magnetic mineral phases in rocks which carry natural remanent magnetism are predominantly iron and iron-titanium oxides. Iron oxyhydroxides and iron sulfides are of significance but not very abundant.

The magnetic phases occur as small grains ranging in sizes from a fraction of a micron to several millimeters in diameter. For the most part, they are found in dilute dispersions in an effectively paramagnetic matrix of silicate minerals. Their total modal abundance typically varies between <1% to about 5%.

The reader is referred to the following texts for comprehensive reviews of the basic physical principles of magnetism, specifically rock magnetism, oxide mineralogy, and paleomagnetism.

Magnetische Mineralphasen mit natürlichem remanenten Magnetismus in Gesteinen sind hauptsächlich Eisen- und Eisen-Titan-Oxide. Eisenhydroxide und Eisensulfide sind auch von Bedeutung, aber nicht sehr häufig.

Die magnetischen Phasen kommen als kleine Körner vor, mit Durchmessern zwischen einem Bruchteil eines Mikrons und mehreren Millimetern. Meistens werden sie in geringer Konzentration in einer paramagnetischen Grundsubstanz von Silikatmineralen gefunden. Ihre totale modale Häufigkeit wechselt im allgemeinen zwischen <1% und etwa 5%.

Der Leser wird auf folgende Texte hingewiesen, in welchen physikalische Grundlagen des Magnetismus, insbesondere des Gesteinsmagnetismus, der Mineralogie der Oxide und des Paläomagnetismus behandelt werden.

Nicholls, G.D.: The mineralogy of rock magnetism. Phil. Mag. Suppl. Adv. Phys. **4** (1955) 113.

Nagata, T.: Rock magnetism. Maruzen Comp., Tokyo (2nd ed.) **1961**.

Stacey, F.D.: The physical theory of rock magnetism. Phil. Mag. Suppl. Adv. Phys. **12** (1963) 46.

Irving, E.: Paleomagnetism and its application to geological and geophysical problems. Wiley-Interscience, New York **1964**.

Fuller, M.D.: Geophysical aspects of paleomagnetism. Crit. Rev. Solid State Phys. **1** (1970) 137.

McElhinny, M.W.: Paleomagnetism and plate tectonics. Cambridge University Press **1973**.

Stacey, F.D. and S.K. Banerjee: The physical principles of rock magnetism. Elsevier Scientific Publishing Comp., Amsterdam **1974**.

Rumble, D. (Ed.): Oxide minerals. Min. Soc. Am. Short Course Notes **3** (1976).
O'Reilly, W.: Magnetic minerals in the crust of the earth. Rep. Prog. Phys. **39** (1976) 857.
Day, R.: TRM and its variation with grain size: A review. J. Geomag. Geoelectr. **29** (1977) 233.

6.1.4.2 The ternary system $FeO-Fe_2O_3-TiO_2$ — Das Dreistoffsystem $FeO-Fe_2O_3-TiO_2$

The chemical composition of the most important iron and iron-titanium oxides is given by the ternary diagram:

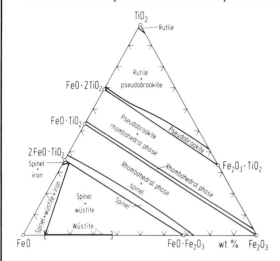

Fig. 11. Solid solutions in the ternary system $FeO-Fe_2O_3-TiO_2$ at $T=1300\,°C$ [Tay64].
The following solid solutions occur:
1) α-series: α-Fe_2O_3—$FeTiO_3$;
2) β-series: Fe_3O_4—Fe_2TiO_4;
3) γ-series: defect correlates of γ-Fe_2O_3;
4) ω-series: Fe_2TiO_5—$FeTi_2O_5$.

6.1.4.3 Ilmenite-hematite solid solution series — Ilmenite-Hämatit-Mischkristallreihe

6.1.4.3.1 Introductory remarks — Einleitende Bemerkungen

The rhombohedral ilmenite-hematite series, defined by the end-members ilmenite, $FeTiO_3$ and hematite α-Fe_2O_3 forms a complete solid solution series at high temperatures ($T>$ about 800 °C).

Pure ilmenite is a rather common mineral in a variety of rock types and essentially paramagnetic.

Hematite which shows a weak ferromagnetism at ambient temperature is a frequent carrier of the natural remanent magnetization in sediments. It is mainly found in specular grains but sometimes also present in the pigment. It may originate either directly from hematite bearing igneous rocks or is produced by chemical changes such as oxidation of magnetite or dehydration of iron oxyhydroxides. Precipitation from iron rich solutions is another possible mechanism of formation.

In igneous rocks, the primary composition of the ilmenite-hematite solid solution series strongly relates to the bulk chemistry of the rock. With decreasing total basicity the content of ilmenite is decreasing [Bud64, Car67, Pet73, Hag67].

Bulk chemistry	Acid	Intermediate	Basic
Rock	Granite Rhyolite	Dacite Trachyte	Gabbro Basalt
Ilmenite-Hematite $x\,FeTiO_3 \cdot (1-x)Fe_2O_3$ $0\leqq x\leqq 1$	$0.5\leqq x\leqq 0.85$	$0.5\leqq x\leqq 0.8$	$0.8\leqq x\leqq 1.0$

Subsolidus reactions, especially the deuteric oxidation of titanomagnetites favour the formation of very ilmenite rich compounds.

Major impurities in the ilmenite-hematite series of igneous rocks are Mg and Mn. Zn, Al, Cr, and V are commonly present in very minor concentrations.

The ilmenite-hematite solid solution series is also found in a wide variety of metamorphic rocks [Rum76]. Due to the limited miscibility at metamorphic temperatures exolution textures of mutual intergrowths of essentially end-member composition are typically observed.

The magnetic phase diagram of the ilmenite-hematite series (Fig. 20) has been found complex in detail, a particular interesting aspect being the self-reversal of thermoremanent magnetization in the compositional range of 50 to 70 mol-% ilmenite [Uye58, Nag61, Wes71a, Hof75].

6.1.4.3.2 Figures and tables — Abbildungen und Tabellen

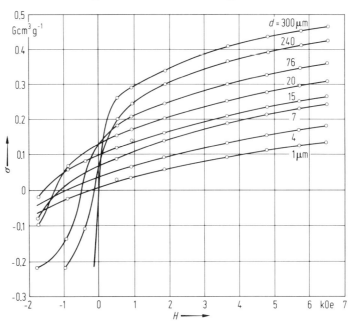

Fig. 12. α-Fe$_2$O$_3$. Specific magnetization, σ, vs. magnetic field, H, dependence upon grain size, d [Che43].

Table 6. Hematite, α-Fe$_2$O$_3$. General properties*). Room temperature unless otherwise specified.

Crystal type	Rhombohedral	Shi59
Structure type	Corundum	
Lattice constants	$a = 5.424$ Å, $\alpha = 55°17'$	Shi59
Molecular weight	(M) $= 159.70$ g mol^{-1}	
X-ray density	$\varrho_X = 5.277$ g cm^{-3}	

*) For further data see Table 5 of Vol. III/12b, p. 8.

Table 7. Hematite, α-Fe$_2$O$_3$. Magnetic properties. Room temperature unless otherwise specified.

Néel temperature	$T_N \simeq 950$ K	Shi59
Low temperature magnetic (Morin) transition	$T_M \simeq 263$ K (impurity sensitive, see e.g. Fig. 30 of Vol. III/4a, p. 11)	Bal74
Spin structure	See Fig. 20 of Vol. III/4a, p. 9	Shu51
Weak (parasitic) ferromagnetism between T_M and T_N	($\sigma \simeq 0.4$ Gauss cm^3 g^{-1})	Fla64
Specific magnetic susceptibility	$\chi_g \simeq 20 \cdot 10^{-6}$ cm^3 g^{-1} (see Fig. 14)	Fla64
Paramagnetic Curie temperature	$\Theta_p = -2940$ K	Gui51
Magnetostriction	$\lambda_{100} \simeq -20 \cdot 10^{-6}$	Miz66
	$\lambda_{111} \simeq 2 \cdot 10^{-6}$	Miz66, Urq56
	$\lambda_{112} \simeq 8 \cdot 10^{-6}$	Urq56
Mössbauer data	See 3.1.2 of Vol. III/12b and 1.1.2 of Vol. III/4a	

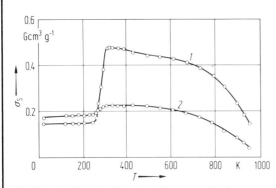

Fig. 13. α-Fe$_2$O$_3$. Specific saturation magnetization, σ_S, as a function of temperature, T, for single crystal hematite. 1, H parallel to a direction in the (111) plane; 2, H parallel to [111] direction [Nee52].

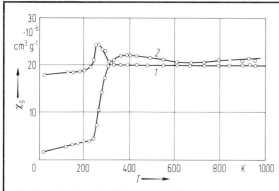

Fig. 14. α-Fe_2O_3. Specific susceptibility, χ_g, as a function of temperature, T, for single crystal. *1*, H parallel to a direction in the (111) plane; *2*, H parallel to [111] direction [Nee52].

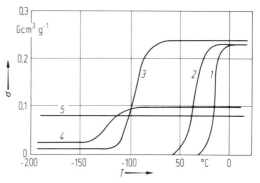

Fig. 15. α-Fe_2O_3. Specific magnetization, σ, as a function of temperature, T, for powders containing particles of size *1*, "bulk"; *2*, 1 micron; *3*, 1000 Å; *4*, 350 Å; *5*, 250 Å [Ban65].

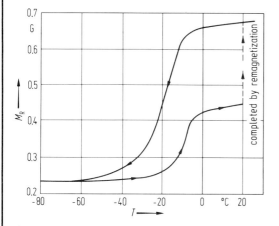

Fig. 16. α-Fe_2O_3. Thermal hysteresis of remanent magnetization, M_R, when the thermal cycle is made across the low temperature magnetic transition in zero magnetic field [Hai57].

Table 8. Ilmenite, $FeTiO_3$. General properties*). Room temperature unless otherwise specified.

Crystal type	Rhombohedral	Bar34
Structure type	$FeTiO_3$, ilmenite	Bar34
Lattice constants	$a = 5.538$ Å, $\alpha = 54°41'$	Nag61
Molecular weight	$(M) = 151.74$ g mol^{-1}	
X-ray density	$\varrho_X = 4.79$ g cm^{-3}	

*) For further data see Tab. 22 of Vol. III/4a, p. 29.

Table 9. Ilmenite, $FeTiO_3$. Magnetic properties. Room temperature unless otherwise specified.

Néel temperature	$T_N = 55\cdots68$ K	Biz56, Ish57
Spin structure	See Fig. 85 of Vol. III/4a, p. 30	Shi59
Magnetic susceptibility	See Fig. 17	
Paramagnetic Curie temperature	$\Theta_p = 23$ K	Ish57
Curie constant per mol	$C_m = 3.74$ cm^3 K mol^{-1}	Ish57
Mössbauer data	See 3.2 of Vol. III/12b and 1.2 of Vol. III/4a	

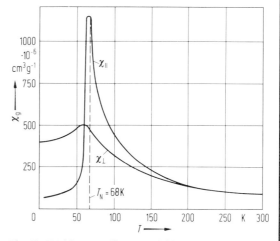

Fig. 17. $FeTiO_3$. Specific susceptibility, χ_g, as a function of temperature, T, with applied magnetic field parallel ($\chi_{\parallel}$) and perpendicular ($\chi_{\perp}$) to the ternary axis [Biz56].

Table 10. Ilmenite-hematite solid solution series, x FeTiO$_3 \cdot (1-x)$Fe$_2$O$_3$, $(0 \leq x \leq 1)$. Crystallographic and magnetic properties. Room temperature unless otherwise specified.

Structure type	Hematite structure, R$\bar{3}$c for Fe$_2$O$_3$-rich compounds	Ish62
	Ilmenite structure, R$\bar{3}$ for FeTiO$_3$-rich compounds	Ish62
Lattice constants	(see Fig. 19 [Nag56, Ish57, Ish58, Uye58, Lin65])	
Magnetic phases	$0 \leq x \lesssim 0.5$ antiferromagnetic with weak ferromagnetism	Ish62
	$0.5 \lesssim x \leq 0.8$ ferrimagnetic	
	$0.8 \leq x < 1.0$ superparamagnetic, ferromagnetic at low temperatures (see Fig. 20)	
Curie temperatures	$T_C[K] = 943 - 1143x$ for $0 \leq x \leq 0.8$	Lin66, Nag56
	$T_C[K] = 895 - 928x$ for $0.35 \leq x \leq 1.0$	Wes71,
Saturation moment at low temperature	[Boz57, Ish57, Ish58a, Ish62], see Fig. 21	see also Ish57, Uye58
Saturation magnetization	[Uye58, Wes71], see Fig. 22	
Initial susceptibility	see Table 11	
Paramagnetic Curie temperature	see Fig. 23	
Curie constant	see Fig. 24	
Mössbauer data	[Shi62, Shi62a]	

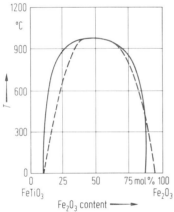

Fig. 18. FeTiO$_3$—Fe$_2$O$_3$ series. Solvus curve. Solid line [Uye58], broken line [Car61]. See also [Lin73].

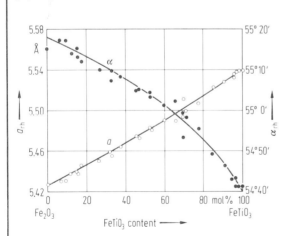

Fig. 19. FeTiO$_3$—Fe$_2$O$_3$ system. Lattice parameters a and α as a function of composition [Ish58].

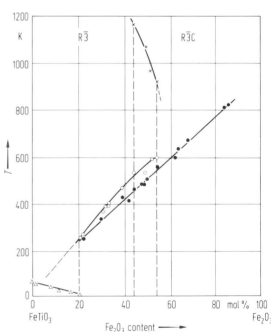

Fig. 20. FeTiO$_3$—Fe$_2$O$_3$ series. Magnetic phase diagram. Crosses: order-disorder transition temperatures; open circles: Curie temperatures of ordered specimens; full circles: Curie temperatures of disordered or partially ordered specimens; triangles: transition temperatures where the second nearest neighbour interaction operates in long range order [Ish62].

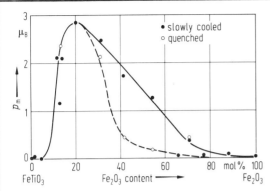

Fig. 21. FeTiO$_3$—Fe$_2$O$_3$ system. Magnetic moment per molecule, p_m, as a function of composition [Boz57].

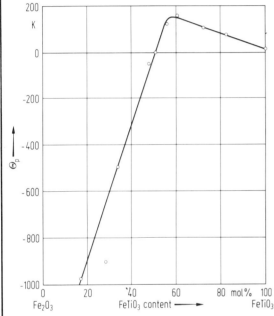

Fig. 23. FeTiO$_3$—Fe$_2$O$_3$ system. Variation of paramagnetic Curie temperature, Θ_p, with composition [Ish57].

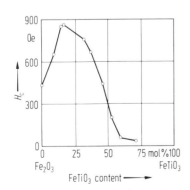

Fig. 25. FeTiO$_3$—Fe$_2$O$_3$ system. Variation of coercive force, H_C, with composition for a grain size of about 15 μm [Nag56].

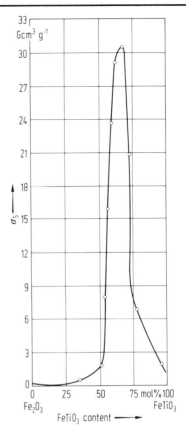

Fig. 22. FeTiO$_3$—Fe$_2$O$_3$ system. Variation of specific saturation magnetization, σ_S, at room temperature with composition for 9.8 μm diameter grains [Wes71].

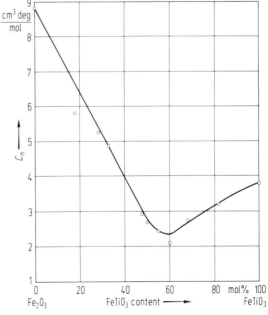

Fig. 24. FeTiO$_3$—Fe$_2$O$_3$ system. Variation of Curie constant per mol, C_m, with composition [Ish57].

Table 11. $x\,\mathrm{FeTiO_3 \cdot (1-x)Fe_2O_3}$ series. Variation of hysteresis magnetic properties, specific saturation magnetization, σ_S, initial mass susceptibility, χ_{gi}, specific saturation remanence, σ_{RS}, coercive force, H_C, and remanence coercivity, H_{RC}, with grain size, d, for various compositions [Wes71].

x	d μm	σ_S Gauss cm^3 g^{-1}	χ_{gi} 10^{-3} cm^3 g^{-1}	σ_{RS} Gauss cm^3 g^{-1}	H_C Oe	H_{RC} Oe	$\tau = H_{RC}/H_C$
0.78	26	6.7	1.5	0.051	38	390	11.0
	16	6.6	1.4	0.068	41	465	11.3
	9.8	6.8	1.3	0.061	47	430	9.7
	4.6	6.7	1.3	0.073	62	466	7.5
	1.5	6.1	1.2	0.141	99	780	7.9
0.73	26	21.0	28.0	0.84	22	63	2.9
	16	20.8	27.6	0.92	26	68	2.6
	9.8	20.7	26.4	1.05	28	71	2.5
	4.6	20.5	23.7	1.46	41	84	2.1
	1.5	18.1	18.6	2.33	65	114	1.8
0.68	26	29.8	19.2	3.4	112	224	2.0
	16	30.2	20.6	3.6	117	230	2.0
	9.8	30.5	16.8	4.0	134	244	1.8
	4.6	29.3	14.1	4.9	182	286	1.6
	1.5	25.3	13.4	6.0	267	390	1.5
0.63	26	30.6	10.0	3.5	215	520	2.4
	16	29.4	9.5	3.9	234	555	2.4
	9.8	29.2	8.1	4.3	295	610	2.1
	4.6	27.7	5.8	5.6	437	810	1.9
	1.5	25.3	4.3	6.6	680	1200	1.8
0.60	26	23.2	4.5	1.55	282	800	2.8
	16	23.0	3.7	1.74	312	1060	3.4
	9.8	23.6	3.9	2.00	410	1170	2.9
	4.6	21.8	3.9	2.62	610	1600	2.6
	1.5	19.2	2.7	2.43	860	2400	2.8
0.57	26	16.6	2.5	0.77	231	1000	4.3
	16	16.3	2.6	0.77	185	1040	5.6
	9.8	15.8	2.3	0.87	277	1300	4.7
	4.6	15.2	2.3	1.34	505	1950	3.9
	1.5	13.2	2.1	1.71	670	3300	4.9
0.54	26	8.3	1.2	0.25	198	2020	10.2
	16	8.2	2.2	0.23			
	9.8	8.0	1.0	0.29	225	2600	11.6
	4.6	7.2	1.0	0.36	276	3100	11.2
	1.5	6.0	1.1	0.50	375	4100	11.0
0.51	26	2.4	0.3	0.075	172	870	5.1
	16	1.9	0.5	0.071	108	830	7.1
	9.8	1.8	0.3	0.075	202	1290	6.4
	4.6	1.7	0.3	0.072	133	2000	15.1
	1.5	1.5	0.3	0.087	185	1820	9.8

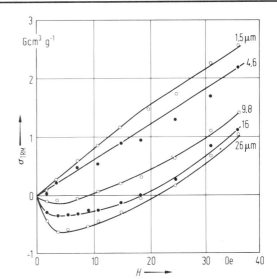

Fig. 26. $FeTiO_3—Fe_2O_3$ system $(Fe_{2-x}Ti_xO_3)$: Intensity of thermoremanent magnetization, σ_{TRM}, as function of applied field, H, for $x = 0.68$ using various grain sizes. Negative values indicate a reverse TRM [Wes71a].

6.1.4.4 Titanomagnetite solid solution series — Titanomagnetit-Mischkristallreihe

6.1.4.4.1 Introductory remarks — Einleitende Bemerkungen

Titanomagnetites, $Fe_{3-x}Ti_xO_4 \cong x\,Fe_2TiO_4 \cdot (1-x)Fe_3O_4$, are the most common magnetic minerals in igneous rocks. At high temperatures ($T \geqq 600\,°C$), there is a continuous solid solution ($0 \leqq x \leqq 1$) in the two-component series magnetite, Fe_3O_4 and ulvöspinel, Fe_2TiO_4 which upon cooling is restricted towards the end-members.

Ulvöspinel is a rare natural crystal in terrestrial rocks, almost always intergrown with magnetite. It is frequently observed in lunar samples [ElG76].

Magnetite occurs in a great variety of igneous, metamorphic, and sedimentary rock types, in certain meteorites, but not in lunar samples. Typically it is formed in various types of subsolidus reactions. As carrier of rock magnetism magnetite is the most abundant and important oxide mineral.

Among other factors, the primary composition as well as the modal abundance of titanomagnetites in igneous rocks depends on the initial bulk chemistry of the rock. As general trend, basic rocks tend to contain larger concentrations of oxides than intermediate or acid suites [Hag76]. In response to falling temperature titanomagnetites show a depletion in Fe_2TiO_4 content in more fractionated magmas leading to magnetite-rich compositions in acid and intermediate rocks, and to ulvöspinel-rich compositions in basic suites [Bud64, Car67, Pet73, Hag76].

Bulk chemistry	Acid	Intermediate	Basic
Rock	Granite Rhyolite	Dacite Trachyte	Gabbro Basalt
Titanomagnetite $x\,Fe_2TiO_4 \cdot (1-x)Fe_3O_4$ $0 \leqq x \leqq 1$	$0 \leqq x \leqq 0.2$	$0.2 \leqq x \leqq 0.4$	$0.4 \leqq x \leqq 0.85$

Titanomagnetites in basaltic lavas of continental margins and island arcs on the avarage contain $30 \cdots 45\%$ ulvöspinel, whereas $60 \cdots 80\%$ are typical for oceanic tholeiites [Car74].

The oxygen fugacity, a further fundamental parameter controlling the oxide abundance and composition of a crystallizing magma, essentially also determines their subsolidus alteration during cooling. Deuteric high temperature oxidation results in rather complex oxide assemblages [Hag76] predominantly found in continental igneous rocks. These processes tend to produce almost pure magnetite as the only major magnetic oxide mineral.

Subsequent oxidation of titanomagnetites at low temperature forming a cation deficient spinel phase (titanomaghemites) will be discussed separately.

Natural titanomagnetites always contain variable amounts of impurities like Al, Mg, Mn and less abundant Ni, Zn, Cr, and V. In basalts typically concentrations of $<1 \cdots 3$ wt% MgO, $<1 \cdots 4$ wt% Al_2O_3 and $<1 \cdots 2$ wt% MnO are observed [Hag76]. In basic lavas titanomagnetites generally have more Al_2O_3 ($3 \cdots 5$ wt%) and MgO ($1 \cdots 3$ wt%) than in tholeiitic lavas ($1 \cdots 2$ wt% Al_2O_3 and $0.5 \cdots 1.5$ wt% MgO) [Car74]. The average primary composition of titanomagnetites in the basaltic oceanic crust is [Pet79]:

$$Fe_{2.27}Ti_{0.58}Al_{0.07}Mg_{0.06}Mn_{0.02}O_4.$$

The amount of impurities critically affects the magnetic properties of titanomagnetites. In addition to the data given in this chapter, the reader is referred to LB Vol. III/4b and Vol. III/12b for magnetic and other properties of various solid solution systems including magnetite which are only of limited general interest in rock magnetism.

6.1.4.4.2 Figures and tables — Abbildungen und Tabellen

Table 12. Magnetite, Fe_3O_4. General properties*). Room temperature unless otherwise specified.

Crystal type	Inverse spinel-cubic above $T \simeq 119$ K	Shu51a
	Orthorhombic below $T \simeq 119$ K	Ver47
Structure type	$MgAl_2O_4$, spinel	
Lattice constant	$a = 8.397$ Å	Sha65
Molecular weight	$(M) = 231.6$ g mol^{-1}	
X-ray density	$\varrho_X = 5.238$ g cm^{-3}	

*) For further data see Table 1 of Vol. III/12b, p. 58.

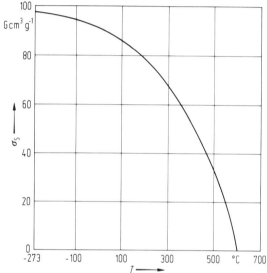

Fig. 27. Fe_3O_4. Specific saturation magnetization, σ_S, vs. temperature, T [Smi59].

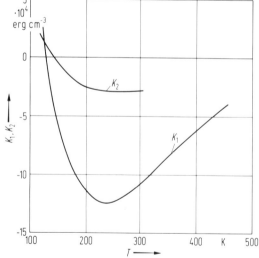

Fig. 28. Fe_3O_4. First- and second-order anisotropy constants, K_1, K_2, vs. temperature, T, as measured by the torque method [Bic57].

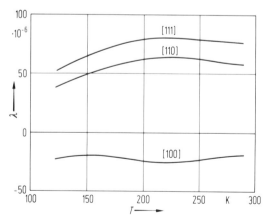

Fig. 29. Fe_3O_4. Magnetostriction constants, λ, as a function of temperature, T, for the three principal directions of a synthetic single crystal [Bic55]. See also [Dom50].

Table 13. Magnetite, Fe_3O_4. Magnetic properties. Room temperature unless otherwise specified.

Curie temperature	$T_C \simeq 850$ K	Sam69
Magnetic ordering temperature (Verwey transition)	$T_V \simeq 119$ K (impurity sensitive)	Ver47
Specific saturation magnetization	$\sigma_S \simeq 92$ Gauss cm^3 g^{-1} (see Fig. 27)	Gor55
Initial mass susceptibility	$\chi_{gi} \simeq 0.2$ cm^3 g^{-1}	Par65, Day76
Anisotropy constant, first order	$K_1 \simeq -1.10 \cdot 10^5$ erg cm^{-3}	Bic57
Anisotropy constant, second order	$K_2 \simeq -0.28 \cdot 10^5$ erg cm^{-3} (see Fig. 28)	Bic57
Magnetostriction constants, polycrystalline single crystal	$\lambda_S \simeq 35 \cdot 10^{-6}$	Bic55
	$\lambda_{100} \simeq -19.5 \cdot 10^{-6}$	Bic55
	$\lambda_{110} \simeq 57.1 \cdot 10^{-6}$ (see Fig. 29)	Bic55
	$\lambda_{111} \simeq 77.6 \cdot 10^{-6}$	Bic55
Mössbauer data	See 4.1.1 of Vol. III/12b and 6.1.1 of Vol. III/4b	

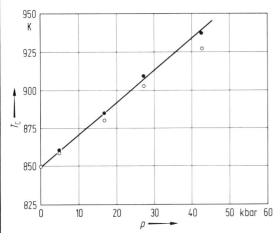

Fig. 30. Fe_3O_4. Curie temperature, T_C, vs. pressure, p, for a cylindrical powder sample of natural crystalline material. Open circles and closed circles are experimental data and data corrected for pressure effect on thermocouple emf, respectively. The slope of the curve, dT_C/dp, is 2.05 K/kbar [Sam69]. See also [Sch68, Sch70].

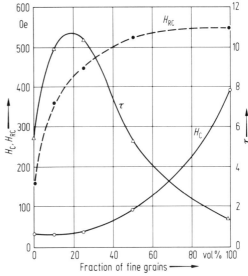

Fig. 32. Fe_3O_4. Variation of coercive force, H_C, remanence coercivity, H_{RC}, and $\tau = H_{RC}/H_C$ with volume fraction of fine grained material for mixtures of fine ($d \simeq 1$ μm) and coarse ($d \geq 150$ μm) magnetite particles [Day76].

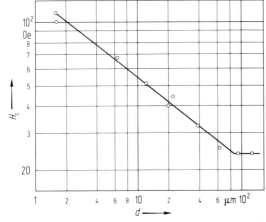

Fig. 31. Fe_3O_4. Coercive force, H_C, as a function of grain size, d, for dispersed magnetite particles [Par65].

Table 14. Ulvöspinel, Fe_2TiO_4. General properties*). Room temperature unless otherwise specified.

Crystal structure	Inverse spinel	Bar32
Cation distribution	$Fe^{2+}[Fe^{2+}Ti^{4+}]O_4$	Bar32
	See also Table 1 of Vol. III/12b, p. 712	
Lattice constant	$a = 8.534$ Å	Pou50
	$a = 8.536$ Å	Lin65, Aki67, Rea78
Molecular weight	$(M) = 223.60$ g mol^{-1}	
X-ray density	$\varrho_X = 4.777$ g cm^{-3}	

*) For further references see Table 1 of Vol. III/12b, p. 712 – 713.

Table 15. Ulvöspinel, Fe_2TiO_4. Magnetic properties. Room temperature unless otherwise specified.

Curie temperature	$T_C \simeq 120$ K	Nag61, Rea68, Sch71
	$T_C \simeq 115$ K	Rea78
Weak ferromagnetism	Between T_C and about $T \simeq 60$ K, maximum at ≈ 100 K	Ish67, Rea68, Rea78
	(paramagnetic at room temperature)	
Specific saturation magnetization at 4.2 K	$\sigma_S \simeq 0.3$ Gauss cm^3 g^{-1}	Ozi71, Rea78
Curie constant per mol	$C_m = 7.038$ cm^3 K mol^{-1}	Rea68
Anisotropy constant, first order at 77 K	$K_1 = 5.8 \cdot 10^5$ erg cm^{-3} *)	Ish71
Magnetostriction constants at 77 K	$\lambda_{100} = 4.7 \cdot 10^{-3}$*)	Ish71, Kat74
	$\lambda_{111} = 1.3 \cdot 10^{-3}$	
Mössbauer data	[Ros65, Ban67, Ono68]	

*) Single crystal with $Fe_{2.05}Ti_{0.95}O_4$ composition.

Table 16. Magnetite-ulvöspinel solid solution series, $x\,Fe_2TiO_4 \cdot (1-x)Fe_3O_4$, $(0 \leqq x \leqq 1)$. Crystallographic and magnetic properties. Room temperature unless otherwise specified.

Crystal type	Spinel cubic	
Cation distribution	theoretical: [Aki54, Che55, Nee55]	
	experimental: [Ish64, O'Re65a, Ste69, Ble71, Jen73, Ble76], (see Fig. 34)	
Lattice constants	a [Å] $= 8.389 + 0.146x$ (see Fig. 35)	Ble76, Aki57a, Lin62, Gho65, O'Re65, Syo65, Ble73, see also Uye58, Kaw59, Aki62, Lin65
Curie temperatures	T_C [K] $= 851 - 580x - 150x^2$	Lin66, Aki57a
	T_C [K] $= 856 - 567x - 186x^2$	Ble76, Aki57a, Uye58,
	(see Fig. 36)	Ozi70, Ble73, see also Kaw59, Aki62, Syo65, Gho66, Ozi71, Rea72, Ste72
Saturation moment at low temperature	[Aki57a, O'Re65a, Ste69, Ble71, Ble76], (see Fig. 38)	
Saturation magnetization	[Aki57a, Gho66, Ozi70] (see Fig. 39)	
Initial mass susceptibility	[Day76, Day77, Day77a], see Table 17	
Anisotropy constants	see Figs. 40, 41	
Magnetostriction constants	see Figs. 42, 43	
Mössbauer data	[Ban67]	

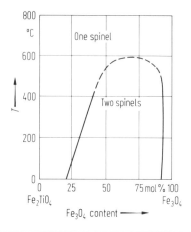

Fig. 33. $Fe_{3-x}Ti_xO_4$ system. Possible form of solvus curve [Bas60]. See also [Phi54, Kaw56, Vin57].

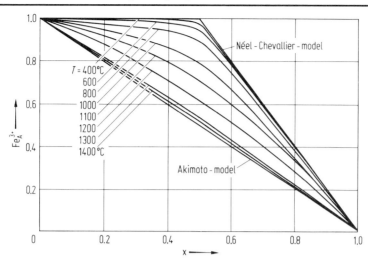

Fig. 34. $Fe_{3-x}TiO_4$ system. Cation distribution: Dependence on composition of the equilibrium number of Fe^{3+} ions on tetrahedral (A) sites per formula unit at different temperatures [Ble76]. See also [Ste69]. Theoretical models indicated are of [Aki54, Che55, Nee55].

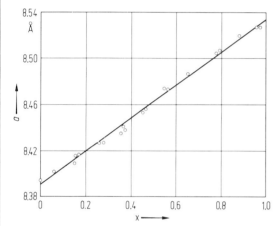

Fig. 35. $Fe_{3-x}Ti_xO_4$ system. Variation of lattice constant with composition [Aki57a].

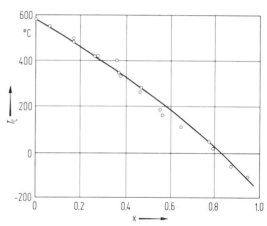

Fig. 36. $Fe_{3-x}Ti_xO_4$ system. Variation of Curie temperature, T_C, with composition [Aki57a].

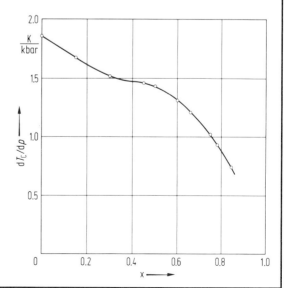

Fig. 37. $Fe_{3-x}Ti_xO_4$ system. Dependence of Curie temperature on pressure, dT_C/dp, vs. composition [Sch70].

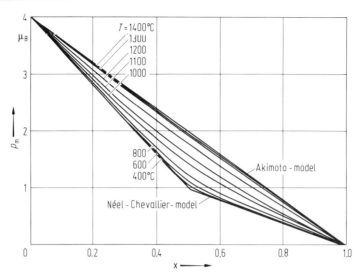

Fig. 38. $Fe_{3-x}Ti_xO_4$ system. Dependence on composition of the molecular saturation moment, p_m, for equilibrium cation distributions at different temperatures [Ble76], based on experimental data of [Aki57a, O'Re65a, Ble71]. See also [Ste69]. Theoretical models indicated are of [Aki54, Che55, Nee55].

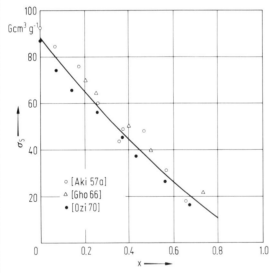

Fig. 39. $Fe_{3-x}Ti_xO_4$ system. Varation of specific saturation magnetization, σ_S, at room temperature with composition [Smi77].

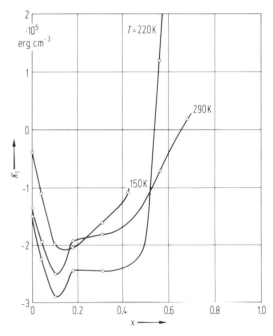

Fig. 40. $Fe_{3-x}Ti_xO_4$ system. Variation of first order magnetocrystalline anisotropy constant, K_1, with composition for different temperatures [Syo65].

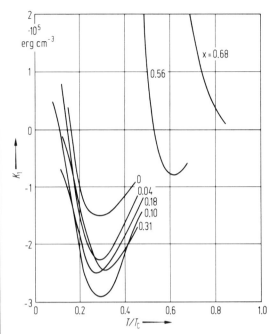

Fig. 41. $Fe_{3-x}Ti_xO_4$ system. Variation of first order magnetocrystalline anisotropy constant, K_1, with temperature reduced by Curie temperature, T/T_C, for various compositions [Syo65].

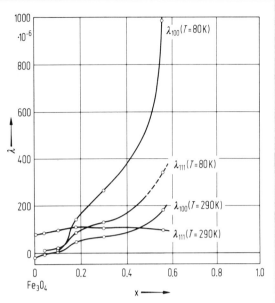

Fig. 42. $Fe_{3-x}Ti_xO_4$ system. Variation of magnetostriction constants, λ, with composition for different temperatures [Syo65].

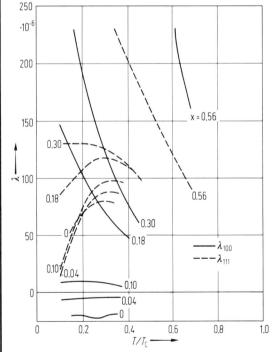

Fig. 43. $Fe_{3-x}Ti_xO_4$ system. Variation of magnetostriction constants, λ, with temperature reduced by Curie temperature, T/T_C, for various compositions [Syo65].

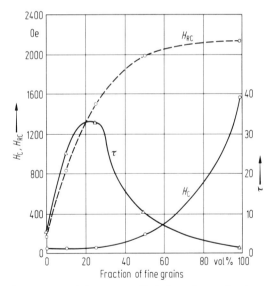

Fig. 44. $Fe_{3-x}Ti_xO_4$ system. Variation of coercive force, H_C, remanence coercivity, H_{RC}, and $\tau = H_{RC}/H_C$ with volume fraction of fine grained material for mixtures of fine $(d \simeq 1\,\mu m)$ and coarse $(d \geq 150\,\mu m)$ titanomagnetite particles of composition x = 0.6 [Day76].

Table 17. Titanomagnetites $Fe_{3-x}Ti_xO_4$. Variation of hysteresis magnetic properties, coercive force, H_C, remanence coercivity, H_{RC}, ratio of specific saturation remanence to specific saturation magnetization, and specific initial susceptibility, χ_{gi}, with grain size, d, for various compositions [Day77].

d μm	Standard deviation	H_C Oe	H_{RC} Oe	$\tau = H_{RC}/H_C$	σ_{RS}/σ_S	χ_{gi} $cm^3 g^{-1}$
x = 0:						
0.8		295	551	1.87	0.133	0.139
0.96		245	501	2.04	0.117	0.144
1.9	1.1	165	432	2.62	0.103	0.159
3.8	1.5	108	359	3.32	0.082	0.118
6.4	2.4	84	300	3.57	0.056	0.224
9.3	2.9	70	280	4.00	0.047	0.224
13.4	4.2	60	250	4.16	0.043	0.192
18.2	5.2	55	209	3.80	0.025	0.218
30	9.0	42	195	4.64	0.021	0.218
57	17.0	37	181	4.89	0.020	0.216
87	13.5	35	169	4.83	0.020	0.203
131	25.1	32	160	5.00	0.019	0.212
x = 0.2:						
0.8		811	998	1.23	0.361	0.083
1.0		620	940	1.52	0.327	0.098
1.8	1.0	360	660	1.83	0.261	0.138
2.9	1.2	240	500	2.08	0.188	0.197
6.1	2.5	125	325	2.60	0.096	0.242
9.8	2.6	92	278	3.02	0.068	0.263
11.1	3.3	85	269	3.16	0.055	0.262
16.5	4.6	66	245	3.71	0.040	0.274
23	7.1	53	225	4.25	0.031	0.291
40	12.0	43	210	4.88	0.021	0.306
75	21.0	40	198	4.95	0.024	0.317
118	19.0	38	195	5.13	0.024	0.301
x = 0.4:						
0.8		1551	1900	1.23	0.449	0.036
0.95		1076	1600	1.49	0.427	0.054
1.64	0.9	594	1030	1.73	0.362	0.088
3.5	1.8	275	580	2.11	0.250	0.137
6.0	2.6	152	385	2.53	0.146	0.182
8.6	2.9	112	320	2.86	0.095	0.208
13.0	3.2	82	270	3.29	0.079	0.224
17.0	5.1	66	235	3.56	0.053	0.236
32	8.7	44	210	4.77	0.034	0.265
61	18	39	185	4.74	0.030	0.279
84	16	36	176	4.89	0.027	0.294
112	21	34	186	4.94	0.033	0.271
x = 0.6:						
0.8		1584	2130	1.34	0.473	0.018
0.94		1115	1815	1.63	0.511	0.030

(continued)

Table 17 (continued)

d µm	Standard deviation	H_C Oe	H_{RC} Oe	$\tau = H_{RC}/H_C$	σ_{RS}/σ_S	χ_{gi} cm^3 g^{-1}
1.7	1.0	581	1115	1.94	0.459	0.048
3.3	1.3	280	625	2.23	0.339	0.074
6.4	2.4	132	330	2.50	0.210	0.105
9.0	3.1	92	260	2.83	0.131	0.118
12.0	3.7	82	235	2.82	0.117	0.176
16.0	5.4	73	201	2.75	0.075	0.217
25.5	7.9	61	185	3.03	0.063	0.237
41	11.2	52	175	3.37	0.058	0.258
78	19	46	160	3.48	0.051	0.266
140	24	39	148	3.79	0.043	0.268

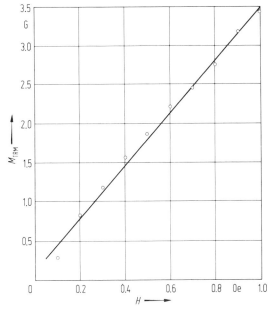

Fig. 45. Fe$_{3-x}$Ti$_x$O$_4$ system. Acquisition of thermoremanent magnetization, M_{TRM}, in fields up to 1 Oe for fine grained ($d \simeq 10^3$ Å) titanomagnetite of x = 0.4 composition [Özd78].

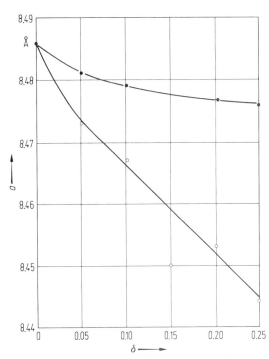

Fig. 46. Fe$_{3-x}$Ti$_x$O$_4$ system with substitutions. Variation of lattice constant, a, with degree of substitution, δ, in the series Fe$_{2.4-\delta}$M$_\delta$Ti$_{0.6}$O$_4$ (M = Mg, solid circles; M = Al, open circles) [Ric73]. See also [Jos75, O'Do77].

For Fig. 47, see next page.

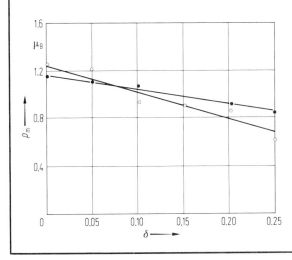

◀

Fig. 48. Fe$_{3-x}$Ti$_x$O$_4$ system with substitutions. Variation of saturation moment, p_m, with degree of substitution, δ, in the series Fe$_{2.4-\delta}$M$_\delta$Ti$_{0.6}$O$_4$ (M = Mg, solid circles; M = Al, open circles) [Ric73]. See also [Jos75].

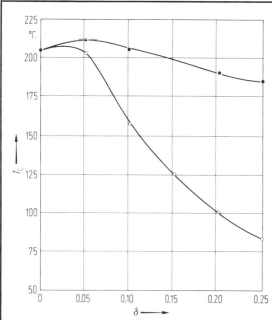

Fig. 47. $Fe_{3-x}Ti_xO_4$ system with substitutions. Variation of Curie temperature, T_C, with degree of substitution, δ, in the series $Fe_{2.4-\delta}M_\delta Ti_{0.6}O_4$ (M = Mg, solid circles; M = Al, open circles) [Ric73]. See also [Jos75, O'Do77].

Table 18. Hysteresis magnetic properties, coercive force, H_C, specific saturation magnetization, σ_S, and specific saturation remanence σ_{RS} for coarse grained ($d \simeq 40\,\mu m$) titanomagnetites $Fe_{2.4-\delta}Al_\delta Ti_{0.6}O_4$ dispersed in a non-magnetic matrix [O'Do77].

δ	H_C Oe	σ_S Gauss cm^3 g^{-1}	σ_{RS} Gauss cm^3 g^{-1}
0.05	40	27.3	3.28
0.1	30	23.5	2.35
0.15	23	17.8	1.60
0.2	22	14.5	1.58
0.25	14	10.4	0.97

Table 19. Hysteresis magnetic properties, specific saturation magnetization, σ_S, specific saturation remanence, σ_{RS}, coercive force, H_C, and remanence coercivity, H_{RC}, for fine grained ($d \simeq 10^3\,\text{Å}$), Al substituted titanomagnetites with $Fe_{3-x-\delta}Ti_xAl_\delta O_4$ compositions dispersed in a non-magnetic matrix [Özd78].

x	δ	σ_S Gauss cm^3 $\cdot$ g^{-1}	σ_{RS} Gauss cm^3 $\cdot$ g^{-1}	H_C Oe	H_{RC} Oe
0.4	0	40.7	22.1	1480	1840
	0.1	34.3	18.7	1660	2040
	0.2	30.9	17.3	1850	2230
0.6	0	22.2	12.8	1790	2270
	0.1	17.1	9.53	1610	2030
	0.2	12.7	6.45	1090	1340

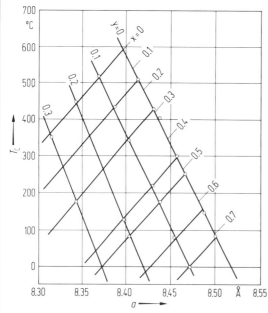

Fig. 49. $Fe_{3-x}Ti_xO_4$ system with substitutions. Variation of lattice constant, a, and Curie temperature, T_C, with composition for titanomagnetite-spinel solid solutions.

$(1-y)(x\,Fe_2TiO_4 \cdot (1-x)Fe_3O_4) \cdot y\,MgAl_2O_4$ [Nis77].

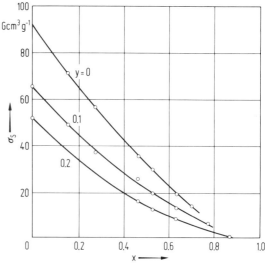

Fig. 50. $Fe_{3-x}Ti_xO_4$ system with substitutions. Variation of specific saturation magnetization, σ_S, at room temperature with composition for titanomagnetite-spinel solid solutions,

$(1-y)(x\,Fe_2TiO_4 \cdot (1-x)Fe_3O_4) \cdot y\,MgAl_2O_4$ [Nis77].

6.1.4.5 Maghemite and titanomaghemites — Maghemit und Titanomaghemite

6.1.4.5.1 Introductory remarks — Einleitende Bemerkungen

Maghemite, γ-Fe_2O_3 has a cubic defect structure, $Fe^{3+}_{8/3}\square_{1/3}O_4$, and is mainly formed by low temperature ($T \leq 250\,°C$) oxidation of magnetite. A complete solid solution $Fe_3O_4 - \gamma$-Fe_2O_3 exists. A member of this series is produced by partial oxidation of Fe_3O_4. The degree of oxidation is usually defined by an oxidation parameter z ($0 \leq z \leq 1$):

$$Fe^{2+} + z/2\,O \;\rightarrow\; (1-z)Fe^{2+} + z\,Fe^{3+} + z/2\,O^{2-}$$

although there is convincing evidence that in nature the process is effectively accomplished by progressive migration of iron out of the original lattice.

At temperatures $T \geq 350\,°C$ maghemite irriversibly converts to hematite. Among other factors the actual transition temperature critically depends on grain size and amount of impurities. It is substantially lowered by increasing hydrostatic pressure (about 3.5° per kbar) [Kus60].

Synthetic maghemites were extensively studied in solid state physics because of their use in magnetic tapes. Occurences of natural maghemite have been reported from certain marine sediments, soils, and intensively altered igneous rocks, specifically laterites. γ-Fe_2O_3 is thought to be responsible for the strong magnetization of 'lodestones' [Nag61].

Titanomaghemites are the non-stoichiometric correlates of the titanomagnetite solid solution series retaining a spinel structure with variable amounts of vacancies in cation positions upon oxidation.

At high temperatures the degree of non-stoichiometry is very limited ($z < 0.1$) [Hau74]. The intrinsic properties of titanomaghemites have been studied using materials synthesized at low temperature. It is now well documented that a continuous increase in Curie temperature and a continuous decrease in lattice constant is related to a progressive degree of oxidation. Although at present considerable discrepancies still exist between the data of different authors, measurements of the saturation magnetization for low initial titanium contents ($x < 0.5$) show a continuous reduction in magnetic intensity with increasing degree of oxidation. For ulvöspinel-rich compounds ($x \geq 0.5$) this trend obviously is not continuous but reverses towards higher magnetization values at an advanced stage of oxidation. This result is frequently interpreted as indicating a potential magnetic self-reversal effect. However, these ideas were not substantiated by studies on natural samples.

Titanomaghemites are the major magnetic constituents in the basaltic oceanic basement. They have also been found in continental igneous rocks and may in fact be much more abundant here than generally anticipated. The presence of water seems to play a decisive rôle in their formation.

6.1.4.5.2 Figures and tables — Abbildungen und Tabellen

Table 20. Maghemite, γ-Fe_2O_3. General properties*).
Room temperature unless otherwise specified.

Crystal type	Inverse spinel cubic	Häg35
	possible tetragonal superstructure	Oos58
Structure type	Al_2O_3 spinel	
Structure formula	$Fe^{3+}_{21.33}\square_{2.67}O^{2-}_{32}$, $\square_{2.67}=2.67$ vacancies (octahedral sites)**)	Häg35, Fer58
Lattice constant	cubic $a = 8.339$ Å	Häg35, for further references, see [Lin76]
Molecular weight	$(M) = 159.70$ g mol^{-1} (Fe_2O_3)	
X-ray density	$\varrho_X = 4.907$ g cm^{-3} ($Fe_{21.33}O_{32}$)	

*) For further data see Table 10 of Vol. III/12b, p. 21
**) See also [Kul67, Web71].

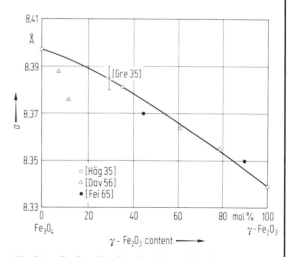

Fig. 51. γ-Fe_2O_3—Fe_3O_4. Variation of lattice constant, a, as function of composition [Lin76]. The data of [Häg35, Fei65] are for samples oxidized at relatively low temperatures ($T < 500\,°C$), the data of [Gre35, Dav56] are for specimens quenched from high temperatures ($T > 1200\,°C$).

Table 21. Maghemit, γ-Fe_2O_3. Magnetic properties. Room temperature unless otherwise specified.

Curie temperature	$T_C \simeq 1020$ K (extrapolated)	Ban70
	$T_C \simeq 1020$ K (calculated, see Fig. 52)	Bro62
Specific saturation magnetization	$\sigma_S = 85$ Gauss $cm^3\,g^{-1}$	Sta74
	$\sigma_S = 75$ Gauss $cm^3\,g^{-1}$ (see Fig. 53)	Ber68
Anisotropy constant, first order	$K_1 \simeq -1.1 \cdot 10^5$ erg cm^{-3}	Sta74
	$K_1 \simeq -2.5 \cdot 10^5$ erg cm^{-3}	Val62
Mössbauer data	See Table 11 of Vol. III/12b, p. 22	

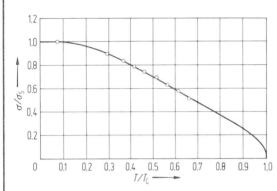

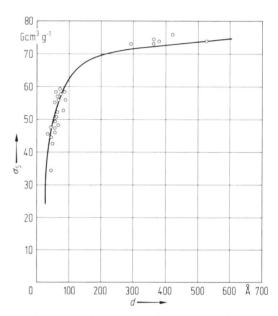

Fig. 52. γ-Fe_2O_3. Reduced magnetization, σ/σ_S, as a function of reduced temperature, T/T_C. Solid curve, theory based on Néel's molecular field model with neglect of A−A and B−B interactions; circles, experimental points with an assumed Curie temperature of $T_C = 1020$ K [Bro62].

Fig. 53. γ-Fe_2O_3. Specific saturation magnetization, σ_S, vs. crystallite size, d, of acicular particles; open circles are measured values, solid line is a theoretical curve [Ber68].

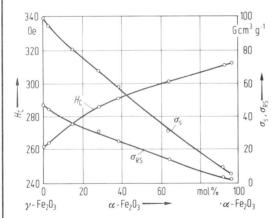

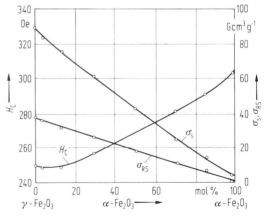

Fig. 54. γ-Fe_2O_3—α-Fe_2O_3. Coercive force, H_C, specific saturation magnetization, σ_S, and remanent magnetization, σ_{RS}, vs. composition for acicular particles [Ima68].

Fig. 55. γ-Fe_2O_3—α-Fe_2O_3. Coercive force, H_C, specific saturation magnetization, σ_S, and remanent magnetization, σ_{RS}, vs. composition for granular particles [Ima68].

Table 22. Titanomaghemites, $Fe_{(1-x)R}Ti_{xR}\square_{3(1-R)}O_4$,
$R = 8/[8 + z(1+x)]$; $(0 \le x \le 1,\ 0 \le z \le 1)$. Crystallographic and magnetic properties.

Crystal structure	Spinel cubic
Cation distribution	[Ver62, O'Re66, O'Re67, Zel67, Zel67a, Rea71, Ste72, O'Do75, O'Do78]
Lattice constant	Synthetic material: [Aki57a, O'Re71, Ozi71, Jos75, O'Do77a, Nis79], (see Figs. 56, 59)
	Natural samples: [Aki59, Bas59, Bas60, Zel65, Mar72, Ozi74, Pet79], (see Fig. 60)
Curie temperature	Synthetic material: [Aki57a, O'Re71, Ozi71, Jos75, O'Do77, Nis79], (see Figs. 57, 59)
	Natural samples: [Aki59, Mar72, Ozi74, Pet79], (see Figs. 58, 60)
Saturation magnetization	Synthetic material: [Aki57a, Rea71, Ozi71, O'Do77], (see Figs. 61···65)
	Natural samples: [Ozi74, Pet79], (see Fig. 66)

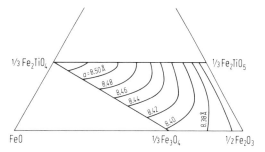

Fig. 56. Titanomaghemites. Variation of spinel structure lattice constant, a, for cation deficient oxidized titanomagnetites [O'Re71].

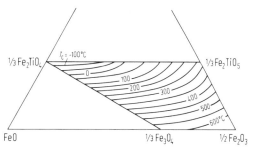

Fig. 57. Titanomaghemites. Variation of Curie temperature, T_C, for cation deficient oxidized titanomagnetites [O'Re71].

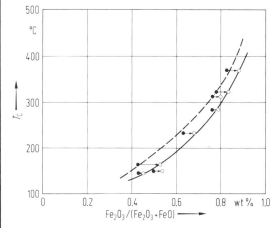

Fig. 58. Titanomaghemites. Curie temperature, T_C, vs. oxidation ratio $Fe_2O_3/Fe_2O_3 + FeO$ for natural oxidized titanomagnetites ($x \simeq 0.6$) extracted from oceanic basalts of the Deep Sea Drilling Project. Dashed curve (full circles) uncorrected, solid curve (open circles) corrected for silicate impurities [Pet79].

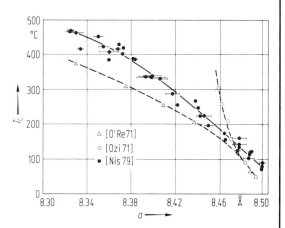

Fig. 59. Titanomaghemites. Variation of lattice constant, a, and Curie temperature, T_C, with increasing degree of oxidation for $x = 0.7$ titanomagnetite. Results of synthetic material: [O'Re71, Ozi71, Nis79].

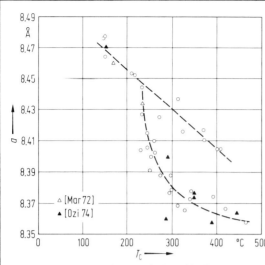

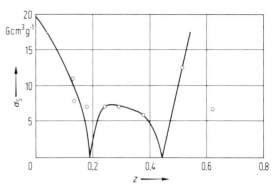

Fig. 61. Titanomaghemites. Variation of specific saturation magnetization, σ_S, measured at $T = 4.2\,\text{K}$ with degree of oxidation, z, for synthetic $x = 0.7$ titanomagnetite [Ozi71]. For $z \geqq 0.3$ the samples are not single phase.

Fig. 60. Titanomaghemites. Variation of lattice constant, a, and Curie temperature, T_C, with increasing degree of oxidation for natural ($x \simeq 0.6$) titanomagnetites extracted from oceanic basalts of various drill sites of the Deep Sea Drilling Projekt [Pet79]. Broken lines indicate two different trends. The differences are not yet understood.

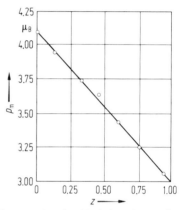

Fig. 63. Titanomaghemites. Variation of saturation moment. p_m, measured at 4.2 K with degree of oxidation, z, for synthetic $x = 0$ titanomagnetite [Rea71].

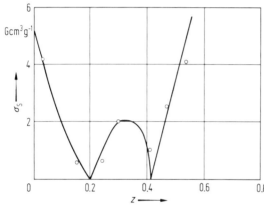

Fig. 62. Titanomaghemites. Variation of specific saturation magnetization, σ_S, measured at $T = 4.2\,\text{K}$ with degree of oxidation, z, for synthetic $x = 0.9$ titanomagnetite [Ozi71]. For $z \geqq 0.3$ the samples are not single phase.

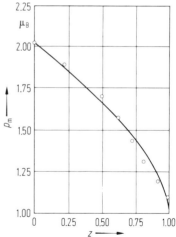

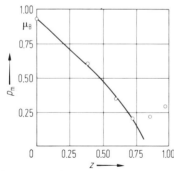

Fig. 65. Titanomaghemites. Variation of saturation moment, p_m, measured at 4.2 K with degree of oxidation, z, for synthetic $x = 0.7$ titanomagnetite [Rea71].

Fig. 64. Titanomaghemites. Variation of saturation moment, p_m, measured at 4.2 K with degree of oxidation, z, for synthetic $x = 0.4$ titanomagnetite [Rea71].

Table 23. Variation of hysteresis magnetic properties, coercive force, H_C, specific saturation magnetization, σ_S, and specific saturation remanence, σ_{RS}, with degree of oxidation, z, for Mg substituted titanomagnetites, $Fe_{2.4-\delta}Ti_{0.6}Mg_{\delta}O_4$. Particle size $d \simeq 10^3$ Å [O'Do77a].

δ	z	H_C Oe	σ_S Gauss cm^3 $\cdot$ g^{-1}	σ_{RS} Gauss cm^3 $\cdot$ g^{-1}
0.05	0.16	2000	22.6	12.9
	0.36	2020	20.3	11.6
	0.59	1890	19.9	9.93
	0.75	720	13.8	7.61
	0.89	290	11.7	6.55
0.15	0.20	1760	19.1	9.74
	0.42	1700	16.8	9.21
	0.57	1560	15.8	8.70
	0.78	645	13.8	6.62
	0.90	260	12.7	6.22
0.25	0.12	1350	17.6	8.96
	0.35	1400	18.5	9.64
	0.62	930	14.7	7.64
	0.74	620	14.4	7.22
	0.95	165		
0.35	0.12	1060	17.6	8.63
	0.30	1170	16.3	8.17
	0.51	955	13.2	7.27
	0.81	270	14.2	6.68
	0.97	155	13.2	6.33

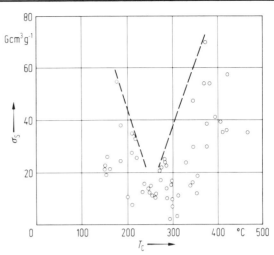

Fig. 66. Titanomaghemites. Variation of specific saturation magnetization, σ_S, measured at room temperature, with Curie temperature, T_C, i.e. with increasing degree of oxidation for natural titanomagnetites in oceanic basalts from various drill sites of the Deep Sea Drilling Project [Pet79]. Broken lines indicate approximate upper limits according to available experimental data.

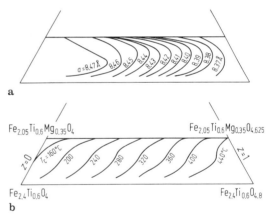

Fig. 67. Titanomaghemtites with substitutions. Variation of a) spinel structure lattice constant, a, and b) Curie temperature, T_C, with degree of oxidation, z, for titanomagnetites containing up to 0.35 Mg^{2+} per formula unit [O'Do77a]. For Al substituted titanomaghemites, see [Jos75].

6.1.4.6 Pseudobrookite solid solution series — Pseudobrookit-Mischkristallreihe

6.1.4.6.1 Introductory remarks — Einleitende Bemerkungen

The pseudobrookite series is defined by the end-members pseudobrookite, Fe_2TiO_5 and ferropseudobrookite, $FeTi_2O_5$. At room temperature it is entirely paramagnetic [Aki57, Che58] and thus not directly relevant to rock magnetism, but important to define the oxidation paragenesis of oxide mineral assemblages [Hag76].

Natural minerals occur in igneous and metamorphic rocks and are a quaternary solid solution containing in minor concentrations karrooite, $MgTi_2O_5$, and tielite, Al_2TiO_5.

Armalcolite, $Fe_{0.5}Mg_{0.5}Ti_2O_5$, has been found in lunar rocks [Lin72, Wec76].

6.1.4.6.2 Tables — Tabellen

Table 24. Pseudobrookite, Fe_2TiO_5. Crystallographic and magnetic properties. Room temperature unless otherwise specified.

Structure	Orthorhombic	
Lattice constants	$a=9.767$ Å, $b=9.947$ Å,	Aki57
	$c=3.717$ Å	
Specific magnetic susceptibility	$\chi_g=43\cdot10^{-6}$ cm^3 g^{-1}	Aki57

Table 25. Ferropseudobrookite, $FeTi_2O_5$. Crystallographic and magnetic properties. Room temperature unless otherwise specified.

Structure	Orthorhombic	
Lattice constants	$a=9.798$ Å, $b=10.041$ Å,	Aki57
	$c=3.741$ Å	
Specific magnetic susceptibility	$\chi_g=53\cdot10^{-6}$ cm^3 g^{-1}	Aki57
Mössbauer data	[Shi62b]	

Table 26. Lattice parameters and specific magnetic susceptibility at room temperature in the pseudo-brookite, $x\,Fe_2TiO_5\cdot(1-x)FeTi_2O_5$ series [Aki57].

x	a Å	b Å	c Å	χ_g $\cdot10^{-6}$ cm^3 g^{-1}
18	9.801	10.031	3.745	48
34	9.800	10.011	3.741	58
50	9.795	9.987	3.739	
74	9.787	9.965	3.734	55

6.1.4.7 Wuestite — Wüstit

6.1.4.7.1 Introductory remarks — Einleitende Bemerkungen

Wuestite, FeO, is an exotic mineral in terrestrial rocks [Wal60]. It has been observed in the melted rims of meteorites [Ram76].

At room temperature wuestite is paramagnetic.

6.1.4.7.2 Tables — Tabellen

Table 27. General properties of wuestite*), Fe_xO. Room temperature unless otherwise specified.

Crystal type	Sodium chloride cubic above $T\simeq198$ K Rhombohedral ($\alpha<60°$) below $T\simeq198$ K	Rot58
Lattice constant	$a=4.332$ Å. See Fig. 3 of Vol. III/12b, p. 3	Hen70
Molecular weight	(M) $=71.85$ g mol^{-1}	
X-ray density	$\varrho_X=5.87$ g cm^{-3}	

*) Wuestite is metastable below $T=843$ K (see Fig. 12 of Vol. III/12b, p. 6) and non-stoichiometric (iron deficient) at low pressure. For further data see Table 1 of Vol. III/12b, p. 2.

Table 28. Magnetic properties of wuestite, Fe_xO. Room temperature unless otherwise specified.

Néel temperature	$T_N=198$ K See also Table 4 of Vol. III/12b, p. 4	Rot58
Spin structure	See Fig. 1 of Vol. III/4a, p. 4	Rot60
Magnetic susceptibility	$\chi_m=8\cdot10^{-3}$ cm^3 mol^{-1} (x<0.9)	Koc67
Paramagnetic Curie temperature	$\Theta_p=-10$ K See also Table 3 of Vol. III/12b, p. 4	Mic70
Curie constant per mole	$C_m=3.30$ cm^3 K mol^{-1} See also Table 3 of Vol. III/12b, p. 4	Mic70
Mössbauer data	See 3.1.1 of Vol. III/12b and 1.1.1 of Vol. III/4a	

6.1.4.8 TiO$_2$ polymorphs — Polymorphe TiO$_2$-Verbindungen

6.1.4.8.1 Introductory remarks — Einleitende Bemerkungen

The three natural polymorphs of TiO$_2$ are rutile, anastase, and brookite. A fourth form has been synthesized at high pressure. They are all paramagnetic at room temperature.

Rutile is generally produced in subsolidus reactions and particular common in metamorphic rocks. Anastase and brookite tend to form in hydrothermal environments and are metastable with respect to rutile.

6.1.4.8.2 Tables — Tabellen

Table 29. Crystallographic and magnetic properties of rutile. Room temperature unless otherwise specified.

Structure	Tetragonal	
Lattice constants	$a = 4.59373$ Å,	Str61
	$c = 2.95812$ Å	
Molar magnetic susceptibility	$\chi_m = 5.4$ $\cdot 10^{-6}$ cm^3 mol^{-1} *)	Sen60

*) Independent of temperature for $T > 55$ K.

Table 30. Crystallographic and magnetic properties of anastase. Room temperature unless otherwise specified.

Structure	Tetragonal	
Lattice constants	$a = 3.785$ Å, $c = 9.514$ Å	Cro55
Molar magnetic susceptibility	$\chi_m = 1.6$ $\cdot 10^{-6}$ cm^3 mol^{-1} *)	Sen60

*) Temperature independent. Experimental temperature dependent part attributed to impurities.

Table 31. Crystallographic and magnetic properties of brookite. Room temperature unless otherwise specified.

Structure	Orthorhombic	
Lattice constants	$a = 9.174$ Å, $b = 5.456$ Å,	Swa64
	$c = 5.138$ Å	
Molar magnetic susceptibility	$\chi_m = 8.0$ $\cdot 10^{-6}$ cm^3 mol^{-1} *)	Sen68

*) Temperature independent. Experimental temperature dependent part attributed to impurities.

6.1.4.9 Natural oxides with spinel structure — Natürliche Oxide mit Spinellstruktur

6.1.4.9.1 Introductory remarks — Einleitende Bemerkungen

Natural magnetites and titanomagnetites are generally impure solid solutions incorporating variable amounts of

chromite	$FeCr_2O_4$
picrochromite	$MgCr_2O_4$
hercynite	$FeAl_2O_4$
spinel	$MgAl_2O_4$
magnesio-ferrite	$MgFe_2O_4$
magnesian-titanite	$TiMg_2O_4$
trevorite	$NiFe_2O_4$
franklinite	$ZnFe_2O_4$
gahnite	$ZnAl_2O_4$
jacobsite	$MnFe_2O_4$
coulsonite	FeV_2O_4

All these oxides have a spinel structure and form extensive mutual and multicomponent solid solutions. They are rarely present in substantial amounts as individual crystals in the mineral assemblage carrying the natural remanent magnetism of common rocks, but have significant influence on crystallographic and magnetic properties of magnetites and titanomagnetites even in minor concentrations.

Most of the oxides listed are diamagnetic or paramagnetic at ambient temperature. Only solid solutions containing trevorite, jacobsite and/or magnesio-ferrite as predominant component are of direct interest for rock magnetism. In a variety of special rock types, namely ores, these were found to be the effective magnetic compounds.

6.1.4.9.2 Figures and tables — Abbildungen und Tabellen

Table 32. Crystallographic and magnetic properties of natural oxides with spinel structure. Room temperature unless otherwise specified.

Mineral	a Å	T_C, T_N K	χ	σ_S Gauss cm^3 g^{-1}
Chromite $FeCr_2O_4$	8.378 [Gil76]	$T_C = 50 \cdots 80$ [Sch76]	$\chi_g \simeq 43 \cdot 10^{-6}$ cm^3 g^{-1} [Lot56]	
Picrochromite $MgCr_2O_4$	8.333 [Gri77]	$T_N = 15$ [Gri77]	$\chi_g \simeq 28 \cdot 10^{-6}$ cm^3 g^{-1} [Bla63]	
Hercynite $FeAl_2O_4$	8.10 [Haf61]	$T_N \simeq 8$ [Sla64]	$\chi_m \simeq 3 \cdot 10^{-3}$ cm^3 mol^{-1} [Sla64]	
Spinel $MgAl_2O_4$	8.083 [Bai71]		$\chi_m = \; \simeq 69.7 \cdot 10^{-6}$ cm^3 mol^{-1} [Ani66]	
Magnesio-ferrite $MgFe_2O_4$	8.375 $\cdots$ 8.397 depending on cation distribution [All66]	$T_C = 690-490\,\eta$ η-fraction of Mg^{2+} ions on tetrahedral sites [Eps58]		$\approx 26 \cdots \approx 45$ depending on cation distribution [Kri56]
Magnesian-titanite Mg_2TiO_4	8.444 [Poi65]		$\chi_m = -48 \cdot 10^{-6}$ cm^3 mol^{-1} [Poi62]	
Trevorite $NiFe_2O_4$	8.337 [Gor54]	$T_C = 858$ [Gor54]		50.4 [Puc71]
Franklinite $ZnFe_2O_4$	8.440 [Rom53]	$T_N \simeq 15$ [Lot66]	$\chi_m \simeq 13 \cdot 10^{-3}$ cm^3 mol^{-1} [Lot66]	
Gahnite $ZnAl_2O_4$	8.088 [Coo72]		$\chi_m = -74.6 \cdot 10^{-6}$ cm^3 mol^{-1} [Ani66]	
Jacobsite $MnFe_2O_4$	8.515 [Tre65]	$T_C = 560$ [Mig61]		63 [Smo67]
Coulsonite FeV_2O_4	8.543 [Reu69]	$T_C = 109$ [Rog63]	$\chi_m \simeq 9 \cdot 10^{-3}$ cm^3 mol^{-1} [Ple71]	

For Fig. 68, see next page.

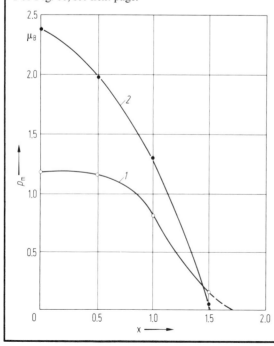

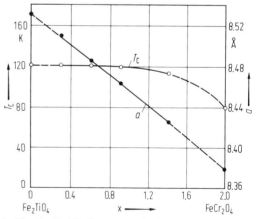

Fig. 70. $(Fe, Cr, Ti)_3O_4$ system. Curie temperature, T_C, and lattice constant, a, vs. composition for $Fe_{2-x/2}Cr_xTi_{1-x/2}O_4$ ($0 \leq x \leq 2$) [Sch71].

◄

Fig. 69. $MgFe_{2-x}Cr_xO_4$ system. Magnetic moment per molecule, p_m, at 0 K vs. composition. 1, annealed samples. 2, quenched samples [McG60].

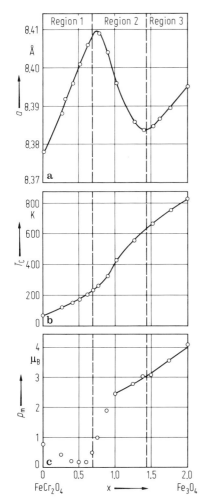

Fig. 68. $Fe_{1+x}Cr_{2-x}O_4$ system.
a) Lattice constant, a,
b) Curie temperature, T_C,
c) magnetic moment per molecule, p_m, $(H = \infty,\ T = 1.5\ K)$, vs. composition, x.

Open circles are data points, the solid line in c) is calculated using 2.28 μ_B per Cr^{3+} and ionic distribution derived from Mössbauer measurements [Rob71].

Table 33. Variation of lattice parameter, a, and Curie temperature, T_C, in substituted magnetite $Fe_2O_3 \cdot MeO$ by divalent ions. [Nag61].

Me	Co		Ni		Mg		Cu	
MeO mol%	T_C °C	a Å	T_C °C	a Å	T_C °C	a Å	T_C °C	a Å
0	570	8.38	570	8.38	570	8.38	570	
10							570	
20					555			
25	555	8.375						
30							560	
40			575	8.36			550	
50	540	8.37						
60			580	8.35	530	8.37	420	
70							445	
75	530	8.36			490			
85							460	
100	520	8.355	595	8.34	315	8.37	480	

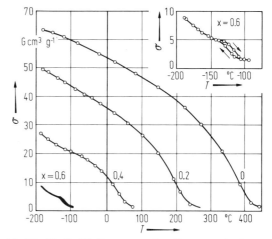

Fig. 72. $(Fe, Cr, Ti)_3O_4$ system. Specific magnetization, σ, vs. temperature, T, for series

$$Fe_{2.4-x}Cr_{0.6}Ti_xO_4 \quad (0 \leq x < 0.7)\ [Sch71].$$

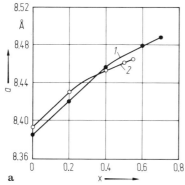

Fig. 71. $(Fe, Cr, Ti)_3O_4$ system. a) Lattice constant, a, b) Curie temperature, T_C, vs. composition for series:

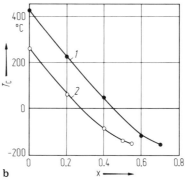

1, $Fe_{2.4-x}Cr_{0.6}Ti_xO_4$, $0 \leq x \leq 0.7$;
2, $Fe_{2.1-x}Cr_{0.9}Ti_xO_4$, $0 \leq x \leq 0.55$ [Sch71].

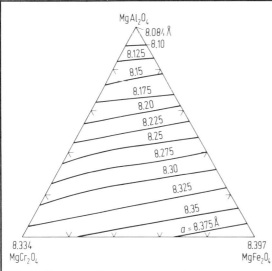

Fig. 73. Variation of lattice constant, a, in the ternary system $MgAl_2O_4 - MgCr_2O_4 - MgFe_2O_4$. All samples were quenched from $T = 1600\,°C$ [All66a].

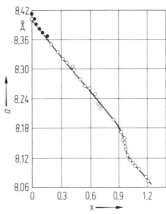

Fig. 75. $Al_xFe_{3-x}O_4$ system. Lattice constant, a, vs. composition [Hof56].

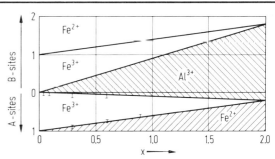

Fig. 74. $Al_xFe_{3-x}O_4$ system. Cation distribution vs. composition [Dek75].

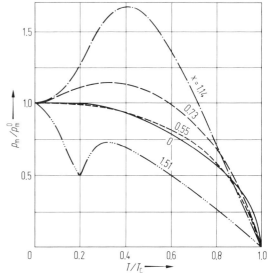

Fig. 77. $Al_xFe_{3-x}O_4$ system. Magnetic moment per molecule reduced by extrapolated moment at $0\,K$, p_m/p_m^0, as function of reduced temperature, T/T_C [Pic59].

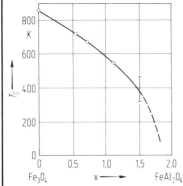

Fig. 76. $Al_xFe_{3-x}O_4$ system. Curie temperature, T_C, vs. composition [Pic59].

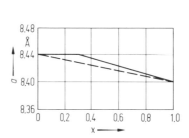

Fig. 78. $Zn_{1-x}Fe_{2+x}O_4$ system. Lattice constant, a, vs. composition. Dashed line calculated from Vegard's law [Pop71].

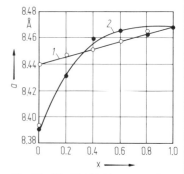

Fig. 79. $(Zn, Ti, Fe)_3O_4$ system. Lattice constant, a, vs. composition. Curve 1 $(Zn_2TiO_4)_x(ZnFe_2O_4)_{1-x}$; curve 2 $(Zn_2TiO_4)_x(Fe_3O_4)_{1-x}$. [Shc69a].

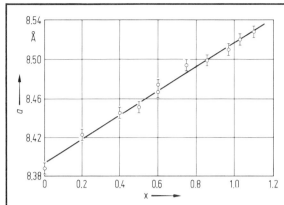

Fig. 80. $Mn_xFe_{3-x}O_4$ system. Lattice constant, a, vs. composition [Fun59].

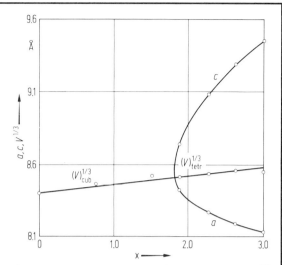

Fig. 81. $Mn_xFe_{3-x}O_4$ system. Lattice parameters, a, c, $V^{1/3}$, vs. composition [Fin57].

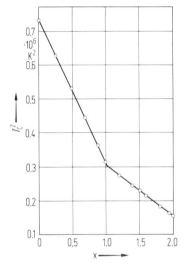

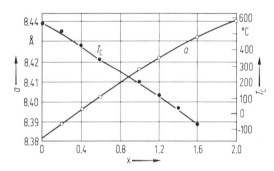

Fig. 83. $V_xFe_{3-x}O_4$ system. Lattice constant, a, and Curie temperature, T_C, vs. composition [Len57].

Fig. 82. $Mn_xFe_{3-x}O_4$. The variation of the Curie temperature with composition in the range $0 \le x \le 2$. In the range $0 \le x \le 1$ the curve exactly fits the theoretically derived relation

$$\left(\frac{T_C}{10^3}\right)^2 = 0.735 - 0.377\,x - 0.042\,x^2.$$

In the range $1.1 \le x \le 2.0$ it is found that the appropriate relation is

$$\left(\frac{T_C}{10^3}\right)^2 = 0.459 - 0.154\,x. \text{ [Esc60]}.$$

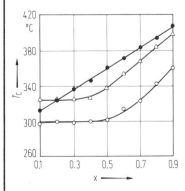

◀

Fig. 84. $Mg_xMn_{1-x}Fe_2O_4$ system. Curie temperature, T_C, vs. composition. Full circles: rapidly cooled samples, triangles: slowly cooled samples, open circles: samples treated in oxygen [Rez74].

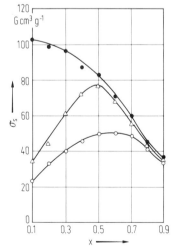

Fig. 85. $Mg_xMn_{1-x}Fe_2O_4$ system. Saturation magnetization, σ_S, (0 K), vs. composition. Full circles: rapidly cooled samples, triangles: slowly cooled samples, open circles: samples treated in oxygen [Rez74].

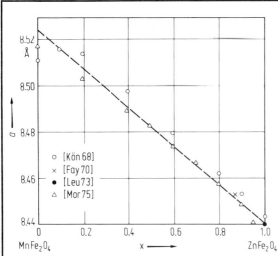

Fig. 86. $Mn_{1-x}Zn_xFe_2O_4$ system. Lattice constant, a, vs. composition.

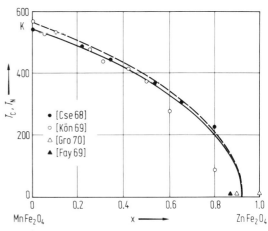

Fig. 87. $Mn_{1-x}Zn_xFe_2O_4$ system. Curie temperature, T_C, and Néel temperature, T_N, vs. composition. Solid and dashed curves are calculated (see original paper [Kön72]).

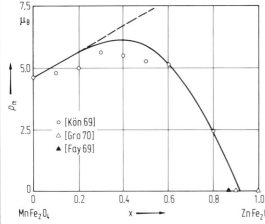

Fig. 88. $Mn_{1-x}Zn_xFe_2O_4$ system. Magnetic moment per molecule at 0 K, p_m, vs. composition. Solid curve: calculated (see original paper [Kön72]), dashed line: Néel theorie.

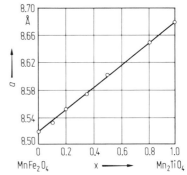

Fig. 89. $(Mn_2TiO_4)_x(MnFe_2O_4)_{1-x}$ system. Lattice constant, a, vs. composition [Shc69].

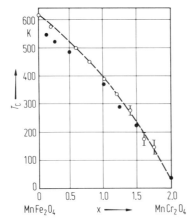

Fig. 90. $MnFe_{2-x}Cr_xO_4$ system. Curie temperature, T_C, vs. composition [Dmi73]. Full circles: data from [Gor54].

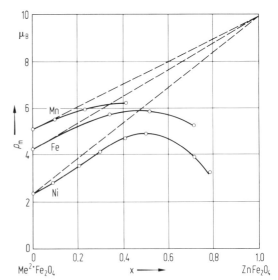

Fig. 91. $(MeFe_2O_4)_{1-x}(ZnFe_2O_4)_x$ systems. Magnetic moment per molecule at 0 K, p_m, vs. composition. Broken lines: theoretical relationship (see original paper [Wen52]).

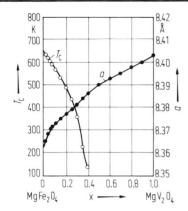

Fig. 92. MgFe$_{2-2x}$V$_{2x}$O$_4$ system. Lattice constant, a, and Curie temperature, T_C, vs. composition [Tel67].

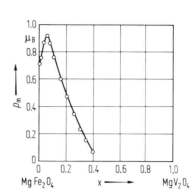

Fig. 93. MgFe$_{2-2x}$V$_{2x}$O$_4$ system. Magnetic moment per molecule at 0 K, p_m, vs. composition [Tel67].

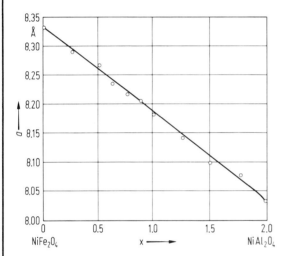

Fig. 94. NiFe$_{2-x}$Al$_x$O$_4$ system. Lattice constant, a, vs. composition [Max53].

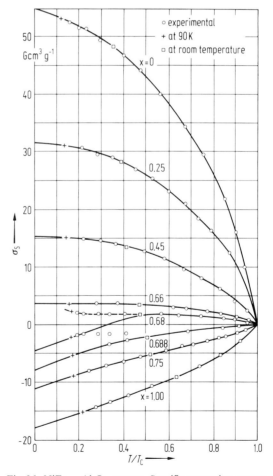

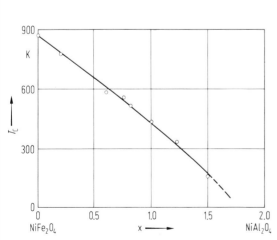

Fig. 95. NiFe$_{2-x}$Al$_x$O$_4$ system. Curie temperature, T_C, vs. composition [Max53].

Fig. 96. NiFe$_{2-x}$Al$_x$O$_4$ system. Specific saturation magnetization, σ_s, vs. reduced temperature, T/T_C, [Bla62].

6.1.4.10 Iron oxyhydroxides — Eisenhydroxide

6.1.4.10.1 Introductory remarks — Einleitende Bemerkungen

There are two forms of naturally occuring iron oxyhydroxides, goethite, α-FeOOH, and much less abundant, lepidocrocite, γ-FeOOH. They dehydrite to hematite, α-Fe$_2$O$_3$ and maghemite, γ-Fe$_2$O$_3$, respectively, at $T = 100 \cdots 300\,°C$. Additionally two synthetic forms are known.

6.1.4.10.2 Figures and tables — Abbildungen und Tabellen

Table 34. Crystallographic and magnetic properties of goethite. Room temperature unless otherwise specified.

Crystal type	Orthorhombic	Gol35
Lattice constants	$a = 4.602\,\text{Å}$, $b = 9.992\,\text{Å}$, $c = 3.021\,\text{Å}$	Sam69a
Néel temperature	$T_{\mathrm{N}} \simeq 385\,\text{K}$	Hed71
Spin structure	[For68]	
Specific magnetic susceptibility	$\chi_{\mathrm{g}} = 42 \cdots 49$ $\cdot 10^{-6}\,\text{cm}^3\,\text{g}^{-1}$ (synthetic material)	Alb29, Hof46
	$\chi_{\mathrm{g}} = 24 \cdots 38$ $\cdot 10^{-6}\,\text{cm}^3\,\text{g}^{-1}$ (natural crystals)	Str68, Hed71
Weak ferro-magnetism	See Fig. 98	
Specific saturation magnetization	$\sigma_{\mathrm{S}} = 0.01 \cdots 1$ Gauss cm^3 g^{-1} *)	Hed71
Mössbauer data	[Tak65, Hry65, Her66, Van66]	

*) Depending on Al content.

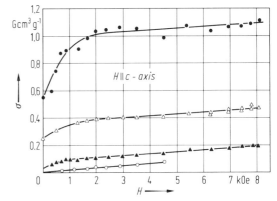

Fig. 97. α-FeOOH. Specific susceptibility, χ_{g}, vs. temperature, T, for a natural single crystal [Hed71].

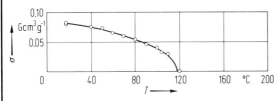

Fig. 98. α-FeOOH. Weak ferromagnetism; specific magnetization, σ, vs. temperature, T, for a natural single crystal parallel to the [001] axis [Hed71].

Fig. 99. α-FeOOH. Specific magnetization, σ, vs. field, H, for various natural goethites, increasing with higher Al concentrations [Hed71].

Table 35. Crystallographic and magnetic properties of lepidocrocite. Room temperature unless otherwise specified.

Crystal type	Orthorhombic	Gol35
Lattice parameters	$a = 3.87\,\text{Å}$, $b = 12.51\,\text{Å}$, $c = 3.06\,\text{Å}$	Her66
Specific magnetic susceptibility	$\chi_{\mathrm{g}} = 32 \cdots 42$ $\cdot 10^{-6}\,\text{cm}^3\,\text{g}^{-1}$	Alb29, Hof46, Str68
Mössbauer data	[Tak65]	

6.1.4.11 Iron-Sulfides — Eisensulfide

6.1.4.11.1 Introductory remarks — Einleitende Bemerkungen

Next to the Fe—Ti oxide system the iron-sulfides are of second importance to rock magnetism. Similarly to the Fe—Ti oxides the iron sulfides are characterized by complicated mineralogical and magnetic properties, which are not fully understood yet.

6.1.4.11.2 Pyrrhotite — Pyrrhotit

From the system $FeS - FeS_2$ it is mainly pyrrhotite $Fe_{1-x}S$ $(0 < x \leqq 0.182)$ which is of interest to rock magnetism [War70, Sch72] (Fig. 100, Table 36). Two types of magnetic pyrrhotites can be distinguished:

(a) Monoclinic "4c" pyrrhotite Fe_7S_8 which can also be written $Fe_{0.875}\square_{0.125}S$, is strongly ferrimagnetic [War70, Sch72] (Fig. 102, Table 36). It has variable stoichiometry and is therefore only nominally Fe_7S_8. It is separated from hexagonal NA- and NC-type pyrrhotites by narrow solvi [Cra74] (Fig. 100, Table 36). The NiAs-based structure consists of a hexagonal close-packed S-lattice with cations in octahedral interstices. The crystal parameters a and b are twice the dimensions of the ortho-hexagonal double-cell of NiAs, while the c-axis is four times larger (4c-pyrrhotite) [Wue74]. The vacant sites are confined to alternate layers of octahedral sites in the NiAs-configuration and form an ordered structure [Ber53, Tok72] (Fig. 103). The ordered arrangement of the vacancies has two effects: (1) the hexagonal crystallographic symmetry is slightly distorted to form a monoclinic structure; and (2) due to the inequality in numbers of ions on the antiferromagnetically coupled sublattices a net ferrimagnetic moment is present (Table 36) [Wei29, Har41, Ber53, Sch72]. The easy directions of magnetization at room temperature lie in the basal plane and exhibit triaxial anisotropy [Wei05] (Fig. 104). Pyrrhotite behaves paramagnetic along the [001]-direction and ferrimagnetic in the (001)-plane [Ben55]. Consequently the strong-field magnetization of polycrystalline material is considered:

$$\sigma(H, T) = 2/3\,\sigma_S\,(\infty, T) + 1/3\,\chi_{c\text{-axis}}(T) \cdot H \quad [\text{Ben55}].$$

The ferrimagnetism disappears at the "β-transition" $(T \approx 310\,°C)$ (Figs. 102, 106, 107) which appears to correspond both to a randomization of vacancy distribution and the Curie temperature [Hir54, Des65]. Above the β-transition the mineral behaves paramagnetically [Sch72].

There exists a second form of Fe_7S_8-pyrrhotite (3c-pyrrhotite) which is antiferromagnetic [Fle71].

(b) Low temperature hexagonal pyrrhotite. Between its melting temperature of 1190 °C and 308 °C, the full width of the pyrrhotite phase field is occupied by a single solid solution, hexagonal $Fe_{1-x}S$ $(0 < x < 0.185)$, in which iron and vacancies are randomly distributed in the cation sites of the NiAs (1c) structure [Cra74] (Fig. 100). This disordered 1c-pyrrhotite cannot be quenched. With decreasing temperature ordering of the vacancies takes place giving rise to various superstructures. These are the 5c, 11c and 6c types found at room temperature and which are stoichiometric phases with compositions

$$Fe_{n-1}S_n \,(n = 10, 11 \text{ and } 12, \text{ respectively}) \quad [\text{Nak71, Cra74, Kis74}].$$

Between the upper stability limit of these stoichiometric types (probably below 100 °C [Cra74]) and 308 °C there exist three other phases, named NA, NC and MC types [Nak71], which also possess subcells of the NiAs structure but have vacancy-supercells of uncertain symmetry [Cra74].

Compositions $Fe_{1-x}S$ with $0 \leqq x \leqq 0.077$ (i.e. Fe-content $\geqq 48$ at %) are antiferromagnetic [Sch72] (Figs. 109a and b). They are characterized by the "α-transition" (T_α), which corresponds to a structural change [And67, Kis74] (Figs. 100 and 102), but also appears to be related to a spin-flop $(T_{\alpha,s})$ with spins parallel to the c-axis for $T \leqq T_{\alpha,s}$, and perpendicular to the c-axis for $T_{\alpha,s} < T < T_\beta$ [Sch72]. However, disagreement exists as to whether the spin-flop occurs at or slightly below T_γ [Sch72] (Fig. 109a) or above T_α (at about 180 °C for FeS) [Tak73, Mol76] (Fig. 108).

The intermediate region between $Fe_{0.923}S$ and $Fe_{0.898}S$ (Fe-content between 48.0 and 47.3 at %) is characterized by increasing susceptibility and the appearance of the "γ-transition" (T_γ) [Sch72] or "anti-Curie point" [Har41, Lot56a] (Figs. 109b and c). At T_γ a peak in susceptibility is observed (Fig. 109b and c). Also the γ-transition appears to correspond to structural change (Fig. 100). The peak in susceptibility may therefore be interpreted in terms of transition from one superstructure to another [Nak71].

6.1.4.11.3 Figures and tables — Abbildungen und Tabellen

Table 36. Minerals and phases of the system FeS—FeS$_2$.

Mineral	Composition	Thermal stability max °C	Thermal stability min °C	Structure	Magnetic properties	Remarks	Ref.
troilite	FeS	140		hexagonal	antiferromagnetic $T_\alpha \approx 150\,°C$ $T_\beta = T_N \approx 330\,°C$ $\chi_g(20\,°C) = 1 \cdot 10^{-5}$ cm^3 g^{-1}	occurs as natural mineral only in meteorites and moon rocks	Hir54, Sch72, Cra74
mackinavite	FeS$_{1-x}$, $0.04 < x < 0.07$, corresponding to 51.8 at % Fe	?		tetragonal	?	occurs in deep sea black mud, hydrothermal deposits, also in moon rocks	Arn67, Cra74
hexagonal pyrrhotite	Fe$_{1-x}$S, $0 < x < 0.185$, 50.0…44.9 at % Fe	1190	≈ 100	hexagonal NiAs-1c-structure	partly antiferromagnetic, partly ferrimagnetic $T_N \approx 330\,°C$	with decreasing temperature ordering of Fe-vacancies gives rise to various superstructures NA, NC, MC-type	Har37, Ber53, Nak71, Sch72, Cra74
6c-pyrrhotite	Fe$_{11}$S$_{12}$, Fe$_{0.917}$S, $x = 0.083$, 47.8 at % Fe	≈ 100		orthorhombic?	antiferromagnetic with weak ferrimagnetism superimposed $\chi_g(20\,°C) \approx 2 \cdot 10^{-5}$ cm^3 g^{-1}	distinct increase of susceptibility above 200 °C: γ-transformation	Har37, Har41, Har41a, Ber53, Sch72, Cra74
11c-pyrrhotite	Fe$_{10}$S$_{11}$, Fe$_{0.909}$S, $x = 0.091$, 47.6 at % Fe	≈ 100		orthorhombic	antiferromagnetic with weak ferrimagnetism superimposed $\chi_g(20\,°C) = 3 \cdot 10^{-5}$ cm^3 g^{-1}	distinct increase of susceptibility above 200 °C: γ-transformation	Har37, Har41, Har41a, Ber53, Sch72, Cra74
5c-pyrrhotite	Fe$_9$S$_{10}$, Fe$_{0.900}$S, $x = 0.100$, 47.4 at % Fe	≈ 100		hexagonal	antiferromagnetic with weak ferrimagnetism superimposed $\chi_g(20\,°C) = 300 \cdot 10^{-5}$ cm^3 g^{-1}	distinct increase of susceptibility above 200 °C: γ-transformation	Har37, Har41, Har41a, Ber53, Sch72, Cra74

(continued)

Table 36 (continued)

Mineral	Composition	Thermal stability max °C	Thermal stability min °C	Structure	Magnetic properties	Remarks	Ref.
metastable pyrrhotite	Fe$_{1-x}$S, 0.03 < x < 0.06, 49.2...48.5 at % Fe	metastable		hexagonal?	antiferromagnetic?	"metastable pyrrhotite" is encountered when iron-rich pyrrhotites are quenched into the two-phase field of troilite+pyrrhotite (Fig. 100)	Cra74
4c-monoclinic pyrrhotite	Fe$_7$S$_8$, Fe$_{0.875}$S, x=0.125, 46.7 at % Fe, interval of non-stoichiometry: Fe$_{7\pm y}$S$_8$, 0 ≤ y ≤ 0.185?, Fe$_{0.852}$S...Fe$_{0.898}$S, 46.0...47.3 at % Fe	254		monoclinic	ferrimagnetic σ_s(20 °C) ≈ 17...20 Gauss cm^3 g^{-1} T_C ≈ 315 °C T_β ≈ 335 °C	most important magnetic mineral of the system FeS—FeS$_2$. The interval of nonstoichiometry is possibly narrower than indicated here (compare Fig. 101)	Wei29, Har37, Har41, Har41a, Ber53, Lot56a, Sch72, Car74
anomalous pyrrhotite	Fe$_{7.075}$S$_8$, Fe$_{0.866}$S, x=0.134, 46.4 at % Fe	?		triclinic?	antiferromagnetic	widespread in low-temperature sedimentary environments	Cla66, Cra74
smythite	Fe$_9$S$_{11}$, Fe$_{0.818}$S, x=0.182, 45.0 at % Fe	≈ 75		pseudorhombohedral, possibly another ordered pyrrhotite	?	occurs as exsolution lamellae in monoclinic pyrrhotite. All natural samples contain 0.4...7.5 wt % Ni	Ben72, Cra74

(continued)

Table 36 (continued)

Mineral	Composition	Thermal Stability max °C	min °C	Structure	Magnetic properties	Remarks	Ref.
greigite	Fe_3S_4, 42.9 at % Fe	metastable?		spinel $a = 9.880$ Å	ferrimagnetic $\sigma_s(20\,°C) = 20$ Gauss cm^3 g^{-1} $T_C = 307\,°C$ [Uda67, Uda68] $\sigma_s(20\,°C) = 34$ Gauss cm^3 g^{-1} $T_C = 300 \ldots 340\,°C$ [Spe72]	found in low temperature environments like lake sediments	Ski64, Uda67, Uda68, Del72, Spe72, Cra74
γ-Fe-sulfide	Fe_2S_3, $Fe_{2.667}S_4$, 40.0 at % Fe	?		spinel, similar to greigite	ferrimagnetic, magnetic characteristics similar to those of greigite	γ-Fe$_2$S$_3$ bears the same relation to greigite as γ-Fe$_2$O$_3$ does to magnetite. It has not yet been encountered as natural mineral	Yam73, Yam73a, Cra74
pyrite	FeS_2, 33.3 at % Fe	743		cubic	diamagnetic with weak temperature independent paramagnetism $\chi_g = 0.4 \ldots$ $1.0 \cdot 10^{-6}$ cm^3 g^{-1}	at temperatures above $T = 743\,°C$ pyrite breaks down to hexagonal pyrrhotite + sulfur	Voi07, Ben55, Kul59, Cra74
marcasite	FeS_2, 33.3 at % Fe, slightly S-deficient?	metastable		orthorhombic	diamagnetic with weak temperature independent paramagnetism $\chi_g = 1 \ldots$ $4.0 \cdot 10^{-6}$ cm^3 g^{-1}		Stu18, Ben55, Pov64, Cra74

Table 37. Magnetic properties of troilite and pyrrhotite phases stable at room temperature.

Mineral	Composition	Magnetic character	σ_s at 20 °C Gauss cm³ g⁻¹	χ_g at 20 °C 10⁻⁵ cm³ g⁻¹	T_α °C	$T_{\alpha,s}$ °C	T_γ °C	T_c °C	T_β °C	Ref.
troilite	FeS	antiferromagnetic		1.3 1.8 $H \perp c$ 1.5 $H \parallel c$	150 155	125 180			330 330	Sch72 Tak73
6 c-pyrrhotite	Fe$_{11}$S$_{12}$	antiferromagnetic with weak ferrimagnetism superimposed	12*)? [Ben55]	≈2			180	210***)	325	Har37, Har41, Sch72
11 c-pyrrhotite	Fe$_{10}$S$_{11}$	antiferromagnetic with weak ferrimagnetism superimposed	15*)? [Ben55]	≈3			210	230***)	330	Har37, Har41, Sch72
5 c-pyrrhotite	Fe$_9$S$_{10}$	antiferromagnetic with weak ferrimagnetism superimposed	1…8**) [Lot56a] 17*)? [Ben55]	≈300			225	270***)	330	Har37, Har41, Lot56a, Sch72
4 c-monoclinic pyrrhotite	Fe$_7$S$_8$	ferrimagnetic	20 [Wei05, Sch72] 17 [Ben55, Lot56a] 19.4 [Wei29]	grain size dependent $\chi_{gi} =$ 5500 cm³ g⁻¹ [Koe34]				305 315 292	325…340 330	Wei05 Ben55 Lot56a, Sch72

*) Magnetization σ measured in $H = 20000$ Oe. The values are distinctly higher than indicated by [Sch72] (compare Figs. 102 and 110).

**) Magnetization σ measured in $H = 12000$ Oe. The wide range of values is obtained from samples of same bulk chemistry but different heat treatment prior to measurement.

***) This may not be the true Curie temperature of the original phase, as the measured T_c is above the upper stability boundary of the original phase.

Bleil / Petersen

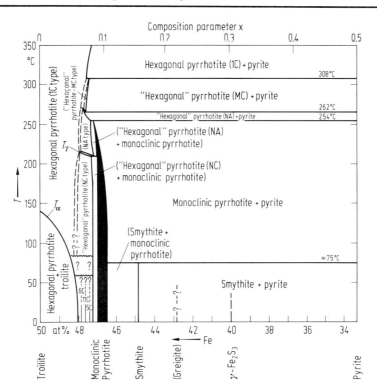

Fig. 100. Relations among the phases of the system FeS (troilite) – FeS$_2$ (pyrite) below 350 °C. The compositional variation is expressed in at % Fe of total composition and also in iron-deficiency x of the formula Fe$_{1-x}$S ($0 \le x \le 0.5$). T_α and T_γ denote the temperature of α- and γ-transition, respectively [Cra74, Kis74].

For Fig. 101, see next page.

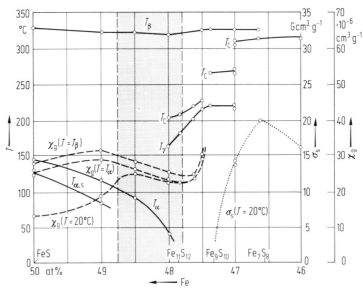

Fig. 102. Magnetic phase diagram for synthetic pyrrhotites annealed at $T = 144$ °C. Initial compositions are plotted along the abscissa. The ordinate gives a temperature, a magnetization, and a susceptibility scale. The dashed lines represent susceptibility, χ_g, in the region where antiferromagnetic behaviour is observed. The solid lines represent the temperatures, T, of α-, β-, γ-transitions, T_α, T_β, T_γ, spin flop, $T_{\alpha,s}$, and Curie point, T_C. The dotted line appears in the ferrimagnetic region and represents the estimated specific saturation magnetization, σ_S, at room temperature. The susceptibility is recorded at $T = 20$ °C and at the α- and β-transition temperatures. The shaded area between the dashed vertical lines shows the range of compositions of primary pyrrhotites in various igneous rocks [Sch72, Car74, Pet80].

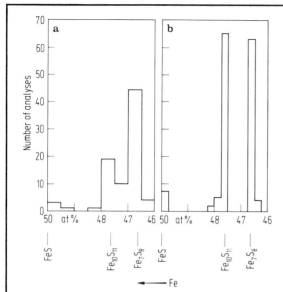

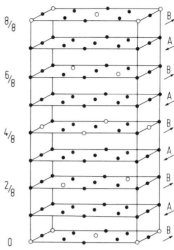

Fig. 101. Frequency distribution of natural terrestrial pyrrhotite compositions. Composition is in at% Fe [Kul67a].
a) Bulk compositions of pyrrhotites
b) Composition of individual pyrrhotite phases in the bulk
material.

Fig. 103. Schematic view of the arrangement of iron-ions
and vacancies in a unit-cell of monoclinic $4c$-pyrrhotite.
Solid circles represent Fe-atoms occupying octahedral sites
in the NiAs-type structure of a hexagonal close-packed
sulfur lattice. Open circles indicate a vacancy. Sulfur atoms
are not shown. The Fe-ions of neighbouring layers are
antiferromagnetically coupled. Due to the inequality in
numbers of Fe-ions on these layers a net ferrimagnetic
moment is present [Ber53].

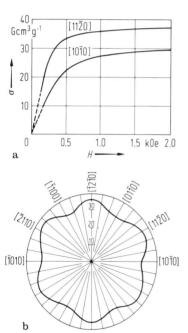

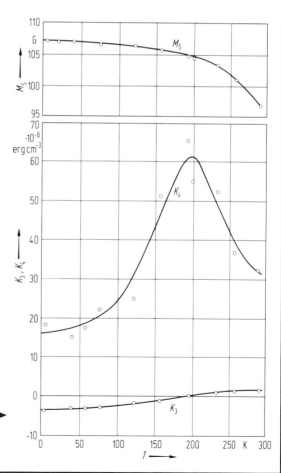

Fig. 104. Magnetization curves of a natural crystal of monoclinic pyrrhotite from Asio, Japan. [Kay39].
a) The [$11\bar{2}0$]-curve represents the magnetization measured
in the plane perpendicular to the c-axis. The [$10\bar{1}0$]-curve
represents the magnetization parallel to the c-axis.
b) Anisotropy of magnetization (measured in a field
$H = 2130$ Oe) in the plane perpendicular to the c-axis.

Fig. 105. Thermal variation of anisotropy constants K_3 and ▶
K_4 and saturation magnetization, M_S, of monoclinic
pyrrhotite Fe_7S_8. K_1 (not plotted here) is nearly zero and
K_2 is approximately constant at $0.35 \cdot 10^{-6}$ erg cm^{-3} [Bin63].

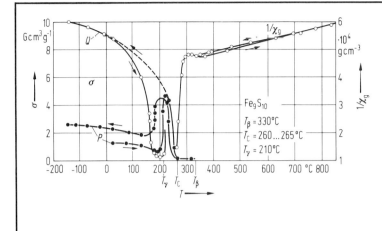

Fig. 106. Specific magnetization, σ, (measured in a field of $H = 8000$ Oe) and reciprocal specific susceptibility, $1/\chi_g$, of hexagonal pyrrhotite F_9S_{10} vs. temperature, T. Curve P: cooled slowly from $T = 1000$ °C, curve Q: annealed at $T = 220$ °C and thereafter quenched to room temperature. T_β and T_γ denote the β- and γ-transition temperatures, respectively, T_C denotes the Curie temperature [Lot56a].

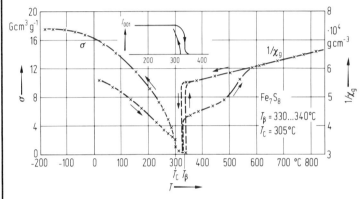

Fig. 107. Specific magnetization, σ, (measured in a field of $H = 8000$ Oe), reciprocal specific susceptibility, $1/\chi_g$, and the intensity of the 001-reflection, I_{001}, of monoclinic pyrrhotite Fe_7S_8 vs. temperature. T_C denotes the Curie temperature, T_β the β-transition temperature [Lot56a].

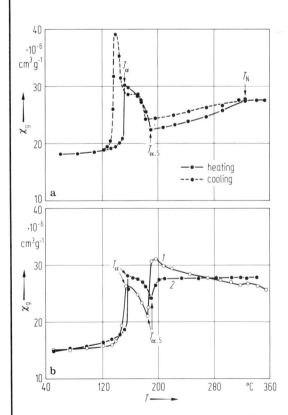

Fig. 108a, b. Magnetic susceptibility, χ_g, vs. temperature, T, for FeS (troilite). Magnetic field $H = 3600$ Oe. T_α denotes the α-transition, $T_{\alpha, s}$ the spin flop, and T_N the Néel temperature [Tak73].
a) Susceptibility measured in the basal plane (H perpendicular to c-axis).
b) Susceptibility measured during heating. Curve 1: H parallel to c-axis, curve 2: angle H to c-axis $= 50°$.

Fig. 109. Specific susceptibility, χ_g, vs. temperature, T, of troilite and pyrrhotite, $Fe_{1-x}S$ ($0 \leqq x \leqq 0.125$, equivalent to $50 \cdots 46.7$ at % Fe).
a) FeS (troilite). T_A denotes annealing temperature prior to measurement. T_α, T_β and $T_{\alpha,s}$ denote α-transition, β-transition and spin flop, respectively. Applied field: $H = 5350$ Oe [Sch72].
b) Hexagonal pyrrhotite of various composition. α, β, γ denote α, β, γ-transition, respectively. Applied field: $H = 1060 \cdots 3640$ Oe. Susceptibilities not dependent on applied magnetic field [Har37].
c) Hexagonal and monoclinic pyrrhotite. γ denotes γ-transition. Applied field: $H = 4116$ Oe [Har41].
d) Hexagonal and monoclinic pyrrhotite. Applied field: $H = 1060$ Oe [Har37].

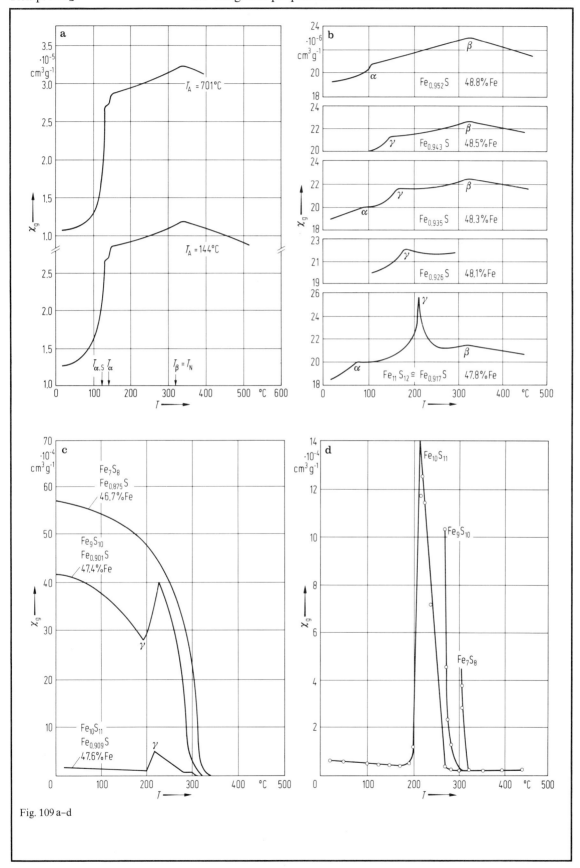

Fig. 109 a–d

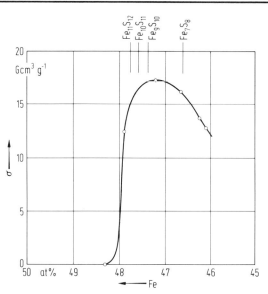

Fig. 110. Specific magnetization, σ, at room temperature of pyrrhotite of various composition. Composition expressed in at % Fe. Applied magnetic field: $H = 20000$ Oe [Ben55].

6.1.5 References for 6.1.1 ⋯ 6.1.4 — Literatur zu 6.1.1 ⋯ 6.1.4

Alb29	Albrecht, W.H.: Chem. Ber. **62** (1929) 1475.
Aki54	Akimoto, S.: J. Geomag. Geoelectr. **6** (1954) 1.
Aki57	Akimoto, S., Nagata, T., Katsura, T.: Nature **179** (1957) 37.
Aki57a	Akimoto, S., Katsura, T., Yoshida, M.: J. Geomag. Geoelectr. **9** (1957) 165.
Aki58	Akimoto, S., Horai, K., Boku, T.: J. Geomag. Geoelectr. **10** (1958) 7.
Aki59	Akimoto, S., Katsura, T.: J. Geomag. Geoelectr. **10** (1958) 69.
Aki62	Akimoto, S.: J. Phys. Soc. Jpn. **17** Suppl. B-1 (1962) 706.
Aki67	Akimoto, S., Syono, Y.: J. Chem. Phys. **47** (1967) 1813.
All66	Allen, W.C.: J. Am. Ceram. Soc. **49** (1966) 251.
All66a	Allen, W.C.: Am. Mineral. **51** (1966) 239.
And67	Andresen, A.F., Torbo, P.: Acta Chem. Scand. **21** (1967) 2841.
Ani66	Anishchenko, R.I., Nikolaev, A.P., Men', A.N.: Sov. Phys. Solid State (English Transl.) **8** (1966) 202.
Arn67	Arnold, R.G.: Can. Mineral. **9** (1967) 31.
Bai71	Bailey, J.T., Russell, R.: J. Am. Ceram. Soc. Bull. **50** (1971) 493.
Bal74	Balko, B., Hoy, G.R.: Phys. Lett. A **47** (1974) 171. ·
Ban65	Bando, Y., Kiyama, M., Yamamoto, N., Takada, T., Shinjo, T., Takaki, H.: J. Phys. Soc. Jpn. **20** (1965) 2086.
Ban67	Banerjee, S.K., O'Reilly, W., Johnson, C.E.: J. Appl. Phys. **38** (1967) 1289.
Ban70	Banerjee, S.K., Bartholin, H.: IEEE Trans. Magn. June **1970**, 299.
Bar32	Barth, T.F.W., Posnjak, E.: Z. Krist. **82** (1932) 325.
Bar34	Barth, T.F.W., Posnjak, E.: Z. Krist. **88** (1934) 265.
Bas59	Basta, E.Z.: Econ. Geol. **54** (1959) 698.
Bas60	Basta, E.Z.: Neues Jahrb. Mineral. Abh. **94** (1960) 1017.
Ben55	Benoit, R.: J. Chim. Phys. **52** (1955) 119.
Ben72	Bennett, C.E.G., Graham, J., Thornber, M.R.: Am. Mineral. **57** (1972) 445.
Ber53	Bertaut, E.F.: Acta Cryst. **6** (1953) 557.

Ber68	Berkowitz, A.E., Schuele, W.J., Flanders, P.J.: J. Appl. Phys. **39** (1968) 1262.
Bic55	Bickford, L.R., Pappis, J., Stull, J.L.: Phys. Rev. **99** (1955) 1211.
Bic57	Bickford, L.R., Brownlow, J.M., Penoyer, R.F.: Proc. Inst. Elec. Eng. (London), Suppl. 5, **B 104** (1957) 238.
Bin63	Bin, M., Pauthenet, R.: J. Appl. Phys. **34** (1963) 1161.
Biz56	Bizette, H., Tsai, B.: C.R. Acad. Sci. **242** (1956) 2124.
Bla62	Blasse, G., Gorter, E.W.: J. Phys. Soc. Jpn. **17** Suppl. B-1 (1962).
Bla63	Blasse, G., Fast, J.F.: Philips Res. Rept. **18** (1963) 1253.
Ble71	Bleil, U.: Z. Geophys. **37** (1971) 305.
Ble73	Bleil, U.: Thesis München **1973**.
Ble76	Bleil, U.: Pageoph. **114** (1976) 165.
Boz57	Bozorth, R.M., Walsh, D.E., Williams, A.J.: Phys. Rev. **108** (1957) 157.
Bro62	Brown, W.F., Johnson, C.E.: J. Appl. Phys. **33** (1962) 2752.
Bud64	Buddington, A.F., Lindsley, D.H.: J. Petrol. **5** (1964) 310.
Car61	Carmichael, C.M.: Thesis London **1961**.
Car67	Carmichael, C.M., Nicholls, J.: J. Geophys. Res. **72** (1967) 4665.
Car74	Carmichael, I.S.E., Turner, F.J., Verhoogen, J.: Igneous Petrology. New York: McGraw-Hill Book Comp. **1974**, p. 305.
Che43	Chevallier, R., Mathieu, S.: Ann. Phys. **18** (1943) 258.
Che55	Chevallier, R., Bolfa, J., Mathieu, S.: Bull. Soc. Franc. Min. Crist. **78** (1955) 307.
Che58	Chevallier, R., Mathieu, S.: Bull. Soc. Chim. France **5** (1958) 726.
Cla66	Clark, A.H.: Inst. Min. Metal. Trans. **75B** (1966) 232.
Coo72	Cooley, R.F., Reed, J.S.: J. Am. Ceram. Soc. **55** (1972) 395.
Cra02	Crane, W.R.: Trans. Am. Inst. Min. Eng. **31** (1902) 405.
Cra74	Craig, J.R., Scott, S.D., in: Min. Soc. Am., Short Course Notes, vol. 1, P.H. Ribbe (Ed.), **1974**, CS-1.
Cro55	Cromer, D.T., Herrington, K.: J. Am. Chem. Soc. **77** (1955) 4708.
Cse68	Cser, L., Dezsi, I., Gladkih, I., Keszthelyi, L., Kulgawczuk, D., Eissa, N.A., Sterk, E.: Phys. Status Solidi (b) **27** (1968) 131.
Dav56	David, I., Welch, A.J.E.: Trans. Faraday Soc. **52** (1956) 1642.
Day76	Day, R., Fuller, M.D., Schmidt, V.A.: J. Geophys. Res. **81** (1976) 873.
Day77	Day, R.: J. Geomag. Geoelectr. **29** (1977) 233.
Day77a	Day, R., Fuller, M.D., Schmidt, V.A.: Phys. Earth Planet. Inter. **13** (1977) 260.
Dek75	Deke, G., Seidel, B., Melzer, K., Michalk, C.: Phys. Status Solidi (a) **31** (1975) 446.
Del72	Dell, C.I.: Am. Mineral. **57** (1972) 1303.
Des65	Desborough, G.A., Carpenter, R.A.: Econ. Geol. **60** (1965) 1431.
Dmi73	Dmitrieva, T.V., Lyubutin, I.S., Pokrobskii, B.I., Bondareva, N.D.: Sov. Phys. JETP (English Transl.) **36** (1973) 709.
Dom50	Domenicalli, C.A.: Phys. Rev. **78** (1950) 461.
Elf65	Elfinger, F.: Dipl.-Arbeit, Inst. f. Angew. Geophysik, Univ. München **1965**, unpublished.
ElG76	El Goresy, A.: Min. Soc. Am. Short Course Notes **3** (1976) EG-1.
Eps58	Epstein, D.J., Frackiewicz, B.: J. Appl. Phys. **29** (1958) 376.
Esc60	Eschenfelder, A.H.: cited in Landolt-Börnstein Vol. III/4b, 211.
Fay69	Fayek, M.K., Leciejewicz, J., Murazik, A., Yamzin, I.I.: Phys. Status Solidi (b) **34** (1969) K 29.
Fay70	Fayek, M.K., Leciejewicz, J., Murasik, A., Yamzin, I.I.: Phys. Status Solidi **37** (1970) 843.
Fei65	Feitknecht, W.: Pure Appl. Chem. **9** (1965) 423.
Fer58	Ferguson, G.A., Hass, M.: Phys. Rev. **112** (1958) 1130.
Fin10	Finke, W.: Ann. Physik **31** (1910) 149.
Fin57	Finch, G.I., Sinha, A.P.B., Sinha, K.P.: Proc. Roy. Soc. (London), Ser. A **242** (1957) 28.
Fla64	Flanders, P.J., Schuele, W.J.: Philos. Mag. **9** (99) (1964) 487.
Flee71	Fleet, M.E.: Acta Cryst. **B27** (1971) 1864.
Fli59	Flinter, B.H.: Am. Mineral. **44** (1959) 738.
For68	Forsyth, J.B., Hedley, I.G., Johnson, C.E.: J. Phys. C **1** (1968) 179.
Fro60	Frost, M.J.: Mineral. Mag. **32** (1960) 573.
Fun59	Funatogawa, Z., Miyata, H., Usami, S.: J. Phys. Soc. Jpn. **14** (1959) 1583.
Gau43	Gaudin, A.M., Spedden, H.R.: Am. Inst. Min. Metall. Eng., Techn. Publ. No. 1549 (1943) 1.
Gho65	Ghorbanian, J., Poix, P.: C.R. Acad. Sci. **261** (1965) 3625.
Gho66	Ghorbanian, J.: Thesis Paris **1966**.

Gil76	Gillot, B., Ferriot, J., Dupré, G., Rousset, A.: Mater. Res. Bull. **11** (1976) 843.
Gol35	Goldsztaub, S.: Bull. Soc. Franc. Min **58** (1935) 6.
Gor54	Gorter, E.W.: Philips Res. Rept. **9** (1954) 295, 403.
Gor55	Gorter, E.W.: Proc. IRE **43** (1955) 245.
Gre30	Grenet, G.: Ann. Phys. (Paris) **13** (1930) 263.
Gre35	Greig, J.W., Posnjak, E., Merwin, H.E., Sosman, R.B.: Am. J. Sci. **30** (1935) 239.
Gri77	Grimes, N.W., Isaac, E.D.: Philos. Mag. **35** (1977) 503.
Gro70	Gros, Y.: Thesis, Grenoble **1970**.
Gui51	Guillaud, C.: J. Phys. Radium **12** (1951) 490.
Haf61	Hafner, S.: Z. Krist. **115** (1961) 343.
Häg35	Hägg, G.: Z. Phys. Chem. **B29** (1935) 95.
Hag76	Haggerty, S.E.: Min. Soc. Am. Short Course Notes **3** (1976) Hg-1.
Hai57	Haigh, G.: Philos. Mag. **2** (1957) 505.
Han66	Hanus, V., Krs, M.: Miner. Deposita **2** (1966) 139.
Har37	Haraldsen, H.: Z. Anorg. Allg. Chem. **231** (1937) 78.
Har41	Haraldsen, H.: Z. Anorg. Allg. Chem. **246** (1941) 169.
Har41a	Haraldsen, H.: Z. Anorg. Allg. Chem. **246** (1941) 195.
Hau74	Hauptman, Z.: Geophys. J. **38** (1974) 29.
Hed68	Hedley, I.G.: Phys. Earth Planet. Inter. **1** (1968) 103.
Hed71	Hedley, I.G.: Z. Geophys. **37** (1971) 409.
Hen70	Hentschel, B.: Z. Naturforsch. **25a** (1970) 1997.
Her66	Herzenberg, C.L., Toms, D.: J. Geophys. Res. **71** (1966) 2661.
Hir54	Hirone, T., Maeda, S., Chiba, S.: J. Phys. Soc. Jpn. **9** (1954) 500.
Hof46	Hofer, L.J.E., Peebles, W.C., Dieter, W.E.: J. Am. Chem. Soc. **68** (1946) 1953.
Hof56	Hoffmann, A., Fischer, W.A.: Z. Physik. Chem. (Frankfurt am Main) **7** (1956) 343.
Hof75	Hoffman, K.A.: Geophys. J. **41** (1975) 65.
Hoy72	Hoye, G.S., O'Reilly, W.: J. Phys. Chem. Solids **33** (1972) 1827.
Hry65	Hrynkiewicz, A.Z., Kulgawczuk, D.S., Tomala, K.: Phys. Lett. **17** (1965) 93.
Ima68	Imaoka, Y.: J. Electrochem. Soc. Jpn. **36** (1968) 17.
Ish57	Ishikawa, Y., Akimoto, S.: J. Phys. Soc. Jpn. **12** (1957) 1083.
Ish58	Ishikawa, Y., Akimoto, S.: J. Phys. Soc. Jpn. **13** (1958) 1110.
Ish58a	Ishikawa, Y., Akimoto, S.: J. Phys. Soc. Jpn. **13** (1958) 1298.
Ish62	Ishikawa, Y.: J. Phys. Soc. Jpn. **17** (1962) 1835.
Ish64	Ishikawa, Y., Syono, Y., Akimoto, S.: Prog. Rept. Rock Magnetism Group Japan **1964**, 14.
Ish67	Ishikawa, Y.: Phys. Lett A**24** (1967) 725.
Ish71	Ishikawa, Y., Syono, Y.: J. Phys. Soc. Jpn. **31** (1971) 461.
Jen73	Jensen, S.D., Shive, P.N.: J. Geophys. Res. **78** (1973) 8474.
Jos75	Joshima, M.: Rock Mag. Paleogeophys. **3** (1975) 5.
Kat74	Kataoka, M.: J. Phys. Soc. Jpn. **36** (1974) 456.
Kaw56	Kawai, N.: Proc. Jpn. Acad. **32** (1956) 464.
Kaw59	Kawai, N.: Congr. Geol. Intern., Mexico, Sect. 11A (1959) 103.
Kay39	Kaya, S., Miyahara, S.: Sci. Rep. Tohoku Univ. **27** (1939) 450.
Kis74	Kissin, S.A.: Ph.D. Thesis, Univ. of Toronto **1974**.
Kle36	Klemm, W.: Magnetochemie, Leipzig: Akad. Verlagsges. **1936**.
Koc67	Koch, F.B., Fine, M.E.: J. Appl. Phys. **38** (1967) 1471.
Koe98	Koenigsberger, J.: Ann. Phys. (Leipzig) **66** (1898) 698.
Koe34	Koenigsberger, J.G.: Beitr. Angew. Geophys. **4** (1934) 385.
Kön68	König, U., Chol, G.: J. Appl. Crystallogr. **1** (1968) 124.
Kön69	König, U., Gros, Y., Chol, G.: Phys. Status Solidi (b) **33** (1969) 811.
Kön72	König, U.: Tech. Mitt. Krupp, Forsch.-Ber. **30** (1972) 1.
Kri56	Kriessman, C.J., Harrison, S.E.: Phys. Rev. **103** (1956) 857.
Kru64	Kruglyakova, G.I., in: Aspects of theoretical mineralogy in the USSR, M.H. Battey and S.I. Tomkeieff (Eds.), Pergamon Press **1964**, 435.
Kul59	Kullerud, G.: Econ. Geol. **54** (1959) 533.
Kul67	Kullerud, G., Donnay, G.: Carnegie Inst. Washington Year Book **65** (1967) 356.
Kul67a	Kullerud, G., in: Researches in Geochemistry, Vol. 2, P.H. Abelson (Ed.), John Wiley and Sons **1967**, p. 286.
Kus60	Kushiro, I.: J. Geomag. Geoelectr. **11** (1960) 148.

Len57	Lensen, M., Michel, A.: Congr. Intern. Chim. Pure Appl., Paris **1957**.
Leu73	Leung, L.K., Evans, B.J., Morrish, A.H.: Phys. Rev. B**8** (1973) 29.
Lew66	Lewis, R.R., Senftle, F.E.: Am. Mineral. **51** (1966) 1467.
Lin62	Lindsley, D.H.: Carnegie Inst. Washington Yearb. **61** (1962) 100.
Lin65	Lindsley, D.H.: Carnegie Inst. Washington Yearb. **64** (1965) 144.
Lin66	Lindsley, D.H., Andreasen, G.E., Balsley, J.R.: Handbook of Physical Constants, Geol. Soc. Am. Memoir **97** (1966) 543.
Lin72	Lind, M.D., Housley, R.M.: Science **175** (1972) 521.
Lin73	Lindsley, D.H.: Geol. Soc. Am. Bull. **84** (1973) 657.
Lin76	Lindsley, D.H.: Min. Soc. Am. Short Course Notes **3** (1976) L-1.
Lot56	Lotgering, F.K.: Philips Res. Rept. **11** (1956) 218.
Lot56a	Lotgering, F.K.: Philips Res. Rept. **11** (1956) 190.
Lot66	Lotgering, F.K.: J. Phys. Chem. Solids **27** (1966) 139.
Mar72	Marshall, M., Cox, A.: J. Geophys. Res. **77** (1972) 6459.
Max53	Maxwell, L.R., Pickart, S.J.: Phys. Rev. **92** (1953) 1120.
McG60	McGuire, T.R., Greenwald, S.W.: Proc. Intern. Conf. Solid State Physics, Brussels **1958**, Vol. 3.
Mic70	Michel, A., Poix, P., Bernier, J.C.: Ann. Chim. **1970**, 263.
Miy61	Miyata, N.: J. Phys. Soc. Jpn. **16** (1961) 206.
Miz66	Mizushima, K., Iida, S.: J. Phys. Soc, Jpn. **21** (1966) 1526.
Mol76	Moldenhauer, W., Brückner, W.: Phys. Status Solidi (a) **34** (1976) 565.
Mor75	Morrish, A.H., Clark, P.E.: Phys. Rev. B**11** (1975) 278.
Nag56	Nagata, T., Akimoto, S.: Geofis. Pura Appl. **34** (1956) 36.
Nag57	Nagata, T., Yukutake, T., Uyeda, S.: J. Geomag. Geoelectr. **9** (1957) 51.
Nag61	Nagata, T.: Rock magnetism. Tokyo: Maruzen Comp. (2nd ed.) **1961**.
Nak71	Nakazawa, H., Morimoto, N.: Mater. Res. Bull. **6** (1971) 345.
Nee52	Néel, L., Pauthenet, R.: C.R. Acad. Sci. **234** (1952) 2172.
Nee55	Néel, L.: Adv. Phys. **4** (1955) 191.
Nil38	Nilakonton, P.: Indian Acad. Sci. **1938**, 39.
Nis77	Nishitani, T., Tanaka, H., Katsura, T.: Rock Mag. Paleogeophys. **4** (1977) 1.
Nis79	Nishitani, T.: Rock Mag. Paleogeophys. **6** (1979) 128.
O'Do75	O'Donovan, J.B.: Thesis, Newcastle upon Tyne **1975**.
O'Do77	O'Donovan, J.B., O'Reilly, W.: Adv. Earth Planet. Sci. **1** (1977) 99.
O'Do77a	O'Donovan, J.B., O'Reilly, W.: Earth Planet. Sci. Lett. **34** (1977) 291.
O'Do78	O'Donovan, J.B., O'Reilly, W.: Phys. Earth Planet. Inter. **16** (1978) 200.
Ono78	Ono, K., Chander, L., Ito, A.: J. Phys. Soc. Jpn. **25** (1968) 174.
Oos58	Oosterhout, C.W. van, Rooijmans, C.J.M.: Nature **181** (1958) 44.
O'Re65	O'Reilly, W.: Thesis, Newcastle upon Tyne **1965**.
O'Re65a	O'Reilly, W., Banerjee, S.K.: Phys. Lett. **17** (1965) 237.
O'Re66	O'Reilly, W., Banerjee, S.K.: Nature **211** (1966) 26.
O'Re67	O'Reilly, W., Banerjee, S.K.: Mineral. Mag. **26** (1967) 29.
O'Re71	O'Reilly, W., Readman, P.W.: Z. Geophys. **37** (1971) 321.
Özd78	Özdemir, Ö., O'Reilly, W.: Phys. Earth Planet. Inter. **16** (1978) 190.
Ozi70	Ozima, M., Larson, E.E.: J. Geophys. Res. **75** (1970) 1003.
Ozi71	Ozima, M., Sakamoto, N.: J. Geophys. Res. **76** (1971) 7035.
Ozi74	Ozima, M., Joshima, M., Kinoshita, H.: J. Geomag. Geoelectr. **36** (1974) 335.
Par65	Parry, L.G.: Philos. Mag. **11** (1965) 303.
Pet73	Petersen, N., Bleil, U.: Z. Geophys. **39** (1973) 965.
Pet79	Petersen, N., Eisenach, P., Bleil, U.: Am. Geophys. Union, M. Ewing Series **2** (1979) 169.
Pet80	Petersen, N.: Unpublished results, Univ. of Munich **1980**.
Phi54	Phillips, R., Vincent, E.A.: Geochim. Cosmochim. Acta **6** (1954) 1.
Pic59	Pickart, S.J., Turnock, A.C.: J. Phys. Chem. Solids **10** (1959) 242.
Ple71	Du Plessis, P.V.: J. Phys. C. **4** (1971) 2919.
Poi62	Poix, P., Michel, A.: Bull. Soc. Chim. France **1962**, 1010.
Poi65	Poix, P.: Ann. Chim. **10** (1965) 49.
Pop71	Popov, G.P., Il'Inova, G.N.: Zh. Fiz. Khim. **45** (1971) 541.
Pou50	Pouillard, E.: Ann. Chim. **5** (1950) 164.
Pov64	Povarennykh, A.S., in: Aspects of theoretical mineralogy in the USSR, M.H. Battey and S.I. Tomkeieff (Eds.), Pergamon Press **1964**, 451.

Puc71	Pucher, R.: Z. Geophys. **37** (1971) 349.
Ram76	Ramdohr, P · Die Erzmineralien und ihre Verwachsungen, Berlin: Akademie-Verlag, 4. Auflg. **1975**, 956.
Rea68	Readman, P.W., O'Reilly, W., Banerjee, S.K.: Phys. Lett. A **25** (1968) 446.
Rea71	Readman, P.W., O'Reilly, W.: Z. Geophys. **37** (1971) 329.
Rea72	Readman, P.W., O'Reilly, W.: J. Geomag. Geoelectr. **24** (1972) 69.
Rea78	Readman, R.W.: Phys. Earth Planet. Inter. **16** (1978) 196.
Reu69	Reuter, B., Riedel, E., Hug, P., Arndt, D., Geisler, U., Behnke, J.: Z. Anorg. Allg. Chem. **369** (1969) 306.
Rez74	Rezlescu, N., Istrate, S., Rezlescu, E., Luca, E.: J. Phys. Chem. Solids **35** (1974) 43.
Ric73	Richards, J.C.W., O'Donovan, J.B., Hauptman, Z., O'Reilly, W., Creer, K.M.: Phys. Earth Planet. Inter. **7** (1973) 437.
Rob71	Robbins, M., Wertheim, G.K., Sherwood, R.C., Buchanan, D.N.E.: J. Phys. Chem. Solids **32** (1971) 717.
Rog63	Rogers, D.B., Arnott, R.J., Wold, A., Goodenough, J.B.: J. Phys. Chem. Solids **24** (1963) 347.
Rom53	Romeijn, F.C.: Philips Res. Rept. **8** (1953) 321.
Ros65	Rossiter, M.J., Clarke, P.T.: Nature **207** (1965) 402.
Rot58	Roth, W.L.: Phys. Rev. **110** (1958) 1333.
Rot60	Roth, W.L.: Acta Cryst. **13** (1960) 146.
Rum76	Rumble, D.: Min. Soc. Am. Short Course Notes **3** (1976) R-1.
Sam69	Samara, G.A., Giardini, A.A.: Phys. Rev. **186** (1969) 577.
Sam69a	Sampson, C.F.: Acta Cryst. B **25** (1969) 1683.
Sch68	Schult, A.: Z. Geophys. **34** (1968) 505.
Sch70	Schult, A.: Earth Planet. Sci. Lett. **10** (1970) 81.
Sch71	Schmidbauer, E.: J. Phys. Chem. Solids **32** (1971) 71.
Sch72	Schwarz, E.J., Vaugham, D.J.: J. Geomag. Geoelectr. **24** (1972) 441.
Sch76	Schmidbauer, E.: Solid State Commun. **18** (1976) 301.
Sen60	Senftle, F.E., Pankey, T., Grant, F.A.: Phys. Rev. **120** (1960) 820.
Sen68	Senftle, F.E., Thorpe, A.N.: Phys. Rev. **175** (1968) 1144.
Sha65	Sharma, V.N.: J. Appl. Phys. **36** (1965) 1452.
Shc69	Shchepetkin, A.A., Zakharov, R.G., Chufarov, G.I.: Izv. Akad. Nauk SSSR, Neorg. Mater. **5** (1969) 1953.
Shc69a	Shchepetkin, A.A., Zakharov, R.G., Zinigrag, M.I., Chufarov, G.I.: Kristallografiya **14** (1969) 889.
Shi59	Shirane, G., Pickart, S.J., Nathans, R., Ishikawa, Y.: J. Phys. Chem. Solids **10** (1959) 35.
Shi62	Shirane, G., Cox, D.E., Takei, W.J., Ruby, S.L.: J. Phys. Soc. Jpn. **17** (1962) 1607.
Shi62a	Shirane, G., Ruby, S.L.: J. Phys. Soc. Jpn. Suppl. B-1 **17** (1962) 133.
Shi62b	Shirane, G., Cox, D.E., Ruby, L.: Phys. Rev. **125** (1962) 1158.
Shu51	Shull, C.G., Strauser, W.A., Wollan, E.O.: Phys. Rev. **83** (1951) 344.
Shu51a	Shull, C.G., Wollan, E.O., Kohler, W.C.: Phys. Rev. **84** (1951) 912.
Ski64	Skinner, B.J., Erd, R.C., Grimaldi, F.S.: Am. Mineral. **49** (1964) 543.
Sla64	Slack, G.A.: Phys. Rev. A **134** (1964) 1268.
Smi59	Smit, J., Wijn, H.P.J.: Ferrites, New York: John Wiley and Sons, **1959**.
Smi77	Smith, B.M., Prévot, M.: Phys. Earth Planet. Inter. **14** (1977) 120.
Smo67	Smolin, R.P., Ryabinkina, L.I.: Zh. Tekhn. Fiz. **37** (1967) 572; Sov. Phys. Tech. Phys. (English Transl.) **12** (1967) 411.
Spe72	Spender, M.R.: Can. J. Phys. **50** (1972) 2313.
Sta74	Stacey, F.D., Banerjee, S.K.: The physical principles of rock magnetism. Amsterdam: Elsevier Scientific Publishing Comp., **1974**.
Ste69	Stephenson, A.: Geophys. J. **18** (1969) 199.
Ste72	Stephenson, A.: Philos. Mag. **25** (1972) 1213.
Str61	Straumanis, M.E., Ejima, T., James, W.J.: Acta Cryst. **14** (1961) 493.
Str68	Strangway, D.W., Honea, R.M., McMahon, B.E., Larson, E.E.: Geophys. J. **15** (1968) 345.
Stu18	Stutzer, F., Groß, W., Bornemann, K.: Met. Erz **15** (1918) 1.
Swa64	Swanson, H.E., Morris, M.C., Evans, E.H., Ulmer, L.: US Natl. Bur. Stand. Mon. **25** (1964) Sec. 3.
Syo60	Syono, Y.: J. Geomag. Geoelectr. **11** (1960) 85.
Syo65	Syono, Y.: Jpn. J. Geophys. **4** (1965) 71.

Tak65	Takada, T., Kiyama, M., Bando, Y., Nakamura, T., Shiga, M., Shinjo, T., Yamamoto, N., Endoh, T., Takaki, H.: J. Phys. Soc. Jpn. **19** (1965) 1744.
Tak73	Takahashi, T.: Solid State Commun. **13** (1973) 1335.
Tay64	Taylor, R.W.: Am. Mineral. **49** (1964) 1016.
Tel67	Tellier, J.C.: Rev. Chem. Minéral. **4** (1967) 325.
Ter62	Teranishi, T.: J. Phys. Soc. Jpn. **17** Suppl. B-1 (1962) 263.
Tok72	Tokonami, M., Nishiguchi, K., Morimoto, N.: Am. Mineral. **57** (1972) 1066.
Tre65	Tret'yakov, Yu.O., Saksonov, Yu.G., Gordeev, I.V.: Izv. Akad. Nauk SSSR, Neorg. Mater. **1** (1965) 413; Inorg. Mater. (English Transl.) **1** (1965) 382.
Uda67	Uda, M.: Z. Anorg. Allg. Chem. **350** (1967) 105.
Uda68	Uda, M.: Sci. Papers I.P.C.R. **62** (1968) 14.
Urq56	Urquhart, H.M.A., Goldman, J.E.: Phys. Rev. **101** (1956) 1448.
Uye58	Uyeda, S.: Jpn. J. Geophys. **2** (1958) 1.
Val62	Valstyn, E.P., Hanton, J.P., Morrish, A.H.: Phys. Rev. **128** (1962) 2078.
Van66	Van der Woude, F., Dekker, A.J.: Phys. Status Solidi **13** (1966) 181.
Ver47	Verwey, E.J.W., Heilmann, E.L.: J. Chem. Phys. **15** (1947) 174.
Ver62	Verhoogen, J.: J. Geol. **70** (1962) 168.
Vin57	Vincent, E.A., Wright, J.B., Chevallier, R.: Mineral. Mag. **31** (1957) 624.
Voi07	Voigt, W., Kinoshita, S.: Ann. Phys. (Leipzig) **24** (1907) 492.
Wal60	Walenta, K.: Neues Jahrb. Mineral. Mh. **1960**, 151.
War70	Ward, J.C.: Rev. Pure Appl. Chem. **20** (1970) 175.
Web71	Weber, H.P., Hafner, S.S.: Z. Krist. **133** (1971) 327.
Wec76	Wechsel, B.A., Prewitt, C.T., Papike, J.J.: Earth Planet. Sci. Lett. **29** (1976) 91.
Wei05	Weiss, P.: J. Phys. Radium **4** (1905) 469.
Wei29	Weiss, P., Forrer, R.: Ann. Phys. (Paris) **12** (1929) 279.
Wen52	Went, J.J., Gorter, E.W.: Philips Tech. Rev. **13** (1952) 185.
Wes71	Westcott-Lewis, M.F., Parry, L.G.: Aust. J. Phys. **24** (1971) 719.
Wes71a	Westcott-Lewis, M.F., Parry, L.G.: Aust. J. Phys. **24** (1971) 735.
Wil20	Wilson, E.: Proc. Roy. Soc. London Ser. A**96** (1920) 429.
Wue74	Wuensch, B.J., in: Sulfide Mineralogy, Min. Soc. Am., Short Course Notes, vol. 1, P.H. Ribbe (Ed.), **1974**, W-21.
Yam73	Yamaguchi, S., Wada, H.: J. Appl. Phys. **44** (1973) 1929.
Yam73a	Yamaguchi, S., Wada, H.: Krist. Tech. **8** (1973) 1017.
Zel65	Zeller, C., Babkine, J.: C.R. Acad. Sci. **250** (1965) 1375.
Zel67	Zeller, C., Hubsch, J., Bolfa, J.: C.R. Acad. Sci. B**265** (1967) 1034.
Zel67a	Zeller, C., Hubsch, J., Reithler, J.C., Bolfa, J.: C.R. Acad. Sci. B**265** (1967) 1335.

6.2 Magnetic properties of rocks — Magnetische Eigenschaften der Gesteine

6.2.1 General remarks — Allgemeine Bemerkungen

The total magnetization M of a rock is the vector sum of "induced magnetization M_i" – dependent on an external field – and "remanent magnetization M_R" – independent of an external field.

Die Gesamtmagnetisierung M eines Gesteines entspricht der Vektorsumme der „induzierten Magnetisierung M_i", die von einem äußeren Feld abhängig ist, und der „remanenten Magnetisierung M_R", die von einem äußeren Feld unabhängig ist.

$$M = M_i + M_R$$

In common rocks most minerals are either paramagnetic or diamagnetic with positive or negative susceptibilities of the order of 10^{-6} Gauss/Oe (see 6.1.3.3 and 6.1.2.3). The ferromagnetic properties of a rock, however, are imposed by the ferrimagnetic minerals contained in the rock although generally low in concentration. The content of ferrimagnetic minerals in major rock types rarely exceeds 10 vol-%.

In den gewöhnlichen Gesteinen sind die meisten Minerale entweder paramagnetisch oder diamagnetisch mit positiven oder negativen Suszeptibilitäten der Größenordnung 10^{-6} Gauss/Oe (siehe 6.1.3.3 und 6.1.2.3). Die im Gestein im allgemeinen in niedriger Konzentration enthaltenen ferrimagnetischen Minerale bestimmen jedoch die ferromagnetischen Eigenschaften eines Gesteins. In den meisten Gesteinstypen übersteigt der Gehalt an ferrimagnetischen Mineralen selten 10 vol-%.

In the overwhelming number of rocks minerals of the Fe–Ti-oxide-group (see 6.1.4.2) and – to a much smaller percentage – minerals of the iron-sulfide-group (6.1.4.11) impose the ferromagnetic properties. In sedimentary rocks also iron-hydroxides (6.1.4.10) play an important role.

In den weitaus meisten Gesteinen sind die Minerale der Fe–Ti-Oxid-Gruppe (siehe 6.1.4.2) und – zu einem viel geringeren Prozentsatz – die Minerale der Eisen-Sulfid-Gruppe (6.1.4.11) ausschlaggebend für die ferromagnetischen Eigenschaften. In sedimentären Gesteinen spielen auch die Eisenhydroxide (6.1.4.10) eine bedeutende Rolle.

For comprehensive reviews of rock magnetism the reader is referred to the following literature:

Zusammenfassende Arbeiten über den Gesteinsmagnetismus:

Nicholls, G.D.: The mineralogy of rock magnetism, Phil. Mag. Suppl. Adv. Phys. **4** (1955) 113.

Néel, L.: Some theoretical aspects of rock magnetism, Phil. Mag. Suppl. Adv. Phys. **4** (1955) 191.

Stacey, F.D.: The physical theory of rock magnetism, Phil. Mag. Suppl. Adv. Phys. **12** (1963) 46.

Angenheister, G. and H. Soffel: Gesteinsmagnetismus und Paläomagnetismus, Borntraeger, Stuttgart, (1972)

Stacey, F.D. and S.K. Banerjee: The physical principles of rock magnetism, Elsevier Sci. Publ. Comp., Amsterdam (1974).

Creer, K.M., I.G. Hedley and W.O'Reilly: Magnetic oxides in geomagnetism, In: Magnetic oxides, part 2, D.J. Craik (Ed.), John Wiley and Sons (1975), 649.

O'Reilly, W.: Magnetic minerals in the crust of the earth, Rep. Prog. Phys. **39** (1976) 857.

6.2.1.1 List of abbreviations and symbols — Liste der Abkürzungen und Symbole

a, b, c [Å]	lattice parameters – Gitterkonstanten
ARM	anhysteretic remanent magnetization – anhysteretische remanente Magnetisierung
CRM	chemical remanent magnetization – chemische remanente Magnetisierung
c [Gauss Oe^{-1}]	coefficient of piezo-remanent magnetization – Koeffizient der Piezo-Remanenz
DRM	depositional remanent magnetization – Sedimentations-Remanenz
DSDP	Deep Sea Drilling Project – Tiefsee-Bohrprojekt
F	emplacement mode indicator for basaltic rocks – Indikator für die Art der Platznahme basaltischer Gesteine
FAMOUS	French-American Midocean Undersea Study
H [Oe]	magnetic field – Magnetfeld
H_C [Oe]	coercive force – Koerzitivkraft
H_d [Oe]	peak alternating field – Spitzenwert der Amplitude des Wechselfeldes
H_{RC} [Oe]	remanence coercivity – Remanenz-Koerzitivkraft
I_H [deg]	inclination of magnetic field – Inklination des Magnetfeldes
I_R [deg]	inclination of remanent magnetization – Inklination der remanenten Magnetisierung

IRM	isothermal remanent magnetization – isothermale remanente Magnetisierung
l [cm]	distance – Länge
M_i [Gauss] ($=[\text{emu cm}^{-3}])*)$	induced magnetization (volume magnetization) – induzierte Magnetisierung (Volumen-Magnetisierung)
M_R [Gauss]	remanent magnetization (volume magnetization) – remanente Magnetisierung (Volumen-Magnetisierung)
M_{RS} [Gauss]	saturation remanent magnetization (volume magnetization) – Sättigungs-Remanenz (Volumen-Magnetisierung)
M_S [Gauss]	saturation magnetization (volume magnetization) – Sättigungsmagnetisierung (Volumen-Magnetisierung)
MD	multi-domain particles – Mehrbereichs-Teilchen
MDF	median destructive field – mittleres Abmagnetisierungsfeld
N	demagnetizing factor – Entmagnetisierungsfaktor
N	number of samples – Anzahl der Proben
NRM	natural remanent magnetization – natürliche remanente Magnetisierung
P	anisotropy factor – Anisotropie-Faktor
$p (0 \leq p \leq 1)$	volume content of ferrimagnetic minerals in a rock – Volumengehalt ferrimagnetischer Minerale im Gestein
p_{ms}	saturation magnetic moment per molecule – magnetisches Moment per Molekül bei Sättigung
PDRM	post-depositional remanent magnetization – Post-Sedimentations-Remanenz
PRM	pressure or piezo-remanent magnetization – Druck- oder Piezo-Remanenz
PSD	pseudo-single-domain particles – Pseudo-Einbereichs-Teilchen
pTRM	partial thermoremanent magnetization – partielle thermoremanente Magnetisierung
Q	Koenigsberger Q-factor – Koenigsberger-Q-Faktor
S [Gauss, Gauss cm³ g⁻¹]	viscosity coefficient – Viskositätskoeffizient
SD	single-domain particles – Einbereichs-Teilchen
SIRM	saturation isothermal remanent magnetization – Sättigungs-Isothermalremanenz
SRM	shock-induced remanent magnetization – Schock-Remanenz
T [°C, K]	temperature – Temperatur
t [s, min, h, d]	time – Zeit
T_B [°C, K]	blocking temperature – Blockungs-Temperatur
T_C [°C, K]	Curie temperature – Curie-Temperatur
T_{comp} [°C, K]	self-reversal compensation temperature – Selbstumkehr-Kompensationstemperatur
TRM	thermoremanent magnetization – thermoremanente Magnetisierung
VRM	viscous remanent magnetization – viskose remanente Magnetisierung
x	titanomagnetite composition ($Fe_{3-x}Ti_xO_4$) – Titanomagnetit-Zusammensetzung
y	ilmenite-hematite composition ($Fe_{2-y}Ti_yO_3$) – Ilmenit-Hämatit-Zusammensetzung
z	titanomagnetite oxidation parameter – Titanomagnetit-Oxidationsparameter
β [cm² kg⁻¹]	stress sensitivity of magnetic susceptibility – Spannungsempfindlichkeit der magnetischen Suszetibilität
δ [deg]	inclination error ($I_H - I_R$) – Inklinations-Fehler
σ_R [Gauss cm³ g⁻¹]	specific remanent magnetization (mass magnetization) – spezifische remanente Magnetisierung
σ_S [Gauss cm³ g⁻¹]	specific saturation magnetization (mass magnetization) – spezifische Sättigungsmagnetisierung
τ [kg cm⁻²]	uniaxial compression – axialer Druck
χ_v [cm³ cm⁻³ = 1] $= [\text{emu cm}^{-3}]*)$ $= [\text{Gauss Oe}^{-1}]$	susceptibility (volume susceptibility) – Suszeptibilität (Volumen-Suszeptibilität)
χ_g [cm³ g⁻¹]	specific susceptibility (mass susceptibility) – spezifische Suszeptibilität
χ_{intr} [cm³ cm⁻¹ = 1]	intrinsic volume susceptibility – Material-Volumensuszeptibilität
$\chi_{a,b,c}$	principal susceptibilities of the susceptibility tensor – Haupt-Suszeptibilitäten des Suszeptibilitäts-Tensors
χ_m [cm³ mol⁻¹]	molar susceptibility – molare Suszeptibilität

*) Note that "emu" is used in two different definitions.

6.2.2 Induced magnetization — Induzierte Magnetisierung

6.2.2.1 Introductory remarks — Einleitende Bemerkungen

The natural induced magnetization of a rock is the reversible part of its total natural magnetization induced by the ambient earth's magnetic field at ambient temperature. Inducing field H and induced magnetization M_i are related by the magnetic susceptibility χ:

Die natürliche induzierte Magnetisierung eines Gesteins ist der reversible Anteil seiner gesamten natürlichen Magnetisierung, der durch das erdmagnetische Feld bei Umgebungstemperatur hervorgerufen wird. Das induzierende Feld H und die induzierte Magnetisierung M_i sind durch die magnetische Suszeptibilität χ verknüpft:

$$M_i = \chi_{ij} H$$

In general the susceptibility is a tensor of second rank, dependent on H, T, τ etc. (see 6.2.2.2). For many practical purposes however (when dealing with the natural induced magnetization of rocks) the susceptibility of common rocks can conveniently be considered as an isotropic constant

Im allgemeinen ist die Suszeptibilität ein Tensor zweiter Stufe, der von H, T, τ usw. abhängig ist (siehe 6.2.2.2). Zu praktischen Zwecken (wenn es sich um die natürliche induzierte Magnetisierung der Gesteine handelt) kann jedoch die Suszeptibilität gewöhnlicher Gesteine einfach als isotrope Konstante betrachtet werden:

$$M_i = \chi H_e$$

where H_e represents a magnetic field of the order of the earth's magnetic field.

wobei H_e ein magnetisches Feld der Größenordnung des magnetischen Feldes der Erde darstellt.

The ratio of natural remanent magnetization NRM to natural induced magnetization is called Koenigsberger Q-factor

Das Verhältnis der natürlichen remanenten Magnetisierung NRM zur natürlichen induzierten Magnetisierung wird Koenigsberger-Q-Faktor genannt:

$$Q = \frac{\text{NRM}}{\chi \cdot H_{\text{earth}}} \quad \text{[Koe38]}$$

where H_{earth} represents the earth's magnetic field at the location of the rock.

wobei H_{earth} das magnetische Feld der Erde am Ort des Gesteins ist.

A plot of total natural magnetization vs. susceptibility shows a proportional relationship for a wide range of rocks. An exception from this linear relationship are basalts and rocks that have been struck by lightnings Fig. 1.

Eine graphische Darstellung der gesamten natürlichen Magnetisierung als Funktion der Suszeptibilität zeigt bei einer großen Anzahl von Gesteinen eine lineare Beziehung. Vom Blitz getroffene Gesteine und Basalte bilden hiervon eine Ausnahme Fig. 1.

The magnetic susceptibility of rocks measured by the usual methods is an apparent susceptibility because of the self-demagnetizing effect of the magnetic minerals contained in the rock. The apparent susceptibility χ of a mineral is related to the intrinsic susceptibility χ_{intr} of this mineral by

Die durch die üblichen Methoden gemessene magnetische Suszeptibilität der Gesteine ist wegen des Eigen-Entmagnetisierungseffekts der im Gestein enthaltenen magnetischen Minerale eine scheinbare Suszeptibilität. Die scheinbare Suszeptibilität χ eines Minerals steht mit der Material-Suszeptibilität χ_{intr} dieses Minerals in folgender Beziehung:

$$\chi^{\text{mineral}} = \frac{\chi_{\text{intr}}}{1 + N \chi_{\text{intr}}}$$

where N is the demagnetizing factor of the mineral.

wobei N der Entmagnetisierungsfaktor des Minerals ist.

The apparent susceptibility of a rock with a volume content p of magnetic minerals is related to the intrinsic susceptibility χ_{intr} of the magnetic minerals contained in the rock by

Die scheinbare Suszeptibilität eines Gesteins mit dem Volumengehalt p an magnetischen Mineralen steht mit der Material-Susceptibilität χ_{intr} der im Gestein enthaltenen magnetischen Minerale in folgender Beziehung:

$$\chi^{\text{rock}} = \frac{p \cdot \chi_{\text{intr}}}{1 + N \chi_{\text{intr}}} \quad p \ll 1 \quad \text{[Nag61]}$$

where N is the demagnetizing factor of the magnetic grains.

wobei N der Entmagnetisierungsfaktor der magnetischen Körner ist.

6.2.2.2 Factors influencing the susceptibility of rocks — Faktoren, welche die Suszeptibilität der Gesteine beeinflussen

6.2.2.2.1 Type and concentration of magnetic minerals contained in the rock — Art und Konzentration magnetischer Minerale im Gestein

The magnetic susceptibility of common rocks is roughly proportional to the content of magnetite (Figs. 2 and 3). Of the other magnetic minerals only pyrrhotite contributes in a significant way to the magnetic susceptibility in certain metamorphic rocks and hydrothermal deposits [Lin66, Moo53]. With higher magnetite concentrations magnetostatic interaction of the magnetite grains causes bulk rock susceptibility to increase more than linearly with increasing concentration (Fig. 4) [Puz30].

Magnetite is an accessory mineral (<10 vol-%) in most rocks. Its content can vary significantly from a magnetic viewpoint without affecting the rock classification. Nevertheless, there is a general trend of variation of susceptibility in igneous rocks, with mafic igneous rocks having higher susceptibilities than more silicic ones (Figs. 5 and 6) [Lin66, Kri73, Kro76, Ref60].

Die magnetische Suszeptibilität der gewöhnlichen Gesteine ist im großen und ganzen proportional zum Magnetitgehalt (Fig. 2 und 3). Von den weiteren magnetischen Mineralen trägt nur Pyrrhotit deutlich zur magnetischen Suszeptibilität in bestimmten metamorphen Gesteinen und hydrothermalen Lagerstätten bei [Lin66, Moo53]. Bei höherem Magnetitgehalt wächst die Suszeptibilität des Gesteins wegen der magnetostatischen Wechselwirkung der Magnetitkörner stärker als linear mit zunehmender Konzentration (Fig. 4) [Puz30].

In den meisten Gesteinen ist Magnetit nur ein akzessorisches Mineral (<10 vol-%). In magnetischer Hinsicht kann seine Konzentration deutlich variieren, ohne die Klassifikation des Gesteins zu ändern. Jedoch haben mafische Eruptivgesteine in der Regel höhere Susceptibilitäten als die Eruptivgesteine mit höherem Siliziumgehalt (Fig. 5 und 6) (Lin66, Kri73, Kro76. Ref60].

6.2.2.2.2 Strength of inducing field — Stärke des induzierenden Feldes

For ferrimagnetic minerals and rocks containing ferrimagnetic minerals the susceptibility is a function of the external field.

Für ferrimagnetische Minerale und Gesteine mit ferrimagnetischen Mineralen ist die Suszeptibilität eine Funktion des äußeren Feldes.

$$\chi = \chi(H) \quad \text{(Fig. 7)}$$

6.2.2.2.3 Grain size of magnetic minerals — Korngröße der magnetischen Minerale

The initial susceptibility of rocks depends strongly on the grain size of the magnetic minerals contained in them. This grain size dependence is particularly evident in the range between lower end of MD-grains and SD-grains [Dun81] (Figs. 8, 9, 10 and Table 6 of section 6.1.4.4.2). Here the susceptibility decreases with decreasing grain diameter. For even smaller grain sizes (superparamagnetic behaviour) the susceptibility increases again.

In the MD-grain-size-range initial susceptibility is determined by factors other than grain size, namely demagnetizing field, inclusions, internal stress etc. [Dun81, Nag61, Nee55].

In the case of magnetite- and titanomagnetite-MD-grains the initial susceptibility is limited by self demagnetization:

Die Anfangs-Suszeptibilität der Gesteine hängt stark von der Korngröße der in ihnen enthaltenen magnetischen Minerale ab. Diese Abhängigkeit von der Korngröße ist besonders ausgeprägt im Bereich zwischen kleinen MD-Körnern und den SD-Körnern [Dun81] (Fig. 8, 9, 10 und Tab. 6 des Abschnitts 6.1.4.4.2). Hier nimmt die Suszeptibilität mit abnehmendem Korndurchmesser ab. Bei noch kleineren Korngrößen (superparamagnetisches Verhalten) nimmt die Suszeptibilität wieder zu.

Für MD-Körner wird die Anfangs-Suszeptibilität durch andere Faktoren als die Korngröße bestimmt, nämlich durch das entmagnetisierende Feld, Einschlüsse, innere Spannung u.s.w. [Dun81, Nag61, Nee55].

Für Magnetit- und Titanomagnetit-MD-Körner wird die Anfangs-Suszeptibilität durch Eigen-Entmagnetisierung begrenzt:

$$\chi = \frac{\chi_{\text{intr}}}{1 + N\chi_{\text{intr}}} \leqq \frac{1}{N} \approx 0.25 \quad \text{[Dun81]}$$

with the intrinsic susceptibility $\chi_{\text{intr}} \approx 10$ Gauss/Oe for magnetite [Bic55].

mit der Material-Suszeptibilität $\chi_{\text{intr}} \approx 10$ Gauss/Oe für Magnetit [Bic55].

6.2.2.2.4 Temperature — Temperatur

Rock samples exposed to temperatures approaching the Curie temperature exhibit enhanced susceptibility due to the Hopkinson effect [Dun74, Rad74] (Figs. 11 to 13). This effect may be important for the interpretation of geomagnetic anomalies of deep-seated sources which are close to the Curie depth.

Gesteinsproben, die Temperaturen nahe der Curie-Temperatur ausgesetzt werden, weisen eine erhöhte Suszeptibilität infolge des Hopkinson-Effekts auf [Dun74, Rad74] (Fig. 11 bis 13). Dieser Effekt könnte wichtig für die Deutung der geomagnetischen Anomalien tiefliegender Quellen sein, die sich nahe der Curie-Tiefe befinden.

6.2.2.2.5 Fabric of the rock — Gesteinsgefüge

In general rocks with anisotropic structures show anisotropic magnetic susceptibilities. The long axes of ferrimagnetic minerals tend to lie in the bedding plane of detrital sediments (Figs. 14, 15a and b, 16), in the foliation planes of metamorphic rocks (Fig. 17), and – less pronounced – along flow lines of magmatic rocks (Figs. 18, 19, 20a and b, 21). This orientation results in a plane (magnetic foliation) or direction (magnetic lineation) of maximum susceptibility parallel to these structures (sediments: [Ful63, Gra58, Isi42, Rad70, Rah75, Ree66]; metamorphic rocks: [Bal60, Hro71, Sta60]; magmatic rocks: [Hal74, Hel73, Kha62, Ell78, Uye63]). A great number of parameters has been introduced in the literature to describe the magnetic anisotropy of rocks (Table 1).

Gesteine mit anisotroper Struktur zeigen im allgemeinen anisotrope magnetische Suszeptibilitäten. Die langen Achsen ferrimagnetischer Minerale findet man in den Ablagerungsebenen von detritischen Sedimenten (Fig. 14, 15a und b, 16), in den Schieferungsebenen der metamorphen Gesteine (Fig. 17) und – weniger ausgeprägt – entlang den Fließlinien magmatischer Gesteine (Fig. 18, 19, 20a und b, 21). Diese Orientierung ergibt eine Ebene (magnetische Foliation) oder Richtung (magnetische Lineation) maximaler Suszeptibilität parallel zu diesen Strukturen (Sedimente: [Ful63, Gra58, Isi42, Rad70, Rah75, Ree66]; metamorphe Gesteine: [Bal60, Hro71, Sta60]; magmatische Gesteine: [Hal74, Hel73, Kha62, Ell78, Uye63]). Eine große Anzahl von Parametern ist in die Literatur eingeführt worden, um die magnetische Anisotropie der Gesteine zu beschreiben (Tab. 1).

Anisotropic susceptibility is a tensor of the second rank χ_{ij}, characterized by three principal susceptibilities $\chi_a \geq \chi_b \geq \chi_c$.

Die anisotrope Suszeptibilität ist ein Tensor zweiter Stufe, χ_{ij}, der durch drei Hauptsuszeptibilitäten $\chi_a \geq \chi_b \geq \chi_c$ charakterisiert ist.

The susceptibility tensor is conveniently represented by the susceptibility ellipsoid (Fig. 22), expressed in rectangular co-ordinates x, y, z as:

Der Suszeptibilitätstensor läßt sich durch das Suszeptibilitätsellipsoid (Fig. 22) darstellen und zwar in rechtwinkligen Koordinaten x, y und z als:

$$\chi_a x^2 + \chi_b y^2 + \chi_c z^2 = 1$$

The length of the semiaxes of the ellipsoid are:

Die Längen der Halbachsen des Ellipsoids sind:

$$1/\sqrt{\chi_a},\ 1/\sqrt{\chi_b},\ 1/\sqrt{\chi_c} \quad [\text{Nag61}]$$

Anisotropy of susceptibility of rocks is caused:

Die Anisotropie des Gesteinssuszeptibilität ist verursacht:

1. by the anisotropic demagnetization factor of the magnetic grains contained in the rock. This is the case in rocks containing grains of magnetite or titanomagnetite with a preferred orientation (Fig. 23). The degree of magnetic anisotropy, the so-called P-factor (Table 1), for a single magnetite or titanomagnetite grain of spheroidal shape is related to the demagnetizing factor N by

1. durch den anisotropen Entmagnetisierungsfaktor der im Gestein enthaltenen magnetischen Körner. Dies ist der Fall für Gesteine, die Magnetit oder Titanomagnetit mit einer bevorzugten Orientierung enthalten (Fig. 23). Der Grad der magnetischen Anisotropie, der sogenannte P-Faktor (Tab. 1), ist mit dem Entmagnetisierungsfaktor N für ein einzelnes Magnetit- oder Titanomagnetitkorn sphäroidischer Form durch folgende Gleichung verbunden:

$$P = \frac{\chi_{\max}}{\chi_{\min}} = \frac{\chi_a}{\chi_c} = \frac{1 + \chi_{\text{intr}} N_c}{1 + \chi_{\text{intr}} N_a} \quad [\text{Uye63}]$$

χ_{intr} = intrinsic susceptibility of magnetite or titanomagnetite, $N_{a,c}$ = demagnetizing factor of a grain in the a and c-direction, respectively (Figs. 24a and b).

wobei χ_{intr} die Material-Suszeptibilität des Magnetits oder Titanomagnetits und $N_{a,c}$ der Entmagnetisierungsfaktor eines Korns in der a- bzw. c-Richtung ist (Fig. 24a und b).

2. crystalline anisotropy of the magnetic minerals. This is the case in rocks containing hematite, ilmenite or pyrrhotite as magnetic minerals. When the intrinsic susceptibility is small, as in the case of hematite, ilmenite and troilite, the degree of anisotropy, the *P*-factor, becomes:

2. durch die Kristall-Anisotropie der magnetischen Minerale. Dies gilt für Gesteine, die magnetische Minerale wie Hämatit, Ilmenit oder Pyrrhotit enthalten. Wenn die Material-Suszeptibilität klein wird, wie bei Hämatit, Ilmenit und Troilit, so ergibt sich der *P*-Faktor, der Grad der Anisotropie, zu:

$$P = \frac{\chi_a}{\chi_c} \approx \frac{\chi_{a,\,intr}}{\chi_{c,\,intr}} \qquad \text{[Har59, Hel73, Uye63]}$$

$\chi_{a,\,intr}$, $\chi_{c,\,intr}$ is the intrinsic susceptibility of the magnetic mineral in the *a* and *c*-direction respectively.

(See section 6.1, Figs. 14 and 17 for hematite and ilmenite and section 6.1, Fig. 108 for troilite)

Hier ist $\chi_{a,\,intr}$, $\chi_{c,\,intr}$ die Material-Suszeptibilität des magnetischen Minerals, in *a*- bzw. *c*-Richtung.

(Siehe Abschnitt 6.1, Fig. 14 und 17 für Hämatit und Ilmenit und Abschnitt 6.1, Fig. 108 für Troilit.)

6.2.2.2.6 Stress — Spannung

Uniaxial compression of rock specimens produces in general a decrease in susceptibility when measured in a magnetic field parallel to the direction of compression. Susceptibility measured perpendicular to the compression axis increases normally with compression [Kap55, Nag66, Nag70, Nag70a, Nag72] (Figs. 25, 26, 27).

Einachsige Kompression von Gesteinsproben ruft normalerweise eine Verminderung der Suszeptibilität hervor, wenn die Messungen in einem magnetischen Feld parallel zur Druckrichtung durchgeführt werden. Wird die Suszeptibilität senkrecht zur Kompressionsachse gemessen, so zeigt sich eine Zunahme der Suszeptibilität mit wachsender Kompression [Kap55, Nag66, Nag70, Nag70a, Nag72] (Fig. 25, 26, 27).

For magnetite or titanomagnetite bearing rocks the stress dependence of initial susceptibility at room temperature ($\tau \lesssim 500$ kg cm^{-2}) is experimentally expressed by:

Für Gesteine, die Magnetit oder Titanomagnetit enthalten, wird die Spannungsabhängigkeit der Anfangs-Suszeptibilität bei Zimmertemperatur ($\tau \lesssim 500$ kg cm^{-2}) durch folgende Gleichung wiedergegeben:

$$\chi^{\|}(\tau) = \frac{\chi_0}{1 + \beta\tau} \qquad \text{[Kap55, Nag70]},$$

when the direction of the applied magnetic field is parallel to the axis of uniaxial compression τ, and by:

wenn die Richtung des angelegten magnetischen Feldes parallel zur Achse der einachsigen Kompression τ ist, und durch die Gleichung:

$$\chi^{\perp}(\tau) = \chi_0(1 + \tfrac{1}{2}\beta\tau) \qquad \text{[Kap55, Nag70]},$$

when the direction of the applied magnetic field is perpendicular to τ.

The observed stress sensitivity β ranges from $0.5 \cdot 10^{-4}$ to $5 \cdot 10^{-4}$ cm^2 kg^{-1} for a wide range of igneous rocks [Nag70a] (Tables 2 and 3).

The stress dependence of susceptibility increases with increasing Ti-content in titanomagnetites (Table 2) [Nag70a, Ohn68] and it decreases with decreasing grain size [Nag66] (Figs. 28 and 29).

The stress sensitivity of magnetic susceptibility of rocks is the main cause for the occurrence of the so-called "seismomagnetic effect" [Nag70, Nag72, Sta64].

wenn das magnetische Feld senkrecht zu τ gerichtet ist.

Die beobachtete Spannungsempfindlichkeit β beträgt $0.5 \cdot 10^{-4}$ bis $5 \cdot 10^{-4}$ cm^2 kg^{-1} für die meisten Eruptivgesteine [Nag70a] (Tab. 2 und 3).

Die Spannungsabhängigkeit der Suszeptibilität nimmt mit der zunehmenden Ti-Konzentration in Titanomagnetiten zu (Tab. 2) [Nag70a, Ohn68] und nimmt mit abnehmender Korngröße ab [Nag66] (Fig. 28 und 29).

Die Spannungsempfindlichkeit der magnetischen Suszeptibilität der Gesteine ist die Hauptursache für den sogenannten „seismomagnetischen Effekt" [Nag70, Nag72, Sta64].

6.2.2.2.7 Figures and tables — Abbildungen und Tabellen

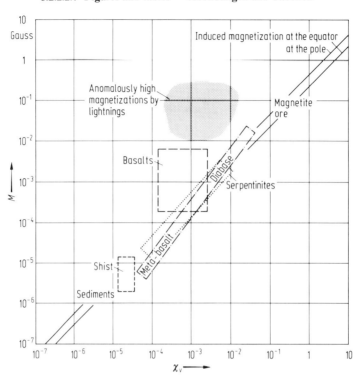

Fig. 1. Total natural magnetization M vs. susceptibility χ_v for a wide range of rock types. Basalts and rocks struck by lightning fall out of the general linear relationship between the two variables [Ref60, Ang81].

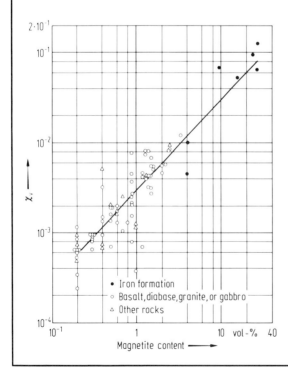

◀ Fig. 2. Relationship between magnetic susceptibility χ_v and magnetite content for a variety of rocks and ores. The empirical formula for susceptibility (full line) $\chi_v = 2.89 \cdot 10^{-3} \cdot V^{1.01}$ ($V =$ vol-% of magnetite) was derived from the data [Moo53].

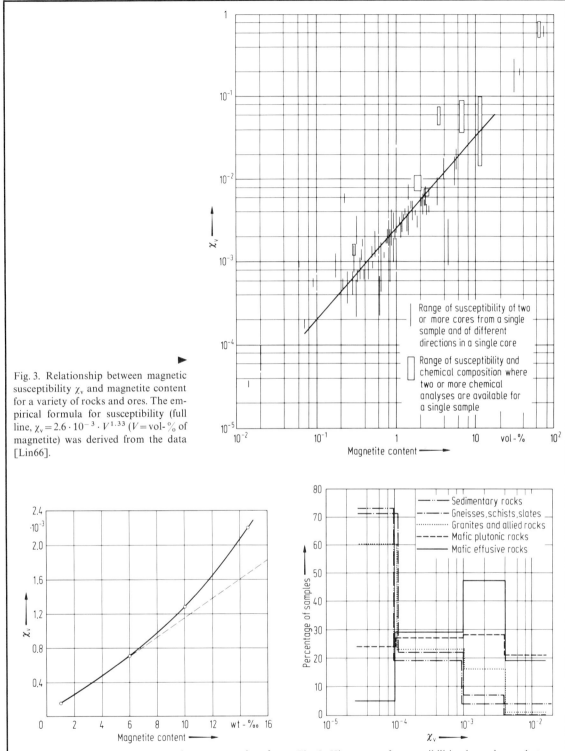

Fig. 3. Relationship between magnetic susceptibility χ_v and magnetite content for a variety of rocks and ores. The empirical formula for susceptibility (full line, $\chi_v = 2.6 \cdot 10^{-3} \cdot V^{1.33}$ ($V =$ vol-% of magnetite) was derived from the data [Lin66].

Range of susceptibility of two or more cores from a single sample and of different directions in a single core

Range of susceptibility and chemical composition where two or more chemical analyses are available for a single sample

Fig. 4. Susceptibility χ_v vs. magnetite concentration for various mixtures of granite and magnetite powders (solid line). The susceptibility has been measured in a field of 15 Oe. The dashed line shows a hypothetical linear relationship between the two variables, that should be expected without the influence of magnetostatic interaction between the dispersed multi-domain magnetite particles [Puz30].

Fig. 5. Histogram of susceptibilities in major rock types [Lin66].

Sedimentary rocks
Gneisses, schists, slates
Granites and allied rocks
Mafic plutonic rocks
Mafic effusive rocks

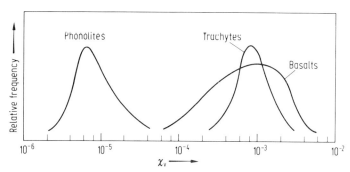

Fig. 6. Frequency distribution of susceptibilities, χ_v, of tertiary continental basalts, trachytes and phonolites from different localities. Basalt data are from [Ref60] (N = 61) and [Kri73] (N = 109), trachytes (N = 25) and phonolites (N = 20) from [Kro76]. (N = number of samples).

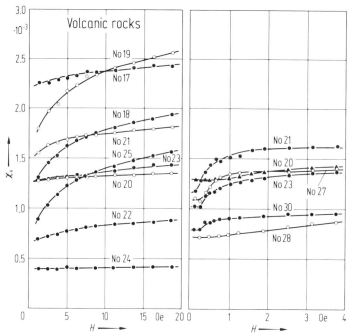

Fig. 7. Magnetic susceptibility, χ_v, of volcanic rocks from Japan as a function of external magnetic field H [Nag61]. No. 17: olivine basalt from Huzi, Hoei crater, No. 18: hypersthene-augite-olivine basalt, Huzi, Makuiwa 1; No. 19: olivine basalt, Huzi, Tawaranotaki; No. 20: olivine basalt, Huzi, Makuiwa 2; No. 21: augite-olivine basalt, Huzi, Makuiwa 3; No. 22: augite-olivine basalt, Huzi, Hoei bomb; No. 23: pyroxene-olivine basalt, Huzi, Makuiwa 4; No. 24: aphanitic andesite, Huzi, Karasuiwa; No. 25: pyroxene-olivine basalt, Huzi, Aokigahara; No. 27: olivine basalt, Amagi, Zizodo; No. 28: olivine basalt, Amagi, Hatikubo-yama; No. 30: pyroxene-olivine basalt, Amagi, Inatori.

It is evident from these curves that considerable error can be introduced to the determination of the Koenigsberger Q-factor if the induced magnetization of a rock sample is measured in a magnetic field not equal to the ambient magnetic field.

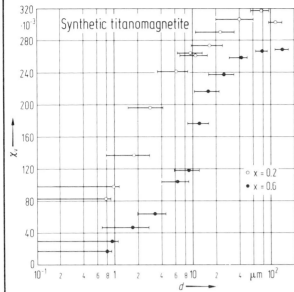

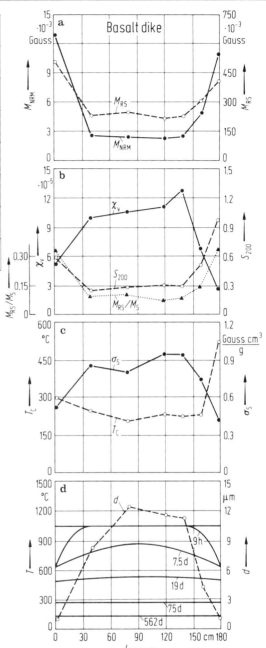

Fig. 8. Grain size dependence of initial susceptibility χ_v of synthetic titanomagnetite (x = 0.2 and 0.6) measured at room temperature. 1 % by weight of magnetic material dispersed in KBr-powder. d = diameter [Day77a].

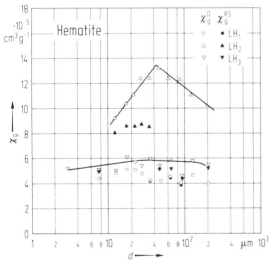

Fig. 9. Grain size dependence of initial susceptibility χ_g of natural hematite. d = diameter.

LH$_1$-samples are from a Kidney ore with radial fibrous structure. The separated grains of all LH$_1$-grain size fractions were platy.

LH$_2$-samples are from hematite ore from Kimberley (South Africa). The grains of the separated fractions were very thin, but well crystallized.

LH$_3$-samples are from hematite ore from Framont (Vosges, France). It was very poorly crystallized. The separated grains were less platy-shaped than those of LH$_1$ and LH$_2$.

χ_g^0 is the natural mass susceptibility and χ_g^{RS} the mass susceptibility after the sample has been given an isothermal saturation remanent magnetization [Dan78].

Fig. 10. Variation of magnetization and related parameters in a basalt dike from Monte Somma (Italy), measured along a horizontal profile across the 180 cm thick dike. l = distance from margin.

a) Natural remanent magnetization M_{NRM} and saturation remanence M_{RS}.

b) Susceptibility χ_v and magnetic stability S_{200} (NRM after alternating field demagnetization in 200 Oe/original NRM).

c) Specific saturation magnetization σ_s and Curie temperature T_C.

d) Mean grain diameter of titanomagnetites d and theoretical temperature profiles for various times after intrusion (h = hours, d = days) [Pet76].

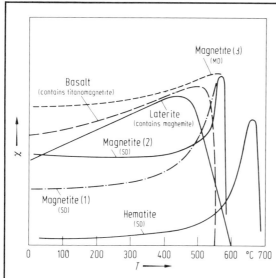

Fig. 11. Temperature dependence of initial susceptibility χ of different materials. Magnetite (1) are single-domain (SD) particles with mean grain size 0.04 µm, magnetite (2) is closer to the multi-domain (MD) threshold with needle-like grains, about 0.2 µm long, magnetite (3) is multi-domain (MD) material. The SD-hematite with grain size of 0.1···1.0 µm shows spectacular enhancement of susceptibility at about 650 °C. The two rock samples show only modest enhancement of susceptibility. The basalt sample contains titano-magnetite of oxidation class V (see [Wil67, Wat67]). Hopkinson peak (enhancement of initial susceptibility close to Curie temperature) is pronounced in SD-material, less pronounced in MD-material [Dun74, Uye63].

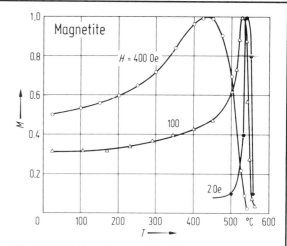

Fig. 12. Field dependence of Hopkinson peak in single-domain magnetite particles. The curves represent the variation of normalized induced magnetization, M, vs. temperature, T, in different magnetic fields, H. The sample consists of synthetic single-domain magnetite dispersed to 0.5 wt-% in aluminum powder [Sug80].

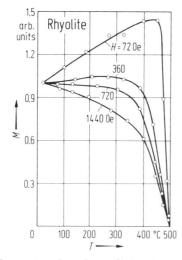

Fig. 13. Temperature dependence of induced magnetization M of rhyolite sample (Niisima, Japan). Magnetic mineral contained in the rock is Ti-poor titanomagnetite. Measurements have been carried out in different magnetic fields, H. Hopkinson peak is suppressed at higher field strengths [Nag61].

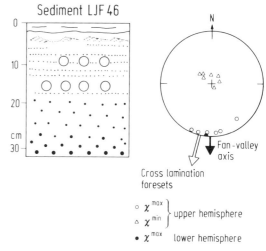

Fig. 14. Sedimentary and magnetic fabric of recent sediments from the La Jolla submarine Fan-Valley (California). Box core LJF 46 is a good example of a current-produced, primary sedimentary and magnetic fabric. The graded sand layer is parallel- and ripple-drift cross-laminated in its top third. In its basal, unlaminated, coarser grained part it is rich of finely dispersed, small mica flakes and plant chips. Cross-lamination foresets trend about 190°, approximately parallel with the Fan-valley axis. The magnetic fabric of the entire graded sand layer is characterized by well-grouped, near horizontal susceptibility maxima with a mean azimuth of 185°. Minima are well grouped and near-vertical. Equal area (Schmidt) net for susceptibility axes (projection plane = bedding plane) [Rad70].

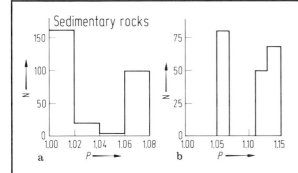

a

b

◄

Fig. 15a. Histogram of degree of anisotropy of susceptibility (*P*-factor) of a wide variety of sedimentary rocks: Red conglomerates, red sandstones, red silty shales [Jan72], carboniferous sandstones [Iow67], artificial sediments [Ham68], ash, limestones, various sandstones and shale [Ful63]. $P = \chi^{max}/\chi^{min}$; N = number of samples. Mean anisotropy degree is 1.03. The low value is influenced by the masking effect of paramagnetic fraction (see Fig. 16).

Fig. 15b. Histogram of degree of anisotropy of susceptibility (*P*-factor) of slightly metamorphosed sedimentary rocks: Slates (highest anisotropy), silty slates (lower anisotropy), graywackes (lowest anisotropy). Data from [Ham67, Jan72]. (N = number of samples). Anisotropy distinctly higher than for non-metamorphosed sediments.

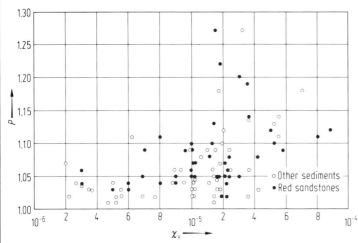

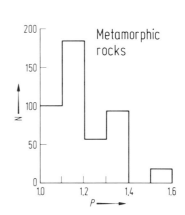

Fig. 16. Variation of the anisotropy of susceptibility (*P*-factor) of sedimentary rocks with bulk volume susceptibility χ_v. The apparent correlation between the degree of susceptibility anisotropy and bulk susceptibility is due to the masking of the anisotropy by an isotropic contribution to the susceptibility of the rock caused by the paramagnetic matrix [Ful63].

Fig. 17. Histogram of the degree of anisotropy of susceptibility (*P*-factor) of metamorphic rocks. The mean anisotropy degree (from 453 samples, including all of the most frequently occurring metamorphic rock types, ranging from slates to gneisses) is 1.24. (N = number of samples). Data from [Ful63, Jan67, Hro70].

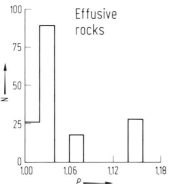

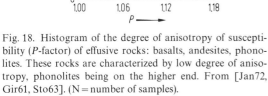

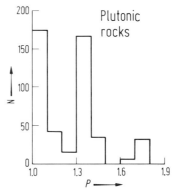

Fig. 18. Histogram of the degree of anisotropy of susceptibility (*P*-factor) of effusive rocks: basalts, andesites, phonolites. These rocks are characterized by low degree of anisotropy, phonolites being on the higher end. From [Jan72, Gir61, Sto63]. (N = number of samples).

Fig. 19. Histogram of the degree of anisotropy of susceptibility (*P*-factor) of plutonic rocks: Gabbros, granodiorites and granites from various localities. Data from [Gir61, Bal60, Hro70]. (N = number of samples).

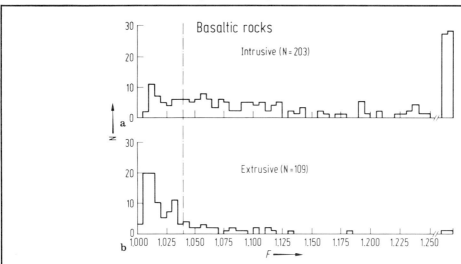

Fig. 20. Histogram of F-values (emplacement mode indicator F, see Table 1) for basaltic rocks.
a) 203 intrusive samples (dikes and sills) from Iceland, Scotland and Bermuda,

b) 109 extrusive samples (lava flows) from Iceland, Oregon, the Azores, Mexico and Bermuda.
Dashed line at $F = 1.040$ represents the line which can be used to distinguish intrusives from extrusives [Ell75].

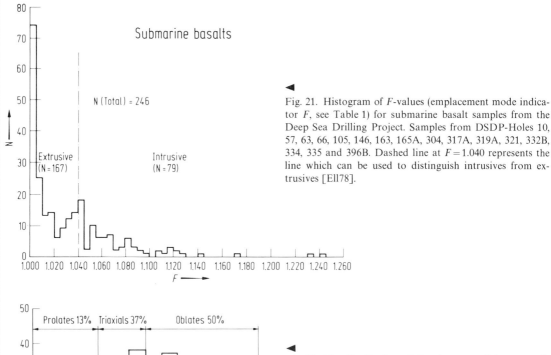

Fig. 21. Histogram of F-values (emplacement mode indicator F, see Table 1) for submarine basalt samples from the Deep Sea Drilling Project. Samples from DSDP-Holes 10, 57, 63, 66, 105, 146, 163, 165A, 304, 317A, 319A, 321, 332B, 334, 335 and 396B. Dashed line at $F = 1.040$ represents the line which can be used to distinguish intrusives from extrusives [Ell78].

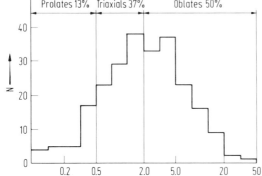

Fig. 22. The distribution of the character of the magnetic susceptibility ellipsoid (oblate versus prolate) of a wide variety of rocks. The susceptibility ellipsoid is here defined as an ellipsoid with semiaxes χ_a, χ_b, χ_c.

For $\dfrac{\chi_b - \chi_c}{\chi_a - \chi_b} > 1$ the ellipsoid becomes oblate,

for $\dfrac{\chi_b - \chi_c}{\chi_a - \chi_b} < 1$ the ellipsoid becomes prolate.

The dominance of rocks with oblate susceptibility ellipsoid shows that on the whole most structures are of the planar rather than of the linear type [Kha62].

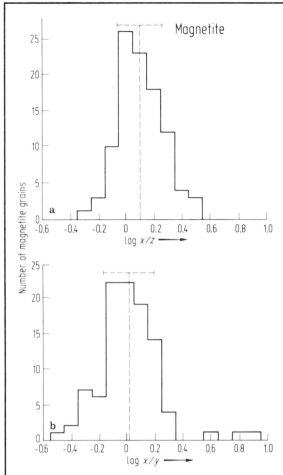

Fig. 23. The frequency distribution of the dimensions (x, y, z) of magnetite grains parallel to the principal directions of the susceptibility ellipsoid of Skaergaard ferrogabbro samples. The susceptibility values have been found to form an oblate spheroid, with the well defined minimum axis close to the vertical (z-direction). Susceptibility ellipsoid defined as in Fig. 22.

a) vertical section: The mean ratio of grain dimensions is 1.19 (x represents horizontal direction, z vertical direction), as indicated by the dashed line. The long axes of the grains, therefore, are preferentially oriented in the layering plane.

b) horizontal section: The mean ratio of grain dimensions is 0.99 (x and y in the horizontal plane). No preferential direction observed [Uye63].

Table 1. Parameters describing magnetic anisotropy in rock fabric studies.

Property	Parameter	Ref.
Degree of anisotropy P-factor	χ_a/χ_c	Nag61
Degree of anisotropy (percentage)	$\left(\dfrac{\chi_a - \chi_c}{\chi_a}\right) \cdot 100$	Gra66
Degree of anisotropy	$\dfrac{\chi_a - \chi_c}{\chi_b}$	Ree66
Foliation F-factor	$\dfrac{\chi_a + \chi_b}{2\chi_c}$	Bal60
Foliation F-factor	χ_b/χ_c	Hro70
Foliation F-factor	$\dfrac{\chi_b - \chi_c}{1/3(\chi_a + \chi_b + \chi_c)}$	Kha62
Lineation L-factor	χ_a/χ_b	Bal60
Lineation L-factor	$\dfrac{\chi_a - \chi_b}{1/3(\chi_a + \chi_b + \chi_c)}$	Kha62
Lineation L-factor	$\dfrac{2\chi_a}{\chi_b + \chi_c}$	Hro71
$R = \dfrac{\text{magnetic lineation}}{\text{magnetic foliation}}$	$\dfrac{N_b - N_a}{N_c - N_b}$	Sta60
Character of susceptibility ellipsoid E-factor	$\dfrac{\chi_b^2}{\chi_a \cdot \chi_c}$	Hro71
Prolateness of susceptibility ellisoid [1]	$\dfrac{\chi_a - \chi_b}{\chi_b - \chi_c}$	Kha62
Oblateness of susceptibility ellipsoid [1]	$\dfrac{\chi_b - \chi_c}{\chi_a - \chi_b}$	Kha62
Emplacement mode indicator F (basalts only)	$\dfrac{\chi_a'}{\sqrt{\chi_b' \cdot \chi_c'}}$ [2]	Ell75
Magnetic excess	$1/2(\chi_a - \chi_b) - \chi_c$	Gra58
Azimuthal anisotropy	$\chi_a - \chi_b$	Gra58
Azimuthal anisotropy quotient	$\dfrac{\chi_a - \chi_b}{1/2(\chi_a - \chi_b) - \chi_c}$	Gra58
"angle V"	$\sin^{-1}\sqrt{\dfrac{\chi_b - \chi_c}{\chi_a - \chi_c}}$	Gra58

[1] Here the susceptibility ellipsoid is defined as an ellipsoid with half axis χ_a, χ_b, χ_c, so being different from the usual definition.

[2] $\chi_a', \chi_b', \chi_c'$ are the susceptibility tensor magnitudes recalculated for a standard volume susceptibility of $1.0 \cdot 10^{-3}$.

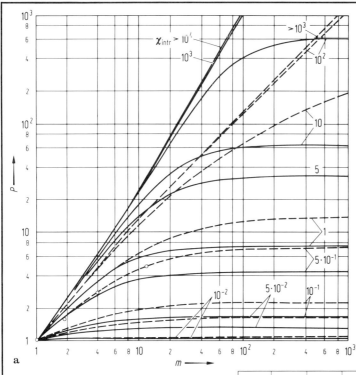

Fig. 24a. Relationship between the degree of anisotropy of susceptibility

$$P = \frac{\chi_{max}}{\chi_{min}} = \frac{1 + \chi_{intr}\, N_c}{1 + \chi_{intr}\, N_a}$$

and the dimension ratio, m, of the axes of a spheroidal magnetic grain for various susceptibility values.

  Solid line = prolate spheroid,
dashed line = oblate spheroid,
  χ_{intr} = intrinsic susceptibility,
    N_a = demagnetization factor of the grain in the direction of maximum susceptibility,
    N_c = demagnetization factor perpendicular to a-direction, in the direction of minimum susceptibility,
  circles = magnetite ore discs.

Fig. 24b. Same as in Fig. 24a with the lower left corner enlarged,
  full circle = average for Skaergaard ferrogabbros,
open circles = individual Skaergaard rocks [Uye63].

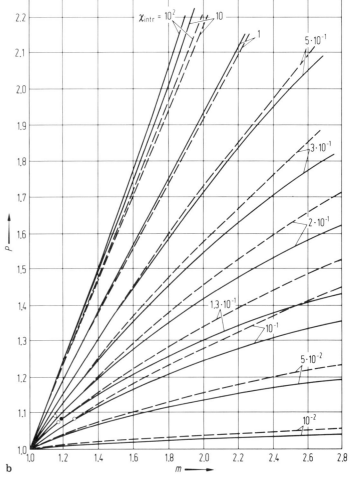

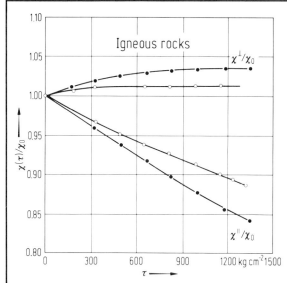

Fig. 25. Examples of the effect of uniaxial compression τ on the reversible initial magnetic susceptibility $\chi(\tau)/\chi_0$ of igneous rocks.

$\chi^{\parallel}$ is the susceptibility measured parallel to the axis of compression.

$\chi^{\perp}$ is the susceptibility measured perpendicular to compression.

χ_0 denotes the isotropic susceptibility for compression $\tau = 0$ [Kap55].

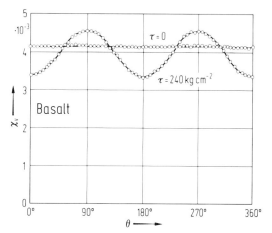

Fig. 26. Example of anisotropic initial magnetic volume susceptibility χ_v of a basalt sample under uniaxial compression $\tau = 240$ kg cm^{-2}, measured along the axis of $\theta = 0°$ to $\theta = 180°$, where θ is the angle between τ and the applied magnetic field.

For $\tau = 0$ the rock has isotropic susceptibility;

for low uniaxial compression $\tau \lesssim 250$ kg cm^{-2} the anisotropic susceptibility can be described by the expression

$$\chi(\theta, \tau) = \chi_0 \left[1 - \frac{\beta \tau}{4} (3 \cos \theta + 1) \right]$$

where β denotes the stress sensitivity [Nag72].

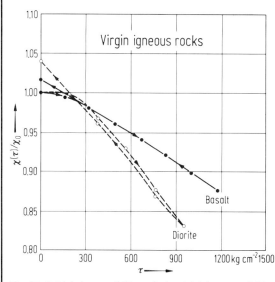

Fig. 27. Initial irreversibility of the initial susceptibility $\chi(\tau)/\chi_0$ with uniaxial compression τ. Examples of the effect of uniaxial compression on the magnetic susceptibility of virgin igneous rocks that have not experienced uniaxial compression prior to the experiment. Susceptibility has been measured parallel to the axis of compression [Nag70].

Table 2. Stress sensitivity β of magnetic susceptibility $\chi = \dfrac{\chi_0}{1 + \beta \tau}$ (χ measured parallel to uniaxial compression τ) for titanomagnetites $Fe_{3-x}Ti_xO_4$ at low pressures ($\tau \lesssim 500$ kg cm^{-2}). χ_0 is the susceptibility at zero pressure.

Titano-magnetite composition x	β [10^{-4} cm^2 kg^{-1}] theoretical values [Sta72]	β [10^{-4} cm^2 kg^{-1}] experimental values extrapolated from [Ohn68]
0	1.11	2.3
0.04	1.28	2.4
0.10	1.5	2.5
0.18	2.4	2.9
0.31	3.2	4.1
0.56	11.5	7.1
0.60	–	7.5
0.70	–	8.0

Table 3. Examples of experimentally observed values of stress sensitivity β of magnetic volume susceptibility $\chi_v = \chi_0/(1 + \beta\tau)$ of igneous rocks (χ_v measured parallel to uniaxial compression). χ_0 is the susceptibility at zero pressure.

Rock type	β 10^{-4} cm² kg⁻¹	χ_0 10^{-3}	Ref.
Olivine basalt	0.42	3.3	Nag70a
Homogeneous basalt	5.6	8.4	Nag70a
Basalt	1.42	5.1	Kap55
Basalt	2.1	0.71	Kap55
Basalt	3.3	3.86	Kap55
Diabase	1.81	7.1	Kap55
Gabbro	0.93	0.14	Kap55
Andesite	2.95	2.5	Kap55
Syenite	0.85	1.75	Kap55
Diorite	0.86	7.1	Kap55
Diorite	1.9	9.9	Nag70a
Porphyrite	1.18	1.51	Kap55

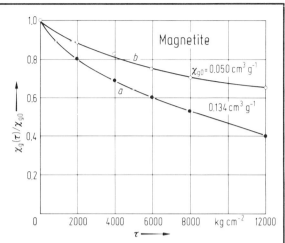

Fig. 28. Stress dependence of initial magnetic susceptibility $\chi_g(\tau)/\chi_{g_0}$ of two magnetite grain assemblages of different grain sizes.
Curve (a) represents a sample of magnetite grains with average grain size 134 μm, curve (b) a sample of magnetite with average grain size 50 μm.
Susceptibility χ_g has been measured parallel to the axis of compression. χ_{g_0} is the mass susceptibility at $\tau = 0$ [Nag66].

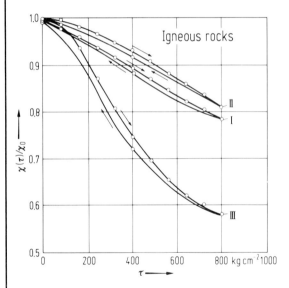

◀

Fig. 29. Stress dependence of the initial magnetic susceptibility $\chi(\tau)/\chi_0$ (measured parallel to the axis of compression) of different igneous rock samples with increasing effective magnetic grain size.
Sample I contains large magnetite grains (2 mm in diameter) subdivided into much smaller forms by small ilmenite lamellae and numerous spinel lenses. Sample II is similar to sample I. Sample III contains large homogeneous magnetite grains.
χ_0 is the susceptibility at $\tau = 0$ [Jel75].

6.2.3 Remanent magnetization — Remanente Magnetisierung

6.2.3.1 Introductory remarks — Einleitende Bemerkungen

The irreversible part of the total natural magnetization of a rock that is left in zero-field is termed natural remanent magnetization NRM. Origin and type of natural remanent magnetization differs greatly in different rock types. It depends on the magnetic minerals contained in the rock on the one hand, and on the individual history of the rock on the other.

Der im Nullfeld gemessene irreversible Teil der natürlichen Gesamtmagnetisierung eines Gesteins wird natürliche remanente Magnetisierung NRM genannt. Ursprung und Art der natürlichen Remanenz ist in verschiedenen Gesteinsarten sehr unterschiedlich. Das hängt einerseits von der Zusammensetzung der im Gestein enthaltenen magnetischen Minerale ab, andererseits von der individuellen Entstehungsgeschichte des Gesteins.

6.2.3.2 The different types of remanent magnetization acquired by rocks — Die verschiedenen Arten der in Gesteinen auftretenden Remanenzen

6.2.3.2.1 Thermoremanent magnetization TRM — Thermoremanente Magnetisierung TRM

Thermoremanent magnetization is the magnetization acquired by a ferromagnetic substance upon cooling at temperatures below the Curie point (more accurately: upon cooling below the "blocking temperature" $T_B < T_C$ [Nee55]) in a magnetic field. This magnetization is usually much more intense than the isothermal remanent magnetization produced by the same field at room temperature, as long as the external magnetic field is small, i.e. as it is of the order of the earth's magnetic field.

Most igneous and high-temperature metamorphic rocks have acquired thermoremanent magnetization TRM upon cooling [Koe38, Nag66, Nee55, Sta74]. The natural TRM of rocks is usually larger than the natural induced magnetization $M_i = \chi \cdot H_{earth}$ i.e. Koenigsberger Q-factor > 1.

This is particularly evident for basalts (Fig. 30) [Mce77, Moo53].

The predominance of TRM over the induced magnetization $M_i = \chi \cdot H_{earth}$ is a function of grain size of the magnetic minerals contained in the rock (Fig. 31) [Day77, Dun81].

For the application in paleomagnetism, in particular for the determination of paleointensities of the earth's magnetic field, the following characteristics of TRM of common rocks are of importance:

1. The direction of thermoremanent magnetization TRM of rocks is in general parallel to the magnetic field generating the TRM (for exceptions, see 6.2.3.2.7) [Che25, Koe36, Lam66, Nag61].

2. The intensity of TRM varies linearly with the magnetic field generating the TRM, in fields of the order of the earth's magnetic field (Figs. 32, 33 and 34) [Day77, Nag61, The38].

3. The additivity of partial TRM (pTRM):
The partial TRM at room temperature T_0, $M_{T_2}^{T_1}(T_0)$ produced by cooling in a weak field H through $\Delta T = T_2 - T_1$ ($T_1 < T_2 < T_C$), is independent of the magnetization produced by corresponding field-cooling through other temperature intervals (Fig. 35). This means that the total TRM is equal to the sum of partial TRMs:

$$M_{T_C}^{T_0}(T_0) = \sum_{T_{i-1}=T_0}^{T_i=T_C} M_{T_i, H}^{T_{i-1}}(T_0); \ (T_0 \leq T_{i-1} < T_i \leq T_C) \qquad \text{(Figs. 36 and 37) [Nag61, Nee55, Sta74, The59]}.$$

The intensity and stability of TRM is critically dependent on grain size of the magnetic mineral that carries the magnetization (Figs. 31, 38, 39, 40) [Day77, Lar69, Dun73b, Dun74a].

Theoretical treatment of thermoremanent magnetization is found in [Day77, Nag61, Nee55, Sta74].

6.2.3.2.1.1 Figures — Abbildungen

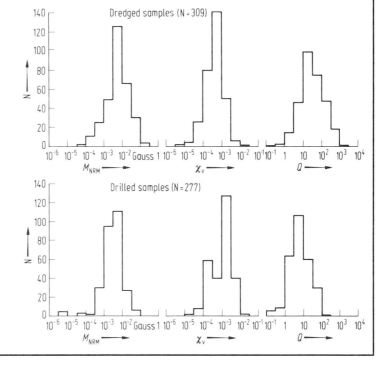

Fig. 30. Histograms of natural remanent magnetization M_{NRM}, volume susceptibility χ_v and Koenigsberger Q-factor of ocean floor basalts. Distinction is made between samples that have been obtained by dredging and samples that have been obtained by deep sea drilling [Mce77]. N = number of samples.

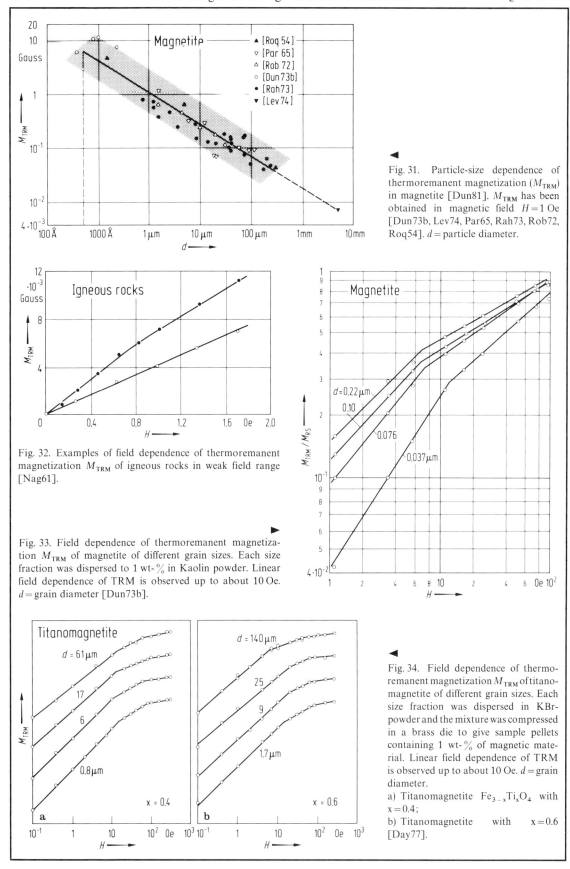

Fig. 31. Particle-size dependence of thermoremanent magnetization (M_{TRM}) in magnetite [Dun81]. M_{TRM} has been obtained in magnetic field $H = 1$ Oe [Dun73b, Lev74, Par65, Rah73, Rob72, Roq54]. d = particle diameter.

Fig. 32. Examples of field dependence of thermoremanent magnetization M_{TRM} of igneous rocks in weak field range [Nag61].

Fig. 33. Field dependence of thermoremanent magnetization M_{TRM} of magnetite of different grain sizes. Each size fraction was dispersed to 1 wt-% in Kaolin powder. Linear field dependence of TRM is observed up to about 10 Oe. d = grain diameter [Dun73b].

Fig. 34. Field dependence of thermoremanent magnetization M_{TRM} of titanomagnetite of different grain sizes. Each size fraction was dispersed in KBr-powder and the mixture was compressed in a brass die to give sample pellets containing 1 wt-% of magnetic material. Linear field dependence of TRM is observed up to about 10 Oe. d = grain diameter.
a) Titanomagnetite $Fe_{3-x}Ti_xO_4$ with $x = 0.4$;
b) Titanomagnetite with $x = 0.6$ [Day77].

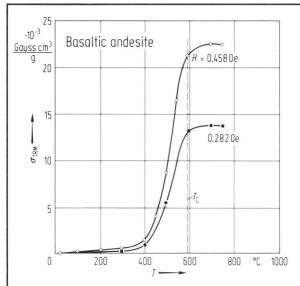

Fig. 35. Acquisition of thermoremanent specific magnetization σ_{TRM} of a sample of basaltic andesite (Miyake-sima, Japan) in two different magnetic fields. The curves represent the TRM measured at room temperature T_0 after weak-field-cooling from temperatures $T > T_0$. The cooling temperature T has been increased stepwise until above the Curie point T_C [Nag53].

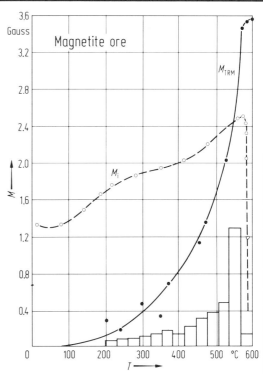

Fig. 37. Acquisition of thermoremanent magnetization M_{TRM} and histogram of partial thermoremanent magnetizations of a sample of magnetite ore. The solid curve represents the cumulative partial TRM measured at room temperature T_0 after weak-field-cooling from temperatures $T_C \geqq T > T_0$. The cooling temperatures T have been increased stepwise up to the Curie temperature T_C. The histogram represents the partial TRM acquired over the respective temperature intervals ΔT. The dashed line represents the temperature dependence of the weak field induced magnetization M_i. Inducing field $H = 0.5$ Oe [Koe32].

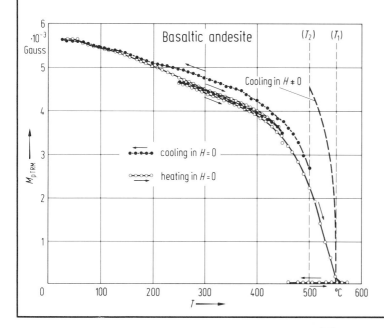

◄

Fig. 36. Acquisition of partial thermo-remanent magnetization M_{pTRM} (same sample as in Fig. 35). The sample has been cooled from temperature T_1 to T_2 $(T_2 < T_1 \leqq T_C)$ in a magnetic field $H \neq 0$ and further cooled in zero field $H = 0$ [Nag53].

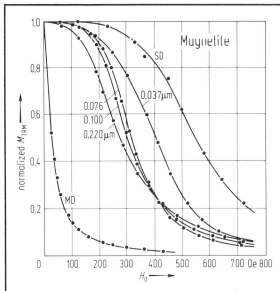

Fig. 38. Grain size dependence of the stability of thermoremanent magnetization of magnetite against alternating-field demagnetization. Magnetite powders were dispersed to 1 wt-% in Kaolin. Grain size range from multi-domain-size MD to single-domain-size SD (see also Fig. 94). Magnetic field inducing TRM was 1 Oe. H_d = peak alternating field [Dun73b].

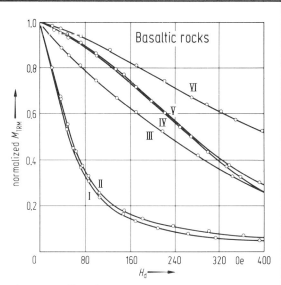

Fig. 39. Stability of thermoremanent magnetization of basaltic samples of different oxidation classes against alternating field demagnetization. Carrier of remanent magnetization are titanomagnetite grains. The oxidation state of the titanomagnetite grains varies from oxidation class I of [Wat67, Wil67] (unoxidized homogeneous grains) to oxidation class VI (highly oxidized grains subdivided by secondary non-magnetic minerals). H_d = peak alternating field [Dun74a].

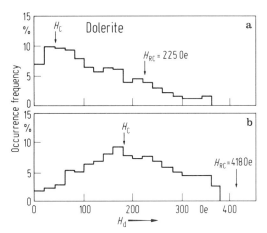

Fig. 40. Stability of thermoremanent magnetization of a dolerite sample against alternating-field demagnetization [Lar69].
a) Coercivity spectrum of the natural sample. Carrier of remanent magnetization are fairly homogeneous grains of titanomagnetite. Average grain size 50 μm.
b) Coercivity spectrum after heating in air to 600 °C. Part of the titanomagnetite grains have broken down to hematite thereby forming numerous tiny lamellae walls, effectively increasing the number of small magnetic grains.
H_C is the bulk coercive force of the sample. H_{RC} the bulk remanence coercivity.

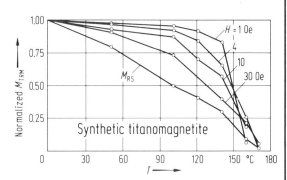

Fig. 41. Stability of thermoremanent magnetization M_{TRM} obtained in different magnetic fields H and of saturation isothermal remanent magnetization M_{RS} in a synthetic titanomagnetite ($Fe_{3-x}Ti_xO_4$ with x = 0.6) powder sample against stepwise thermal demagnetization. The titanomagnetite powder (average grain size 6.4 μm) was dispersed to 1 wt-% in KBr [Day77].

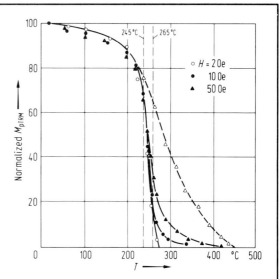

Fig. 42. Field dependence of blocking temperatures T_B of single-domain particles. The solid curves represent heating curves for partial thermoremanent magnetization pTRM produced in different magnetic fields H applied between 265 °C to 245 °C.

The dashed curve (open triangles) represents the heating curve for a pTRM produced by cooling in 50 Oe from 430···245 °C. The sample used was laterite containing single-domain-hematite- and maghemite-particles as magnetic minerals [Eve61].

6.2.3.2.2 Chemical remanent magnetization CRM — Chemische remanente Magnetisierung CRM

When a ferromagnetic material is produced by a chemical process or phase-change and grows slowly from tiny nuclei at a temperature below its Curie temperature T_C, the remanence which it thus acquires in the ambient magnetic field is referred to as chemical remanent magnetization CRM (Figs. 43 and 44). Intensity and stability of CRM is comparable to that of a thermoremanent magnetization TRM produced in the same magnetic field [Hai58, Joh72, Kob59, Nee55, Sta74].

Chemical remanent magnetization may be acquired by rocks upon oxidation or reduction, recrystallization, chemical precipitation, or exsolution of ferrimagnetic minerals at temperatures below the respective Curie temperatures. Consequently a chemical remanent magnetization is involved to a certain percentage in almost all rocks carrying natural remanent magnetization. CRM plays an important role in ocean floor basalts [Hal77, Joh78, Ozi74, Pet79, Pre79] (Figs. 49, 50 and 51).

For a theoretical treatment of chemical remanent magnetization the reader is referred to [Nee55, Sta74].

6.2.3.2.2.1 Figures   and   tables   —   Abbildungen   und   Tabellen

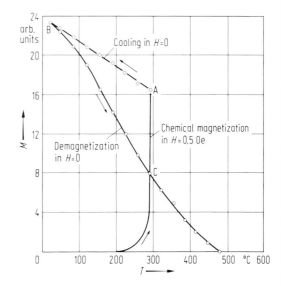

Fig. 43. Growth and thermal demagnetization of chemical remanent magnetization formed by the reduction process α-$Fe_2O_3 \rightarrow Fe_3O_4$ in $H = 0.5$ Oe. The demagnetization curve lies distinctly below the cooling curve (non-coincidence of points A and C). This is due to a time decay of the chemical remanent magnetization [Hai58].

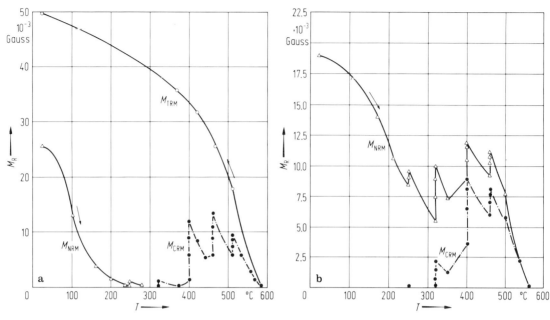

Fig. 44. Growth of chemical remanent magnetization CRM in samples of ocean floor pillow basalt from the Juan de Fuca Ridge during heating experiments in air. Carrier of remanent magnetization is titanomagnetite [Mar71].

a) Oxidation of titanomagnetite and associated growth of CRM occurs at temperatures above the Curie temperature of the original titanomagnetite. The temperature was held constant for 30 min at 250, 320, 400, 460 and 510 °C, whereby the sample was kept in a magnetic field of 0.5 Oe. CRM was produced parallel to the applied field.

triangles: thermal demagnetization of NRM;

full circles: growth of CRM (for the moment of measurement the inducing field $H = 0.5$ Oe was switched off);

open circles: acquisition of TRM ($H = 0.5$ Oe) by the oxidized rock.

b) Oxidation of titanomagnetite and associated growth of CRM occurs at temperatures below the Curie temperature of the original magnetic mineral. During the growth experiment the direction of original NRM and of inducing magnetic field was kept perpendicular to each other. CRM was produced parallel to the applied field. The original NRM was not destroyed by this process, but rather increased in intensity during the oxidation, thereby retaining its original direction of NRM.

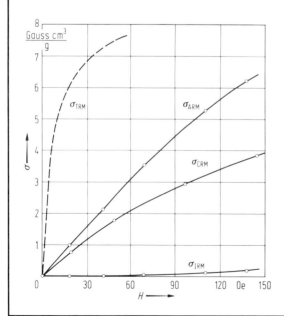

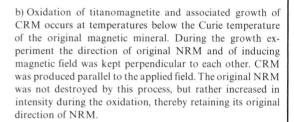

Fig. 45. Dependence of different kinds of specific remanent magnetization on the strength of the inducing magnetic field [Hai58]. Comparison of chemical remanent magnetization CRM with isothermal remanent magnetization IRM, anhysteretic remanent magnetization ARM and thermoremanent magnetization TRM in magnetite specimens. CRM, ARM and IRM curves have been measured on samples produced by the reduction process

$$\alpha\text{-Fe}_2\text{O}_3 \rightarrow \text{Fe}_3\text{O}_4.$$

The dashed curve for TRM has been taken from [The42].

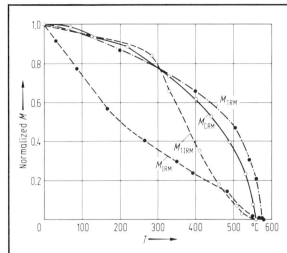

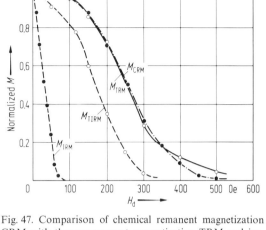

Fig. 46. Comparison of chemical remanent magnetization CRM with thermoremanent magnetization TRM and isothermal remanent magnetization IRM in magnetite: Stability against thermal demagnetization.
The samples have been produced by the reduction process

$$\alpha\text{-Fe}_2\text{O}_3 \rightarrow \text{Fe}_3\text{O}_4.$$

Inducing magnetic field was $H = 3$ Oe in the case of CRM and TRM, $H = 200$ Oe in the case of IRM. TIRM denotes an isothermal magnetization obtained at $T = 340\,°\text{C}$ in $H = 20$ Oe and then cooled to room temperature in $H = 0$ [Kob59].

Fig. 47. Comparison of chemical remanent magnetization CRM with thermoremanent magnetization TRM and isothermal remanent magnetization IRM in magnetite: Stability against alternating field demagnetization.
The samples have been produced by the reduction process

$$\alpha\text{-Fe}_2\text{O}_3 \rightarrow \text{Fe}_3\text{O}_4.$$

Inducing field was $H = 10$ Oe for CRM, $H = 0.5$ Oe for TRM and $H = 30$ Oe for IRM. TIRM denotes an isothermal magnetization obtained at $T = 340\,°\text{C}$ in $H = 10$ Oe and then cooled to room temperature in $H = 0$ [Kob59].

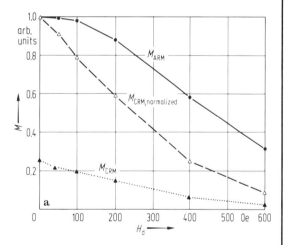

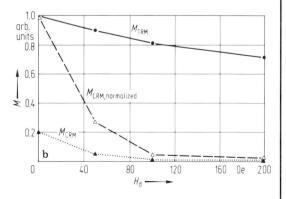

Fig. 48. Stability of chemical remanent magnetization CRM against alternating field demagnetization. CRM was formed by oxidation of natural magnetite particles with average diameter less than 44 µm. Prior to the oxidation process the magnetite sample was given an anhysteretic remanent magnetization ARM in a direct field of $H = 10$ Oe. The CRM formed in an inducing field of $H = 6$ Oe which was oriented perpendicular to the original ARM [Joh72].
a) Magnetite sample partly oxidized to maghemite ($\gamma\text{-Fe}_2\text{O}_3$). CRM is oriented parallel to the inducing field, perpendicular to the ARM, suggesting that no exchange interaction between the two magnetic phases takes place. Oxidation temperature was $T = 150\,°\text{C}$.
b) Magnetite sample partly oxidized to hematite ($\alpha\text{-Fe}_2\text{O}_3$). CRM is also oriented parallel to the inducing field, perpendicular to the ARM, suggesting that no exchange interaction between the two phases takes place. Oxidation temperature was $T = 200\,°\text{C}$.

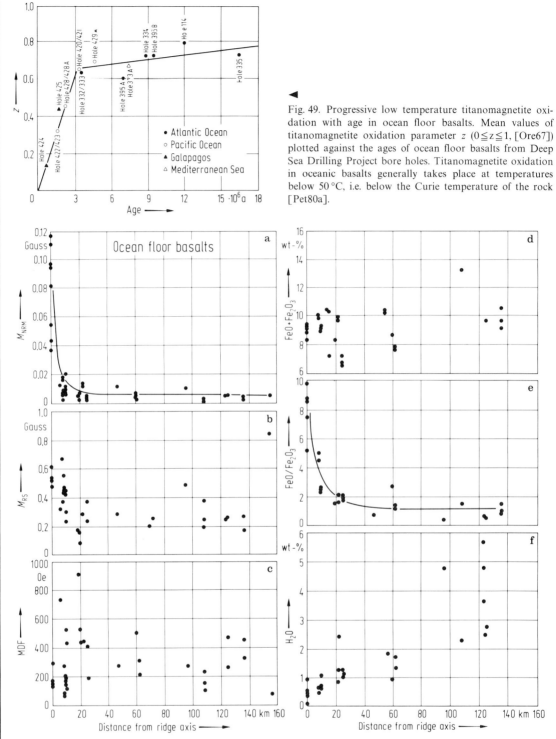

Fig. 49. Progressive low temperature titanomagnetite oxidation with age in ocean floor basalts. Mean values of titanomagnetite oxidation parameter z ($0 \leq z \leq 1$, [Ore67]) plotted against the ages of ocean floor basalts from Deep Sea Drilling Project bore holes. Titanomagnetite oxidation in oceanic basalts generally takes place at temperatures below $50\,°C$, i.e. below the Curie temperature of the rock [Pet80a].

Fig. 50. Variation of magnetic and chemical properties of ocean floor basalts dredged along a traverse across the Mid-Atlantic Ridge at $45°$ N with increasing distance l from the ridge axis (i.e. increasing age of the rock). The variation of the magnetic properties is caused primarily by low temperature oxidation of the titanomagnetites contained in the rocks (see also Fig. 49) [Irv70].

a) Natural remanent magnetization, M_{NRM}.

b) Saturation remanence, M_{RS}.

c) Stability of remanent magnetization. The medium destructive field MDF of a rock is the magnetic field necessary to erase half of the original remanent magnetization by alternating field demagnetization.

d) Bulk rock total iron content (expressed in $FeO + Fe_2O_3$).

e) Bulk rock oxidation ratio FeO/Fe_2O_3.

f) Bulk rock water content.

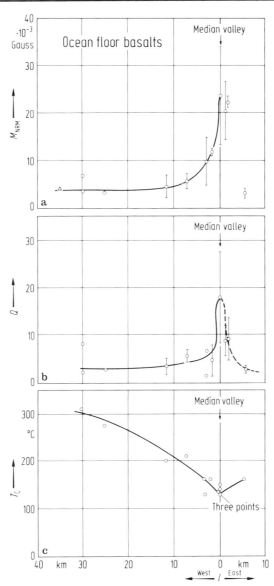

Fig. 51. Variation of magnetic properties in ocean floor basalts dredged and drilled along a traverse across the Mid-Atlantic Ridge near 37° N with increasing distance l from the ridge axis (i.e. increasing age). The cause for this variation of magnetic properties is primarily low temperature oxidation ($T < 50$ °C) of the titanomagnetites contained in the basalts (see also Fig. 49) [Joh77].

a) Natural remanent magnetization M_{NRM}. Points with error bars (one standard deviation) are site averages and points without bars are values of individual rocks. Triangle is the average value for the top unit of DSDP Site 332.

b) Koenigsberger Q-factor ($M_{NRM}/\chi \cdot H_{earth}$).

c) Curie temperature T_C. Samples from the inner valley floor are clustered around $T_C = 140$ °C, very close to the value for unoxidized titanomagnetite in fresh ocean floor basalts. [Pet79] give a Curie temperature of 125 °C for unoxidized ocean floor basalts.

Table 4. Variation of magnetic properties of ocean floor basalts from the Atlantic with age. Comparison of pillow basalts from the FAMOUS axial rift valley and Deep Sea Drilling Project Site 332 [Pre79].

	FAMOUS $(0 \cdots 0.1 \cdot 10^6 \text{ a})$		DSDP Site 332 $(3.5 \cdot 10^6 \text{ a})$	
T_C [°C]	159 ± 9	(N=86)	294 ± 14	(N=69)
z	$\simeq 0.2$		0.7 ± 0.04	(N=69)
M_S [Gauss]	0.87 ± 0.11	(N=53)	0.40 ± 0.04	(N=85)
M_{NRM} [10^{-2} Gauss]	$1.44 \begin{cases} 1.23 \\ 1.68 \end{cases}$	(N=103)	$0.366 \begin{cases} 0.312 \\ 0.428 \end{cases}$	(N=85)
χ_v [10^{-4}]	2.99 ± 0.34	(N=103)	2.41 ± 0.22	(N=84)
Q	$151 \begin{cases} 129 \\ 178 \end{cases}$	(N=103)	$39.5 \begin{cases} 32.9 \\ 47.6 \end{cases}$	(N=84)
M_{RS}/M_S	0.41 ± 0.03	(N=53)	0.54 ± 0.12	(N=8)
H_{RC}/H_C	1.43 ± 0.07	(N=51)	1.41 ± 0.14	(N=20)
H_C [Oe]	297 ± 42	(N=53)	253 ± 31	(N=20)
H_{RC} [Oe]	407 ± 74	(N=51)	353 ± 50	(N=20)
MDF [Oe]	359 ± 51	(N=53)	303 ± 25	(N=85)

N: number of samples analyzed; z: oxidation parameter of titanomagnetites contained in the basalts ($0 \leq z \leq 1$; [Ore67]); MDF: median destructive field (magnetic field necessary during alternating field demagnetization to erase half of the original NRM).

6.2.3.2.3 Depositional remanent magnitization DRM — Sedimentations-Remanenz DRM

During deposition of sediments, previously magnetized particles tend to become aligned parallel to an external magnetic field. In a low-turbulence depositional environment this alignment may be retained in the deposited sediments and produce a remanent magnetization. Varved clays provide the best example of such a DRM [Col65, Gri60, Isi42, Joh48, Khr68] (Figs. 52, 53a, 54, 55). Although the declination record preserved in the sediment is generally a good representation of the earth's magnetic field declination, the recorded inclinations are often too shallow dependent on the mean shape of the grains or on the physical nature of the sediment/water interface [Kin55] (Fig. 56 and 57).

However, analyses of both natural and laboratory deposited sediments have shown that the inclination error is often less than predicted or not observed at all, the latter generally being the case in deep-sea sediments [Ken73, Opd73] (Figs. 58 and 60). Post-depositional models in which grain re-alignment occurs after the initial deposition and leads to a post-depositional remanent magnetization PDRM, have been invoked to account for these features [Løv76, Tuc80, Tuc80a, Ver77].

DRM and/or PDRM is an important but not the only cause of the natural remanence of most sedimentary rocks. Another important contribution may come from a chemical remanent magnetization CRM. This has been well documented for red sandstones [Col65, Col74], partly also for limestone [Hel78] and loess [Tuc76].

6.2.3.2.3.1 Figures and tables − Abbildungen und Tabellen

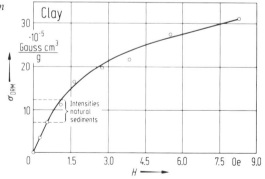

Fig. 52. Intensity of depositional remanent specific magnetization σ_{DRM} as function of the magnetic field H applied during laboratory redeposition of varved clay samples from New England (USA). Range of intensities for the natural sediments prior to redeposition is indicated separately. The magnetization of the sediments is carried by fine grained magnetite with grain sizes $d = 0.3\ \mu m$ and smaller [Joh48].

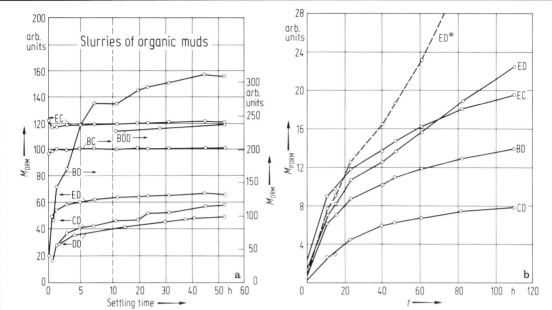

Fig. 53a. Acquisition of depositional remanent magnetization M_{DRM} of slurries of very fine grained organic muds from two volcanic crater lakes, Lake Bullenmerri and Lake Keilambete, Australia. The muds were dispersed in a kitchen blender, poured into glass tubes, vibrated in an ultrasonic bath at 30 kHz for 10 min and then placed into magnetic field for settling. For slurry characteristics and local field parameters, see Table 5. The magnetization was determined by moving the glass tube through a cryogenic magnetometer (zero field) and measuring the peak signal. Magnetization intensities are not corrected for compaction of the sediment column. The time scale between 0 and 10 hours is expanded [Bar80].

Fig. 53b. Acquisition of post-depositional remanent magnetization M_{PDRM} of slurries of very fine grained muds (same material as in Fig. 53a). The slurries settled in zero field for 24.5 hours and were then exposed to the ambient earth's magnetic field. As in Fig. 53a, the magnetization intensities were recorded at the peak signal level. Only ED showed significant settling after the initial 24.5 hour. ED* is uncorrected for compaction, ED is corrected for compaction [Bar80].

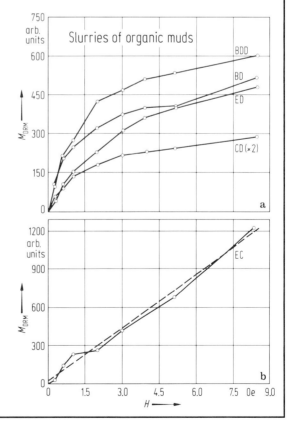

Fig. 54. Intensity of depositional remanent magnetization M_{DRM} as a function of applied magnetic field during laboratory redeposition of slurries of very fine grained organic muds (same material as in Fig. 53a, b). Each sample settled for 30 hours in a uniform magnetic field before measurement [Bar80].
a) Dilute slurries. Intensities for CD have been plotted at twice their true value to improve clarity. The dilute slurries show a slow trend towards a limiting value of DRM in fields above about 4 Oe.
b) Concentrated slurry. The sample shows an approximately linear dependence of DRM on the applied magnetic field.

Table 5. Slurry characteristics and local earth's magnetic field parameters [Bar80] (see Figs. 53a and b, 54).

Slurry	Tube	Water [wt-%]	Field inclination [°]	Field intensity H [Oe]
Concentrated Bullenmerri Mud	BC	87.5	−65.4	0.589
Dilute strong Bullenmerri Mud	BD	97.5	−65.7	0.593
Less dilute strong Bullenmerri Mud with the same composition as BC	BDD	≈95	−67.2	0.551
Dilute weak Bullenmerri Mud	CD	97.3	−66.1	0.596
Dilute upper Keilambete Mud	DD	95.9	−63.4	0.593
Concentrated lower Keilambete Mud	EC	88.4	−66.5	0.605
Dilute lower Keilambete Mud	ED	97.9	−66.7	0.610

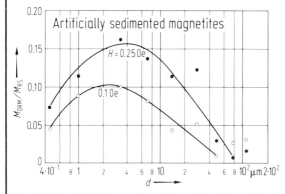

Fig. 55. Particle-size dependence of normalized depositional remanent magnetization (M_{DRM}/M_{RS}) for artificially sedimented magnetites. At large grain sizes alignment efficiency drops because of mechanical settling effects. The decrease in submicron magnetites is caused by thermal fluctuations perturbing perfect alignment of the moments of settling particles. d = grain diameter [Dun81].

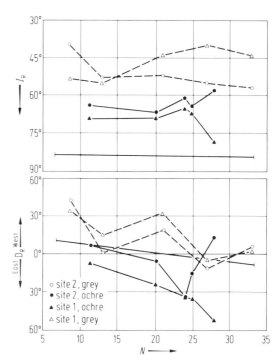

Fig. 56. Measurements of inclination I_R and declination D_R of the natural remanent magnetization NRM of varves from Hagavatn, Iceland. The varves were laid down within the period A.D. 1700 to 1800. The declination record preserved in the sediment is a good representation of the earth's magnetic field declination, the recorded inclinations, however, are too shallow.

Solid line: direction of the earth's magnetic field at the time of deposition of the varves. The two sites were about 80 m apart. N = layer number of varves [Gri60].

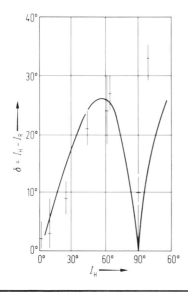

◄ Fig. 57. Inclination error $\delta = I_H - I_R$ plotted against the inclination of the applied magnetic field I_H. I_R is the inclination of depositional remanent magnetization DRM. Solid curve represents the empirical expression

$$\tan I_R = 0.4 \tan I_H.$$

Crosses show DRM inclination values observed in sediment redeposition experiments [Kin55].

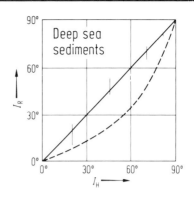

Fig. 58. Post-depositional remanent magnetization PDRM of reconstituted deep sea sediments from the Indian Ocean. Inclination of PDRM, I_R, plotted against the inclination of the applied magnetic field, I_H. Error bars represent range of remanent inclinations for each sample. Straight line is the line of perfect agreement; dashed curve represents the empirical expression $\tan I_R = 0.4 \tan I_H$ which approximates the variation of inclination error (Fig. 57) with applied magnetic field inclination observed in depositional remanent magnetization DRM of redeposited sediments [Kin55]. Magnetic minerals in the sediment are titanomagnetite and magnetite [Opd73]. For redeposition the sediment was dispersed in distilled water to form a thick slurry, similar in consistancy to that normally found near the tops of deep sea piston cores. The slurry was placed into plastic cylinders and then thoroughly stirred in the presence of a known field. It was then allowed to dry in this same field. After drying (4···7 days) the remanent magnetization was measured [Ken73].

For Fig. 59, see next page.

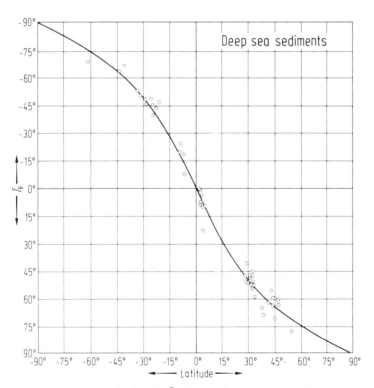

Fig. 60. Mean inclination $\bar{I}_R$ of natural remanent magnetization NRM of deep sea sediments from different oceans, plotted against the latitude of the sampling site. The sediments (piston cores) were taken from geographic latitudes between 55° N and 62° S. The age of the sediments ranges between recent and 700000 years. The data points represent mean inclination values averaged over the whole length of each sediment core (most of them about 5 meters long). The solid curve shows the latitudinal dependence of the inclination of the axial earth magnetic dipole field [Opd69].

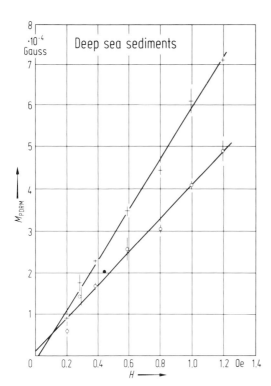

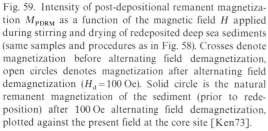

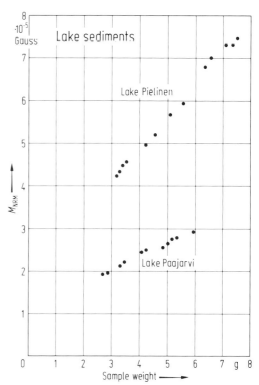

Fig. 59. Intensity of post-depositional remanent magnetization M_{PDRM} as a function of the magnetic field H applied during stirring and drying of redeposited deep sea sediments (same samples and procedures as in Fig. 58). Crosses denote magnetization before alternating field demagnetization, open circles denotes magnetization after alternating field demagnetization ($H_d = 100$ Oe). Solid circle is the natural remanent magnetization of the sediment (prior to redeposition) after 100 Oe alternating field demagnetization, plotted against the present field at the core site [Ken73].

Fig. 61. Effect of dehydration on the natural remanent magnetization M_{NRM} of lake sediments. Remanent magnetization intensity plotted against sample weight during drying of the sediment samples. Remanence intensity is approximately linearly related to the weight loss due to drying. The intensity decrease was not accompanied by any change in remanence direction. Sediments from Lake Pielinen and Lake Paajarvi, Finnland [Sto79].

6.2.3.2.4 Isothermal remanent magnetization IRM — Isothermale remanente Magnetisierung IRM

Isothermal remanent magnetization is acquired by a rock at constant ambient temperature when placed in a magnetic field greater than the smallest coercive force of any of its magnetic mineral grains. IRM acquired in the earth's magnetic field is generally very weak and unstable compared to a TRM, CRM or DRM acquired in the same field (Figs. 45, 46, 47). However, lightning strokes are accompanied by intense magnetic fields and they can easily produce strong IRM [Mat54, Poc01, Roe63] (Fig. 62a···g).

6.2.3.2.4.1 Figures — Abbildungen

▶

Fig. 62. Distribution of the natural remanent magnetization NRM in a tertiary basalt at a site hit by a lightning stroke. The locality is Backenberg near Göttingen, Germany [Roe63].
a) Vertical cross-section of sampling site. Sampling was carried out in different horizontal levels marked I, II, etc.
b) Declination and horizontal intensity of NRM of the rock in level II (horizontal cross-section of sampling site). The cross marks the supposed center of the lightning stroke.

c) NRM inclination of the rock in level II (horizontal cross-section of sampling site).
d) Declination and horizontal intensity of NRM in level IV.
e) NRM inclination in level IV.
f) Variation of M_{NRM} with distance r from supposed center of lightning stroke (marked by a cross in Fig. 62b). Samples taken from all different levels.
g) Stability of M_{NRM} against alternating field demagnetization of six samples taken at different distances r.

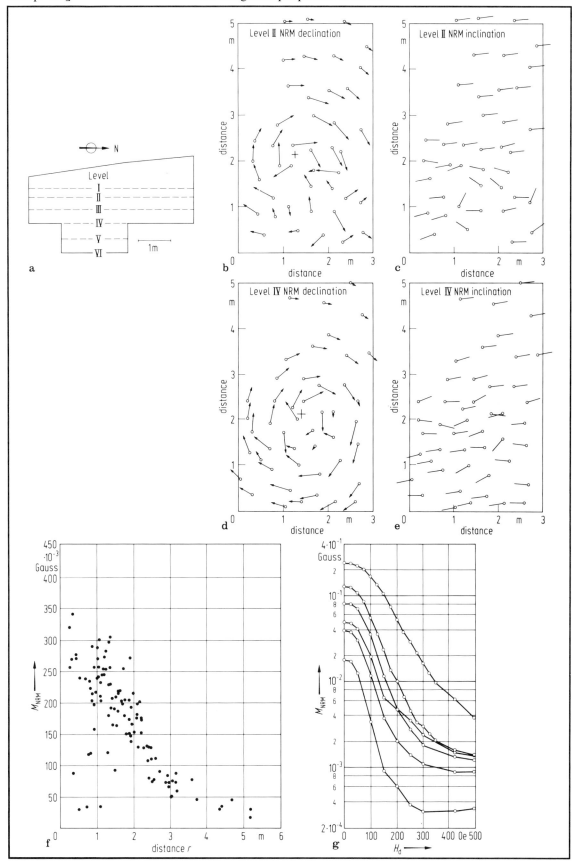

Petersen/Bleil

6.2.3.2.5 Viscous remanent magnetization VRM — Viskose remanente Magnetisierung VRM

If a specimen is kept in a constant magnetic field during an appreciably long time, the resultant remanent magnetization becomes larger than the ordinary IRM of the specimen adopted in a short time [Rim56, Rim59] (Fig. 63). This effect of time, caused by thermal agitation, is called magnetic viscosity [Nee51a]. The increase in remanence is often a logarithmic function of time (Figs. 64, 65 and 68):

$$M(t) = M_0 + S \cdot \ln(t/t_0)$$

S is called viscosity coefficient. However, in basalts a simple logarithmic law is rarely obeyed. Two or three stages are usually observed, in each of which a simple logarithmic relationship can be assumed (Figs. 69 and 70).

Different parameters are used in the literature to describe the acquisition of VRM in rocks (Table 6).

Rocks containing very small grains of minerals, on the verge of superparamagnetism, are most sensitive to VRM acquisition. Viscous magnetization is strongly size-dependent [Dun81] (Figs. 66, 67, 68, 72 and 73).

VRM plays also an important role in rocks containing large and homogeneous magnetic mineral grains (multi-domain grains) [Dun81, Nee55, Shi60] (Fig. 72).

The stability of VRM is a function of inducing field and time of acquisition [Rim56, Rim59] (Figs. 74 and 75). VRM acquired in the earth's magnetic field can be very stable with respect to alternating field demagnetization, particularly in hematite-bearing sedimentary rocks [Biq70, Biq71, Rim56] (Fig. 75).

The stress dependence of the acquisition of VRM has been described by [Poz70]. VRM acquisition is greatly enhanced under stress.

Theoretical treatment of VRM is found in [Dun73a, Nee51a, Rim 56, Sta74].

6.2.3.2.5.1 Figures and tables — Abbildungen und Tabellen

Table 6. Parameters used to describe acquisition of viscous remanent magnetization in rocks.

Parameter	Defined by	Ref.
Viscosity coefficient S	$M(t) = M_0 + S \cdot \ln(t/t_0)$	Nee51a, Rim56
Thellier's viscosity coefficient	VRM(14 days)/NRM VRM acquired in the ambient earth's magnetic field in the laboratory over a period of 14 days ($= 336$ h) in the presence of the original NRM.	The59
VRM(1000)/NRM	VRM acquired in a constant magnetic field of usually 1 Oe over a period of 1000 hours. VRM acquisition usually after alternating field demagnetization of the original NRM.	Low73, Low78

Table 7. Relative increase of viscous remanent magnetization of a basalt from the Auvergne (France) in different magnetic fields: Ratio of remanent magnetization measured after a time t to remanent magnetization measured after 1 s (data taken from Fig. 63) [Rim59].

t	H [Oe]							
	2	5	10	20	100	150	200	300
1 s	1	1	1	1	1	1	1	1
1 min	3	3	2	1.3	1.1	1.2	1.1	1.1
15 min	9	6	3	2	1.3	1.3	1.1	1.1
1 h	13	10	4	2.3	1.3	1.3	1.1	1.2
1 d	22	17	7	3	1.5	1.5	1.2	1.2

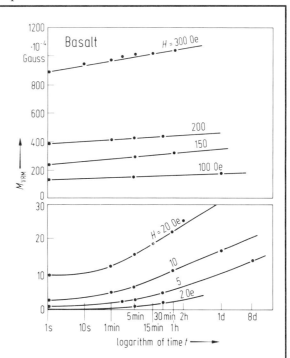

Fig. 63. Time dependence of viscous remanent magnetization of a basalt sample (Auvergne, France) induced in a constant magnetic field at room temperature. Prior to the experiment, the sample has been thermally demagnetized. The sample was then placed in a magnetic field for a certain time t. For measurement of remanent magnetization the constant field was switched off and the sample was left in zero field for 1 min before the actual measurement was carried out. The remanent magnetization measured after $t = 1$ s represents the ordinary isothermal remanent magnetization IRM. The relative increase of viscous remanent magnetization with time becomes smaller with increasing constant field H (Table 7) [Rim59].

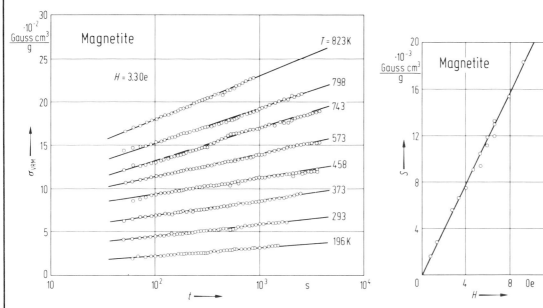

Fig. 64. Time dependence of viscous remanent specific magnetization (σ_{VRM}) of magnetite powder (grain size $d = 2$ μm), measured at different temperatures. Applied magnetic field was $H = 3.3$ Oe. The zero point of the ordinate scale is arbitrary, relative displacements of the graphs being arranged for convenience in the drawing. The slopes of the curves give the magnetic viscosity coefficient at the respective temperatures [Shi60].

Fig. 65. Variation of magnetic viscosity coefficient S as a function of the applied magnetic field H. The sample is magnetite powder with a grain size $d = 100$ μm. Temperature of measurement is room temperature [Shi60].

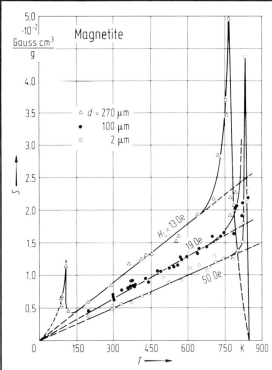

Fig. 66. Variation of magnetic viscosity coefficient S as a function of absolute temperature for magnetite powders of different grain size, d. Coercive force H_C measured at room temperature of each specimen is indicated on the curve [Shi60].

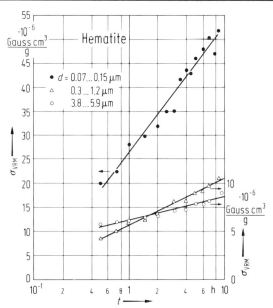

Fig. 68. Growth of viscous remanent specific magnetization σ_{VRM} of hematites of various particle sizes in a magnetic field of 5 Oe measured at room temperature. The left hand scale corresponds to the data for $d = 0.07\cdots0.15$ μm, the right hand scale to the other data [Dun77].

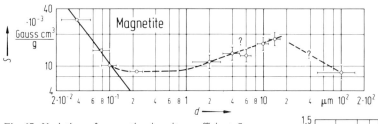

Fig. 67. Variation of magnetic viscosity coefficient S as a function of grain diameter d of magnetite at room temperature. The trend of increasing S with increasing particle size in the range $5\cdots15$ μm is doubtful as it contradicts other observations [Zhi65] [Dun81].

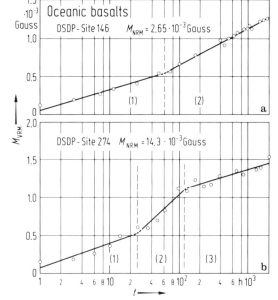

Fig. 69. Growth of viscous remanent magnetization M_{VRM} in two oceanic basalts from Deep Sea Drilling Project Leg 15, Caribbean Sea, (a), and Leg 28, South Pacific Ocean, (b). Prior to the VRM measurements the original NRM has been erased by alternating field demagnetization. Applied magnetic field during VRM growth was $H = 1$ Oe at room temperature. No simple logarithmic VRM growth is observed but a two-stage (a) and three-stage (b) acquisition [Low78].

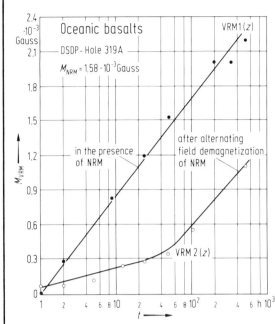

Fig. 70. Growth of viscous remanent magnetization VRM in medium-grained oceanic basalt from Deep Sea Drilling Project Leg 34, Nazca plate. Acquisition of VRM in a constant magnetic field $H = 10$ Oe before (VRM 1) and after (VRM 2) alternating field demagnetization of the original NRM. Direction of the applied magnetic field (z-direction) was perpendicular to the NRM [Low78].

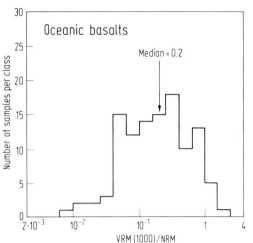

Fig. 71. Histogram of the ratio VRM (1000)/NRM in 111 oceanic basalts from all oceans. VRM (1000) was acquired after alternating field demagnetization of the original NRM in a constant magnetic field $H = 10$ Oe over 1000 hours. The distribution is possibly not truly representative of oceanic basalts in general since VRM investigations are usually carried out deliberately on those rocks which display unstable characteristics during NRM measurements [Ken77, Low73, Low73a, Low73b, Low74, Low75, Low76, Low78, Pei74, Tar76].

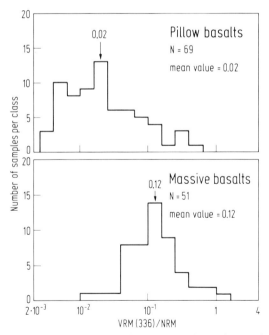

Fig. 72. Histogram of the ratio VRM (336)/NRM in oceanic basalts from Deep Sea Drilling Project Leg 51, Atlantic ocean. VRM (336) was acquired in the ambient earth's magnetic field in the laboratory ($H \simeq 0.5$ Oe) over 336 hours = 14 days (Thellier's viscosity coefficient [The59]) [Smi79].

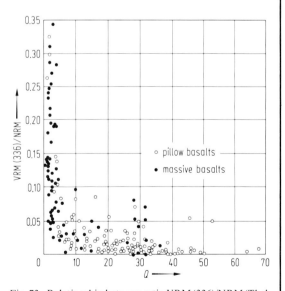

Fig. 73. Relationship between ratio VRM (336)/NRM (Thellier's viscosity coefficient [The59]) and Koenigsberger Q-factor for oceanic basalts from Deep Sea Drilling Project Leg 51, Atlantic ocean [Smi59].

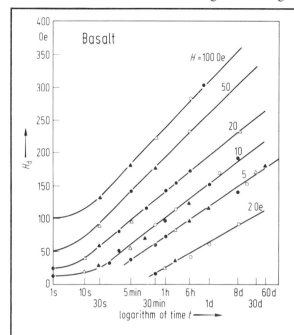

◄

Fig. 74. Relationship between peak alternating field H_d necessary to erase (to less than one hundredth of its original intensity) a VRM acquired in a constant magnetic field H over a time t for different basalts from the Auvergne (France) [Rim56].

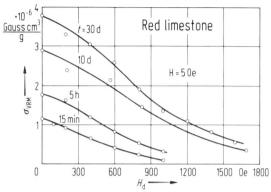

Fig. 75. Alternating field demagnetization of viscous remanent specific magnetization σ_{VRM} in red limestone. VRM has been acquired in a constant magnetic field $H = 5$ Oe over various periods of time [Biq71a].

6.2.3.2.6 Pressure or piezo-remanent magnetization PRM — Druck- oder Piezo-Remanenz PRM

A geometrical deformation of a magnetized specimen caused by mechanical stress generally results in a change of intensity and direction of magnetization. The irreversible part of this change has been called pressure or piezo-remanent magnetization *(in sensu lato)* [Dom62, Mar78, Nag70, Poz77]. This remanent magnetization is extremely sensitive to the sequence of the application and release of uniaxial compression τ and inducing magnetic field H.

Piezo-remanent magnetization PRM *in sensu stricto* is defined as

$$M_R(H_+ \, \tau_+ \, \tau_0 \, H_0), \quad [\text{Nag65}], (\text{Fig. 76})$$

where ($+$) denotes the application and (0) the removal of the magnetic field H or uniaxial compression τ. The order of the symbols in parentheses indicates the time sequence of the procedure: M_R is the remanent magnetization left in a rock sample when first the magnetic field H and then the uniaxial compression τ was applied, then τ and finally H was removed again (notation according to [Nag65]). In this notation ordinary isothermal remanence without any pressure treatment is denoted as

$$M_R(H_+ \, H_0).$$

The following characteristics of piezo-remanent magnetization *(in sensu lato)* of igneous rocks in small magnetic fields $H = 0 \cdots 10$ Oe and under uniaxial compression comparable to the earth's crust stress $0 \cdots 100 \, \text{kg} \cdot \text{cm}^{-2}$ have been demonstrated [Nag68, Nag69]:

a) Linear dependence of piezo-remanent magnetization *(in sensu stricto)* on inducing magnetic field H,

$$M_R(H_+ \, \tau_+ \, \tau_0 \, H_0) = c(\tau) \cdot H$$

where $c(\tau)$ is a constant, the coefficient of piezo-remanent magnetization, depending on τ. The following relation holds:

$$M_R^{\perp}(H_+ \, \tau_+ \, \tau_0 \, H_0) < M_R^{\parallel}(H_+ \, \tau_+ \, \tau_0 \, H_0)$$

where $M_R^{\perp}$ denotes the PRM acquired in a magnetic field H perpendicular to τ and $M_R^{\parallel}$ the PRM acquired in a field H parallel to τ (Fig. 77 and Table 8).

b) Non-commutativity of τ and H,

$$M_R(H_+ \, \tau_+ \, \tau_0 \, H_0) \geqq M_R(\tau_+ \, H_+ \, \tau_0 \, H_0) > M_R(H_+ \, \tau_+ \, H_0 \, \tau_0) \geqq M_R(\tau_+ \, H_+ \, H_0 \, \tau_0)$$

This relationship holds in both cases $\tau \parallel H$ and $\tau \perp H$ (Fig. 78).

c) "After-effect" of uniaxial compression:

$$M_R(H_+ \, H_0) < M_R(\tau_+ \, \tau_0 \, H_+ \, H_0) \quad \text{for} \ \tau > \tau_{cr}.$$

After a rock sample is uniaxially compressed in a non-magnetic space by a pressure τ larger than a critical value τ_{cr}, its isothermal remanence becomes larger than the ordinary IRM without such a pressure history. This effect takes place in both cases $\tau \parallel H$ and $\tau \perp H$ (Fig. 79).

d) "Pressure-demagnetization" effect,

$$M_R(H_+ H_0) > M_R(H_+ H_0 \tau_+ \tau_0)$$

After uniaxially compressing a rock sample having an isothermal remanence $M_R(H_+ H_0)$ in a non-magnetic space, the residual magnetization becomes smaller than the original IRM. This effect takes place in both cases $\tau \parallel H$ and $\tau \perp H$ (Fig. 80 and 81).

Most susceptible to PRM effects are rocks containing relatively large homogeneous (multi-domain) grains of magnetite or titanomagnetite. The smaller the grain size, the smaller the pressure dependence of magnetization. Subdivision of grains by exsolution, very common in titanomagnetites in basalts, also reduces strongly the pressure dependence [Ohn68, Sch68].

A special kind of piezo-remanent magnetization is the shock-induced remanent magnetization SRM which is generated in a rock when hit by a shock wave while exposed to a magnetic field [Poh75, Sha67]. SRM may be of importance in rocks in and around impact craters [Har69] Figs. 82 and 83).

6.2.3.2.6.1 Figures and tables — Abbildungen und Tabellen

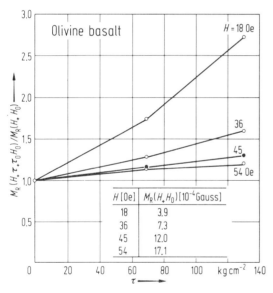

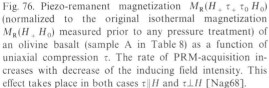

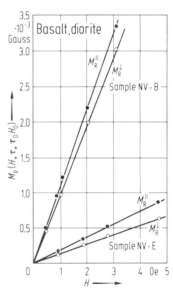

Fig. 76. Piezo-remanent magnetization $M_R(H_+ \tau_+ \tau_0 H_0)$ (normalized to the original isothermal magnetization $M_R(H_+ H_0)$ measured prior to any pressure treatment) of an olivine basalt (sample A in Table 8) as a function of uniaxial compression τ. The rate of PRM-acquisition increases with decrease of the inducing field intensity. This effect takes place in both cases $\tau \parallel H$ and $\tau \perp H$ [Nag68].

Fig. 77. Relationship between piezo-remanent magnetization $M_R(H_+ \tau_+ \tau_0 H_0)$ and inducing magnetic field H for a basalt (NV-B) and a diorite sample (NV-E, same sample code as in Table 8). Uniaxial compression τ was 118 kg·cm² in both cases [Nag69].

Table 8. Coefficient of piezo-remanent magnetization $c(\tau)$ and other magnetic parameters of various igneous rocks. $c(\tau)$ defined by: $M_{\kappa}(H_{+}\tau_{+}\tau_{0}H_{0})-c(\tau)\cdot H$, [Nag68, Nag69]. $c^{\parallel}$ denotes the PRM acquired in a magnetic field parallel to τ, $c^{\perp}$ in a magnetic field perpendicular to τ.

Sample code	Rock type	$c^{\parallel}(\tau)$ $\tau=118\,\mathrm{kg\,cm^{-2}}$ $[10^{-4}\,\mathrm{Gauss\,Oe^{-1}}]$	$c^{\perp}(\tau)$ $\tau=118\,\mathrm{kg\,cm^{-2}}$ $[10^{-4}\,\mathrm{Gauss\,Oe^{-1}}]$	M_S [Gauss]	H_C [Oe]	χ_v $[10^{-3}]$	M_{RS} [Gauss]	M_{RS}/M_S
NV-A	Olivine basalt	0.20	–	0.75	293	0.48	0.382	0.51
NV-B	Homogeneous basalt	10.7	9.5	1.10	68	2.61	0.320	0.29
NV-C	Basalt	0.57	0.4	0.71	152	0.82	0.141	0.20
NV-E	Diorite intrusion	1.8	1.4	1.08	72	0.89	0.127	0.14
NV-H	Basalt intrusion	0.34	–	0.81	90	0.93	0.166	0.17
NV-K	Andesite intrusion	1.9	1.7	2.99	97	3.09	0.395	0.12
NV-L	Homogeneous basalt	7.2	7.0	0.71	85	1.34	0.234	0.35
A	Olivine basalt	0.2	–	–	245	1.9	–	–
B	Continental basalt	13.0	10.2	–	71	5.3	–	–
D	Oceanic basalt	0.35	–	–	–	0.3	–	–
E	Diorite	2.1	1.5	–	–	5.9	–	–

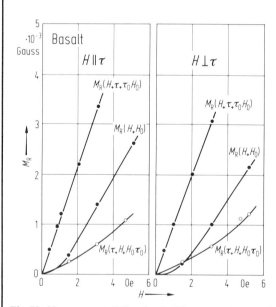

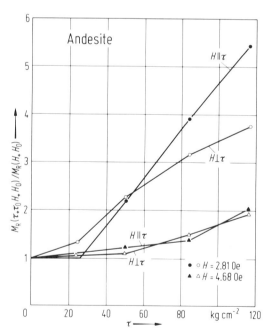

Fig. 78. Non-commutativity of uniaxial compression τ and inducing magnetic field H in PRM experiments. Comparison of $M_R(H_+\tau_+\tau_0 H_0)$, ordinary isothermal remanence $M_R(H_+ H_0)$ and $M_R(\tau_+ H_+ H_0 \tau_0)$ in a homogeneous basalt (sample NV-B, Table 8). The effect of non-commutativity of τ and H takes place in both cases $H\|\tau$ and $H\perp\tau$ [Nag69].

Fig. 79. "After effect" of uniaxial compression τ on isothermal remanence $M_R(\tau_+ \tau_0 H_+ H_0)$ (normalized to the original isothermal remanence $M_R(H_+ H_0)$ measured prior to any pressure treatment) of an andesite (sample NV-K, Table 8) [Nag69].

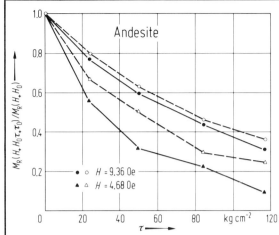

Fig. 80. Pressure demagnetization of isothermal remanent magnetization $M_R(H_+ H_0 \tau_+ \tau_0)$ (normalized to the original isothermal remanence $M_R(H_+ H_0)$ measured prior to any pressure treatment) of an andesite (sample NV-K, Table 8). Isothermal remanence was acquired in inducing fields $H = 4.68$ Oe and $H = 9.36$ Oe. Remanent magnetization decreases with increase of τ regardless of whether $\tau \| M_R$ (solid curves) or $\tau \perp M_R$ (dashed curves) [Nag69].

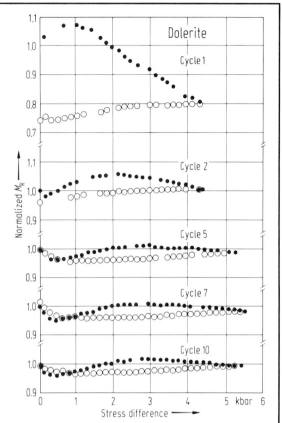

Fig. 81. Pressure demagnetization of thermoremanent magnetization and the effect of recycling for a dolerite (Ralston intrusive, Golden, Colorado, USA). Magnetic mineral component are large grains of titanomagnetite (mean grain diameter 0.7 mm) showing a fine texture of ilmenite exsolution lamellae. Intensity of remanent magnetization plotted as a function of differential stress at confining pressure of 500 bar.
Solid circles: Remanence measured during increasing stress; Open circles: Remanence measured during decreasing stress [Mar78].

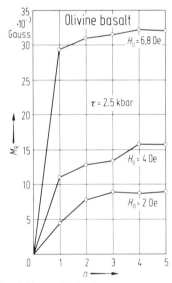

Fig. 82. Acquisition of shock remanent magnetization $M_R(H_+ S_+^1, S_+^2 \dots S_+^n H_0)$ in different magnetic fields. Dependence on the number n of impacts $S_1, S_2 \dots S_n$ ($S =$ stress wave amplitude). Sample: Olivine basalt containing homogeneous titanomagnetite (average grain diameter 30 µm) as magnetic mineral component [Poh75].

▶

Fig. 83. Shock remanent magnetization (open circles) $M_R(H_+ S_n H_0)$ as dependent on the stress wave amplitude S and piezo-remanent magnetization (full circles) $M_R(H_+ \tau_+ \tau_0 H_0)$ as dependent on the uniaxial compression τ. Sample: Olivine basalt (same sample as in Fig. 82) [Poh75].

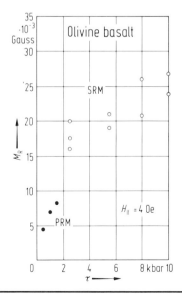

6.2.3.2.7 Inverse remanent magnetization — Inverse remanente Magnetisierung

In certain cases, rocks occur in nature which acquire a remanent magnetization whose direction is just opposite to that of the external magnetic field acting on it while the rock acquires its magnetization. This "inverse" magnetization is either a thermoremanent or a chemical remanent magnetization, or a combination of both, a so-called thermochemical remanence [Cre69, Nag59, Nee51, Uye55].

The mechanisms causing self-reversal can be devided in two groups according to the presence of only one magnetic phase or two magnetically interacting phases (Table 9).

Complete self-reversal of natural remanent magnetization of rocks seems to be a rare phenomenon (Figs. 84, 85, 86). However, partial self-reversal occurs frequently, particularly in basalts [Cre69, Hav65, Hel79, Hel80], (Fig. 87).

Theoretical treatment of inverse remanent magnetization is found in [Nee51, Uye55, Ver56].

6.2.3.2.7.1 Figures and tables — Abbildungen und Tabellen

For Table 9, see next page.

Table 10a. Lattice constant (a), spontaneous molecular moment (p_{ms}), saturation volume magnetization (M_s), Curie temperature (T_C) and compensation temperature ($T_{comp.}$) of titanomagnetites contained in the alkaline basalt from Steinberg (near Meensen, Germany) which shows Néel N-type self-reversal of the natural remanent magnetization (Fig. 84) [Sch76].

Sample number (from Fig. 84)	$p_{ms}(0\,K)$ [μ_B]	$M_s(20\,°C)$ [Gauss]	T_C [°C]	$T_{comp.}$ [°C]	a [Å]
1	0.10	26.6	157 ± 8	−180	8.449 ± 0.003
2	0.12	19.5	170 ± 10	−165	8.431 ± 0.002
3b	0.14	17.3	190 ± 8	−120	8.417 ± 0.003
4	0.18	13.0	192 ± 5	− 66	8.411 ± 0.002
5	0.22	8.9	205 ± 5	− 14	8.397 ± 0.003
6	0.26	10.1	215 ± 5	0	8.386 ± 0.002
7	0.30	15.0	222 ± 8	+ 50	8.382 ± 0.002
B	0.32	17.3	225 ± 10	+ 70	8.380 ± 0.003

Table 10b. Chemical composition of titanomagnetites contained in the alkaline basalt from Steinberg (near Meensen, Germany) which shows Néel N-type self-reversal of the natural remanent magnetization. [Sch76a].

Sample number (from Fig. 84)	1	7
Components	wt-%	wt-%
Fe as Fe_2O_3	77.3	74.06
TiO_2	24.3	22.6
Al_2O_3	0.55	2.68
MgO	1.82	1.92
MnO	0.27	0.3
Cr_2O_3	<0.2	<0.2
Ca	<0.2	<0.2
Zn	—	—
Ni	<0.2	<0.2
Zr	—	—
Nb	—	—

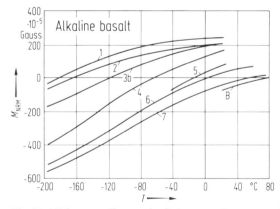

Fig. 84. Néel type self-reversal of the natural remanent magnetization NRM of the alkaline basalt from Steinberg (near Meensen, Germany). The NRM has been partially demagnetized in an alternating field of 150 Oe prior to the measurement. The curves are reversible within the measured range of temperatures. Carrier of remanent magnetization is cation-deficient titanomagnetite with relatively high Mg and Al content (see Table 10b for curve numbers) [Sch76].

Table 9. Self-reversal of remanent magnetization in natural and synthetic samples.

		Ilmenite-hematite $Fe_{2-y}Ti_yO_3$	$T_{comp.}$ °C	reversal mechanism*)	Titano-magnetite $Fe_{3-x}Ti_xO_4$	$T_{comp.}$ °C	reversal mechanism*)	Pyrrhotite $Fe_{7+\delta}S_8$	$T_{comp.}$ °C	reversal mechanism*)
One-phase systems		$0.45 \leqq y \leqq 0.73$			$0.40 \leqq x \leqq 0.65$			$0 \leqq \delta \leqq 0.185$		
	synthetic samples	−			Mg, Al-doped titanomagnetite [Ver56]	?	IO?	−		
	natural samples	−			alkaline basalt containing Néel N-type titanomagnetite [Sch68a, Sch76, Nis74]	−180 ···+70	N	−		
Two-phase systems		$0.51 \leqq y \leqq 0.73$			$0.40 \leqq x \leqq 0.60$					
	synthetic samples	complete self-reversal for grains > 5 µm [Wes71]	200	SE	complete self-reversal for grains > 50 µm [Hav65, Pet73, Pet74]	100 ···170	M?	−		
	natural samples	granite (not reproducible in the laboratory) [Hel71]	?	SE?	continental basalt (complete self-reversal) [Mei63, Uye55]	150	M?	carboniferous shale [Eve62]	100 ···320 and 0	M
		diorite (not reproducible in the laboratory) [Mer69]	?	SE?	continental basalt (partial self-reversal) [Cre69, Hav65]	−	M?			
		dacite (Haruna volcano, Japan) [Nag52, Nag59, Uye55]	100 ···200	SE	submarine basalt (from seamounts) [Ozi67, Ozi68]	200 ···330	SE?			
		crystals from Allard Lake, Canada [Car61]	100 ···160	SE	lava (complete self-reversal) [Hel80]	50	M?			
					lava (partial self-reversal) [Hel79]	−	M?			
					magnetite sand (partial self-reversal) [Nag55]	−	M?			

*) SE = superexchange interaction;
   M = magnetostatic interaction;
   N = ferrimagnetic Néel N-type;
   IO = ionic ordering.

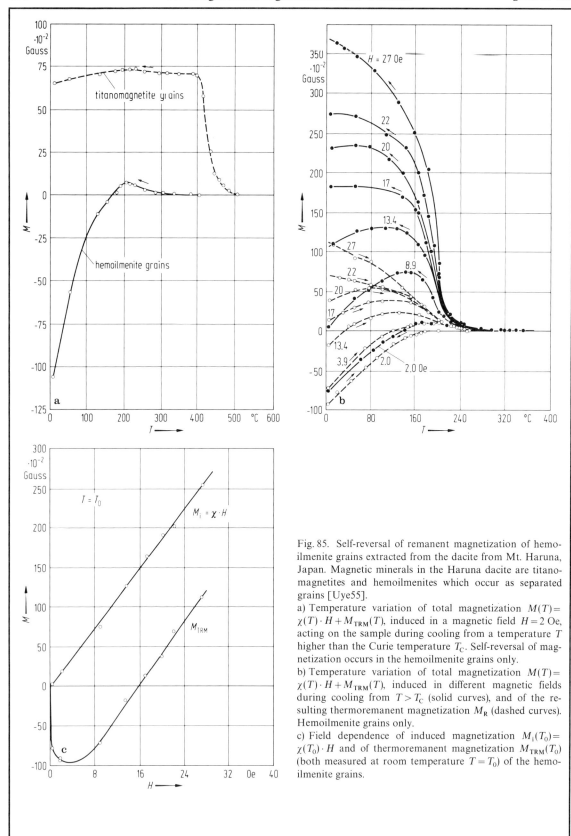

Fig. 85. Self-reversal of remanent magnetization of hemoilmenite grains extracted from the dacite from Mt. Haruna,
Japan. Magnetic minerals in the Haruna dacite are titanomagnetites and hemoilmenites which occur as separated
grains [Uye55].
a) Temperature variation of total magnetization $M(T) =$
$\chi(T) \cdot H + M_{\text{TRM}}(T)$, induced in a magnetic field $H = 2$ Oe,
acting on the sample during cooling from a temperature T
higher than the Curie temperature T_{C}. Self-reversal of magnetization occurs in the hemoilmenite grains only.
b) Temperature variation of total magnetization $M(T) =$
$\chi(T) \cdot H + M_{\text{TRM}}(T)$, induced in different magnetic fields
during cooling from $T > T_{\text{C}}$ (solid curves), and of the resulting thermoremanent magnetization M_{R} (dashed curves).
Hemoilmenite grains only.
c) Field dependence of induced magnetization $M_{\text{i}}(T_0) =$
$\chi(T_0) \cdot H$ and of thermoremanent magnetization $M_{\text{TRM}}(T_0)$
(both measured at room temperature $T = T_0$) of the hemoilmenite grains.

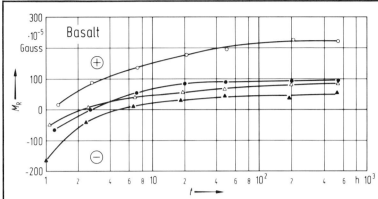

Fig. 86. Time dependence of inverse thermochemical remanent magnetization of different samples from the basalt Backenberg (near Göttingen, Germany). Previous to the time measurements the samples were heated in air at 250 °C for 40.2 hours in a magnetic field $H = 0.186$ Oe (during which time a thermochemical remanence was acquired) and then cooled to room temperature in zero field. After this procedure the samples were left for various times (plotted on the abscissa) in zero field before the remanent magnetization (plotted on the ordinate) was measured. The magnetic mineral component prior to the heating is slightly inhomogeneous titanomagnetite (mean grain diameter 12 μm) with occasional exsolution lamellae of ilmenite. Similar instability of artificial inverse thermochemical remanence has also been observed in other basalts [Cre70] [Mei63].

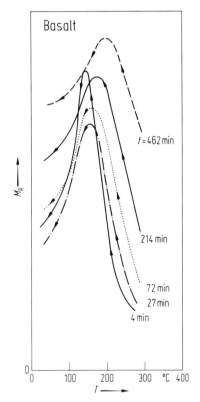

Fig. 87. Thermochemical remanent magnetization of different samples from the basalt Rauher Kulm (near Kemnath, Germany) vs. temperature. The samples were heated up to 285 °C in a magnetic field $H = 1$ Oe and were kept at this temperature for a certain time t (maximum period: 462 min) during which a thermochemical remanence was acquired. The applied field $H = 1$ Oe was then switched off and the sample cooled down to room temperature in zero field. The magnetization of the samples during cooling is shown. Only partial self-reversal of remanent magnetization is observed. The magnetic mineral component prior to the heating experiments is homogeneous titanomagnetite (mean grain diameter 30 μm) with a Curie temperature $T_C = 220$ °C [Cre69].

6.2.4 Magnetic hysteresis curves and related parameters — Magnetische Hysterese-Kurven und verwandte Parameter

6.2.4.1 Introductory remarks — Einleitende Bemerkungen

The shape of hysteresis curves of rocks varies widely [Nag61, Puz30, Rad71, Was73] (Figs. 88, 89, 90) depending upon the intrinsic properties of the ferrimagnetic grains contained in them as well as grain size [Day77] (Figs. 91 and 92) and internal stress [Jel75, Nag70, Sch68].

Grain size and magnetic domain structure of magnetic minerals are related to each other [Dun81] (Figs. 93 and 94). Consequently it is possible to deduce information on the domain structure from magnetic

Die Form der Hysterese-Kurven von Gesteinen variiert erheblich [Nag61, Puz30, Rad71, Was73] (Fig. 88, 89, 90) und hängt sowohl von den Materialeigenschaften der ferrimagnetischen Körner im Gestein, als auch von der Korngröße [Day77] (Fig. 91 und 92) und der inneren Spannung [Jel75, Nag70, Sch68] ab.

Die Korngröße und die magnetische Bereichsstruktur magnetischer Minerale stehen in Beziehung zueinander [Dun81] (Fig. 93 und 94). Daher ist es möglich, Informationen über die Bereichsstruktur

hysteresis parameters [Day77, Par65]. Particularly the ratios of remanence coercivity to coercive force H_{RC}/H_C and saturation remanence to saturation magnetization M_{RS}/M_S are used in this context (Figs. 95, 96, 97, 98).

Due to the low concentration of ferrimagnetic minerals in most major rock types (<10 vol-%) the magnetostatic interaction between the magnetic mineral grains is small. Consequently the external demagnetization factor N (shape of rock sample) can be neglected with respect to the internal demagnetization factor N_{int} (shape of the individual magnetic mineral grain) [Dav76, Kra56, Sch80].

aus magnetischen Hysterese-Parametern zu gewinnen [Day77, Par65]. Speziell das Verhältnis von Remanenz-Koerzivkraft zu Koerzivkraft, H_{RC}/H_C, und das von Sättigungsremanenz zu Sättigungsmagnetisierung, M_{RS}/M_S, wird in diesem Zusammenhang benutzt (Fig. 95, 96, 97, 98).

Da ferrimagnetische Minerale in den Haupt-Gesteinstypen nur in geringer Konzentration enthalten sind (<10 vol-%), ist die magnetostatische Wechselwirkung zwischen den magnetischen Mineralkörnern klein. Der äußere Entmagnetisierungsfaktor N (Form der Gesteinsprobe) kann deshalb gegenüber dem inneren Entmagnetisierungsfaktor N_{int} (Form der einzelnen magnetischen Mineralkörner) vernachlässigt werden [Dav76, Kra56, Sch80].

6.2.4.2 Figures — Abbildungen

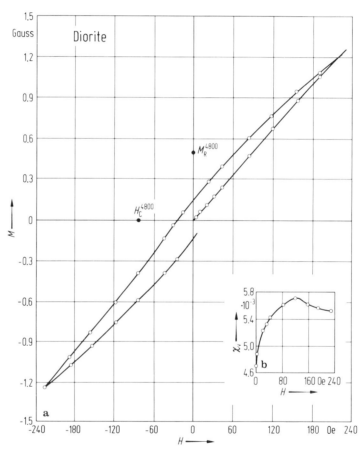

Fig. 88. Example of magnetic hysteresis curve (a) and field dependence of volume susceptibility (b) of rocks at room temperature. The sample has been demagnetized in alternating magnetic fields prior to measurement. The curve is unsheared. Diorite from Ahris, Sweden [Puz30]. The two points M_R^{4800} and H_C^{4800} have been measured after magnetizing the sample in a field $H = 4800$ Oe.

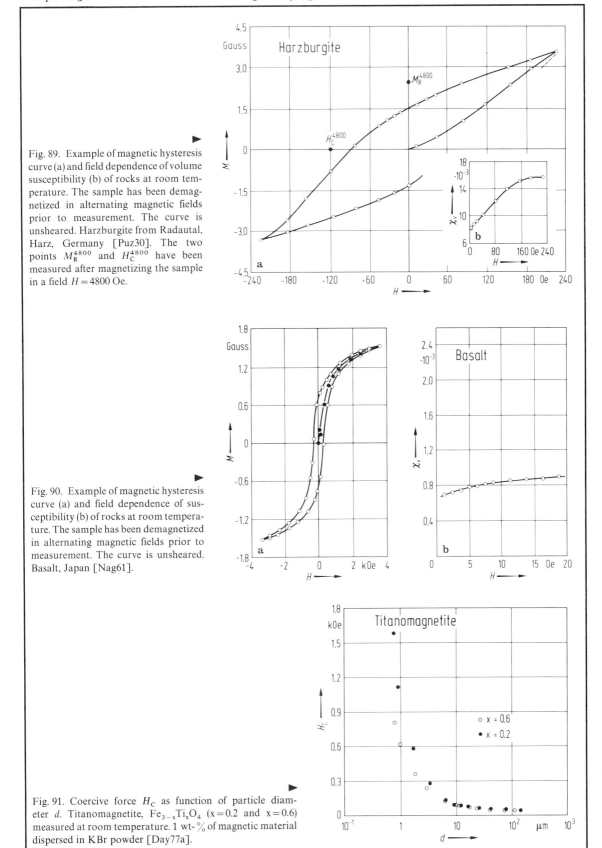

Fig. 89. Example of magnetic hysteresis curve (a) and field dependence of volume susceptibility (b) of rocks at room temperature. The sample has been demagnetized in alternating magnetic fields prior to measurement. The curve is unsheared. Harzburgite from Radautal, Harz, Germany [Puz30]. The two points M_R^{4800} and H_C^{4800} have been measured after magnetizing the sample in a field $H = 4800$ Oe.

Fig. 90. Example of magnetic hysteresis curve (a) and field dependence of susceptibility (b) of rocks at room temperature. The sample has been demagnetized in alternating magnetic fields prior to measurement. The curve is unsheared. Basalt, Japan [Nag61].

Fig. 91. Coercive force H_C as function of particle diameter d. Titanomagnetite, $Fe_{3-x}Ti_xO_4$ ($x = 0.2$ and $x = 0.6$) measured at room temperature. 1 wt-% of magnetic material dispersed in KBr powder [Day77a].

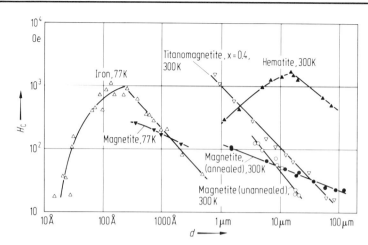

Fig. 92. Coercive force H_C as function of particle diameter d. Iron measured at 77 K [Lub61, Mei53], magnetite at 77 K [Dun73] and 300 K [Got35, Par65], titanomagnetite ($x = 0.4$) at 300 K [Day77], hematite at 300 K [Ban71, Che43].

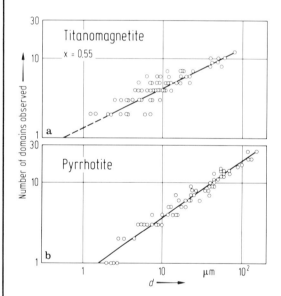

Fig. 93. Observation of number of domains in natural titanomagnetite ($x = 0.55$) [Sof71] (a) and pyrrhotite [Sof77] (b) particles of various sizes (d = grain diameter). The single domain (SD) – two domain transition is directly observed in pyrrhotite and estimated by extrapolation in titanomagnetite.

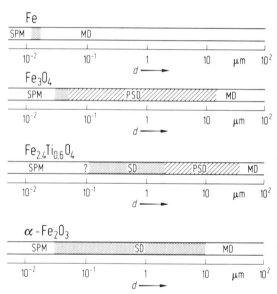

Fig. 94. Domain structure as a function of grain size in common magnetic minerals at room temperature. Fe from [But75], magnetite from [Dic66, Dun73], titanomagnetite ($x = 0.6$) from [Day77], hematite from [Ban71]. SPM = superparamagnetic, SD = single-domain structure, PSD = pseudo-single domain structure, MD = multi-domain structure.

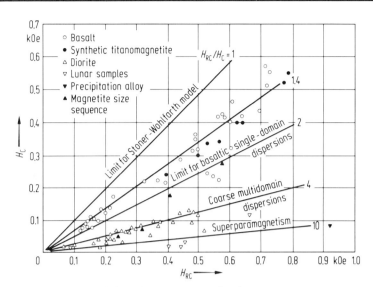

Fig. 95. Relationship between coercive force H_C and remanence coercive force H_{RC} at room temperature for basalts, diorites, lunar samples, synthetic titanomagnetite, magnetite and precipitation alloys. The ratio H_{RC}/H_C varies in a characteristic way depending on the magnetic domain structure [Was73].

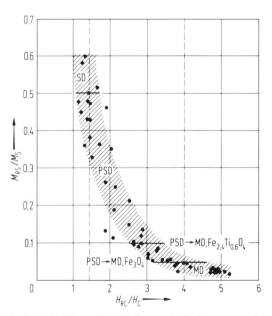

Fig. 96. M_{RS}/M_S vs. H_{RC}/H_C for synthetic titanomagnetites of varying composition ($x=0$, 0.2, 0.4, 0.6) measured at room temperature [Day77]. This graph can be used for a diagnosis of the magnetic domain structure. The solid horizontal lines indicate the single-domain (SD) – pseudo-single-domain (PSD) transition and PSD – multi-domain (MD) transition, respectively. The dashed vertical lines are taken from [Was73] and give the critical values of H_{RC}/H_C for the SD–PSD and PSD–MD transitions, respectively.

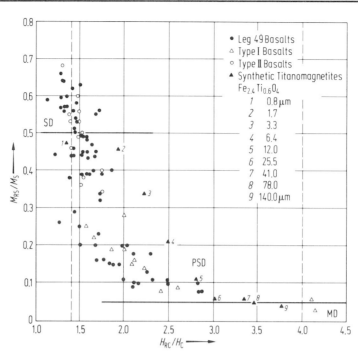

Fig. 97. Magnetic domain structure diagnosis using M_{RS}/M_S and H_{RC}/H_C. The critical values for SD–PSD and PSD–MD transitions have been taken from Fig. 96. Ocean floor basalts from Deep Sea Drilling Project Leg 49 (North Atlantic) together with synthetic titanomagnetite (x = 0.6).

Type I basalts show reversible thermomagnetic curves, type II basalts irreversible thermomagnetic curves. Apart from two samples of type I basalt all samples show either PSD or SD structure [Day78].

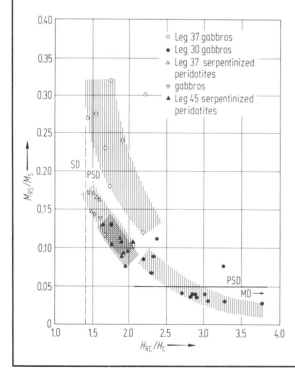

Fig. 98. Magnetic domain structure diagnosis using M_{RS}/M_S and H_{RC}/H_C. The critical values for SD–PSD and PSD–MD transitions have been taken from Fig. 96. Ocean floor intrusive rocks from Deep Dea Drilling Project Legs 30, 37, and 45 [Dun81].

6.2.5 Curie temperature — Curie-Temperatur
6.2.5.1 Introductory remarks — Einleitende Bemerkungen

The Curie temperature T_C of rocks is caused by the ferrimagnetic minerals contained in the rock. Frequently rocks do not show one well defined Curie temperature but rather a range of Curie temperatures reflecting the spread of compositions of the ferrimagnetic mineral component.

As the Curie temperature is a physical constant of material it is better suited than most other magnetic parameters to serve in estimating the composition of the ferrimagnetic minerals contained in rocks (for Curie temperatures of minerals, see Chapter 6.1.).

In the case of igneous rocks it is useful to distinguish between primary (original) and secondary (subsequent) Curie temperatures [Car67, Har71]. The primary Curie temperature is expression of the primary magnetic mineral composition that is present in a rock prior to any subsequent alteration and is relatively restricted in the case of common igneous rocks (Tables 11, 12, 13, Figs. 99, 100, 101, and 102).

Secondary Curie temperatures are acquired in the course of transformation of the primary magnetic mineral component (oxidation, reduction or inversion). A comparison of the measured Curie temperature of a rock with the theoretical Curie temperature deduced from the expected primary composition of the Fe—Ti oxides contained in the rock, provides a simple method to estimate the degree of alteration of the magnetic minerals [Car67] (Figs. 103 and 110c).

The by far most frequent type of alteration of the magnetic minerals in common rocks under present terrestrial surface conditions is oxidation. The effect of oxidation is – generally speaking – the increase of the Curie temperature of a rock (Figs. 103, 104, 105, 107, 108, and 110) and, conversely, the effect of reduction is the decrease of the Curie temperature, apart from the rare case where extreme reduction results in the formation of native iron with Curie temperatures around 770 °C [Den77].

For basaltic rocks (in sensu lato) the following types of magnetic mineral oxidation are distinguished:

Die Curie-Temperatur von Gesteinen wird durch die ferrimagnetischen Minerale des Gesteins bestimmt. Häufig zeigen Gesteine keine genau definierte Curie-Temperatur, sondern vielmehr einen Bereich von Curie-Temperaturen, der die Variationsbreite der Zusammensetzung der ferrimagnetischen Mineralkomponente widerspiegelt.

Die Curie-Temperatur, als eine physikalische Materialkonstante, ist besser als die meisten anderen magnetischen Größen für eine Abschätzung der Zusammensetzung der im Gestein enthaltenen ferrimagnetischen Minerale geeignet. (Zur Curie-Temperatur in Mineralen siehe Abschnitt 6.1).

Im Fall magmatischer Gesteine ist es sinnvoll, zwischen primären (ursprünglichen) und sekundären (subsequenten) Curie-Temperaturen zu unterscheiden [Car67, Har71]. Die primäre Curie-Temperatur ist Ausdruck der ursprünglichen Zusammensetzung der magnetischen Minerale, die in einem Gestein vor irgendeiner subsequenten Alteration vorhanden ist. Der Bereich der primären Curie-Temperaturen ist im Fall gewöhnlicher magmatischer Gesteine relativ beschränkt (Tab. 11, 12, 13, Fig. 99, 100, 101 und 102).

Sekundäre Curie-Temperaturen entstehen im Lauf der Umformung (Alteration) der ursprünglichen magnetischen Mineralkomponenten (durch Oxidation, Reduktion oder Inversion). Ein Vergleich zwischen der gemessenen Curie-Temperatur eines Gesteins und der theoretischen Curie-Temperatur, die aus der für das Gestein zu erwartenden ursprünglichen Zusammensetzung der Fe—Ti-Oxide abgeleitet wird, ergibt eine einfache Methode, den Grad der Veränderung (Alteration) der magnetischen Minerale abzuschätzen [Car67] (Fig. 103 und 110c).

Die bei weitem häufigste Art der Alteration der magnetischen Minerale in gewöhnlichen Gesteinen unter den gegebenen Bedingungen der Erdoberfläche ist die Oxidation. Der Effekt der Oxidation besteht – allgemein gesprochen – im Ansteigen der Curie-Temperatur eines Gesteins (Fig. 103, 104, 105, 107, 108 und 110), und umgekehrt, der der Reduktion im Fallen der Curie-Temperatur, abgesehen von den seltenen Fällen, in denen durch extreme Reduktion gediegenes Eisen mit Curie-Temperaturen um 770 °C entsteht [Den77].

Bei basaltischen Gesteinen (im weiten Sinn) unterscheidet man folgende Arten der Oxidation magnetischer Minerale:

a) high temperature oxidation ($T > 350\,°C$) [Hag76, Pet76]

a) Hochtemperatur-Oxidation ($T > 350\,°C$) [Hag76, Pet76].

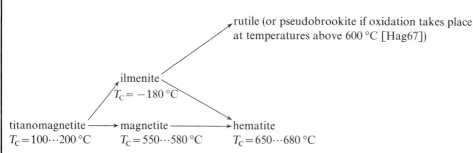

The formation of magnetite causes an increase of Curie temperature and magnetization (Fig. 105), further oxidation to hematite causes further increase of Curie temperature and decrease of magnetization [Cre69, Ore67, Pet66].

High temperature oxidation is common in subaerial lava flows (Fig. 106) [Ade68, Law71, Pet76, Wat67] and gabbros (Fig. 104) [Bud64, Har71] but is only rarely observed in ocean floor basalt [Joh78] (Table 15). The high temperature oxidation of the magnetic minerals in gabbros is also termed "internal oxidation" [Har71]. The term "deuteric oxidation" is frequently used for a high temperature oxidation of the primary Fe—Ti oxide assemblage taking place during the initial cooling of an igneous rock [Ade68, Hol20].

Die Bildung von Magnetit erhöht die Curie-Temperatur und die Magnetisierung (Fig. 105); bei der weiteren Oxidation zu Hämatit erhöht sich die Curie-Temperatur weiter, die Magnetisierung aber nimmt ab [Cre69, Ore67, Pet66].

Hochtemperatur-Oxidation tritt hauptsächlich in subaerischen Laven (Fig. 106) [Ade68, Law71, Pet76, Wat67] und in Gabbros auf (Fig. 104) [Bud64, Har71], ist aber nur selten in Meeresbodenbasalten beobachtet worden [Joh78] (Tab. 15). Die Hochtemperatur-Oxidation von magnetischen Mineralen in Gabbros wird auch als „interne Oxidation" bezeichnet [Har71]. Der Ausdruck „deuterische Oxidation" wird häufig für die Hochtemperatur-Oxidation von primären Fe—Ti-Oxiden gebraucht, die bei der Abkühlung eines magmatischen Gesteins stattfindet [Ade68, Hol20].

b) low temperature oxidation ($T < 300\,°C$) [Ore67, Pet76, Pet79, Rea72]

b) Tieftemperatur-Oxidation ($T < 300\,°C$) [Ore67, Pet76, Pet79, Rea72].

 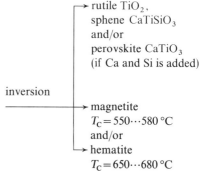

Low temperature oxidation is the common form of titanomagnetite alteration in ocean floor basalts where it proceeds at ambient temperatures (Fig 107 a and b). This process is frequently termed "sea floor weathering" [Joh78, Pet79].

Low temperature oxidation is also common in subaerial basalts where it often grades into hydrothermal alteration.

Tieftemperatur-Oxidation ist die übliche Form der Titanomagnetit-Alteration in Ozean-Basalten, wo sie bei Umgebungstemperatur stattfindet (Fig. 107 a und b). Dieser Prozeß wird „Meeresboden-Verwitterung" genannt [Joh78, Pet79].

Tieftemperatur-Oxidation tritt aber auch häufig in subaerischen Basalten auf, wobei sie oft in hydrothermale Alteration übergeht.

c) hydrothermal alteration ($50 < T < 400$ °C)

This form of oxidation is important in buried volcanic sequences [Ade71] (Fig. 110) and ophiolite complexes [Lev77, Vin72].

In natural samples it may be difficult to make clear distinction between the above named three cases as the different types of oxidation frequently grade continuously into each other. Therefore, the temperature limits given here may serve only as rough guidelines.

c) Hydrothermale Alteration ($50 < T < 400$ °C)

Diese Art der Oxidation ist in abgesunkenen vulkanischen Formationen [Ade71] (Fig. 110) und in Ophiolith-Komplexen von Bedeutung [Lev77, Vin72].

In natürlichen Proben kann es schwierig sein, zwischen den drei obengenannten Oxidationsarten klar zu unterscheiden, da sie oft kontinuierlich ineinander übergehen. Die hier angegebenen Temperaturgrenzen sind deshalb nur als ungefähre Richtwerte zu verstehen.

6.2.5.2 Figures and tables — Abbildungen und Tabellen

Table 11. Range of compositions and Curie temperatures of primary Fe—Ti oxides in acid to basic rocks*) [Pet73].

Bulk chemistry Rock type	acid granite rhyolite	intermediate dacite trachyte	basic gabbro basalt
Titanomagnetite composition $Fe_{3-x}Ti_xO_4$ Curie temperature T_C [°C]	$0 \leq x \leq 0.2$ $580 \cdots 463$	$0.2 \leq x \leq 0.6$ $463 \cdots 165$	$0.5 \leq x \leq 0.85$ $244 \cdots -41$
Ilmenite-hematite composition $Fe_{2-y}Ti_yO_3$ Curie temperature T_C [°C]	$0.5 \leq y \leq 0.9$ $220 \cdots -130$	$0.5 \leq y \leq 0.9$ $220 \cdots -130$	$0.8 \leq y \leq 1.0$ $-40 \cdots -200$

*) Fe-Ti oxide composition (and corresponding theoretical Curie temperature) at the temperature of precipitation of the oxides from the silicate melt, prior to any alteration or internal equilibration (see also Fig. 104).

Table 12. Primary Fe—Ti oxide composition and calculated Curie temperatures in basalts and andesites (the numbers in parentheses give the number of analyses).

Rock type	Titanomagnetites[2]					Hemoilmenites	
	$Fe_{3-x-y-z}Ti_xAl_yMg_zO_4$			T_C [°C]		$Fe_{2-y}Ti_yO_3$	T_C [°C]
	x	y	z	Ti only[3]	Ti, Mg, Al[4]	y	
All basalts[1]	0.61 (237)	0.05 (37)	0.11 (102)	168	128	0.89 (51)	−121
Tholeiites	0.64 (49)	0.03 (13)	0.04 (13)	144	123		
Alkaline basalts	0.52 (57)	0.08 (12)	0.12 (20)	253	181		
Andesites	0.38 (32)			341			

[1] The number of analyses of 'all basalts' is larger than the sum of 'tholeiites' and 'alkaline basalts' as for many titanomagnetite analyses the host rock is described as 'basalt' only.

[2] In many titanomagnetite analyses the content of Mg and Al has not been determined. In cases where the x-value has then been determined from the Fe/Ti ratio, this neglection leads to slightly higher values than correct. The above listed x-values may therefore be slightly too high (particularly for alkali basalt as there the Mg and Al content is higher).

[3] Curie temperature was calculated under the assumption that the titanomagnetite did not contain any impurities of Mg and Al. Calculation according to [Ric73].

[4] Curie temperature calculation including the influence of Mg and Al impurities according to [Ric73].

Table 13. Compilation of data of the primary titanomagnetite composition of ocean floor basalts. (The mean primary titanomagnetite composition of continental tholeiites is included for comparison) N − number of analyses.

Rock type locality	Ref.	$Fe_{3-x-y-z-a}Ti_xAl_yMg_zMn_aO_4$				T_C °C [1]	Technique Used	N
		x	y	z	a			
Continental tholeiites	Pet76	0.64	0.03	0.04	−	123	Microprobe	49
Ocean floor tholeiites Atlantic Ocean DSDP Leg 45	Joh78a	0.63 ± 0.05	0.03 ± 0.01	0.05 ± 0.02	0.02 ± 0.002	110	Microprobe	6
Ocean floor tholeiites Atlantic Ocean DSDP Leg 37	Ble77	0.62 ± 0.02	0.09 ± 0.01	0.03 ± 0.01	0.01	89	Microprobe	4
Ocean floor tholeiites Atlantic Ocean FAMOUS Area	Pre76	0.61 ± 0.02	0.11 ± 0.01	0.03 ± 0.01	0.02 ± 0.01	89	Microprobe	10
Ocean floor tholeiites Atlantic and Pacific Ocean, DSDP Legs 16, 34, 38 and MOHOLE EM-7	Joh78	0.65 ± 0.02	−	−	−	117	Microprobe	60
Ocean floor basalts Atlantic and Pacific Ocean, dredged and MOHOLE EM-7	Ozi74	0.73 ± 0.11	−	−	−	55	Wet chemical	5
Ocean floor tholeiites Pacific Ocean DSDP Leg 34	Maz76	0.68 ± 0.05	0.07 ± 0.02	0.03 ± 0.01	0.02 ± 0.002	55	Microprobe	14
Ocean floor tholeiites Pacific Ocean DSDP Leg 34	Ade76	0.62 ± 0.02	−	−	−	140 138 [2]	Curie temperature	8
Ocean floor tholeiites Atlantic, Pacific, Mediterranean Sea	Pet79	0.58 ± 0.02	0.07 ± 0.02	0.06 ± 0.02	0.02 ± 0.005	131 125 [3]	Microprobe	24

[1]) Curie temperature of synthetic equivalent calculated according to [Ric73].
[2]) Curie temperature measured on natural samples.
[3]) Curie temperature measured on synthetic equivalent.

Fig. 99

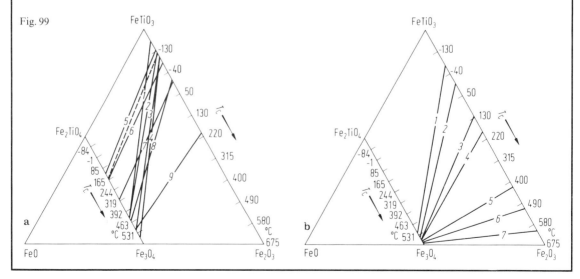

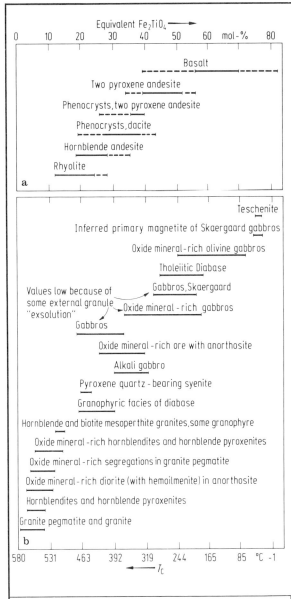

◀
Fig. 100. Range of compositions and corresponding Curie temperatures T_C of primary titanomagnetites (Fe$_2$TiO$_4$—Fe$_3$O$_4$) in igneous rocks.

a) Volcanic rocks. 70% or more of the analyses lie in the range shown by the heavy lines. Modified from [Bud64].

b) Plutonic rocks. The titanomagnetite compositions given here represent the composition immediately after precipitation from the silicate melt prior to any subsequent internal equilibration (as shown in Fig. 104a).

The partly relatively low Ti values in titanomagnetites of gabbros are due to subsequent "granule exsolution" which results in a partition of the original titanomagnetite grains into separate ilmenite and Ti-poor titanomagnetite grains. Modified from [Bud64].

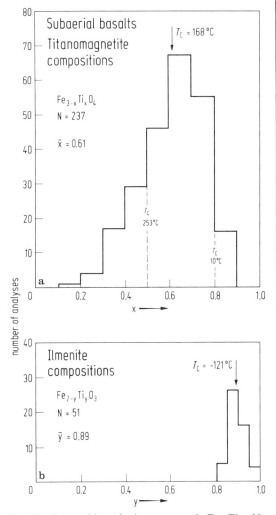

Fig. 101. Composition of primary magnetic Fe—Ti oxides (titanomagnetites Fe$_{3-x}$Ti$_x$O$_4$ and ferrian ilmenites Fe$_{2-y}$Ti$_y$O$_3$) in subaerial basalts (*in sensu lato*) [Pet76].

a) Histogram of the titanomagnetite compositions. Also given is the Curie temperature T_C corresponding to the mean x value. N is the total number of analyses.

b) Histogram of the ilmenite compositions. Also given is the Curie temperature T_C corresponding to the mean y value. N is the total number of analyses.

◀
Fig. 99 Composition and Curie temperatures T_C of primary Fe-Ti oxides for a wide variety of rocks represented in the ternary system FeO—FeTiO$_3$—Fe$_2$O$_3$. Coexisting members of the titanomagnetite (Fe$_2$TiO$_4$—Fe$_3$O$_4$) and ilmenite-hematite (FeTiO$_3$—Fe$_2$O$_3$) solid solution series are connected by tie lines as representative for different rock types.

a) Igneous rocks. The dotted line gives the subaerial basalt mean taken from Fig. 101 [Pet76]. *1, 2:* quartz-bearing pyroxene syenite [Bud64]; *3:* biotite granite [Bud64]; *4:* granite pegmatite [Bud64]; *5:* Heimaey mugearite, Iceland [Jak73]; *6:* Kilauea Lava, Hawaii [And72]; *7:* andesite [And68]; *8:* rhyolite [Low70]; *9:* Haruna dacite, Japan [Uye55].

b) Metamorphic rocks. *1, 2, 3, 4, 5, 6:* metasomatized paragneisses; *7:* ore replacement of paragneisses [Bud64].

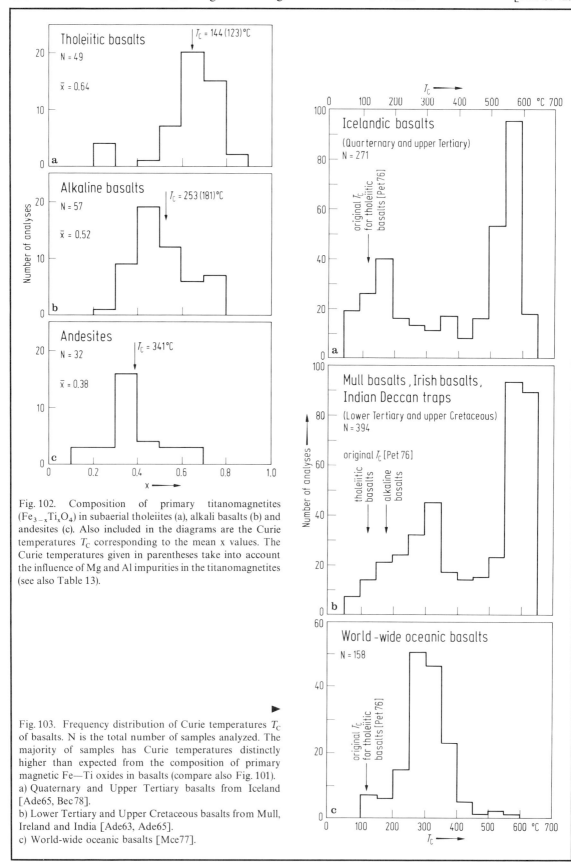

Fig. 102. Composition of primary titanomagnetites $(Fe_{3-x}Ti_xO_4)$ in subaerial tholeiites (a), alkali basalts (b) and andesites (c). Also included in the diagrams are the Curie temperatures T_C corresponding to the mean x values. The Curie temperatures given in parentheses take into account the influence of Mg and Al impurities in the titanomagnetites (see also Table 13).

Fig. 103. Frequency distribution of Curie temperatures T_C of basalts. N is the total number of samples analyzed. The majority of samples has Curie temperatures distinctly higher than expected from the composition of primary magnetic Fe—Ti oxides in basalts (compare also Fig. 101).
a) Quaternary and Upper Tertiary basalts from Iceland [Ade65, Bec78].
b) Lower Tertiary and Upper Cretaceous basalts from Mull, Ireland and India [Ade63, Ade65].
c) World-wide oceanic basalts [Mce77].

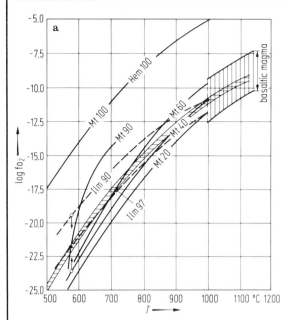

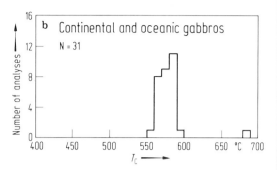

Fig. 104b. Frequency distribution of Curie temperatures T_C of gabbro from the Skaergaard intrusion [Che54] and of gabbro samples dredged from the North Atlantic ocean floor [Ken78].

Fig. 104a. Internal oxidation/reduction of the primary Fe—Ti oxide assemblage in basaltic magmas, cooling very slowly in a closed system, as it is commonly exemplified in gabbros and thick intrusions (see also Table 15). Oxygen fugacity (log f_{O_2}) plotted against temperature T. The vertically dashed area represents the range of oxygen fugacities in basaltic magmas [Car67]. The horizontally dashed area gives the temperature variation of oxygen fugacity of the quartz-fayalite-magnetite (QFM) buffer which represents a reasonable approximation of the natural conditions in a closed-system cooling basaltic magma [Hag76]. The curves labelled Mt and Ilm are equilibrium oxygen fugacities of different members of the titanomagnetite and ilmenite-hematite solid solution series [Bud64]. When precipitating from the silicate melt at about 1050 °C, titano-magnetite, $Fe_{3-x}Ti_xO_4$, with x = 0.60 (Mt40) and ferrian ilmenite, $Fe_{2-y}Ti_yO_3$, with y = 0.90 (Ilm90) are in equilibrium with the environmental oxygen fugacity (full circle at 1000 °C). After cooling down to 600 °C, titanomagnetite with x = 0.10 (Mt90) coexisting with ilmenite y = 0.97 (Ilm97) are now the phases in equilibrium with the environment, represented by the QFM-buffer (full circle at 600 °C). During cooling from 1050 to 600 °C the primary titanomagnetite with x = 0.60 (open circle) gradually oxidized to Ti-poor magnetite with x = 0.10, simultaneously forming exsolution lamellae of Ilm97. The primary ferrian ilmenite with y = 0.90 (other open circle) was gradually reduced to ilmenite with y = 0.97, becoming enriched in Ti and simultaneously forming exsolution lamellae of magnetite with x = 0.10. The whole process results in magnetic phases with Curie temperatures close to that of magnetite, i.e. between 550 and 600 °C [Har71].

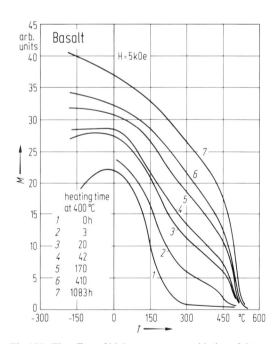

Fig. 105. The effect of high temperature oxidation of titano-magnetite in basalt. Strong field magnetization M vs. temperature T of basalt samples from Rauher Kulm (near Kemnath, Germany). The samples have been heated in air for various times prior to the measurement. The magnetization curves evolve in a characteristic manner: gradual increase of Curie temperature and magnetization intensity [Cre69].

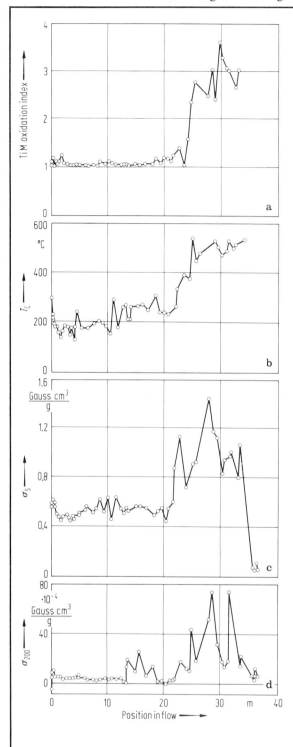

◄

Fig. 106a···d. Spatial variation of magnetic properties in an Antrim (Northern Ireland) tholeiitic lava flow, measured along a vertical profile from bottom to top of the flow, High temperature oxidation dominates the upper half of the flow, resulting in increased magnetization intensity and Curie temperature [Law71].

a) Variation of the magnetic mineral (titanomagnetite) oxidation state, expressed by the titanomagnetite (TiM) high temperature oxidation index [Ade68, Hag76]. The oxidation index runs continuously from 1.0 (= no oxidation) to 6.0 (= maximum oxidation).

b) Curie temperature T_C.

c) Specific saturation magnetization σ_S measured at room temperature.

d) σ_{200}-specific natural remanent magnetization after demagnetization in alternating field, $H_d = 200$ Oe.

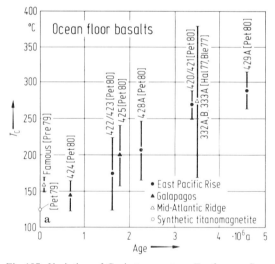

Fig. 107. Variation of Curie temperature T_C of ocean floor basalts with age. The basalt samples have been recovered by FAMOUS and Deep Sea Drilling Project, DSDP. The numbers in the diagrams are the respective DSDP drill hole numbers. The error bars give the spread of Curie temperatures (one standard deviation) within the drill hole, the symbols give the mean value. The open circle represents the Curie temperature of a synthetic titanomagnetite with a composition equivalent to the primary titanomagnetite composition in ocean floor basalts as deduced from electron microprobe analyses [Pet79]. The variation of Curie temperature with age is the effect of low temperature oxidation of the titanomagnetites contained in the ocean floor basalts [Joh78, Ozi74, Pet79].

a) ocean floor basalts with ages up to $5 \cdot 10^6$ a.

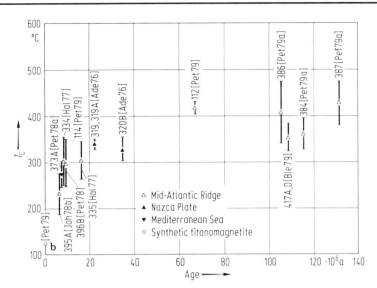

Fig. 107. b) ocean floor basalts with ages between 7 and 131 · 10⁶ a (see also Table 14).

Table 14. Mean Curie temperatures of ocean floor basalts recovered by the Deep Sea Drilling Project (DSDP) and FAMOUS.

Locality	DSDP Hole number	Rock type[1])	Age 10⁶ a	T_C °C	Number of analyses	Ref.
Mid-Atlantic Ridge	FAMOUS	P	0.1···0.5	159 ± 9	86	Pre79
	332A, B, 333A	P	3.5	274 ± 112	359	Ble77, Hal77
	395A	P	7.0	234 ± 42	131	Joh78b
	334	P	8.9	300 ± 51	18	Hal77
	396B	P	9.5	297 ± 56	68	Pet78
	114	P	12	340	1	Pet79
	335	P	16.5	306 ± 40	43	Hal77
	112	P	67	417 ± 13	4	Pet79
	386	P	105	408 ± 70	4	Pet79a
	417A, D	P	108	353 ± 34	44	Ble79
		M	108	281 ± 42	31	Ble79
	384	P	115	362 ± 32	11	Pet79a
	387	P	131	428 ± 49	6	Pet79a
East Pacific Rise	422, 423	P	1.6	173 ± 55	10	Pet80
	428A	P	2.25	210 ± 44	14	Pet80
	420, 421	P	3.4	268 ± 20	2	Pet80
	429A	P	4.6	286 ± 26	4	Pet80
Galapagos Spreading Center	424A, B, C	P	0.7	143 ± 24	17	Pet80
	425	P	1.8	201 ± 65	12	Pet80
Nazca Plate	319, 319A	P	23	340 ± 10	13	Ade76
	319A	M	23	246 ± 89	7	Ade76
	320B	P	30···40	330 ± 23	6	Ade76
	321	M	40	185 ± 69	14	Ade76
Mediterranean Sea	373A	P	7.5	274 ± 28	8	Pet78a

[1]) P – mainly pillow basalts and/or thin flows,
   M – mainly massive basalt.

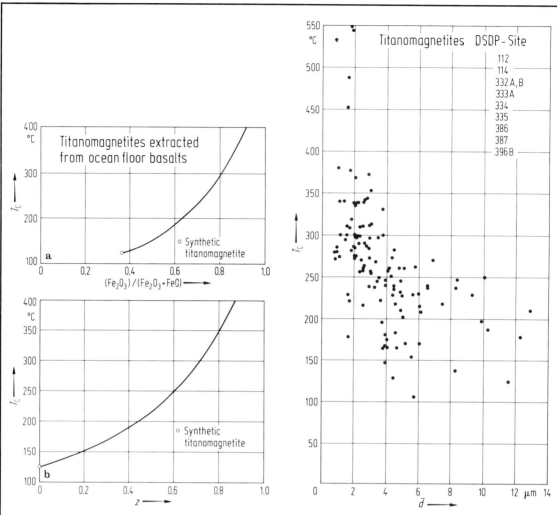

Fig. 108. The effect of low temperature oxidation of titano-magnetites in ocean floor basalts on Curie temperature [Pet 79].

a) Curie temperature T_C vs. weight ratio $Fe_2O_3/(Fe_2O_3+FeO)$ of titanomagnetites extracted from ocean floor basalts from different oceans and various ages. The curve has been corrected for silicate impurities adhering to the extracted titanomagnetite grains. The open circle represents a synthetic titanomagnetite with a composition equivalent to the primary titanomagnetite composition in ocean floor basalts as deduced from electron microprobe analyses.

b) Curie temperature T_C vs. oxidation parameter z (for definition of z, see text). The curve has been derived from the curve in Fig. 108a by conversion of the $Fe_2O_3/(Fe_2O_3+FeO)$ values into z values assuming Fe-migration out of the titanomagnetite crystals. For the open circle, see Fig. 108a.

Fig. 109. Curie temperature T_C vs. mean grain diameter of titanomagnetites, $\bar{d}$, contained in ocean floor basalts from various Deep Sea Drilling Project (DSDP) drill sites. The obvious inverse relationship between the two variables is a consequence of a grain size dependence of low temperature oxidation of titanomagnetites [Pet 79].

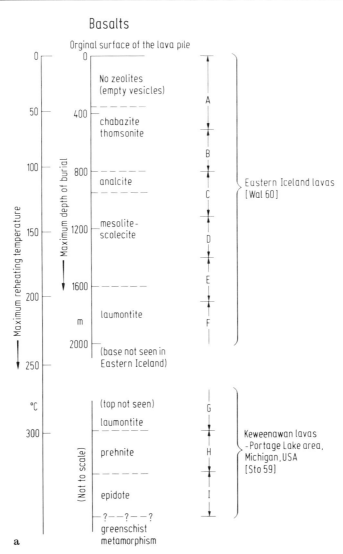

Fig. 110. The effect of hydrothermal alteration of basalts on Curie temperature [Ade71].

a) Standard succession of regional secondary mineral zones, based on the Eastern Iceland [Wal60] and Keweenawan (Michigan, USA) [Sto59] lava piles. The succession of zones A to I is characterized by certain zeolite minerals and represents increasing temperature and pressure, i.e. increasing hydrothermal alteration.

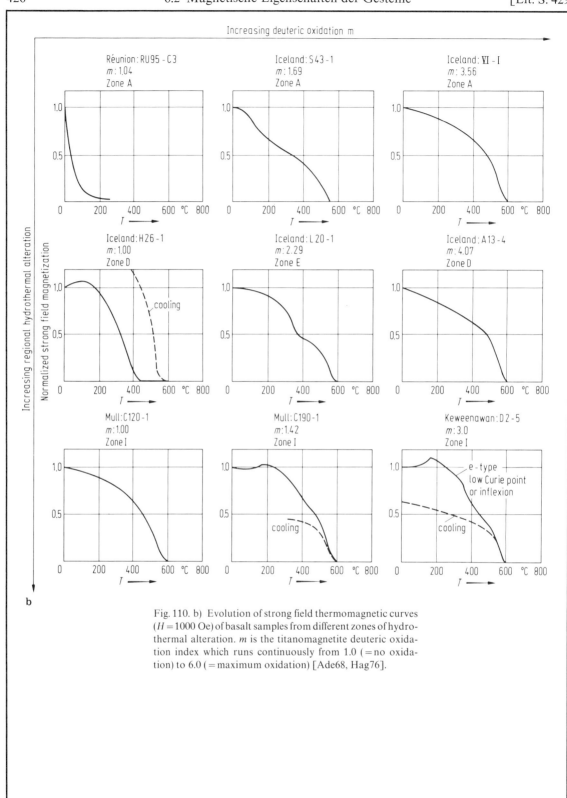

Fig. 110. b) Evolution of strong field thermomagnetic curves ($H = 1000$ Oe) of basalt samples from different zones of hydrothermal alteration. m is the titanomagnetite deuteric oxidation index which runs continuously from 1.0 ($=$ no oxidation) to 6.0 ($=$ maximum oxidation) [Ade68, Hag76].

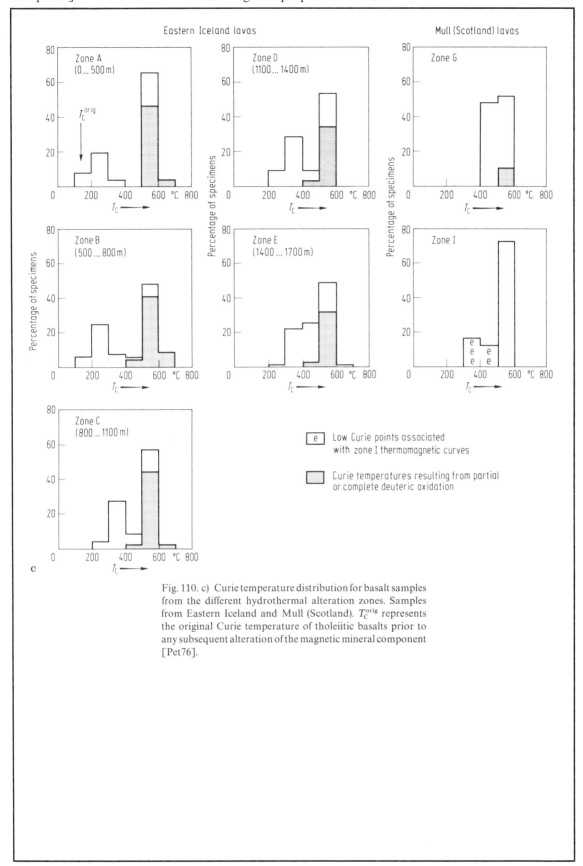

Fig. 110. c) Curie temperature distribution for basalt samples from the different hydrothermal alteration zones. Samples from Eastern Iceland and Mull (Scotland). T_C^{orig} represents the original Curie temperature of tholeiitic basalts prior to any subsequent alteration of the magnetic mineral component [Pet76].

Table 15. Correlation between petrology, Curie temperature and Koenigsberger Q-factor of basaltic rocks. The effect of low temperature oxidation and hydrothermal alteration has not been taken into account, only the effect of initial cooling history. Subsequent low temperature oxidation may effect all different cases listed. Modified from [Har71].

	Gabbros	Dolerite dikes and sills		Basalts		Thick flows	
		Border	Center	Pillows	Thin flows	Border	Center
Cooling rate	slow	fast	slow	quench	quench	fast	intermediate
Volatile compounds	juvenile, closed system	probably juvenile	juvenile, closed system	contamination, water	contamination, water/air	contamination, water/air	possibly contaminated. f_{O_2} high by H_2-diffusion [Gro69, Wat67]
Subsolidus Fe-Ti oxide relations	equilibrated down to 600 °C "internal oxidation"	may be quenched from 1000 °C	varying equilibration down to 600 °C, "internal oxidation"	primary assemblage preserved	primary assemblage preserved. possibly some high T oxidation	primary assemblage, partly preserved, some high T oxidation	varying oxidation within zones of high f_{O_2} due to H_2-diffusion
Resulting Fe—Ti oxide phases	Ti-poor magnetite, ilmenite mostly intergrown, possibly minor exsolved ulvöspinel, iron sulfides	Ti-rich titanomagnetite, ferrian ilmenite, iron sulfides	Ti-poor magnetite, ilmenite, possibly minor exsolved ulvöspinel, iron sulfides	Ti-rich titanomagnetite, ferrian ilmenite, iron sulfides	Ti-rich titanomagnetite, ferrian ilmenite	partly magnetite/ilmenite intergrowths, partly Ti-rich titanomagnetite, ferrian ilmenite	magnetite/ilmenite intergrowths, iron sulfides
Dominant T_C [°C]	550···600	0···550	550···600	100···200	100···300	100···670	100···600
Q-factor	0.5···5	5···15	0.5···5	10···100	10···100	10···100	5···20

6.2.6 References for 6.2.1 ··· 6.2.5 — Literatur zu 6.2.1 ··· 6.2.5

Ade63	Ade-Hall, J.M., Wilson, R.L.: Nature **198** (1963) 659.
Ade65	Ade-Hall, J.M., Wilson, R.L., Smith, P.J.: Geophys. J. **9** (1965) 323.
Ade68	Ade-Hall, J.M., Khan, M.A., Dagley, P., Wilson, R.L.: Geophys. J. **16** (1968) 375.
Ade71	Ade-Hall, J.M., Palmer, H.C., Hubbard, T.P.: Geophys. J. **24** (1971) 137.
Ade76	Ade-Hall, J.M., Johnson, H.P., Ryall, J.L.: In. Rep. DSDP **34** (1976) 459.
And68	Anderson, A.T.: Am. J. Sci. **266** (1968) 704.
And72	Anderson, A.T., Wright, T.L.: Am. Mineral. **57** (1972) 188.
Ang81	Angenheister, G., Bleil, U., Petersen, N.: unpubl. data (1981).
Bal60	Balsley, J.R., Buddington, A.F.: Am. J. Sci. **258 A** (1960) 6.
Ban71	Banerjee, S.K.: Nature Phys. Sci. **232** (1971) 15.
Bar80	Barton, C.E., McElhinny, M.W., Edwards, D.J.: Geophys. J. **61** (1980) 355.
Bec78	Becker, H.: Ph.D. Thesis Univ. München (1978).
Bic55	Bickford, L.R., Pappis, J., Stull, J.L.: Phys. Rev. **99** (1955) 1211.
Biq70	Biquand, D., Prévot, M.: C.R. Acad. Sci. (Paris) **270** (1970) 362.
Biq71	Biquand, D., Prévot, M.: Z. Geophys. **37** (1971) 471.
Biq71a	Biquand, D., Prévot, M., Dunlop, D.J.: J. Physique C-1, **32** (1971) 1043.
Ble77	Bleil, U., Petersen, N.: In. Rep. DSDP **37** (1977) 446.
Ble79	Bleil, U., Smith, B.M.: In. Rep. DSDP **51–53** (2) (1979) 1411.
Bud64	Buddington, A.F., Lindsley, D.H.: J. Petrol. **5** (1964) 310.
But75	Butler, R.F., Banerjee, S.K.: J. Geophys. Res. **80** (1975) 252.
Car61	Carmichael, C.W.: Proc. Roy. Soc. London, Ser. A **263** (1961) 508.
Car67	Carmichael, I.S.E., Nicholls, J.: J. Geophys. Res. **72** (1967) 4665.
Che25	Chevallier, R.: Ann. Phys. **4** (1925) 5.
Che43	Chevallier, R., Mathieu, S.: Ann. Phys. **18** (1943) 258.
Che54	Chevallier, R., Mathicu, S., Vincent, E.A.: Geochim. Cosmochim. Acta **6** (1954) 27.
Col65	Collinson, D.W.: Geophys. J. **9** (1965) 203.
Col74	Collinson, D.W.: Geophys. J. **38** (1974) 253.
Cre69	Creer, K.M., Petersen, N.: Z. Geophys. **35** (1969) 501.
Cre70	Creer, K.M., Petersen, N., Petherbridge, J.: Geophys. J. **21** (1970) 471.
Dan78	Dankers, P.H.M.: Ph. D. Thesis Univ. Utrecht (1978).
Day77	Day, R.: J. Geomag. Geoelectr. **29** (1977) 233.
Day77a	Day, R., Fuller, M., Schmidt, V.A.: Phys. Earth Planet. Int. **13** (1977) 260.
Day78	Day, R., Halgedahl, S., Steiner, M., Kobayashi, K., Furuta, T., Ishii, T., Faller, A.: In. Rep. DSDP **49** (1978) 781.
Dav76	Davis, P.M., Evans, M.E.: J. Geophys. Res. **81** (1976) 989.
Deu77	Deutsch, E.R., Rao, K.V., Laurent, R., Seguin, M.K.: Nature **269** (1977) 684.
Dic66	Dickson, G.O., Everitt, C.W.F., Parry, L.G.: Earth Planet. Sci. Lett. **1** (1966) 222.
Dom62	Domen, H.: J. Geomag. Geoelectr. **13** (1962) 66.
Dun73	Dunlop, D.J.: J. Geophys. Res. **78** (1973) 1780.
Dun73a	Dunlop, D.J.: Rev. Geophys. Space Phys. **11** (1973) 855.
Dun73b	Dunlop, D.J.: J. Geophys. Res. **78** (1973) 7602.
Dun74	Dunlop, D.J.: J. Geophys. **40** (1974) 439.
Dun74a	Dunlop, D.J.: Proc. T. Nagata Conf., Univ. Pittsburgh (1974) 58.
Dun77	Dunlop, D.J., Stirling, J.M.: Geophys. Res. Lett. **4** (1977) 163.
Dun81	Dunlop, D.J.: Phys. Earth Planet. Int. **26** (1981) 1.
Ell75	Ellwood, B.B.: J. Geophys. Res. **80** (1977) 4805.
Ell78	Ellwood, B.B., Watkins, N.D.: In. Rep. DSDP **46** (1978) 363.
Eve61	Everitt, C.W.F.: Phil. Mag. **6** (1961) 713.
Eve62	Everitt, C.W.F.: Phil. Mag. **7** (1962) 831.
Ful63	Fuller, M.D.: J. Geophys. Res. **68** (1963) 293.
Gir61	Girdler, R.W.: Geophys. J. **5** (1961) 197.
Got35	Gottschalk, V.H.: Physics **6** (1935) 127.
Gra66	Graham, J.W.: Am. Geophys. Union, Geophys. Monogr. **10** (1966) 627.
Gra58	Granar, L.: Ark. Geofysik **3** (1958) 1.
Gri60	Griffiths, D.H., King, R.F., Rees, A.I., Wright, A.E.: Proc. Roy. Soc. London Ser. A **256** (1960) 359.

Hag76	Haggerty, S.E.: Min. Soc. Am. Short Course Notes **3** (1976).
Hai58	Haigh, G.: Phil. Mag. **3** (1958) 267.
Hal74	Halvorsen, E.: Earth Planet. Sci. Lett. **21** (1974) 127.
Hal77	Hall, J.M.: J. Geomag. Geoelectr. **29** (1977) 411.
Hal77a	Hall, J.M., Ryall, P.: In. Rep. DSDP **37** (1977) 489.
Ham67	Hamilton, N., Loveland, P.J.: Brit. Antarct. Bull. No. 11 (1967) 59.
Ham68	Hamilton, N., Owens, W.H., Rees, A.I.: J. Geology **76** (1968) 465.
Har59	Hargraves, R.B.: J. Geophys. Res. **64** (1959) 1565.
Har69	Hargraves, R.B., Perkins, W.E.: J. Geophys. Res. **74** (1969) 2576.
Har71	Hargraves, R.B., Petersen, N.: Z. Geophys. **37** (1971) 367.
Hav65	Havard, A.D., Lewis, M.: Geophys. J. **10** (1965) 59.
Hel71	Heller, F.: Z. Geophys. **37** (1971) 557.
Hel73	Heller, F.: Earth Planet. Sci. Lett. **20** (1973) 180.
Hel78	Heller, F.: J. Geophys. **44** (1978) 525.
Hel79	Heller, F., Markert, H., Schmidbauer, E.: J. Geophys. **45** (1979) 235.
Hel80	Heller, F.: Nature **284** (1980) 334.
Hol20	Holmes, A.: Nomenclature of petrology. Th. Murby & Co., London (1920).
Hro70	Hrouda, F.: Vestnik UUG **45** (1970) 147.
Hro71	Hrouda, F., Janak, F., Rejl, L., Weiss, J.: Geol. Rundschau **60** (1971) 1124.
Iow67	Iowerth, H., Hamilton, N., Loveland, P.J.: Geol. Mag. **104** (1967) 564.
Irv70	Irving, E.: Can. J. Earth Sci. **7** (1970) 1528.
Isi42	Ising, G.: Ark. Mat. Fys. **29A** (1942) 1.
Jak73	Jakobsson, S.P., Pedersen, A.K., Rönsbo, J.G., Melchior-Larson, L.: Lithos **6** (1973) 203.
Jan67	Janak, F., Stovickova, N.: Sbor. geol. ved. **6** (1967) 169.
Jan72	Janak, F.: Geophys. Prosp. **20** (1972) 375.
Jel75	Jelenska, M.: Pageoph **113** (1975) 635.
Joh48	Johnson, E.A., Murphy. T., Torreson, O.W.: Terr. Mag. Atm. Electr. **53** (1948) 349.
Joh72	Johnson, H.P., Merrill, R.T.: J. Geophys. Res. **77** (1972) 334.
Joh77	Johnson, H.P., Atwater, T.: Geol. Soc. Am. Bull. **88** (1977) 637.
Joh78	Johnson, H.P., Hall, J.M.: Geophys. J. **52** (1978) 45.
Joh78a	Johnson, H.P., Melson, W.G.: In. Rep. DSDP **45** (1978) 575.
Joh78b	Johnson, H.P.: In. Rep. DSDP **45** (1978) 397.
Kap55	Kapitsa, S.P.: Izv. Akad. Nauk USSR, Geophys. Ser. (English Transl.) **6** (1955) 489.
Ken73	Kent, D.V.: Nature **246** (1973) 32.
Ken77	Kent, D.V., Lowrie, W.: In. Rep. DSDP **37** (1977) 525.
Ken78	Kent, D.V., Honnorez, B.M., Opdyke, N.D., Fox, P.J.: Geophys. J. **55** (1978) 513.
Kha62	Khan, M.A.: J. Geophys. Res. **67** (1962) 2873.
Khr68	Khramov, A.N.: Phys. Solid Earth (1968) (1) 63.
Kin55	King, R.F.: Mon. Not. R. astr. Soc. Geophys. Suppl. **7** (1955) 115.
Kob59	Kobayashi, K.: J. Geomag. Geoelectr. **10** (1959) 99.
Koe32	Koenigsberger, J.: Phys. Zeitschr. **33** (1932) 468.
Koe36	Koenigsberger, J.: Beitr. Angew. Geophysik **5** (1936) 193.
Koe38	Koenigsberger, J.: Terr. Mag. Atm. Electr. **43** (1938) 119, 299.
Kra56	Kranz, J.: In: Beiträge zur Theorie des Ferromagnetismus und der Magnetisierungskurve, Köster, W. (Hrsg.) Springer-Verlag, Berlin, (1956) 180.
Kri73	Kristjansson, L.G., Deutsch, E.R.: Geol. Surv. Can. Paper 71-23 (1973) 545.
Kro76	Kropacek, V.: Publ. Inst. Geoph. Pol. Ac. Sci. **102** C-1 (1976) 65.
Lam66	Lamoureux, C., Klerkx, J.: Ann. Soc. Geol. Belgique **90** (1966) 261.
Lar69	Larson, E., Ozima, M., Ozima, M., Nagata, T., Strangway, D.: Geophys. J. **17** (1969) 263.
Law71	Lawley, A., Ade-Hall, J.M.: Earth Planet. Sci. Lett. **11** (1971) 113.
Lev74	Levi, S.: Ph.D. Thesis Univ. Washington, Seattle (1974).
Lev77	Levi, S., Banerjee, S.K.: J. Geomag. Geoelectr. **29** (1977) 241.
Lin66	Lindsley, D.H., Andreasen, G.E., Balsley, J.R.: In: Handbook of physical constants, Geol. Soc. Am. Mem. **97** (1966) 543.
Løv76	Løvlie, R.: Earth Planet. Sci. Lett. **30** (1976) 209.
Low73	Lowrie, W.: Nature **243** (1973) 27.
Low73a	Lowrie, W., Opdyke, N.D.: In. Rep. DSDP **15** (1973) 1017.
Low73b	Lowrie, W., Løvlie, R., Opdyke, N.D.: J. Geophys. Res. **78** (1973) 7647.

Low74	Lowrie, W.: J. Geophys. **40** (1974) 513.
Low75	Lowrie, W., Hayes, D.E.: In. Rep. DSDP **28** (1975) 869.
Low76	Lowrie, W., Kent, D.V.: In. Rep. DSDP **34** (1976) 479.
Low78	Lowrie, W., Kent, V.D.: J. Geophys. **44** (1978) 297.
Lub61	Luborsky, F.E.: J. Appl. Phys. **32** (1961) 171.
Mar71	Marshall, M., Cox, A.: Nature **230** (1971) 28.
Mar78	Martin, R.J., Habermann, R.E., Wyss, M.: J. Geophys. Res. **83** (1978) 3485.
Mat54	Matsuzaki, H., Kobayashi, K., Momose, K.: J. Geomag. Geoelectr. **6** (1954) 53.
Maz76	Mazullo, L.J., Bence, A.E.: J. Geophys. Res. **81** (1976) 4327.
Mce77	McElhinny, M.W.: In: Indian Ocean Geology and Biostratigraphy, Am. Geophys. Union, Washington (1977) 301.
Mei53	Meiklejohn, W.E.: Rev. Mod. Phys. **25** (1953) 302.
Mei63	Meitzner, W.: Beitr. Mineral. Petrol. **9** (1963) 320.
Mer69	Merrill, R.T., Grommé, C.S.: J. Geophys. Res. **74** (1969) 2014.
Moo53	Mooney, H.M., Bleifuss, R.: Geophysics **18** (1953) 383.
Nag52	Nagata, T., Uyeda, S., Akimoto, S.: J. Geomag. Geoelectr. **4** (1952) 22.
Nag53	Nagata, T.: Rock magnetism, Maruzen Comp., Tokyo, (1. ed.) (1953).
Nag55	Nagata, T., Ozima, M.: J. Geomag. Geoelectr. **7** (1955) 105.
Nag59	Nagata, T., Uyeda, S.: Nature **184** (1959) 890.
Nag61	Nagata, T.: Rock Magnetism, Maruzen Comp., Tokyo, (2. ed.) (1961).
Nag65	Nagata, T., Kinoshita, H.: J. Geomag. Geoelectr. **17** (1965) 121.
Nag66	Nagata, T.: In: Handbuch der Physik, Flügge (Ed.), Springer-Verlag, (1966).
Nag68	Nagata, T., Carleton, B.J.: J. Geomag. Geoelectr. **20** (1968) 115.
Nag69	Nagata, T., Carleton, B.J.: J. Geomag. Geoelectr. **21** (1969) 427.
Nag70	Nagata, T.: Tectonophysics **9** (1970) 167.
Nag70a	Nagata, T.: Pageoph **78** (1970) 110.
Nag72	Nagata, T.: Tectonophysics **14** (1972) 263.
Nee51	Néel, L.: Ann. Geophys. **7** (1951) 90.
Nee51a	Néel, L.: J. Phys. Rad. **12** (1951) 339.
Nee55	Néel, L.: Adv. Phys. **4** (1955) 191.
Nis74	Nishida, S., Sasajima, S.: Geophys. J. **37** (1974) 453.
Ohn68	Ohnaka, M., Kinoshita, H.: J. Geomag. Geoelectr. **20** (1968) 93.
Opd69	Opdyke, N.D., Henry, K.W.: Earth Planet. Sci. Lett. **6** (1969) 139.
Opd73	Opdyke, N.D., Kent, D.V., Lowrie, W.: Earth Planet. Sci. Lett. **20** (1973) 315.
Ore67	O'Reilly, W., Banerjee, S.K.: Min. Mag. **36** (1967) 29.
Ozi67	Ozima, M., Ozima, M.: Earth Planet. Sci. Lett. **3** (1967) 213.
Ozi68	Ozima, M., Larson, E.E.: J. Geomag. Geoelectr. **20** (1968) 337.
Ozi74	Ozima, M., Joshima, M., Kinoshita, H.: J. Geomag. Geoelectr. **26** (1974) 335.
Par65	Parry, L.G.: Phil. Mag. **11** (1965) 303.
Pei74	Peirce, J.W., Denham, C.R., Luyendyk, B.P.: In. Rep. DSDP **26** (1974) 517.
Pet66	Petersen, N.: J. Geomag. Geoelectr. **18** (1966) 463.
Pet73	Petersen, N., Bleil, U.: Z. Geophys. **39** (1973) 965.
Pet74	Petherbridge, J., Campbell, A.L., Hauptman, Z.: Nature **250** (1974) 479.
Pet76	Petersen, N.: Pageoph **114** (1976) 117.
Pet78	Petersen, N.: In. Rep. DSDP **46** (1978) 357.
Pet78a	Petersen, N., Bleil, U., Eisenach, P.: In. Rep. DSDP **42** (1) (1978) 881.
Pet79	Petersen, N., Eisenach, P., Bleil, U.: Am. Geophys. Union, Maurice Ewing Series **2** (1979) 169.
Pet79a	Petersen, N., Bleil, U., Eisenach, P.: In. Rep. DSDP **43** (1979) 773.
Pet80	Petersen, N., Roggenthen, W.M.: In. Rep. DSDP **54** (1980) 865.
Poc01	Pockels, F.: Phys. Zeitschr. **3** (1901) 22.
Poh75	Pohl, J., Bleil, U., Hornemann, U.: J. Geophys. **41** (1975) 23.
Poz70	Pozzi, J.P.: C.R. Acad. Sci. (Paris) **B 271** (1970) 819.
Poz77	Pozzi, J.P.: Phys. Earth Planet. Int. **14** (1977) 77.
Pre76	Prévot, M., Lecaille, A.: Bull. Soc. Géol. France **18** (1976) 903.
Pre79	Prévot, M., Lecaille, A., Hekinian, R.: Am. Geophys. Union, Maurice Ewing Series **2** (1979) 210.
Puz30	Puzicha, K.: Z. Prakt. Geol. **38** (1930) 161.

Rad70	Rad v., U.: Geol. Rundschau **60** (1970) 331.
Rad71	Radhakrishnamurty, C., Raja, P.K.S., Likhite, S.D.: Curr. Sci. **40** (1971) 1.
Rad74	Radhakrishnamurty, C., Deutsch, E.R.: J. Geophys. **40** (1974) 453.
Rah73	Rahman, A.A., Duncan, A.D., Parry, L.G.: Riv. It. Geofis. **22** (1973) 259.
Rah75	Rahman, A.U., Gough, D.I., Evans, M.E.: Can. J. Sci. **12** (1975) 1465.
Rea72	Readman, P.W., O'Reilly, W.: J. Geomag. Geoelectr. **24** (1972) 69.
Ree66	Rees, A.I.: J. Geol. **74** (1966) 856.
Ref60	Refai, E.: Ph.D. Thesis Univ. München (1960).
Ric73	Richards, J.C.W., O'Donovan, J.B., Hauptman, Z., O'Reilly, W.: Phys. Earth Planet. Int. **7** (1973) 437.
Rim56	Rimbert, F.: C.R. Acad. Sci. (Paris) **242** (1956) 2536.
Rim59	Rimbert, F.: Rev. Inst. Franc. Pétrole **14** (1959) 17.
Rob72	Robins, B.W.: Ph.D. Thesis, Univ. New South Wales, Sydney (1972).
Roe63	Roeser, H.A.: Diplomarbeit, Geophysik, Univ. Göttingen (1963).
Roq54	Roquet, J.: Ann. Géophys. **10** (1954) 226, 282.
Sch68	Schmidbauer, E., Petersen, N.: J. Geomag. Geoelectr. **20** (1968) 169.
Sch68a	Schult, A.: Earth Planet. Sci. Lett. **4** (1968) 57.
Sch76	Schult, A.: J. Geophys. **42** (1976) 81.
Sch76a	Schult, A.: Habil.-Schrift, Fak. Geowiss., Univ. München (1976).
Sch80	Schmidbauer, E., Veitch, R.J.: J. Geophys. **48** (1980) 148.
Sha67	Shapiro, V.A., Ivanov, N.A.: Dokl. Akad. Nauk. USSR (English Transl.) **173** (1967) 1065.
Shi60	Shimizu, Y.: J. Geomag. Geoelectr. **11** (1960) 125.
Smi79	Smith, B.M., Bleil, U.: In. Rep. DSDP **51–53** (1979) 1379.
Sof71	Soffel, H.: Z. Geophys. **37** (1971) 451.
Sof77	Soffel, H.: J. Geophys. **42** (1977) 351.
Sta60	Stacey, F.D., Jophin, G., Linsay, J.: Geof. pura e appl. **47** (1960) 30.
Sta64	Stacey, F.D.: Pageoph **58** (1964) 5.
Sta72	Stacey, F.D., Johnston, M.J.S.: Pageoph **97** (1972) 146.
Sta74	Stacey, F.D., Banerjee, S.K.: The physical principles of rock magnetism, Elevier Sci. Publ. Comp., Amsterdam (1974).
Sto59	Stoiber, R.E., Davidson, E.S.: Econ. Geol. **54** (1959) 1250.
Sto63	Stone, D.B.: Geophys. J. **7** (1963) 375.
Sto79	Stober, J.C., Thompson, R.: Geophys. J. **57** (1979) 727.
Sug80	Sugiura, N.: Earth Planet. Sci. Lett. **46** (1980) 438.
Tar76	Tarasiewicz, G., Tarasiewics, E., Harrison, C.G.A.: In. Rep. DSDP **34** (1976) 473.
The38	Thellier, E.: Ann. Inst. Phys. du Globe, Paris **16** (1938) 157.
The42	Thellier, E., Thellier, O.: C.R. Acad. Sci. (Paris) **214** (1942) 382.
The59	Thellier, E., Thellier, O.: Ann. Géophys. **15** (1959) 285.
Tuc76	Tucholka, P.: Publ. Inst. Geophys. Pol. Ac. Sci. **102C-1** (1976) 127.
Tuc80	Tucker, P.: J. Geophys. **48** (1980) 153.
Tuc80a	Tucker, P.: Geophys. J. **63** (1980) 149.
Uye55	Uyeda, S.: J. Geomag. Geoelectr. **7** (1955) 9.
Uye63	Uyeda, S., Fuller, M.D., Belshé, J.C., Girdler, R.W.: J. Geophys. Res. **68** (1963) 293.
Ver56	Verhoogen, J.: J. Geophys. Res. **61** (1956) 201.
Ver77	Verosub, K.L.: Rev. Geophys. Space Phys. **15** (1977) 129.
Vin72	Vine, F.J., Moores, E.M.: Geol. Soc. Am. Mem. **132** (1972) 195.
Wal60	Walker, G.P.L.: J. Geol. **68** (1960) 515.
Wat67	Watkins, N.D., Haggerty, S.E.: Contr. Mineral. Petrol. **15** (1967) 251.
Wil67	Wilson, R.L., Watkins, N.D.: Geophys. J. **12** (1967) 405.
Was73	Wasilewski, P.J.: Earth Planet. Sci. Lett. **20** (1973) 67.
Wes71	Westcott-Lewis, M.F., Parry, L.G.: Earth Planet. Sci. Lett. **12** (1971) 124.
Zhi65	Zhilyaeva, V.A., Minibaev, R.A.: Izv. Akad. Nauk. USSR, Ser. Fiz. Zemli **4** (1965) 91.

7 Radioactivity of rocks — Radioaktivität der Gesteine

7.1 Radioactive isotopes in rocks — Radioaktive Isotope in den Gesteinen

7.1.1 Introductory remarks — Vorbemerkungen

Unless otherwise stated, the concentration is given in % for K and in ppm (10^{-6} g/g) for Th and U.

Soweit nicht anders vermerkt, werden Konzentrationsangaben für K in %, für Th und U in ppm (10^{-6} g/g) gemacht.

$\text{AM} = \text{arithmetic mean} = \dfrac{1}{N} \sum\limits_{i=1}^{N} x_i$ — arithmetisches Mittel

s = standard deviation of the individual value — Standardabweichung des Einzelwertes

$\text{GM} = \text{geometric mean} = \exp\left(\dfrac{1}{N} \sum \ln x_i\right)$ — geometrisches Mittel

$a = \exp\left(\dfrac{1}{N} \sum \ln x_i + s'\right) - \text{GM}$

$b = -\exp\left(\dfrac{1}{N} \sum \ln x_i - s'\right) + \text{GM}$

s' = standard deviation for $\ln x_i$ of $\left(\dfrac{1}{N} \sum \ln x_i\right)$ — Standardabweichung für $\ln x_i$ von $\dfrac{1}{N} \sum \ln x_i$

min = smallest measured value — kleinster vorkommender Wert
max = highest measured value — größter vorkommender Wert
n = number of measurements taken as basis — Anzahl der zugrundegelegten Messungen
N = number of values used for calculation, with $n \geq N$ — Anzahl der zur Berechnung verwendeten Werte, wobei $n \geq N$

The following discussion, figures and calculations are based on the AM. The use of the GM is indicated in a special note.

It was intended to avoid granting too much weight to those rocks for which many measurements were made. Therefore only the AM of the individual measurements given by one author for one occurrence were used; i.e. for the Rotondo granite [Ki78] N=1, but n=151. n always represents the minimum number of measurements; n=1 was assumed in each case in which it was not clear on how many measurements a value for a mean concentration is based.

In der Diskussion, in Darstellungen oder bei Berechnungen sind die AM zugrundegelegt. Die Verwendung von GM wird besonders angegeben.

Es sollte vermieden werden, daß bei der Berechnung der durchschnittlichen Zusammensetzungen diejenigen Vorkommen ein besonderes Gewicht erhalten, für die viele Messungen vorliegen. In die Berechnung wurden deshalb die arithmetischen Mittel der Meßwerte eines Autors für die jeweiligen Vorkommen eingesetzt. Für den Rotondo-Granit [Ki78] ist z.B. N=1, aber n=151. n ist dabei immer die Mindestanzahl von Messungen; immer dann, wenn nicht klar zu erkennen war, auf wie vielen Werten eine Konzentrationsangabe beruht, wurde n=1 gesetzt.

The heat production rate in rocks is called A and is calculated by means of the formula

Die Wärmeproduktionsrate in den Gesteinen wurde A genannt; sie wird berechnet nach der Formel

$$A = \frac{\varrho}{100} (8.5 \cdot \text{K} + 6.3 \cdot \text{Th} + 23.4 \cdot \text{U}) \quad [\text{HGU}]$$

1 HGU (heat generation unit) = $0.418 \cdot 10^{-6}$ W m^{-3}

The density of rocks ϱ was assumed uniformly as $2 \cdot 67$ g/cm^3; the units used for this calculation are % for K and ppm for Th and U.

Die Dichte der Gesteine ϱ wurde einheitlich zu 2,67 g/cm^3 angenommen; K wird in %, Th und U werden in ppm in die Formel eingesetzt.

Haack

7.1.2 Isotopic composition of K, Th and U — Isotopenzusammensetzung von K, Th, U

The heat generating radioactive isotopes in the earth are ^{40}K, ^{232}Th, ^{235}U, ^{238}U. Their decay constants λ and relative abundances are [IUGS78]:

Die für die Wärmeerzeugung in der Erde verantwortlichen radioaktiven Isotope sind ^{40}K, ^{232}Th, ^{235}U, ^{238}U. Ihre Zerfallskonstanten λ und relativen Häufigkeiten sind [IUGS77]:

$$^{40}K \quad \beta: \quad \lambda = 4.962 \cdot 10^{-10}\,a^{-1}$$
$$e: \quad \lambda = 0.581 \cdot 10^{-10}\,a^{-1}$$
$$^{232}Th \quad \lambda = 4.9475 \cdot 10^{-11}\,a^{-1}$$
$$^{235}U \quad \lambda = 9.8485 \cdot 10^{-10}\,a^{-1}$$
$$^{238}U \quad \lambda = 1.55125 \cdot 10^{-10}\,a^{-1}$$
$$^{238}U/^{235}U = 137.88$$
$$^{232}Th = 100\,\%$$
$$^{40}K = 0.01167\ at\text{-}\%.$$

For geochronological determinations the ratio $^{238}U/^{235}U = 137.88 \pm 0.14$ has been recommended for natural uranium [IUGS77]. However, some deviations from this ratio are known [Co76a]: the uranium deposits of the Colorado Plateau are distinguished by a statistically significantly higher value of 139.706. The highest known deviations are those in the fossil, natural reactors of the uranium deposit in Oklo/Gabon where about $1.8 \cdot 10^9$ years ago part of ^{235}U was consumed during chain reactions. The ratio $^{238}U/^{235}U$ is as high as 227 in some ores of Oklo [Bo72, Ne72], but in others it diminishes to 134 because of the formation of new ^{235}U in the reaction

Für Zwecke der Geochronologie wurde für natürliches Uran das Verhältnis $^{238}U/^{235}U = 137,88 \pm 0,14$ empfohlen [IUGS77]. Es gibt aber Abweichungen davon: So sind nach [Co76a] die Uranlagerstätten des Colorado-Plateaus durch einen statistisch signifikant höheren Wert von 139,706 ausgezeichnet. Die größten Abweichungen sind bekannt aus den fossilen, natürlichen Reaktoren der U-Lagerstätte von Oklo in Gabun, in der vor etwa 1,8 Mrd. Jahren ein Teil des ^{235}U bei Kettenreaktionen verbraucht wurde. Das Verhältnis von $^{238}U/^{235}U$ steigt in den Erzen von Oklo bis 227 [Bo72, Ne72], sinkt aber auch auf 134 wegen Neubildung von ^{235}U nach der Reaktion

$$^{238}U\,(n, \gamma)\,^{239}U \rightarrow {}^{239}Np \rightarrow {}^{239}Pu \rightarrow {}^{235}U.$$

Some of the variation of the ratio $^{238}U/^{235}U$ between 137.5 and 138.4 in deep-reaching fault zones is apparently caused by diffusion [Ma75].

Anscheinend durch Diffusion verursachte geringe Schwankungen im Verhältnis $^{238}U/^{235}U$ zwischen 137,5 und 138,4 in tiefreichenden Störungszonen werden von [Ma75] mitgeteilt.

While the U-isotope ratio of the moon corresponds to that of the earth, this ratio varies for some meteorites between 106.8 and 137.5. These variations are due to the fact that diverse quantities of a different uranium with isotope ratios down to 40.2 are mixed with the main quantity of uranium with a nearly terrestrial ratio of $^{238}U/^{235}U$. This unusual uranium is contained in an almost insoluble component which has already been isolated but not yet identified [Ar77].

Während das U-Isotopenverhältnis auf dem Mond demjenigen der Erde entspricht, schwankt es in einigen Meteoriten zwischen 106,8 und 137,5. Diese Schwankungen beruhen darauf, daß zu der Hauptmenge Uran mit nahezu terrestrischem Verhältnis $^{238}U/^{235}U$ unterschiedliche Mengen eines andersartigen Urans mit Isotopenverhältnissen bis herab zu 40,2 zugemischt sind. Dieses besondere Uran ist in einer zwar schon isolierten, aber noch nicht näher identifizierten schwer löslichen Komponente enthalten [Ar77].

Large deviations from the equilibrium value of the ratio $^{234}U/^{238}U = 57 \cdot 10^{-6}$ are to be found (factor: $0.4\cdots15$).

Große Abweichungen des Verhältnisses $^{234}U/^{238}U$, das im Gleichgewicht $57 \cdot 10^{-6}$ betragen müßte, kommen vor (Faktor: $0,4\cdots15$).

The ratio is considerably higher in formation waters and in their young precipitated compounds [Ma75]. ^{234}U is more easily removed from its compounds than ^{238}U because - as a consequence of the α-recoil - it resides in a disturbed site of elevated chemical potential which is dissolved more easily than the undisturbed rest (etching reveals α-recoil tracks in

Vor allem in Formationswässern und aus diesen ausgefällten Verbindungen [Ma75] ist das Verhältnis stark erhöht; ^{234}U wird offenbar leichter als ^{238}U aus seinen Verbindungen gelöst, weil es sich infolge des α-Zerfalls und des damit verbundenen Rückstoßes an einer gestörten, der Auflösung leichter anheimfallenden Stelle befindet (durch Ätzen können α-

many minerals). In river-water, the ratio of $^{234}U/^{235}U$ is more than 1.15 times the equilibrium value. The factor for seawater is nearly constant at 1.14 [Du72].

Rückstoßspuren in vielen Mineralen sichtbar gemacht werden). Fast alle Flußwässer haben $^{234}U/^{235}U$-Verhältnisse, die um mehr als den Faktor 1,15 über dem Gleichgewichtswert liegen. Der Faktor 1,14 ist für Meerwasser nahezu konstant [Du72].

7.1.3 Natural transuranic elements — Transurane in der Natur

The only natural transuranic elements identified up to now are plutonium isotopes. ^{239}Pu is produced in uranium ores from ^{238}U by the reaction with fission neutrons according to: $^{238}U(n, \gamma)^{239}U \rightarrow ^{239}Np \rightarrow ^{239}Pu$. It is also considered possible that part of the ^{239}Pu is the decay product of a superheavy transuranic element (Eka-Os) [Ch68, Me74]. Up to now, natural ^{244}Pu has been found only in bastnaesite, a rare-earth mineral from Mountain Pass, California [Ho71]. The spectrum of the noble gases [Al71] produced during spontaneous fission as well as fission track excesses have revealed the presence of ^{244}Pu at the formation of numerous meteorites [Wa69].

Up to now, the search for natural superheavy transuranic elements with atomic numbers around 114 whose existence has been predicted to be theoretically possible was unsuccessful [Wy70, Gr71, Th72, Ha73a, He79a].

Die einzigen bisher sicher nachgewiesenen natürlichen Transurane sind Plutonium-Isotope. ^{239}Pu kommt in Uranerzen vor, wo es aus der Reaktion von ^{238}U mit Spaltneutronen entsteht nach der Reaktion $^{238}U(n, \gamma)^{239}U \rightarrow ^{239}Np \rightarrow ^{239}Pu$. Auch die Abkunft von einem überschweren Transuranelement (Eka-Os) wird für einen Teil davon für möglich gehalten [Ch68, Me74]. Natürliches ^{244}Pu wurde bisher nur in dem Seltenerdmineral Bastnaesit von Mountain Pass, Californien [Ho71] gefunden. Für zahlreiche Meteorite wurde sowohl durch das Spektrum der aus Spontanspaltung hervorgegangenen Edelgase [Al71] als auch durch Überschüsse an Spaltspuren nachgewiesen [Wa69], daß bei der Entstehung dieser Himmelskörper ^{244}Pu anwesend war.

Die Suche nach den theoretisch als möglich vorausgesagten überschweren Transuranelementen mit Ordnungszahlen von ungefähr 114 verlief bisher ergebnislos [Wy70, Gr71, Th72, Ha73a, He79a].

7.1.4 Radioactivity of rocks — Radioaktivität der Gesteine

The decay of ^{40}K, ^{232}Th, ^{235}U and ^{238}U provides more than 99% of the heat produced by radioactivity in rocks. The importance of these isotopes cannot be overestimated, for there would be no source of energy for the continuous geological transformation of the earth without them.

There is - with some exceptions - a positive correlation between the elements K, Rb, Cs, Tl, Pb, Th, U in eruptive and in many sedimentary rocks. On the other hand, these elements often correlate negatively with the elements Mg, Ca and Fe, which are typical of mafic rocks. This is a typical characteristic of these elements which are known as the group of the so-called "incompatible" elements. Their name indicates that they are not easily accomodated in the lattice positions coordinated with 6 oxygen atoms in the minerals of the earth's mantle. Other elements with large ions, among them the rare earths, belong to this group. During the melting process, they are enriched in the melt and then extracted from the mantle. A further enrichment takes place in the earth's crust towards the surface. Table 1 gives the mean K, Th and U concentrations in numerous rocks and waters.

Der Zerfall von ^{40}K, ^{232}Th, ^{235}U und ^{238}U liefert mehr als 99% der durch Radioaktivität in den Gesteinen entstehenden Wärme. Diese Isotope sind von gar nicht zu unterschätzender geologischer Bedeutung, denn ohne sie wäre keine Energiequelle für die ständige Umgestaltung der Erde vorhanden.

In den Eruptiv- und vielen Sedimentgesteinen besteht - mit Ausnahmen - eine positive Korrelation der Elemente K, Rb, Cs, Tl, Pb, Th, U untereinander, und diese Elemente wiederum sind oft negativ korreliert mit den für mafische Gesteine typischen Elementen Mg, Ca, Fe. Dies ist ein charakteristisches Kennzeichen dieser Elementgruppe, die unter dem Namen „inkompatible" Elemente zusammengefaßt wird, weil ihnen gemeinsam ist, daß sie alle nur sehr schlecht in die 6-fach mit Sauerstoff koordinierten Gitterpositionen der Minerale des Erdmantels eingebaut werden. Zu dieser Gruppe gehören noch andere großionige Elemente, u.a. die Seltenen Erden. Sie reichern sich bei Schmelzprozessen in der Schmelze an und werden aus dem Mantel extrahiert. In der Erdkruste findet eine weitere Anreicherung nach oben statt. Tab. 1 gibt die mittleren Gehalte an K, Th, U in zahlreichen Gesteinen und in Gewässern.

Table 1. K, Th, U content of rocks (and waters).

No.		K [in %] AM N; n; s min···max	Th [in ppm] AM N; n; s min···max	GM a; b	U [in ppm] AM N; n; s min···max	GM a; b	GM(Th)/GM(U) AM(Th)/AM(U)
Mantle rocks							
00	ordinary Chondrites	0.083 5; 57; 0.007 0.073···0.0910	0.041 4; 27; 0.005 0.034···0.046	0.040 0.005; 0.005	0.013 4; 35; 0.002; 0.010···0.015	0.013 0.002; 0.002	3.1 3.2
01	Dunites	0.002 8; 21; 0.002 0.00016···0.006 GM = 0.0012 a = 0.0026; b = 0.0008	0.020 3; 5; 0.008 0.013···0.028	0.019 0.009; 0.006	0.018 27; 47; 0.023 0.0004···0.098	0.008 0.024; 0.006	2.3 1.1
02	Harzburgites [1])	0.087 9; 101; 0.109 0.004···0.36	(0.048) 1; 1; −		0.051 16; 16; 0.079 0.007···0.318	0.028 0.049; 0.018	−
03	Lherzolites	0.017 59; 79; 0.027 0.0005···0.180 GM = 0.0069 a = 0.0215, b = 0.0052	0.060 21; 24; 0.052 0.004···0.202	0.040 0.069; 0.025	0.026 72; 143; 0.017 0.001···0.082	0.020 0.027; 0.011	2.0 2.3
04	Lherzolites from kimberlites	0.124 2; 8; − 0.124···0.126 GM = 0.052 a = 0.098, b = 0.034	0.81 2; 2; − 0.65···0.97	0.79 0.26; 0.20	0.21 8; 27; 0.13 0.068···0.45	0.17 0.16; 0.08	
05	Pyroxenites, websterites	0.035 17; 54; 0.075 <0.001···0.260 GM = 0.0087 a = 0.0314, b = 0.0068	0.110 2; 2; − 0.011···0.210	0.05 0.34; 0.04	0.068 17; 19; 0.123 0.0011···0.520	0.026 0.096; 0.021	
06	terrestrial ultramafites (Nos. 01 + 02 + 03 + 05 + 07)	0.0224 114; 259; 0.0515 0.00016···0.360 GM = 0.0055 a = 0.024; b = 0.0045	0.095 29; 34; 0.159 0.004···0.210	0.044 0.103; 0.031	0.050 144; 235; 0.090 0.0004···0.520	0.022 0.059; 0.016	2.0 1.9

No.	References
00	Ma79
01	Au71, Be68a, Fi70, Gl71, Ha65, Ho53, Ko69, Ko74, Lo63a, Mo66, Mo66a, Ni72, Ti63, Wa67a
02	Ab74, Au71, Da68, Do75a, Ko74, Ku80, Ni72, Wa67a
03	Be68a, Do75a, Do76a, Fi70, Gl71, Gr68, Ha65, Ha73, Ha75a, Hen71, Ho53, Kl69, Ko69, Ko74, Ku80, Ma71, Mo66, Mo66a, Mo69a, Mo75, Na68, Ni72, Pa70, Ti63, Wa67a
04	Ca80, Fi70, Ma71, Wa67a
05	Ab74, Fi70, Ha65, Ha75a, Ho53, Ko69, Ko74, Ku80, Mo66a, Wa67a
06	see Nos. 01, 02, 03, 05, 07

(continued)

[1]) Harzburgite from Midatlantic Ridge not included [Au71].

Haack

Table 1 (continued)

No.		K [in %] AM N; n; s min⋯max	Th [in ppm] AM N; n; s min⋯max	GM a; b	U [in ppm] AM N; n; s min⋯max	GM a; b	GM(Th)/GM(U) AM(Th)/AM(U)
07	Wehrlites	0.048 4; 4; 0.075 0.002⋯0.160 GM=0.016 a=0.082; b=0.013	0.107 2; 2; − 0.018⋯0.197	0.059 0.266; 0.048	0.062 10; 10; 0.079 0.008⋯0.250	0.032 0.074; 0.022	1.8 1.7

Volcanites (oceanic, oceanic islands, island arcs)

No.		K [in %]	Th [in ppm]	GM	U [in ppm]	GM	GM(Th)/GM(U) AM(Th)/AM(U)
08	Alkali and alkali-olivine basalts, oceanic islands	1.13 15; 50; 0.44	2.9 20; 33; 1.8 1.3⋯7.1	2.5 1.8; 1.1	0.91 28, 56, 0.60 0.29⋯2.79	0.75 0.64, 0.35	3.3 3.2
09	Alkali and alkali-olivine basalts, island arcs	1.13 6; 21; 0.36 0.72⋯1.71	4.6 8; 19; 1.8 3.3⋯8.9	4.4 1.6; 1.2	1.0 14; 25; 0.45 0.53⋯2.25	0.96 0.46; 0.31	4.6 4.4
10	Andesites, oceanic islands	1.86 6; 52; 0.51 1.51⋯2.66	5.0 8; 12; 2.0 2.1⋯7.9	4.6 2.6; 1.7	1.6 10; 12; 0.6 0.71⋯2.50	1.5 0.76; 0.51	3.1 3.1
11	Andesites, island arcs	1.23 22; 144; 0.64 0.47⋯2.46	4.1 12; 60; 4.4 0.51⋯12.0	2.3 4.9; 1.6	0.98 22; 54; 0.67 0.21⋯2.60	0.78 0.77; 0.39	2.9 4.2
12	Dacites, island arcs	1.53 9; 18; 0.33 1.03⋯2.03	6.1 14; 18; 5.1 0.86⋯13.0	4.2 10; 12; 0.9	1.5 10; 12; 0.9 0.38⋯2.9	1.2 1.3; 0.6	3.6 4.1
13	Rhyolites, island arcs	2.4 5; 10; 1.0 0.75⋯3.56	10.9 6; 29; 6.0 1.4⋯20.2	8.6 13.1; 5.2	2.8 6; 29; 1.6 1.2⋯6.0	2.5 1.7; 1.0	3.4 3.9
14	Tholeiites[2]), oceanic	0.17 355; 355; 0.10 0.025⋯0.45 GM=0.15 a=0.11; b=0.06	0.52 107; 107; 0.29 0.04⋯0.12	0.42 0.47; 0.22	0.14 119; 119; 0.08 0.016⋯0.34	0.11 0.11; 0.06	3.7 3.7
15	Tholeiites, oceanic islands	0.34 6; 201; 0.10 0.17⋯0.48	0.92 11; 24; 0.44 0.41⋯1.60	0.84 0.49; 0.31	0.26 10; 25; 0.13 0.06⋯0.44	0.23 0.19; 0.10	3.7 3.5

No.	References
07	Ha75a, Ho53, Ko74, Ni72, Wa67a
08	Fe75b, He64, Ka70, La60, Ma64, Ov68, Ov70, Ov72, So66, Ta66a, Ta66b, Ta78, Za73, Zi75
09	He63, Ni72, Pr73, Pr80, So66, Ta66a, Za73
10	Fe75b, He64, La60, Ma64, Po75, Ta66a, We68, Zi75
11	Ch70, Di79, Do71, Ew68a, Ew68c, Go62a, Ja70, Ni80, Ta69a, Wh79
12	Do71, Du71, Ew68a, Go62a, Ja70, Na68, Sm79, Ta69a, Wh79
13	Do71, Ew68a, Ew68b, Ho75a
14	Au71a, Bl76, Bo74, Ca70, Ca70a, Ca73, Ch75, Cu77, Do78a, En63, En65, En65a, En75, Er74, Fi79, Fl77, Fo77, Fr74, He68, Ka70a, Ke76, La77, Lam77, Ma77, Me68, Me71, Mi69, Mi77, Mu64, Mu66, Sh71, Sh75, Su79, Ta66b, Ta78, Un76, Wo79
15	He63, He64, Ma64, So66, Ta66a, Ta66b, Ta78, We68

(continued)

[2]) Extreme values K > 0.5 %, Th > 1.3 ppm, U > 0.36 ppm not included.

Haack

Table 1 (continued)

No.		K [in %] AM N; n; s min⋯max	Th [in ppm] AM N; n; s min⋯max	GM a; b	U [in ppm] AM N; n; s min⋯max	GM a; b	$\dfrac{GM(Th)}{GM(U)}$ $\dfrac{AM(Th)}{AM(U)}$
16	Tholeiites, island arcs	0.32 2; 13; −	0.59 13; 30; 0.44 0.20⋯1.68	0.47 0.47; 0.23	0.36 13; 30; 0.31 0.11⋯1.1	0.28 0.27; 0.14	1.7 1.6
17	Trachytes, oceanic islands	2.95 5; 9; 0.87 2.15⋯4.18	7.5 8; 8; 1.6 5.93⋯11.0	7.4 1.7; 1.4	2.2 10; 10; 0.6 1.4⋯3.15	2.1 0.7; 0.5	3.5 3.4
18	Trachytes, island arcs	4.83 6; 20; 0.77 3.99⋯5.81	16.0 6; 20; 2.9	15.8 3.4; 2.8	3.4 6; 16; 1.0	3.3 1.2; 0.9	4.8 4.7

Volcanites (continental margins)

No.		K	Th	GM	U	GM	
19	Alkali and alkali-olivine-basalts	1.37 2; 10; 0.11 1.29⋯1.45	3.7 3; 11; 1.5 2.15⋯5.10	3.5 1.9; 1.2	1.2 3; 11; 0.7 0.6⋯1.9	1.1 0.9; 5	3.2 3.1
20	Andesites	1.73 22; 111; 0.58 0.34⋯2.62	4.3 7; 40; 3.4 1.5⋯9.7	3.4 3; 7; 1.8	1.9 31; 139; 1.2 0.36⋯4.6	1.5 1.5; 0.8	2.3 2.3
21	Dacites	2.21 14; 41; 0.34 1.64⋯2.55	4.2 5; 12; 1.6 2.1⋯5.7	3.9 2.2; 1.4	2.6 17; 50; 0.9 1.1⋯4.1	2.4 1.1; 0.75	1.6 1.6
22	Rhyolites[3]	3.05 4.14; 0.99 1.6⋯3.7	8.2 3; 4; 3.2 6.1⋯12.0	7.7 3.6; 2.4	5.6 23; 109; 2.8 1.8⋯12.6	5.0 3.3; 2.0	1.5 1.5
23	Tholeiites	0.54 2; 41; − 0.53⋯0.54	1.3 4; 44; 0.3 0.91⋯1.70	1.2 0.4; 0.3	0.54 4; 44; 0.20 0.31⋯0.79	0.51 0.25; 0.17	2.4 2.4

Volcanites (intracontinental)

No.		K	Th	GM	U	GM	
24	Alkali and alkali olivine basalts	1.17 9; 44; 0.29 0.90⋯1.59	6.2 13; 63; 3.1 1.6⋯8.0	5.5 3.5; 2.1	1.1 13; 56; 0.5 0.2⋯1.9	0.97 0.81; 0.44	5.7 5.6
25	Andesites	2.52 8; 70; 0.70 1.62⋯3.80	8.7 7; 12; 2.5 5.5⋯12.2	8.4 2.9; 2.1	5.5 10; 15; 2.8 1.6⋯9.4	4.8 3.7; 2.1	1.8 1.6
26	Basanites	1.78 5; 28; 0.46 1.43⋯2.57	10.7 7; 14; 6.4 5.6⋯23.5	9.4 6.5; 3.8	2.2 7; 14; 1.0 1.3⋯4.2	2.0 1.0; 0.7	4.7 4.9

No.	References
16	Go62a, He63, So66, Ta66a
17	He64, Ma64, Ov70, So66, Ta66a, Zi75
18	Pr72, Sm79
19	Ba70, Li73, Za73
20	Ad54, Ad55, Al76, Ba70, Du75, Go62a, La56, La80, Li73, McN79, No80, Si69, Ze74, Zi76
21	Ad54, Du71, Go62a, Ja70, Kl74, La56, McN79
22	Ad54, Ba70, Go62a, La56, La58, Sh70, Wh59, Zi76, Zi78
23	Go62a, Li73
24	Ci74, Co68, Fr78, Gh76, Gl71, Ka78, Lo76, Me80
25	Ch63, Ci74, Do76, Ec79, On73, Po71
26	Co68, Fr78, Gl71, Gr68, Ku80, We78

(continued)

[3]) U: value 35.8 ppm not included.

Table 1 (continued)

No.		K [in %] AM N; n; s min···max	Th [in ppm] AM N; n; s min···max	GM a; b	U [in ppm] AM N; n; s min···max	GM a; b	$\dfrac{GM(Th)}{GM(U)}$ $\dfrac{AM(Th)}{AM(U)}$
27	Carbonatites		20.0 20; 341; 36 0.12···200	6.2 31.1; 5.2	8.3 24; 345; 13.5 0.05···94.6	2.4 2.7; 2.0	2.6 2.4
28	Kimberlites	0.52 32; 93; 0.53 0.02···1.67	16.0 61; 123; 11.3 4.3···56.5	12.9 11.9; 6.2	2.6 70; 134; 1.3 0.75···8.0	2.3 1.6; 0.9	5.6 6.2
29	Nephelinites, melilitites	2.34 10; 52; 1.10 1.19···4.49	12.1 19; 61; 4.4 5.0···21.0	11.3 5.0; 3.5	3.1 17; 59; 2.3 1.5···10.4	2.7 1.7; 1.1	4.2 3.9
30	Phonolites	4.97 9; 31; 0.90 3.86···6.3	18.5 9; 30; 4.7 11.8···25.4	17.9 5.3; 4.1	7.8 10; 45; 5.5 2.3···18.0	6.1 6.9; 3.2	2.9 2.4
31	Pikrites	0.89 8; 22; 0.38 0.28···1.55	7.5 7; 21; 3.2 4.0···11.4	6.9 3.9; 2.5	2.9 11; 27; 1.6 1.2···5.7	2.5 1.9; 1.1	2.8 2.6
32	Rhyolites	4.43 11; 122; 1.20 2.9···6.89	17.5 15; 159; 6.1 7.6···34.0	16.5 7.0; 4.9	5.3 19; 374; 3.1 2.1···13.5	4.7 3.2; 1.8	3.5 3.3
33	Tholeiites	0.63 11; 189; 0.14 0.50···0.96	3.1 10; 193; 1.4 1.5···5.6	2.8 1.5; 1.0	0.75 9; 180; 0.31 0.36···1.3	0.68 0.42; 0.26	4.1 4.1
34	Trachytes[4])	3.10 8; 70; 1.13 1.26···4.40	12.7 9; 74; 5.2 6.0···4.0	11.8 6.0; 4.0	5.0 10; 75; 3.9	4.1 1.7; 1.9	2.9 2.5

Volcanites (Italy)

No.		K [in %] AM N; n; s min···max	Th [in ppm] AM N; n; s min···max	GM a; b	U [in ppm] AM N; n; s min···max	GM a; b	
35	Alkali basalts	0.73 3; 4; 0.21 0.65···0.97	4.1 3; 4; 1.7 2.3···5.6	3.8 2.2; 1.4	1.5 4; 4; 0.15 1.3···1.7	1.5 0.16; 0.14	2.6 2.7
36	Andesites	1.96 12; 37; 0.88 0.76···3.33	10.0 13; 40; 4.1 4.2···16.0	9.2 5.1; 3.3	3.9 13; 40; 2.6 1.4···10.3	3.3 2.8; 1.6	2.8 2.6
37	Latites	3.81 10; 27; 1.33 1.97···6.9	25.8 10; 27; 20.9 7.7···76.0	20.1 21.4; 10.4	8.5 10; 27; 4.7 2.3···16.0	7.1 7.1; 3.5	2.8 3.0

No.	References
27	Be68, Be68a, Bu73, Da70, He62, La74, Po71, Wa64, Zv76
28	Fe75a, Ko75, Kr77, Lo68, Ma71, Mo71a, Pa77, Zv76
29	Da70, Fr78, Gl71, Gl79, Lo68, Me80, Po71, We78
30	Ch63, Da70, Gh76, Gl71, Gl79, La56, Po71, Wa64
31	Bu73, Ci74, Gl71, Ko75, Po71
32	Ab59, Ba70, Be70, Ch63, Ec79, Fo75, Gh76, Ha81, Ho75, Ku74, Po71, Pu72, Ta77, Ti74, Tu57, Si74, Vl76
33	Co68, Fr78, Gh76, He65, Ka78, Me80, Lo76
34	Ch63, Ch73, Ci74, Co76, Gh76, Gl71, Ho68, Po71
35	Do75, Ci72
36	Ch63, Ci72, Ci73, Co80, Kl74
37	Ca71, Ci65, Ci72, Fe75, Ga63, Kl74

(continued)

[4]) U: two values, 11.9 and 12.5 ppm, not included.

Haack

Table 1 (continued)

No.		K [in %] AM N; n; s min···max	Th [in ppm] AM N; n; s min···max	GM a; b	U [in ppm] AM N; n; s min···max	GM a; b	GM(Th)/GM(U) AM(Th)/AM(U)
38	Leucitites (tephritic)	6.45 8; 132; 0.90 4.98···7.47	39.8 7; 111; 16.7 20.0···58.9	36.2 23.0; 14.1	17.0 7; 111; 6.9 6.0···26.2	15.5 10.2; 6.1	2.3 2.3
39	Phonolites	6.60 4; 72; 1.72 5.17···9.10	105.4 3; 69; 59.3 55.0···171.6	94.8 71.8; 40.9	26.3 3; 69; 10.5 20.0···38.4	25.0 11.2; 7.8	3.8 4.0
40	Rhyolites Obsidians	4.21 12; 33; 1.30 2.02···6.8	36.2 12; 30; 15.7 14.6···70.3	33.2 18.6; 11.9	14.0 16; 34; 8.7 7.1···35.2	11.9 9.5; 5.3	2.8 2.6
41	Trachytes	5.64 13; 121; 1.33 3.96···7.47	58.6 11; 112; 59.2 16.0···208.0	40.6 55.0; 23.4	17.0 11; 112; 12.1 6.0···47.0	14.1 12.4; 6.6	2.9 3.4
Intrusive rocks							
42	Anorthosites	0.34 18; 117; 0.21 0.05···0.91	1.6 5; 19; 2.4 0.07···5.9	0.60 2.5; 0.48	0.64 7; 26; 0.62 0.01···1.6	0.28 1.4; 0.24	2.1 2.5
43	Diorites, quartz diorites	1.83 20; 44; 0.67 0.5···2.72	9.2 23; 70; 5.2 0.9···23.0	7.4 8.9; 4.0	2.6 31; 82; 2.2 0.37···9.4	1.9 2.4; 1.1	3.9 3.5
44	Gabbros from ophiolites				0.026 15; 15; 0.011 0.013···0.054	0.024 0.011; 0.007	
45	Gabbros (excluding No. 44)	0.60 16; 86; 0.54 0.18···2.0	3.1 19; 26; 4.0 0.5···16.0	1.7 3.1; 1.1	0.59 32; 182; 0.74 0.005···3.5	0.33 0.91; 0.24	5.2 5.3
46	Granites all (Nos. 47 + 50)	3.73 210; 2180; 0.95 0.78···6.40	24.6 265; 2798; 17.7 2.2···139.1	20.0 18.4; 9.6	5.4 332; 4447; 3.8 0.63···35.2	4.4 0.41; 2.1	4.5 4.6
47	only Granites, above active sub-duction zones	3.15 46; 303; 1.18 0.78···5.77	16.5 54; 268; 9.4 2.60···58.7	14.2 10.9; 6.2	4.1 61; 391; 2.9 0.92···13.6	3.3 3.1; 1.6	4.4 4.0
48	only Granites, Japan	2.93 18; 18; 1.37 1.50···5.77	16.0 18; 18; 5.5 5.4···26.9	15.0 7.1; 4.8	3.7 18; 18; 1.8 1.2···8.3	3.3 2.1; 1.3	4.5 4.3

No.	References
38	Ca71, Ci65, Ci72, Lu63, Lu64, Luo65, Sa67
39	Ba67, Ci65, Ci72, Ga63
40	Ba67, Be70, Ca71, Ch63, Ci72, Do75, Fe75, Kl74
41	Ba67, Ca71, Ci65, Ci72, Fe75, Ga63, Lu64, Lu65, Sa67
42	At78, Gr69, He71, Ho53, Sw74, Wi73, Zh75
43	Ab59, Br70, Do71, He66, Ko71, La56, Le63, Lo79, Ly61, McN72, Mi72, Mo69a, On73, Ro61, Ry78, Sw77, Tu57, Vl76
44	Do75a
45	Ab59, Ar72, Au71, Bu61, Ch77, Da68, Do71, Ge71, Go62a, He63, He75, Hen71, Ko69, La56, La60, Le63, Mo69a, Ni72, Ro61, Sw77, Tu57, Wh59, Zh65a
46	See Nos. 47, 50
47	At67, Bu73a, Do71, Ed78, La56, La60, Le76, Lo79, Mi72, Mi76, Ni76, Ro61, Sw77
48	Mi72

(continued)

Table 1 (continued)

No.		K [in %] AM N; n; s min⋯max	Th [in ppm] AM N; n; s min⋯max	GM a; b	U [in ppm] AM N; n; s min⋯max	GM a; b	GM(Th)/GM(U) AM(Th)/AM(U)
49	only Granites, Damara Orogen SWA/Namibia	4.30 24; 137; 0.61 2.95⋯5.0	33.3 24; 137; 19.9 9.0⋯88.3	29.0 20.0; 11.8	5.2 24; 137; 2.7 1.5⋯12.5	4.6 3.0; 1.8	6.3 6.4
50	only Granites, not over active subduction zones	3.89 164; 1878; 0.83 1.40⋯6.4	26.0 213; 2419; 17.7 2.23⋯139.1	21.5 19.0; 10.1	5.9 273; 3934; 4.3 0.63⋯35.2	4.7 4.4; 2.6	4.5 4.4
51	aplitic Granites	3.45 4; 16; 0.25 3.2⋯3.7	25.9 10; 70; 9.1 12.6⋯39.2	25.4 11.1; 7.6	10.5 12; 72; 6.4 3.0⋯21.1	8.6 8; 8; 4.3	2.8 2.5
52	Granodiorites, all (Nos. 53 + 55)	2.34 59; 134; 0.84 0.43⋯4.68	12.4 67; 237; 9.4 1.3⋯53.0	9.9 9.9; 4.9	3.1 82; 312; 2.4 0.53⋯14.4	2.4 2.3; 1.2	4.1 4.0
53	only Granodiorites, above active subduction zones	2.22 41; 73; 0.80 0.43⋯4.15	11.1 41; 97; 8.2 1.3⋯53.0	9.2 7.8; 4.2	2.8 52; 162; 2.1 0.35⋯14.0	2.3 2.1; 1.1	4.0 4.0
54	only Granodiorites, Japan	2.04 23; 23; 0.85 0.43⋯3.49	9.4 24; 24; 4.9 1.3⋯22.9	8.1 7.1; 3.8	2.4 24; 24; 1.4 0.35⋯7.3	2.0 1.8; 1.0	4.0 3.9
55	only Granodiorites, not above active subductions zones	2.61 18; 79; 0.88 1.22⋯4.68	14.6 26; 138; 10.8 2.7⋯36.0	10.9 13.7; 6.1	3.6 29; 148; 2.9 0.74⋯14.4	2.8 3.0; 1.4	4.0 4.1
56	Syenites	4.75 17; 351; 1.02 2.87⋯6.71	27.8 37; 451; 26.3 0.55⋯96.0	16.3 36.8; 11.3	8.6 48; 830; 10.8 0.09⋯51	4.6 11.5; 3.3	3.5 3.2
57	Tonalites	1.50 31; 60; 0.64 0.33⋯3.3	6.4 32; 73; 5.2 0.8⋯24.0	4.7 6.2; 2.7	2.0 36; 99; 1.9 0.08⋯11.0	1.3 2.3; 0.8	3.5 3.2

Metamorphic rocks

No.		K [in %] AM N; n; s min⋯max	Th [in ppm] AM N; n; s min⋯max	GM a; b	U [in ppm] AM N; n; s min⋯max	GM a; b	GM(Th)/GM(U) AM(Th)/AM(U)
58	Amphibolites	0.78 21; 138; 0.85 0.11⋯3.8	2.3 19; 88; 3.1 0.01⋯11.2	0.89 3.8; 0.7	1.2 27; 301; 1.3 0.01⋯5.3	0.56 1.9; 0.43	1.6 1.9

No.	References
49	Ha81
50	Ab59, Ba61 b, Bar70, Br69, Br70, Br79, Bu61, Bu73 a, Ca71, Ch77, Co59, Co76, Di62, Ev74, Fa76, Fa78, Fe75, Fi59, Gl76, Go62 b, Ha64, Ha66, Ha81, He66, Ho75, Ki75, Ki78, Ko66a, Kr58, Ma71 a, Mah71, Mi76, Mo66, Mo69a, Mo71, Na69, On73, Pu72, Ra67, Ra79, Ro61, Ro69a, Ro78, Ry73, Ry78, Sw74, Sw77, Ta56, Ta75, Tu57, Vl76, Wh59, Zh75
51	Ba61 b, Ki78, La76, Le63, Le64 b, Ly61, Ma71 a, Mo71, Vl76
52	See Nos. 53, 55
53	Do71, Ed78, La56, La60, Le76, Lo74, Lo79, McN79, Mi72, Mi76, Pi62 b, Pl62, Ro61
54	Mi72
55	Br70, De79, Do75, Fa76, Fe75, He63, Ho75, Ko66a, Ko71, La76, Le63, Ly61, Ma71 a, On73, Ry78, Tu57, Vl76, Wh59, Zh65 a
56	Ba67, Be68, Bu61, Da70, Er63, Ge56, Go62 b, Ha64, He62, Ko71, La56, La71, Le61, Le64 a, Mi76, Po64, Po71, Ro61, Sa59, Sm73, Sw74, Ti74, Tu57, Ye62, Zh65 a, Zh65 b, Zh75
57	Con80, Do71, Gl76, Ha81, Ho75, La56, La60, Lo74, Lu79, Ly61, Ro61, Ry78, Wh59
58	At78, Bi62, Co80, Do78, Dr73, Ho75, Ry73, Ry78, Sm74, Sw74, Vl76, Zh75

(continued)

Table 1 (continued)

No.		K [in %] AM; N; n; s; min···max	Th [in ppm] AM; N; n; s; min···max	GM a; b	U [in ppm] AM; N; n; s; min···max	GM a; b	GM(Th)/GM(U) AM(Th)/AM(U)
59	Eclogites from pipes	0.20; 11; 20; 0.21; 0.02···0.69; GM=0.12; a=0.27; b=0.08	0.27; 14; 25; 0.19; 0.10···0.62	0.22; 0.21; 0.11	0.11; 17; 30; 0.11; 0.03···0.41	0.078; 0.091; 0.042	2.8; 2.5
60	Eclogites from metamorphic terrains [5])	0.21; 16; 16; 0.19; 0.009···0.75; GM=0.14; a=0.26; b=0.09	0.28; 9; 10; 0.23; 0.003···0.60	0.11; 0.67; 0.10	0.29; 18; 64; 0.25; 0.018···0.91	0.17; 0.39; 0.12	0.64; 1.0
61	Gneisses	3.10; 30; 165; 0.92; 0.32···4.33	12.5; 29; 149; 4.9; 0.98···24.8	11.2; 9.0; 5.0	4.1; 44; 589; 2.6; 0.50···13.1	3.4; 3.3; 1.7	3.3; 3.0
62	Granulites, acid	2.73; 42; 576; 1.12; 0.90···5.52	9.3; 36; 512; 8.5; 0.10···35.0	5.5; 13.3; 3.9	0.97; 41; 586; 0.86; 0.02···4.5	0.66; 1.4; 0.45	8.3; 9.6
63	Granulites, basic	0.86; 11; 158; 0.42; 0.3···1.51	6.8; 6; 76; 13.0; 1.0···33.2	2.3; 6.7; 1.7	0.30; 7; 76; 0.14; 0.11···0.50	0.27; 0.19; 0.11	8.5; 22.7
64	all Granulites (Nos. 62+63)	2.34; 53; 734; 1.28; 0.30···5.52	9.0; 42; 588; 9.1; 0.1···3.5	4.9; 12.5; 3.5	0.87; 48; 662; 0.83; 0.023···2.5	0.56; 1.1; 0.36	8.8; 10.3
65	Schists (metamorphic pelites and psammites, various grades)	2.97; 17; 344; 0.63; 1.84···3.84	12.9; 13; 297; 2.7; 8.2···21.9	12.6; 3.0; 2.4	3.0; 20; 383; 1.3; 1.6···5.7	2.8; 1.4; 0.9	4.6; 4.3

Sediments (old)

No.		K [in %]	Th [in ppm]	GM	U [in ppm]	GM	
66	Dolomites	0.68; 19; 19; 0.79; 0.14···3.5; GM=0.39; a=0.94, b=0.28	1.8; 5; 5; 1.1; 0.8···2.7	1.5; 1.4; 0.7			
67	Greywackes	1.33; 15; 176; 0.49; 0.4···2.1	6.9; 17; 182; 2.8; 1.5···12.3	6.2; 4.4; 2.6	2.0; 17; 182; 1.1; 0.5···4.6	1.8; 1.3; 0.8	3.5; 3.5

No.	References
59	Al69a, He63a, Lo63, Lo63a, Lo68, Ma71, Mo66a, Ti63, Wa67a
60	Ah54, Al69a, Be68a, Fe74, He63a, Lo63, Mo66a, Na68
61	Ar72, Bi62, Ch77, Gr69, He71, Ho75, Lam76, Nu74, Ry73, Zh75
62	At78, Ch73, De79, Do78, Dr73, Du79, Ga79, Gr69, Gr72, Gr77, He71, Lam68, Le79, Lo68, Mo69, Mo69a, Mo79, Or78, Si77, Sk70, Sm71, Zh75
63	At78, Do78, Dr73, Du79, Gr77, Le79, Mo69, Pa66, Sh69, Si78
64	See Nos. 62, 63
65	At79, Bi62, Bu73b, Ch77, Do78, Ha81, Ho75, Ri76, Ro70a, Ry73, Sm74, Zh75
66	Ba77, Be56, Kh64, La67, We64
67	Ga75, Nan76, Ro64, Ro70a, Sc80, Vl76, Wo67

(continued)

[5]) Th: value 1.7 ppm not included.

Haack

Table 1 (continued)

No.		K [in %] AM N; n; s min···max	Th [in ppm] AM N; n; s min···max	GM a; b	U [in ppm] AM N; n; s min···max	GM a; b	GM(Th)/GM(U) AM(Th)/AM(U)
68	Limestones	0.66 6; 70; 0.51 0.2···1.5 GM=0.50 a=0.64, b=0.28	2.0 62; 5632; 1.8 0.01···8.3	1.3 3.0; 0.9	2.5 72; 5687; 2.8 0.3···15.0	1.8 1.9; 0.9	0.7 0.8
69	Sandstones	0.74 13; 273; 0.35 0.2···1.2	5.5 23; 293; 4.5 1.0···21.4	4.4 4.3; 2.2	2.0 26; 320; 1.5 0.3···6.5	1.5 1.8; 0.8	2.8 2.8
70	Shales and silt-stones, (excluding black shales)	2.29 28; 994; 1.37 0.64···8.5	11.5 90; 5576; 4.8 3.35···33.5	10.7 5.2; 3.5	3.4 90; 5634; 1.5 1.0···9.0	3.1 1.7; 1.1	3.5 3.4
71	black Shales[6])	2.60 13; 144; 0.62 1.3···3.41	10.9 31; 407; 6.1 1.6···28.0	9.3 7.8; 4.3	20.2 56; 770; 21.1 1.4···80.0	12.1 22.8; 7.9	0.8 0.5
Sediments (young to recent)							
72	Bauxites		42.0 38; 38; 31.5 5.0···132	31.1 40.6; 17.6	7.5 38; 38; 6.2 1.6···26.7	5.4 7.1; 3.1	5.7 5.6
73	Bentonites	4.14 64; 64; 1.70 0.5···6.7	24.0 69; 69; 7.3 7.2···42.0	22.9 8.7; 6.3	5.5 69; 69; 4.6	4.5 2.8; 1.7	5.1 4.4
74	Clays (anaerobic, marine)		7.5 2; 13; 6.9 2.6···12.4	5.7 11.4; 3.8	14.2 30; 286; 11.3 2.48···48	10.8 11.9; 5.7	0.5 0.5
75	Clays and silts (continental shelves)		5.7 2; 18; − 5.4···6.0		2.6 34; 1092; 1.7 1.0···5.8	2.3 1.5; 0.9	
76	Clays and silts (deep sea)		9.5 40; 357; 3.8 2.24···20.6	8.6 5.2; 3.2	2.8 79; 620; 1.4 0.5···8.6	1.7 1.5; 0.8	5.1 3.4
77	Clay (windborne dust, Caribbean)		− 2; 18; − 1.7···12.4		− 2; 18; − 3.6···5.5		

No.	References
68	Ad58a, Ba56, Ba77, Be55, Be56, Fa68, Ga75, Kh64, Ku74, La67, Nan76, Sc80, Ta59, Vl76
69	Ad58a, Al69, Ba77, Ga75, Ka75, Kh64, Ku74, Mu58, Nan76, Pl62a, Ro64, Sc80, Wo67, Wo73
70	Ad58, Ad58a, Al59, Ba56, Ga75, Ka75, Kh64, La67, McL80, Pl62a, Pl71b, Pl75, Sc80, Vi56, Vl76, Wo73
71	Ad58, Ad58a, Ba58, Ba77, Bl64, Co61, Ed74, Ma56, Pl71b, Pl75, Po55, Sc80, Sch80, Se63, Sw61, Vl76
72	Ad58a
73	Ad58, Ad58a
74	Ba68, Ba73c, Bert70, De77, Ko57, Ma78, Mo73, Ni77, Sw61, Vl67, Vl74
75	Ba73, Ba73a, Ba73b, Ba73c, Ba76a, Mo73, St58
76	Ba57, Ba71, Ba73, Be74, Be74a, Ber70, Bert70, Bo71, Bo76, Ch63a, Dy69, Dy75, Go63, Go73, Im76, Kh69, Ko73, Ku65, Ku68a, Ku68b, Ku68c, Ma78, Mo73, Ro61a, So66a, St58
77	Ry72

(continued)

[6]) U: six values, 168, 209, 288, 1244, 4960, and 5010 ppm, not included.

Haack

Table 1 (continued)

No.		K [in %] AM N; n; s min···max	Th [in ppm] AM N; n; s min···max	Th [in ppm] GM a; b	U [in ppm] AM N; n; s min···max	U [in ppm] GM a; b	$\frac{GM(Th)}{GM(U)}$ $\frac{AM(Th)}{AM(U)}$
78	Limestones (corals)		<0.1		2.6 43; 262; 0.7 1.2···4.3	2.5 0.8; 0.6	
79	Limestones (molluscs)		0.1 21; 161; 0.2 0···1.0	0.05 0.2; 0.04	1.2 34; 271; 1.5 0.06···6.0	0.6 1.5; 0.4	
80	Limestones (ooids)				3.0 4; 12; 0.8 2.4···4.2	2.9 0.8; 0.6	
81	Limestones (stalagmites)		0 1; 3; −		0.30 4; 21; 0.28 0.07···0.45	0.18 0.44; 0.13	
82	Manganese nodules (marine)		25.0 16; 151; 33.8 0.1···137.0	9.3 53.0; 7.9	6.8 34; 169; 4.1 0.3···13.9	5.2 7.5; 3.1	1.8 3.7
83	Phosphorites (including cretaceous phosphorites)	0.3 1; 5; −	27.0 15; 80; 56.8 0.06···179.0	3.8 35.6; 3.4	91.9 51; 532; 83.6 0.1···485.0	59.9 146.2; 42.5	
84	Sands[7]) (continental shelves)		4.7 4; 31; 3.8 2.2···10.4	3.8 3.8; 1.9	1.7 18; 193; 1.0 0.5···3.8	1.4 1.2; 0.6	2.7 2.8
85	Soils		9.0 9; 551; 3.1 5.1···14.1	8.6 3.4; 2.5	2.7 10; 552; 1.2 1.7···5.5	2.5 1.2; 0.8	3.4 3.3

Organic sediments

No.						
86	Asphalts				range 10···3760 (normal 100···500)	
87	Coals (including lignites)				range <10··· <6000	

No.　References
78　Ba56a, Ben73, Br65, Ch70, Ku68d, Sa63, Sz71, Ta59, Th70, Ve66, Vl68
79　Ba71, Fa65, Ka71, Og74, Sa63, Ta59
80　Br65
81　Th75, Va69
82　Bo75, Gl72, Ku67, Ku69, Ku76, Mo73, So71, Ta59
83　Al58, Ba71, Be74a, Bu77, Ch70, Co77, Ha70a, Ko70, Ko75a, Ko78, McK58, Na76, Pr79, Sa70, Sz69, Tu69, Vl74
84　Ba68, Ba71, Ba73, Ba73b, Ba73c, Ba76, Jo76, Jo77, Ko73, Tr73
85　Ba76, Dob76, Ro66, Sc76
86　Ha56 cited from [Ro67a]
87　Vi56a cited from [Ro67a]　　　　　　　　　　　　　　　　　　　　　　(continued)

[7]) Not included beach sands near monazite-bearing areas in India with contents of Th: 33.9 and 92.0 ppm, and of U: 1.4 and 8.1 ppm [Jo77].

Haack

Table 1 (continued)

No.	K [in %] AM N; n; s min···max	Th [in ppm] AM N; n; s min···max	GM a; b	U [in ppm] AM N; n; s min···max	GM a; b	GM(Th)/GM(U) AM(Th)/AM(U)
88a Oil				range 0.00001··· 0.414		
88b Oil ashes				range 4.5···77		
89 Peat		4.1 1; 8; −		8.1 1; 8; −		
Waters						
90 Sea water, dissolved		$0.68 \cdot 10^{-6}$ 5; 32; $0.88 \cdot 10^{-6}$ $0.07 \cdot 10^{-6}$ $\cdots 2.2 \cdot 10^{-6}$	$0.35 \cdot 10^{-6}$ $0.95 \cdot 10^{-6}$; $0.26 \cdot 10^{-6}$	$2.4 \cdot 10^{-3}$ 12; 22; $1.2 \cdot 10^{-3}$ $0.2 \cdot 10^{-3}$ $\cdots 3.4 \cdot 10^{-3}$	$1.9 \cdot 10^{-3}$ $2.9 \cdot 10^{-3}$; $1.2 \cdot 10^{-3}$	
91 Sea water, pore sulution in sediments				$22 \cdot 10^{-3}$ 2; 13; $10 \cdots 65 \cdot 10^{-3}$		
92 Sweet water: rivers, dissolved		$5.6 \cdot 10^{-5}$ 3; 20; $4.0 \cdot 10^{-5}$ $2.7 \cdots 9.6 \cdot 10^{-5}$	$4.8 \cdot 10^{-5}$ $4.3 \cdot 10^{-5}$; $2.3 \cdot 10^{-5}$	$1.4 \cdot 10^{-3}$ 61; 132; 0.0020	$0.43 \cdot 10^{-3}$ $2.3 \cdot 10^{-3}$; $0.36 \cdot 10^{-3}$	
93 Sweet water: rivers, dissolved, (only Russian rivers)				$2 \cdot 10^{-3}$ 15; 15; $1 \cdot 10^{-3}$ $3 \cdot 10^{-4} \cdots$ $4 \cdot 10^{-3}$	$1.7 \cdot 10^{-3}$ $1.7 \cdot 10^{-3}$; $0.9 \cdot 10^{-3}$	
94 Sweet water: rivers, dissolved (Mississippi, St. Lawrence, Amazon, Congo)				$0.21 \cdot 10^{-3}$ 8; 8; $0.34 \cdot 10^{-3}$ $0.016 \cdot 10^{-3} \cdots$ $1 \cdot 10^{-3}$	$0.082 \cdot 10^{-3}$ $0.25 \cdot 10^{-3}$; $0.062 \cdot 10^{-3}$	
95 Sweet water: rivers, suspended		8.9 2; 2; − 7.8···10		2.8 19; 19; 3.5 0.2···16.0	1.8 2.6; 1.1	
96 Sweet water: springs				$4.2 \cdot 10^{-3}$ 32; 119; $2.8 \cdot 10^{-3}$ 0.2···11.8	$3.1 \cdot 10^{-3}$ $4.8 \cdot 10^{-3}$; $1.9 \cdot 10^{-3}$	

No.	References
88	Hy56 cited from [Ro67a] and [Al59, Li60, Tv74]
89	Ba61a
90	Ba59, Ba73c, Bh69a, Ka69, Ko57, Ni77, Pa81, Ro52, Sm54, So66a
91	Ba73c, Bo76
92	Ba73c, Bert70, Bh69, Ch76, He58, He59, Ju55, Ko57, Lo67, Mi73, Mo67, Ni77, Ro72
93	Ba73c, Ch76, Ni77
94	Bert70, Mo67, Ro62
95	Ba73c, Bert70, Ju55, Mo67
96	Ch76, He58, Ju57

7.1.4.1 K, Th and U in ultramafic rocks — K, Th, U in Ultramafiten

(Nos. 01···07, Table 1)

Ultramafic rocks are samples of the earth's mantle. Most of them have been brought up to the surface as inclusions in basalts. They probably don't represent the primary mantle but are remnants of one or more partial melting processes. This is indicated by their low Th/U ratio of 1.9, which shows a considerable deviation from the ratio for the whole earth which is probably 3.9 in the light of Pb-isotope-data [Ma79a]. Ultramafic rocks are not in chemical equilibrium with the melts which brought them to the surface; they are xenoliths trapped in rising magmas.

A very frequent group are peridotites averaging 54% olivine, 25% orthopyroxene, 18% clinopyroxene and about 1% Cr-spinel. Garnet-peridotites often contain $\approx 8\%$ garnet [We75].

The values shown in Table 1 are only an indication of the order of magnitude of the concentrations of K, Th and U to be found in terrestrial ultramafic rocks. In fact, there are many doubts about the accurate numbers because there is a high risk of contamination by geological processes and incorrect laboratory treatment. Because of the very different methods used, the results may systematically differ by an order of magnitude from author to author. Actually, one should differentiate between ultramafic rocks of the mid-ocean ridges, the oceanic islands, the continental margins and the continents. Such a classification is not reasonably possible yet, because of the considerable differences in the results given by the various authors. In view of these restrictions, it is difficult to give well-founded mean values.

On the basis of the element-content of the minerals given in Table 2, the upper mantle contains an average of $(5···15)\cdot10^{-3}$ ppm U according to whether the arithmetic or geometric means are considered more relevant. Histograms of the U-concentration in olivine, orthopyroxene and clinopyroxene are given in Fig. 1a–c. Information on Th in minerals is so rare that it is not possible to make a statement about the content of this element in minerals.

If Th/U $\approx 3···4$ is assumed for the upper mantle, either in analogy to the cosmic abundance, or based on the histograms in Fig. 2 or on Pb-isotope data (e.g. [We68]), then the Th concentration should be $(15···60)\cdot 10^{-3}$ ppm. However, Table 1 shows values about twice as high for the average of the peridotites (No. 03). The discrepancy cannot be cleared up yet.

Die Ultramafite sind Proben des Erdmantels. Die meisten von ihnen wurden als Einschlüsse in Basalten gefördert. Wahrscheinlich stellen sie nicht den unveränderten Mantel dar, sondern sind Reste einer oder mehrerer partieller Aufschmelzungen. Ein Hinweis darauf ist ihr niedriges Th/U-Verhältnis von 1,9, das stark von dem Wert 3,9 für die Gesamterde abweicht, der von [Ma79a] aufgrund von Pb-Isotopen-Daten für wahrscheinlich gehalten wird. Die Ultramafite stehen auch nicht im chemischen Gleichgewicht mit den Schmelzen, die sie – als knollenförmige Einschlüsse – gefördert haben. Sie wurden vielmehr während des Aufstieges der Magmen auf dem Wege nach oben mitgerissen.

Ein sehr häufiger Typ sind Peridotite mit durchschnittlich 54% Olivin, 25% Orthopyroxen, 18% Klinopyroxen und etwa 1% Cr-Spinell. Granat-Peridotite enthalten oft $\approx 8\%$ Granat [We75].

Die in Tab. 1 enthaltenen Werte geben Hinweise auf die Größenordnung der in irdischen Ultramafiten anzutreffenden Konzentrationen an K, Th und U. Tatsächlich bestehen aber große Unsicherheiten über die richtigen Gehalte, weil die Gefahr der Kontamination durch geologische Prozesse und durch fehlerhafte Behandlung im Labor besonders groß ist. Außerdem bestehen systematische Unterschiede bis zu einer Größenordnung zwischen den Angaben verschiedener Autoren. Eigentlich müßte auch unterschieden werden zwischen Ultramafiten der mittelozeanischen Rücken, der ozeanischen Inseln, Kontinentalränder und Kontinente. Wegen der stark unterschiedlichen Ergebnisse verschiedener Autoren läßt sich diese Unterteilung aber nicht sinnvoll durchführen. Angesichts dieser Einschränkungen ist es schwierig, gut begründete Mittelwerte anzugeben.

Aus den in Tab. 2 mitgeteilten Elementgehalten der Minerale ergibt sich ein durchschnittlicher U-Gehalt des oberen Erdmantels von $(5···15)\cdot10^{-3}$ ppm, je nachdem, ob man die arithmetischen oder die geometrischen Mittel für aussagekräftiger hält. Histogramme der U-Gehalte in Olivin, Ortho- und Klinopyroxen sind in Fig. 1a–c gegeben. Angaben für Th in Mineralen sind so spärlich, daß über dieses Element in Mineralen keine Aussage möglich ist.

Nimmt man für den oberen Erdmantel Th/U $\approx 3···4$ an, sei es aus Analogie zur kosmischen Häufigkeit, aufgrund des Histogrammes Fig. 2 oder aufgrund von Pb-Isotopen-Daten (z.B. [We68]), so folgt ein Th-Gehalt von $(15···60)\cdot10^{-3}$ ppm. Die Tab. 1 weist dagegen etwa doppelt so hohe Werte für den Durchschnitt der Peridotite (No. 03) aus. Einstweilen läßt sich diese Diskrepanz nicht aufklären.

Table 2. K, Th, U content in minerals of the mantle.

	K [in 10^{-3} %]		Th [in 10^{-3} ppm]		U [in 10^{-3} ppm]		Ref.
	AM n; s min···max	GM a; b	AM n; s min···max	GM a; b	AM n; s min···max	GM a; b	Remarks
Olivine			4.1 3; 2.1 0.2···19	3.8 2.3; 1.4	3.6 87; 6.2	1.5 3.8; 1.1	Au71, Ca80, Do75a, Ha73, Ha75a, He79, Hen75, Kl69 Ko74, Mi74, Mo75, Na68a, Ni72, Se73, Ta78 excluded 10 values $U = (30 \cdots 100) \cdot 10^{-3}$ ppm
Ortho- pyroxene			16 1; —		6.1 65; 9.9 <1···13	3.4 6.2; 2.2	Au71, Ca80, Do75a, Do76a, Gi74, Ha73, Kl69, Ko74, Mi74, Na68a, Ni72, Se73, Ta78 excluded 8 values $U = (19 \cdots 180) \cdot 10^{-3}$ ppm
Clino- pyroxene	30 14; 18 13···63	26 19; 11	161 3; 188 18···374	84 302; 66	28 112; 75 0.6···84	16 39; 11	Au71, Ca80, Do75a, Do76, Ha73, Hen71, Kl69, Ko74, Kr77, Ki79, Mi74, Mo75, Na68a, Ni72, Se73, Ta78 excluded 18 values $U = (116 \cdots 950) \cdot 10^{-3}$ ppm
Garnet					100 6; 100 40···300	70 97; 41	Ca80, Mi74
Spinel					25 25; 23 0.2···71	14 32; 10	Do75a, Do76a, Ha73, Ko74, Kl69, Ta78 excluded 2 values $U = (96 \text{ and } 150) \cdot 10^{-3}$ ppm
Phlogopite			130 1; —		0.5 1; —		Ha73 Kl69

Fig. 3 gives a histogram of the logarithms of the uranium concentrations in ultramafic rocks. Here the geometric mean for U, $(22 \cdot 10^{-3})$ ppm seems to be the most relevant estimate. The corresponding value for Th should be $\approx 80 \cdot 10^{-3}$ (for Th/U = 3···4).

Fig. 3 gibt ein Histogramm der Logarithmen der Urangehalte in Ultramafiten. Danach scheint das geometrische Mittel für U, $(22 \cdot 10^{-3})$ ppm, der aussagekräftigste Schätzwert zu sein. Entsprechend müßte Th $\approx (80 \cdot 10^{-3})$ ppm sein (bei Th/U = 3···4).

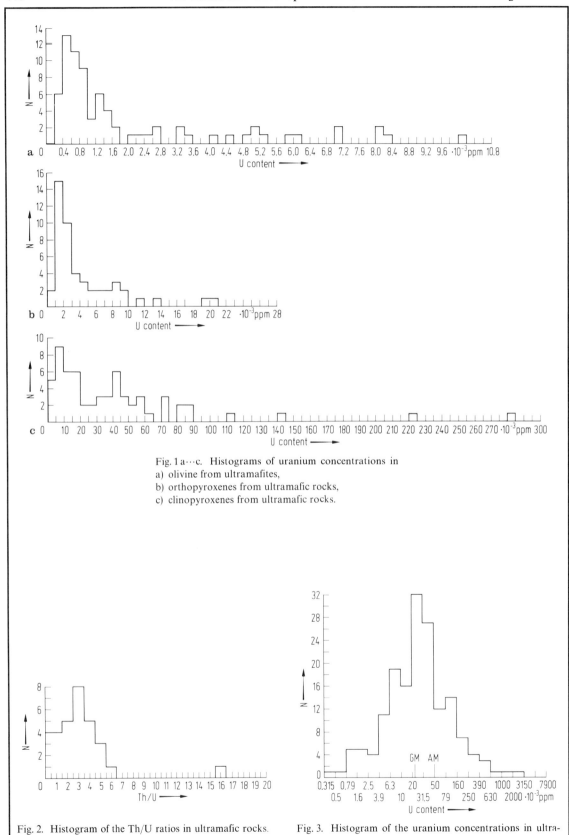

Fig. 1 a···c. Histograms of uranium concentrations in
a) olivine from ultramafites,
b) orthopyroxenes from ultramafic rocks,
c) clinopyroxenes from ultramafic rocks.

Fig. 2. Histogram of the Th/U ratios in ultramafic rocks.

Fig. 3. Histogram of the uranium concentrations in ultramafic rocks.

Haack

7.1.4.2 K, Th and U in volcanic rocks — K, Th, U in vulkanischen Gesteinen

The type and composition of volcanic rocks shows characteristic differences according to the geotectonic environment in which they were brought to the surface. This is also demonstrated by the trace element concentration. For the purpose of this article, the volcanites have therefore been classified into those of the

Art und Zusammensetzung der vulkanischen Gesteine unterscheiden sich charakteristisch voneinander je nachdem, in welcher geotektonischen Situation sie gefördert wurden. Dieses prägt sich auch im Gehalt an Spurenelementen aus. Deshalb wurden die Vulkanite für die Zwecke dieses Artikels je nach ihrer Herkunft eingeteilt in solche

> deep-sea (ocean floor tholeiitic basalts)
> oceanic islands
> island arcs
> subduction zones at continental margins
> interior of the continents (intracontinental volcanites)
> Italian volcanic province.

> aus der Tiefsee (ocean floor tholeiitic basalts)
> von ozeanischen Inseln
> von Inselbögen
> von Subduktionszonen an Kontinentalrändern
> aus dem Inneren von Kontinenten (intrakontinentale Vulkanite)
> aus der italienischen Vulkanprovinz

The following compilation of especially frequent volcanites depicts these characteristic differences very clearly.

Die folgende Zusammenstellung für besonders häufige Vulkanite zeigt diese charakteristischen Unterschiede sehr deutlich.

Table 3. K, Th, U content in volcanites of different geotectonic settings.

		Ocean floor	Oceanic islands	Island arcs	Continental margins	Intra-continental	Italy
Alkali and	K [in %]		1.13	1.13	1.37	1.17	0.73
alkali olivine	Th [in ppm]		2.9	4.6	3.7	6.2	4.1
basalts	U [in ppm]		0.91	1.0	1.2	1.1	1.5
Andesites	K [in %]		1.86	1.23	1.73	2.52	1.96
	Th [in ppm]		5.0	4.1	4.3	8.7	10.0
	U [in ppm]		1.6	0.98	1.9	5.5	3.9
Dacites	K [in %]			1.53	2.21		
	Th [in ppm]			6.1	4.2		
	U [in ppm]			1.5	2.6		
Rhyolites	K [in %]			2.4	3.05	4.43	4.21
	Th [in ppm]			10.9	8.2	17.5	36.3
	U [in ppm]			2.8	5.6	5.3	14.0
Tholeiitic	K [in %]	0.17	0.34	0.32	0.54	0.63	
basalts	Th [in ppm]	0.52	0.92	0.59	1.3	3.1	
	U [in ppm]	0.14	0.26	0.36	0.54	0.75	
Trachytes	K [in %]		2.95	4.83		3.10	5.64
	Th [in ppm]		7.5	16.0		12.7	58.6
	K [in ppm]		2.2	3.4		5.0	17.0

It follows from Figs. 4, 5, 7···12 that certain ratios of K/Th, K/U and also of Th/U are typical of volcanites of different geotectonic position.

Aus den Fig. 4, 5, 7···12 folgt außerdem, daß ganz bestimmte K/Th und K/U-Verhältnisse und damit auch Th/U-Verhältnisse für die Vulkanite geotektonisch unterschiedlicher Herkunft typisch sind.

Table 4 Element ratios in volcanites of different geotectonic settings, chondrites and ultramafites.

	K/Th	K/U	Th/U
Volcanites from:			
Ocean floor	$3.3 \cdot 10^3$	$12.4 \cdot 10^3$	3.7
Ocean islands	$3.9 \cdot 10^3$	$12.5 \cdot 10^3$	3.2
Island arcs	$2.4 \cdot 10^3$	$9 \cdot 10^3$	3.8
Continental margins	$4.5 \cdot 10^3$	$10 \cdot 10^3$	2.2
Intracontinental	$2.8 \cdot 10^3$ and $2 \cdot 10^3$	$8.4 \cdot 10^3$	3.0 and 4.2
Italy	$1.7 \cdot 10^3$	$4.8 \cdot 10^3$	2.8
Ordinary chondrites	$20.2 \cdot 10^3$	$63.9 \cdot 10^3$	3.2
Ultramafites	$2.4 \cdot 10^3$	$4.5 \cdot 10^3$	1.9

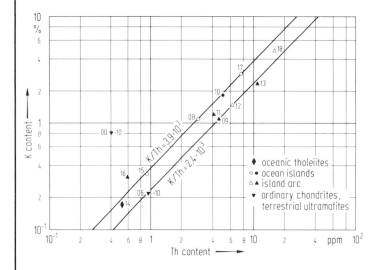

◄

Fig. 4. K/Th ratio in different volcanites from the ocean floor, ocean islands and island arcs. Numbers refer to those of Table 1. Full symbols: rocks of large volume; open symbols: rocks of small volume. Concentrations in chondrites and ultramafites: times 10.

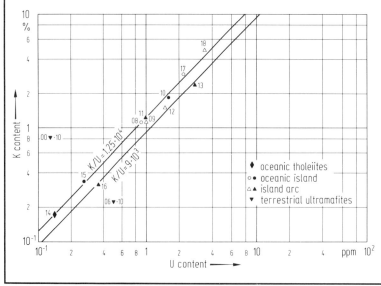

◄

Fig. 5. K/U ratio in different volcanites from the ocean floor, ocean islands and island arcs. Numbers refer to those of Table 1. Full symbols: rocks of large volume; open symbols: rocks of small volume. Concentrations in chondrites and ultramafites: times 10.

7.1.4.2.1 Ocean floor tholeiitic basalts — Tiefseebasalte
(No. 14, Table 1)

The ocean floor represents the largest volume of volcanic rocks on earth. Most of these are tholeiitic basalts with very little chemical variation and with a particularly low content of incompatible elements such as K, Th, U. Their origin is due to a high degree of partial melting ($10 \cdots 20\%$) of a mantle which had already been depleted in incompatible elements before. At the oceanic ridges – where these basalts erupt – the earth's mantle provides the crust with most of the juvenile K, Th and U. Therefore, it is particularly important to know the exact concentration of these elements in the oceanic tholeiites. However, the primary K and U contents are still subject of dispute for there are hardly any fresh samples. The mineral and chemical composition of practically all basalts are changed by reactions with seawater. The values given in Table 1 (No. 14) represent good averages of the K, Th and U concentrations in the oceanic crust after these changes. However, fresh ocean floor basalts have an average K concentration of only 0.13%, as opposed to 0.44% after strang alteration [Ha70]. The concentrations of K [Ha70] and U [Au71] augment with increasing distance from the oceanic ridges, i.e. with increasing age [Ha70]; they also grow with increasing total water content and increasing ratio Fe_2O_3/FeO, both of which are considered as indicators for the degree of secondary alteration [Ha70].

In the following Table 5, some analyses of particularly fresh ocean floor basalts are compared with the mean values. Glass crusts over submarine effusions show the smallest secondary changes.

Except for rocks near the continents, the concentrations of the three elements K, Th and U in these unusually fresh, selected samples are lower than the mean; for K and U they correspond to the mode of the histograms of Fig. 12a$\cdots$c. However, the element ratios ($K/Th \approx 8 \cdot 10^3$, $K/U = (12$ and $19) \cdot 10^3$, $Th/U = (1.5$ and $2.3)$ similar to those of ultramafic rocks) deviate so much from the usual ratios in volcanites that, in view of the small number of such samples, it should remain doubtful how typical these low values are. Undeniably, K and U concentrations can increase during secondary changes. But, on the other hand, it is also known that the lavas of some sections of the oceanic ridges are richer in incompatible elements than others, and can be compared with those of the oceanic islands. This is taken as reflecting inhomogeneities in the mantle [We81].

Die ozeanische Kruste stellt das größte Volumen vulkanischer Gesteine dar. Es handelt sich fast immer um wenig variierende tholeiitische Basalte mit besonders niedrigen Gehalten an inkompatiblen Elementen, unter ihnen K, Th, U. Ihre Herkunft wird auf hohe partielle ($10 \cdots 20\%$) Aufschmelzung eines an inkompatiblen Elementen bereits vorher verarmten Erdmantels zurückgeführt. An den ozeanischen Rücken, wo diese Basalte empordringen, erhält die Erdkruste das meiste neue K, Th und U aus dem Erdmantel zugeführt. Deshalb ist eine genaue Kenntnis der Gehalte der ozeanischen Tholeiite an diesen Elementen besonders wichtig. Indessen ist trotz einer großen Zahl von Analysen besonders der primäre K- und U-Gehalt noch strittig, weil es kaum frische Proben gibt. Nahezu alle Basalte sind durch Reaktionen mit dem Meerwasser in ihrem Mineralbestand und ihrer chemischen Zusammensetzung stark verändert. Die in Tab. 1 aufgeführten Werte (No. 14) geben einen guten Durchschnitt über die K-, Th- und U-Gehalte der ozeanischen Kruste, wie sie nach dieser Veränderung vorliegen. Nach [Ha70] enthalten jedoch frische Tiefseebasalte im Mittel nur 0,13% K, stark veränderte hingegen 0,44% K. Die Gehalte an K [Ha70] und U [Au71] nehmen mit der Entfernung von den ozeanischen Rücken, also mit wachsendem Alter, zu [Ha70]; desgleichen steigen sie mit wachsendem Gehalt an Gesamt-H_2O und wachsendem Verhältnis Fe_2O_3/FeO, welche als Indikatoren für das Ausmaß der sekundären Veränderung gelten [Ha70].

Im folgenden werden einige Analysen von ganz besonders frischen Tiefseebasalten den Durchschnittswerten gegenübergestellt (Tab. 5). Glaskrusten auf untermeerischen Ergüssen zeigen die geringsten sekundären Veränderungen.

Mit Ausnahme der Gesteine nahe den Kontinenten liegen für alle drei Elemente die Konzentrationen in diesen ausgesucht frischen Proben deutlich unter den Mittelwerten für die Gesamtheit der Proben. Sie fallen außerdem bei K und U in das Häufigkeitsmaximum der Histogramme der Fig. 12a$\cdots$c. Allerdings weichen die Elementverhältnisse (($K/Th \approx 8 \cdot 10^3$, $K/U = (12$ und $19) \cdot 10^3$, $Th/U = 1,5$ und $2,3$, ähnlich wie in Ultramafiten)) so weit von dem bei Vulkaniten üblichen ab, daß es bei der geringen Anzahl dieser Proben zweifelhaft bleiben muß, wieweit diese niedrigen Werte als typisch anzusehen sind. Es ist zwar unstrittig, daß die K- und U-Gehalte durch sekundäre Veränderung anwachsen können. Andererseits ist aber auch bekannt, daß manche Abschnitte der ozeanischen Rücken Laven fördern, die ähnlich wie die Basalte der ozeanischen Inseln an inkompatiblen Elementen reicher sind als diejenigen anderer Abschnitte [We81]. Daraus wird auf Inhomogenitäten im Erdmantel geschlossen.

Table 5. Selected radioelement contents in ocean floor tholeiitic basalts.

	K [in %] (n; s)	Th [in ppm] (n; s)	U [in ppm] (n; s)	Th/U (n; s)	Ref.
Atlantic and Pacific ocean [1]	0.17 (355; 0.10)	0.52 (107; 0.29)	0.14 (119; 0.08)	3.7	see Table 1, No. 14
Atlantic ocean [2]	0.142	0.18 (7; 0.07)	0.12 (7; 0.04)	1.6 (7; 0.9)	Ta65
Atlantic ocean [3]		0.14 (4; 0.07)	0.099 (4; 0.03)	1.5	Ta78
Midatlantic ridge [4]			0.065 (6; 0.056)		Fi79
Pacific ocean [2]	0.146	0.18 (10; 0.06)	0.078 (10; 0.016)	2.3	Ta65
NE Pacific ocean [4,5]			0.098 (4; 0.027)		Ma77
Pacific ocean [4,6]			0.042 (19; 0.010)		Ma77
near continental margins [4,7]			0.410 (12; 0.0)		Ma77

[1]) Compilation not subdivided as to origin or degree of secondary alteration, No. 14, Table 1.
[2]) Relatively fresh basalts.
[3]) Basalts expressively called "fresh".
[4]) Glassy crusts; fission track method.
[5]) DSDP-sites 32, 36, 37, 39.
[6]) S. Indonesia, Somali basin, Red Sea, Bauer Deep, Peru basin, DSDP-sites 163, 211, 213, 231, 236, 238, 240, 319, 320.
[7]) DSDP-sites 10, 215, 220.

7.1.4.2.2 Volcanites of the oceanic islands — Vulkanite der ozeanischen Inseln

(Nos. 08, 10, 15, 17, Table 1)

These rocks are formed in isolated volcanoes or in volcanic chains in the interior of oceanic plates. These basalts are probably produced by stationary "hot spots" in the upper mantle over which the plates move. Their distinctly higher content of the incompatible elements Th and U (compare the ocean floor tholeiites with those of oceanic islands) indicates that they have melted out of a mantle which was less depleted in incompatible elements than that from which the oceanic tholeiites originate. The mean values of the element concentrations correlate very well with one another (Fig. 4 and Fig. 5). Their element ratios $K/Th = 3.9 \cdot 10^3$ and $K/U = 12.5 \cdot 10^3$ contrast clearly with those of the island arcs.

Diese Gesteine entstehen in isolierten Vulkanen oder Vulkanketten inmitten von ozeanischen Platten. Man nimmt an, daß stationäre "hot spots" im oberen Mantel, über die die Platten hinwegwandern, diese Basalte fördern. Ihr deutlich höherer Gehalt an den inkompatiblen Elementen Th, U (vgl. die Tholeiite der Tiefsee mit denjenigen der ozeanischen Inseln) deutet darauf hin, daß sie aus einem Mantel erschmolzen wurden, der noch nicht so stark an inkompatiblen Elementen verarmt ist wie derjenige, dem die ozeanischen Tholeiite entstammen. Die Mittelwerte der Elementgehalte korrelieren sehr gut miteinander (Fig. 4 und Fig. 5). Sie heben sich in ihren Elementverhältnissen $K/Th = 3,9 \cdot 10^3$ und $K/U = 12,5 \cdot 10^3$ deutlich von denen der Inselbögen ab.

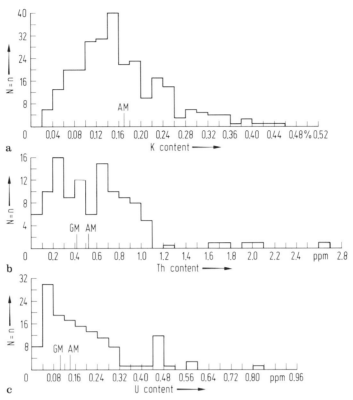

Fig. 6 a···c. Histogram of K (a), Th (b), U (c) concentrations in basalts of the deep sea.

7.1.4.2.3 Volcanites of the island arcs — Vulkanite der Inselbögen

(Nos. 09, 11, 12, 13, 16, 18, Table 1)

These rocks originate where oceanic crust is subducted under oceanic crust and remelted. The arcs contain $\gtrsim 80\%$ tholeiites and basaltic andesites, as well as significant quantities of rhyolites, dacites and andesites. $K/Th = 2.4 \cdot 10^3$ and $K/U = 9 \cdot 10^3$ are the characteristic ratios. The most important deviations herefrom concern the andesites with $K/Th = 3.0 \cdot 10^3$ and $K/U = 12.5 \cdot 10^3$ (Figs. 4, 5).

Diese Gesteine entstehen dort, wo ozeanische Kruste unter ozeanische Kruste subduziert und erneut aufgeschmolzen wird. Es werden $\gtrsim 80\%$ Tholeiite und basaltische Andesite gefördert, bedeutende Mengen Rhyolite und Dacite sowie Andesite. Charakteristisch sind $K/Th = 2,4 \cdot 10^3$ und $K/U = 9 \cdot 10^3$. Die wichtigste Abweichung davon betrifft die Andesite mit $K/Th = 3,0 \cdot 10^3$ und $K/U = 12,5 \cdot 10^3$ (Fig. 4, 5).

7.1.4.2.4 Volcanites of continental margins — Vulkanite der Kontinentalränder

(Nos. 19, 20, 21, 22, 23, Table 1)

These rocks originate from the subduction of oceanic under continental crust. Andesites, dacites and rhyolites are the main magma types. Apparently, the

Diese Gesteine entstehen, wo ozeanische unter kontinentale Kruste subduziert wird. Der Hauptmagmentyp sind Andesite, Dacite und Rhyolite. An-

actual growth of continents is caused by this type of volcanism. It differs from the preceding types by higher concentration of K, Th and U and higher K/Th ratios of $4.5 \cdot 10^3$ and lower Th/U ratios of 2.2 (Figs. 7, 8).

scheinend wachsen gegenwärtig die Kontinente vor allem durch diesen Typ von Vulkanismus. Er unterscheidet sich hinsichtlich K, Th und U von den vorigen durch seine höheren Konzentrationen und höheren K/Th-Verhältnisse von $4,5 \cdot 10^3$, die aber einhergehen mit besonders niedrigen Th/U-Verhältnissen von 2,2 (Fig. 7, 8).

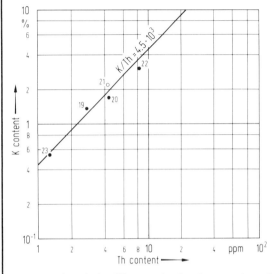
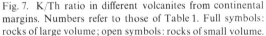
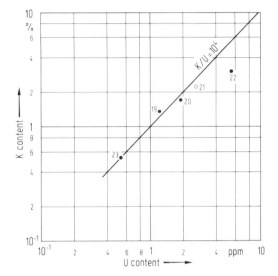

Fig. 7. K/Th ratio in different volcanites from continental margins. Numbers refer to those of Table 1. Full symbols: rocks of large volume; open symbols: rocks of small volume.

Fig. 8. K/U ratio in different volcanites from continental margins. Numbers refer to those of Table 1. Full symbols: rocks of large volume; open symbols: rocks of small volume.

7.1.4.2.5 Volcanites of the continents — Vulkanite der Kontinente

(Nos. 24···34, Table 1)

These rocks are brought to the surface in the large fracture zones of the continents. The East African Rift – still under formation – is the most important example. The large regions of plateau basalt (Deccan, W-Siberia, Paranà, Columbia Plateau) are typical continental basalts. The ratios K/Th (Fig. 9) show two different values, one for andesites, rhyolites and phonolites with $K/Th = 2.8 \cdot 10^3$ and one for basalts (alkali-olivine basalts and tholeiites) with $K/Th = 2 \cdot 10^3$. The ratio $K/U = 8.4 \cdot 10^3$ (Fig. 10) is rather uniform for all basaltic rocks and rhyolite, but it tends to lower values for andesites and phonolites. Pikrite, a rock formed in most cases as an olivine cumulate deviates clearly from these correlations. Kimberlites occupy a particular position with regard to their occurrence in explosion pipes. They are characterized by very variable and exceptional K, Th and U contents and element ratios.

Diese Gesteine werden an den großen Bruchzonen der Kontinente gefördert; das wichtigste Beispiel – noch in Bildung begriffen – ist der Ostafrikanische Graben. Die großen Plateaubasaltgebiete (Deccan, W-Sibirien, Paranà, Columbia-Plateau) sind typische kontinentale Bildungen. Die Verhältnisse K/Th (Fig. 9) spalten sich auf in einen Wert $2,8 \cdot 10^3$ für Andesite, Rhyolite und Phonolite und einen Wert für die Basalte (Alkali-Olivinbasalte und Tholeiite) von $2 \cdot 10^3$. Das K/U-Verhältnis $= 8,4 \cdot 10^3$ (Fig. 10) ist recht einheitlich für die basaltischen Gesteine und Rhyolit, streut aber zu niedrigeren Werten für Andesite und Phonolite. Überhaupt nicht zu diesen Korrelationen gehört Pikrit; das ist verständlich, weil diese Gesteine meist Kumulate von Olivin sind. Eine Sonderstellung hinsichtlich ihres Vorkommens in Explosionsschloten nehmen auch die Kimberlite ein, die sich durch sehr variable und vom üblichen abweichende K-, Th-, U-Gehalte und Elementverhältnisse auszeichnen.

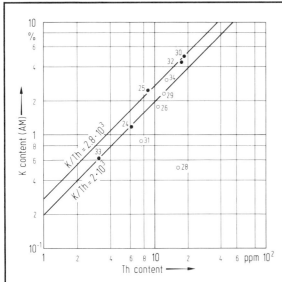

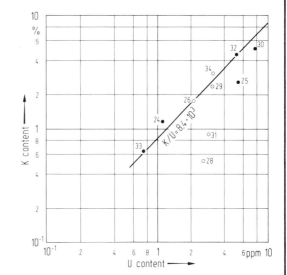

Fig. 9. K/Th ratio in different intracontinental volcanites. Numbers refer to those of Table 1. Full symbols: rocks of large volumes; open symbols: rocks of small volumes.

Fig. 10. K/U ratio in different intracontinental volcanites. Numbers refer to those of Table 1. Full symbols: rocks of large volumes; open symbols: rocks of small volumes.

7.1.4.2.6 Volcanites of Italy — Vulkanite Italiens

(Nos. 35···41, Table 1)

Sr and O isotope studies [Ta76] have proved that the volcanites in the complicated mediterranean plate mosaic originate under participation of crustal material. They are characterized by high SiO_2 contents and often erupt explosively. The ratios $K/Th = 1.7 \cdot 10^3$ and $K/U = 4.8 \cdot 10^3$ are typical for alkali basalts and andesites but not for phonolites, rhyolites and trachytes, which are 2···6 times richer in Th and 2···3 times richer in U compared with the usual intracontinental rocks. They are therefore an exception (Figs. 11, 12).

Im komplizierten Plattenmosaik des Mittelmeeres entstehen Vulkanite unter Beteiligung von Krustenmaterial, wie durch Sr- und O-Isotopenuntersuchungen nachgewiesen wurde [Ta76]. Sie zeichnen sich aus durch hohe SiO_2-Gehalte und oft explosive Förderung. Für Alkalibasalte und Andesite gilt $K/Th = 1,7 \cdot 10^3$ und $K/U = 4,8 \cdot 10^3$, für Phonolite, Rhyolite und Trachyte gelten diese Verhältnisse nicht, weil sie im Vergleich zu üblichen intrakontinentalen Gesteinen um das 2···6fache an Th und das 2···3fache an U angereichert sind. Sie nehmen hierin eine Sonderstellung ein (Fig. 11, 12).

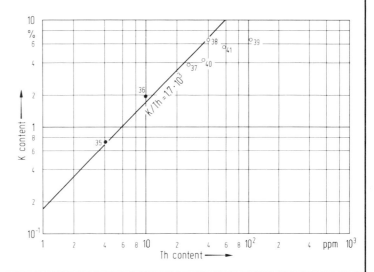

Fig. 11. K/Th ratio in different volcanites from Italy. Numbers refer to those of Table 1. Full symbols: rocks of large volumes; open symbols: rocks of small volumes.

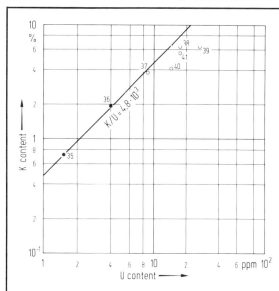

◄ Fig. 12. K/U ratio in different volcanites from Italy. Numbers refer to those of Table 1. Full symbols: rocks of large volumes; open symbols: rocks of small volumes.

7.1.4.3 K, Th and U in intrusive rocks — K, Th, U in Intrusivgesteinen

(Nos. 42···57, Table 1)

The same ratio $K/Th = 1.9 \cdot 10^3$ is valid for all the intrusive rocks from anorthosites (No. 42), over gabbro (No. 45) and diorites (No. 43) to granodiorites (No. 53) and granites of subduction zones (No. 47) as well as the ratio $K/U = 7.4 \cdot 10^3$, except for gabbros and anorthosites. Thus the ratio $Th/U = 3.9$ is valid for all these rocks. Similar ratios also characterize the most important clastic sediments: greywacke (No. 67) and slate (No. 70), so that they must be considered as characteristic of a large part of the upper continental crust. However, all metamorphites and a large part of the granites – those not from active subduction zones – (see Fig. 13 and Fig. 14) have different ratios.

Für die Intrusivgesteine von Anorthosit (No. 42) über Gabbro (No. 45), Diorit (No. 43) zu Granodiorit (No. 53) und den Graniten aus Subduktionszonen (No. 47) gilt das gleiche Verhältnis $K/Th = 1,9 \cdot 10^3$, und – mit Ausnahme von Gabbro und Anorthosit – das gleiche Verhältnis $K/U = 7,4 \cdot 10^3$. Damit gilt für diese Gesteine auch: $Th/U = 3,9$. Ähnliche Verhältnisse charakterisieren ebenfalls die wichtigsten klastischen Sedimente: Grauwacken (No. 67) und Tonschiefer (No. 70), so daß sie als ein durchgängiges Kennzeichen eines großen Teiles der obersten kontinentalen Kruste angesehen werden müssen. Allerdings weichen alle Metamorphite davon ab und ein sehr großer Teil der Granite (siehe Fig. 13 und Fig. 14), nämlich diejenigen, die nicht über aktiven Subduktionszonen intrudierten.

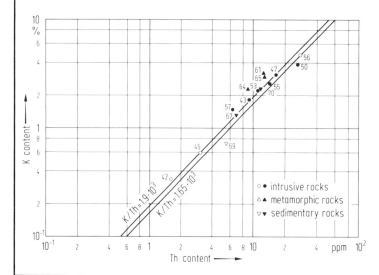

◄ Fig. 13. K/Th ratio in different rocks. Numbers refer to those of Table 1. Full symbols: rocks of large volumes; open symbols: rocks of small volumes.

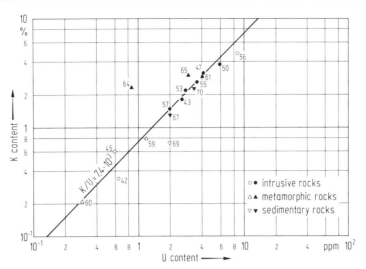

Fig. 14. K/U ratio in different rocks. Numbers refer to those of Table 1. Full symbols: rocks of large volumes; open symbols: rocks of small volumes.

7.1.4.3.1 Granites — Granite

(Nos. 46···51, Table 1)

As shown in the histograms (Fig. 15a···d), there are wide fluctuations in the K, Th and U-content of granites. Accordingly, the heat production rate of granites varies between 0.6 and $17.8 \cdot 10^{-6}$ W/m^3, with a broad maximum between 1.7 and $3.3 \cdot 10^{-6}$ W/m^3 (see Fig. 16). The transition to the granodiorites is frequently diffuse, which can falsify the picture, especially at low concentrations (e.g. see granites with <2% K).

One must distinguish clearly between granites above (No. 47) and not above (No. 50) active subduction zones, the latter showing the higher K, Th and K concentrations. Granites from Japan (No. 48) and from the Damara Orogen in South-West Africa (No. 49), respectively, are particularly clear examples of both types. It is known from Sr and O isotope studies [Ta78] that primary mantle material contributed to the material of granites above subduction zones. This is not the case for granites from the Damara Orogen [Ha81]; these are completely intracrustal formations, exceptionally rich in K and Th and with Th/U ratios mostly >6. Some parts of the earth's crust seem to be much enriched in Th and U. This is expressed by the fact that granites of these areas, although differing much in age, all have elevated contents of the radioactive elements. One of these so-called Th- and U provinces which often contain rich U ore deposits is the Erzgebirge [Kl56].

Die Zusammensetzung der Granite hinsichtlich K, Th, U schwankt in weiten Grenzen, wie die Histogramme (Fig. 15a···d) ausweisen. Dementsprechend reicht auch die Wärmeproduktionsrate der Granite von 0,6 bis $17,8 \cdot 10^{-6}$ W/m^3, mit einem breiten Maximum zwischen 1,7 und $3,3 \cdot 10^{-6}$ W/m^3 (siehe Fig. 16). Die Trennung zu den Granodioriten ist häufig unscharf, was besonders bei den niedrigen Konzentrationen zu einer Verfälschung des Bildes beitragen kann (siehe z.B. Granit mit <2% K).

Bei den Graniten ist ganz klar zu unterscheiden zwischen denen der aktiven Subduktionszonen (No. 47), deren Gehalte an K, Th, U bedeutend unter denjenigen der Granite liegen, die nicht über aktiven Subduktionszonen intrudiert sind (No. 50). Als besonders klare Beispiele für beide Arten sind die japanischen Granite (No. 48) und die Granite des Damara-Orogens in Südwest-Afrika herausgegriffen (No. 49). Von den Graniten über Subduktionszonen ist aufgrund von Sr- und O-Isotopen-Untersuchungen [Ta78a] bekannt, daß primäres Mantelmaterial an ihrem Aufbau beteiligt ist, während dies für diejenigen des Damara-Orogens [Ha81] ausgeschlossen werden kann. Letztere sind gänzlich intrakrustale Bildungen, überdurchschnittlich reich an K und Th und mit Th/U-Verhältnissen von meist >6. In manchen Teilen der Erdkruste scheinen ausgesprochen starke Anreicherungen von Th und U vorzukommen, was sich darin ausprägt, daß alle Granite solcher Gebiete trotz ganz unterschiedlichen Alters erhöhte Gehalte an diesen Elementen aufweisen. Eine dieser sogenannten Th-, U-Provinzen, die oft besonders ergiebige U-Lagerstätten enthalten, ist z.B. das Erzgebirge [Kl56].

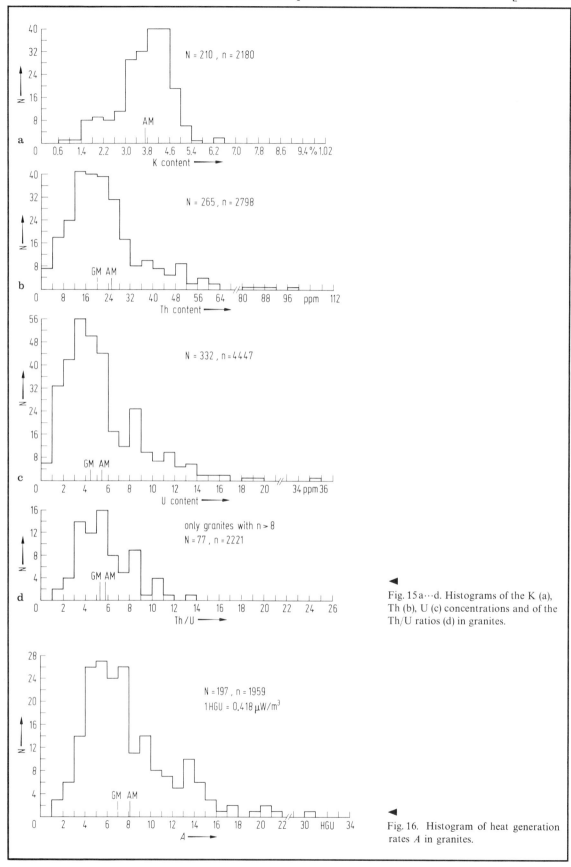

◀

Fig. 15a···d. Histograms of the K (a), Th (b), U (c) concentrations and of the Th/U ratios (d) in granites.

◀

Fig. 16. Histogram of heat generation rates A in granites.

Haack

7.1.4.3.2 Granodiorites — Granodiorite

(Nos. 52···55, Table 1)

Just as granites, granodiorites contain more K, Th and U when they have not been intruded above active subduction zones (No. 55) than when they have been formed above such zones (No. 53). Generally the K, Th and U content is lower in granodiorites than in granites. Some high K values could be due to a faulty petrographic distinction from granites. The histograms (Figs. 17a···c, 18) give more information about the scatter of the element concentrations and of the heat generation rates.

Ebenso wie die Granite, enthalten die Granodiorite mehr K, Th, U, wenn sie nicht über aktiven Subduktionszonen intrudiert sind (No. 55), als wenn sie über solchen Subduktionszonen entstanden (No. 53). Generell ist der K, Th, U-Gehalt der Granodiorite geringer als derjenige der Granite. Einige hohe K-Werte dürften auf mangelhafte petrographische Abgrenzung zu den Graniten zurückzuführen sein. Über die Streubreite der Elementgehalte und der Wärmeproduktionsraten geben die Histogramme Auskunft (Fig. 17a···c, 18).

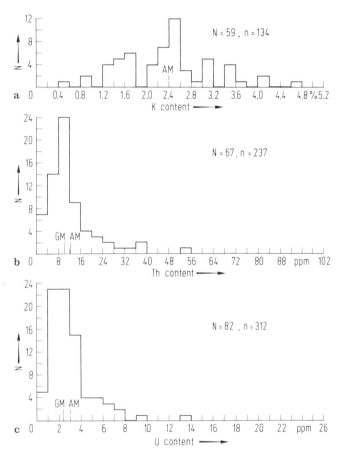

Fig. 17a···c. Histograms of the K (a), Th (b) and U (c) concentrations in granodiorites.

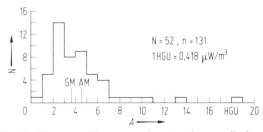

Fig. 18. Histogram of heat generation rate A in granodiorites.

Haack

7.1.4.4 K, Th and U in metamorphic rocks — K, Th, U in metamorphen Gesteinen

(Nos. 58···65, Table 1)

The Th/U ratios of the means of the metamorphic rocks deviate clearly from values between 3 and 4 which are typical of many intrusive rocks. In this respect, they also differ from their non-metamorphic equivalents. This is due to unequal losses of K, Th and U during metamorphic processes in which the mobility of uranium plays a great part (uranium being easily soluble in its hexavalent form as the uranyl ion).

The pair amphibolite-eclogite (No. 58 and No. 60), both rocks chemically equivalent to basalts, should be particularly pointed out. Compared to basalt, amphibolite lost but a small amount of K and Th. By contrast, the concentrations of K, Th and U in eclogite have dropped drastically. This is a general characteristic of the rocks of the deeper earth's crust which have been metamorphosed under extremely high pressure and temperature. This applies especially to the granulites (Nos. 62···64). Their main components are often comparable to those of granites and gneisses of highest metamorphic grade, but often they also represent the refractory residues of magmatic processes. The latter applies particularly to the "basic" granulites (No. 63).

The histograms in Fig. 19a···d show the wide range of Th and U concentrations as well as of the Th/U ratios in granulites. Fig. 20 shows the ratio Th/U versus the Th content. The Th/U ratio approaches 2 as the Th content decreases to 1 ppm, but this graph demonstrates also that Th/U ratios up to 92 do occur. These data indicate that during highest grade metamorphism U is preferentially extracted – probably together with H_2O and CO_2 – and that often Th only follows when hardly any uranium is left. A reason for the better mobility of U is that, in contrast to Th, it is partly loosely bound on grain boundaries and internal surfaces (e.g. [Do78]).

K, Th and U analyses of granulites are very important because they demonstrate to what degree the lower crust has been depleted in radioactive elements. Realistic models for the mean content of K, Th and U in the earth's crust – also leading to accordance with heat flow measurements – are only possible when the low concentrations in granulites are taken into account. Compared with the upper crust, granulites are much depleted also in the other incompatible elements.

Die Metamorphite weichen hinsichtlich der Th/U-Verhältnisse ihrer Mittelwerte deutlich von den Werten zwischen 3 und 4 ab, die für viele der Intrusiva (außer den Graniten) und eine Reihe vulkanischer Gesteine typisch sind. Desgleichen unterscheiden sie sich hierin von ihren unmetamorphen Äquivalenten. Dies ist auf den unterschiedlichen Verlust von K, Th, U bei metamorphen Prozessen zurückzuführen, wobei vor allem die Mobilität des Urans eine Rolle spielt, das in seiner sechswertigen Form als Uranylion leicht löslich ist.

Besonders sei hingewiesen auf das Paar Amphibolit-Eklogit (No. 58 und No. 60), Gesteine, die beide Basalten chemisch äquivalent sind. Amphibolit hat gegenüber Basalt nur eine leichte Verarmung an K und Th erlebt. In Eklogit hingegen sind die Gehalte an K, Th, U drastisch abgesenkt, ein ganz allgemeines Kennzeichen für die Gesteine der tieferen Erdkruste, die unter höchsten Drucken und Temperaturen umgebildet werden. Dies trifft in ganz besonderem Maße für die Granulite zu (No. 62···64). Diese Gesteine entsprechen in ihren Hauptbestandteilen einerseits Graniten und Gneisen in höchstmetamorphem Zustand, stellen aber andererseits vermutlich oft auch den nicht aufschmelzbaren Rest bei magmatischen Prozessen dar. Letzteres gilt insbesondere für die sogenannten „basischen" Granulite (No. 63).

Die Histogramme der Fig. 19a···d zeigen den weiten Streubereich der Th- und U-Gehalte sowie der Th/U-Verhältnisse in Granuliten. In Fig. 20 ist das Verhältnis Th/U gegen Th aufgetragen. Aus dieser Darstellung wird ersichtlich, daß das Verhältnis Th/U gegen 2 strebt, wenn der Th-Gehalt auf 1 ppm sinkt. Andererseits zeigt diese Darstellung auch, daß Th/U-Verhältnisse bis 92 vorkommen. Aus diesen Befunden läßt sich ablesen, daß bei der höchstgradigen Metamorphose zunächst vor allem das Uran – vermutlich zusammen mit H_2O und CO_2 – weggeführt wird, während das Th oft erst dann folgt, wenn kaum noch U vorhanden ist. Ein Grund für die leichtere Mobilisierbarkeit des U liegt darin, daß es im Gegensatz zu Th zum Teil auf Korngrenzen und inneren Oberflächen locker gebunden ist (z.B. [Do78]).

Die Bedeutung der Analysen von K, Th, U in Graniten liegt vor allem darin, daß sie belegen, in welchem Ausmaß die untere Kruste an radioaktiven Elementen gegenüber der oberen Kruste verarmt ist. Erst die Kenntnis der niedrigen Konzentrationen in Granuliten ermöglicht realistische Modelle über die mittleren Gehalte an K, Th, U in der Erdkruste, die auch zu Übereinstimmung mit Wärmeflußmessungen führen. Die Granulite sind auch an den anderen inkompatiblen Elementen im Vergleich zur oberen Kruste stark verarmt.

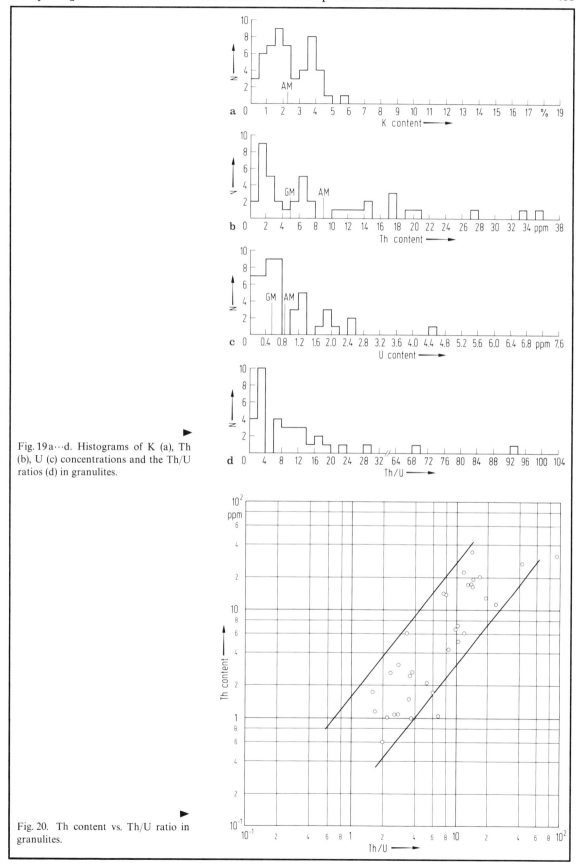

Fig. 19 a···d. Histograms of K (a), Th (b), U (c) concentrations and the Th/U ratios (d) in granulites.

Fig. 20. Th content vs. Th/U ratio in granulites.

Haack

7.1.4.5 K, Th and U in sediments — K, Th, U in Sedimenten

(Nos. 66···89, Table 1)

[Sc80] has determined a mean heat production rate of $1.38 \cdot 10^{-6}$ W/m³ for the Rheinisches Schiefergebirge taking into account the relative frequency of the rocks. [Ro76] calculated an identical value for platform-sediments. This value is probably typical for a wide range of the continental sedimentary cover. Estimates of the concentration of radioactive elements in sediments of continental geosynclines and platforms as well as in oceanic sediments – mainly based on Russian studies – have been published by [Ro76] (see the following Table 6). Some important differences to Table 1 can be explained by another classification of the rocks (Example: greywackes and sandstones have apparently been put together by [Ro76]).

[Sc80] hat die mittlere Wärmeproduktionsrate des Rheinischen Schiefergebirges unter Berücksichtigung der Gesteinshäufigkeiten zu $1,38 \cdot 10^{-6}$ W/m³ bestimmt. Zum gleichen Wert gelangte [Ro76] für Plattform-Sedimente. Vermutlich ist dies ein Wert, der für weite Bereiche der kontinentalen Sedimenthülle typisch ist. Abschätzungen über Konzentrationen an Radioelementen in Sedimenten von kontinentalen Geosynklinen und Plattformen sowie von ozeanischen Sedimenten – basierend vor allem auf russischen Arbeiten – sind von [Ro76] veröffentlicht (siehe folgende Tab. 6). Einige wichtige Unterschiede zu Tab. 1 sind durch andersartige Abgrenzungen der Gesteine voneinander zu erklären. (Beispiel: Grauwacken und Sandsteine sind bei [Ro76] offenbar zusammengefaßt).

Table 6. K, Th, U content in sediments [Ro76].

	K %	Th ppm	U ppm
Continental platforms:			
Sands	1.69	8.0	2.0
Clays	2.89	12.8	3.0
Carbonates	0.31	1.9	1.6
Evaporites	0.30	0.7	0.8
Lavas	0.90	3.3	1.0
mean composition of sedimentary layers in platforms	1.84	8.3	2.3
Continental geosynclines:			
Sands	1.60	9.2	1.9
Clays	2.33	7.8	2.9
Carbonates	0.60	1.8	1.1
Evaporites	0.14	0.7	0.8
Lavas	1.34	4.3	1.1
mean composition of continental geosynclines	1.59	6.0	1.9
Oceanic surface sediments in first seismic layer:			
terrigenous	2.55	3.1	2.6
calcareous	0.86	1.0	1.3
siliceous	2.24	2.1	1.3
red deep-water clays	2.37	6.6	2.7
mean composition of surface sediments	1.69	2.9	1.8
sedimentarey shell as a whole, excluding lavas	1.59	6.1	2.1

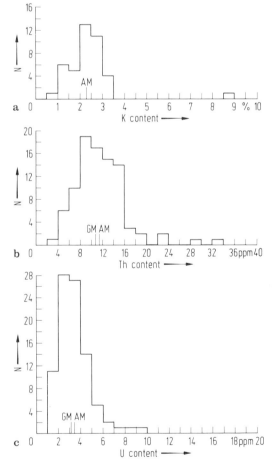

Fig. 21 a···c. Histograms of K (a), Th (b), U (c) concentrations in shales and siltstones (excluding black shales).

Histograms of element concentrations in some of the most common sediments are given in Figs. 21···23. The most abundant sedimentary rocks (greywackes 11 %, clays and siltstones 74 % according to [Ga72a]) show Th/U ratios of ≈3.5, very similar to those of most of the intrusive rocks and of many volcanites. The sediments are a representative mixture of material from all parts of the earth's crust. Therefore, the element ratios have great importance. A Th/U ratio of 3.4···3.9 probably corresponds to the mean value for the continental crust. By contrast, the mean Th/U ratio of the crystalline continental crust is 7.3 (see the following section). Most of the uranium must have been extracted from this crystalline crust and either been returned to the mantle during subduction processes or deposited in the sediments. From the distribution coefficient of U in crystal-melt equilibria we know that it is not possible to bring U (and Th) back into the mantle in significant quantities. It must therefore be deposited in the continental margin volcanites (low Th/U ratios) and the sedimentary cover. The quantity in question is $\approx 8 \cdot 10^{12}$ t U, a large part of which is probably contained in black shales. According to [Sc80] and [Ro68a], the proportion of black shales in platform-sediments amounts to about 5 %. By generalizing this number and assuming the total quantity of sediments to be $3.2 \cdot 10^{18}$ t [Ga72a], we have about $3.2 \cdot 10^{12}$ t U bound in black shales. With Th/U = 3.9 in the total crust, $2.8 \cdot 10^{12}$ t are uranium "excess". It is possible that the black shales alone contain half or more of the uranium extracted from the lower crust, when one considers that the mean uranium content of the black shales rises to 30.4 ppm U if the extreme values between 100 and 300 ppm U are not excluded as was done for the calculation of the mean values for Table 1; it could even be many times higher than that if the extreme values >1000 ppm were also taken into account. Recent anaerobic sediments, being early stages of black shales, prove that the black shales must be important sinks of U. Additional $\approx 1.2 \cdot 10^{12}$ t "excess" uranium are bound in limestones and sandstones (15 % and 11 % of the total sediments according to [Ga72b]). The uranium quantities bound in recent limestones (Nos. 78···80, Table 1), phosphorites (No. 83), coals (No. 87), soils (No. 85) and ore deposits could not be considered for this calculation because the numerical data is still rather incomplete. Nevertheless, it seems that the excess uranium in the sedimentary cover and in continental margin volcanites may equal the amount of uranium extracted from the crystalline crust.

Histogramme der Elementhäufigkeiten in einigen der verbreitetsten Sedimente werden in Fig. 21···23 gegeben. Die mengenmäßig wichtigsten Sedimentgesteine (Grauwacken 11 %, Tonschiefer und Siltsteine 74 % nach Angaben von [Ga72a]) weisen Th/U-Verhältnisse von ≈3,5 auf, sehr ähnlich denen der meisten Intrusiva und sehr vieler Vulkanite. Vermutlich stellen diese Sedimente eine repräsentative Mischung von Material aus allen Bereichen der Erdkruste dar. Deshalb ist auch den Elementverhältnissen eine besondere Bedeutung zuzumessen. Wahrscheinlich entspricht ein Th/U-Verhältnis von 3,4···3,9 dem mittleren Wert für die kontinentale Kruste. Dem steht gegenüber ein mittleres Th/U-Verhältnis von 7,3 in der kristallinen kontinentalen Kruste (siehe folgenden Abschnitt). Das Uran muß zu großen Teilen aus dieser kristallinen Kruste extrahiert und entweder bei Subduktionsprozessen in den Mantel zurückgebracht oder in den Sedimenten deponiert worden sein. – Aus den Verteilungskoeffizienten für U bei Gleichgewichten zwischen Schmelze und Kristallen wissen wir, daß die Möglichkeit, Uran (und Th) in nennenswerter Menge in den Mantel zurückzubringen, ausgeschlossen ist. Es muß deshalb in den Vulkaniten der Kontinentalränder (niedrige Th/U-Verhältnisse) und in der Sedimenthülle deponiert sein. Es handelt sich um $\approx 8 \cdot 10^{12}$ t U, von denen ein großer Teil vermutlich in den Schwarzschiefern fixiert ist. Nach [Sc80] und [Ro68a] beträgt der Anteil der Schwarzschiefer an Plattformsedimenten etwa 5 %. Verallgemeinert man diese Zahl, so sind bei einer Gesamtmasse der Sedimente von $3,2 \cdot 10^{18}$ t [Ga72a] etwa $3,2 \cdot 10^{12}$ t U in Schwarzschiefern gebunden. Davon sind – bei Th/U=3,9 in der gesamten Kruste – $2,8 \cdot 10^{12}$ t „Überschuß"-Uran. Bedenkt man, daß der mittlere Urangehalt der Schwarzschiefer unter Berücksichtigung der Extremwerte zwischen 100 und 300 ppm U (die bei der für Tab. 1 erfolgten Mittelwertbildung weggelassen wurden) auf 30,4 ppm U steigt und wenn, man die Extremwerte >1000 ppm hinzunimmt, um ein Vielfaches höher liegt, so ist es leicht möglich, daß die Schwarzschiefer alleine die Hälfte oder mehr des Urans enthalten, das aus der unteren Kruste extrahiert wurde. Daß die Schwarzschiefer ein bedeutendes Reservoir für U sein müssen, zeigt sich auch an den rezenten anaeroben Sedimenten (No. 74, Tab. 1), welche die Vorstadien für Schwarzschiefer darstellen. In den Kalksteinen und Sanden (15 bzw. 11 % Anteil an den Sedimenten nach [Ga72b]) sind weitere $\approx 1,2 \cdot 10^{12}$ t „Überschuß"-Uran gebunden. Für die Rechnung nicht berücksichtigt werden konnten die Mengen Uran, die in rezenten Kalksteinen (No. 78···80, Tab. 1), Phosphoriten (No. 83), Kohlen (No. 87), Böden (No. 85) oder Lagerstätten gebunden sind, weil das Zahlenmaterial zu unvollständig ist. Trotz alledem zeichnet sich aber ab, daß die Bilanz, „Überschuß"-Uran in der Sedimenthülle und in den Vulkaniten der Kontinentalränder = aus der kristallinen Kruste extrahiertes Uran, aufgeht.

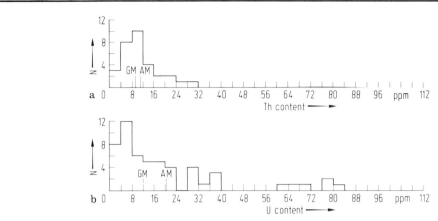

Fig. 22 a, b. Histograms of Th (a) and U (b) concentrations in black shales. U concentrations > 100 ppm not shown.

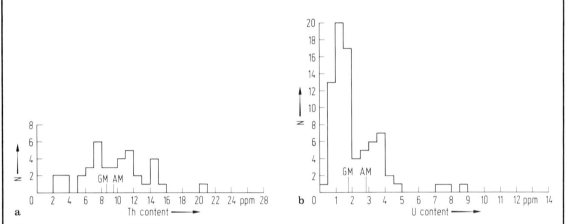

Fig. 23 a, b. Histograms of Th (a) und U (b) concentrations in clays of the deep sea.

7.1.5 K, Th and U content of the earth's crust — Gehalte an K, Th, U in der Erdkruste

The average values for the intrusive rocks given in Table 1 and estimated relative abundances in the continental upper earth's crust (44% granites, 34% granodiorites, 9% diorites and quartz diorites, 13% gabbros [We69]) yield the mean values for the contents in the upper crust which are very similar to those typical of granodiorites (V.M. Goldschmidt [Go54] already pointed out this fact).

Benutzt man die in Tab. 1 angegebenen Durchschnittswerte der K-, Th-, U-Konzentrationen für die Intrusiva und die von [We69] abgeschätzten Gesteinshäufigkeiten in der oberen kontinentalen Erdkruste (44% Granite, 34% Granodiorite, 9% Diorite und Quarzdiorite, 13% Gabbros), so erhält man Mittelwerte der Gehalte für die obere Kruste, die sehr nahe bei denjenigen liegen, die für Granodiorite typisch sind (worauf schon V.M. Goldschmidt [Go54] hingewiesen hatte).

Weighted mean K, Th and U concentrations of intrusive rocks of the upper continental crust

K	Th	U	K/Th	K/U	Th/U	heat generation rate A
2.74%	16 ppm	3.8 ppm	$1.7 \cdot 10^3$	$7.2 \cdot 10^3$	4.2	$2.38 \cdot 10^{-6} \, \text{W/m}^3$

When comparing these values to those given as average composition of precambrian shields – see following table – it is noteworthy that many K concentrations and nearly all Th and U concentrations, and, consequently, the heat generation in the crust are considerably lower than the average for intrusive rocks. The element ratios are generally different as well, e.g. Th/U is very often ≥ 6 and not ≈ 4. These differences are probably caused by the deep erosion of old shields and are an indication for the enrichment of the upper crust with radioactive elements during metamorphism and formation of magmatic melts. The degree of the relative enrichment is $U > Th \gg K$.

Vergleicht man diese Werte mit denjenigen, die als Durchschnittszusammensetzung der präkambrischen Schilde – siehe folgende Tabelle – angegeben werden, so fällt auf, daß dort viele der K- und fast alle der Th- und U-Gehalte und dementsprechend die Wärmeproduktion in der Kruste bedeutend niedriger sind als der Durchschnitt der Intrusivgesteine. Auch die Elementverhältnisse sind generell andere, z.B. ist Th/U sehr oft ≥ 6 und nicht ≈ 4. Diese Unterschiede hängen vermutlich mit der tieferen Erosion der alten Schilde zusammen und signalisieren, daß die radioaktiven Elemente im Zuge der Metamorphose und bei der Bildung magmatischer Schmelzen in der obersten Erdkruste angereichert werden. Der Grad der relativen Anreicherung ist $U > Th \gg K$.

Table 7. Various estimates of radioactivity in upper continental crust. A = heat generation.

No.		K in %	Th in ppm	U in ppm	K/Th [10^3]	K/U [10^3]	Th/U	A $10^{-6}\,\mathrm{W\,m^{-3}}$	Ref.
	Canadian shield,								
1	archaic	2.21	9.7	1.2	2.3	18.4	8.1	1.19	Fa68
2	proterozoic	2.80	13.6	2.2	2.1	12.7	6.2	1.78	Fa68
3	weighted average of 1 and 2	2.35	10.8	1.5	2.2	15.7	7.2	1.36	Fa68
4	N. Quebec	2.20	6.8	2.0	3.2	11.0	3.4	1.20	Sh67
5	SW. Quebec	3.46	6.7	1.4	5.2	24.7	4.8	1.15	Sh67
6	Baffin Island	2.78	14.7	2.1	1.9	13.2	7.0	1.83	Sh67
7	S. Saskatchewan	3.02	13.0	4.3	2.3	7.0	3.0	2.31	Sh67
8	grand mean of 4$\cdots$7	2.87	10.3	2.45	2.8	11.7	4.2	1.62	Sh67
	Canadian shield,								
9	W. Canada Basement		21.1	4.13			5.1		Bu76
10	Bear Province		35.7	8.1			4.4		Bu76
11	Slave Province		8.4	1.7			4.9		Bu76
12	Churchill Province		15.5	2.6			6.0		Bu76
13	average of 9$\cdots$12 weighted according to size of province		13.0	2.1			6.2		Bu76
14	W. Australian shield	2.6	20	3	1.30	8.7	6.7	2.42	Hy68
15	Snowy Mountains, Australia	1.6	5.6	1.1	2.9	14.5	5.1	0.82	Hy68
16	Greenstone Area, Australia	2.9	16	4.1	1.8	7.1	3.9	2.45	Hy68
17	Baltic and Ukrainian shields and basement of Russian platform	2.56	18	2.8	1.4	9.1	6.2	2.26	Ro70
18	mean composition of shields (including Canadian and Australian)	2.52	14.3	2.4	1.8	10.5	6.0	1.86	Ro70
19	continents as a whole	1.64	6.3	1.6	2.6	10.1	3.9	1.01	Ro76
20	subcontinental crust as a whole	1.58	5.7	1.5	2.7	10.4	3.8	0.93	Ro76
21	crust Archean shield SW. Australia as a whole	1.5	4.5	0.7 or 1.1[1]	3.3	21.4 or 13.6[1]	6.4 or 4.1[1]	0.63 or 0.74[1]	La68
22	Paleozoic crust E. Australia as a whole	1.7	7	1.3 or 1.7[1]	2.4	13.1 or 10.0[1]	5.4 or 4.1[1]	0.98 or 1.09[1]	La68
23	upper post-Archean crust	2.74	10	2.5	2.7	11.0	4	1.60	Ta79
24	lower post-Archean crust	0.50		0.25		20.0			Ta79
25	total post-Archean crust	1.25	2.5	1.0	5.0	12.5	2.5	0.55	Ta79

[1]) Keeping Th/U = 4.

Haack

An attempt to estimate the mean K, Th and U concentration of the crust can be based on the following assumptions: granites and granulites are the extreme end members of a process of crustal differentiation, the granites being the enriched, the granulites the depleted parts (Nos. 46 and 64, Table 1). According to [La68a] and [Ro68] the change of the heat generation with depth is best described by the relation $A = A_0 e^{-d/b}$. Inserting $A = 0.713 \cdot 10^{-6}$ W/m³ (based on geometric means for Th and U) as the heat generation rate of granulites, $A_0 = 3.49 \cdot 10^{-6}$ W/m³ that of granites and choosing an intermediate depth parameter $b = 10$ km (values between 3 and 14 km occur [Je78]) the depth $d = 15.9$ km results. This, in turn, can be used to calculate the depth parameters of the three elements (assuming an analogous depth distribution):

$$b_K = 34.4 \text{ km}$$
$$b_{Th} = 9.9 \text{ km}$$
$$b_U = 7.1 \text{ km}.$$

The mean concentrations of a 45 km thick crust which has not yet lost its granite layer is then:

$$\overline{K} = 2.1\%$$
$$\overline{Th} = 5.4 \text{ ppm}$$
$$\overline{U} = 0.85 \text{ ppm}$$
$$\text{and} \quad Th/U = 6.3.$$

Such a crust can, in first approximation, be identified as the phanerozoic fold belts which have a mass of $0.53 \cdot 10^{19}$ t [Ga72]. On the basis of the mean composition of the shields (Table 7) [Ro70] and with the same depth parameters, the mean concentrations of K, Th and U in the shields (thickness ≈ 37 km, mass = $1.08 \cdot 10^{19}$ t [Ga72]) is:

$$\overline{K} = 1.6\%$$
$$\overline{Th} = 3.8 \text{ ppm}$$
$$\overline{U} = 0.46 \text{ ppm}$$
$$\text{and} \quad Th/U = 8.2.$$

If the masses of the shields and fold belts are also considered, the mean concentration of K, Th and U in the earth's continental crust (excluding the sedimentary cover) are as follows:

$$\overline{K} = 1.7\%$$
$$\overline{Th} = 4.3 \text{ ppm}$$
$$\overline{U} = 0.59 \text{ ppm}$$
$$\text{and} \quad Th/U = 7.3.$$

This result is intended to supplement other estimates. It differs from them in its low Th and U contents and the high Th/U ratio of 7.3. However, it is well compatible with the continental heat flow values.

The uranium extracted from the lower crust must now reside elsewhere. If the total mass of the continental crust amounts to $1.61 \cdot 10^{19}$ t [Ga72] and has an average Th content of 4.3 ppm and the Th/U ratio

Eine Abschätzung der mittleren Konzentration dieser Elemente in der Kruste ist möglich unter der Voraussetzung, daß die Granite die besonders an radioaktiven Elementen angereicherten und die Granulite die besonders verarmten Endglieder dieser – kontinuierlichen – Entwicklung darstellen. Ferner gelte nach [La68a] und [Ro68] $A = A_0 e^{-d/b}$ mit den Wärmeproduktionsraten A_0 und A an der Ober- und Unterseite einer Schicht der Dicke d, und dem Tiefenparameter b. Mit $A_0 = 3.49 \cdot 10^{-6}$ W/m³ für Granite und $A = 0.713$ W/m³ für Granulite (geometrische Mittel für Th und U) sowie $b = 10$ km (Werte zwischen 3 und 14 km, meist 6 km, kommen vor [Je78]) ergibt sich $d = 15.9$ km. Analoge Tiefenverteilung für die Elemente vorausgesetzt, lassen sich damit deren Tiefenparameter berechnen:

$$b_K = 34,4 \text{ km}$$
$$b_{Th} = 9,9 \text{ km}$$
$$b_U = 7,1 \text{ km}.$$

Für eine 45 km dicke, ihrer Granite noch nicht beraubte Kruste ergeben sich damit folgende mittlere Konzentrationen

$$\overline{K} = 2,1\%$$
$$\overline{Th} = 5,4 \text{ ppm}$$
$$\overline{U} = 0,85 \text{ ppm}$$
$$\text{und} \quad Th/U = 6,3.$$

Eine solche Kruste kann in erster Näherung mit den phanerozoischen Faltengürteln gleichgesetzt werden, deren Masse $0,53 \cdot 10^{19}$ t beträgt [Ga72]. Ausgehend von der mittleren Zusammensetzung der Schilde nach [Ro70] (Tab. 7) ergeben sich mit den gleichen Tiefenparametern als mittlere Konzentrationen für die Schilde (Dicke 37 km, Masse $1,08 \cdot 10^{19}$ t [Ga72]).

$$\overline{K} = 1,6\%$$
$$\overline{Th} = 3,8 \text{ ppm}$$
$$\overline{U} = 0,46 \text{ ppm}$$
$$\text{und} \quad Th/U = 8,2.$$

Unter Berücksichtigung der Massen der Schilde und der Faltengürtel sind die mittleren Konzentrationen der kontinentalen Erdkruste (ohne Sedimenthülle):

$$\overline{K} = 1,7\%$$
$$\overline{Th} = 4,3 \text{ ppm}$$
$$\overline{U} = 0,59 \text{ ppm}$$
$$\text{und} \quad Th/U = 7,3.$$

Dieses Ergebnis ist als Ergänzung zu den sonstigen Schätzungen gedacht. Es unterscheidet sich von ihnen vor allem durch seine niedrigen Th- und U-Konzentrationen, die mit einem hohen Th/U-Verhältnis von 7,3 einhergehen, ist aber dafür gut verträglich mit den Werten für kontinentale Wärmeflüsse.

Das aus der unteren Kruste extrahierte Uran muß sich jetzt woanders befinden. Wenn die Gesamtmasse der kontinentalen Kruste $1,61 \cdot 10^{19}$ t beträgt [Ga72], einen mittleren Th-Gehalt von 4,3 ppm aufweist und

equals 3.9 for the total earth [Ma79] as well as for the whole crust (most of intrusive rocks, except granites), then the quantity of uranium in question is $\approx 8 \cdot 10^{12}$ t. A large part of it is bound in the black shales, which are only $\lesssim 5\%$ of the total sedimentary mass [Sc80]. Limestones are also an important sedimentary reservoir of such "excess" uranium as are also many other sediments [Ro76] (see Table 6). Volcanites of the continental margins with their low Th/U ratios may also contain appreciable amounts of this uranium.

The K, Th and U concentrations in the oceanic crust are probably very similar to the values for oceanic tholeiites (No. 14, Table 1).

das sowohl für die Gesamterde [Ma79] als auch die Gesamtkruste (die meisten Intrusiva außer Graniten) zutreffende Th/U-Verhältnis = 3,9 ist, so läßt sich diese infrage stehende Uranmenge zu $\approx 8 \cdot 10^{12}$ t bestimmen. Ein großer Teil davon ist in den Schwarzschiefern gebunden, die selber nur $\lesssim 5\%$ der Gesamtmasse der Sedimente ausmachen [Sc80]. Aber auch die Kalke sind ein wichtiges Reservoir von solchem „Überschuß"-Uran ebenso wie viele andere Sedimente [Ro76] (siehe Tab. 6). Auch die Vulkanite der Kontinentalränder mit ihren niedrigen Th/U-Verhältnissen könnten solches Uran enthalten.

Der Gehalt an K, Th, U in der ozeanischen Kruste dürfte sehr gut durch die Werte für ozeanische Tiefsee-Tholeiite (No. 14, Tab. 1) wiedergegeben werden.

7.1.6 K, Th and U in minerals — K, Th, U in Mineralen

Table 8. Th and U content in some accessory and rock forming minerals. Data for minerals of ultramafic rocks are given in Table 2.

Mineral	Th [in ppm]		U [in ppm]		Ref.
	AM N; n; s min···max	GM a; b	AM N; n; s min···max	GM a; b	Remarks
Anatase			10 2; 2		Ye77
Baryte, deep sea, authigenic	37 1; 13; —		6 1; 11; —		Ch72
Biotite, igneous rocks	4.5 3; 3; 0.6 3.8···4.9	4.5 0.7; 0.6	0.23 7; 7; 0.07 0.098···0.29	0.22 0.10; 0.07	Be74, Do75, Do76, Ko67, Na66
Biotite, metamorphic rocks			0.031 8; 8; 0.008 0.022···0.044	0.030 0.009; 0.007	Do78
Chromite			0.025 1; 1; — 0.0006···0.03		Gi74 Se73
Clinopyroxene, andesites, basalts, dacites	0.028 5; 5; 0.017 0.011···0.056	0.024 0.020; 0.011	0.021 14; 14; 0.043 0.0067···0.070	0.0046 0.018; 0.0037	Co80, Do75, Do76, Na68a (not included Th > 1.0 and U > 0.3 ppm)
Clinopyroxene, gabbros			0.0078 8; 8; 0.0047 0.0045···0.018	0.0069 0.0043; 0.0026	Do75a, Hen71
Clinopyroxene, granulites			0.038 5; 9; 0.021 0.016···0.064	0.033 0.027; 0.015	Do78
Garnet, andradite rich			0.3···3.5 12; —; —		Ha72
Garnet, metamorphic rocks			0.001···0.05		Do78, Ha75, Ko67

(continued)

Haack

Table 8 (continued)

Mineral	Th [in ppm]		U [in ppm]		Ref.
	AM	GM	AM	GM	Remarks
	N; n; s	a; b	N; n; s	a; b	
	min···max		min···max		
Hornblende, igneous and metamorphic	≈0.5		0.0005···0.28		Be74, Do78, Ha73, Kl69, Lu80
Magnetite, igneous rocks	4.8 7; 11; 3.3 1.7···9.9	3.9 3.9; 1.9	0.95 9; 13; 0.58 0.29···1.85	0.78 0.79; 0.39	Al76, Be74, Co80, Ye77
Muscovite			0.0003···0.05		Ko66
Olivine, basalts			0.0055 6; 6; 0.0016 0.0040···0.0082	0.0054 0.0016; 0.0012	Do75, Do76, Ni72
Orthopyroxene, volcanites	0.12···1.93 7; 8; −		0.02···0.7		Al76, Do75, Do76, Lu80
Orthopyroxene, granulites			0.018 8; 8; 0.0073 <0.015···0.030	0.016 0.0079; 0.0053	Do78
Phillipsite, authigenic, deep sea sediments	3.6 4.36; 0.24; − 1.4···6.7	3.0 3.0; 1.5	0.51 3; 24; 0.11 0.39···0.61	0.50 0.13; 010	Be69, Be74a, Ber70
Plagioclase, andesites, basalts	0.018 7; 7; 0.011 0.01···0.025	0.015 0.012; 0.007	0.015 9; 9; 0.006 0.0055···0.025	0.013 0.008; 0.005	Al76, Co80, Do75, Do75a, Do76, Do76a, Lu80 (not included Th > 0.05, U > 0.05 ppm)
Plagioclase, dacites	0.37 6; 6; 0.20 0.16···0.64	0.33 0.25; 0.14	0.014 1; 1; −		Do76, Du71
Plagioclase, gabbros			0.041 7; 13; 0.021 0.0025···0.053	0.037 0.021; 0.013	Do75a, Gi74
Plagioclase, ignimbrites from Italy			0.077 4; 4; 0.002 0.074···0.079	0.077 0.0023; 0.0022	Do75
Plagioclase, metamorphic rocks (amphibolite facies)			0.020 7; 7; 0.006 0.010···0.028	0.019 0.008; 0.006	Do78
Plagioclase, metamorphic rocks (basic granulites)			0.013 13; 13; 0.0068 0.007···0.015	0.012 0.007; 0.005	Do78
Potassium feldspar (igneous rocks from Italy)			0.064 5; 5; 0.003 0.059···0.066	0.064 0.003; 0.003	Do75
Rutile			5···194 6; 9; −		Be74, Do78, Ye77
Vesuvianite (idiocrase)			0.7···4.7 11; −; −		Ha76

Table 9. Th and U abundances in common accessory minerals which concentrate radioactive elements [Ro67, Ro67a].

Mineral	Th [in ppm]	U [in ppm]
Allanite (Orthit)	$10^3 \cdots 2 \cdot 10^4$	$30 \cdots 1000$
Apatite	$50 \cdots 250$	$1 \cdots 200$
Monazite	$2 \cdot 10^4 \cdots 2 \cdot 10^5$	$500 \cdots 3000$
Sphene (Titanit)	$100 \cdots 1000$	$10 \cdots 700$
Zircon	$100 \cdots 10^4$	$10 \cdots 6000$

As shown in Table 8, only small quantities of Th and U are contained in the normal rock forming minerals. The main quantity of Th is to be found in accessory minerals (see Table 9). These minerals bear a large part of the uranium, too. However, the main part of the uranium occurs finely dispersed as submicroscopic, ill-defined microinclusions, loosely bound on grain boundaries, fractures and internal surfaces. It can be easily leached by weak acids [La56a, Ba67a, Sz69a]. This can explain the high mobility of uranium during the processes of metamorphism and anatexis and probably is the main reason for the strong enrichment of uranium in the upper kilometers of the earth's crust, a fact of great geological importance. Truly isomorphically bound uranium occurs only in very few rock forming minerals and only in low concentrations of some ppb (spessartine) or some ppm (andradite, epidote, vesuvianite). They are – without exception – Ca-minerals with the Ca in lattice positions coordinated with 8 O-ions [Ha75]. Feldspar does not contain such positions (only Ca coordinated with 6 O) and is, for this reason, very poor in uranium. Even the value of 0.015 ppm U is probably too high. In the past a continuous revision towards lower values for the isomorphically bound U content of minerals could be observed; this development is still going on.

Th und U werden in die normalen gesteinsbildenden Minerale nur in sehr geringen Mengen eingebaut, wie die Tab. 8 ausweist. Die Hauptmenge des Th befindet sich in akzessorischen Mineralen (siehe Tab. 9). In diesen Mineralen wird auch ein großer Teil des Urans aufgenommen. Der größere Teil des Urans kommt jedoch in fein verteilter Form als submikroskopische Mikroeinschlüsse nicht näher bekannter Art auf Korngrenzen, Rissen und inneren Oberflächen lose gebunden vor und kann durch Behandlung mit schwachen Säuren herausgelöst werden [La56a, Ba67a, Sz69a]. So erklärt sich auch die gute Mobilisierbarkeit des Urans bei Prozessen der Metamorphose und Anatexis. Sie dürfte der Hauptgrund für die starke Anreicherung des Urans in den obersten Kilometern der Erdkruste sein. Damit ist sie von großer geologischer Bedeutung. Echt isomorph eingebaut kommt das Uran nur in wenigen gesteinsbildenden Mineralen in Konzentrationen von einigen ppb (Spessartin), oder einigen ppm (Andradit, Epidot, Vesuvian) vor. Es handelt sich ausnahmslos um Ca-Minerale, in denen 8-fach mit O koordinierte Gitterpositionen des Ca vorhanden sind [Ha75]. Feldspat enthält keine solchen Positionen (nur 6-fach koordiniertes Ca) und ist deshalb sehr uranarm. Der Wert von 0,015 ppm U dürfte wahrscheinlich eher zu hoch sein. In der Vergangenheit war jedenfalls eine ständige Revision der U-Werte von Mineralen nach unten zu beobachten, eine Entwicklung, die noch nicht abgeschlossen ist.

7.1.7 Distribution coefficients Th and U — Verteilungskoeffizienten von Th und U

Distribution coefficients of minerals in equilibrium with melts are very important in order to understand the geochemistry of Th and U, especially for enrichment processes during partial melting of rocks of the earth's mantle and of the crust. The quality of Th and U as incompatible elements – fitting only badly into octahedrally coordinated lattice positions – is clearly illustrated by very low distribution coefficients for silicates. Therefore, partial melting as well as crystallization differentiation cause strong enrichment of these elements in the melt. Some of these distribution coefficients are listed in Table 10; for uranium, they have been determined by the extremely sensible fission track method. The distribution coefficients for

Verteilungskoeffizienten von Mineralen im Gleichgewicht mit Schmelzen sind von großer Bedeutung für das Verständnis der Geochemie von Th und U, insbesondere auch für die Anreicherungsprozesse beim partiellen Aufschmelzen von Gesteinen des Erdmantels und der Kruste. Ganz allgemein ist festzustellen, daß der Charakter von Th und U als inkompatible Elemente, die nur sehr schlecht in 6-fach koordinierte Gitterpositionen eingebaut werden, sehr deutlich durch die immer sehr niedrigen Verteilungskoeffizienten der Silikate belegt wird. Es kommt deshalb beim partiellen Schmelzen ebenso wie bei der Kristallisationsdifferentiation zu starken Anreicherungen in der Schmelze. Die Tab. 10 führt einige dieser Verteilungskoeffizienten

Table 10. Distribution coefficient D of Th and U

$$D_{Th, U} = \frac{\text{concentration(Th, U) in mineral}}{\text{concentration (Th, U) in ground mass}}.$$

Mineral	Rock	D_{Th} n; s min$\cdots$max	D_U n; s min$\cdots$max	Ref.
Apatite	andesite	2.56 1; −	1.96 1; −	Al76
			3.38 1; −	Do76
Biotite	dacite	0.31 1; −		Hi69
	ignimbrites and 1 andesite		0.021 5; 0.002 0.019$\cdots$0.023	Do75, Do76
Clinopyroxene	andesites, basalts, dacites	0.022 6; 0.019 0.0033$\cdots$0.053	0.032 14; 0.025 0.011$\cdots$0.084 GM = 0.023 a = 0.031, b = 0.014	Do75, Do76, Na68a, On68
	experimental melt	0.002 1; −	0.002 1; −	Be80
		0.0065 2; −	0.00151 12; −	Se72
Hornblende	andesite	0.18 4; 0.05 0.14$\cdots$0.25		Lu80
Magnetite	andesite	0.069 1; −	0.063 1; −	Al76
Olivine	basalts		0.0030 4; 0.0007 0.0024$\cdots$0.004	Do75, Do76
	gabbro		0.015 6; 0.004 0.011$\cdots$0.021	He79
Orthopyroxene	2 andesites, 1 ignimbrite		0.0065 3; 0.0005 0.006$\cdots$0.007	Do75, Do76
	andesite	0.074 1; −	0.074 1; −	Al76
Plagioclase	andesites and dacites	0.056 12; 0.018 0.032$\cdots$0.080	0.0029 4; 0.0010 0.002$\cdots$0.004	Al76, Do76, Do76a, Du71, Lu80[1])
	basalts and ignimbrites		0.0080 8; 0.0019 0.0064$\cdots$0.0120	Do75, Do76a
Potassium feldspar	3 ignimbrites, 1 granodiorite		0.0058 4; 0.0006 0.0054$\cdots$0.0064	Do75
Whitlockite	experimental melt	1.2 1; − 2.13 1; −	0.5 1; −	Be80 Se72

[1]) Not included D_{Th} = 0.22, D_U = 0.04.

Haack

olivine and clinopyroxene are especially important for processes during the melting of the earth's mantle. However, as shown in Table 10, unresolved differences between experimentally determined distribution coefficients and those deduced from natural occurrences of the order of the factor 20 exist in the case of clinopyroxene. Generally, great uncertainty still prevails about the distribution coefficients of Th and U.

auf, die im Falle des U meist mit der außerordentlich empfindlichen Spaltspurenmethode bestimmt wurden. Besonders wichtig für Vorgänge beim Aufschmelzen des Erdmantels sind die Verteilungskoeffizienten für Olivin und Klinopyroxen. Wie die Tab. 10 ausweist, bestehen für Klinopyroxen noch Unterschiede um den Faktor 20 zwischen experimentell gewonnenem Verteilungskoeffizienten und an natürlichen Vorkommen bestimmten. Überhaupt herrscht auf dem Gebiet der Verteilungskoeffizienten noch große Unsicherheit.

7.1.8 References for 7.1 — Literatur zu 7.1

Ab59	Abramovich, I.I.: Geochemistry (USSR) (English Transl.) **1959**, 442–450.
Ab74	Abramovich, I.I., Nemtsovich, N.M., Paradeyeva, L.N.: Geochem. Int. **11** (1974) 727–733.
Ad54	Adams, J.A.S., in: H. Faul (ed): Nuclear Geology, New York: John Wiley **1954**, p. 89–98.
Ad55	Adams, J.A.S.: Geochim. Cosmochim. Acta **8** (1955) 74–85.
Ad58	Adams, J.A.S., Richardson, J.E., Templeton, C.C.: Geochim. Cosmochim. Acta **13** (1958) 270–279.
Ad58a	Adams, J.A.S., Weaver, C.E.: Bull. Am. Assoc. Petroleum Geologists **42** (1958) 387–430.
Ad60	Adams, J.A.S., Richardson, K.A.: Econ. Geol. **55** (1960) 1653–1675.
Ad62	Adams, J.A.S., Kline, M.C., Richardson, K.A., Rogers, J.J.W.: Proc. Natl. Acad. Sci. U.S.A. **48** (1962) 1898–1905.
Ah54	Ahrens, L.H., in: H. Faul (ed): Nuclear Geology, New York: John Wiley **1954**, p. 128–133.
Al58	Altschuler, Z.S., Clarke jr., R.S., Jong, E.J.: U. S. Geol. Survey Prof. Paper **314-D** (1958), 90 pp.
Al59	Alekperov, R.A., Efendiev, G.H.: Geochemistry (USSR) (English Transl.) **1959**, 621–627.
Al69	Alekseyev, F.A., Gottikh, R.P., Vorb'yeva, V.Y., Murav'yeva, L.V.: Geochem. Int. **6** (1969) 963–970.
Al69a	Allsopp, H.L., Nicolaysen, L.O.: Earth Planet. Sci. Lett. **5** (1969) 231–244.
Al71	Alexander jr., E.C., Lewis, R.S., Reynolds, J.H., Michel, M.C.: Science **172** (1971) 837–840.
Al76	Allègre, C.J., Condomines, M.: Earth Planet. Sci. Lett. **28** (1976) 395–406.
Al77	Alexander, P.O., Gibson, I.L.: Lithos **10** (1977) 143–147.
Ar72	Arshavskaya, N.F., Berzina, I.G., Lubimova, E.A.: Geothermics **1** (1972) 25–30.
Ar77	Arden, J.W.: Nature (London) **269** (1977) 788–789.
At67	Atrashenok, L.Ya., Krylov, A.Ya.: Geochem. Int. **4** (1967) 825.
At78	Atal, B.S., Bhalla, N.S., Lall, Y., Mahadevan, T.M., Udas, G.R., in: B.F. Windley, S.M. Naqvi (eds): Archean Geochemistry. Developments in Precambrian Geology Vol. 1, Elsevier **1978**, p. 205–220.
At79	Atherton, M.P., Brotherton, M.S.: Chem. Geol. **27** (1979) 329–342.
Au71	Aumento, F., Hyndman, R.D.: Earth Planet. Sci. Lett. **12** (1971) 373–384.
Au71a	Aumento, F.: Earth Planet. Sci. Lett. **11** (1971) 90–94.
Ba56	Baranov, V.I., Ronov, A.B., Kunashova, K.G.: Geochemistry (USSR) (English Transl.) **1956**, 227–235.
Ba56a	Barnes, J.W., Lang, E.J., Potratz, H.A.: Science **124** (1956) 175–176.
Ba57	Baranov, V.I., Kuzmina, L.A.: Geochemistry (USSR) (English Transl.) **1957**, 25–36.
Ba58	Bates, T.F., Strahl, E.O.: Proc. Intern. Conf. Peaceful Uses At. Energy, 2nd, Geneva **1958**, Vol. 2, p. 404–411.
Ba59	Baranov, V.I., Khristianova, L.A.: Geochemistry (USSR) (English Transl.) **1959**, 765–769.
Ba61a	Baranov, V.I., Titayeva, N.A.: Geochemistry (USSR) (English Transl.) **1961**, 121–126.
Ba61b	Baranov, V.I., Du Lieh T'ien: Geochemistry (USSR) (English Transl.) **1961**, 1180–1191.
Ba67	Barberi, F., Borsi, S., Ferrara, G., Innocenti, F.: Mem. Soc. Geol. Ital. **6** (1967) 581–606.
Ba67a	Barbier, J., Carrat, H.G., Ranchin, G.: C.R. Acad. Sci. Paris **D264** (1967) 2436–2439.
Ba68	Baturin, G.N.: Geochem. Int. **5** (1968) 344–348.
Ba70	Baker, I., Ridley, W.I.: Earth Planet. Sci. Lett. **10** (1970) 106–114.
Ba71	Baturin, G.N., Kochenov, A.V., Senin, Yu.M.: Geochem. Int. **8** (1971) 281–284.
Ba73	Baturin, G.N.: Geochem. Int. **10** (1973) 1031–1041.
Ba73a	Baturin, G.N., Lisitsyn, A.P.: Acad. Sci. USSR, Oceanology (English Transl.) **13** (1973) 876–881.
Ba73b	Baturin, G.N., Yemel'yanov, Ye.M.: Acad. Sci. USSR, Oceanology (English Transl.) **13** (1973) 674–679.

Ba73c	Baturin, G.N.: Lithol. Mineral Resources (English Transl.) **8** (1973) 540–549.
Ba76	Baltakmens, T · N.Z, J, Sci. **19** (1976) 375–382.
Ba76a	Baturin, G.N., Lebedev, L.I., Mayev, Ye.G.: Acad. Sci. USSR, Oceanology (English Transl.) **15** (1976) 87–91.
Ba77	Bareja, E.: Kwartal. Geol. (Polska) **21** (1977) 37–49.
Bar70	Barthel, F., Mehnert, K.R.: Neues Jahrb. Mineral. Abhandl. **114** (1970) 18–47.
Be55	Bell, K.G.: U.S. Geol. Surv. Prof. Paper **474-A** (1955) A1–A29.
Be56	Bell, K.G.: U.S. Geol. Surv. Prof. Paper **300** (1956) 381–386.
Be68	Berezina, L.A., Bagdasarov, Yu.A.: Geochem. Int. **5** (1968) 714–721.
Be68a	Becker, V., Bennett, J.H., Manuel, O.K.: Earth Planet. Sci. Lett. **4** (1968) 357–367.
Be69	Bernat, M., Goldberg, E.D.: Earth Planet. Sci. Lett. **5** (1969) 308–312.
Be70	Belluomini, G., Tadeucci, A.: Period. Mineral. **XXXIX** (1970) 387–395.
Be74	Berzina, I.G., Yeliseyeva, O.P., Popenko, D.P.: Int. Geol. Rev. **16** (1974) 1191–1204.
Be74a	Bernat, M., Allègre, C.J.: Earth Planet. Sci. Lett. **21** (1974) 310–314.
Be80	Benjamin, T., Heuser, W.R., Burnett, D.S., Seitz, M.G.: Geochim. Cosmochim. Acta **44** (1980) 1251–1264.
Ben73	Bender, M.L.: Geochim. Cosmochim. Acta **37** (1973) 1229–1247.
Ber70	Bernat, M., Bieri, H., Koide, M., Griffin, J.J., Goldberg, E.D.: Geochim. Cosmochim. Acta **34** (1970) 1053–1071.
Bert70	Bertine, K.K., Chan, L.H., Turekian, K.K.: Geochim. Cosmochim. Acta **34** (1970) 641–648.
Bh69	Bhat, S.G., Krishnaswamy, S.: Indian Acad. Sci. Proc. **A70** (1969) 1–17.
Bh69a	Bhat, S.G., Krishnaswamy, S., Lal, D., Rama, Moore, W.S.: Earth Planet. Sci. Lett. **5** (1969) 483–491.
Bi62	Billings, G.K.: Tex. J. Sci. **14** (1962) 328–351.
Bl64	Bloxam, T.W.: Geochim. Cosmochim. Acta **28** (1964) 1177–1185.
Bl76	Blanchard, D.P., Rhodes, J.M., Dungan, M.A., Rodgers, K.V., Donaldson, C.H., Brannon, J.C., Jacobs, J.W., Gibson, E.K.: J. Geophys. Res. **81** (1976) 4231–4246.
Bo71	Bonatti, E., Fisher, D.E., Joenssuu, O., Rydell, H.S.: Geochim. Cosmochim. Acta **35** (1971) 189–201.
Bo72	Bodu, R., Bonzigues, H., Morin, N., Pfiffelmann, J.-P.: C.R. Acad. Sci. Paris **D275** (1972) 1731–1732.
Bo74	Bougault, H., Hekinian, R.: Earth Planet. Sci. Lett. **24** (1974) 249–261.
Bo75	Boulard, A.P., Condomines, M., Bernat, M., Michard, G., Allègre, C.J.: C.R. Acad. Sci. Paris **D280** (1975) 2425–2428.
Bo76	Boulard, A.P., Michard, G.: Earth Planet. Sci. Lett. **32** (1976) 77–83.
Br65	Broecker, W.S., Thurber, D.L.: Science **149** (1965) 58–60.
Br69	Brimhall, W.H., Adams, J.A.S.: Geochim. Cosmochim. Acta **33** (1969) 1308–1311.
Br70	Bräuer, H.: Freiberg. Forschungsh. **C259** (1970) 83–139.
Br79	Brown, G.C., Plant, J., Lee, M.K.: Nature (London) **280** (1979) 129–130.
Bu61	Butler, A.P.: U.S., Geol. Surv. Prof. Paper **424-B** (1961) 67–69.
Bu73	Bulakh, A.G., Mazalov, A.A., Saturin, A.A., Bakhtiarov, A.V.: Geochem. Int. **10** (1973) 1063–1065.
Bu73a	Bunker, C.M., Bush, C.A.: J. Res. U.S. Geol. Surv. **1** (1973) 289–292.
Bu73b	Bunker, C.M., Bush, C.A., Forbes, R.B.: J. Res. U.S. Geol. Surv. **1** (1973) 659–663.
Bu76	Burwash, R.A., Cumming, G.L.: Can. J. Earth Sci. **13** (1976) 284–293.
Bu77	Burnett, W.C., Veeh, H.H.: Geochim. Cosmochim. Acta **41** (1977) 755–764.
Ca70	Caun, J.C.: Deep Sea Res. **17** (1970) 477.
Ca70a	Caun, J.R.: Earth Planet. Sci. Lett. **10** (1970) 7.
Ca71	Carrat, H.G.: Miner. Deposita **6** (1971) 1–22.
Ca71a	Capaldi, G., Civetta, L., Gasparini, P.: Geochim. Cosmochim. Acta **35** (1971) 1067–1072.
Ca73	Campsie, J., Bailey, J.C. Rasmussen, M., Dittmer, F.: Nature (London) Phys. Sci. **244** (1973) 71.
Ca80	Carswell, D.A., Rice, C.M.: Mineral. Mag. **43** (1980) 689–693.
Ch68	Cherdyntsev, V.V., Zverev, V.L., Kuptsov, V.M., Kislitsina, G.I.: Geochem. Int. **5** (1968) 355–361.
Ch70	Cherdyntsev, V.V., Senina, N.I.: Geochem. Int. **7** (1970) 652–663.
Ch72	Church, T.M., Bernat, M.: Earth Planet. Sci. Lett. **14** (1972) 139–144.
Ch73	Cheminée, J.-L.: Rev. Géogr. Phys. Géol. Dyn. **15** (1973) 353–372.
Ch73a	Church, S.E.: Contrib. Miner. Petrol. **39** (1973) 17–32.

Ch75	Church, S.E., Tatsumoto, M.: Contrib. Miner. Petrol. **53** (1975) 253.
Ch76	Chalov, P.I., Tuzova, T.V., Alekhina, V.M., Merkulova, K.I., Sveilichnaya, N.A.: Izv. Akad. Nauk SSSR, Fiz. Zemli **12** (1977) 77–86.
Ch77	Chapman, D.S., Pollack, H.N.: Tectonophysics **41** (1977) 79–100.
Ci65	Civetta, L., Gasparini, P., Rapolla, A.: Ann. Oss. Vesuviano, Ser. 6, **7** (1965) 77–105.
Ci72	Civetta, L., Gasparini, P.: Int. Symp. Natural Radiation Environment, Vol. 2, **1972**, p. 483–515.
Ci74	Civetta, L., De Fino, M., La Volpe, L., Lirer, L.: Chem. Geol. **13** (1974) 149–162.
Cl73	Clifford, T.N.: Spec. Publ. Geol. Soc. S. Afr. **3** (1973) 17–23.
Co59	Coulomb, R.: Rapport C.E.A. No. 1173, **1959.**
Co61	Cobb, J.C., Kulp, J.L.: Geochim. Cosmochim. Acta **24** (1961) 226–249.
Co68	Compton, W., McDougall, I., Heier, K.S.: Geochim. Cosmochim. Acta **32** (1968) 129–149.
Co76	Coulon, M.: C.R. Acad. Sci. Paris **D283** (1976) 1691–1694.
Co76a	Cowan, G.A., Adler, H.II.: Geochim. Cosmochim. Acta **40** (1976) 1487–1490.
Co77	Coppens, R., Bashir, S., Richard, P.: Miner. Deposita **12** (1977) 189–196.
Co80	Condomines, M., Allègre, C.J.: Nature (London) **288** (1980) 354–357.
Con80	Condie, K.C., Allen, P.: Contrib. Miner. Petrol. **74** (1980) 35–43.
Cu77	Cummings, G.L.: Can. J. Earth Sci. **14** (1977) 768–770.
Da68	Danchin, R.V.: Earth Planet. Sci. Lett. **5** (1968) 41–44.
Da70	Dawson, J.B., Gale, N.H.: Chem. Geol. **6** (1970) 221–231.
De77	Degens, E.T., Khoo, F., Michaelis, W.: Nature (London) **269** (1977) 566–569.
De79	Demaiffe, D., Duchesne, J.C., Hertogen, J., in: L.H. Ahrens (ed): Physics and Chemistry of the Earth, Pergamon Press, Vol. 11, **1979**, p. 361–366.
Di62	Dimitriyev, L.V., Leonova, L.L.: Geochemistry (USSR) (English Transl.) **1962**, 769–777.
Di79	Dixon, T.H., Batiza, R.: Contrib. Miner. Petrol. **70** (1979) 167–181.
Do71	Donnelly, T.W., Rogers, J.J.W., Pushkar, P., Armstrong, R.L.: Geol. Soc. Am. Mem. **130** (1971) 181–224.
Do75	Dostal, J., Capedri, S.: Chem. Geol. **15** (1975) 285–294.
Do75a	Dostal, J., Capedri, S., Aumento, F.: Earth Planet. Sci. Lett. **26** (1975) 345–352.
Do76	Dostal, J., Capedri, S., Dupuy, C.: Lithos **9** (1976) 179–183.
Do76a	Dostal, J., Capedri, S.: Contrib. Miner. Petrol. **54** (1976) 245–254.
Do78	Dostal, J., Capedri, S.: Contrib. Miner. Petrol. **66** (1978) 409–414.
Do78a	Dostal, J., Muecke, G.K.: Earth Planet. Sci. Lett. **40** (1978) 415–422.
Dob76	Dobbs, J.E., Matthews, K.M.: N.Z. J. Sci. **19** (1976) 243–247.
Dr73	Drury, S.A.: Chem. Geol. **11** (1973) 167–188.
Du71	Dudas, M.J., Schmidt, R.A., Harward, M.E.: Earth Planet. Sci. Lett **11** (1971) 440–446.
Du72	Duursma, E.K.: Oceanogr. Mar. Biol. Ann. Rev. **10** (1972) 137–223.
Du75	Dupuy, C., Dostal, J., Capedri, S., Lefevre, C.: Bull. Volcanol. **39** (1975) 363–370.
Du79	Dupuy, C., Leyreloup, A., Vernières, J., in: L.H. Ahrens (ed): Physics and Chemistry of the Earth, Pergamon Press, Vol. 11, **1979**, p. 401–415.
Dy69	Dymond, J.: Earth Planet. Sci. Lett. **6** (1969) 9–14.
Dy75	Dymond, J., Veeh, H.H.: Earth Planet. Sci. Lett. **28** (1975) 13–22.
Ec79	Eckardt, F.J., in: L.H. Ahrens (ed): Physics and Chemistry of the Earth, Pergamon Press, Vol. 11, **1979**, p. 527–532.
Ed70	Edgington, D.N., Callender, E.: Earth Planet. Sci. Lett. **8** (1970) 97–100.
Ed74	Edling, B.: Publications from the Paleontological Institution of the University of Uppsala, Special Volume **2** (1974) p. 118.
Ed78	Edwards, C.L., Reiter, M., Shearer, C., Young, W.: Geol. Soc. Am. Bull. **89** (1978) 1341–1350.
En63	Engel, C.G., Engel, A.E.J.: Science **140** (1963) 1321.
En65	Engel, A.E.J., Engel, C.G., Havens, R.G.: Geol. Soc. Am. Bull. **76** (1965) 719–734.
En65a	Engel, C.G., Fisher, R.L., Engel, A.E.J.: Science **150** (1965) 605–610.
En75	Engel, C.G., Fisher, R.L.: Geol. Soc. Am. Bull. **86** (1975) 1553–1578.
Er63	Erickson, R.L., Blade, L.V.: U.S. Geol. Surv. Prof. Paper **425** (1963).
Er74	Erlank, A.E., Reid, D.L.: Initial Reports DSDP **25** (1974) 543–551.
Ev74	Evans, T.R., Tammemagi, H.Y.: Earth Planet. Sci. Lett. **23** (1974) 349–356.
Ew68a	Ewart, A., Stipp, J.J.: Geochim. Cosmochim. Acta **32** (1968) 699–736.
Ew68b	Ewart, A., Taylor, S.R., Capp, A.C.: Contrib. Miner. Petrol. **17** (1968) 116–140.
Ew68c	Ewart, A., Taylor, S.R., Capp, A.C.: Contrib. Miner. Petrol. **18** (1968) 76–104.

Fa65	Fanale, F.P., Schaeffer, O.A.: Science **149** (1965) 312–317.
Fa68	Fahrig, W.F., Eade, K.F.: Can. J. Earth Sci. **5** (1968) 1247–1252.
Fa76	Farquharson, R.B.: Can. J. Earth Sci. **13** (1976) 993–997.
Fa78	Falkum, T., Rose-Hansen, J.: Chem. Geol. **23** (1978) 73–86.
Fe74	Feld'man, V.I., Stupnikova, N.I., Kovalenko, V.I.: Geochem. Int. **11** (1974) 661–666.
Fe75	Ferrara, G., Macera, P., Valentinetti, R.: Rend. Soc. Ital. Mineral. Petrol. **31** (1975) 209–219.
Fe75a	Fesq, H.W., Kable, E.J.D., Gurney, J.J., in: L.H. Ahrens (ed): Physics and Chemistry of the Earth, Pergamon Press, Vol. 9, **1975**, p. 687–707.
Fe75b	Ferreira, M.P., Macedo, R., Costa, V., Reynolds, J.H., Riley, J.E., Rowe, H.W.: Earth Planet. Sci. Lett. **25** (1975) 142–150.
Fi59	Filippov, M.S., Komlev, L.V.: Geochemistry (USSR) (English Transl.) **1959**, 535–549.
Fi70	Fisher, D.E.: Geochim. Cosmochim. Acta **34** (1970) 630–634.
Fi79	Fisher, D.E.: Geochim. Cosmochim. Acta **43** (1979) 709–716.
Fl77	Flower, M.F.J., Robinson, P.T., Schmincke, H.U., Ohnmacht, W.: Contrib. Miner. Petrol. **64** (1977) 167–195.
Fo75	Forgac, J.: Geol. Zb. (Bratislava) **26** (1975) 309–322.
Fo77	Fodor, R.V., Husler, J.W., Keil, K.: Initial Reports DSDP **39** (1977) 513–523.
Fr74	Frey, F.A.: J. Geophys. Res. **79** (1974) 5507–5527.
Fr78	Frey, F.A., Green, D.H., Roy, S.D.: J. Petrol. **19** (1978) 463–513.
Ga63	Gasparini, P.: Ann. Oss. Vesuviano, Ser. 6, **5** (1963) 185–218.
Ga66	Gasparini, P., Lirer, L., Luongo, G.: Ann. Oss. Vesuviano, Ser. 6, **8** (1966) 37–52.
Ga72	Gast, P.W., in: E.C. Robertson (ed): The Nature of the Solid Earth, McGraw Hill, Inc. **1972**, p. 19–40.
Ga72a	Garrels, R.M., Mackenzie, F.T., Silver, R., in: E.C. Robertson (ed): The Nature of the Solid Earth, McGraw Hill, Inc., **1972**, p. 93–124.
Ga75	Gavrilenko, V.A.: Geochem. Int. **12** (1975) 53–61.
Ga79	Gasparini, P., Mantovani, M.S.M.: Earth Planet. Sci. Lett. **42** (1979) 311–320.
Ge56	Gerasimovskii, V.I.: Geochemistry (USSR) (English Transl.) **1956**, 494–510.
Ge71	Gerling, E.K., Mamyrin, B.A., Tolstikhin, I.N., Yakoleva, S.S.: Geochem. Int. **8** (1971) 755–762.
Ge73	Geisler, F.H., Philips, P.R., Walker, R.M.: Nature (London) **244** (1973) 428–429.
Ge79	Gerasimovsky, V.I., in: L.H. Ahrens (ed): Physics and Chemistry of the Earth, Pergamon Press, **1979**, Vol. 11, p. 361–366.
Gh76	Ghose, N.C.: Lithos **9** (1976) 65–73.
Gi74	Gijbels, R.H., Millard jr., H.T., Desborough, G.A., Bartel, A.J.: Geochim. Cosmochim. Acta **38** (1974) 319–337.
Gl71	Gladkikh, V.S., Lebedev-Zinov'ev, A.A.: Geochem. Int. **8** (1971) 813–820.
Gl72	Glasby, G.P.: Mar. Chem. **1** (1972) 105–125.
Gl76	Glikson, A.Y.: Geochim. Cosmochim. Acta **40** (1976) 1261–1280.
Go54	Goldschmidt, V.M.: Geochemistry, Oxford: Clarendon Press **1954**.
Go62a	Gottfried, D., Moore, R., Campbell, E.: U.S. Geol. Surv. Prof. Paper **450-E** (1962) 85–89.
Go62b	Gottfried, D., Moore R., Caemmerer, A.: U.S. Geol. Surv. Prof. Paper **450** (1962/1963) B70–B72.
Go63	Goldberg, E.D., Koide, M., in: J. Geiss, E.D. Goldberg (eds): Earth Science and Meteoritics, Amsterdam: North Holland Publ. Comp. **1963**, p. 90–102.
Gr68	Green, D.H., Morgan, J.W., Hue, K.S.: Earth Planet. Sci. Lett. **4** (1968) 155–166.
Gr69	Green, T.H., Brunfeld, A.O., Heier, K.S.: Earth Planet. Sci. Lett. **7** (1969) 93–98.
Gr71	Grimm, W., Herrmann, G., Schüssler, H.D.: Phys. Rev. Lett. **26** (1971) 1040–1043.
Gr72	Gray, C.M., Oversby, V.M.: Geochim. Cosmochim. Acta **36** (1972) 939–952.
Gr77	Gray, C.M.: Contrib. Miner. Petrol. **65** (1977) 79–89.
Gu73	Gurney, J.J., Hobbs, J.B.M., in: Int. Conf. Kimberlites, Extended Abstr. Pap: Cape Town **1973**, p. 143–146.
Gv73	Gvirtzman, G., Friedman, G.M., Miller, D.S.: J. Sediment. Petrol. **43** (1973) 985–997.
Ha56	Hail jr., W.J., Myers, A.T., How, C.A.: U.S. Geol. Surv. Prof. Papers **300** (1956) 521–526.
Ha65	Hamilton, W., Mountjoy, W.: Geochim. Cosmochim. Acta **29** (1965) 661–671.
Ha66	Hamilton, E.I.: Earth Planet. Sci. Lett. **1** (1966) 317–318.
Ha70	Hart, R.: Earth Planet. Sci. Lett. **9** (1970) 269–279.
Ha70a	Hansen, R.O., Begg, E.L.: Earth Planet. Sci. Lett. **8** (1970) 411–419.

Ha72	Haack, U.K., Gramse, M.: Contrib. Miner. Petrol. **34** (1972) 258–260.
Ha73	Haines, E.L., Zartmann, R.E.: Earth Planet. Sci. Lett. **20** (1973) 45–53.
Ha73a	Haack, U.: Naturwissenschaften **60** (1973) 65–70.
Ha75	Haack, U.K.: Habilitationsschrift, Universität Göttingen **1975**.
Ha75a	Hamilton, E.I.: Chem. Geol. **16** (1975) 221–231.
Ha76	Haack, U.: Neues Jahrb. Miner. Abhandl. **129** (1976) 160–170.
Ha81	Haack, U., Gohn, E., Hartmann, O.: Spec. Publ. Geol. Soc. S. Afr. (1982) in press.
Ha81a	Haack, U., Hoefs, J., Gohn, E.: Final Report Sonderforschungsbereich 48, Universität Göttingen; Stuttgart: Schweizerbart **1982** (in press).
He58	Hecht, F., Kupper, H., Petrascheck, W.E.: Proc. Intern. Conf. Peaceful Uses At. Energy, 2nd., Vol. 2, **1958**, p. 158–160.
He59	Heide, F., Proft, G.: Naturwissenschaften **46** (1959) 352–353.
He62	Heier, K.S.: Nor. Geol. Tidsskr. **42** (1962) 287–392.
He63	Heier, K.S., Rodgers, J.J.W.: Geochim. Cosmochim. Acta **27** (1963) 137–154.
He63a	Heier, K.S.: Geochim. Cosmochim. Acta **27** (1963) 849–860.
He64	Heier, K.S., McDougall, I., Adams, J.A.S.: Nature (London) **201** (1964) 254–256.
He65	Heier, K.S., Compston, W., McDougall, I.: Geochim. Cosmochim. Acta **29** (1965) 643–659.
He66	Heier, K.S., Rhodes, J.M.: Econ. Geol. **61** (1966) 563–571.
He68	Hekinian, R.: Deep Sea Res. **15** (1968) 195–213.
He71	Heier, K.S., Thoresen, K.: Geochim. Cosmochim. Acta **35** (1971) 89–99.
He75	Henderson, P.: Mineral. Mag. **40** (1975) 285–291.
He79	Henderson, P., Williams, C.T., in: L.H. Ahrens (ed): Physics and Chemistry of the Earth, Pergamon Press, Vol. 11, **1979**, p. 191–198.
He79a	Herrmann, G.: Nature (London) **280** (1979) 543–549.
Hen71	Henderson, P., Mackinnon, A., Gale, N.H.: Geochim. Cosmochim. Acta **35** (1971) 917–925.
Hi69	Higuchi, H., Nagasawa, H.: Earth Planet. Sci. Lett. **7** (1969) 281–287.
Ho53	Holyk, W., Ahrens, L.H.: Geochim. Cosmochim. Acta **4** (1953) 241–250.
Ho68	Hoefs, J., Wedepohl, K.H.: Contrib. Miner. Petrol. **19** (1968) 328–338.
Ho71	Hoffman, D.C., Lawrence, F.O., Mewherter, J.L., Rourke, F.M.: Nature (London) **234** (1971) 132–134.
Ho75	Höhndorf, A.: Schweiz. Mineral. Petrogr. Mitt. **55** (1975) 89–102.
Ho75a	Howorth, R., Rankin, P.C.: Chem. Geol. **15** (1975) 239–250.
Hy56	Hyden, H.J.: U.S. Geol. Surv. Prof. Paper **300** (1956) 511–515.
Hy68	Hyndman, R.D., Lambert, I.B., Heier, K.S., Jaeger, J.C., Ringwood, A.E.: Phys. Earth Planet. Interiors **1** (1968) 129–135.
Im76	Immel, R., Osmond, J.K.: Chem. Geol. **18** (1976) 263–272.
Is69	Ishizaka, K., Yamaguchi, M.: Earth Planet. Sci. Lett. **6** (1969) 179–185.
IUGS77	Steiger, R.H., Jäger, E.: Earth Planet. Sci. Lett. **36** (1977) 359–362.
IUGS78	Streckeisen, A.: IUGS subcommission on the systematics of igneous rocks: Classification and nomenclature of volcanic rocks, lamprophyres, carbonatites and melilitic rocks. Neues Jahrb. Miner. Abhandl. **134** (1978) 1–14.
Ja70	Jakeš, P., White, A.J.R.: Geochim. Cosmochim. Acta **34** (1970) 849–856.
Ja71	Jakeš, P., White, A.J.R.: Earth Planet. Sci. Lett. **12** (1971) 224–230.
Ja72	Jakeš, P., White, A.J.R.: Bull. Geol. Soc. Am. **83** (1972) 29–40.
Je78	Jessop, A.M., Lewis, T.: Tectonophysics **50** (1978) 55–77.
Jo76	Joshi, L.V., Ganguly, A.K.: J. Radioanal. Chem. **34** (1976) 299–308.
Jo77	Joshi, L.V., Ganguly, A.K.: Proc. Indian Acad. Sci. **85 B** (1977) 173–256.
Ju55	Judson, S., Osmond, J.K.: Am. J. Sci. **253** (1955) 104–116.
Ju57	Jurain, G.: C.R. Acad. Sci. Paris **245** (1957) 1071–1074.
Ka69	Kaufmann, A.: Geochim. Cosmochim. Acta **33** (1969) 717–724.
Ka70	Kable, E.J.D., Erlank, A.J., Cherry, R.D.: cited in I. Baker et al.: Earth Planet. Sci. Lett **10** (1970) 106–114.
Ka70a	Kay, R., Hubbard, N.J., Gast, P.W.: J. Geophys. Res. **75** (1970) 1585–1613.
Ka71	Kaufmann, A., Broecker, W.S., Ku. T.L., Thurber, D.L.: Geochim. Cosmochim. Acta **35** (1971) 1155–1183.

Ka75	Kanaya, H., Katada, M.: J. Japan. Assoc. Mineralogists Petrologists Econ. Geologists **70** (1975) 286–294.
Ka78	Kanegaonkar, N.B., Powar, K.B.: Neues Jahrb. Mineral. Monatsh. **1978**, 506–511.
Ke76	Keays, R.R., Scott, R.B.: Econ. Geol. **71** (1976) 705–720.
Kh64	Kharitonova, R.Sh.: Geochem. Int. **1** (1964) 816–821.
Kh69	Kharkar, D.P., Turekian, K.K., Scott, M.R.: Earth Planet. Sci. Lett. **6** (1969) 61–68.
Ki75	Killeen, P.G., Heier, K.S.: Nor. Vidensk. Akad. I. Mat.-Naturv. Klasse, Skr. N. Ser. No **35** (1975) 1–32.
Ki78	Kissling, E., Labhart, T.P., Rybach, L.: Schweiz. Mineral. Petrogr. Mitt. **58** (1978) 357–388.
Kl56	Klepper, M.R., Wyant, D.G.: U.S. Geol. Surv. Prof. Pap. **300** (1956) 17–25.
Kl69	Kleemann, J.-D., Green, D.H., Lovering, J.F.: Earth Planet. Sci. Lett. **5** (1969) 449–458.
Kl74	Klerkx, J., Deutsch, S., Hertogen, J., De Winter, J., Gijbels, R., Pichler, H.: Earth Planet. Sci. Lett. **23** (1974) 297–303.
Kn78	Knauss, K.G., Ku, T.L., Moore, W.S.: Earth Planet. Sci. Lett. **39** (1978) 235–249.
Ko57	Koczy, F.F., Tomic, E., Hecht, F.: Geochim. Cosmochim. Acta **11** (1957) 86–102.
Ko66	Komarov, A.N., Shukolyukov, Ya.A.: Geochem. Int. **3** (1966) 1065.
Ko66a	Kolbe, P., Taylor, S.R.: Contrib. Miner. Petrol. **12** (1966) 202–222.
Ko67	Komarov, A.N., Shukolyukov, Ya.A., Skovorodkin, N.V.: Geochem. Int. **4** (1967) 647–659.
Ko69	Komarov, A.N., Skovorodkin, N.V.: Geochem. Int. **6** (1969) 127–133.
Ko70	Kolodny, Y., Kaplan, I.R.: Geochim. Cosmochim. Acta **34** (1970) 3–24.
Ko71	Kovalev, V.P., Malyasova, Z.V.: Geochem. Int. **8** (1971) 541–549.
Ko73	Koide, M., Bruland, K.W., Goldberg, E.D.: Geochim. Cosmochim. Acta **37** (1973) 1171–1187.
Ko74	Komarov, A.N., Zhitkov, A.S.: Int. Geol. Rev. **16** (1974) 971–977.
Ko75	Komarov, A.N., Ilupin, I.P.: Dokl. Akad. Nauk SSSR **222** (1975) 1210–1212; Dokl. Earth Sci. Sect. (English Transl.) **222** (1975) 243–245.
Ko75a	Kozlov, A.A.: Geochem. Int. **12** (1975) 136–138.
Ko78	Kohler, E.E., Haeussler, H.: Geol. Jahrb. **A 46** (1978) 69–91.
Kr58	Krylov, A.Ya.: Geochemistry (USSR) (English Transl.) **1958**, 240–247.
Kr77	Kramers, J.D.: Earth Planet. Sci. Lett. **34** (1977) 419–431.
Kr79	Kramers, J.D.: Earth Planet. Sci. Lett. **42** (1979) 58–70.
Ku65	Ku, T.L.: J. Geophys. Res. **70** (1965) 3457–3474.
Ku67	Ku, T.L., Broecker, W.S.: Earth Planet. Sci. Lett. **2** (1967) 317–320.
Ku68a	Kuznetsov, Yu.V., Simonyak, Z.N., Lisityn, A.P., Frenklikh, M.S.: Geochem. Int. **5** (1968) 169–177.
Ku68b	Kuznetsov, Yu.A., Simonyak, Z.N., Lisityn, A.P., Frenklikh, M.S.: Geochem. Int. **5** (1968) 306–313.
Ku68c	Ku, T.L., Broecker, W.S., Opdyke, N.: Earth Planet. Sci. Lett. **4** (1968) 1–16.
Ku68d	Ku, T.L.: J. Geophys. Res. **73** (1968) 2271–2276.
Ku69	Ku, T.L., Broecker, W.S.: Deep Sea Res. **16** (1969) 625–637.
Ku74	Kurat, G., Niedermayr, G., Korkisch, J., Seemann, R.: Carinthia 2, **84** (1974) 87–98.
Ku76	Kunzendorf, H., Friedrich, G.H.: Trans. Inst. Min. Metall. Sect. B **85** (1976) 284–288.
Ku80	Kurat, G., Palme, H., Spettel, B., Baddenhausen, H., Hofmeister, H., Palme, C., Wänke, H.: Geochim. Cosmochim. Acta **44** (1980) 45–60.
La56	Larsen, E.S., Phair, G., Gottfried, D., Smith, W.L.: U.S. Geol. Surv. Prof. Paper **300** (1956) 65–74.
La56a	Larsen jr., E.S., Phair, G., Gottfried, D., Smith, W.S.: Int. Conf. Peaceful Uses At. Energy Vol. 6, **1956**, 240–247.
La58	Larsen, E.S., Gottfried, D., Mollog, M.: Proc. Conf. Peaceful Uses At. Energy, 2nd. Vol. 2, **1958**, 509–514.
La60	Larsen, E.S., Gottfried, D.: Am. J. Sci. **258-A** (1960) 151–169.
La61	Larsen, E.S., Gottfried, D.: U.S. Geol. Surv. Bull. **1070-C** (1961) 63–103.
La67	Larionov, V.V., Shvartsman, M.D.: Geochem. Int. **4** (1967) 151–157.
La68	Lambert, I.B., Heier, K.S.: Chem. Geol. **3** (1968) 233–238.
La68a	Lachenbruch, A.H.: J. Geophys. Res. **73** (1968) 6977–6989.
La71	Labhart, T.P., Rybach, L.: Chem. Geol. **7** (1971) 237–251.
La74	Lancelot, J.R., Allègre, C.J.: Earth Planet. Sci. Lett. **22** (1974) 233–238.
La76	Labhart, T.P., Rybach, L.: Schweiz. Mineral. Petrogr. Mitt. **56** (1976) 669–673.

La77	Langmuir, C.H., Bender, J.F., Bence, A.E., Hanson, G.N.: Earth Planet. Sci. Lett. **36** (1977) 133–156.
Lal76	Lal, N., Nagpaul, K.K., Sharma, K.K.: Geol. Soc. Am. Bull. **87** (1976) 687–690.
Lam68	Lambert, L.B., Heier, K.S.: Lithos **1** (1968) 30–53.
Lam76	Lambert, R.St. John, Holland, J.G., in: B.F. Windley (ed): The early history of the earth, John Wiley **1976**, p. 191–201.
Lam77	Lambert, R.St. John, Holland, J.G.: Can. J. Earth Sci. **14** (1977) 809–836.
Le61	Leonova, L.L., Gavrilin, R.D., Bagreyev, V.V.: Geochemistry (USSR) (English Trans.) **1961**, 1173–1179.
Le63	Leonova, L.L., Balashov, Yu.A.: Geochemistry (USSR) (English Transl.) **1963**, 1047–1055.
Le64a	Leonova, L.L.: Geochem. Int. **1** (1964) 925–929.
Le64b	Leonova, L.L., Renne, O.S.: Geochem. Int. **1** (1964) 775–781.
Le76	Lewis, T.S.: Can. J. Earth Sci. **13** (1976) 1634–1642.
Le79	Leeman, W.P.: Nature (London) **281** (1979) 365–366.
Li60	Lisitin, A.K.: Geochemistry (USSR) (English Transl.) **1960**, 761–768.
Li73	Lipman, P.W., Bunker, C.M., Bush, C.A.: J. Res. U.S. Geol. Surv. **1** (1973) 387–401.
Lo63	Lovering, J.F., Morgan, J.W.: Nature (London) **197** (1963) 138–140.
Lo63a	Lovering, J.F., Morgan, J.W.: Nature (London) **199** (1963) 479–480.
Lo67	Lopatkina, A.P.: Geochem. Int. **4** (1967) 577–588.
Lo68	Lovering, J.F., Tatsumoto, M.: Earth Planet. Sci. Lett. **4** (1968) 350–356.
Lo74	López-Escobar, L., Oyarzún, J.M.: Pac. Geol. **8** (1974) 47–50.
Lo76	Lo, H.H., Goldes, G.G.: Lithos **9** (1976) 149–159.
Lo79	López-Escobar, L., Frey, F.A., Oyarzún, J.: Contrib. Miner. Petrol. **70** (1979) 439–450.
Lu63	Luongo, G.: Ann. Oss. Vesuviano, Ser. 6, **5** (1963) 219–228.
Lu64	Luongo, G., Rapolla, A.: Ann. Oss. Vesuviano, Ser. 6, **6** (1964) 45–66.
Lu65	Luongo, G., Rapolla, A.: Ann. Oss. Vesuviano, Ser. 6, **7** (1965) 215–225.
Lu73	Lutts, B.G., Mineyeva, I.G.: Geochem. Int. **10** (1973) 1278–1281.
Lu78	Ludwig, K.R., Stuckless, J.S.: Contrib. Miner. Petrol. **65** (1978) 243–254.
Lu80	Luhr, J.F., Carmichael, I.S.E.: Contrib. Miner. Petrol. **71** (1980) 343–372.
Luo65	Luongo, G., Rapolla, A.: Ann. Oss. Vesuviano, Ser. 6, **7** (1965) 58–76.
Ly61	Lyons, J.B.: U.S. Geol. Surv. Prof. Paper **424-B** (1961) 69–71.
Ma56	Mapel, W.J.: U.S. Geol. Surv. Prof. Paper **300** (1956) 469–476.
Ma64	Macdonald, G.A., Katsura, T.: J. Petrol. **5** (1964) 82–133.
Ma67	Manson, V., in: H.H. Hess, A. Poldervaart: The Poldervaart treatise on rocks of basaltic composition, Vol. 1, **1967**, p. 215–269.
Ma69	Manton, W.I., Tatsumoto, M.: EOS, Trans. Am. Geophys. Union **50** (1969) 343.
Ma71	Manton, W.I., Tatsumoto, M.: Earth Planet. Sci. Lett. **10** (1971) 217–226.
Ma71a	Magarovskiy, V.V., Mel'nichenko, A.K., Kozyrev, V.I.: Geochem. Int. **8** (1971) 259–267.
Ma75	Malystev, V.I., Melkov, V.G., Yakubovich, A.L., Sokolova, Z.A., Sumin, L.V., Shiryayeva, M.B., Kotsen, M.Ye., Salmin, Yu.P., Kharlamov, V.I., Dunayev, V.V., Marmilova, L.I.: Dokl. Akad. Nauk SSSR **223** (1975) 212–214.
Ma77	MacDougall, J.D.: Earth Planet. Sci. Lett. **35** (1977) 65–70.
Ma78	Mangini, A.: Meteor-Forschungsergeb., Reihe C, Nr. 29, **1978**, 1–5.
Ma79	Mason, B.: U.S. Geol. Surv. Prof. Paper **440-B-1**, (1979).
Ma79a	Manhès, G., Allègre, C.J., Dupré, B., Hamelin, B.: Earth Planet. Sci. Lett. **44** (1979) 91–104.
Mah71	Mahfouz, S.: Chem. Erde **29** (1971) 347.
McK56	McKelvey, V.E., Carswell, L.D.: U.S. Geol. Surv. Prof. Paper **300** (1956) 483–487.
McL80	McLennan, S.M., Taylor, S.R.: Nature (London) **285** (1980) 621–624.
McN79	McNutt, R.H., Clark, A.H., Zentilli, M.: Econ. Geol. **74** (1979) 827–837.
McN81	McNeal, J.M., Lee, D.E., Millard jr., H.T.: J. Geochem. Explor. **14** (1981) 25–40.
Me68	Melson, W.G., Thompson, G., van Andel, T.H.: J. Geophys. Res. **73** (1968) 5925–5941.
Me71	Melson, W.G., Thompson, G.: Phil. Trans. R. Soc. London **A268** (1971) 423–441.
Me74	Meier, H., Bösche, D., Zeitler, G., Albrecht, W., Hecker, W., Menge, D., Unger, E., Zimmerhackl, E.: Radiochim. Acta **21** (1974) 110–116.
Me80	Mengel, K., Pourmoafi, M.: Neues Jahrb. Miner. Monatsh. **1980**, 66–74.
Mi69	Miyashiro, A., Shido, F., Ewing, M.: Contrib. Miner. Petrol. **23** (1969) 38–52.
Mi72	Miyake, Y., Sugimura, Y., Hirao, Y.: Int. Symp. Natural Radiation Environment **2** (1972) 535–558.

Mi73	Miyake, Y., Sugimura, Y., Yasujima, T.: Papers Metereolog. Geophys. **24** (1973) 67–73.
Mi74	Mitchell, W.S., Aumento, F. J. Geophys. Res. **79** (1974) 5529–5532.
Mi76	Miller, T.P., Bunker, C.M.: J. Res. U.S. Geol. Surv. **4** (1976) 367–377.
Mi77	Mitchell, W.S., Aumento, F.: Can. J. Earth Sci. **14** (1977) 794–808.
Mo66	Morgan, J.W., Heier, K.S.: Earth Planet. Sci. Lett. **1** (1966) 158–160.
Mo66a	Morgan, J.W., Goode, A.D.T.: Earth Planet. Sci. Lett. **1** (1966) 110–112.
Mo67	Moore, W.S.: Earth Planet. Sci. Lett. **2** (1967) 231–234.
Mo69	Moorbath, S., Welke, H., Gale, N.H.: Earth Planet. Sci. Lett. **6** (1969) 245–256.
Mo69a	Moorbath, S., Welke, H.: Earth Planet. Sci. Lett. **5** (1969) 217–230.
Mo69b	Morgan, J.W.: Earth Planet. Sci. Lett. **7** (1969) 53–63.
Mo71	Mogarovskiy, V.V., Mel'nichenko, A.K., Kozyrev, V.I.: Geochem. Int. **8** (1971) 259–267.
Mo71a	Morgan, J.W., Lovering, J.F., in: A.O. Brunfelt, E. Steinnes (eds): Activation Analysis in Geochemistry and Cosmochemistry. Proc. NATO Advanced Study Institute. Oslo: Universitetsforlaget **1971**, 445–454.
Mo73	Mo, T, Suttle, A.D., Sackett, W.M.: Geochim. Cosmochim. Acta **37** (1973) 35–51.
Mo75	Morioka, M., Kigoshi, K.: Earth Planet. Sci. Lett. **25** (1975) 116–120.
Mo79	Montgomery, C.W.: Contrib. Miner. Petrol. **69** (1979) 167–176.
Mu58	Murray, E.G., Adams, J.A.S.: Geochim. Cosmochim. Acta **13** (1958) 260–269.
Mu64	Muir, I.D., Tilley, C.E., Scoon, J.H.: J. Petrol. **5** (1964) 409–434.
Mu66	Muir, I.D., Tilley, C.E., Scoon, J.H.: J. Petrol. **7** (1966) 193–201.
Na66	Naumov, G.B., Polyakov, A.I., Sergeyev-Bobr, A.A.: Geochem. Int. **3** (1966) 1066–1074.
Na68	Naydenov, B.M., Yefimov, I.A.: Geochem. Int. **5** (1968) 504–510.
Na68a	Nagasawa, H., Wakita, H.: Geochim. Cosmochim. Acta **32** (1968) 917–921.
Na69	Narayamaswamy, R., Venkatasubramanian, V.S.: Geochim. Cosmochim. Acta **33** (1969) 1007–1009.
Na76	Nathan, Y., Shiloni, Y.: Proc. Symp. Exploration Uranium Ores, Vienna **1976**, pp. 645–655, IAEA (1976).
Nan76	Nance, W.B., Taylor, S.R.: Geochim. Cosmochim. Acta **40** (1976) 1539–1551.
Ne72	Neuilly, M., Bussac, J., Frèjaques, C., Nief, G., Vendryes, G., Yvon, J.: C.R. Acad. Sci. Paris **275-D** (1972) 1847–1853.
Ni72	Nishimura, S.: Chem. Geol. **10** (1972) 211–221.
Ni76	Nicolli, H.B., Lucero-Michaut, H.N., Gamba, M.A.: Bol. Acad. Nac. Cienc. Argent. **51** (1976) 225–242.
Ni77	Nikolaev, D.S., Lazarev, K.F., Dozhzhin, V.M.: Geochem. Int. **14** (1977) 141–146.
Ni80	Nishimura, S., Ikeda, T., Ishizaka, K.: Phys. Geol. Indonesian Island Arcs **1980**, 109–113.
No80	Noble, D.C., Rose jr., W. I., Zielinski, R.A.: Econ. Geol. **75** (1980) 127–130.
Nu74	Nunes, P.D., Steiger, R.H.: Contrib. Miner. Petrol. **47** (1974) 255–280.
Og74	Ogloblin, K.F., Khalifa-Zade, Ch.M.: Geochem. Int. **11** (1974) 239–244.
On68	Onuma, N., Higuchi, H., Wakita, H., Nagasawa, H.: Earth Planet. Sci. Lett. **5** (1968) 47–51.
On72	Ondra, P.: Cas. Mineral. Geol. **17** (1972) 413–428.
On73	Ondra, P.: Geol. Zb. (Bratislava) **24** (1973) 315–324.
Or78	Ormaasen, D.E., Raade, G.: Earth Planet. Sci. Lett. **39** (1978) 145–150.
Ov68	Oversby, V.M., Gast, P.W.: Earth Planet. Sci. Lett. **5** (1968) 199–206.
Ov70	Oversby, V.M., Gast, P.W.: J. Geophys. Res. **75** (1970) 2097–2114.
Ov72	Oversby, V.M.: Geochim. Cosmochim. Acta **36** (1972) 1167–1179.
Pa66	Park, R.G.: Scott. J. Geol. **2** (1966) 179–199.
Pa70	Paul, D.K., Hutchinson, R.: Geochim. Cosmochim. Acta **34** (1970) 1249–1251.
Pa76	Paul, D.K., Potts, P.J.: Chem. Geol. **18** (1976) 161–162.
Pa77	Paul, D.K., Gale, N.H., Harris, P.G.: Geochim. Cosmochim. Acta **41** (1977) 335–339.
Pa81	Pattenden, N.J., Cambray, R.S., Playford, K.: Geochim. Cosmochim. Acta **45** (1981) 93–100.
Pe80	Pe-Piper, G.: Contrib. Miner. Petrol. **72** (1980) 387–396.
Pl62	Pliler, R., Adams, J.A.S.: Geochim. Cosmochim. Acta **26** (1962) 1137–1146.
Pl62a	Pliler, R., Adams, J.A.S.: Geochim. Cosmochim. Acta **26** (1962) 1115–1135.
Pl71b	Pluman, I.I.: Geochem. Int. **8** (1971) 716–721.
Pl75	Pluman, I.I.: Geochem. Int. **12** (1975) 97–107.

Po55	Ponsford, D.R.A.: Bull. Geol. Surv. G.B. **10** (1955) 24–44.
Po64	Polyakov, A.I., Kot, G.A.: Geochem. Int. **1** (1964) 479–488.
Po71	Polyakov, A.I., Sobornov, O.P.: Geochem. Int. **8** (1971) 697–707.
Po75	Polyakov, A.I., Sobornov, O.P.: Geochem. Int. **12** (1975) 20–27.
Pr73	Price, R.C., Taylor, S.R.: Contrib. Miner. Petrol. **40** (1973) 195–205.
Pr79	Prévôt, L., Lucas, J., Nathan, Y., Shiloni, Y., in: L.H. Ahrens (ed): Physics and Chemistry of the Earth, Pergamon Press **1979**, Vol. 11, p. 293–304.
Pr80	Price, R.C., Taylor, S.R.: Contrib. Miner. Petrol. **72** (1980) 1–18.
Pu72	Putinstsev, V.K., Ditmar, G.V., Maksimovskiy, V.A., Selivanov, V.A.: Geochem. Int. **9** (1972) 583–588.
Ra67	Ragland, P.C., Billings, G.K., Adams, J.A.S.: Geochim. Cosmochim. Acta **31** (1967) 17–33.
Ra76	Rao, R.U.M., Rao, G.V., Narain, H.: Earth Planet. Sci. Lett. **30** (1976) 57–64.
Ra79	Raade, G., cited in: K.S. Heier, Phil. Trans. R. Soc. London **A291** (1979) 413–421.
Ri76	Richardson, S.W., Powell, R.: Scott. J. Geol. **12** (1976) 237–268.
Ro52	Rona, E., Urry, W.D.: Am. J. Sci. **250** (1952) 241–262.
Ro56	Rona, E., Gilpatrick, L.O., Jeffrey, L.M.: Trans. Am. Geophys. Union **37** (1956) 697–701.
Ro61	Rogers, J.J.W., Ragland, P.C.: Geochim. Cosmochim. Acta **25** (1961) 99–109.
Ro61a	Rosholt, J.N., Emiliani, C., Geiss, J., Koczy, F.F., Wangersby, P.J.: J. Geol. **69** (1961) 162–185.
Ro64	Rogers, J.J.W., Richardson, K.A.: Geochim. Cosmochim. Acta **28** (1964) 2005–2011.
Ro65	Rogers, J.J.W., Adams, J.A.S., Gatlin, B.: Am. J. Sci. **263** (1965) 817–822.
Ro66	Rosholt, J.N., Doe, B.R., Tatsumoto, M.: Bull. Geol. Soc. Am. **77** (1966) 987–1004.
Ro67	Rogers, J.J.W., Adams, J.A.S., in: K.H. Wedepohl (ed): Handbook of Geochemistry, Berlin-Heidelberg-New York: Springer **1967**, II-5, 90-C-1 ··· 90-C-3.
Ro67a	Rogers, J.J.W., Adams, J.A.S., in: K.H. Wedepohl (ed): Handbook of Geochemistry, Berlin-Heidelberg-New York: Springer **1967**, II-5, 92-C-1 ··· 92-C-4.
Ro68	Roy, R.F., Blackwell, D.D., Birch, F.: Earth Planet. Sci. Lett. **5** (1968) 1–12.
Ro68a	Ronov, A.B.: Sedimentology **10** (1968) 25–43.
Ro69	Rosholt, J.N., Noble, D.C.: Earth Planet. Sci. Lett. **6** (1969) 268–270.
Ro69a	Rosholt, J.N., Bartel, A.J.: Earth Planet. Sci. Lett. **7** (1969) 141–147.
Ro70	Ronov, A.B., Migdisov, A.A.: Geochem. Int. **7** (1970) 294–325.
Ro70a	Rodgers, J.J.W., Condie, K., Mahan, S.: Chem. Geol. **5** (1970) 207–213.
Ro71	Rosholt, J.N., Prijana, Noble, D.C.: Econ. Geol. **66** (1971) 1061–1069.
Ro76	Ronov, A.B., Yaroshevskiy, A.A.: Geochem. Int. **13** (1976) 89–121.
Ro78	Rogers, J.J.W., Ghuma, M.A., Nagy, R.M., Greenberg, J.K., Fullagar, P.D.: Earth Planet. Sci. Lett. **39** (1978) 109–117.
Ry70	Rybach, L., Grauert, B., Labhart, T.P.: Eclogae Geol. Helv. **63** (1970) 291–298.
Ry72	Rydell, H.S., Prospero, J.M.: Earth Planet. Sci. Lett. **14** (1972) 397–402.
Ry73	Rybach, L.: Beitr. Geol. Schweiz, Geotech. Ser. **51** (1973) 1–43.
Ry78	Rye, D.M., Roy, R.F.: Am. J. Sci. **278** (1978) 354–378.
Sa59	Saprykina, T.V.: Geochemistry (USSR) (English Transl.) **1959**, 565–570.
Sa63	Sackett, W.M., Potratz, H.A.: U.S. Geol. Surv. Prof. Paper **260-BB** (1963) 1053–1066.
Sa67	Savelli, C.: Contrib. Miner. Petrol. **16** (1967) 328–353.
Sa70	Sarswat, A.C., Varada Raju, H.N., Taneja, P.C., Bargaja, V.B., Sankaran, A.V.: Econ Comm. Asia Far East, Miner. Resources Development Series **38** (1970) 179–187.
Sc76	Scharpenseel, H.W., Pietig, F., Kruse, E., in: J.O. Nriagu (ed): Env. Biochem. Proc. 2nd Int. Symp. Env. Biochem., Hamilton **1975**, Vol. 2 (1976) 597–607.
Sc80	Schulz-Dobrick, B.: Fortschr. Mineral. **58**, Beih. 1, (1980) 121–123.
Sch80	Schmid, H.: Erzmetall **33** (1980) 540–545.
Se63	Serikov, Yu.I.: Geochemistry (USSR) (English Transl.) **1963**, 535–539.
Se72	Seitz, M.G.: Carnegie Inst. Washington, Yearb. **72** (1972/1973) 581–586.
Se73	Seitz, M.G., Hart, S.R.: Earth Planet. Sci. Lett. **21** (1973) 97–107.
Sh69	Sheraton, J.W.: unpubl. Ph.D. Thesis, University of Birmingham **1969**.
Sh70	Shatkov, G.A., Shatkova, L.N., Gushchin, Ye.N.: Geochem. Int. **7** (1970) 1051–1063.
Sh71	Shido, F., Miyashiro, A., Ewing, M.: Contrib. Mineral. Petrol. **31** (1971) 251–266.
Sh75	Shibata, T., Fox, P.J.: Earth Planet. Sci. Lett. **27** (1975) 62–72.
Sh76	Shaw, D.M., Dostal, J., Keays, R.R.: Geochim. Cosmochim. Acta **40** (1976) 73–83.

Sh80	Sheraton, J.W., Cundari, A.: Contrib. Miner. Petrol. **71** (1980) 417–427.
Si69	Siegers, A., Pichler, H., Zeil, W.: Geochim. Cosmochim. Acta **33** (1969) 882–887.
Si74	Sighinolfi, G.P., Sakai, T.: Chem. Geol. **14** (1974) 23–30.
Si77	Sighinolfi, G.P., Sakai, T.: Geochem. J. **11** (1977) 33–39.
Si78	Sighinolfi, G.P., Gorgoni, C.: Chem. Geol. **22** (1978) 157–176.
Sk73	Skinner, A.C.: unpubl. Ph.D. Thesis University of Birmingham **1970**; cited in [Dr73].
Sm54	Smith, A.P., Grimaldi, F.S.: U.S. Geol. Surv. Bull. **1006** (1954) 125–131.
Sm71	Smithson, S.B., Heier, K.S.: Earth Planet. Sci. Lett. **12** (1971) 325–326.
Sm73	Smithson, S.B., Decker, E.R.: Earth Planet. Sci. Lett. **19** (1973) 131–134.
Sm74	Smithson, S.B., Decker, E.R.: Earth Planet. Sci. Lett. **22** (1974) 215–225.
Sm79	Smith, I.E.M., Taylor, S.R., Johnson, R.W.: Contrib. Miner. Petrol. **69** (1979) 227–233.
So66	Somayajulu, B.L.K., Tatsumoto, M., Rosholt, J.N., Knight, R.J.: Earth Planet. Sci. Lett. **1** (1966) 387–391.
So66a	Somayajulu, B.L.K., Goldberg, E.D.: Earth Planet. Sci. Lett. **1** (1966) 102–106.
So71	Somayajulu, B.L.K., Heath, G.R., Moore jr., T.C., Cronon, D.S.: Geochim. Cosmochim. Acta **35** (1971) 621–624.
St58	Starik, I.E., Kuznetsov, Yu.V., Grashchenko, S.M., Frenklikh, M.S.: Geochemistry (USSR) (English Transl.) **1958**, 1–15.
St78	Stuckless, J.S., Nkomo, I.T.: Econ. Geol. **73** (1978) 427–441.
Su79	Sun, S.S., Nesbitt, R.W.: Earth Planet. Sci. Lett. **44** (1979) 119–138.
Sw61	Swanson, V.E.: U.S. Geol. Surv. Prof. Paper **356-C** (1961) 67–112.
Sw72	Swanberg, C.A.: J. Geophys. Res. **77** (1972) 2508–2513.
Sw74	Swanberg, C.A., Chessman, M.D., Simmons, G., Smithson, S.B., Grønlie, G., Heier, K.S.: Tectonophysics **23** (1974) 31–48.
Sy68	Syromyatnikov, N.G., Ivanova, E.I.: Geochem. Int. **5** (1968) 299–305.
Sz69	Szabo, B.J., Mulde, H.-E., Irwin-Williams, C.: Earth Planet. Sci. Lett. **6** (1969) 237–244.
Sz69a	Szalay, S., Samson, Z.: Geochem. Int. **6** (1969) 613–623.
Sz71	Szabo, B.J., Vedder, J.G.: Earth Planet. Sci. Lett. **11** (1971) 283–290.
Ta56	Tauson, L.V.: Geochemistry (USSR) (English Transl.) **1956**, 236–245.
Ta59	Tatsumoto, M., Goldberg, E.D.: Geochim. Cosmochim. Acta **17** (1959) 201–208.
Ta65	Tatsumoto, M., Hedge, C.E., Engel, A.E.J.: Science **150** (1965) 886–888.
Ta66a	Tatsumoto, M.: J. Geophys. Res. **71** (1966) 1721–1733.
Ta66b	Tatsumoto, M.: Science **153** (1966) 1094–1101.
Ta69	Tatsumoto, M.: Earth Planet. Sci. Lett. **6** (1969) 369–376.
Ta69a	Taylor, S.R., Capp, A.C., Graham, A.L.: Contrib. Miner. Petrol. **23** (1969) 1–26.
Ta75	Tammemagi, H.Y., Smith, N.L.: J. Geol. Soc. (London) **131** (1975) 415–427.
Ta76	Taylor jr., H.P., Turi, B.: Contrib. Miner. Petrol. **55** (1976) 33–54.
Ta77	Taylor, S.R., Hallberg, J.A.: Geochim. Cosmochim. Acta **41** (1977) 1125–1129.
Ta78	Tatsumoto, M.: Earth Planet. Sci. Lett. **38** (1978) 63–87.
Ta78a	Taylor, H.P., Silver, L.T.: U.S. Geol. Surv. Open-File Report **78-701** (1978) 423–426.
Ta79	Taylor, S.R., in: M.W. McElhinny (ed): The Earth: Its Origin, Structure and Evolution. London-New York-San Francisco: Academic Press **1979**, p. 353–377.
Te75	Tenyakov, V.A., Vinokurov, P.K., Zheleznova, E.I., Smirnova, A.I.: Geochem. Int. **12** (1975) 173–177.
Th70	Thompson, G., Livingston, H.D.: Earth Planet. Sci. Lett. **8** (1970) 439–442.
Th72	Thompson, S.G., Tsang, C.F.: Science **178** (1972) 1047–1055.
Th75	Thompson, G.M., Lunsden, D.N., Walker, R.L., Carter, J.A.: Geochim. Cosmochim. Acta **39** (1975) 1211–1218.
Ti63	Tilton, G.R., Reed, G.W., in: J. Geiss, E.D. Goldberg (eds): Earth Science and Meteoritics, Amsterdam: North Holland Publ. Comp. **1963**, p. 31–43.
Ti74	Tiek, T.T., Nelson, R.A., Eggler, D.H.: Tex. J. Sci. **25** (1974) 67–80.
Tr73	Triulzi, C.: Thalassia Jugosl. **9** (1973) 119–125.
Tu57	Turovskii, S.T.: Geochemistry (USSR) (English Transl.) **1957**, 199–215.
Tu69	Turekian, K.K., Kharkar, D.P., Funkhouser, J., Schaeffer, O.A.: Earth Planet. Sci. Lett. **7** (1969) 420–424.
Tv74	Tverdova, R.A., Fedina, V.V.: Geochem. Int. **11** (1974) 722–726.
Un76	Unruh, D.M., Tatsumoto, M.: Initial Reports DSDP **34** (1976) 341.

Va69	Van, N.H., Lalou, C.: C.R. Acad. Sci. Paris **D269** (1969) 560–563.
Ve66	Veeh, H.H.: J. Geophys. Res. **71** (1966) 3379–3386.
Ve67	Veeh, H.H.: Earth Planet. Sci. Lett. **3** (1967) 145–150.
Ve68	Veeh, H.H., Turekian, K.K.: Limnol. Ozeanogr. **13** (1968) 304–308.
Ve74	Veeh, H.H., Calvert, S.E., Price, N.B.: Mar. Chem. **2** (1974) 189–202.
Vi56	Vinogradov, V.R., Ronov, A.B.: Geochemistry (USSR) (English Transl.) **1956**, 123–139.
Vl76	Vlašimský, P.: Vestn. Ustred. Ustava Geol. **51** (1976) 85–98.
Vo76	Vollmer, R.: Geochim. Cosmochim. Acta **40** (1976) 283–295.
Wa64	van Wambeke, L., in: L. van Wambeke, J.W. Brinck, W. Deutzmann, R. Gontiantini, A. Hubaux, D. Métais, P. Omenetto, E. Tongiorgi, G. Verfaillié, K. Weber, W. Wimmenauer: Report EUR 1827. d, f, e, Communauté Européenne de l'Énergie atomique – EURATOM, Brüssel **1964**, p. 93–196.
Wa67a	Wakita, H., Nagasawa, H., Uyeda, S., Kuno, H.: Geochem. Int. **1** (1967) 183–198.
Wa69	Wasserburg, G.J., Huneke, J.C., Burnett, D.S.: J. Geophys. Res. **74** (1969) 4221–4232.
We64	Weber, J.N.: Geochim. Cosmochim. Acta **28** (1964) 1817–1868.
We67	Wedepohl, K.H.: Geochemie. Sammlung Göschen, Berlin: Walter de Gruyter+Co. **1967**, p. 49.
We68	Welke, H., Moorbath, S., Cumming, G.L., Sigurdsson, H.: Earth Planet. Sci. Lett. **4** (1968) 221–231.
We69	Wedepohl, K.H., in: K.H. Wedepohl (ed): Handbook of Geochemistry I, Heidelberg-Berlin-New York: Springer **1969**, p. 227–249.
We75	Wedepohl, K.H.: Fortschr. Miner. **52** (1975) 141–172.
We78	Wedepohl, K.H.: Aufschluß, Sonderband **28** (1978) 156–167.
We81	Wedepohl, K.H.: Naturwissenschaften **68** (1981) 110–119.
Wh59	Whittfield, J.M., Rogers, J.J.W., Adams, J.A.S.: Geochim. Cosmochim. Acta **17** (1959) 248–271.
Wh79	Whitford, D.J., Nicholls, I.A., Taylor, S.R.: Contrib. Miner. Petrol. **70** (1979) 341–356.
Wi60	Wilson, J.D., Webster, R.K., Milner, G.W., Barnett, G.A., Smales, A.A.: Anal. Chim. Acta **23** (1960) 505.
Wi73	Windley, B.F.: Spec. Publ. Geol. Soc. S. Afr. **3** (1973) 319–332.
Wo67	Wollenberg, H.A., Smith, A.R.: J. Geophys. Res. **72** (1967) 4139–4150.
Wo73	Wollenberg, H.A., Dodge, F.C.W.: Geol. Surv. U.S.A. Bull. **Nr. 1382** (1973) 1–17.
Wo79	Wood, D.A., Tarney, J., Varet, J., Saunders, A.D., Bougault, H., Joron, J.L., Treuil, M., Cann, J.R.: Earth Planet. Sci. Lett. **42** (1979) 77–97.
Wy70	Wyttenbach, A.: Z. Naturforsch. **25a** (1970) 307–308.
Ye62	Yes'kova, E.M., Mineyev, D.A., Mineyeva, I.G.: Geochemistry (USSR) (English Transl.) **1962**, 885–894.
Ye77	Yeliseeyeva, O.P.: Geochem. Int. **14** (1977) 37–49.
Za73	Zartmann, R.E., Tera, F.: Earth Planet. Sci. Lett. **20** (1973) 54–66.
Ze77	Zentilli, M., Dostal, J.: J. Volcanol. Geotherm. Res. **2** (1977) 251–258.
Zh65a	Zhuralev, R.S., Osipov, D.K.: Geochem. Int. **2** (1965) 308–312.
Zh65b	Zhuravlev, R.S., Osipov, D.K., Gladikh, Z.V.: Geochem. Int. **2** (1965) 582–586.
Zh75	Zhukova, A.M., Bergman, I.A., Zhukov, G.V.: Geochem. Int. **12** (1975) 103–106.
Zi75	Zielinski, R.A.: Geochim. Cosmochim. Acta **39** (1975) 713–734.
Zi76	Zielinski, R.A., Lipman, P.W.: Geol. Soc. Am. Bull. **87** (1976) 1477–1485.
Zi78	Zielinski, R.A.: Geol. Soc. Am. Bull. **89** (1978) 409–414.
Zn76	Zverev, V.L., Semenov, G.S., Spiridonov, A.I., Cheshko, A.L.: Geochem. Int. **13** (1976) 174–176.

7.2 The age of rocks — Das Alter der Gesteine

See p. 561

8 Physical properties of ice — Physikalische Eigenschaften von Eis

8.0 Introduction — Einleitung

Ice covers about 10% of the earth's land surface and, varying with the seasons up to 12% of the surface of the oceans. Therefore it must be regarded as a very common naturally occuring rock; it plays a very important role in global heat balance and surface processes. It is also the largest resource of fresh water. Its singularity in composition – it is a mono-mineralic rock – justifies a special treatment in this volume.

Altogether there are nine crystalline forms of water as is shown in Fig. 1. As the high pressure polymorphs of ice do not occur naturally, we shall discuss here ice Ih only.

Eis bedeckt etwa 10% der Landoberfläche der Erde und mit jahreszeitlichen Veränderungen bis zu 12% der Oberfläche der Ozeane. Eis ist daher als weitverbreitetes natürliches Gestein anzusehen und ist von großer Bedeutung für die globale Wärmebilanz, für das Klima und für Prozesse, die an der Erdoberfläche stattfinden. Eis ist auch das größte Süßwasservorkommen. Da Eis ein monomineralisches Gestein ist und dazu noch aus einem sehr einfachen Molekül aufgebaut ist, wird es in diesem Band gesondert behandelt.

Insgesamt sind 9 verschiedene kristalline Phasen des Wassers bekannt (Fig. 1). Da die Hochdruckphasen von Eis nicht in der Natur vorkommen, wird im folgenden immer von Eis Ih gesprochen.

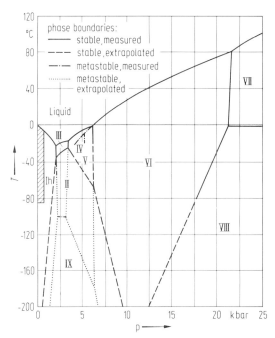

Fig. 1. Phase diagram of the solid phases of water. Ice IV is metastable in the region of ice V. Hatched area: approximate region for naturally occurring ice.

The clathrate hydrates will not be discussed here although they potentially form economic hydrocarbon deposits in permafrost-regions. They are solid phases of water with large holes which are filled with inert molecules such as ethylene chloride.

Ice Ih is a member of the hexagonal crystal system with the cell dimensions

Die Klathrat-Hydrate, die in Perma-Frost-Regionen potentiell nutzbare Kohlenwasserstoff-Lagerstätten bilden können, werden hier nicht behandelt. Es handelt sich dabei um Eis, dessen Kristallstruktur offen ist; in den relativ großen Löchern finden sich neutrale Moleküle wie z.B. Äthylenchlorid.

Eis Ih bildet Kristalle des hexagonalen Kristallsystems mit den Zell-Parametern

$$a_0 = 4.5 \cdot 10^{-8} \text{ m}$$

$$c_0 = 7.32 \cdot 10^{-8} \text{ m}$$

The density of pure ice is 0.931 Mg m^{-3}.

Die Dichte von reinem Eis bei 0 °C ist 0.931 Mg m^{-3}.

Natural fresh-water ice occurs on the earth's surface in the form of Glaciers or Ice-sheets or within the uppermost few hundred meters of the earth's crust as Permafrost; it is always an agglomerate of single crystal grains. In greater amounts, naturally occurring ice is almost never free of impurities and imperfections which are incorporated during the freezing process in the atmosphere or during the metamorphism from snow to ice. Thus physical properties vary whenever the pure one-phase-system ice is contaminated. This holds much more so for sea ice, which is frozen from sea water and incorporates not only ions of the dissolved salts, but also brine pockets during the freezing process.

Natürlich vorkommendes Süßwasser-Eis ist immer ein Agglomerat von einzelnen Eiskristallkörnern. Über ein größeres Volumen betrachtet, ist natürliches Eis niemals frei von Verunreinigungen und Fehlstellen, die bereits während des Gefriervorganges in der Atmosphäre oder während der Metamorphose von Schnee zu Eis in die Kristalle eingebaut werden. Entsprechend sind die physikalischen Eigenschaften beeinflußt und ändern sich, wenn das reine Einphasensystem verunreinigt ist. Dies wird beim Meer-Eis besonders deutlich, da dieses aus Meerwasser gefriert und während des Gefriervorganges nicht nur gelöste Salzionen aufnimmt, sondern auch Laugenbläschen in das Gefüge eingeschlossen werden.

List of symbols – Symbolliste

a	$[\text{cm}^{-1}]$	absorption coefficient in the infrared – Absorptionskoeffizient im Infraroten
a_0, c_0	$[10^{-8}\,\text{m}]$	hexagonal lattice parameters – hexagonale Gitterkonstanten
c	$[\text{m s}^{-1}]$	velocity of light – Lichtgeschwindigkeit
C_{ij}	$[\text{bar}]$	matrix elements of stiffness constant – Matrixelemente der Elastizitätsmoduln
c_p	$[\text{kJ kg}^{-1}\,\text{K}^{-1}]$	specific heat at constant pressure – spezifische Wärme bei konstantem Druck
E	$[\text{bar}]$	modulus of elasticity (Young's modulus) – Elastizitätsmodul
E	$[\text{eV}]$	activation energy – Aktivierungsenergie
E_τ	$[\text{eV}]$	electrical activation energy – elektrische Aktivierungsenergie
E_m	$[\text{eV}]$	activation energy for mechanical relaxation – Aktivierungsenergie für mechanische Relaxation
f	$[\text{Hz}]$	frequency of seismic wave – Frequenz der seismischen Welle
G	$[\text{bar}]$	modulus of rigidity – Scherungsmodul
H	$[\text{J}]$	enthalpy – Enthalpie
i	$[\text{deg}]$	angle of incidence – Einfallwinkel
I_0	$[\text{cd}]$	total incident intensity – totale Einfallsintensität
I_z	$[\text{cd}]$	intensity at distance z below the surface – Intensität in einer Entfernung z unterhalb der Oberfläche
K	$[\text{bar}]$	bulk modulus – Kompressionsmodul
k	$[\text{bar}^{-1}]$	modulus of compressibility – Kompressibilitätsmodul
k	$[\text{eV K}^{-1}]$	Boltzman constant – Boltzmankonstante
k	$[\text{W m}^{-1}\,\text{K}^{-1}]$	thermal conductivity – Wärmeleitfähigkeit
L_F	$[\text{kJ kg}^{-1}]$	heat of fusion – Schmelzwärme
L_S	$[\text{kJ kg}^{-1}]$	heat of sublimation – Sublimationswärme
n_E		extraordinary index of refraction – außerordentlicher Brechungsindex
n_O		ordinary index of refraction – ordentlicher Brechungsindex
p	$[\text{bar}]$	pressure – Druck
Q		quality factor – Qualitätsfaktor
R	$[\%]$	surface reflection coefficient – Reflexionskoeffizient an der Oberfläche
S_{ij}	$[\text{bar}^{-1}]$	matrix elements of compliance constant – Matrixelemente der Elastizitätskoeffizienten
T	$[°\text{C}, \text{K}]$	temperature – Temperatur
t	$[s, h]$	time – Zeit
U	$[\text{J}]$	internal energy – innere Energie
V	$[\text{m}^3]$	volume – Volumen
v	$[\text{m s}^{-1}]$	velocity of seismic waves – Geschwindigkeit der seismischen Wellen
v_E	$[\text{m s}^{-1}]$	velocity of extraordinary wave – Geschwindigkeit der außerordentlichen Welle
v_O	$[\text{m s}^{-1}]$	velocity of ordinary wave – Geschwindigkeit der ordentlichen Welle
v_P	$[\text{m s}^{-1}]$	longitudinal wave velocity – Geschwindigkeit der longitudinalen Welle
v_S	$[\text{m s}^{-1}]$	transversal wave velocity – Geschwindigkeit der transversalen Welle

α	$[\text{m}^{-1}]$	absorption coefficient – Absorptionskoeffizient
α	$[°\text{C}^{-1}, \text{K}^{-1}]$	coefficient of linear expansion – linearer Ausdehnungskoeffizient
β	$[°\text{C}^{-1}, \text{K}^{-1}]$	coefficient of cubic expansion – kubischer Ausdehnungskoeffizient
γ		Poisson's ratio – Poissonsche Zahl
δ	$[\text{deg}]$	loss angle – Verlustwinkel
Δ		Laplace operator – Laplace-Operator
ε		relative permittivity – relative Dielektrizitätskonstante
ε	$[\%]$	strain – Verformung
$\dot\varepsilon$	$[\text{s}^{-1}]$	strain rate – Verzerrungsgeschwindigkeit
ε'		ordinary relative permittivity – Realteil der relativen Dielektrizitätskonstante
ε''		dielectric loss factor – dielektrischer Verlustfaktor
ε_∞		high frequency relative permittivity – relative Dielektrizitätskonstante für hohe Frequenzen
ε_0	$[\%]$	instantaneous strain – plötzliche Verzerrung
ε_c	$[\%]$	compressive strain – Druckverformung
ε_i	$[\%]$	strain component – Verformungskomponente
ε_s		static relative permittivity – statische relative Dielektrizitätskonstante
$\dot\varepsilon_\text{st}$	$[\% \, \text{s}^{-1}]$	steady-state creep rate – stationäre Kriechrate
ε_t	$[\%]$	tensile strain – Dehnungsverformung
ε_shear	$[\%]$	shear strain – Scherungsverformung
κ	$[\text{m}^2 \, \text{s}^{-1}]$	thermal diffusivity – Temperaturleitfähigkeit
λ	$[\mu\text{m}]$	wave length – Wellenlänge
ν	$[\text{Hz}]$	frequency – Frequenz
$\tilde\nu$	$[\text{cm}^{-1}]$	wave number – Wellenzahl
ν_b	$[\%]$	brine volume (content), relative volume of brine in ice to total volume – Laugengehalt, Laugengehalt im Eis relativ zum Eisvolumen
ϱ	$[\Omega \, \text{m}]$	resistivity – Widerstand
ϱ	$[\text{g cm}^{-3}, \text{Mg m}^{-3}]$	density – Dichte
σ	$[\Omega^{-1} \, \text{m}^{-1}]$	conductivity – Leitfähigkeit
σ	$[\text{bar}]$	stress – Spannung (1 bar $= 10^5$ N m^{-2})
σ_∞	$[\Omega^{-1} \, \text{m}^{-1}]$	high frequency conductivity – Leitfähigkeit für hochfrequenten Wechselstrom
σ_c	$[\text{bar}]$	compressive strength – Druckfestigkeit
σ_i	$[\text{bar}]$	stress component – Spannungskomponente
σ_max	$[\text{bar}]$	yield stress for constant strain rate – Nachgebespannung bei konstanter Deformationsrate
σ_s	$[\Omega^{-1} \, \text{m}^{-1}]$	dc conductivity – Gleichstromleitfähigkeit
σ_shear	$[\text{bar}]$	shear stress – Scherungsfestigkeit
σ_t	$[\text{bar}]$	tensile stress – Zugfestigkeit
τ	$[\text{s}]$	dielectric relaxation time – dielektrische Relaxationszeit
τ_m	$[\text{s}]$	mechanical relaxation time – mechanische Relaxationszeit
ϕ	$[\%]$	porosity – Porosität
ω	$[\text{s}^{-1}]$	angular frequency – Kreisfrequenz

8.1 Thermal properties — Thermische Eigenschaften

The differential equation describing the conduction of heat is

Die Wärmeleitung wird durch folgende Differentialgleichung beschrieben:

$$\frac{\partial T}{\partial t} = \kappa \, \Delta T,$$

where

κ is the thermal diffusivity,

Δ the Laplace operator,

$\dfrac{\partial T}{\partial t}$ the time derivative of the temperature.

mit

κ Temperaturleitfähigkeit

Δ Laplace-Operator

$\dfrac{\partial T}{\partial t}$ zeitliche Änderung der Temperatur.

The thermal conductivity is related to the thermal diffusivity by the equation

$$k = \kappa \cdot c_p \cdot \varrho \qquad [\mathrm{W\,m^{-1}\,K^{-1}}] \qquad (1)$$

where c_p is the specific heat at constant pressure and ϱ the density.

The temperature dependence of the thermal conductivity of ice can be expressed by the equation

$$k = \frac{488 \cdot 19}{T} + 0.4685 \qquad (2)$$

with T in [K] [Ho74].

In natural samples containing impurities and gas or liquid filled bubbles the effective values for the thermal properties may vary widely. This is especially the case for bubbly glacier ice in the melting stage.

Latent heat of fusion and sublimation

The latent heat is defined as the change in enthalpy

$$dH = dU + p\,dV,$$

where

dH   is change in enthalpy,
dU   is change in internal energy,
dV   is change in volume.

When ice is converted isothermally and reversibly into water or water vapour, the heat of fusion is:

$$L_F = 333.6 \text{ kJ kg}^{-1},$$

and the heat of sublimation:

$$L_S = 2838 \text{ kJ kg}^{-1} \text{ [Ho74]}.$$

Die Wärmeleitfähigkeit ist über folgende Gleichung mit der Temperaturleitfähigkeit verknüpft:

mit: c_p spezifische Wärme bei konstantem Druck, ϱ Dichte.

Die Temperaturabhängigkeit der Wärmeleitfähigkeit wird durch die Gleichung

beschrieben [Ho74], wobei T in [K].

In natürlichen Proben, die Verunreinigungen enthalten, können die effektiven thermischen Eigenschaften in weiten Bereichen schwanken. Dies trifft besonders dann zu, wenn blasenhaltiges Gletschereis in den Schmelzbereich kommt.

Schmelz- und Sublimationswärme

Die latente Wärme wird definiert als die Änderung der Enthalpie

mit

dH   Änderung der Enthalpie,
dU   Änderung der inneren Energie,
dV   Änderung des Volumens.

Wenn Eis isotherm und reversibel in Wasser oder Wasserdampf übergeführt wird, ist die Schmelzwärme:

und die Sublimationswärme:

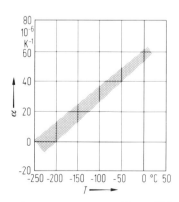

Fig. 2. Dependence on temperature of the coefficient of linear expansion α of bulk ice at atmospheric pressure. Measured values within hatched area [Ho74, Da62].

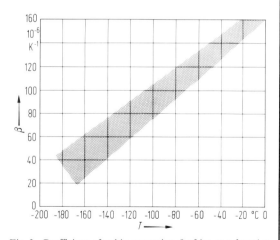

Fig. 3. Coefficient of cubic expansion β of ice as a function of temperature at atmospheric pressure. Values calculated from different measurements [Ho74, Da62, Br67, La60] within hatched area.

Table 1. Specific heat, thermal diffusivity and thermal conductivity for some types of ice at different temperatures.

T °C	Type of ice	ϱ g cm^{-3}	κ m^2 s^{-1}	k W$_m^{-1}$ K^{-1}	c_p kJ kg^{-1} K^{-1}	Ref.
-10	Pure ice	0.9187	$1.24 \cdot 10^{-6}$	2.32	2.04	Ho74
-0.5	Dry, bubbly glacier ice	0.82	$1.1 \cdot 10^{-6}$	1.90	2.115	Ho74
-20	Antarctic ice	0.86	$1.54 \cdot 10^{-6}$	1.59	1.963	Ho74, 63Sc
0	Wet, bubbly glacier ice	0.9	$2.2 \cdot 10^{-10}$ $-1.1 \cdot 10^{-6}$		$200 \cdots 10000$	Ra76

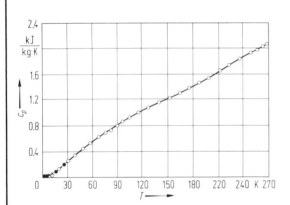

Fig. 4. Heat capacity of ice at constant pressure c_p as a function of temperature [Ho74].

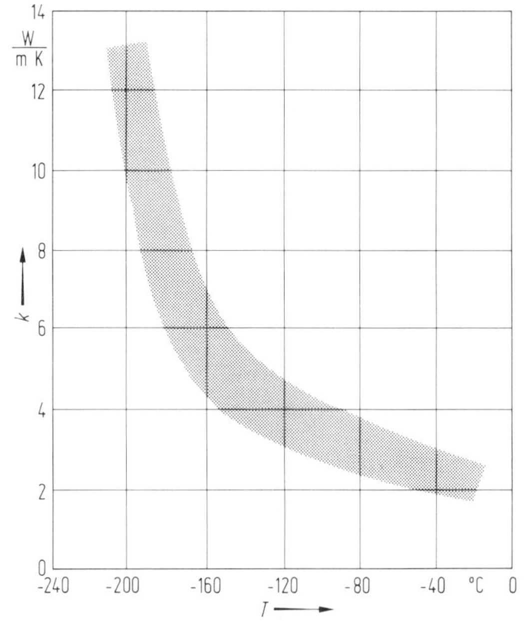

Fig. 5. Variation of thermal conductivity of polycrystalline ice with temperature. Range of values indicated by hatched area [Ra62, Ho74].

Table 2. Specific heat of sea ice as function of temperature and salinity [Po65].

Salinity ‰	T [°C]										
	-2	-4	-6	-8	-10	-12	-14	-16	-18	-20	-22
	c_p [kJ kg^{-1} K^{-1}]										
0	2.1	2.1	2.05	2.05	2.05	2.01	2.01	2.01	1.96	1.96	1.92
1	6.62	3.22	2.55	2.34	2.22	2.14	2.1	2.1	2.01	2.01	1.96
2	11.14	4.35	3.05	2.59	2.43	2.26	2.17	2.14	2.1	2.05	2.01
4	20.15	6.62	4.06	3.18	2.76	2.51	2.34	2.30	2.18	2.13	2.1
6	28.45	8.88	5.06	3.73	3.14	2.76	2.55	2.43	2.30	2.22	2.13
8	38.25	11.10	6.07	4.31	3.47	3.02	2.72	2.55	2.38	2.30	2.22
10	47.34	13.31	7.03	4.86	3.85	3.27	2.89	2.72	2.51	2.39	2.30

8.2 Electrical properties — Elektrische Eigenschaften

Many of the electrical properties of ice are unusual and have been the subject of intense experimental and theoretical investigations.

Both the dielectric and the conduction properties of ice may be investigated by freezing electrodes into a block of ice and measuring the complex impedance Z of the ice as a function of frequency. During measurement the electrodes should be in true ohmic contact with the ice. Different experimental procedures are given in [Bu69, En69, Ho74, Wo69]. Surface effects may be significantly reduced by using guard ring electrodes or four-probe-measurements.

The Debye dispersion formula which describes the electrical behaviour of ice below the optic frequencies lets us define the complex relative permittivity

Viele der elektrischen Eigenschaften von Eis sind ungewöhnlich und sind ausgiebig experimentell und theoretisch untersucht worden.

Die Leitfähigkeit wie auch die dielektrischen Eigenschaften von Eis können mit Hilfe von Messungen der komplexen Impedanz in Abhängigkeit von der Frequenz über ins Eis eingefrorene Elektroden untersucht werden. Details über experimentelle Verfahren bei [Bu69, En69, Ho74, Wo69]. Oberflächeneffekte können durch Vier-Pol-Messungen oder Schutzring-Elektroden deutlich verringert werden.

Die Debye'sche Dispersions-Beziehung, die die elektrischen Eigenschaften von Eis unterhalb der optischen Frequenz beschreibt, definiert die komplexe relative Dielektrizitätskonstante

$$\varepsilon = \varepsilon_\infty + \frac{\varepsilon_S - \varepsilon_\infty}{1 + i\omega\tau} = \varepsilon' - i\varepsilon'' \tag{3}$$

where

τ is the dielectric relaxation time
ε' the ordinary relative permittivity
ε'' the dielectric loss factor
ε_S the static relative permittivity
ε_∞ the high frequency relative permittivity

mit

τ dielektrische Relaxationszeit
ε' Realteil der relativen Dielektrizitätskonstante
ε'' dielektrischer Verlustfaktor
ε_S statische relative Dielektrizitätskonstante
ε_∞ relative Dielektrizitätskonstante für hohe Frequenzen

τ is a function of temperature and its dependence can be represented by:

τ ist abhängig von der Temperatur

$$\tau = C_\tau \exp\left(\frac{E_\tau}{kT}\right) \tag{4}$$

where

T in [K]:   absolute temperature
k in [eV K^{-1}]: Boltzmann constant $(= 0.86166 \cdot 10^{-4}$ eV K$^{-1})$
E_τ in [eV]:   activation-energy
C_τ in [s]:   constant

mit

T absolute Temperatur
k Boltzmann-Konstante $(0.86166 \cdot 10^{-4}$ eV K$^{-1})$

E_τ Aktivierungsenergie in [eV]
C_τ Zeitkonstante

The phase angle between the displacement current and the total current in an alternating electric field is called the loss angle δ and the loss tangent is given by

Der Verlustwinkel δ ist definiert als der Phasenwinkel zwischen dem Verschiebungsstrom und dem Gesamtstrom in einem elektrischen Wechselfeld. Er ist definiert durch

$$\operatorname{tg}\delta = \frac{\varepsilon''}{\varepsilon'} = \frac{\varepsilon_s - \varepsilon_\infty}{\varepsilon_s + \varepsilon_\infty \omega^2\tau^2}\omega\tau \tag{5}$$

Conductivity

The conductivity is also treated as a complex quantity

Leitfähigkeit

Die Leitfähigkeit wird ebenfalls als komplexe Größe behandelt

$$\sigma = \sigma' + i\sigma'' \tag{6}$$

A semi-circle results if the real and imaginary parts are plotted against each other (Cole-Cole Plot). The points where this plot touches the real axis correspond to the low-frequency and high-frequency limits of the

Wird der Realteil gegen den Imaginärteil aufgetragen (Cole-Cole-Diagramm), erhält man einen Halbkreis. Die Punkte, an denen dieser Halbkreis die Abszisse schneidet, entsprechen Grenzwerten der Leit-

conductivity and are called σ_s (static or dc conductivity) and σ_∞ (high-frequency conductivity).

fähigkeit im niederfrequenten und hochfrequenten Bereich. Diese Grenzwerte sind σ_s (Gleichstrom-Leitfähigkeit) und σ_∞ (Leitfähigkeit für hochfrequenten Wechselstrom).

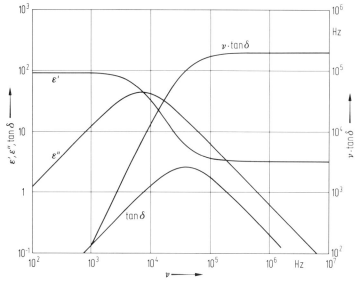

Fig. 6. Plot of the ordinary relative permittivity ε', the dielectric loss factor ε'', the loss tangent $\tan\delta$, and $v\cdot\tan\delta$ against frequency for ice of 0 °C, calculated using equation (3). Data is taken from [Ho74, Wo69].

Table 3. High frequency relative permittivity ε_∞ of ice [Ho74].

Ref.	Type of ice	T °C	v Hz	ε_∞ at -40 °C	$\frac{1}{\varepsilon_\infty}\cdot\frac{\partial\varepsilon_\infty}{\partial T}$ $10^{-4}\,\mathrm{K}^{-1}$
La49	Polycrystalline ice grown in laboratory	$0\cdots-200$	$2.4\cdot10^{10}$	3.17	0.95
Cu52	Polycrystalline ice grown in laboratory	$0\cdots-20$		3.15	
Ma65	Polycrystalline ice grown in laboratory	$-60\cdots-180$	10^4	3.07	5.1
Hi69a	Single crystals grown in laboratory. Field parallel to c-axis	$-4\cdots-130$	$>10^5$	3.23	1
Hi69a	Single crystals grown in laboratory. Field perpendicular to c-axis	$-4\cdots-80$	$>10^5$	3.30	3
Pa70	Single crystals from Mendenhall Glacier. Field parallel to c-axis	$-10\cdots-80$	10^5	3.20	4.0 5.9 3.7
Pa70	Polycrystalline commercial ice	$-20\cdots-80$			4.2
Pa70	Polycrystalline commercial ice	$-20\cdots-60$	10^5	3.20	4.3
Pa70	Polycrystalline commercial ice	$-10\cdots-50$			4.0
Pa70	Polycrystalline glacier ice from Greenland	$-35\cdots-55$	10^5	3.21	4.8
Pa70	Polycrystalline ice from Ward Hunt ice shelf. Ellesmere Island				3.6
Pa70	Polycrystalline ice from Ward Hunt Sea	$-10\cdots-60$	$1.5\cdot10^8$	3.17	5.5
Pa70	Polycrystalline ice from Tuto tunnel, Greenland				3.2
Pa70	Polycrystalline ice from Little America, Antarctica				4.2

Table 4. Static relative permittivity ε_s of ice [Ho74].

Ref.	Type of ice	T °C	A_c[1] 10^4 K	T_c[1] K	Deduced value of ε_s at 0 °C	Deduced value of ε_s at −40 °C	Remarks
Au52	Polycrystalline ice grown in laboratory	0···−44.7	3.75	−146	92.6	102.0	Possible errors from unknown amounts of dissolved gases in ice, polarization and surface effects
Hu53	Single crystals grown in laboratory. Field perpendicular to c-axis	0···−40	2.74	−38	91.6	104.4	Possible errors from unknown amounts of dissolved gases in ice, polarization and surface effects
Hu53	Single crystals grown in laboratory. Field parallel to c-axis	0···−40	2.29	51	106.4	129.0	Possible errors from unknown amounts of dissolved gases in ice, polarization and surface effects
Ru69	Single crystals grown in laboratory. Field parallel to c-axis (?)	0···−70	2.80	0	105.7	123.3	Two and three terminal guard rings used
Wo69	Polycrystalline ice but consisting of parallel single crystals. Field perpendicular to c-axis	0···−80	2.07	38	91.2	109.3	Three terminal guard ring and two terminal with potential probes used
Go70	Polycrystalline ice grown in laboratory	−10···−98	2.2	32	95	115	Guard rings used
Hi71	Annealed single crystals grown in laboratory. Field perpendicular to c-axis	−20.2 and −78.1 °C	–	–	(96.4 at −20.2 °C)	125.9 at −78.1 °C	Three terminal guard ring used. Values given are for inherent volume polarization
	Annealed single crystals grown in laboratory. Field parallel to c-axis	−20 °C			(112.3 at −20 °C)		

---

[1]) The static relative permittivity increases with decreasing temperature as predicted by the Curie-Weiss law:

$$\varepsilon_s - \varepsilon_\infty = \frac{A_c}{T - T_c}$$

The values given for A_c and T_c are the values best fitting this equation.

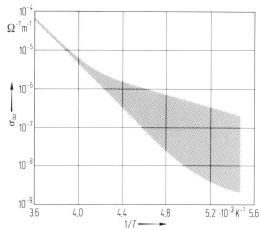

Fig. 7. Temperature dependence of the high frequency conductivity, σ_∞. Laboratory measurements of natural and artificial polycrystalline and single crystal samples. After [Gl75].

Miller

Table 5. Dielectric relaxation time τ of ice [Ho74]. E_τ = activation energy.

Ref	Type of ice	T °C	E_τ eV	τ at 0 °C 10^{-4} s
Au52	Polycrystalline ice grown in laboratory	0⋯−66	0.571	2.08
Hu53	Single crystals grown in laboratory. Field parallel to c-axis	−5⋯−30	0.571	2.12
Hu53	Single crystals grown in laboratory. Field perpendicular to c-axis	−5⋯−30	0.571	2.24
Yo68	Polycrystalline ice grown in laboratory	−8⋯−27	0.574±0.004	2.19±0.10
Co69	Single crystals grown in laboratory	0⋯−40	0.537	2.1
Ca69	Single crystals grown in laboratory	−5⋯−30	0.570	2.18
Ru69	Single crystals grown in laboratory	−4⋯−53 −53⋯−73	0.62±0.01 0.45±0.02	1.86±0.10
Hi71	Single crystals grown in laboratory	0⋯−94	0.59	1.8
Jo78	Zone refined . Single crystals		0.53±0.02	3.6

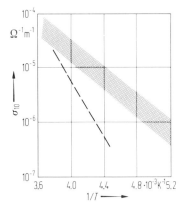

Fig. 8. Temperature dependence of the conductivity at 10 kHz, σ_{10}, for samples from different depths in the Camp Century and Site 2 (Greenland) ice cores (hatched region). The dashed line was obtained at 100 kHz for very pure ice crystals from Mendenhall Glacier (Alaska). After [Fi75].

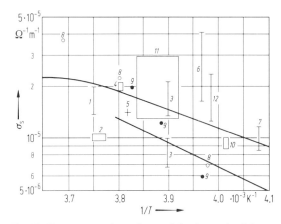

Fig. 10. Temperature dependence of the dc conductivity σ_s for polar ice samples. From [Gl75] with changes. Laboratory measurements on borehole samples are indicated by continuous lines. Numbers refer to following locations

1 Observatory Glacier, Baffin Island;
2 Penny Ice Cap, Baffin Island
3 Meighen Island ice cap;
4 Tuto, Greenland;
5 Paris Gletscher, Greenland;
6 Point Nord, Greenland;
7 Station Centrale, Greenland;
8 Ross Ice Shelf, Antarctica;
9 Roosevelt Island, Antarctica;
10 Camp Century, Greenland;
11 Sverdrup Glacier, Devon Island;
12 "Ice cap station", Devon Island

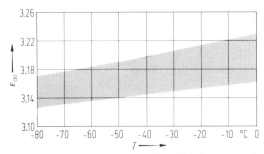

Fig. 9. Temperature dependence of the high frequency relative permittivity ε_∞ of ice. Hatched region indicates range of possible values. Data from [Go70].

Table 6. Bulk (σ_s) and surface (σ_s') dc-conductivities of ice [Ho74].

Ref.	Type of ice	T	σ_s $\Omega^{-1}\cdot m^{-1}$	σ_s' Ω^{-1}	$\dfrac{\sigma_s'}{\sigma_s}$ [mm]	E_s eV	E_s eV	Method
Ei64	Polycrystalline ice grown in laboratory	−10	$(1.0\pm0.15)\cdot10^{-7}$			0.47		Two platinum probes
Ja67	Single crystals grown in laboratory	−11	$6.6\cdot10^{-8}$	$1.6\cdot10^{-10}$	2.4			Four-point
Ja67	Single crystals grown in laboratory	−11	$4.5\cdot10^{-8}$	$1.4\cdot10^{-10}$	3.4			Four-point
Ja67	Single crystals grown in laboratory	−11	$6.6\cdot10^{-8}$	$1.6\cdot10^{-10}$	2.4			Four-point
Ja67	Single crystals grown in laboratory	−11	$1.9\cdot10^{-7}$	$6.3\cdot10^{-10}$	3.4			Four-point
Bu66	Single crystals grown in laboratory	−11	$2.2\cdot10^{-7}$	$3.9\cdot10^{-10}$	1.8	0.37 (−3···−20°C)	1.29 (−3···−20°C)	Guard rings Platinum electrodes
Bu69	Single crystals grown in laboratory. Electric field parallel to c-axis	−10	$(1.1\pm0.5)\cdot10^{-8}$			0.34 ± 0.02 (above −60°C)	0.86···1.29	Guarded palladium probes
Ka69	Single crystals grown in laboratory. Electric field parallel to c-axis					0.39 (−20···−70°C) 0.56 (−70···−100°C)		Ion-exchange membranes
Co69	Single crystals grown in laboratory		$1\cdot10^{-7}$ $6\cdot10^{-8}$			0.08 (at −10°C) 0.58 (0°C)		Guard rings Guard rings
Ca69	Polycrystalline ice grown in laboratory	−10	$\approx5\cdot10^{-2}$ Independent of temperature down to −65°C			≈0.53 (below −65°C)		Four probes
Ru69	Single crystals grown in laboratory	−10	$1.5\cdot10^{-8}$	$1.5\cdot10^{-10}$	10	0.30 (−1.3···−10°C) 0.09 (−20···−73°C)	1.29	Guard rings
Ma70	Single crystals grown in laboratory	0···−10°C				0.52	1.4	Comparison of two-terminal with three-terminal holders

Table 7. Relative permittivity ε' of natural ice as determined from wide-angle radar reflection measurements.

Location Type of ice	ε'	v MHz	Ref.
Roosevelt Island Dome	3.00 ± 0.05	30	Be72
McMurdo Ice Shelf			
Station 203	3.00 ± 0.08	30	Be72
Station 204	2.95 ± 0.05	30	
Skelton Glacier	3.17 ± 0.04	30	Be72
Tuto East	3.26 ± 0.17	35	Ro69
Barnes Ice Cap	3.07 ± 0.03	35	Cl70
Molodezhnaya	3.23	213	Be72
	3.70	213	
	3.60	213	
	3.52	213	
	3.52	213	

Table 8. dc resistivity, ϱ, of temperate glacier ice as determined by 4-electrode-configuration measurements.

Location	ϱ [10^6 Ωm]	Ref.
Glacier de St. Sorlin	59$\cdots$170	Ro67
Athabaska Glacier	10$\cdots$ 20	Ro67
Lednik Tuyuksy	>100	Ro67
Iles Kerguelen	40$\cdots$ 60	Ro67
Mer de Glace	> 40	Ro67
Unteraargletscher	30$\cdots$ 80	Ro67
Aletschgletscher	30$\cdots$ 70	Ro67
Vernagtferner	10$\cdots$ 60	Mi71
Talku Glacier	20$\cdots$ 65	Mi73

Table 9. dc resistivity ϱ, of polar glacier ice as determined by 4-electrode-configuration measurements.

Location	Electrode configuration	ϱ [10^4 Ωm]	Ref.
Ice cap station, Devon Island	Schlumberger	4$\cdots$18	Vo67
Sverdrup glacier	Schlumberger	3$\cdots$12	Vo67
Moltke glacier	Wenner	5$\cdots$15	Me62
Camp Century	Wenner	10$\cdots$50	Me62
Station Centrale, Greenland	Schlumberger	8.5$\cdots$11.5	Be72
Ross Ice Shelf, Antarctica	Schlumberger	3 $\cdots$14	Be72
Roosevelt Island, Antarctica	Schlumberger	2 $\cdots$16.5	Be72
Tuto, Greenland	Schlumberger	4 $\cdots$ 6	An67

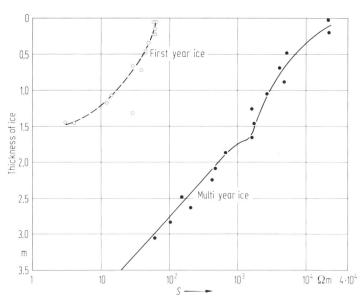

Fig. 11. Vertical resistivity profiles of typical first year and multi-year sea ice measured at 18.6 kHz [Mc73].

Miller

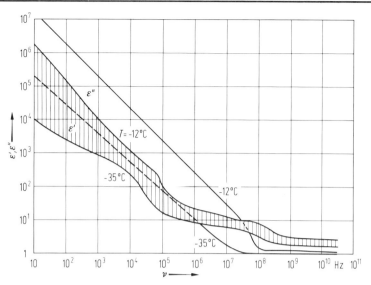

Fig. 12. Variation of the ordinary relative permittivity ε' and the dielectric loss factor ε'' as a function of frequency ν in sea ice [Sc77].

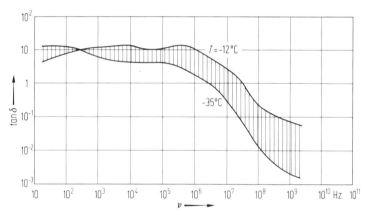

Fig. 13. Variation of the loss tangent $\tan \delta$ of sea ice as a function of frequency [Sc77].

8.3 Optical properties — Optische Eigenschaften

Ice is a double refractive, uniaxial and optically positive crystal. The c-axis coincides with the optic axis. Hence it has to be characterized by two indices of refraction,

Eis ist ein doppelbrechender, uniaxialer und optisch positiver Kristall. Die c-Achse stimmt mit der optischen Achse überein. Daher genügen zwei Brechungsindices zur Beschreibung der optischen Eigenschaften.

$$n_E = \frac{c}{v_E} \quad \text{and} \quad n_O = \frac{c}{v_O} \tag{7}$$

where

v_E = minimum normal velocity of extraordinary wave

v_O = velocity of ordinary wave
c = velocity of light = $2.997925 \cdot 10^8$ ms^{-1}.

mit

v_E = minimale Geschwindigkeit der außerordentlichen Welle

v_O = Geschwindigkeit der ordentlichen Welle
c = Lichtgeschwindigkeit = $2.997925 \cdot 10^8$ ms^{-1}

The absorption coefficient a for light passing through an ice sample is defined by the equation

Der Absorptionskoeffizient a für Licht, das durch einen Eiskristall dringt, wird durch die folgende Gleichung definiert

$$I_Z = (1 - R) I_0 \exp(-az) \tag{8}$$

where

mit

I_Z = the intensity at distance z below the surface

I_Z = Intensität in einer Entfernung z unterhalb der Oberfläche

I_0 = total incident intensity

R = surface reflection coefficient

a = absorption coefficient.

I_0 = totale einfallende Intensität

R = Reflexionskoeffizient an der Oberfläche

a = Absorptionskoeffizient

Table 10. Indices of refraction n_O, n_E of ice at $-3\,°C$ with respect to vacuum [Ho74].

$\tilde{v}$ cm^{-1}	λ µm	n_O	n_E
24716	0.4046	1.3183	1.3198
22946	0.4358	1.3159	1.3174
20572	0.4861	1.3129	1.3143
20342	0.4916	1.3126	1.3140
18312	0.5461	1.3104	1.3118
17301	0.5780	1.3093	1.3107
16969	0.5893	1.3090	1.3104
16041	0.6234	1.3079	1.3093
15237	0.6563	1.3070	1.3084
14476	0.6908	1.3063	1.3077
14154	0.7065	1.3060	1.3074

Table 11. Values for the absorption coefficient a of ice in the infrared [Ir68].

$\tilde{v}$ cm^{-1}	a cm^{-1}	$\tilde{v}$ cm^{-1}	a cm^{-1}	$\tilde{v}$ cm^{-1}	a cm^{-1}
10526	0.11	3333	9522	1538	1065
10000	0.25	3279	13093	1515	1065
9709	0.34	3252	14008	1504	1065
9524	0.30	3226	13184	1493	1039
9091	0.21	3175	9979	1471	988
8696	0.32	3125	6135	1449	925
8333	0.96	3077	3479	1429	881
8000	1.42	3030	2381	1408	837
7692	1.20	2985	1649	1389	790
7143	1.84	2941	1135	1333	669
6897	10.2	2899	824	1250	580
6667	46.8	2857	586	1176	520
6579	53.0	2817	458	1111	510
6452	47.4	2778	367	1053	410
6250	29.9	2632	270	1000	520
6061	23.1	2564	335	952	720
5882	14.2	2500	389	909	1090
5714	10.2	2439	459	870	1240
5556	7.9	2381	525	833	1260
5405	4.3	2326	636	800	1190
5263	25.9	2273	806	769	1040
5128	71.7	2222	921	741	870
5000	01.0	2174	783	667	640
4878	85.8	2128	576	571	250
4762	50.1	2083	454	500	160
4651	39.5	2041	377	400	150
4545	17.6	2000	334	333	220
4444	11.9	1961	321	286	400
4348	10.8	1923	322	250	560
4255	18.7	1887	360	230	930
4167	29.9	1852	408	223	1630
4082	36.2	1786	548	215	1170
4000	4.01	1754	690	192	650
3922	41.8	1724	918	179	340
3846	39.0	1695	1120	161	580
3810	39.8	1667	1293	120	330
3774	67.8	1653	1341	100	100
3571	550	1639	1325	85	32
3509	1557	1613	1248	67	25
3448	4394	1587	1164	66	25
3390	7690	1563	1090		

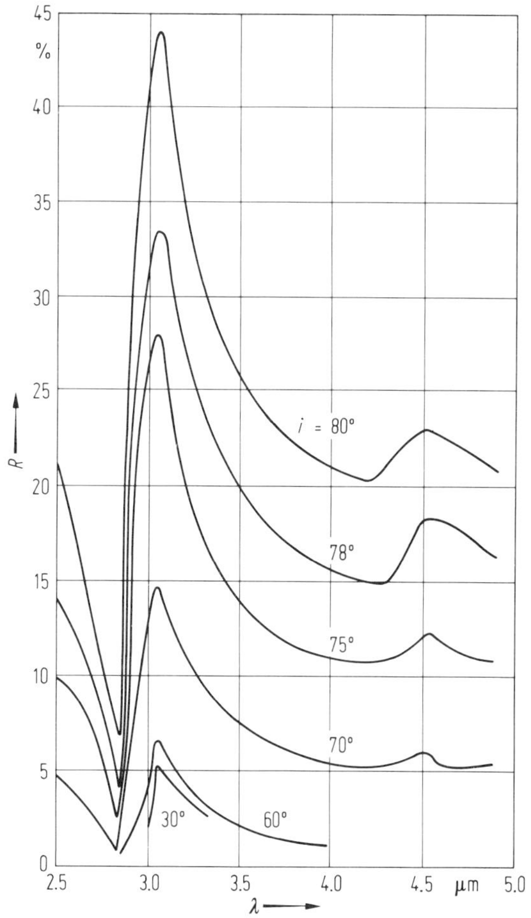

Table 12. Values of the absorption coefficient a of different naturally occuring ice types in the visible and near infrared part of the spectrum.

Ref.	$\tilde{v}$ cm^{-1}	a cm^{-1}	Remarks
Ka35	38120	0.048	Brown, bubbly pond
	28902	0.049	ice. 10 cm thick
	27322	0.049	
	25510	0.048	
	24038	0.042	
	22831	0.044	
	22422	0.048	
Sa38	26316	0.040	Clear ice with small
	22989	0.030	air bubbles. 2.5 cm thick
	19048	0.029	
	16949	0.031	
	15873	0.023	
	14286	0.023	
	13605	0.023	
	13158	0.029	
Ly59	18182	0.047	Average measurements
	16978	0.178	on artificial ice. First sample 2 % bubbles, second sample 4 % bubbles. Average diameter of bubbles 0.36 mm
Am62	25000	0.025	Homogeneous glacier
	20000	0.027	ice
	16666	0.032	
	14286	0.048	
	12500	0.089	
	11111	0.156	
	10000	0.224	
Sa38	26316	0.052	Snow ice. 4 cm thick.
	22989	0.038	Reflected light for this
	19048	0.021	sample was 45 to 49 %
	16949	0.038	
	15873	0.028	
	15152	0.021	
	14286	0.031	
	13605	0.038	
	13158	0.038	
	12346	0.061	
Pi47	Sunlight	0.031	Bubbly lake ice. 32 cm thick
Po50	White	0.015	4 cm of bubbly ice over
	Violet	0.046	4 cm of clear ice.
	Azure	0.023	Diameter of bubbles
	Green	0.015	2–6 mm
	Clear red	0.037	
	Dark red	0.049	

Fig. 14. The reflection coefficient R of clear polycrystalline ice at $-7\,°C$ as a function of wavelength λ and the angle of incidence i [Al66].

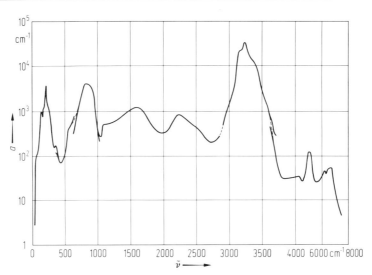

Fig. 15. Infrared absorption spectrum of ice in the range
30 to 7000 cm^{-1} [Be69].

8.4 Mechanical properties — Mechanische Eigenschaften

8.4.1 Elastic properties — Elastische Eigenschaften

Generally, the elastic behaviour of crystals is defined by:

Im allgemeinen wird das elastische Verhalten von Kristallen definiert durch

$$\varepsilon_i = S_{ij}\sigma_j \quad \text{or} \quad \sigma_i = C_{ij}\varepsilon_j \quad i, j = 1\cdots 6$$

where S_{ij} and C_{ij} are matrix elements representing compliance and stiffness constants respectively, σ_i and ε_i are the stress and strain components

wobei die S_{ij} und C_{ij} Elemente einer Matrix sind, und die Elastizitätskoeffizienten bzw. Elastizitätsmoduln darstellen. σ_i und ε_i stellen die Komponenten der Spannung und Verformung dar.

For ice Ih (hexagonal) there are only 5 non zero independent compliance constants, namely S_{11}, S_{12}, S_{13}, S_{33} and S_{44} or the corresponding C_{ij}.

Für Eis Ih (hexagonal) gibt es nur 5 von Null verschiedene unabhängige Elastizitätskoeffizienten, nämlich $S_{11}, S_{12}, S_{13}, S_{33}$ und S_{44}, bzw. die entsprechenden C_{ij}.

The following relationships hold:

Es gelten die folgenden Beziehungen:

$$C_{11} + C_{12} = \frac{S_{33}}{S}, \quad C_{11} - C_{12} = \frac{1}{S_{11} - S_{12}}$$

$$C_{13} = \frac{S_{13}}{S}; \quad C_{33} = \frac{S_{11} + S_{12}}{S}, \quad C_{44} = \frac{1}{S_{44}} \tag{9}$$

with $S = S_{33}(S_{11} + S_{12}) - 2S_{13}^2$.

mit $S = S_{33}(S_{11} + S_{12}) - 2S_{13}^2$

The bulk elastic behaviour can be described by the conventional elastic moduli, namely

Das elastische Verhalten des Eises als Gestein wird durch die üblichen elastischen Moduln beschrieben.

E: modulus of elasticity or Young's modulus
G: modulus of rigidity (also called μ)
γ: Poisson's ratio (also called σ)
K: bulk modulus
k: modulus of compressibility

E: Elastizitätsmodul oder Young'scher Modul
G: Scherungsmodul (häufig auch mit μ bezeichnet)
γ: Poisson Zahl (häufig auch mit σ bezeichnet)
K: Kompressionsmodul
k: Kompressibilitätsmodul

Since ice is slightly anisotropic, the bulk values given are dependent on the crystal orientation in polycrystalline natural ice and may therefore vary widely.

The same holds for the velocities of elastic waves, which for larger ice masses are bulk values dependent on temperature, density and water content of the ice. Natural ice in glaciers and ice sheets can be isotropic or anisotropic depending upon its stress history. In the Byrd station borehole (Antarctica), strong anisotropy was found within a certain depth range [Ko79].

Properties of isotropic polycrystalline ice can be derived from single crystal data by using apropriate averaging methods. Here use is made of the Voigt average [Vo28, Ro72].

The fundamental loss mechanism in single crystal ice can be described in terms of motion of water molecules in the crystal lattice which results in a Debye-type relaxation spectrum. In polycrystalline ice losses associated with grain boundary phenomena which predominate over the relaxation spectrum at temperatures above about $-10\,°C$ also exist. Both these processes are affected by impurities in the ice, the former by impurities in the crystal lattice, the latter by grain boundary impurities.

The Debye-type relaxation spectrum can be expressed by the equation

$$\text{tg}\,\sigma_m = (\text{tg}\,\sigma_m)_{max}\frac{2\omega\tau_m}{1+\omega^2\tau_m^2} \tag{10}$$

where ω is angular frequency
and τ_m the mechanical relaxation time which depends on temperature

$$\tau_m = \tau_{m_0}\cdot\exp\left(\frac{E_m}{kT}\right) \tag{11}$$

with E_m activation energy for mechanical relaxation given in eV and

$$\tau_{m_0} = 6.9\cdot10^{-16}\,\text{s}.$$

$$E_m = 0.57\,\text{eV}$$

which incidentally is the same as for dielectric relaxation [Fl70, Ku64].

The loss tangent for mechanical relaxation is equal to Q^{-1}, a quantity usually determined from the absorption of seismic waves by using the relation

$$Q^{-1} = \frac{\alpha v}{\pi f}$$

where
α: absorption coefficient
v: velocity of seismic wave
f: frequency of seismic wave.

Da Eis in seinem elastischen Verhalten anisotrop ist, sind die Eigenschaften von natürlichem polykristallinem Eis abhängig von der Orientierung der einzelnen Eiskörner und streuen dementsprechend. Dies gilt insbesondere für die Geschwindigkeiten elastischer Wellen, die im weiteren noch von der Temperatur, der Dichte und dem Wassergehalt abhängen. Natürliches Eis in Gletschern und Eisschilden kann isotrop oder anisotrop sein, je nach Beanspruchungsgeschichte. In der Bohrung an der Byrd-Station (Antarktis) konnte starke Anisotropie für ein bestimmtes Tiefenintervall nachgewiesen werden [Ko79].

Die Eigenschaften für polykristallines isotropes Eis können aus den Daten für Einkristalle bei geeigneter Mittelung abgeleitet werden. Hier wird die Mittelung nach Voigt durchgeführt [Vo28, Ro72].

Der Dämpfungsmechanismus in Eis-Einkristallen kann durch die Bewegung von Wassermolekülen im Kristallgitter beschrieben werden. Dies führt auf ein Relaxationsspektrum vom Debye-Typ. In polykristallinem Eis treten zusätzliche Dämpfungsverluste auf, die auf inelastische Vorgänge an den Korngrenzen zurückgeführt werden und bei Temperaturen oberhalb $-10\,°C$ gegenüber dem Relaxationsspektrum überwiegen. Beide Vorgänge werden durch Verunreinigungen im Eis beeinflußt, der erste durch Verunreinigungen im Kristallgitter, der zweite durch Verunreinigungen an den Korngrenzen.

Das Relaxationsspektrum vom Debye-Typ wird durch die Gleichung:

mit: ω Kreisfrequenz
τ_m mechanische Relaxationszeit, die selbst temperaturabhängig ist

mit: $E_m = 0.57\,\text{eV}$ Aktivierungsenergie für mechanische Relaxation

$$\tau_{m_0} = 6.9\cdot10^{-16}\,\text{s}$$

beschrieben. Die Aktivierungsenergie für die mechanische Relaxation ist dieselbe wie für die dielektrische Relaxation [Fl70, Ku64].

Der Tangens des Verlustwinkels ist gleich Q^{-1} und diese Größe wird im allgemeinen aus der Absorption seismischer Wellen bestimmt nach der Beziehung:

mit
α: Absorptionskoeffizient
v: Geschwindigkeit der seismischen Welle
f: Frequenz der seismischen Welle

Table 13. Polynomial constants A_i and B_i, giving the temperature dependence of the stiffness constants C_{ij} in [10^4 bar] and compliance S_{ij} constants in [10^{-6} bar] (T in °C) as

$$C_{ij} = A_0 + A_1 T + A_2 T^2, \quad S_{ij} = B_0 = B_1 T + B_2 T^2$$

in the temperature range $0 \cdots -140$ °C [Da69].

C_{ij}	A_0 10^4 bar	A_1 10^2 bar centigrade^{-1}	A_2 10^{-1} bar centigrade^{-2}
C_{11}	12.904	-1.921	-2.387
C_{12}	6.487	-1.344	-2.348
C_{13}	5.622	-1.053	0
C_{33}	14.075	-2.293	-4.124
C_{44}	2.819	-0.451	-1.02

S_{ij}	B_0 10^{-6} bar^{-1}	B_1 10^{-8} bar^{-1} centigrade^{-1}	B_2 10^{-11} bar^{-1} centigrade^{-2}
S_{11}	10.40	1.113	1.945
S_{12}	-4.42	-0.204	0.910
S_{13}	-1.89	-0.285	-1.162
S_{33}	8.48	1.191	3.9
S_{44}	33.42	5.029	13.5

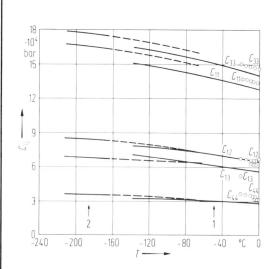

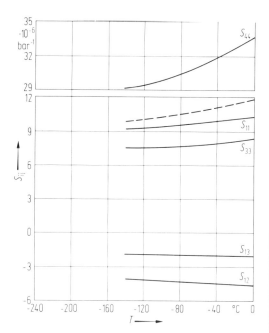

Fig. 16. Elastic stiffness constants C_{ij} of ice I vs. temperature 1 [Da69]; 2 [Pr66]; circles [Br64]. Dashed line is extrapolation of data from [Pr66] to higher temperatures. Anomalous behaviour between -200 and -120 °C is reported in [He69]. This is not shown in the graph.

Fig. 17. Compliance constants S_{ij} of ice I vs. temperature [Da69]. Dashed curve is adiabatic compressibility.

Table 14. Bulk elastic properties and seismic velocities of bubble free polycrystalline ice with randomly oriented crystals as computed from the elastic properties of single crystals as determined in the laboratory.

Ref.	T °C	ϱ Mg m^{-3}	Young's modulus E 10^4 bar	Shear modulus G 10^4 bar	Poisson's ratio γ	P-wave velocity v_p m s^{-1}	S-wave velocity v_s m s^{-1}
Jo52	−16	0.919	9.40	3.55	0.324	3851	1965
Ba57	−16	0.919	9.43	3.62	0.304	3736	1985
Br64	−16	0.919	9.424	3.585	0.3143	3796	1975
Be72	−16	0.918	9.403	3.545	0.3263	3870	1965
Da69	−16	0.919	8.82	3.32	0.329	3766	1900
	−5	0.917	8.702	3.27	0.328	3737	1889
	−25	0.920	8.920	3.354	0.330	3788	1909
	−35	0.921	9.022	3.391	0.330	3812	1918

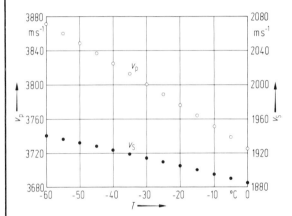

Fig. 18. P-wave and S-wave velocities of dry bubbly polycrystalline ice vs. temperature. Values calculated from elastic moduli [Da69] of ice.

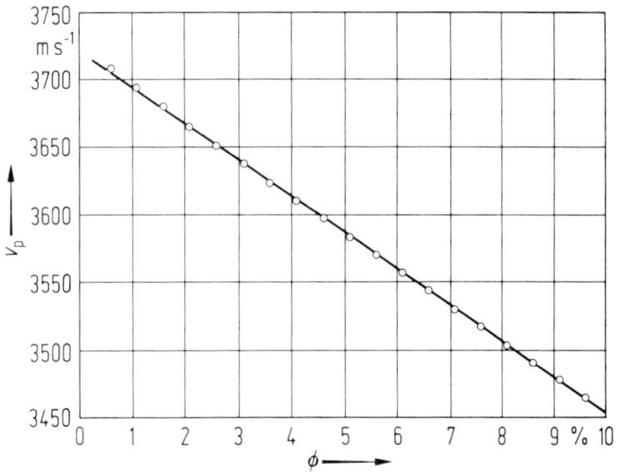

Fig. 19. Dependence of P-wave velocity of dry bubbly isotropic polycrystalline ice on porosity ϕ. Curve is calculated from elastic moduli of ice [Da69].

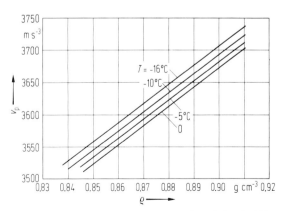

Fig. 20. P-wave velocity vs. bulk density of dry bubbly polycrystalline ice. Curves calculated from elastic moduli of ice [Da69].

Miller

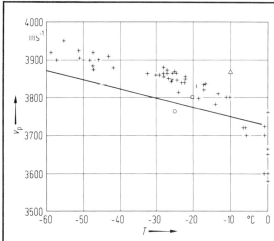

Fig. 21. P-wave velocity vs. temperature in ice sheets, cold and temperate glaciers [Ko74, Th67, Ro72, Ro58, Mi71]. Laboratory data from [Be72]: triangle; [Br64]: square; [Ba57]: circle. Solid curve calculated from data on elastic moduli of ice [Da69].

At temperatures below $-10\,°C$ the experimental data (crosses) yield the relationship
$v_p = 3795 - (2.3 \pm 0.17) \cdot T, \; [m/s^{-1}]$
$v_s = 1915 - (1.2 \pm 0.58) \cdot T, \; [m/s^{-1}], \; T \text{ in } °C \; [Ko74].$
These temperature gradients agree well within error limits with theoretical values from data on single crystals [Da69].

Table 15. Activation energies E_m of mechanical relaxation for various types of ice [Ho74].

Ice type	E_m eV	Cause of internal friction
pure H_2O ice, single crystal	0.57	proton reorientation
pure ice, polycrystalline	2.6···3.0	grain boundary relaxation
NaCl doped, polycrystalline ice	1.3···2.6	relaxation
natural glacier ice, polycrystalline	1.6	relaxation
natural ice, single crystal	0.22···0.59	relaxation

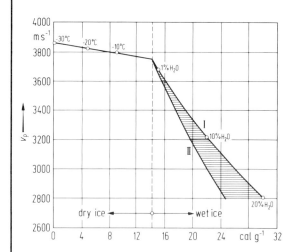

Fig. 22. P-wave velocity of ice vs. amount of heat introduced into the sample [Th67]. For wet ice the hatched region is bounded by curve I [Wy56] and II [La63]. (1 cal g^{-1} = 4.19 J g^{-1}.)

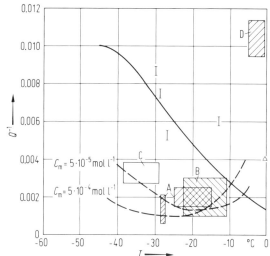

Fig. 23. Field determinations of Q^{-1} as function of temperature, T.
Areas A: data from Greenland [Br65];
area B: data from West Antarctica [Be76];
area C: data from East Antarctica [Be76, Ro69];
bars: data from Greenland [Ko69];
triangle: temperate ice (Blue Glacier) [We65];
area D: sea ice [La63];
solid line: single crystal [Ho74, Ku64];
dashed line: polycrystalline ice with impurities, C_m [Be76, Ku64].

8.4.2 Creep of ice — Kriechen von Eis

1. Single crystal data

Single crystal creep behaviour is strongly anisotropic with a pronounced direction of "easy glide" along the basal plane. During deformation along the basal plane work softening is observed, in other directions work hardening. For all types of deformation a strong temperature dependence is observed.

The general, relationship between steady state strain rate $\dot{\varepsilon}$ and stress σ can be expressed by

1. Daten für Einkristalle

Das Kriechverhalten von Einkristallen zeigt eine ausgeprägte Anisotropie mit einer bevorzugten Gleitebene parallel zur Basisfläche. Für Verzerrungen parallel der Basisfläche wird im Anfangsstadium des Kriechens Verformungserweichung, für Verzerrungen parallel zur *c*-Achse Verformungsverfestigung beobachtet. Alle Arten der Verformung sind stark temperaturabhängig.

Die allgemeine Beziehung zwischen stationärer Verzerrungsgeschwindigkeit $\dot{\varepsilon}$ und Spannung σ lautet:

$$\dot{\varepsilon} = A \cdot \exp\left(-\frac{E}{kT}\right) \cdot \sigma^n \tag{12}$$

where A and n are constants and
E activation energy.

The yield stress σ_{max} for constant strain rate deformation is given by

mit: A und n Konstanten
E Aktivierungsenergie

Die Nachgebespannung σ_{max} für Verzerrung bei konstanter Deformationsrate wird durch die folgende Gleichung beschrieben:

$$\sigma_{max} = B \exp - \frac{E}{kT} \dot{\varepsilon}^{1/m} \tag{13}$$

with: $m = 1.53$ and $E = 0.45$ eV [Hi69].

It appears that surface conditions of single crystals affect values of n and E.

The creep curves of single crystals of ice oriented for easy glide are qualitatively different from those of polycrystalline ice.

mit: $m = 1.53$ und $E = 0.45$ eV [Hi69].

Es gibt Hinweise, daß der Zustand der Kristalloberfläche die Werte von n und E beeinflußt.

Die Kriechkurven für Eis-Einkristalle für Verzerrung in der bevorzugten Gleitebene unterscheiden sich in der Form deutlich von denen für polykristallines Eis.

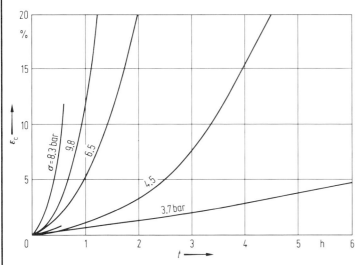

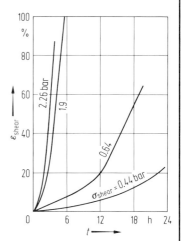

Fig. 24. Creep of single crystals of ice at temperatures between −10 and −11 °C (compressive strain ε_c vs. time t). Parameter: compressive stress σ. Stress was oriented at 45° with respect to *c*-axis. [Ho74, Gr54].

Fig. 25. Creep of single crystals of ice at −23 °C under shear stress in the basal plane (shear strain ε_{shear} vs. time t). Parameter: shear stress σ_{shear} [Ho74, St54].

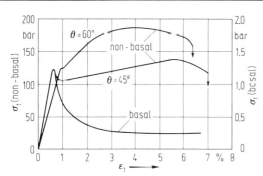

Fig. 26. Stress-strain curves for single crystal of ice for basal and non basal glide (tensile stress σ_t vs. tensile strain ε_t). θ is the angle between the a-axis and the direction of tension. Temperature $-19\,°C$. Strain rate $\dot\varepsilon = 3 \cdot 10^{-6}\,s^{-1}$ [Hi69].

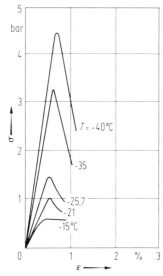

Fig. 28. Stress-strain (σ vs. ε) curves at various temperatures for single crystals of ice deformed by basal shear at a constant strain rate $\dot\varepsilon = 1.3 \cdot 10^{-7}\,s^{-1}$. Temperature range: $-15\,°C$ to $-40\,°C$ [Ho74, Hi69].

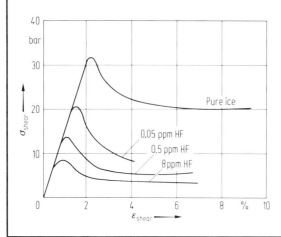

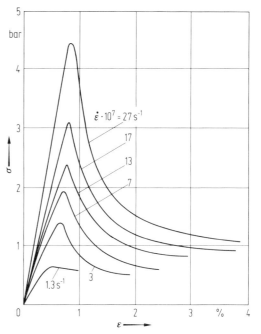

Fig. 27. Stress-strain (σ vs. ε) curves at $-15\,°C$ for single crystals of ice deformed by basal shear at constant strain rates $\dot\varepsilon$ given as parameter in $[10^{-7}\,s^{-1}]$. [Ho74, Hi69].

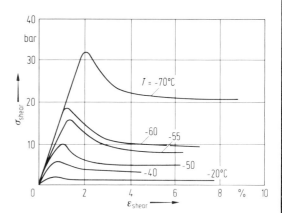

Fig. 29. Stress-strain (σ_{shear} vs. ε_{shear}) curves at different temperatures for single crystals of ice deformed by basal shear. Strain rate $\dot\varepsilon = 2.7 \cdot 10^{-7}\,s^{-1}$; temperature range: $-20\,°C$ to $-70\,°C$ [Jo68].

Fig. 30. Stress-strain (σ_{shear} vs. ε_{shear}) curves for single crystals of ice in easy glide at $-70\,°C$ for various concentrations of HF dissolved in the ice. Constant strain rate $\dot\varepsilon = 2.7 \cdot 10^{-7}\,s^{-1}$ [Jo69].

2. *Polycrystalline Ice*

The creep behaviour of polycrystalline ice is shown schematically in Fig. 31 and in Fig. 32 experimental data for the first part of the schematic strain vs. time diagram are given.

For moderate stresses ($1 < \sigma < 50$ bar) four different types of creep as function of time exist. They are usually named:

1. instantaneous stress upon loading
2. transient creep
3. secondary creep
4. tertiary creep

At high stresses (> 50 bar) secondary creep is not observed, while at low stresses the creep behaviour is dominated by transient processes and secondary creep is not reached even after several months [Ho74, We73].

The deformation can be described by

$$\varepsilon = \varepsilon_0 + \alpha t^{\frac{1}{3}} + \dot{\varepsilon}_{st} \cdot t \qquad (14)$$

where

ε_0 is the instantaneous strain
α a suitable constant
$\alpha t^{\frac{1}{3}}$ describes transient creep

$\dot{\varepsilon}_{st}$ steady-state creep rate.
t time [Ba71]

For the steady-state creep rate the relation

$$\dot{\varepsilon}_{St} = A' (\sinh \alpha' \sigma)^n \exp - \left(\frac{E}{kT} \right) \qquad (15)$$

is used with

A' and α' constants
E activation energy
k Boltzmann constant
T absolute Temperature.

For moderate stresses $\alpha' \sigma \leqq 0.8$ this equation is reduced to

$$\dot{\varepsilon}_{St} = A \sigma^n \exp \left(-\frac{E}{kT} \right) \qquad (16)$$

with $A = A'(\alpha')^n$.

The activation energy is multiple valued in the temperature range $0 \cdots -45$ °C [Ho74], and is higher in the temperature range $0 \cdots -11$ °C.

2. *Polykristallines Eis*

Das Kriechverhalten von polykristallinem Eis ist in Fig. 31 schematisch dargestellt und in Fig. 32 sind experimentelle Ergebnisse für den Anfangsteil des schematischen Verzerrungs-Zeit-Diagramms.

Für mittlere Spannungen ($1 < \sigma < 50$ bar) können vier verschiedene Arten des Kriechverhaltens unterschieden werden, die sich zeitabhängig gliedern lassen. Sie werden i. a. wie folgt bezeichnet:

1. Momentaner Anteil der Verformung
2. Übergangs-Kriechen
3. Sekundäres Kriechen
3. Tertiäres Kriechen

Bei hohen Spannungen (> 50 bar) wird sekundäres Kriechen nicht mehr beobachtet, während bei niedrigen Spannungen Übergangs-Kriechen vorherrscht und das Stadium des sekundären Kriechens auch nach mehreren Monaten nicht erreicht wird [Ho74, We73].

Die Verzerrung kann beschrieben werden durch:

mit:

ε_0 momentane Verzerrung
α geeignete Konstante
$\alpha t^{\frac{1}{3}}$ Ausdruck, der das Übergangs-Kriechen beschreibt

$\dot{\varepsilon}_{st}$ stationäre Kriechrate.
t Zeit [Ba71]

für die stationäre Kriechrate gilt die Beziehung:

mit:

A' und α' Konstanten
E Aktivierungsenergie
k Boltzmannkonstante
T absolute Temperatur.

Für mittlere Spannungen ($\alpha' \sigma \leqq 0.8$) kann diese Gleichung vereinfacht werden zu:

und $A = A' \cdot (\alpha')^n$.

Die Aktivierungsenergie ist im Temperaturbereich von $0 \cdots -45$ °C mehrdeutig [Ho74] und ist im Bereich von $0 \cdots -11$ °C höher.

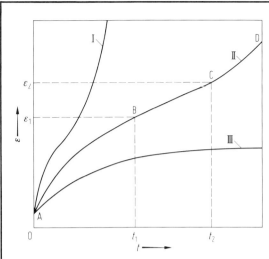

Fig. 31. Schematic representation of creep of polycrystalline ice at various stresses at constant temperature.
I: Flow under high stresses;
II: Flow under moderate stresses;
III: Flow under low stresses [Ho74].
Curve II shows 4 distinct regions,
0A: instantaneous stress which occurs on loading;
AB: transient creep;
BC: secondary creep;
CD: tertiary creep.

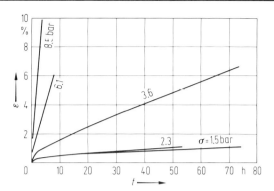

Fig. 32. Creep curves (strain ε vs. time t) of ice at $-0.02\,°C$ for various stresses σ in [bar] [Gl55].

Table 16. Values of n and Q in eg. (14) for steady state creep of single crystals from constant strain rate tests [We73, Jo69, Hi69].

n	Q	Glide direction	Remarks
2	0.62	parallel basal plane	
1.3	0.48	parallel basal plane	chemically polished
1.7	0.65	parallel basal plane	mechanically
2…3	0.43… 0.78	parallel basal plane	polished
1.58	0.68	parallel basal plane	
6.5	0.72	parallel c-axis	

Table 17. Values of constants A', α', n and E in eq. (15) [Ho74, Ba71].

A' $\%\,s^{-1}$	α' bar^{-1}	n	E eV	T °C
$4.6 \cdot 10^{20}$	$2.79 \cdot 10^{-2}$	1.3	1.3	$-2…-8$
$3.14 \cdot 10^{12}$	$2.54 \cdot 10^{-2}$	3.08	0.8	$-8…-14$
$1.88 \cdot 10^{12}$	$2.82 \cdot 10^{-2}$	2.92	0.8	$-14…-22$
$2.70 \cdot 10^{12}$	$2.62 \cdot 10^{-2}$	3.15	0.8	$-8…-45$

Table 18. Values of constants A, n and E in eq. (16) as analyzed in [Ba71] from data of [Gl55, St58]. Numerical values of constant A to be used when steady state strain rate $\dot\varepsilon_{st}$ is given in $\%\,s^{-1}$ and stress σ in MPa.

A	n	E eV	T °C	Remarks
$1.02 \cdot 10^{19}$	2.97	1.2	$-1.5…-12.8$	
$2.21 \cdot 10^{21}$	2.77	1.4	$-1.9…-11.5$	data from tensile and compressive experiments
$4.91 \cdot 10^{11}$	3.12	0.9	$-11.5…-21.5$	
$1.52 \cdot 10^{21}$	3.25	1.4	$-1.9…-11.5$	compression data only
$1.47 \cdot 10^{11}$	3.60	0.9	$-11.5…-21.5$	

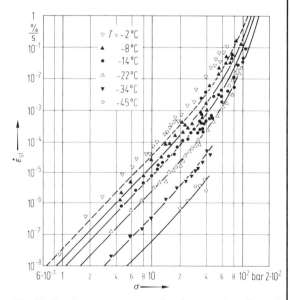

Fig. 33. Steady-state creep rate for the secondary flow of polycrystalline ice in uniaxial compression against applied stress, $\dot\varepsilon_{st}$ vs. σ. Curves are creep rates calculated with equation (15) [Ba71].

8.4.3 Strength of ice — Festigkeit von Eis

Failure of ice during loading at high strain rates is preceded by or due to the formation of cracks either at the surface or in the interior. Initiation and propagation of cracks are dependent on stress, temperature, strain, strain rate, grain size and structure. Therefore there are large variations in reported values. This is especially true for sea ice where brine volume (content) plays a very important role.

Bei rascher Verformung von Eis kommt es zum spröden Bruch. Vor dem eigentlichen Bruch kommt es zur Ausbildung von Rissen an der Oberfläche oder auch im Innern der Probe. Beginn und Ausbreitung der Risse hängen von der Spannung, der Temperatur, der Verzerrung, der Verzerrungsrate und der Größe und Struktur der Eiskörner ab. Daher streuen die beobachteten Werte erheblich. Dies trifft besonders für Meereis zu, wo der Laugengehalt eine entscheidende Rolle spielt.

Table. 19. Values of strength, σ, of different types of ice for different types of stress. $\dot{\varepsilon}$ = strain rate.

σ bar	Type	T °C	$\dot{\varepsilon}$ s^{-1}	Remarks	Ref.
10···20	tensile	−10	10^{-3}	bubbly polycrystalline ice	Go77
2.5···16	tensile	−10		sea ice with varying brine volume	Sc77
50···80	compressive	−10	10^{-3}	bubbly polycrystalline ice	Ha77
30···60	compressive	−10	10^{-3}	sea ice with salinity 2.7‰	Sc77
2.8···15	flexural	−3···0		lake ice	Go77a
1···10	flexural	< −1		sea ice with varying brine volume	Sc77

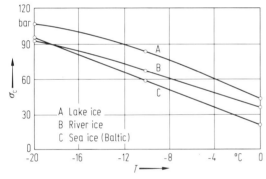

Fig. 34. Compressive strength σ_c of freshwater and sea ice as a function of temperature T. Load applied perpendicular to the growth direction [Sc77]. Strain rate $\dot{\varepsilon} = 3 \cdot 10^{-3}\,\text{s}^{-1}$.

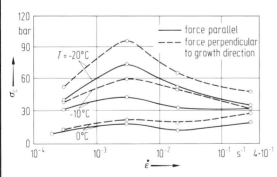

Fig. 35. Compressive strength σ_c of sea ice (salinity 2.7‰) as function of strain rate $\dot{\varepsilon}$. Parameter: ice temperature and orientation of stress [Sc77].

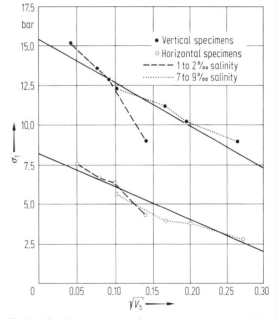

Fig. 36. Tensile strength σ_t of sea ice vs. square root of the brine volume V_b [Sc77].

8.5 References for 8 — Literatur zu 8

Al66	Alkezweeny, A.J., Hobbs, P.V.: J. Geophys. Res. **71** (1966) 1083–1086.
Am62	Ambach, W., Habicht, H.L.: Arch. Met. Geophys. Bioklim. Ser. B**11** (1962) 512–531.
An67	Andrieux, G.: Thesis Diplome, IPG Strasbourg, **1967**.
Au52	Auty, R.P., Cole, R.H.: J. Chem. Phys. **20** (1952) 1309–1314.
Ba57	Bass, R., Rossberg, D., Ziegler, G.: Z. Phys. **149** (1957) 199–203.
Ba71	Barnes, P., Tabor, D., Walker, J.C.F.: Proc. R. Soc. London A**324** (1971) 127–155.
Be69	Bertie, J.E., Labbé, J.H., Whalley, E.: J. Chem. Phys. **50** (1969) 4501–4520.
Be72	Bennett, H.F.: CRREL Res. Rep. 237, Hanover, N.H., **1972**.
Be76	Bentley, C.R., Kohnen, H.: J. Geophys. Res. **81** (1976) 1519–1526.
Br64	Brockamp, B., Querfurth, H.: Polarforschung V**34** (1964) 253–262.
Br65	Brockamp, B., Kohnen, H.: Polarforschung V**35** (1965) 2–12.
Br67	Brill, R.: Acta Crystallog. **23** (1967) 343–345.
Bu66	Bullemer, B., Riehl, N.: Solid State Commun. **4** (1966) 447–448.
Bu69	Bullemer, B., Engelhardt, H., Riehl, N., in: Physics of Ice, New York: Plenum Press, **1969**, p. 416–429.
Ca69	Camp, P.R., Kiscenick, W., Arnold, D.A., in: Physics of Ice, New York: Plenum Press, **1969**, p. 450–470.
Cl70	Clough, J.W., Bentley, C.R.: IASH Publ. 68, Proc. ISAGE **1970**, p. 115–127.
Co69	Cole, R.H., Wörz, O., in: Physics of Ice, New York: Plenum Press, **1969**, p. 546–554.
Cu52	Cumming, W.A.: J. Appl. Phys. **23** (1952) 768–773.
Da62	Dantl, G.: Z. Phys. **166** (1962) 115–118.
Da69	Dantl, G., in: Physics of Ice, New York: Plenum Press, **1969**, p. 223–230.
Ei64	Eigen, M., De Mayer, L., Spatz, H.: Z. Elektrochem. **68** (1964) 19–29.
En69	Engelhardt, H., Bullemer, B., Riehl, N., in: Physics of Ice, New York: Plenum Press, **1969**, p. 430–442.
Fi75	Fitzgerald, W.J., Paren, J.G.: J. Glaciol. **15** (1975) 39–48.
Fl70	Fletcher, N.H.: The chemical physics of ice, Cambridge: Cambridge University Press **1970**.
Gl55	Glen, J.W.: Proc. R. Soc. London A**228** (1955) 519–538.
Gl75	Glen, J.W., Paren, J.G.: J. Glaciol. **15** (1975) 15–38.
Go70	Gough, S.R., Davidson, D.W.: J. Chem. Phys. **52** (1970) 5442–5449.
Go77	Gold, L.W.: J. Glaciol. **19** (1977) 197–212.
Go77a	Gow, A.J.: J. Glaciol. **19** (1977) 247–256.
Gr54	Griggs, D.T., Coles, N.E.: SIPRE Tech. Rep. 11, 24, **1954**.
Ha77	Haynes, F.D., Mellor, M.: J. Glaciol. **19** (1977) 213–223.
He69	Helmreich, D., in: Physics of Ice, New York: Plenum Press, **1969**, p. 231–238.
Hi69	Higashi, A., in: Physics of Ice, New York: Plenum Press, **1969**, p. 197–212.
Hi69a	von Hippel, A., Westphal, W.B., Knoll, D.B., Maidique, M.A., Mykolajewycz, R.: U.S. Clearinghouse Fed. Sci. Tech. Inform. AD-689379, 59 pp, **1969**.
Hi71	von Hippel, A., Mykolajewycz, R., Runck, A.H., Westphal, W.B.: Tech. Rep. 10, Lab. Insulation Res. M.I.T., **1971**.
Ho74	Hobbs, P.V.: Ice Physics, Oxford: Clarendon Press, **1974**.
Ho79	Hooke, R.LeB., Mellor, M., Jones, S.J., Martin, R.T., Meier, M.F., Weertman, J.: Rep. ICSI/NSF WG ice mechanics, **1979**, 16pp.
Hu53	Humbel, F., Jona, F., Scherrer, P.: Helv. Phys. Acta **26** (1953) 17–32.
Ir68	Irvine, W.M., Pollack, J.B.: Icarus **8** (1968) 324–360.
Ja67	Jaccard, C., in: Physics of snow and ice, Inst. Low Temp. Sci. Hokkaido Univ. Sapporo, **1967**, p. 173–179.
Jo52	Jona, F., Scherrer, P.: Helv. Phys. Acta **25** (1952) 35–54.
Jo68	Jones, S.J., Glen, J.W.: IASH Publ. 79, Gentbrugge **1968**, p. 326–340.
Jo69	Jones, S.J., Glen, J.W.: Phil. Mag. **19** (1969) 13–24.
Jo78	Johari, G.P., Jones, S.J.: J. Glaciol. **21** (1978) 259–300.
Ka35	Kalitin, N.N.: C.R. Acad. Sci. URSS **2** (1935) 216–221.
Ka69	Kahane, A., in: Physics of Ice, New York: Plenum Press **1969**, p. 443–449.
Ko69	Kohnen, H.: Polarforschung VI, **1** (1969) 269–275.
Ko74	Kohnen, H.: J. Glaciol. **13** (1974) 144–147.
Ko79	Kohnen, H., Gow, A.J.: Geophys. Res. **84** (1979) 4865–4874.
Ku64	Kuroiwa, D.: Contrib. Inst. Low Temp. Sci., Hokkaido Univ. A**18** (1964) 1.62.
La49	Lamb, J., Turney, A.: Proc. Phys. Soc. B**62** (1949) 272–273.
La60	La Placa, S.J., Post, B.: Acta Crystallog. **13** (1960) 503–505.
La63	Langleben, M.P., Pounder, E.R., in: Ice and Snow, Cambridge: MIT Press **1963**, p. 69–78.

Ly59	Lyons, J.B., Stoiber, R.E.: Sci. Rep. 3, Dartmouth College, Hanover, N.H., **1959**, 13 pp.
Ma65	Mathes, K.N., in: Annual Report: Conf. Electr. Insulation, Natl. Res. Council (U.S.A.) Pub. 1356, **1965**, p. 20–103.
Ma70	Maidique, M.A., Von Hippel, A., Westphal, W.B.: J. Chem. Phys. **54** (1970) 150–160.
Me62	Meyer, A.U., Röthlisberger, H.: CRREL Tech. Rep. 87, Hanover N.H., **1962**.
Mi71	Miller, H.: Dissertation, Univ. München **1971**.
Mi73	Miller, H.: Abstr. Tagung DGG, Karlsruhe **1973**.
Mc73	McNeil, D., Hoekstra, P.: Radio Sci. **8** (1973) 23–30.
Pa70	Paren, J.G.: Ph. D. Thesis, Univ. of Cambridge, 233 pp, **1970**.
Pi47	Pisiakova, N.M.: Met. Gidrol. Nabl. Transp. Palkhtusov **6** (1947) 65.
Po50	Polli, S.: Ann. Geofis. **3** (1950) 371–377.
Po65	Pounder, E.R.: Physics of Ice, Oxford: Pergamon Press **1965**.
Pr66	Proctor, T.M.: J. Acoust. Soc. Am. **39** (1966) 972–977.
Ra62	Ratcliffe, E.H.: Phil. Mag. **7** (1962) 1197–1203.
Ra76	Raymond, C.F.: J. Glaciol. **16** (1976) 159–171.
Ro58	Robin, G. de Q.: Sci. Results Norw. Brit. Swed. Antarctic Expedition 1949–52, Vol. 5, **1958**.
Ro67	Röthlisberger, H., Vögtli, K.: J. Glaciol. **47** (1967) 607–621.
Ro69	Robin, G. de Q., Evans, S., Bailey, J.T.: Phil. Trans. R. Soc. London A **265** (1969) 437–505.
Ro72	Röthlisberger, H.: CRREL Monograph II-A2a, Hanover, N.H. **1972**.
Ru69	Runnels, L.K., in: Physics of Ice, New York: Plenum Press **1969**, p. 514–526.
Sa38	Sauberer, F.: Met. Z. **55** (1938) 250–255.
Sc63	Schwerdtfeger, P.: J. Glaciol. **4** (1963) 789–807.
Sc77	Schwarz, J., Weeks, W.F.: J. Glaciol. **19** (1977) 499–531.
St54	Steinemann, S.: J. Glaciol. **2** (1954) 404–412.
St58	Steinemann, S.: Beitr. Geol. Karte Schweiz, Geopech. Ser. Hydr. **10** (1958) 1–72.
Th67	Thyssen, F.: Z. Geophys. **33** (1967) 65–79.
Vo28	Voigt, W.: Lehrbuch der Kristallphysik, Leipzig: Teubner **1928**.
Vo67	Vögtli, K.: J. Glaciol. **6** (1967) 635–642.
We65	Westphal, J.A.: J. Geophys. Res. **70** (1965) 1849–1853.
We73	Weertman, J., in: Physics and Chemistry of Ice, Ottawa: Royal Soc. Canada **1973**, p. 320–337.
Wo69	Wörz, O., Cole, R.H.: J. Chem. Phys. **51** (1969) 1546–1550.
Wy56	Wyllie, M.R.J., Gregory, A.R., Gardner, L.W.: Geophys. **21** (1956) 41–70.
Yo68	Young, I.G., Salomon, R.E.: J. Chem. Phys. **48** (1968) 1635–1644.

9 Physical properties of lunar rocks —
Physikalische Eigenschaften von Gesteinen des Mondes

9.0 Introduction — Einleitung

Lunar rock samples were brought to the earth by 6 manned Apollo missions and by three unmanned Luna missions (Table 1). Table 1 also shows the identification numbers for lunar rocks of the different Apollo missions. In the five-digit numbering system the first two digits give the mission for the Apollo 11, 12, 14, and 15 samples (10, 12, 14, 15). For the Apollo 16 and 17 missions the first digit is 6 and 7, respectively, and the second digit indicates the collecting station. Subsamples are identified by additional numbers separated by a comma from the sample number (e.g. 14310,72).

Gesteine des Mondes wurden bei 6 bemannten Apollo-Missionen und bei 3 unbemannten Luna-Missionen zur Erde gebracht (Tabelle 1). In Tabelle 1 sind auch die Kennzahlen zur Identifizierung der Gesteinsproben der verschiedenen Apollo-Missionen angegeben. Die Probennummern sind fünfstellige Zahlen bei denen für Apollo 11, 12, 14 und 15 die zwei ersten Ziffern die Mission bezeichnen (10, 12, 14, 15) und für Apollo 16 und 17 als erste Ziffer eine 6 respektive eine 7 verwendet wurde. Die zweite Ziffer kennzeichnet bei diesen zwei Missionen den Fundort im Landegebiet. Teile von Proben werden durch zusätzliche Ziffern bezeichnet, die durch ein Komma von der Probennummer getrennt sind (z.B. 14310,72).

Table 1. Mission landing sites.

Mission	Site	Latitude	Longitude	Date	Sample number prefix
Apollo landing sites					
Apollo 11	Mare Tranquillitatis (Mare)	0° 67′ N	23° 49′ E	20.07.1969	10
Apollo 12	Oceanus Procellarum (Mare)	3° 12′ S	23° 23′ W	19.11.1969	12
Apollo 14	Fra Mauro (Ejecta blanket formation from Imbrium basin)	3° 40′ S	17° 28′ E	31.01.1971	14
Apollo 15	Hadley-Appenines (Highlands bordering mare-filled Imbrium basin)	26° 06′ N	3° 39′ E	30.07.1971	15
Apollo 16	Descartes (Highlands)	8° 60′ S	15° 31′ E	21.04.1972	6
Apollo 17	Taurus-Littrow (Highlands bordering mare-filled Serenitatis basin)	20° 10′ N	30° 46′ E	11.12.1972	7
Luna landing sites					
Luna 16	Mare Fecunditatis	0° 41′ S	56° 18′ E	19.09.1970	
Luna 20	Apollonius highlands	3° 32′ N	56° 33′ E	21.02.1972	
Luna 24	Mare Crisium	12° 45′ N	62° 12′ E	18.08.1976	

9.0.1 Classification of lunar rocks — Klassifizierung der Gesteine des Mondes

Lunar rock samples were collected from the regolith, which forms the uppermost layer on the moon. The regolith consists of debris ranging in size from finest dust up to blocks several tens of meters in diameter. The average thickness of the regolith in the maria is about 5 m and in the highlands about 10 m, but may reach several tens of m in some places.

The formation of the regolith is attributed predominantly to the continuous bombardment of the lunar surface by cosmic bodies of various sizes. Whereas micrometeorites have mainly an erosional effect, larger impacting bodies can excavate the lunar surface deeply, bring bedrock samples to the surface and form extended ejecta blankets. Multiple excavation and ejection result in mixing of various rock types all over the lunar surface. Shock metamorphic effects associated with impacts are leading to evaporation, melting, comminution and brecciation of rocks and to characteristical physical properties.

A preliminary classification established since the Apollo 11 landing distinguishes crystalline rocks, breccias and loose soil material. Crystalline rocks are predominantly igneous rocks. In some cases they are considered to result from impact melts. Breccias consist of various rock, mineral, and glass fragments and glass droplets embedded in a finer matrix material which sometimes also contains a large amount of glass. Breccias can be monomict or polymict. Breccias in breccias indicate a complex multigeneration impact history. Some breccias have a very weak cohesion whereas others are strongly consolidated, depending on the shock and temperature conditions during their formation.

The fine-grained loose lunar soil contains local and remote rock and mineral fragments, glass droplets and fragments, agglutinates welded together by glass and meteoritic material. In many cases, the fraction with sizes <1 mm is used for measurements of physical properties.

Die zur Erde gebrachten Proben stammen alle aus dem Regolith, der die oberste Schicht des Mondes bildet. Der Regolith ist eine Schuttschicht aus unsortierten Bestandteilen, von feinstem Staub bis zu mehreren m großen Blöcken. Die mittlere Mächtigkeit des Regoliths beträgt in den Maregebieten rund 5 m und in den Hochländern rund 10 m, kann aber an manchen Stellen bis zu mehreren zehner m reichen.

Die Bildung des Regoliths wird hauptsächlich durch das andauernde Bombardement der Mondoberfläche mit kosmischen Körpern erklärt. Mikrometeoriten haben vorwiegend eine erodierende Wirkung. Größere Einschlagskörper können tiefe Krater erzeugen, Proben des im Untergrunde anstehenden Gesteins zur Oberfläche bringen und ausgedehnte Auswurfdecken bilden. Mehrfache Ausgrabung und Auswurf durch Einschläge haben eine starke Durchmischung der oberflächennahen Gesteine des Mondes zur Folge. Mit den Einschlägen verbunden sind die Prozesse der Stoßwellenmetamorphose, die zu Verdampfung, Schmelzen, feiner Zertrümmerung und Breccierung der Gesteine und damit zu besonderen physikalischen Eigenschaften führen.

Seit Apollo 11 werden die zur Erde gebrachten Proben in kristalline Gesteine, Breccien und Mondboden eingeteilt. Kristalline Gesteine sind vorwiegend magmatische Gesteine. In einigen Fällen sind sie aus Impaktschmelzen entstanden. Die Breccien bestehen aus verschiedenen Gesteins-, Mineral- und Glasbruchstücken und Glaströpfchen, die in einer feineren Matrix eingebettet sind, die ebenfalls viel Glas enthalten kann. Breccien können monomikt oder polymikt sein. Breccien in Breccien zeigen eine Genese durch multiple Einschläge an. Manche Breccien sind leicht zerreiblich, während andere stark verfestigt sind, je nach Druck- und Temperaturbedingungen, die bei ihrer Bildung herrschten.

Der feinkörnige, lockere Mondboden enthält Gesteins-, Mineral- und Glasbruchstücke, Glaströpfchen, Agglutinate, die durch Glas zusammengeschweißt sind, und meteoritisches Material. Häufig wurde bei Untersuchungen der physikalischen Eigenschaften des Mondbodens die Fraktion <1 mm verwendet.

9.0.2 Chemical composition and petrology of igneous lunar rocks — Chemische Zusammensetzung und Petrologie der magmatischen Gesteine

According to the general division of the lunar surface into the lighter, densely cratered highlands or Terrae and the darker Maria, the lunar igneous rocks are divided into highland rock types and mare basalts. Tables 2 and 3 ([Ta75]) give the major element concentration in various highland and mare rocks.

Entsprechend der Einteilung der Oberfläche des Mondes in die helleren Hochländer oder Terrae und die dunkleren Maria werden bei den magmatischen Gesteinen des Mondes Hochlandgesteine und Marebasalte unterschieden. Die Tabellen 2 und 3 geben die chemische Zusammensetzung der wichtigsten Hochlandgesteine und Marebasalte an (nach [Ta75]).

Table 2. Composition of lunar highland rocks [Ta75].

Element oxides in weight %. Composition of average highlands estimated from orbital measurements and observed inter-element ratios [Ta75].

(wt %)	Anorthosite	Gabbroic anorthosite	Anorthositic gabbro	Troctolite	Low-K Fra Mauro basalt	Medium-K Fra Mauro basalt	Average highlands
SiO_2	44.3	44.5	44.5	43.7	46.6	48.0	45.0
TiO_2	0.06	0.35	0.39	0.17	1.25	2.1	0.56
Al_2O_3	35.1	31.0	26.0	22.7	18.8	17.6	24.6
FeO	0.67	3.46	5.77	4.9	9.7	10.9	6.6
MnO				0.07			
MgO	0.80	3.38	8.05	14.7	11.0	8.70	8.6
CaO	18.7	17.3	14.9	13.1	11.6	10.7	14.2
Na_2O	0.80	0.12	0.25	0.39	0.37	0.70	0.45
K_2O					0.12	0.54	0.075
Cr_2O_3	0.02	0.04	0.06	0.09	0.26	0.18	0.10
Σ	100.5	100.2	99.9	99.9	99.6	99.4	99.3

Table 3. Composition of lunar mare basalts [Ta75]. Element oxides in weight %.

(wt %)	Olivine basalt Apollo 12	Olivine basalt Apollo 15	Quartz basalt Apollo 15	Quartz basalt Apollo 12	High K basalt Apollo 11	Low K basalt Apollo 11	High Ti basalt Apollo 17	Aluminous mare basalts Apollo 12	Aluminous mare basalts Luna 16
SiO_2	45.0	44.2	48.8	46.1	40.5	40.5	37.6	46.6	45.5
TiO_2	2.90	2.26	1.46	3.35	11.8	10.5	12.1	3.31	4.1
Al_2O_3	8.59	8.48	9.30	9.95	8.7	10.4	8.74	12.5	13.9
FeO	21.0	22.5	18.6	20.7	19.0	18.5	21.5	18.0	17.8
MnO	0.28	0.29	0.27	0.28	0.25	0.28	0.22	0.27	0.26
MgO	11.6	11.2	9.46	8.1	7.6	7.0	8.21	6.71	5.95
CaO	9.42	9.45	10.8	10.9	10.2	11.6	10.3	11.82	12.0
Na_2O	0.23	0.24	0.26	0.26	0.50	0.41	0.39	0.66	0.63
K_2O	0.064	0.03	0.03	0.071	0.29	0.096	0.08	0.07	0.21
P_2O_5	0.07	0.06	0.03	0.08	0.18	0.11	0.05	0.14	0.15
S	0.06	0.05	0.03	0.07			0.15	0.06	
Cr_2O_3	0.55	0.70	0.66	0.46	0.37	0.25	0.42	0.37	
Σ	99.77	99.46	99.08	100.23	99.67	99.85	99.58	100.2	100.42
Sample	12009	15555	15076	12052	Average	Average	71055	12038	B-1 A-35

In the following sections, generally no distinction is made between the various basalt types for the description of physical properties; rocks containing more than $\approx 20\%$ Al_2O_3 are referred to as anorthositic rocks or anorthosites.

Bei der Beschreibung der physikalischen Eigenschaften der Gesteine wird in der Regel nicht zwischen den verschiedenen Basaltarten unterschieden, und Gesteine, die mehr als 20% Al_2O_3 enthalten, werden allgemein als anorthositische Gesteine oder Anorthosite bezeichnet.

9.0.3 Composition of the regolith — Die Zusammensetzung des Regoliths

The overall composition of the regolith is dominated by the local rock types forming the geological formations in the vicinity of the landing sites: Basaltic mare material at Apollo 11, 12, 15 and 17 sites; highland material at Apollo 15, 16 and 17 sites; ejecta

Die allgemeine Zusammensetzung des Regoliths ist vorwiegend bestimmt durch die lokalen Gesteinsformationen in der näheren Umgebung der Landeplätze: Durch basaltisches Marematerial bei Apollo 11, 12, 15 und 17, durch Hochlandgesteine bei Apollo

blanket formation of large basins (predominantly high-land crustal rocks) at Apollo 14 site (see Table 1). A minor part of the regolith, for example anorthositic components of the highlands in the mare regolith at the Apollo 11 site, originates in distant sources which may be many hundreds of km away from the sampling site.

15, 16 und 17 und durch Auswurfformationen eines großen Beckens bei Apollo 14, die allerdings auch hauptsächlich aus anorthositischen Krustengesteinen bestehen. Ein kleinerer Teil des Regoliths besteht aus Material, das von bis zu mehreren hundert km entfernten Orten stammt, wie z.B. anorthositische Bestandteile aus dem Hochland im Mare-Regolith von Apollo 11.

9.0.4 Physical conditions at the lunar surface — Die physikalischen Bedingungen an der Oberfläche des Mondes

Many of the investigated physical properties discussed in the following sections are determined – to a large extent – by the physical conditions near the lunar surface. The Moon has an extremely tenuous atmosphere with a pressure in the order of 10^{-12} Torr and the environment of the lunar surface is characterized by the absence of volatiles, especially H_2O, and of oxidizing factors. Material at the surface is exposed directly to solar electromagnetic radiation, to solar and galactic cosmic rays and to the solar wind. Temperature variations near the surface, where most samples were collected, are extremely large, as shown in Fig. 1. However, at a depth of about 1 m the temperature is nearly constant at about $-50\,°C$.

Für viele der untersuchten physikalischen Eigenschaften der Gesteine des Mondes sind die physikalischen Bedingungen nahe der Oberfläche des Mondes von großer Bedeutung. Der Mond hat mit einem Druck von etwa 10^{-12} Torr eine extrem dünne Atmosphäre, und flüchtige Stoffe, wie zum Beispiel Wasserdampf, fehlen ebenso wie oxidierende. Das Material an der Oberfläche des Mondes ist der elektromagnetischen Strahlung der Sonne, der solaren und galaktischen kosmischen Strahlung und dem Sonnenwind direkt ausgesetzt. Die Temperaturschwankungen nahe der Oberfläche, wo die meisten Proben gesammelt wurden, sind sehr groß (Fig. 1). In einer Tiefe von 1 m herrscht jedoch eine annähernd konstante Temperatur von rund $-50\,°C$.

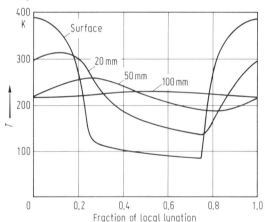

◄

Fig. 1. Calculated temperature variation near the lunar surface [Cr74].

Measurements of physical properties of lunar rocks should be carried out in conditions identical to those at the lunar surface in order to obtain characteristic values. Many measurements have been made in conditions approximating the lunar surface environment relatively well, especially in the later phases of the investigations. At the beginning of the investigations the importance of several factors, such as atmospheric humidity, was underestimated in some cases. Most samples from the Apollo 11, 12 and 14 missions were exposed to normal atmospheric conditions before measurement and irreversible or only partly reversible changes of some physical properties took place. Samples from the later missions stored in dry, non-oxidizing atmosphere did not show such changes. Measurements at elevated temperature, especially of electrical and

Im Idealfall sollten die physikalischen Eigenschaften der Gesteine des Mondes unter denselben Bedingungen gemessen werden, die an der Oberfläche des Mondes herrschen, um repräsentative Werte zu erhalten. In vielen Fällen wurde diese Forderung relativ gut erfüllt, besonders in den späteren Untersuchungsphasen. Zu Beginn der Untersuchungen wurde die Bedeutung mancher Faktoren, wie z.B. der Luftfeuchtigkeit, etwas unterschätzt und die meisten Proben von Apollo 11, 12 und 14 waren vor den Messungen über längere Zeit den normalen atmosphärischen Bedingungen ausgesetzt. Hierbei traten irreversible oder nur zum Teil reversible Veränderungen auf. Solche Veränderungen wurden bei Proben der späteren Flüge, die unter trockenen, nicht-oxidierenden Bedingungen gelagert wurden, nicht beobachtet. Für Messungen

magnetic properties, require an extremely careful control of the oxygen fugacity in order not to modify the reduced state of lunar material. Some of these problems are discussed briefly in the following sections.

bei höheren Temperaturen, besonders der elektrischen und magnetischen Eigenschaften, muß sehr sorgfältig auf die Einstellung der richtigen Sauerstoff-Fugazität geachtet werden, um den stark reduzierten Zustand der Mondgesteine nicht zu verändern.

9.0.5 List of symbols — Symbolliste

A^* $[W\,kg^{-1}]$	radiogenic heat — radiogene Wärme
C $[Pa]$	dynamic elastic modulus — dynamischer Elastizitätsmodul
c	content — Gehalt
c_p $[J\,kg^{-1}\,K^{-1}]$	specific heat at constant pressure — spezifische Wärme bei konstantem Druck
D $(=\tan\delta)$	dissipation factor — Verlustfaktor
E	elastic wave energy — Energie elastischer Wellen
f $[Hz]$	frequency — Frequenz
H $[Oe]$	magnetic field — Magnetfeld
H_c $[Oe]$	coercive force — Koerzitivkraft
H_{rc} $[Oe]$	remanent coercive force — remanente Koerzitivkraft
I $[G\,cm^3\,g^{-1}]$	specific magnetization — spezifische Magnetisierung
I_r $[G\,cm^3\,g^{-1}]$	specific remanent magnetization — spezifische remanente Magnetisierung
I_{rs} $[G\,cm^3\,g^{-1}]$	specific saturation remanent magnetization — spezifische remanente Sättigungsmagnetisierung
I_s $[G\,cm^3\,g^{-1}]$	specific saturation magnetization — spezifische Sättigungsmagnetisierung
K $[Pa]$	static bulk modulus — statischer Kompressionsmodul
K'	dielectric constant — Dielektrizitätskonstante
K_{dyn} $[Pa]$	dynamic bulk modulus — dynamischer Kompressionsmodul
k $[W\,m^{-1}\,K^{-1}]$	thermal conductivity — Wärmeleitfähigkeit
L $[m]$	length — Länge
m	abundance — Häufigkeit
NRM	natural remanent magnetization — natürliche remanente Magnetisierung
P $[Pa]$	pressure — Druck
p $[Torr, atm]$	gaseous pressure - Gasdruck
Q	quality factor — Qualitätsfaktor
S	linear strain — lineare Deformation
T $[K]$	temperature — Temperatur
T_C $[K]$	Curie temperature — Curie-Temperatur
$T_{\alpha\gamma}$ $[K]$	$\alpha\rightleftharpoons\gamma$ transition temperature — $\alpha\rightleftharpoons\gamma$-Übergangstemperatur
V $[m^3]$	volume — Volumen
$V_{specific}$ $[m^3\,kg^{-1}]$	specific volume — spezifisches Volumen
v_p $[km\,s^{-1}]$	compressional wave velocity (P wave velocity) — Kompressionswellengeschwindigkeit (P-Wellen-Geschwindigkeit)
v_s $[km\,s^{-1}]$	shear wave velocity (S wave velocity) — Scherwellen-Geschwindigkeit (S-Wellen-Geschwindigkeit)
α $[K^{-1}]$	thermal volume expansion coefficient — thermischer Volumen-Ausdehnungskoeffizient
β $[Pa^{-1}]$	compressibility — Kompressibilität
$\tan\delta$	loss tangent — Verlustwinkel
ε_0 $[A\,s\,V^{-1}\,m^{-1}]$	vacuum permittivity — absolute Dielektrizitätskonstante des Vakuums
κ $[m^2\,s^{-1}]$	thermal diffusivity — thermische Diffusivität
ϱ $[kg\,m^{-3}]$	density — Dichte
ϱ_{bulk} $[kg\,m^{-3}]$	bulk density — Rohdichte
ϱ_{intr} $[kg\,m^{-3}]$	intrinsic density — Matrixdichte
σ' $[\Omega^{-1}\,m^{-1}]$	dielectric conductivity — dielektrische Leitfähigkeit
$\sigma'_{d.c.}$ $[\Omega^{-1}\,m^{-1}]$	d.c. electrical conductivity — Gleichstrom-Leitfähigkeit
ϕ $[\%]$	porosity — Porosität
χ_g $[G\,cm^3\,g^{-1}\,Oe^{-1}]$	specific susceptibility — spezifische Suszeptibilität
χ_{ga} $[G\,cm^3\,g^{-1}\,Oe^{-1}]$	specific paramagnetic susceptibility — spezifische paramagnetische Suszeptibilität
χ_{g0} $[G\,cm^3\,g^{-1}\,Oe^{-1}]$	specific initial susceptibility — spezifische Anfangssuszeptibilität

9.1 Density and porosity of lunar rocks and soil — Dichte und Porosität der Gesteine des Mondes und des Mondbodens

Density and porosity data of lunar rocks are presented in Table 4 and in Fig. 2. They have been compiled mainly from values published in connection with measurements of elastic and thermal properties. Data and references are given in Tables 6 and 13. Additional references are [Wo70, St71, O'K72, Ca75]. Intrinsic density (density at zero porosity, also matrix density, ideal density) values were determined by different methods (e.g. from powdered material, from direct porosity measurements or calculated from modal mineral compositions), so that they represent a somewhat heterogeneous data set. Therefore the intrinsic density and the porosity can only be considered as approximate values in some cases. Density variation with applied stress can be estimated from the compressibility measurements on lunar rocks given in section 9.3.

The density of lunar soil as a particulate material is strongly dependent on handling of the samples. In Table 5 minimum and maximum values measured on soil samples are given together with the intrinsic density [Ca73]. Minimum values represent the loosest state and maximum values the densest state after tapping, but with no applied stress or overburden pressure. Values near to *in situ* bulk density values are won by measurements on lunar core tube samples and drill samples ([Ho74], Table 5). Density of lunar soil as a function of applied stress is given in Fig. 7 and in section 9.3.

Dichtewerte und Porositätswerte von Gesteinen des Mondes sind in Tab. 4 und in Fig. 2 dargestellt. Die Daten stammen vorwiegend von Messungen, die in Zusammenhang mit der Bestimmung von elastischen und thermischen Eigenschaften durchgeführt wurden (siehe Tab. 6 und 13, mit Literaturangaben). Weitere Daten stammen aus [Wo70, St71, O'K72, Ca75]. Die Matrixdichten (Porosität $\phi = 0$) wurden mit verschiedenen Methoden bestimmt, z.B. mit Hilfe von pulverisiertem Material, aus direkten Porositätsmessungen, oder aber sie wurden aus der Modalzusammensetzung der Gesteine berechnet. Die Werte sind infolgedessen von unterschiedlicher Güte und können in manchen Fällen, ebenso wie die angegebenen Porositäten, nur als Näherungswerte angesehen werden. Änderungen der Dichte unter Druck können aus den Messungen der Kompressibilität von Gesteinen des Mondes entnommen werden, die in 9.3 dargestellt sind.

Die Dichte des feinkörnigen Mondbodens hängt stark von der Behandlung der Proben ab. In Tab. 5 sind Minimal- und Maximalwerte sowie die Matrixdichte von Mondbodenproben angegeben [Ca73]. Minimalwerte entsprechen dem lockersten Zustand und Maximalwerte dem dichtesten Zustand, der durch wiederholtes Klopfen, jedoch ohne zusätzlichen Druck erreicht werden konnten. Dichtewerte, die den *in situ*-Werten relativ gut entsprechen, wurden an Kernen gemessen, die mit Stechrohren sowie durch Bohren aus dem Mondboden gewonnen wurden ([Ho74], Tab. 5). Die Druckabhängigkeit der Dichte des Mondbodens ist in Fig. 7 und im Abschnitt 9.3 dargestellt.

Table 4. Density and porosity of lunar rocks. Compilation basing on Tables 6 and 13 and some additional references (see text). In parentheses: number of samples

Rock type	Bulk density ϱ		Intrinsic density ϱ_{intr}		Porosity ϕ	
	mean	ϱ_{max} ϱ_{min}	mean	$\varrho_{intr, max}$ $\varrho_{intr, min}$	mean	ϕ_{max} ϕ_{min}
	kg m^{-3}		kg m^{-3}			
Basalts	3070 (28)	3370 2560	3280 (13)	3510 3040	0.086 (13)	0.226 0.002
Anorthosites	2830 (8)	2970 2700				
Breccias	2520 (17)	2930 2210	2995 (11)	3120 2790	0.171 (11)	0.260 0.049

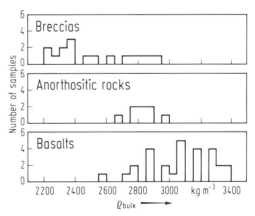

Fig. 2. Bulk density ϱ_{bulk} of lunar rocks. For data, see text and Tables 6 and 13.

Table 5. Density of lunar soil.

(A) [Ca73], ϱ_{min} = loosest state, ϱ_{max} = densest state after tapping.

(B) [Ho74], relative density = $\dfrac{\varrho_{max}}{\varrho_{bulk}} \cdot \dfrac{\varrho_{bulk} - \varrho_{min}}{\varrho_{max} - \varrho_{min}}$

Sample	Bulk density		Intrinsic density
	ϱ_{min}	ϱ_{max}	ϱ_{intr}
	$kg\,m^{-3}$		$kg\,m^{-3}$
10084	1360	1800	3010
10084,68	1260		3010
12001,19	1300		
12029,3	1150	1930	
14163,133	1100		2900
14163,148	890	1550	2900
14259,3	870	1510	2930
15031,38	1300		
15601,82	1100	1890	3240

	Depth range m	Bulk density $kg\,m^{-3}$	Relative density
Core tube samples (Apollo 15, 16, 17)	0···0.15	1500 ± 50	0.64
	0···0.30	1580 ± 50	0.74
	0.30···0.60	1740 ± 50	0.92
	0···0.60	1660 ± 50	0.83
Lunar drill samples (all drill cores)	0···0.30	1690 ± 80	
	0.30···0.60	1770 ± 80	

9.2 Compressional waves and shear waves in lunar rocks and soil — Kompressionswellen und Scherwellen in Gesteinen des Mondes und im Mondboden

9.2.1 Compressional and shear wave velocities of lunar rocks — Die Geschwindigkeiten der Kompressionswellen und der Scherwellen in Gesteinen des Mondes

For lunar rocks compressional and shear wave velocities have been measured mainly as a function of confining hydrostatic pressure in the range 0.1··· 1000 MPa (1 bar···10 kbar). Measurements with no confining pressure were usually made under normal laboratory atmospheric conditions. In most cases a Birch-type pulse method was used in the frequency range from 0.5···5 MHz. As this technique involves a certain alteration of the samples through bonding agents and compression, the measured velocities may in some cases differ somewhat from the values under lunar environment conditions, especially for brecciated material. An extremely low compressional wave velocity ($\bar{v}_p = 0.69\ km\,s^{-1}$), for example, was measured on a highly fractured anorthosite (60025,174) by [So79] with a new technique where the sample is kept in conditions equivalent to the lunar surface environment, i.e. without compression and under high vacuum. Notable differences in the velocity-pressure relationship measured on well outgassed terrestrial rocks as lunar analogues in vacuum and on rocks exposed to normal laboratory conditions were also observed by [Ti76] and [Ti77]. The relative change in velocity with pressure is much less for dry, outgassed rocks than for rocks exposed to laboratory humidity. The difference is attributed to the presence of adsorbed volatiles, mainly H_2O, in cracks.

Die Geschwindigkeiten der seismischen Wellen wurden an Gesteinen des Mondes vorwiegend in Abhängigkeit von hydrostatischem Druck im Bereich 0.1···1000 MPa (1 bar···10 kbar) gemessen. In den meisten Fällen wurde eine Impulsmethode nach Birch verwendet mit Frequenzen zwischen 0.5 und 5 MHz. Bei Messungen ohne zusätzlichen hydrostatischen Druck, also unter Normalbedingungen, unterliegen jedoch bei dieser Methode die Proben gewissen Veränderungen durch die erforderliche Imprägnierung mit Bindemitteln und Andrücken der Impulsgeber, so daß die gemessenen Geschwindigkeiten z.T. von denjenigen abweichen, die unter lunaren Bedingungen gemessen würden. Dies gilt insbesondere für brecciierte Gesteine. Mit einer Methode, bei der die Gesteinsproben sich unter Bedingungen befanden, die denen an der Mondoberfläche entsprechen – d.h. in hohem Vakuum und völlig unbelastet – haben [So79] an an einer stark brecciierten anorthositischen Probe (60025,174) eine extrem niedrige Kompressionswellengeschwindigkeit ($\bar{v}_p = 0.69\ km\,s^{-1}$) erhalten. [Ti76] und [Ti77] stellten ebenfalls erhebliche Unterschiede bezüglich der Abhängigkeit der Geschwindigkeit vom Druck zwischen stark entgasten terrestrischen Gesteinen, die im Vakuum vermessen wurden, und Gesteinen, die den normalen atmosphärischen Bedingungen im Labor ausgesetzt waren, fest. Die Geschwindigkeit nahm mit zunehmendem Druck bei den entgasten, trockenen Gesteinen relativ weniger

Velocity data as a function of confining hydrostatic pressure are compiled in Table 6 together with density measurements. Figs. 3 and 4 show some representative measurement curves. The pressure dependence of compressional wave velocities is given in a logarithmic scale in Figs. 12, 13 and 14. Fig. 5 shows the density-velocity relationship of lunar rocks.

zu als bei den nicht entgasten Proben. Das unterschiedliche Verhalten wird vorwiegend auf das Vorhandensein von adsorbierten flüchtigen Stoffen, besonders H_2O, in Mikrorissen zurückgeführt.

Gemessene Geschwindigkeiten in Abhängigkeit vom hydrostatischen Druck sind in Tabelle 6 zusammengestellt. Die Fig. 3 und 4 zeigen Ergebnisse für einige typische Gesteine. Eine weitere graphische Darstellung der Druckabhängigkeit der Kompressionswellengeschwindigkeit ist in logarithmischer Form in den Fig. 12, 13 und 14 gegeben. Fig. 5 zeigt die Geschwindigkeit-Dichte-Beziehungen von Gesteinen des Mondes.

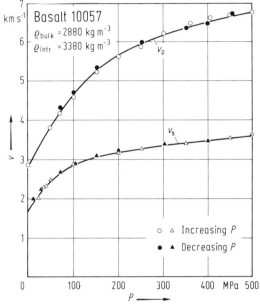

Fig. 3. Compressional wave velocity v_p and shear wave velocity v_s as a function of hydrostatic pressure P (experimental data) for lunar basalt [Ka70]. ϱ_{bulk} is bulk density and ϱ_{intr} is estimated intrinsic density.

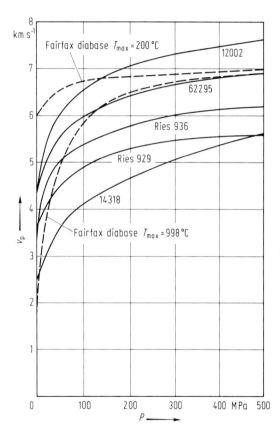

Fig. 4. Compressional wave velocity v_p as a function of pressure P for lunar basalt (12002), anorthositic gabbro (62295) and breccia (14318), shocked granitic rock (936) and high shock glassy breccia (929) from the Ries crater (Germany), and thermally cycled Fairfax diabase [To73].

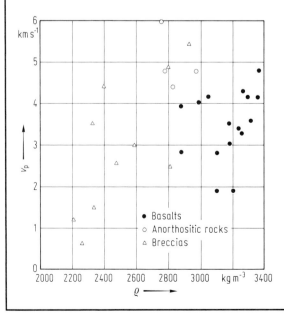

Fig. 5. Velocity-density relationships for lunar rocks. Data from Table 6.

Temperature dependence of compressional wave velocity in the temperature range 27···200 °C under confining pressure in the range 200···900 MPa was measured on lunar breccia 14321,93 (Table 6) by [Mi73] and showed a decrease of velocity with increasing temperature. Prior to the measurements, the samples were exposed to normal laboratory conditions. In the absence of any confining pressure and in high vacuum (10^{-5} Pa), [Ti76] found an increase of velocity in a well outgassed terrestrial basalt – as a lunar analogue – with increasing temperature.

Elastic properties of lunar surface rocks differ basically from those of unshocked terrestrial rocks of similar composition, but they are similar to those of shocked terrestrial rocks. Fig. 6 [To73] shows compressional velocity normalized to the velocity at 1 GPa (10 kbar) as a function of crack porosity for different rocks. Shocked rocks occupy a field which is distinctly different from those of typical igneous rocks and thermally cycled rocks. Therefore shock effects are considered to be the main cause for the difference between measured elastic properties of lunar rocks and their intrinsic values. Thermal effects, such as e.g. thermal cycling under the temperature fluctuations at the lunar surface are thought to be less important [To73, Si75].

Einige Messungen der Temperaturabhängigkeit der Geschwindigkeit der Kompressionswellen im Temperaturbereich von 27···200 °C und unter hydrostatischem Druck (200···900 MPa) an der Breccie 14321,93 (siehe Tab. 6) durch [Mi73] ergaben eine Abnahme der Geschwindigkeit mit zunehmender Temperatur. Die Probe war vor Beginn der Messungen normalen Laborbedingungen ausgesetzt. Eine Zunahme der Geschwindigkeit mit steigender Temperatur wurde hingegen an einem stark entgasten terrestrischen Basalt im Hochvakuum von [Ti76] gemessen.

Die Messungen der elastischen Eigenschaften der oberflächennahen Gesteine des Mondes zeigen, daß diese sich grundsätzlich von denen nicht-geschockter terrestrischer Gesteine unterscheiden. Sie sind jedoch denjenigen geschockter terrestrischer Gesteine in vieler Hinsicht vergleichbar. In Fig. 6 [To73] ist das Verhältnis der Geschwindigkeit der Kompressionswellen bei Normaldruck zur Geschwindigkeit bei 1 GPa (10 kbar) in Abhängigkeit von der Rißporosität für verschiedene Gesteine dargestellt. In diesem Diagramm fallen alle Werte für geschockte Gesteine in ein Gebiet, das sich von den Gebieten für typische terrestrische Gesteine und thermisch behandelte Gesteine charakteristisch unterscheidet. Schockeffekte werden daher als Hauptgrund für die starken Abweichungen der elastischen Eigenschaften der Gesteine des Mondes von ihren Idealwerten angesehen [To73, Si75].

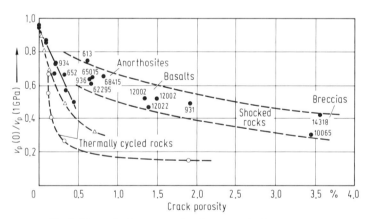

Fig. 6. Normalized compressional wave velocity $v_p(0)/v_p$ (1 GPa) as a function of crack porosity for shocked lunar and terrestrial rocks, terrestrial igneous rocks and thermally cycled rocks [To73]. 62295, 65015 and 68415 are lunar anorthosites, 12002 and 12022 are lunar basalts, 931, 934 and 936 are shocked granitic rocks from the Ries crater (Germany), 613 and 652 are terrestrial breccias. Solid line with data points is for terrestrial igneous rocks which suffered no shock [Nu70]. Thermally cycled rocks are Fairfax diabase and Westerly granite.

Table 6. Compressional and shear wave velocities of lunar rocks as a function of hydrostatic pressure. v_p = compressional wave velocity; v_s = shear wave velocity; ϱ_{bulk} = bulk density; ϱ_{intr} = intrinsic density (at zero porosity, ideal density); * measured value after pressure cycling to several hundred MPa; + average value; ( ) extrapolated values.

Sample	Density ϱ kg m^{-3}	Velocity km s^{-1}	Pressure [MPa] 0.1	0.1*	10	20	25	50	75	100	150	200	300	400	500	600	650	700	800	900	1000	Ref.
Basalts																						
10017	ϱ_{bulk} 3100	v_p	1.84	1.9		4.36		5.26		5.54	5.68	5.86										An70
			1.85	1.9																		
			1.95	2.0																		
		v_s	1.05	1.1																		
			1.03	1.2																		
			—	—																		
10020	ϱ_{bulk} 3180 ϱ_{intr} 3250	v_p v_s	3.50 2.20					4.80 2.88		5.55 3.25		6.30 3.67	6.80 3.87	7.00 4.00	7.20 4.08							Ka70
10057	ϱ_{bulk} 2880 ϱ_{intr} 3380	v_p v_s	2.82 1.70					3.80 2.45		4.65 2.82		5.62 3.20	6.18 3.39	6.52 3.50	6.78 3.62							Ka70
12002,54	ϱ_{bulk} 3300 ϱ_{intr} 3310	v_p v_s	3.60 2.10		4.45 2.30			5.65 2.85		6.25 3.10	6.60 3.25	6.90 3.40	7.20 3.60		7.55 3.85		7.70 3.95					Wa71
12002,58	ϱ_{bulk} 3300	v_p	4.50		5.10		5.65	6.10	6.40	6.55	6.85	7.05	7.30		7.60		7.70					Wa71
			4.05		4.90		5.50	6.00	6.30	6.55	6.85	7.05	7.30		7.60		7.80					
			3.85		4.50		5.05	5.65	6.00	6.30	6.65	6.85	7.10		7.40		7.55					
		v_s	2.45		2.70		2.90	3.20	3.40	3.55	3.75	3.90	4.00		4.10		4.15					
			2.30		2.70		2.95	3.20	3.35	3.50	3.70	3.80	3.95		4.10		4.15					
			2.45		2.55		2.70	2.95	3.10	3.20	3.40	3.55	3.75		3.90		4.00					
12018	ϱ_{bulk} 3200	v_p	1.79																			Wa71a
			2.04																			
			1.81																			
		v_s	1.07																			
			1.17																			
			1.10																			

(continued)

Table 6 (continued)

Sample	Density ρ kg m⁻³	Velocity km s⁻¹	0.1	0.1*	10	20	25	50	75	100	150	200	300	400	500	600	650	700	800	900	1000	Ref.
12022,60	ϱ_{bulk} 3320	v_p	3.55		3.95		4.50	5.25	5.80	6.15	6.60	6.90	7.10		7.50	7.60						Wa71
			3.60		3.85		4.25	4.80	5.20	5.50	6.15	6.55	7.05		7.35	7.40						
			3.55		3.80		4.10	4.65	5.10	5.55	6.20	6.55	6.90		7.30	7.40						
		v_s	1.70		1.85		2.05	2.30	3.00	2.70	3.00	3.25	3.50		3.85	3.95						
			1.55		1.70		1.90	2.20	2.50	2.70	3.05	3.25	3.60		3.90	4.00						
			1.75		1.85		2.05	2.35	2.50	2.75	3.05	3.25	3.50		3.75	3.85						
12022,95	ϱ_{bulk} 3320	v_p	3.65		4.05		4.60	5.20	5.70	6.00	6.40	6.70	7.05		7.35	7.40						Wa71
		v_s	2.25		2.35		2.55	2.75	2.90	3.10	3.35	3.50	3.70		3.85	3.90						
12052,35	ϱ_{bulk} 3270	v_p	4.30			4.90		5.55		5.93		6.32	6.55		6.80			6.90			7.01	Ka71
		v_s	2.59			2.70		2.84		3.03		3.34	3.55		3.74			3.82			3.88	
12063	ϱ_{bulk} 3100	v_p	2.93																			Wa71a
			3.0⁺																			
			2.51																			
		v_s	1.71																			
			1.63⁺																			
			1.37																			
12063,97		v_p	4.02					4.33		4.56	4.71	4.85	5.06	5.21	5.31							Wa73a
12065,68	ϱ_{bulk} 3260	v_p	3.27			4.44		5.21		5.80		6.24	6.47	6.50	6.74			6.86			6.96	Ka71
		v_s	2.14			2.42		2.73		3.04		3.38	3.54	3.59	3.72			3.82			3.86	
14053,32	ϱ_{bulk} 3180 ϱ_{intr} 3190	v_p	3.02					4.54		5.32		6.02	6.35		6.71					6.93			Mi73
14310,72	ϱ_{bulk} 2880 ϱ_{intr} 3040	v_p	3.93		4.21		4.55	4.96	5.23	5.51	5.86	6.08	6.33	6.50	6.68							To72
			3.84		4.12		4.52	4.91	5.26	5.55	5.92	6.14	6.39	6.59	6.79							
		v_s	2.08		2.23		2.40	2.63	2.80	2.95	3.15	3.30	3.46	3.57	3.66							
			2.07		2.19		2.37	2.60	2.79	2.94	3.11	3.26	3.42	3.54	3.63							
14310,82		v_p						4.70		5.42	5.75	5.88	5.95	6.02	6.05	6.07						Tr74
		v_s						2.63		2.90	3.14	3.29	3.53	3.61	3.65	3.67						

(continued)

Table 6 (continued)

Sample	Density ρ [kg m⁻³]	Velocity [km s⁻¹]	Pressure [MPa]																			Ref.
			0.1	0.1*	10	20	25	50	75	100	150	200	300	400	500	600	650	700	800	900	1000	
15058,57	ϱ_{bulk} 2990 ϱ_{intr} 3270	v_p	4.03					5.33		5.54		5.85	6.12		6.49			6.65		6.73		Mi73
15065,27	ϱ_{bulk} 2860	v_p						3.90		4.70	5.25	5.62	6.20	6.52	6.76		6.84	6.98			(7.10)	Ch73
		v_s						2.50		2.80	3.06	3.29	3.50	3.68	3.75		3.86	3.90			(3.97)	
15545,24	ϱ_{bulk} 2560 ϱ_{intr} 3250	v_p		5.60				6.10		6.37		6.63	6.76		6.90			6.98		7.02		Mi73
15555,88	ϱ_{bulk} 3100	v_p						5.20		6.10	6.45	6.66	6.90	7.02	7.14		7.25	7.30			(7.42)	Ch73
		v_s						2.60		3.00	3.24	3.45	3.66	3.76	3.87		3.94	4.01			(4.12)	
60015,29	ϱ_{bulk} 2760	v_p						5.50		6.00	6.27	6.52	6.75	6.86	6.90		6.94	6.97			(7.02)	Ch73
		v_s						2.60		2.90	3.21	3.40	3.58	3.68	3.74		3.86	3.88			(3.91)	
60315,33	ϱ_{bulk} 3050 ϱ_{intr} 3060	v_p	4.15					4.81		5.40		6.13	6.38		6.88			7.03		7.11		Mi73
		v_s	2.00					2.48		2.79		3.10	3.26		3.50			3.68		3.83		
61016,34	ϱ_{bulk} 2790	v_p						5.60		6.20	6.30	6.60	6.77	6.87	6.91		6.96	6.99			(7.02)	Ch73
		v_s						2.40		3.10	3.22	3.36	3.58	3.69	3.74		3.86	3.88			(3.90)	
70215,29	ϱ_{bulk} 3240	v_p	3.38																			Wa74
		v_s	2.22																			
70215,30	ϱ_{bulk} 3370 ϱ_{intr} 3410	v_p	4.79					5.77		6.23		6.70	6.91		7.08			7.14		7.19		Mi74
		v_s	2.48					3.10		3.33		3.56	3.67		3.77			3.80		3.81		
71055,15		v_p					3.19	3.86		4.52	5.00	5.30	5.66	5.90	6.11							Tr74
		v_s					2.11	2.43		2.80	3.00	3.13	3.27	3.39	3.47							
74275,25	ϱ_{bulk} 3360	v_p	4.14					5.20		5.98		6.61	6.88		7.13			7.24		7.28		Mi74
		v_s								2.98		3.66	3.82		4.04			4.09		4.11		

Anorthosites

Sample	Density ρ [kg m⁻³]	Velocity [km s⁻¹]	0.1	0.1*	10	20	25	50	75	100	150	200	300	400	500	600	650	700	800	900	1000	Ref.
15415,57	ϱ_{bulk} 2700	v_p						5.00		5.60	6.02	6.40	6.65	6.70	6.78		6.83	6.85			(6.87)	Ch73
		v_s						2.00		2.50	2.90	3.26	3.42	3.54	3.56		3.58	3.61			(3.69)	
15415,96	ϱ_{bulk} 2760 ϱ_{intr} 2760	v_p	5.98					6.20		6.28		6.43	6.56		6.75			6.89		7.01		Mi73

(continued)

Table 6 (continued)

Sample	Density ρ kg m⁻³	Velocity km s⁻¹	0.1	0.1*	10	20	25	50	75	100	150	200	300	400	500	600	650	700	800	900	1000	Ref.
							Pressure [MPa]															
62295,18	ϱ_{bulk} 2830	v_p	4.39	4.73			5.11	5.54	5.81	5.98	6.23	6.42	6.66	6.82	6.92							To73
		v_s	2.00	2.40			2.61	2.83	2.98	3.11	3.28	3.41	3.56	3.68	3.75							
65015,9	ϱ_{bulk} 2970	v_p	4.77	4.99			5.25	5.58	5.80	5.98	6.17	6.33	6.53	6.72	6.90							To73
		v_s	2.38	2.49			2.63	2.79	2.91	3.02	3.19	3.29	3.45	3.54	3.63							
68415,54	ϱ_{bulk} 2780	v_p	4.70	5.02			5.29	5.63	5.89	6.09	6.37	6.54	6.76	6.85	6.94							To73
			4.95	5.25			5.57	5.92	6.11	6.27	6.49	6.64	6.80	6.92	7.04							
		v_s	2.59	2.69			2.80	2.94	3.05	3.13	3.26	3.35	3.43	3.47	3.54							
			2.48	2.60			2.73	2.88	3.00	3.09	3.23	3.31	3.41	3.46	3.54							
Breccias																						
10046	ϱ_{bulk} 2210 ϱ_{intr} 2990	v_p	1.25 1.19 1.20	2.06				4.89		5.50		6.12										An70
		v_s	0.74 0.76 0.79	1.29																		
10065	ϱ_{bulk} 2340 ϱ_{intr} 3115	v_p	1.50					2.90		3.50		4.05	4.30	4.30	4.50							Ka70
		v_s	1.05					1.70		2.00		2.28	2.42	2.65	2.78							
14311,50	ϱ_{bulk} 2860 ϱ_{intr} 3050	v_p	5.65					6.02		6.18		6.36	6.47		6.56			6.59			6.62	Mi72
		v_s	3.04					3.24		3.35		3.48	3.54		3.60			3.63			3.65	
14313,27	ϱ_{bulk} 2390 ϱ_{intr} (3030)	v_p	2.25					2.61		2.88		3.39	3.82		4.41			4.81			5.16	Mi72
		v_s	1.35					1.60		1.80		2.09	2.32		2.55			2.68			2.79	
14318,30	ϱ_{bulk} 2810	v_p	2.47	2.79			3.14	3.62	3.96	4.16	4.42	4.71	5.11	5.45	5.67							To73
14321,93	ϱ_{bulk} 2400 ϱ_{intr} 3090	v_p ($T=$ 27 °C)	4.41					4.68		4.86		5.10	5.27	5.53				5.69		5.84		Mi73
		v_p ($T=$ 100 °C)										4.82	5.06		5.23			5.59		5.79		
		v_p ($T=$ 200 °C)										4.46	4.71		5.09			5.37		5.68		

(continued)

Pohl

Table 6 (continued)

Sample	Density ρ kg m⁻³	Velocity km s⁻¹	Pressure [MPa] 0.1	0.1*	10	20	25	50	75	100	150	200	300	400	500	600	650	700	800	900	1000	Ref.
15015,18	ρ_bulk 2330	v_p	3.50		3.68		3.90	4.13	4.27	4.38	4.49	4.54	4.64	4.74	4.85							To72
15418,43	ρ_bulk 2800	v_p	4.85		5.00		5.20	5.50	5.77	6.02	6.33	6.50	6.64	6.69	6.75							To72
		v_s	2.82		2.88		2.97	3.08	3.19	3.28	3.42	3.50	3.58	3.63	3.69							
15498,23		v_p					3.07	3.37		3.77	4.00	4.19	4.49	4.70	4.88							Wa73a
60025,174	ρ_bulk 2260, ρ_intr 2790	v_p	0.53 / 0.70 / 0.78																			So79
60335,20	ρ_bulk 2590, ρ_intr 2910	v_p	3.0					4.14		4.70	5.02	5.23	5.49	5.70								Wa73a
		v_s						2.42		2.70	2.86	3.02	3.19	3.28	3.35		3.40					
61175,22	ρ_bulk 2250	v_p						2.94		3.46		4.03	4.38		4.94			5.13		5.46		Mi74
		v_s								1.92		2.19	2.37		2.64			2.91		3.09		
73235,18	ρ_bulk 2930	v_p	5.42					6.02		6.39		6.72	6.88		7.08			7.12		7.14		Mi74
		v_s	2.95					3.32		3.48		3.66	3.77		3.86			3.90		3.92		
77017,24	ρ_bulk 2480	v_p	2.56					2.69		3.20		4.25	4.80		5.51			5.95		6.20		Mi74
		v_s	1.78					1.86		2.01		2.30	2.60		3.03			3.33		3.45		

9.2.2 Compressional and shear wave velocity of lunar soil —
Die Geschwindigkeiten der Kompressionswellen und der Scherwellen im Mondboden

Compressional wave velocities of lunar soil as a function of uniaxial pressure and density are shown in Fig. 7 [Wa71a]. The velocity was measured in the compressed state, then in the decompressed state, and then the pressure was increased, etc. There is good agreement of the velocity-density values of compressed soil at the zero isobar with two breccias from the same site. Velocities measured at the lowest experimental densities ($\approx 2100 \, \mathrm{kg \, m^{-3}}$) are higher than the velocities near the lunar surface, where the *in situ* bulk density varies from 1500 at the surface to $1740 \, \mathrm{kg \, m^{-3}}$ at a depth of 0.6 m (average values, see Table 5).

Die Abhängigkeit der Kompressionswellen-Geschwindigkeit von einaxialem Druck und von der Dichte ist in Fig. 7 [Wa71a] gezeigt. Bei diesen Untersuchungen wurde die Geschwindigkeit zunächst unter Druck gemessen, dann nach Entlastung, und dann wurde der Druck erhöht usw. Die Geschwindigkeiten und Dichten, die im entlasteten Zustand gemessen wurden ($P = 0$-Isobare), stimmen recht gut mit denen zweier Breccien vom selben Fundort überein. Die Geschwindigkeiten, die bei den niedrigsten Dichten in diesem Experiment ($\approx 2100 \, \mathrm{kg \, m^{-3}}$) gemessen wurden, sind höher als die Geschwindigkeiten in der Nähe der Mondoberfläche, wo die *in situ*-Dichte an der Oberfläche etwa $1500 \, \mathrm{kg \, m^{-3}}$ und in einer Tiefe von 0.6 m etwa $1740 \, \mathrm{kg \, m^{-3}}$ beträgt (siehe Tab. 5).

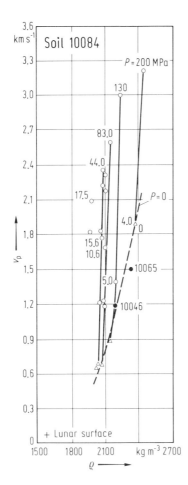

◀

Fig. 7. Compressional wave velocity v_{p} of lunar soil as a function of uniaxial pressure P and density ϱ [Wa71a]. Values on the dashed line ($P = 0$) were obtained after compression to the above indicated pressures (MPa). Full circles are for lunar breccias 10046 [An70] and 10065 [Ka70].

Compressional and shear wave velocities under hydrostatic confining pressure are given in Table 7 [Ta74]. The modified pulse transmission technique used for these measurements is described in [Ta73]. All measurements were carried out with samples exposed to normal laboratory atmospheric and moisture conditions.

Geschwindigkeiten von Kompressions- und Schwerwellen, die unter hydrostatischem Druck gemessen wurden, sind in Tab. 7 zusammengestellt [Ta74]. Eine modifizierte Impulsmethode, die bei diesen Messungen verwendet wurde, ist in [Ta73] beschrieben. Alle Messungen wurden mit Proben, die den normalen Feuchtigkeitsbedingungen im Labor ausgesetzt waren, durchgeführt.

Table 7. Compressional and shear wave velocities v_p and v_s in lunar soil as a function of hydrostatic pressure [Ta74]. Smoothed values obtained from the measurements, which were reversible with pressure cycling. Densities are not reported. Measurements were made under atmospheric conditions.

Sample	Velocity $km\,s^{-1}$	Confining pressure [MPa]								
		0.1	5	25	50	75	100	150	200	250
70051	v_p	0.25	0.95	1.85	2.38	2.64	2.79	3.00	3.17	3.30
	v_s	0.12	0.42	0.80	1.05	1.19	1.30	1.48	1.62	1.73
72161	v_p	0.25	1.20	2.17	2.53	2.74	2.91	3.14	3.31	3.42
72701	v_p	0.25	1.14	1.87	2.22	2.44	2.58	2.74	2.84	2.92
	v_s	0.10	0.40	0.72	0.85	0.93	1.10	1.15	1.25	1.32
75081	v_p	0.17	0.74	1.50	1.81	1.97	2.04	2.13	2.20	2.22
	v_s	0.11	0.33	0.73	0.88	0.98	1.03	1.12	1.18	1.29

Experimental compressional wave velocity measurements under conditions approaching the lunar surface conditions, i.e. in high vacuum and with well outgassed material, have been made under uniaxial compression in the pressure range 0.0015···0.5 MPa (0.015···5 bar) with finely crushed olivine diabase as a lunar analogue soil by [St77] and [St78]. During initial compaction the mean compressional wave velocity increased from 0.165 km s^{-1} at 0.0015 MPa to 0.620 km s^{-1} at 0.5 MPa. For precompacted powders the mean velocity increased form 0.205 km s^{-1} at 0.0015 MPa to 0.635 km s^{-1} at 0.5 MPa.

Untersuchungen unter Bedingungen, die denen an der Mondoberfläche näherkommen, d.h. in hohem Vakuum und mit stark entgasten Proben, wurden mit einaxialem Druck im Bereich 0.0015···0.5 MPa (0.015··· 5 bar) an fein zerkleinertem terrestrischen Olivindiabas von [St77] und [St78] durchgeführt. Bei der Anfangskompression wurde bei einem Druck von 0.0015 MPa eine mittlere Geschwindigkeit von 0.165 km s^{-1} und bei 0.5 MPa eine Geschwindigkeit von 0.620 km s^{-1} gemessen. Bei vorkomprimiertem Material stieg die Geschwindigkeit von 0.205 km s^{-1} bei 0.0015 MPa auf 0.635 km s^{-1} bei 0.5 MPa an.

9.2.3 Quality factor Q — Qualitätsfaktor Q

The quality factor is defined by

Der Qualitätsfaktor Q wird definiert durch

$$Q = 2\pi E/\Delta E$$

where E is the elastic wave energy per unit volume and ΔE is the energy dissipated as heat per cycle and unit volume (e.g. [Kn64]). Very high near surface Q-values in the range 3000···5000 have been found by seismic *in situ* measurements just under the moon's surface (e.g. [La70]), indicating an extremely low absorption under lunar surface conditions. However, laboratory determinations of Q on lunar samples carried out under terrestrial atmospheric conditions yielded only low values for Q in the order 50···100 [Ka70, Wa71a, Wa71, He74]. These values are comparable to values measured on terrestrial rocks. Higher Q-values were measured in vacuum conditions [Ti73, Wa75], but values comparable to lunar *in situ* values were obtained only after strong outgasing at elevated temperatures in ultra-high vacuum. Since measured Q-values strongly depend on the measurement conditions and previous treatments of the samples, a short description of several experiments conducted on lunar rocks by [Ti75, Ti76] and [Sch77] is given in Tables 8, 9 and 10.

wobei E die elastische Wellenenergiedichte und ΔE die pro Wellenlänge als Wärme dissipierte Energiedichte ist (z.B. [Kn64]). Auf dem Mond wurden in den oberflächennahen Schichten durch seismische Messungen sehr hohe *in situ*-Werte für Q bestimmt (3000···5000, z.B. [La70]). Dies bedeutet, daß unter den Druck- und Temperaturbedingungen nahe der Mondoberfläche die seismischen Wellen sehr wenig gedämpft werden. Labormessungen von Q an Mondgesteinen, die zunächst unter terrestrischen atmosphärischen Bedingungen durchgeführt wurden, ergaben sehr niedrige Werte für Q (50···100), wie sie normalerweise auch an terrestrischen Gesteinen gemessen werden [Ka70, Wa71a, Wa71, He74]. Höhere Werte für Q wurden unter Vakuumbedingungen gemessen [Ti73, Wa75]. Werte, die mit den *in situ* bestimmten Werten vergleichbar sind, wurden jedoch erst nach starker Entgasung der Proben bei höheren Temperaturen und im Ultrahochvakuum gemessen. Da die gemessenen Q-Werte sehr stark von der Vorbehandlung der Proben und von den Meßbedingungen abhängen, werden in den Tab. 8, 9 und 10 einige Versuchsreihen, die mit Mondgesteinen durchgeführt wurden, kurz beschrieben ([Ti75, Ti76] und [Sch77]).

Table 8. Quality factor Q in lunar basalt 70215,85 [Ti75]. Vibrating bar technique in longitudinal mode with resonance frequencies around 20 kHz.

Measurement conditions	Q
As received in laboratory air.	60
At 10^{-3} Torr and room temperature.	340
After first heating run at 10^{-3} Torr followed by slow cooling.	400
After second and third heating runs at 10^{-3} Torr followed by rapid cooling.	800
After fourth heating run followed by rapid cooling at 10^{-6} Torr.	2420
After continued pumping at room temperature and at 10^{-7} Torr (1 week).	3130
After long-term re-exposure to laboratory air at room temperature.	142

Vibrating bar technique in longitudinal mode with resonance frequencies around 20 kHz.

Table 9. Effect of ultra-high vacuum on Q for lunar basalt 70215,85 [Ti76].

Pressure Pa	Exposure time hours	Q	Resonance frequency increase %
10^{-5}	12	1851	0
10^{-6}	12	2381	0.06
10^{-8}	14	3330	0.21

Table 10. Quality factor Q in lunar basalt 70215,135 [Sch77]. Sphere resonance technique, $_2T_1$ mode of vibration.

Measurement conditions	Q
Specimen cleaned and equilibrated to air pressure of 1 atm	75
Held in vacuum 24 hr, 10^{-3} Torr	328
After 72 hr, 10^{-3} Torr	no change
Pressure reduced to 10^{-5} Torr	700
Specimen exposed to laboratory air, 1 atm	70
Specimen heated to 175 °C, 10^{-3} Torr	550
Pressure reduced to 10^{-5} Torr	878
Introduced He gas	5 % decrease
Heated to 100 °C, pressure to 10^{-5} Torr	866
Introduced CO_2 gas	11 % decrease
Heated to 100 °C, pressure to 10^{-5} Torr	825
Sample exposed to H_2O vapor	95
Pressure reduced to 10^{-5} Torr	280
Heated to 175 °C, pressure to 10^{-5} Torr	908
Same as above, after 10 days	956
Heated to 225 °C, pressure to $7 \cdot 10^{-6}$ Torr	1185
After 15 days	1093

Sphere resonance technique, $_2T_1$ mode of vibration.

Table 11. Temperature dependence of Q of lunar basalt 70215,85 [Ti78]. Low frequency flexural technique at 56 Hz.

T °C	Q	Conditions
100	740	In vacuum of 10^{-5} Pa
50	950	after strong outgassing
0	1330	
−50	1430	

The low Q-values measured under terrestrial atmospheric humidity conditions are attributed mainly to adsorption of volatiles, especially H_2O, at the crack and grain boundaries. Investigations of the role of various volatiles on the damping mechanism are discussed in [Ti75, Ti76, Sch77, Ti77, Ti78, Ti79, Sp80, Ti80].

Temperature and pressure dependence of Q under lunar environment conditions have been investigated mainly on terrestrial rocks as lunar analogues. Temperature dependence of Q is given in Table 11 for a lunar basalt [Ti78]. Measurements on lunar analogues are reported by [Ti75, Ti76, Ti77, Ti78, Ti79, Ti80].

Die sehr niedrigen Q-Werte, die unter terrestrischen atmosphärischen Feuchtigkeitsbedingungen gemessen wurden, werden vorwiegend einer Adsorption von H_2O an den Rißflächen und Korngrenzen zugeschrieben. Zahlreiche Untersuchungen über den Einfluß der Adsorption von verschiedenen flüchtigen Stoffen auf den Dämpfungsmechanismus sind in [Ti75, Ti76, Sch77, Ti77, Ti78, Ti79, Sp80, Ti80] beschrieben.

Die Temperatur- und Druckabhängigkeit von Q unter lunaren Bedingungen wurde vorwiegend an terrestrischen Gesteinen als simulierten Mondgesteinen untersucht. Die Temperaturabhängigkeit von Q eines Mondbasaltes ist in Tab. 11 [Ti78] angegeben. Messungen an terrestrischen Gesteinen sind bei [Ti75, Ti76, Ti77, Ti78, Ti79, Ti80] beschrieben.

9.3 Elastic moduli — Elastizitätsmoduln

Definitions:

Static bulk modulus

Für die angegebenen Elastizitätsmoduln gelten folgende Definitionen:

Kompressionsmodul (statisch)

$$K^{-1} = \beta = -\frac{1}{V} \cdot \frac{\Delta V}{\Delta P} \left(\simeq -3\frac{\Delta S}{\Delta P} \right)$$

with V = volume, P = pressure, S = linear strain, β = compressibility

mit V = Volumen, P = Druck, S = lineare Deformation und β = Kompressibilität

Dynamic elastic modulus

Dynamischer Elastizitätsmodul

$$C = \varrho\, v_p^2$$

with ϱ = density, v_p = compressional wave velocity

mit ϱ = Dichte, v_p = Geschwindigkeit der Kompressionswelle

Dynamic bulk modulus

Dynamischer Kompressionsmodul

$$K_{dyn} = \varrho\,(v_p^2 - \tfrac{4}{3}v_s^2)$$

with v_s = shear wave velocity

mit v_s = Geschwindigkeit der Scherwelle

Static bulk moduli of a number of lunar rock and soil samples have been determined mainly as a function of hydrostatic compression. Measurements of representative rock types and soil are shown in Figs. 8, 9 and 10. A compilation of static bulk modulus values as a function of pressure is given in a logarithmic scale in Fig. 11 [Wa75].

Die statische Kompressibilität wurde an einer Reihe von Mondgesteinen und an Mondböden in Abhängigkeit vom hydrostatischen Druck gemessen. Meßergebnisse für typische Proben werden in Fig. 8, 9 und 10 gezeigt. Die Kompressionsmoduln in Abhängigkeit vom hydrostatischen Druck sind in Fig. 11 in einer logarithmischen Skala dargestellt [Wa75].

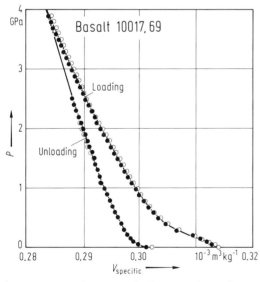

Fig. 8. Pressure-volume curve for lunar basalt [St70].

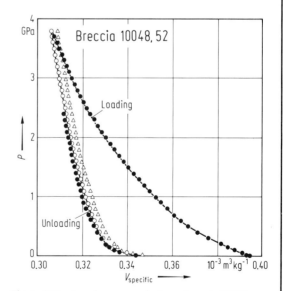

Fig. 9. Pressure-volume curve for lunar breccia [St70].

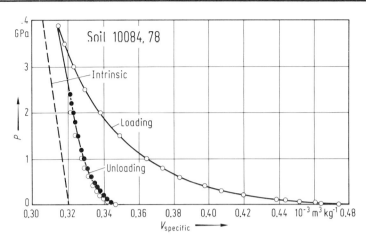

Fig. 10. Pressure-volume curve for lunar soil [St70].

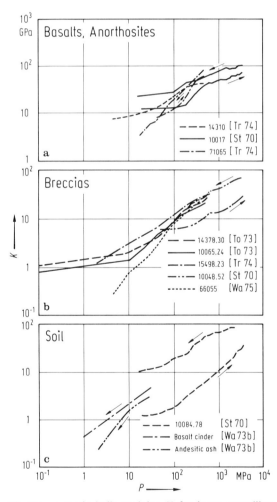

Fig. 11a···c. Static bulk modulus K for lunar crystalline rocks (basalts and anorthositic rocks), breccias and soil (after [Wa75]). Basalt cinder and andesitic ash are terrestrial material.

Pohl

Dynamic elastic moduli C calculated from compressional wave velocity and density data (see Tables 3 and 4) as a function of hydrostatic compression are given in a logarithmic scale in Figs. 12, 13 and 14 [Wa75]. The compilations in Figs. 11···14 represent the correlations of the pressure dependence of elastic moduli (or velocity) with lithological properties. They show characteristic differences for different rock types [Wa75]. A comparison of static and dynamic moduli for several rocks is given in Fig. 15 [Wa75].

Dynamische Elastizitätsmodule C, die aus den gemessenen Geschwindigkeiten der Kompressionswellen berechnet wurden (Tab. 3 und 4) sind als Funktion des hydrostatischen Druckes in einer logarithmischen Skala in Fig. 12, 13 und 14 dargestellt [Wa75]. Das unterschiedliche Verhalten des dynamischen Elastizitätsmoduls (oder der Kompressionswellen-Geschwindigkeit) kann korreliert werden mit den lithologischen Eigenschaften der untersuchten Proben [Wa75]. In Fig. 15 werden statische und dynamische Modeln in Abhängigkeit vom hydrostatischen Druck für verschiedene Gesteinsproben miteinander verglichen.

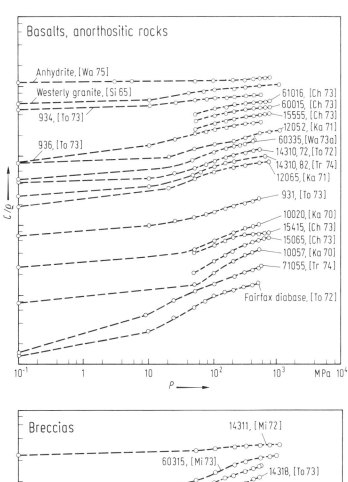

◄

Fig. 12. Dynamic elastic modulus $C/\varrho = v_p^2$ vs. pressure P for lunar basalts and anorthositic rocks, and several terrestrial rocks [Wa75]. Data are not given as absolute values but in stacked form in order to avoid interference. Each logarithmic decade signifies a change by a factor of ten. 931 is a moderately shocked hornblende diorite, 934 and 936 are weakly shocked granitic rocks from the Ries crater (Germany) [To73]. Other sample numbers are for lunar rocks (see Table 6). Fairfax diabase is thermally cycled [To73].

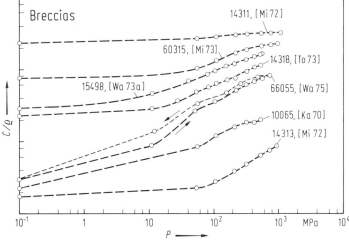

◄

Fig. 13. Dynamic elastic modulus $C/\varrho = v_p^2$ vs. pressure P for lunar breccias [Wa75]. Data are not given as absolute values but in stacked form in order to avoid interference. Each logarithmic decade signifies a change by a factor of ten. Numbers are lunar rock numbers (see Table 6).

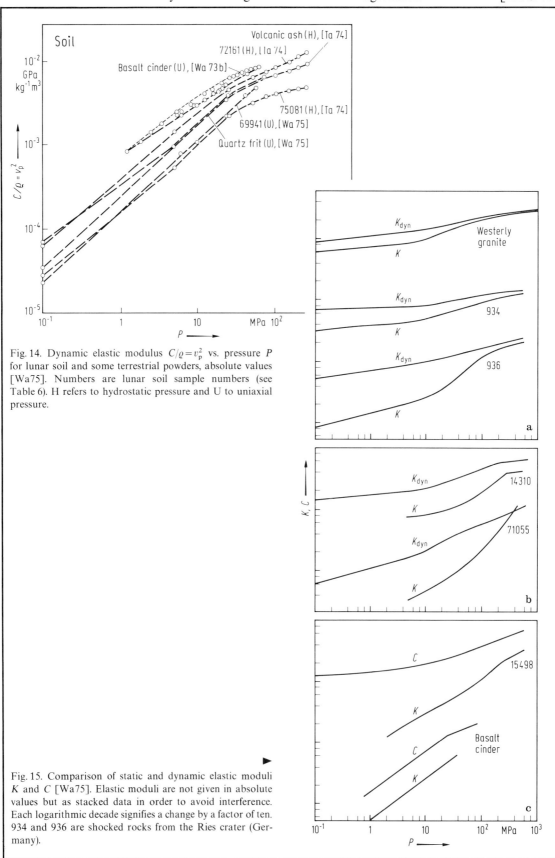

Fig. 14. Dynamic elastic modulus $C/\varrho = v_p^2$ vs. pressure P for lunar soil and some terrestrial powders, absolute values [Wa75]. Numbers are lunar soil sample numbers (see Table 6). H refers to hydrostatic pressure and U to uniaxial pressure.

Fig. 15. Comparison of static and dynamic elastic moduli K and C [Wa75]. Elastic moduli are not given in absolute values but as stacked data in order to avoid interference. Each logarithmic decade signifies a change by a factor of ten. 934 and 936 are shocked rocks from the Ries crater (Germany).

9.4 Thermal properties of lunar rocks — Thermische Eigenschaften der Gesteine des Mondes

9.4.1 Thermal conductivity k — Wärmeleitfähigkeit k

Thermal conductivity measurements were performed on lunar soil of different densities mostly over the range of lunar environment conditions near to the lunar surface, i.e. in the temperature range from about 100 K to 400 K and at vacuum conditions. In lunar surface materials heat transfer is achieved by solid conduction through grains and grain contact and by radiation. Gaseous or liquid conduction or convection can be neglected because of the extremely low pressures in the order of 10^{-12} Torr. Thermal conductivity data is presented as a function of temperature T by the equation $k = A + BT^3$ (e.g. [Cr74]). The constants A and B are obtained from the measured data by least squares fits (Table 12 and Figs. 16, 17, 18).

Die Wärmeleitfähigkeit wurde an Mondboden-Proben (soil) für verschiedene Dichten und annähernd unter den gleichen physikalischen Bedingungen, die an der Mondoberfläche herrschen, gemessen, d.h. bei Temperaturen von etwa 100 K bis etwa 400 K und im Vakuum. In den Gesteinen der Mondoberfläche erfolgt bei diesen Bedingungen der Wärmetransport durch Wärmeleitung im Festkörper und durch Strahlung. Wärmeleitung in Flüssigkeiten und Gasen, sowie konvektiver Wärmetransport können wegen des sehr hohen Vakuums von etwa 10^{-12} Torr an der Mondoberfläche vernachlässigt werden. Meßergebnisse als Funktion der Temperatur T sind in Tab. 12 in Form der Koeffizienten der Gleichung $k = A + BT^3$ [Cr74] dargestellt, wobei die A und B aus den Meßdaten mit der Methode der kleinsten Quadrate bestimmt wurden (s.a. Fig. 16, 17, 18).

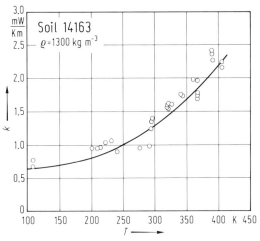

Fig. 16. Experimental thermal conductivity k vs. temperature T for lunar soil 14163, density $\varrho = 1300$ kg m^{-3}. Solid line is least squares fit to the data with the relation $k = A + B T^3$ (see Table 12) [Cr72].

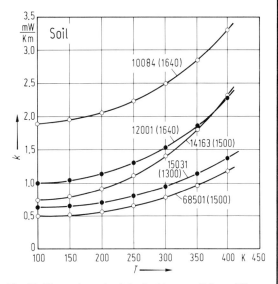

Fig. 17. Thermal conductivity k of lunar soil from different Apollo landing sites vs. temperature T. Lines are least square fits to the data. Measurement density is given in parentheses in kg m^{-3}. See Table 12 for references.

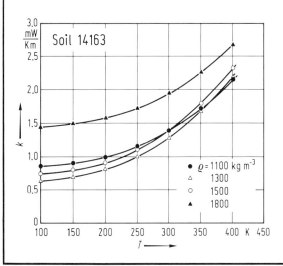

◄

Fig. 18. Thermal conductivity k of lunar soil 14163 at different densities. Solid lines are least squares fits to the data (see Table 12) [Cr74, Cr75].

The ratio BT^3/A of radiative heat transfer to conductive heat transfer is given for 225 K and 300 K. The temperature of 225 K is approximately the steady state temperature near the lunar surface at a depth of 0.2···0.3 m, where solar radiation causes no diurnal variation. Thermal conductivity values are also given for $T = 225$ K and $T = 300$ K.

The determination of the effect of air pressure on thermal conductivity of lunar soil proves that a vacuum of $10^{-2}···10^{-3}$ Torr is low enough to simulate the environment conditions at the lunar surface (Fig. 19, see also Fig. 23).

Das Verhältnis BT^3/A von Wärmetransport durch Strahlung zu Wärmeleitung im Festkörper ist für die Temperaturen 225 K und 300 K angegeben. Die Temperatur von 225 K entspricht in etwa der stationären Temperatur im Mondboden in einer Tiefe von 0.2···0.3 m, in der die Sonneneinstrahlung keine tägliche Variation mehr bewirkt. Die Wärmeleitfähigkeit ist ebenfalls für die Temperaturen 225 K und 300 K angegeben.

Aus Messungen der Wärmeleitfähigkeit in Abhängigkeit vom Luftdruck ergibt sich, daß ein Vakuum von $\approx 10^{-2}···10^{-3}$ Torr ausreicht, um die Vakuumbedingungen an der Mondoberfläche zu simulieren (Fig. 19, s.a. Fig. 23).

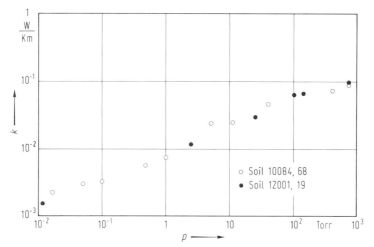

Fig. 19. Effect of interstitial gaseous pressure of air, p, on thermal conductivity k of two samples of lunar soil [Cr74]. 1 Torr = 133 Pa.

Soil conductivity was measured mostly with the line heat source technique [Ca59, Cr71]. The samples were very small (5···10 g) and in a loose and stress-free state. The fact that the samples were not held under pressure during the measurements may explain the lower thermal conductivity values (up to an order of magnitude) in comparison to *in situ* determinations in the regolith with an enhanced intergranular thermal contact (e.g. [Ho77]).

Different thermal conductivity values and temperature dependences for particulate materials at varying densities are due to effects of particle size distribution, particle shape and surface structure, to the fraction and nature of glassy particles, to the material composition and the stress state; they are not easy to explain. Data on powdered terrestrial rocks as an analogue to lunar soil can be found in [We65, Fo70, We72, Fo73, Cr74, Ho78].

Die Wärmeleitfähigkeit wurde mit einer Linien-Wärmequelle-Methode gemessen [Ca59, Cr71]. Während der Messungen waren die Proben, die sehr klein waren (5···10 g), in einem lockeren, spannungsfreien Zustand. Dies mag die im Vergleich zu *in-situ*-Messungen, wo sicher ein wesentlich besserer intergranularer Kontakt herrscht, bis zu einer Größenordnung geringeren Werte für die Wärmeleitfähigkeit erklären [Ho77].

Die Wärmeleitfähigkeit von fein zerkleinertem Material wird durch eine ganze Reihe von Faktoren beeinflußt, wie z.B. die Korngrößenverteilung, die Form und die Oberflächenstruktur der Körner, der Anteil und die Art von Glasteilchen, die Art des Materials und der Spannungszustand. Meßergebnisse an fein zerkleinertem terrestrischen Gesteinen als künstlicher Mondboden sind in [We65, Fo70, We72, Fo73, Cr74, Ho78] zu finden.

Table 12. Thermal conductivity k of lunar soil. p = interstitial gaseous pressure; ϱ = density; T_{min} and T_{max} = approximate temperature limits of experimental data; A and B = coefficients obtained by least squares fits to the data using the equation $k = A + BT^3$; BT^3/A = ratio of radiative heat transfer to conductive heat transfer for $T = 225$ K and $T = 300$ K; k_{225} and k_{300} = thermal conductivity at 225 K and 300 K.

Sample	p Torr	ϱ kg m^{-3}	T_{min} K	T_{max} K	A mW m^{-1} K^{-1}	B pW m^{-1} K^{-4}	BT^3/A 225 K	BT^3/A 300 K	k_{225} mW m^{-1} K^{-1}	k_{300} mW m^{-1} K^{-1}	Ref.
10084,68	10^{-2} ...10^{-7}	1300	200	400	1.425	17.2	0.14	0.33	1.62	1.89	Cr70
	10^{-5}	1640			1.868	22.9	0.14	0.33	2.13	2.49	Cr74
	10^{-5}	1950			1.793	14.7	0.093	0.22	1.96	2.19	Cr74
12001,19	10^{-5}	1300	170	380	0.922	31.9	0.394	0.934	1.29	1.78	Cr71a
	10^{-5}	1640	220	400	0.985	20.6	0.258	0.565	1.22	1.54	Cr74
	10^{-5}	1970	110	400	1.15	15.9	0.157	0.373	1.33	1.58	Cr74
14163,133	10^{-5}	1100	120	350	0.836	20.9	0.285	0.675	1.074	1.40	Cr74
	10^{-5}	1300	110	400	0.619	24.9	0.458	1.086	0.903	1.29	Cr74
14163	10^{-6}	1500	100	400	0.716	25.4	0.404	0.96	1.01	1.40	Cr75
	10^{-6}	1800	100	400	1.43	19.7	0.157	0.37	1.65	1.96	Cr75
15031,38	10^{-6}	1300	95	406	0.625	11.92	0.22	0.52	0.76	0.95	Cr73
68501	10^{-6}	1500	100	390	0.484	11.1	0.26	0.62	0.61	0.78	Cr74a

9.4.2 Thermal diffusivity κ — Thermische Diffusivität κ

The thermal diffusivity $\kappa = k/\varrho c_p$ (where k = thermal conductivity, ϱ = density, c_p = specific heat at constant pressure) was measured for lunar basaltic and anorthositic rocks and for lunar breccias. Measurements were made in the temperature range from 100 K to 600 K at different pressures from 1 atm to 10^{-6} Torr and with different gas compositions.

Heat transfer in solid lunar rocks is achieved by solid conduction and by radiation just as for the lunar soil. The thermal diffusivity data in Table 13 is presented as smoothed values obtained by polynomial least squares fitting to the data. Part of the data was fitted to the equation $\kappa = A + BT^{-1} + CT^3$ ([Fo73, Mi74, Ho75], Fig. 21). [Ho76] and [Ho78] used the equation $\kappa = A + BT + CT^{-2} + DT^2$ (Figs. 20 and 22, for discussion see [Ho76]).

The effect of interstitial gaseous pressure on the thermal diffusivity of a lunar basalt is shown in Fig. 23. A vacuum of 10^{-2}...10^{-3} Torr seems to be sufficient to obtain thermal diffusivity values similar to those at lunar surface conditions.

Thermal diffusivity measurements were made with the modified Ångstrom method [Ka68, Ka69, Ho75]. The accuracy is estimated to about 10% [Mi74].

Die thermische Diffusivität $\kappa = k/\varrho c_p$ (k = Wärmeleitfähigkeit, ϱ = Dichte, c_p = spezifische Wärme bei konstantem Druck) wurde an basaltischen und anorthositischen Gesteinen des Mondes und an Mondbreccien gemessen. Die Messungen wurden im Temperaturbereich von 100 K bis 600 K bei verschiedenen Gasdrucken von 1 atm bis 10^{-6} Torr und mit verschiedenen Gasen durchgeführt.

Wie beim Mondboden erfolgt auch in den festen Mondgesteinen der Wärmetransport durch Wärmeleitung im Festkörper und durch Strahlung. Die Meßergebnisse sind in Tab. 13 als geglättete Werte in Abhängigkeit von der Temperatur angegeben. Sie wurden mit der Methode der kleinsten Quadrate aus den Meßdaten gewonnen, wobei verschiedene Ausgleichspolynome verwendet wurden. Die Gleichung $\kappa = A + BT^{-1} + CT^3$ wurde von [Fo73, Mi74, Ho75], Fig. 21 verwendet. [Ho76] und [Ho78] verwenden die Gleichung $\kappa = A + BT + CT^{-2} + DT^2$ (Fig. 20 und 22).

Der Einfluß des Gasdrucks auf die thermische Diffusivität ist in Fig. 23 dargestellt. Auch hier zeigt sich, daß ein Vakuum von 10^{-2}...10^{-3} Torr ausreicht um Meßwerte, die den Bedingungen an der Mondoberfläche entsprechen, zu erhalten.

Die Messungen der thermischen Diffusivität wurden mit einer modifizierten Ångstrom-Methode durchgeführt [Ka68, Ka69, Ho75]. Die Genauigkeit wird mit etwa 10% angegeben [Mi74].

Table 13. Thermal diffusivity κ of lunar rocks. p = interstitial gaseous pressure; ϱ_{bulk} = bulk density; ϱ_{intr} = intrinsic density (at zero porosity); ϕ = porosity. Thermal diffusivity values are smoothed values obtained by polynomial least squares fitting to the data (see introduction).

Sample	p Torr	ϱ_{bulk} kg m^{-3}	ϱ_{intr} kg m^{-3}	ϕ	100	150	200	250	300	350	400	450	500	550	600 K	Ref.
					\multicolumn{11}{c}{κ [10^{-6} m^2 s^{-1}]}											
Basalts																
10020,44	1 atm, N$_2$	3060			0.907	0.643	0.537	0.482	0.447	0.424	0.406	0.392				Ho78
	10^{-6}, N$_2$		3251	0.059	0.655	0.528	0.462	0.423	0.397	0.380	0.369	0.361				
10049,42	1 atm, air	3070			1.287	0.968	0.803	0.707	0.639	0.591	0.551	0.521	0.492	0.467	0.443	Fo73
	$10^{-3}...10^{-5}$, air		3250	0.055	0.703	0.572	0.505	0.467	0.441	0.424	0.412	0.405	0.400	0.399	0.398	
10069,41	1 atm, air	2900			1.219	0.929	0.783	0.693	0.631	0.584	0.546	0.512	0.482			Fo73
	$10^{-3}...10^{-5}$, air		3260	0.11	0.396	0.311	0.266	0.244	0.230	0.225	0.224	0.229	0.237	0.250	0.268	
12002,85	1 atm, air	2961	3310	0.105	0.631	0.514	0.453	0.412	0.381	0.352	0.324	0.293				Ho75
	10^{-5}, air				0.425	0.343	0.302	0.276	0.259	0.245	0.233	0.223				
	1 atm, He				0.745	0.573	0.487	0.434	0.398	0.371	0.350	0.332				
	1 atm, Ar				0.530	0.397	0.329	0.288	0.258	0.235	0.216	0.198				
70017,77	1 atm, air	2716	3514	0.227	0.915	0.680	0.551	0.473	0.424	0.393	0.375	0.367				Ho76
	10^{-6}, air				0.339	0.295	0.274	0.263	0.257	0.254	0.253	0.253				
	1 atm, CO$_2$						0.468	0.388	0.346	0.327	0.322	0.328				
	5, CO$_2$						0.320	0.308	0.299	0.292	0.289	0.287				
70215,18	1 atm, air	3262	3444	0.053	0.912	0.688	0.561	0.481	0.428	0.392	0.366	0.348				Ho76
	10^{-6}, air				0.284	0.266	0.268	0.275	0.282	0.290	0.299	0.309				
	1 atm, CO$_2$						0.438	0.401	0.369	0.348	0.338	0.339				
	5, CO$_2$						0.296	0.308	0.312	0.314	0.316	0.320				

(continued)

Table 13 (continued)

Sample	p Torr	ϱ_{bulk} kg m⁻³	ϱ_{intr} kg m⁻³	ϕ	100	150	200	250	300	350	400	450	500	550	600 K	Ref.
					\multicolumn{11}{} $\kappa\ [10^{-6}\ \mathrm{m^2\,s^{-1}}]$											
Basalts																
70215,30	$<10^{-2}$, air	3370		0.002?	1.12	0.831	0.681	0.594	0.534	0.494	0.463	0.440	0.422	0.405	0.396	Mi74
Anorthosites																
77017,24	$<10^{-2}$, air	2480		high	0.209	0.157	0.131	0.115	0.108	0.103	0.100	0.995	0.101	0.103	0.108	Mi74
Breccias																
10065,23	1 atm, N₂	2367	3115	0.24	0.366	0.273	0.242	0.227	0.220	0.215	0.211	0.208				Ho78
	10^{-6}, N₂				0.147	0.114	0.104	0.102	0.105	0.111	0.119	0.129				
14311,50	1 atm, air	2710	2850	0.049	1.214	0.937	0.800	0.712	0.652	0.603	0.563	0.524	0.488	0.449	0.409	Fo73
	$10^{-3}\dots10^{-5}$, air				1.015	0.749	0.618	0.539	0.490	0.457	0.435	0.421	0.414	0.413	0.416	
72395,14	1 atm, air	2539	3073	0.174	0.782	0.567	0.465	0.404	0.361	0.329	0.302	0.279				Ho76
	10^{-6}, air				0.141	0.121	0.124	0.132	0.141	0.150	0.160	0.171				
	1 atm, CO₂						0.326	0.255	0.238	0.235	0.233	0.228				
	5, CO₂						0.190	0.190	0.189	0.186	0.183	0.179				
77035,44	1 atm, air	2617	3046	0.141	0.636	0.510	0.429	0.378	0.344	0.323	0.310	0.303				Ho76
	10^{-6}, air				0.266	0.213	0.197	0.194	0.197	0.204	0.214	0.226				
	1 atm, CO₂						0.373	0.322	0.296	0.284	0.281	0.284				
	5, CO₂						0.270	0.262	0.255	0.250	0.246	0.242				

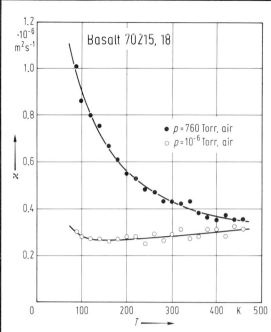

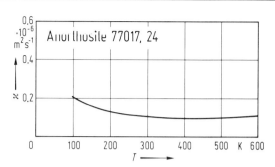

Fig. 21. Thermal diffusivity κ of lunar anorthosite 77017,24 as a function of temperature T, $p < 10^{-2}$ Torr, air. Smoothed curve obtained by least squares fitting to the data with the relation $\kappa = A + BT^{-1} + CT^3$ [Mi74].

Fig. 20. Thermal diffusivity κ as a function of temperature T of lunar basalt 70215,18. Solid lines obtained by least squares fitting to the data with the relation $\kappa = A + BT + CT^{-2} + DT^2$ [Ho76].

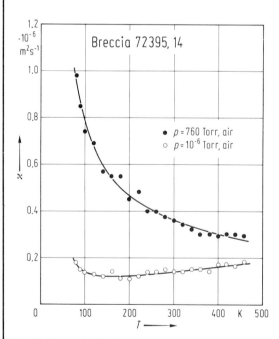

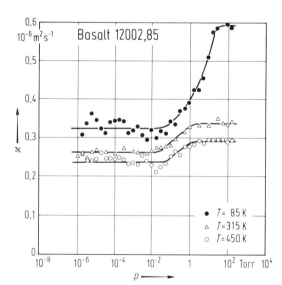

Fig. 22. Thermal diffusivity κ as a function of temperature T of lunar breccia 72395,14. Solid lines obtained by least squares fitting to the data with the relation $\kappa = A + BT + CT^{-2} + DT^2$ [Ho76].

Fig. 23. Thermal diffusivity κ of lunar basalt 12002,85 as a function of interstitial gaseous pressure of air, p, at different temperatures [Ho75].

9.4.3 Specific heat c_p — Spezifische Wärme c_p

Specific heat measurements were made for lunar rocks and lunar soil in the temperature range from about 100 K to 350 K using a low temperature adiabatic calorimeter [Ro70]. Results are presented in Table 14 as smoothed values obtained from polynomial fitting of experimental data. Empirical fitting of measured data and theoretical calculation of the specific heat using the known mineral composition of the sample is discussed in [Cr74].

Experimental data of specific heat at very low temperature is given in Table 15.

Die spezifische Wärme von Mondgesteinen und Mondboden (soil) wurde im Temperaturbereich von 100 K bis 350 K mit adiabatischen Kalorimetern für tiefe Temperaturen gemessen [Ro70]. Ergebnisse sind in Tab. 14 als geglättete Werte von Polynomen, die mit der Methode der kleinsten Quadrate aus den Meßdaten bestimmt wurden, dargestellt. Eine Diskussion von Anpassungsformeln, sowie auch der theoretischen Berechnung der spezifischen Wärme aus der bekannten Mineralzusammensetzung von Gesteinen findet man bei [Cr74].

Einige Meßergebnisse bei sehr tiefen Temperaturen sind in Tab. 15 angegeben.

Table 14. Specific heat c_p of lunar rocks and lunar soil ($J\ kg^{-1}\ K^{-1}$). Smoothed values obtained by polynomial fitting of experimental data. In parentheses: extrapolated values.

Sample	90	100	120	140	160	180	200	220	240	260	280	300	320	340	360 K	Ref.
	$c_p[J\ kg^{-1}\ K^{-1}]$															
Basalts																
10057	(239)	265	323	386	450	509	562	607	647	683	710	747	775	802	(830)	Ro70
12018,84	(208)	247	321	385	445	499	551	602	648	689	727	763	796	(827)		Ro71
15555,159	229	292	355	415	472	526	576	623	667	708	747	783	817	850	886	He73
Breccias																
10021,41	(226)	264	334	396	485	517	564	607	647	683	717	744	780	809	(824)	Ro71
14321,153	(223)	254	323	395	459	512	559	602	644	681	715	749	783	817	(851)	He73
Soil																
10084	(257)	278	336	400	464	517	564	605	642	677	710	741	772	802	(824)	Ro70
14163,186	(234)	274	349	414	472	530	581	625	663	700	738	771	799	826	859	He73
15301,20	241	272	336	397	439	507	557	605	649	691	730	767	804	839	872	He73
60601,31	242	274	339	401	459	512	562	610	656	699	741	780	817	853	887	He73

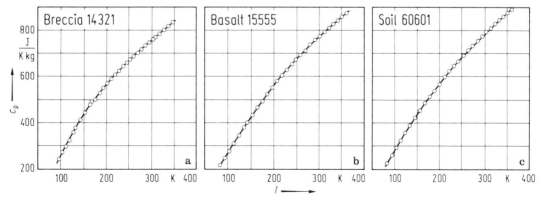

Fig. 24a···c. Experimental specific heat measurements for lunar rocks. Solid lines are polynomial least squares fits to the data [He73].

Table 15. Experimental specific heat measurements on lunar rocks at low temperatures [Wa71].

Basalt 10017		Breccia 10046	
T K	c_p J kg^{-1} K^{-1}	T K	c_p J kg^{-1} K^{-1}
2.344	1.934	3.08	1.172
2.393	1.950	3.26	1.758
2.472	1.921	3.32	1.465
2.713	2.022	3.54	1.884
2.876	2.034	3.71	1.59
3.028	2.085	4.05	1.005
3.399	2.214		
3.483	2.223		
3.819	2.340		
4.036	2.394		
4.27	2.545		
4.43	2.553		
4.52	2.679		
4.53	2.721		
4.54	2.637		
4.70	2.470		
4.97	2.805		

9.4.4 Thermal expansion — Thermische Ausdehnung

Thermal expansion of several lunar igneous rocks and breccias was measured by [Ba71] and [Ba72]. The thermal volume expansion coefficient α is given by

$$\alpha = 3\, \Delta L / L_0\, \Delta T$$

where L = length, T = temperature. The thermal expansion was measured in one direction only and the expansion coefficient was determined from the linear expansion according to the formula given above.

Experimental data for two typical rock types is shown in Fig. 25 and the data for all samples is summarized in Table 16 [Ba72]. Table 16 also includes approximate theoretical values for the thermal expansion coefficient for aggregates calculated according to

$$\alpha_{\text{theor.}} = \sum \alpha_i K_i V_i / \sum K_i V_i$$

where α_i, K_i and V_i are the volume thermal expansion coefficient, the bulk modulus and the volume fraction of the i-th phase of the aggregate, respectively [Ba72].

Comparison with measured data shows that the thermal expansion of most lunar rocks is much lower than the theoretical values suggest. The effect of microfractures on the thermal expansion is discussed by [Ba72] and [To73].

Die thermische Ausdehnung wurde von [Ba71] und [Ba72] an Basalten und Breccien gemessen. Der thermische Volumen-Ausdehnungskoeffizient α ist definiert durch

$$\alpha = 3\, \Delta L / L_0\, \Delta T$$

wobei L die Länge und T die Temperatur ist. Die thermische Ausdehnung wurde jeweils nur in einer Richtung bestimmt und der Ausdehnungskoeffizient dann nach der oben angegebenen Formel berechnet.

Meßergebnisse für zwei typische Proben sind in Fig. 25 dargestellt. Die thermischen Ausdehnungskoeffizienten für alle gemessenen Proben sind in Tab. 16 [Ba72] zusammengefaßt. Tab. 16 enthält ebenfalls theoretische Werte die nach folgender Formel für Mineralaggregate berechnet wurden

$$\alpha_{\text{theor.}} = \sum \alpha_i K_i V_i / \sum K_i V_i$$

Hierbei sind α_i, K_i und V_i der thermische Ausdehnungskoeffizient, der Kompressionsmodul und der Volumenanteil der i-ten Phase in dem Mineralaggregat [Ba72].

Ein Vergleich der gemessenen mit den theoretisch berechneten Werten zeigt, daß für die meisten der untersuchten Gesteine des Mondes die gemessenen Werte viel kleiner sind als die theoretischen. Als Ursache hierfür wird vor allem das Vorhandensein zahlreicher Mikrorisse, die vorwiegend durch Stoßwellen beim·Einschlag von Meteoriten erzeugt wurden, angesehen [Ba72, To73].

Table 16. Volume coefficient of thermal expansion α.

Sample		α [10^{-6} K^{-1}]		$\alpha_{theor.}$ [10^{-6} K^{-1}]
		$-100\cdots$ 25 °C	$25\cdots$ 200 °C	$25\cdots$ 200 °C
10020	Basalt	7.2	16.2	22.5
10046	Breccia (20% glass)	15.0	22.2	22.1
10057	Basalt	8.1	14.7	21.6
12022	Basalt	5.7	15.9	23.7
14318	Breccia	11.7	18.0	–
15015	Breccia	12.6	–	–
15418	Breccia	7.5	12.3	–

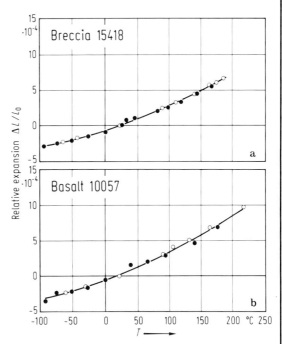

Fig. 25 a, b. Thermal expansion of lunar rocks. Open and solid circles indicate measurements at increasing and decreasing temperature, respectively [Ba72].

9.4.5 Radiogenic heat — Radiogene Wärme

Heat generation in lunar rocks by radioactive nuclides is calculated from the content of ^{40}K, ^{232}Th, ^{235}U and ^{238}U. If it is assumed that the isotopic abundances of K, Th and U are the same as on the earth, the radiogenic heat A^* per s and per kg is given by

Die Wärmeproduktion durch radioaktive Isotope wird aus dem Gehalt an ^{40}K, ^{232}Th, ^{235}U und ^{238}U berechnet. Unter der Annahme, daß die Isotopenverhältnisse in den Gesteinen des Mondes dieselben sind wie in terrestrischen Gesteinen ergibt sich die radiogene Wärme A^* pro s und kg durch

$$A^* = 0.0348\,c_K + 0.0256\,c_{Th} + 0.0955\,c_U$$

with A^* in 10^{-9} W kg^{-1}, c_K = potassium content in weight %, c_{Th} = thorium content in ppm and c_U = uranium content in ppm and using the revised constants of [Ry73].

mit A^* in 10^{-9} W kg^{-1}, c_K = Kaliumgehalt in Gewichtsprozent, c_{Th} = Thoriumgehalt in ppm, c_U = Urangehalt in ppm (revidierte Konstanten nach [Ry73]).

Tables 17 and 18 show the K, Th and U content and the heat generation of typical lunar highland rocks and mare basalts (see also [Ho72]). The major element composition of these rocks is given in Tables 2 and 3 (after [Ta75]). Mare basalts form only about 0.5% of the lunar crust and the heat production is therefore dominated by the anorthositic rock types. The K-rich rock types are not very frequent. Recent discussions of the content of K, Th and U in lunar highland rocks can be found in [Ta78] and [Ta79].

In den Tabellen 17 und 18 sind die Gehalte an K, Th und U sowie die radiogene Wärme für typische Hochlandgesteine und für Mare-Basalte zusammengestellt (s.a. [Ho72]). Die chemische Zusammensetzung dieser Gesteine ist in Tab. 2 und 3 angegeben [Ta75]. Das Volumen der Mare-Basalte beträgt nur rund 0.5% des Volumens der Kruste des Mondes, so daß die radiogene Wärme der Kruste überwiegend durch die anorthositischen Gesteinsarten bestimmt wird. Dabei ist zu berücksichtigen, daß die K-reichen Gesteine relativ selten sind. Rezente Diskussionen der Häufigkeiten von K, Th und U findet man bei [Ta78] und [Ta79].

Table 17. Radiogenic heat in lunar highland rocks. Chemical composition after [Ta75] (see also Table 2).

	Gabbroic anorthosite	Anorthositic gabbro	Troctolite	Low K Fra Mauro basalt	Medium K Fra Mauro basalt	Average *) highlands crust
K [wt.-%]	–	–	–	0.10	0.45	0.055
Th [ppm]	0.73	0.23	1.0	5.3	12	0.80
U [ppm]	0.17	0.05	0.27	1.37	3.2	0.21
A^* [10^{-9} W kg^{-1}]	0.035	0.010	0.052	0.270	0.629	0.043

*) Composition after [Ta78].

Table 18. Radiogenic heat in lunar mare basalts. Chemical composition after [Ta75] (see also Table 3).

	Olivine basalt A 12	Olivine basalt A 15	Quartz basalt A 15	Quartz basalt A 12	High K basalt A 11	Low K basalt A 11	High Ti basalt A 17	Aluminous mare basalt A 12
K [wt.-%]	0.053	0.025	0.025	0.059	0.24	0.08	0.06	0.06
Th [ppm]	0.88	0.05	0.59	1.15	3.4	1.0	0.54	0.90
U [ppm]	0.24	0.16	0.13	0.30	0.8	0.25	0.17	0.25
A^* [10^{-9} W kg^{-1}]	0.047	0.016	0.028	0.060	0.172	0.052	0.032	0.013

9.5 Electrical properties of lunar rocks — Elektrische Eigenschaften der Gesteine des Mondes

9.5.1 Dielectric constant, dissipation factor and conductivity — Dielektrische Eigenschaften

Measurements of dielectric properties usually give values for the dielectric constant K', which is the real part of the complex relative permittivity

Bei Messungen der dielektrischen Eigenschaften werden in der Regel die relative Dielektrizitätskonstante (DK, Dielektrizitätszahl) K' und der Verlustfaktor D bestimmt. K' ist der Realteil der komplexen relativen Dielektrizitätskonstante

$$K^* = K' - j K'', \quad j = \sqrt{-1} \quad (K' = \varepsilon_r)$$

and for the dissipation factor D or loss tangent $\tan \delta$ given by

Der Verlustfaktor D ist gegeben durch

$$D = \tan \delta = K''/K'$$

The real conductivity σ', which is the real part of the complex conductivity

wobei δ der Verlustwinkel ist. Die Leitfähigkeit σ' ist der Realteil der komplexen Leitfähigkeit in Dielektrika:

$$\sigma^* = \sigma' + j \sigma''$$

can be obtained from K' and D by

σ' wird bestimmt durch

$$\sigma' = \omega K' \varepsilon_0 D$$

where $\omega = 2 \pi f$, f = frequency, ε_0 = vacuum permittivity (As/Vm) (see e.g. [Co73, Ka73a]).

Figs. 26···31 give the dielectric constant, loss tangent and conductivity σ' of different lunar rocks as a function of frequency and temperature. Measurements shown in these figures were made under conditions approximating the lunar surface environment conditions, i.e. absence of moisture and high vacuum.

mit $\omega = 2 \pi f$, f = Frequenz, ε_0 = absolute Dielektrizitätskonstante des Vakuums (siehe z.B. [Co73, Ka73a]).

Fig. 26···31 zeigen die Dielektrizitätskonstante, den Verlustfaktor und die Leitfähigkeit σ' von verschiedenen Gesteinen des Mondes in Abhängigkeit von Frequenz und Temperatur. Die in diesen Abbildungen dargestellten Ergebnisse wurden unter Bedingungen gemessen, die denjenigen an der Mondoberfläche ver-

Samples from Apollo 11, 12 and 14 were exposed over a long time to atmospheric humidity before the measurements, and outgasing at elevated temperatures was necessary to minimize the important effect of moisture on dielectric properties (e.g. [Ol73, Ol75a]). Residual moisture is indicated by a frequency dispersion of the dielectric constant and the loss tangent at low frequencies. Table 19 is a compilation of dielectric properties at high frequency ([Ol75]). Fig. 32 shows the density dependence of the dielectric constant at high frequencies and room temperature ([Ol75] and Table 19). A regression analysis of the dielectric constant K' vs. density ϱ data yields the following relation:

gleichbar sind, d.h. in hohem Vakuum und in Abwesenheit von Feuchtigkeit. Die Proben von Apollo 11, 12 und 14 waren vor den Messungen über einen längeren Zeitraum den atmosphärischen Feuchtigkeitsbedingungen ausgesetzt, und es erwies sich als notwendig sie bei höheren Temperaturen zu entgasen, um den starken Einfluß der Feuchtigkeit auf die elektrischen Eigenschaften möglichst gering zu machen [Ol73, Ol75a]. Verbleibender adsorbierter Wasserdampf macht sich durch eine Frequenzabhängigkeit der Dielektrizitätskonstante und des Verlustfaktors bei niedrigen Frequenzen bemerkbar. In Tab. 19 sind die dielektrischen Eigenschaften bei höheren Frequenzen zusammengestellt (nach [Ol75]). Fig. 32 zeigt die Abhängigkeit der Dielektrizitätskonstante bei höheren Frequenzen und Zimmertemperatur in Abhängigkeit von der Dichte ([Ol75] und Tab. 19). Eine Regressionsanalyse ergibt folgenden Zusammenhang zwischen der Dielektrizitätskonstante K' und der Dichte ϱ:

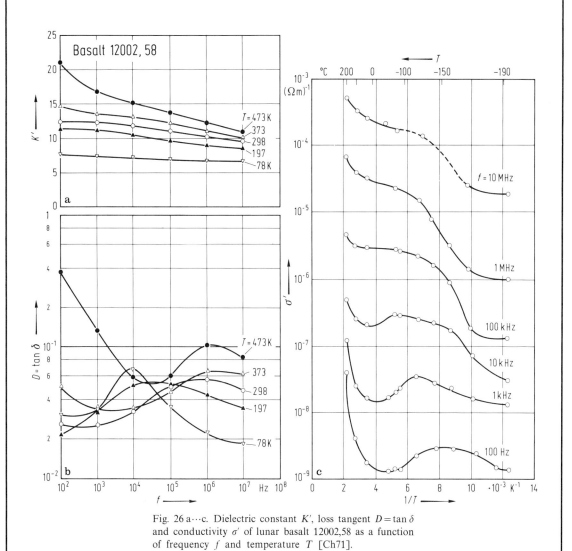

Fig. 26 a⋯c. Dielectric constant K', loss tangent $D = \tan\delta$ and conductivity σ' of lunar basalt 12002,58 as a function of frequency f and temperature T [Ch71].

$$K' = (1.93 \pm 0.17)^{10^{-3} \cdot \varrho}$$

with ϱ in kg m^{-3} and the requirement that $K' - 1.0$ at $\varrho = 0.0$ [Ol75]. A regression analysis of the dissipation factor D data vs. density ϱ and the content c of FeO + TiO$_2$ in % for Apollo 14, 15, 16 and 17 samples, which were uncontaminated by moisture, gives the relation:

wobei ϱ in kg m^{-3} angegeben wird und die Bedingung, daß $K' = 0$ ist bei $\varrho = 0.0$ erfüllt sein muß [Ol75]. Für die Abhängigkeit des Verlustfaktors D von der Dichte und dem Gehalt c an FeO + TiO$_2$ (in %) ergibt sich für die Proben von Apollo 14, 15, 16 und 17, die der Luftfeuchte nicht ausgesetzt waren, die Beziehung:

$$D = ((0.53 \pm 0.56) + (0.25 \pm 0.09)\,c) \cdot 10^{-6} \cdot \varrho$$

[Ol75]. Measurements for the Apollo 15 and 16 deep drill cores are given in [Go77]. Values are comparable to surface soil values.

Messungen der dielektrischen Eigenschaften an den Bohrkernen von Apollo 15 und 16 [Go77] ergeben ähnliche Werte, wie für die Proben von der Oberfläche.

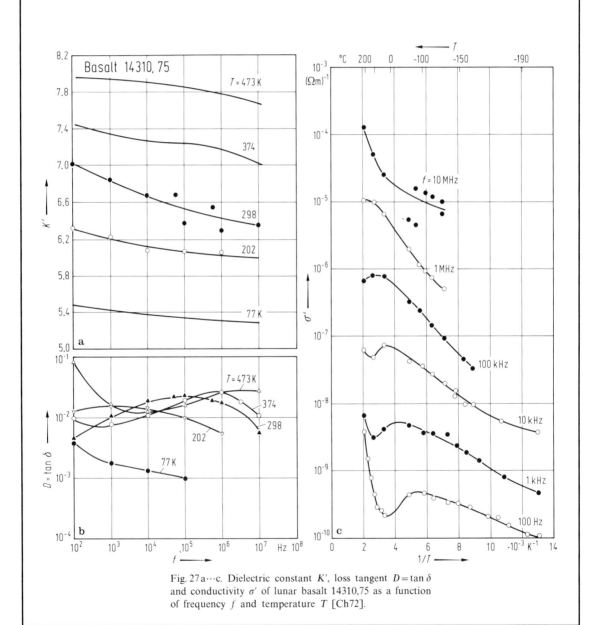

Fig. 27a···c. Dielectric constant K', loss tangent $D = \tan \delta$ and conductivity σ' of lunar basalt 14310,75 as a function of frequency f and temperature T [Ch72].

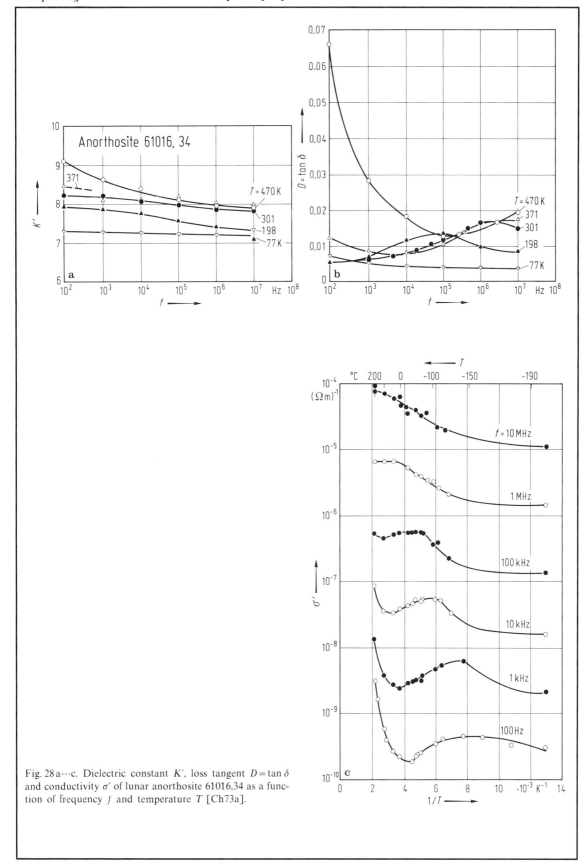

Fig. 28 a···c. Dielectric constant K', loss tangent $D = \tan \delta$ and conductivity σ' of lunar anorthosite 61016,34 as a function of frequency f and temperature T [Ch73a].

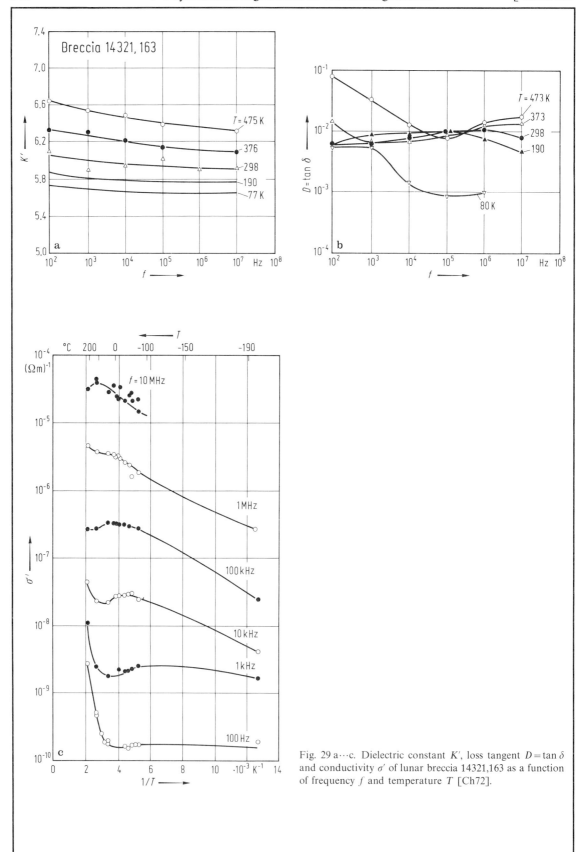

Fig. 29 a···c. Dielectric constant K', loss tangent $D = \tan \delta$ and conductivity σ' of lunar breccia 14321,163 as a function of frequency f and temperature T [Ch72].

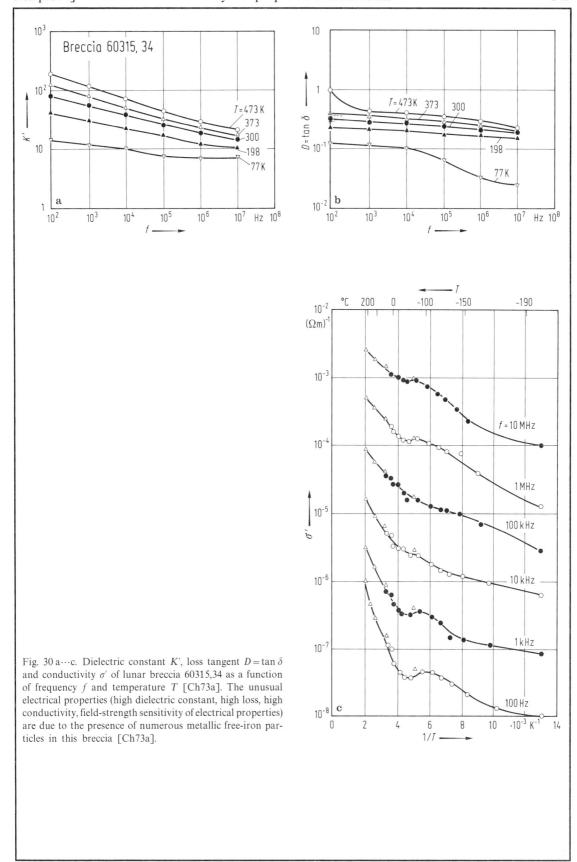

Fig. 30 a⋯c. Dielectric constant K', loss tangent $D = \tan \delta$ and conductivity σ' of lunar breccia 60315,34 as a function of frequency f and temperature T [Ch73a]. The unusual electrical properties (high dielectric constant, high loss, high conductivity, field-strength sensitivity of electrical properties) are due to the presence of numerous metallic free-iron particles in this breccia [Ch73a].

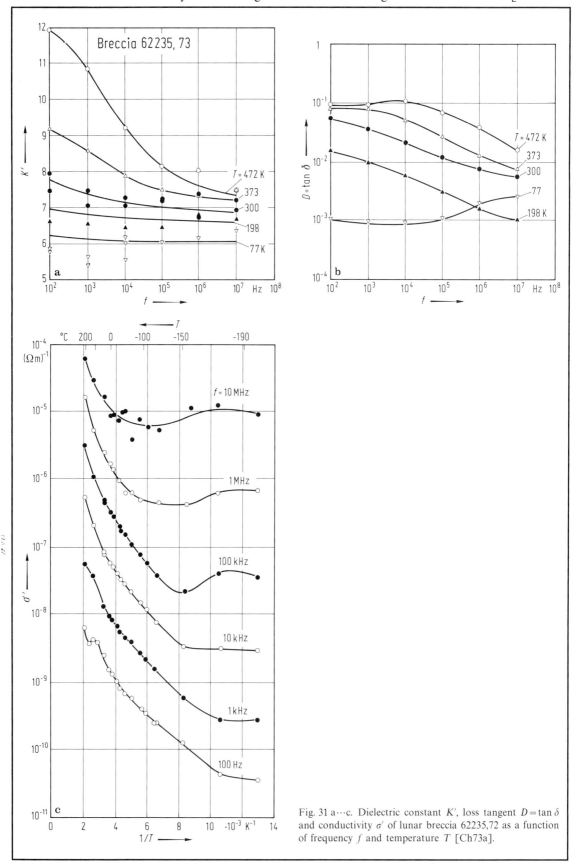

Fig. 31 a⋯c. Dielectric constant K', loss tangent $D = \tan \delta$ and conductivity σ' of lunar breccia 62235,72 as a function of frequency f and temperature T [Ch73a].

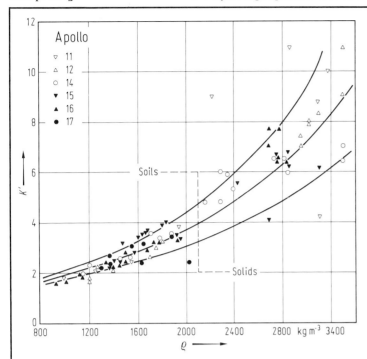

Fig. 32. Dielectric constant K' vs. density ϱ. Solid curves obtained from regression analysis (plus, minus one standard deviation) (after [Ol75] and Table 19).

9.5.2 Low frequency a.c. and d.c. conductivity — Gleichstromleitfähigkeit

Measurements of low frequency a.c. (5 Hz) and d.c. conductivity are shown in Fig. 33 as a function of temperature. In the early phase of these investigations very large irreversible changes of conductivity were observed on heating and cooling under non-appropriate oxygen fugacity conditions [Sch71, Sch73, Sch74]. The results presented in Fig. 33 were obtained in carefully controlled environment avoiding oxidation and giving reproducible results [Sch74] and are thought to be representative for lunar conditions. Additional d.c. conductivity data can be found in [Ol73, Ol74, Ol74a].

Messungen der elektrischen Leitfähigkeit bei sehr niedrigen Frequenzen (5 Hz) und bei Gleichstrom in Abhängigkeit von der Temperatur sind in Fig. 33 dargestellt. Zu Beginn dieser Untersuchungen wurden beim Aufheizen und Abkühlen sehr starke irreversible Veränderungen der Leitfähigkeit beobachtet, was auf ungeeignete Sauerstoff-Fugazitäts-Bedingungen zurückzuführen war [Sch71, Sch73, Sch74]. Die Ergebnisse in Fig. 33 wurden unter Bedingungen gemessen, die eine Oxidation verhinderten und reproduzierbare Meßwerte lieferten [Sch74]. Es wird angenommen, daß sie den wirklichen Werten unter lunaren Bedingungen vergleichbar sind. Weitere Messungen der Gleichstromleitfähigkeit finden sich bei [Ol73, Ol74, Ol74a].

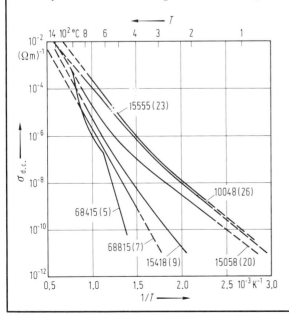

Fig. 33. Temperature dependence of d.c. electrical conductivity $\sigma_{d.c.}$ of lunar rocks. Numbers in parentheses give FeO content in mass %. 10048 = breccia, 15058 = porous basalt, 15418 = anorthositic breccia, 15555 = basalt, 68415 = anorthositic gabbro, 68815 = glass-welded breccia (after [Fi77]).

Table 19. Dielectric properties of lunar rocks and soil at high frequencies at 300 K (after [Ol75]).

Sample	ϱ kg m^{-3}	f 10^6 Hz	Environment	K'	D ($=\tan\delta$)	Ref.	% TiO$_2$	% FeO
Basalts								
10017,30	3100	1.0	N	8.8	0.075	Ka71a	11.74	19.82
10020	3180	1.0	AN	10.0	0.13	Ch70	10.72	19.35
10022	3100	450	A	4.2	0.06	Go71	12.20	18.90
10057	2880	1.0	AN	11.0		Ch70	11.44	19.35
12002,58	3300	1.0	AN	9.00	0.05	Ch71	2.76	19.38
12002,84	3100	1.0	N	8.30	0.051	Ka71a		
12002,85	3040	1.0	N	7.80	0.056	Ka71a		
12002,85	3040	1.0	V	8.00	0.065	Ol74		
12022,60	3320	1.0	AN	11.0	0.18	Ol73a	4.90	21.70
12063,89	2950	450	A	7.0	0.069	Go71	5.00	21.26
12065	2950	450	A	7.30		Go71	3.80	22.00
14310,72	2860	1.0	AN	6.00	0.02	Ch72	1.30	7.70
14310,74	2814	9000	AN	6.46	0.0075	Ba72a		
14310,75	3300	1.0	AN	6.40	0.02	Ch72		
14310,87	3300	1.0	AN	7.0	0.012	Ch72		
14310,161	2750	450	A	6.5	0.00454	Go72		
15065,27	2860	1.0	N	6.70	0.01	Ch73a	1.48	19.18
15555,88	3100	1.0	N	6.15	0.0252	Ch73a	2.26	22.47
15597,30	2850	450	A	6.20	0.0022	Go73	1.87	20.17
62235,17	2780	1.0	N	6.52	0.0066	Ch73a	1.21	9.45
Anorthosites								
15415,57	2700	1.0	N	4.2		Ch73a	0.02	0.23
60015,29	2760	1.0	N	6.60	0.0002	Ch73a	0.06	0.35
61016,34	2790	1.0	N	7.82	0.016	Ch73a	0.69	4.97
Breccias								
10046	2210	1.0	AN	9.0	0.05	Ch70	10.35	19.22
14301,37	2170	1.0	AN	4.80	0.05	Ch72		
14301,41	2300	1.0	AN	4.80	0.05	Ch72	1.70	9.80
14318,30	2300	1.0	AN	5.97	0.0082	Ch72	1.46	9.50
14321,163	2400	1.0	AN	5.28	0.0123	Ch72	2.40	13.00
14321,228	2350	1.0	AN	5.90	0.01	Ch72		
15459,62	2760	1.0	N	6.62	0.005	Ch73a	0.91	9.40
15498,39	2420	450	A	5.45	0.008	Go73	1.60	17.30
60017,45	2850	450	A	6.30	0.0024	Go73	0.30	2.97
65015,6	2700	1.0	V	7.70	0.008	Ol73		
65015,7	2700	1.0	V	7.00	0.008	Si73	1.26	8.59
Soil								
10084,83	1940	1.0	N	3.8	0.0175	Ka71a	7.56	15.94
10084	1000	450	A	1.80		Go70		
	1250	450	A	2.06		Go70		
	1560	450	A	2.45		Go70		
12033	1200	450	A	1.80		Go71	2.48	14.20
	1400	450	A	2.20		Go71		
	1700	450	A	2.60		Go71		
12070,107	1740	1.0	N	3.0	0.025	Ka71a	2.81	16.40
12070	1200	450	A	1.80		Go71		
	1400	450	A	2.20		Go71		
	1810	450	A	3.20		Go71	(continued)	

Table 19 (continued)

Sample	ϱ kg m^{-3}	f 10^6 Hz	Environ-ment	K'	D ($=\tan\delta$)	Ref.	%TiO$_2$	%FeO
14003	1160	450	A	1.95		Go72	1.77	10.45
	1550	450	A	2.55		Go72		
14163	1450	450	A	2.55		Go72	1.77	10.41
	1800	450	A	3.25		Go72		
14163,131	1200	1.0	V	2.3	0.0006	St72	1.79	10.35
14163,164	1710	9000	AN	3.59	0.015	Ba72a		
	1880	9000	AN	3.59	0.015	Ba72a		
15021,144	1303	450	A	2.20	0.00418	Go73	1.80	15.00
15041,81	1451	450	A	2.50	0.00486	Go73	1.70	14.20
15211,38	1359	450	A	2.45	0.00389	Go73	1.34	11.66
15221,59	1529	450	A	2.55	0.00274	Go73	1.27	11.32
15301,38	1470	1.0	V	3.20	0.0008	Ol74	1.17	14.05
15301,38	1600	0.1	V	3.42	0.0122	Fr75		
	1620	0.1	V	3.49	0.0124	Fr75		
	1640	0.1	V	3.51	0.0125	Fr75		
	1680	0.1	V	3.58	0.0122	Fr75		
	1800	0.1	V	3.89	0.0125	Fr75		
	1830	0.1	V	3.98	0.0122	Fr75		
15301,43	1576	450	A	2.80	0.00438	Go73		
15601,105	1945	450	A	3.30	0.00251	Go73	1.98	19.79
61500,7	1143	450	A	1.96	0.00277	Go73	0.56	5.31
	1489	450	A	2.50	0.00347	Go73		
	1906	450	A	3.55	0.00503	Go73		
62240,5	1383	450	A	2.40	0.00364	Go73	0.56	5.49
	1713	450	A	3.15	0.00416	Go73		
	1906	450	A	3.30	0.00482	Go73		
62241,21	1340	1.0	V	2.40	0.001	St72		
63501,25	1014	450	A	1.70	0.00161	Go73	0.53	4.72
	1420	450	A	2.40	0.00253	Go73		
	1788	450	A	3.20	0.00341	Go73		
66041,8	1500	1.0	N	2.70	0.002	Ka73	0.63	5.80
66041,13	932	450	A	1.60	0.00225	Go73		
	1279	450	A	2.20	0.00300	Go73		
	1531	450	A	2.70	0.00388	Go73		
66081,20	1490	1.0	V	2.80	0.001	Si73	0.67	5.85
67601,22	1151	450	A	1.90	0.00216	Go73	0.42	4.09
	1429	450	A	2.40	0.00259	Go73		
	1675	450	A	2.95	0.00290	Go73		
72441,12	1560	0.1	V	3.05	0.005	Fr75	1.53	8.68
	1650	0.1	V	3.12	0.005	Fr75		
	1800	0.1	V	3.26	0.005	Fr75		
74220,24	1370	1.0	V	2.60	0.019	Ol73a	8.81	22.04
74241,2	1380	0.1	V	2.20	0.010	Al73	8.61	15.84
	1610	0.1	V	2.38	0.010	Al73		
75081,3	1900	1.0	N	3.50	0.010	Ol75	9.52	17.41
75081,27	2080	1.0	V	2.40	0.018	Ol73a		

Remarks: Subsample number is not generally the same as for chemical analysis.
Environment: A = air, N = nitrogen, V = vacuum.
References are for electrical measurements. References for chemical analyses are given in [Ol75].

9.6 Magnetic properties of lunar rocks — Magnetische Eigenschaften der Gesteine des Mondes

For conversion table of quantities (magnetic field, magnetisation, susceptibility), see 6.1.1.2

9.6.1 Ferromagnetic phases — Ferromagnetische Phasen

The temperature dependence of saturation magnetization above room temperature demonstrates that pure metallic iron and/or relatively Ni-poor kamacite are the dominant ferromagnetic phases in lunar rocks (Figs. 34, 35 and 36). Pure iron has a Curie-point of 770 °C and most samples have Curie-points in the vicinity of this temperature (Fig. 37). The ratio m_k/m (where m_k is the kamacite content of various kamacite phases in wt-% and $m = m_{Fe} + m_k$ with m_{Fe} = abundance of metallic iron in wt-%) and the average Ni content in the kamacite phases were estimated from the magnetization versus temperature curves and from the transition temperatures $T_{\alpha\gamma}$ and $T_{\gamma\alpha}$ (Figs. 35 and 37) of the kamacite phases ([Na75], Fig. 38). Histograms of saturation magnetization of lunar rocks caused by the ferromagnetic iron and kamacite phases are shown in Fig. 39 (see Fig. 42 for definition).

Messungen der Abhängigkeit der Sättigungsmagnetisierung von der Temperatur zeigen, daß die ferromagnetischen Phasen in den Gesteinen des Mondes überwiegend aus metallischem Eisen und aus Kamazit mit einem relativ geringen Ni-Gehalt bestehen (Fig. 34, 35 und 36). Die Curie-Temperatur der meisten Proben liegt in der Nähe der Curie-Temperatur von reinem Eisen (Fig. 37). Der Kamazitanteil am Gesamtmetallgehalt $m_k/m = m_k/m_{Fe} + m_k$ (m_k = Kamazitgehalt in Gew.-%, m_{Fe} = Eisengehalt in Gew.-%), sowie der mittlere Ni-Gehalt in den Kamazitphasen können aus der Temperaturabhängigkeit der Sättigungsmagnetisierung und aus den Übergangstemperaturen $T_{\alpha\gamma}$ und $T_{\gamma\alpha}$ (Fig. 35 und 37) abgeschätzt werden ([Na75], Fig. 38). Fig. 39 zeigt Häufigkeitsverteilungen für die Sättigungsmagnetisierung von verschiedenen Arten von Mondgesteinen.

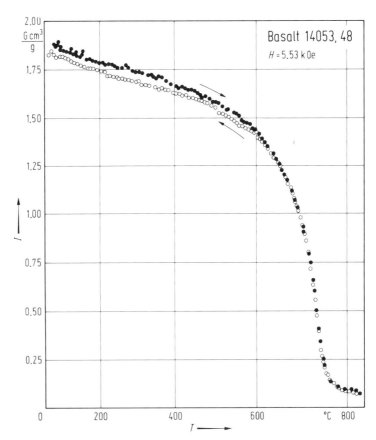

Fig. 34. Specific magnetization I vs. temperature T of lunar basalt 14053,48 measured in a magnetic field $H = 5.53$ kOe [Na72]. Full circles: heating; open circles: cooling. The measurements indicate the presence of nearly pure iron.

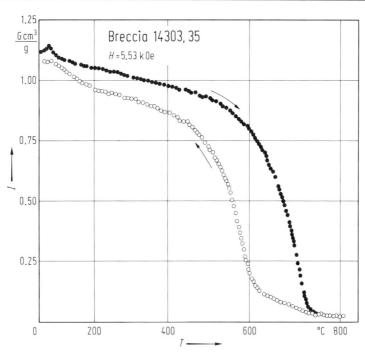

Fig. 35. Specific magnetization I vs. temperature T (first run) of lunar breccia 14303,35 measured in a magnetic field $H = 5.53$ kOe [Na72]. Full circles: heating; open circles: cooling. The thermal hysteresis is explained by the $\alpha \rightleftharpoons \gamma$ transition of the nickel-iron alloy.

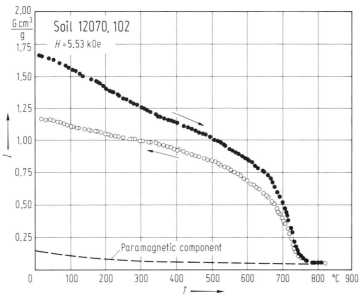

Fig. 36. Specific magnetization I vs. temperature T of lunar soil 12070,102 measured in a magnetic field $H = 5.53$ kOe [Na72]. Full circles: heating; open circles: cooling. The irreversible decrease of magnetization, which is characteristic for lunar soil and many breccias, is explained by oxidation of the very fine metallic iron grains in a vacuum of about 10^{-5} Torr.

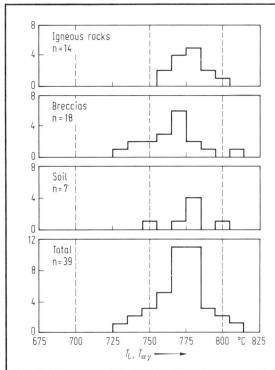

Fig. 37. Histograms of Curie points T_C and $\alpha \rightarrow \gamma$ transition temperatures $T_{\alpha\gamma}$ of lunar igneous rocks, breccias and soil (Apollo 11···17) [Na75]. For samples with $T_C < T_{\alpha\gamma}$ the T_C value and for samples with $T_C > T_{\alpha\gamma}$ the $T_{\alpha\gamma}$ value is included in the histograms.

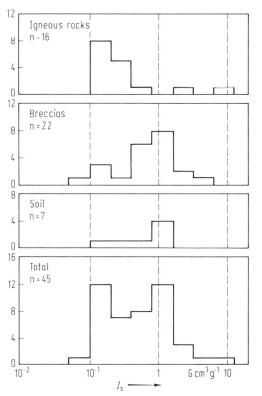

Fig. 39. Histograms of specific saturation magnetization I_s at $T = 300$ K of lunar rocks (Apollo 11···17) [Na75].

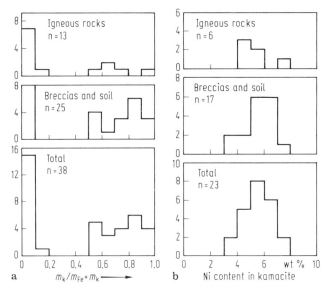

Fig. 38a, b. (a) Histograms of the ratio $m_k/m_{Fe} + m_k$, where m_k is the abundance of kamacite in wt% and m_{Fe} is the abundance of metallic iron in wt%, in lunar igneous rocks, breccias and soil (Apollo 11···17) [Na75].
(b) Histograms of average Ni content in kamacite [Na75].

The ferromagnetic phases in lunar surface rocks, which were identified by thermomagnetic analysis and which determine the saturation magnetization I_s (Fig. 39) include original metal phases in lunar rocks, meteoritic metal phases added to lunar rocks through impacts, and pure iron produced by shock metamorphic effects from Fe-bearing silicates. Most lunar igneous rocks have saturation magnetizations between 0.1 and 0.2 G cm^3 g^{-1}, whereas breccias and soil generally have higher saturation magnetizations. From this it is concluded that any saturation magnetization greater than about 0.15 G cm^3 g^{-1} (corresponding to 0.2 wt-% of metal phases) must be considered to be a result of meteorite impacts on the lunar surface.

Die ferromagnetischen Phasen, die die Sättigungsmagnetisierung der an der Mondoberfläche gesammelten Gesteine bestimmen, bestehen aus primären Metallphasen in den kristallinen Gesteinen, aus Beimengungen von Meteoriten und aus reinem metallischem Eisen, das durch Stoßwelleneffekte aus Fe-haltigen Silikaten gebildet wurde. Die meisten kristallinen Gesteine haben eine Sättigungsmagnetisierung von 0,1 bis 0,2 G cm^3 g^{-1}, die Breccien und der Mondboden hingegen in der Regel 5 bis 10mal höhere Werte. Hieraus wird geschlossen, daß Impaktprozesse die Bildung des höheren Metallgehaltes in Proben mit einer Sättigungsmagnetisierung größer als rund 0,15 G cm^3 g^{-1} (was einem Metallgehalt von 0,2 Gew.-% entspricht) verursacht haben.

9.6.2 Antiferromagnetic and paramagnetic phases — Antiferromagnetische und paramagnetische Phasen

The paramagnetic susceptibility at room temperature mainly includes contributions from pure paramagnetic phases and from antiferromagnetic phases with Néel temperatures below room temperature (Fig. 40). Low temperature measurements permit estimates of the contribution of various components (Fig. 41). A more refined method of analysis consisting in numerical differentiation of the magnetization versus temperature curves with respect to $1/T$ has been used by [Na72].

Zur paramagnetischen Suszeptibilität bei Zimmertemperatur tragen vorwiegend rein paramagnetische Minerale sowie antiferromagnetische Phasen, die Néel-Temperaturen unterhalb der Zimmertemperatur haben, bei (Fig. 40). Aus Messungen bei tiefen Temperaturen läßt sich der Beitrag verschiedener Phasen abschätzen (Fig. 41). Eine verfeinerte Analysenmethode, bei der die Ableitung der Temperaturabhängigkeit der Sättigungsmagnetisierung nach $1/T$ verwendet wird, wurde von [Na72] beschrieben.

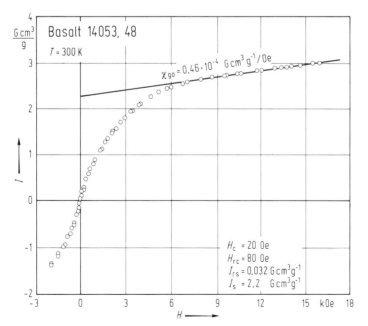

Fig. 40. Magnetization curve of lunar basalt 14053,48 [Na72] (see also Fig. 42).

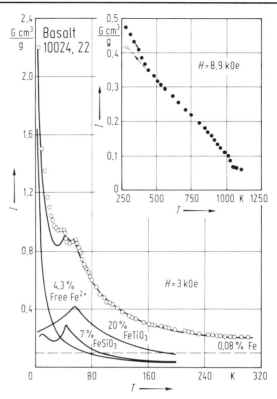

Fig. 41. Specific magnetization I vs. temperature T of lunar basalt 10024,22. Circles are observed values. Solid lines are calculated values [Na70].

9.6.3 Magnetic classification of lunar rocks —
Klassifizierung der Gesteine des Mondes nach magnetischen Parametern

Crossplots of various magnetic parameters have frequently been used for classification of lunar igneous rocks, breccias and soil. These classifications yield information on the physical state of the ferromagnetic phases (multi-domain, single-domain, superparamagnetic) and permit conclusions on the origin and the shock history of the rocks. Fig. 42 shows the definition of different parameters which can be obtained from hysteresis measurements. Various crossplots of magnetic parameters are shown in Figs. 43···48.

Es hat sich gezeigt, daß magnetische Parameter für eine Klassifizierung der kristallinen Gesteine, der Breccien und des Mondbodens sehr geeignet sind. Sie ermöglichen Aussagen über die Natur der ferromagnetischen Phasen (Mehrbereichsteilchen, Einbereichsteilchen, superparamagnetische Teilchen) und damit über den Ursprung und die Stoßwellenmetamorphose der Gesteine. Fig. 42 erläutert verschiedene Parameter, die aus Messungen der Hysteresekurve gewonnen werden. Klassifizierungen von Mondgesteinen mit Hilfe der magnetischen Parameter sind in Fig. 43···48 dargestellt.

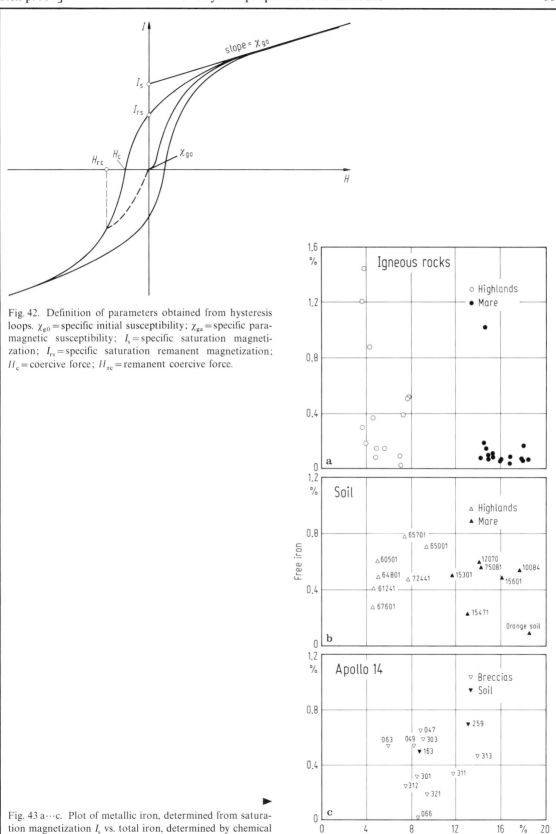

Fig. 42. Definition of parameters obtained from hysteresis loops. χ_{g0} = specific initial susceptibility; χ_{ga} = specific paramagnetic susceptibility; I_s = specific saturation magnetization; I_{rs} = specific saturation remanent magnetization; H_c = coercive force; H_{rc} = remanent coercive force.

Fig. 43 a···c. Plot of metallic iron, determined from saturation magnetization I_s vs. total iron, determined by chemical analysis [Fu75].

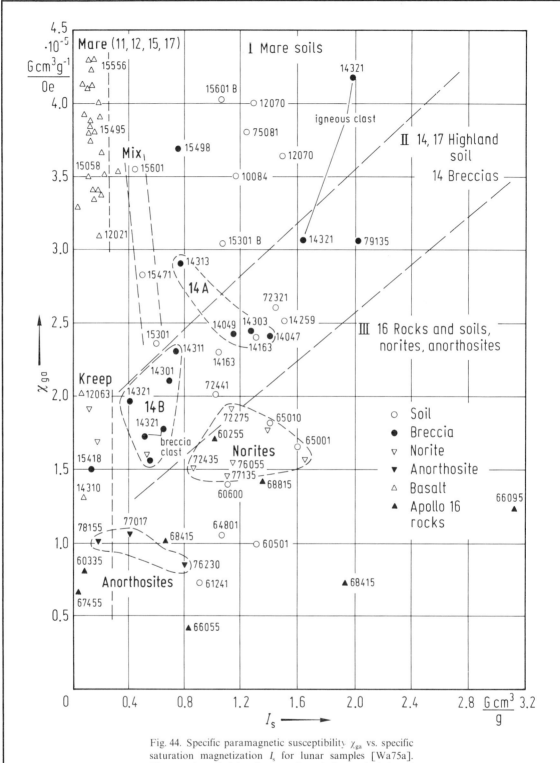

Fig. 44. Specific paramagnetic susceptibility χ_{ga} vs. specific saturation magnetization I_s for lunar samples [Wa75a]. The area bounded by the χ_{ga}-axis and the vertical dotted line at 0.25 G cm³ g⁻¹ is defined as the igneous rock region. Mix = soil fraction < 150 μm. B = bulk soil. Kreep = lunar basaltic rock enriched in potassium **K**alium, **r**are **e**arth **e**lements, **p**hosphorus and other incompatible elements (see e.g. [Ta75]).

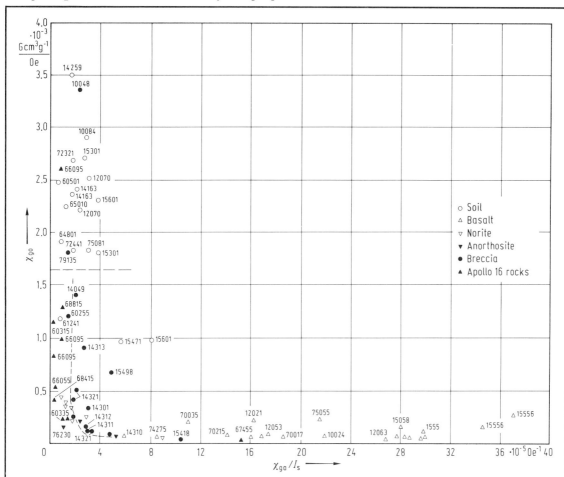

Fig. 45. Specific initial susceptibility χ_{g0} vs. the ratio of specific paramagnetic susceptibility χ_{ga} to specific saturation magnetization I_s for lunar samples [Wa75a].

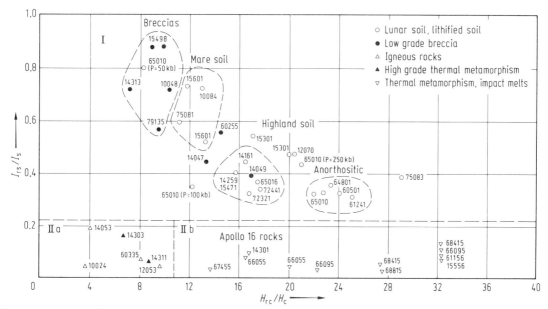

Fig. 46. Ratio of specific saturation remanent magnetization I_{rs} to specific saturation magnetization I_s vs. the ratio of remanent coercive force H_{rc} to coercive force H_c for lunar samples [Wa75a].

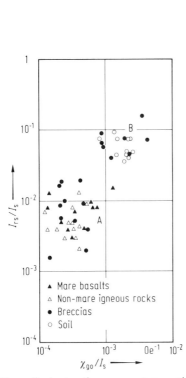

Fig. 47. Normalized saturation remanent magnetization I_{rs} vs. normalized initial susceptibility χ_{g0} [Fu75]. $I_s =$ specific saturation magnetization. A: multidomain hysteretic behaviour; B: single domain and superparamagnetic behaviour.

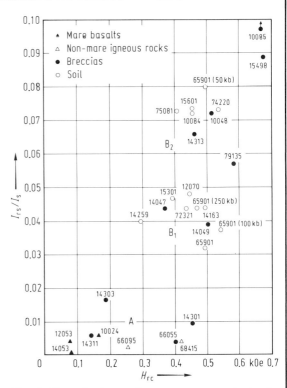

Fig. 48. Normalized saturation remanent magnetization I_{rs} vs. remanent coercive force H_{rc} [Fu75]. $I_s =$ saturation magnetization. A: igneous rocks and recrystallized breccias; B_1: mainly highland soils; B_2: mainly mare soils.

9.6.4 Natural remanent magnetization (NRM) — Natürliche remanente Magnetisierung (NRM)

Fig. 49 shows histograms of natural remanent magnetization (NRM) and of remanent magnetization after demagnetization in an alternating field with peak amplitudes of 100 Oe. The stability of the NRM in lunar rocks is extremely variable and the reader is referred to review discussions (e.g. [Fu74]) for more details.

The magnetizing processes by which lunar rocks acquired their natural remanent magnetization are not yet definitely known. Under discussion are mainly thermoremanent magnetizations, chemical remanent magnetizations and shock remanent magnetizations. In addition also the origin of the magnetizing fields, which range up to 1 Oe, is still obscure (see Soffel, Volume V/2).

Fig. 49 zeigt Häufigkeitsverteilungen der natürlichen remanenten Magnetisierung (NRM) und der Restmagnetisierung nach Wechselfeldentmagnetisierung mit 100 Oe. Die Stabilität der natürlichen remanenten Magnetisierung ist sehr unterschiedlich. Für eine Diskussion einzelner Proben wird auf Übersichtartikel verwiesen (z.B. [Fu74]).

Die Entstehung der natürlichen remanenten Magnetisierung der Gesteine des Mondes ist bisher nicht geklärt. In Frage kommen vor allem thermoremanente Magnetisierungen, chemische remanente Magnetisierungen und schockremanente Magnetisierungen. Weiterhin ist auch noch die Ursache des magnetisierenden Feldes nicht bekannt (vgl. Beitrag von Soffel in Band V/2).

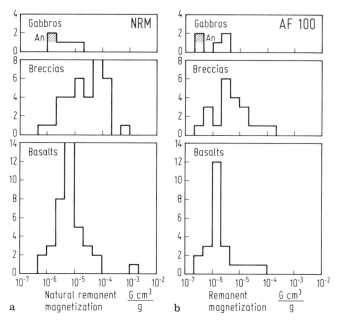

Fig. 49. Histograms of natural remanent magnetization (NRM) of lunar rocks and of remanent magnetization after demagnetization with an alternating field of 100 Oe (AF 100). (See also [St77b]. An = Anorthosite.

9.7 References for 9.0···9.6 — Literatur zu 9.0···9.6

A Introductory literature

Short, N.M.: Planetary Geology. New Jersey: Prentice-Hall, Englewood Cliffs, **1975**, 361 p.
Guest, J.E., Greeley, R.: Geology on the Moon. London: Wykeham, **1977**.
Taylor, S.R.: Lunar Science: A Post-Apollo View. New York: Pergamon Press, **1975**, 372 p.

B General literature and data sources

Proc. Apollo 11 Lunar Sci. Conf. (A.A. Levinson, ed.), **1970**, New York: Pergamon Press, 2492 p.
Proc. Lunar Sci. Conf. 2nd (A.A. Levinson, ed.), **1971**, Cambridge: MIT Press, 2818 p.
Proc. Lunar Sci. Conf. 3rd (D.R. Criswell, ed.), **1972**, Cambridge: MIT Press, 3263 p.
Proc. Lunar Sci. Conf. 4th (W.A. Gose, ed.), **1973**, New York: Pergamon Press, 3290 p.
Proc. Lunar Sci. Conf. 5th (W.A. Gose, ed.), **1974**, New York: Pergamon Press, 3134 p.
Proc. Lunar Sci. Conf. 6th (R.B. Merrill, ed.), **1975**, New York: Pergamon Press, 3637 p.
Proc. Lunar Sci. Conf. 7th (R.B. Merrill, ed.), **1976**, New York: Pergamon Press, 3651 p.
Proc. Lunar Sci. Conf. 8th (R.B. Merrill, ed.), **1977**, New York: Pergamon Press, 3965 p.
Proc. Lunar Planet. Sci. Conf. 9th (R.B. Merrill, ed.), **1978**, New York: Pergamon Press, 3973 p.
Proc. Lunar Planet. Sci. Conf. 10th (R.B. Merrill, ed.), **1979**, New York: Pergamon Press, 3077 p.
Proc. Lunar Planet. Sci. Conf. 11th (R.B. Merrill, ed.), **1980**, New York: Pergamon Press, 2502 p.

Lunar Science Conferences extended abstracts published by:
    The Lunar and Planetary Science Institute (formerly The Lunar Science Institute), Houston, Texas.

    Lunar Science III, **1972**.
    Lunar Science IV, **1973**.
    Lunar Science V, **1974**.
    Lunar Science VI, **1975**.
    Lunar Science VII, **1976**.
    Lunar Science VIII, **1977**.
    Lunar Planet. Science IX, **1978**.
    Lunar Planet. Science X, **1979**.
    Lunar Planet. Science XI, **1980**.

Lunar Sample Preliminary Examination Team Reports (LSPET):

Science **165** (1969) 211, Apollo 11.
Science **167** (1970) 1325, Apollo 12.
Science **173** (1971) 681, Apollo 14.
Science **175** (1972) 363, Apollo 15.
Science **179** (1973) 23, Apollo 16.
Science **182** (1973) 659, Apollo 17.

Preliminary Science Reports

Apollo 11, NASA SP 214, 209 p., **1969**.
Apollo 12, NASA SP 235, 235 p., **1970**.
Apollo 14, NASA SP 272, 313 p., **1971**.
Apollo 15, NASA SP 289, 502 p., **1972**.
Apollo 16, NASA SP 315, 636 p., **1972**.
Apollo 17, NASA SP 330, 710 p., **1973**.

Lunar Sample Information Catalogs (NASA, Houston, Texas):

Apollo 11, **1969**.
Apollo 12, MSC-S-243, **1970**.
Apollo 14, TM X-58062, **1971**.
Apollo 15, MSC-03209, **1971**.
Apollo 16, MSC-03210, **1972**.
Apollo 17, MSC-03211, **1973**.

Lunar Sample Data Bank (analytical data):
NASA, Johnson Space Center, Data System, Houston, Texas.

C Cited literature

Al73   Alvarez, R.: Trans. Am. Geophys. Union **54** (1973) 1129.
An70   Anderson, O.L., Scholz, C., Soga, N., Warren, N., Schreiber, E.: Proc. Apollo 11 Lunar Sci. Conf., **1970**, 1959–1973.
Ba71   Baldridge, W.S., Simmons, G.: Proc. Lunar Sci. Conf. 2nd, **1971**, 2317–2321.
Ba72   Baldridge, W.S., Miller, F., Wang, H., Simmons, G.: Proc. Lunar Sci. Conf. 3rd, **1972**, 2599–2609.
Ba72a   Bassett, H.L., Shackelford, R.G.: Proc. Lunar Sci. Conf. 3rd, **1972**, 3157–3160.
Ca59   Carslaw, H.S., Jaeger, J.S.: Conduction of heat in solids. London and New York: Oxford Univ. Press, **1959**.
Ca73   Carrier, W.D., Mitchell, J.K., Mahmood, A.: Proc. Lunar Sci. Conf. 4th, **1973**, 2403–2411.
Ca75   Cadenhead, D.A., Stetter, J.R.: Proc. Lunar Sci. Conf. 6th, **1975**, 3199–3206.
Ch70   Chung, D.H., Westphal, W.B., Simmons, G.: J. Geophys. Res. **75** (1970) 6524–6531.
Ch71   Chung, D.H., Westphal, W.B., Simmons, G.: Proc. Lunar Sci. Conf. 2nd, **1971**, 2381–2390.
Ch72   Chung, D.H., Westphal, W.B., Olhoeft, G.R.: Proc. Lunar Sci. Conf. 3rd, **1972**, 3161–3172.
Ch73   Chung, D.H.: Proc. Lunar Sci. Conf. 4th, **1973**, 2591–2600.
Ch73a   Chung, D.H., Westphal, W.B.: Proc. Lunar Sci. Conf. 4th, **1973**, 3077–3091.
Co73   Collett, L.S., Katsube, T.J.: Geophysics **38** (1973) 76–91.
Cr70   Cremers, C.J., Birkebak, R.C., Dawson, J.P.: Proc. Apollo 11 Lunar Sci. Conf., **1970**, 2045–2050.
Cr71   Cremers, C.J.: Rev. Sci. Instruments **42** (1971) 1694.
Cr71a   Cremers, C.J., Birkebak, R.C.: Proc. Lunar Sci. Conf. 2nd, **1971**, 2311–2315.
Cr72   Cremers, C.J.: Proc. Lunar Sci. Conf. 3rd, **1972**, 2611–2617.
Cr73   Cremers, C.J., Hsia, H.S.: Proc. Lunar Sci. Conf. 4th, **1973**, 2459–2464.
Cr74   Cremers, C.J.: Advances in Heat Transfer **10** (1974) 39–83.
Cr74a   Cremers, C.J., Hsia, H.S.: Proc. Lunar Sci. Conf. 5th, **1974**, 2703–2708.
Cr75   Cremers, C.J.: J. Geophys. Res. **80** (1975) 4466–4470.
Fi77   Fisher, R.M., Huffmann, G.P., Nagata, T., Schwerer, F.C.: Phil. Trans. R. Soc. London Ser. A **285** (1977) 517–521.
Fo70   Fountain, J.A., West, E.A.: J. Geophys. Res. **75** (1970) 4063–4069.

Fo73	Fountain, J.A., Scott, R.W.: NASA Tech Memo, X-64759, **1973**.
Fr75	Frisillo, A.L., Olhoeft, G.R., Strangway, D.W.: Earth Planet. Sci. Lett. **24** (1975) 345–356.
Fu73	Fujii, N., Osako, M.: Earth Planet. Sci. Lett. **18** (1973) 65–71.
Fu74	Fuller, M.: Rev. Geophys. Space Phys. **12** (1974) 23–70.
Fu75	Fuller, M., Wu, Y., Wasilewski, P.J.: The Moon **13** (1975) 327–338
Go70	Gold, T., Campbell, M.J., O'Leary, B.T.: Proc. Apollo 11 Lunar Sci. Conf., **1970**, 2149–2154.
Go71	Gold, T., O'Leary, B.T., Campbell, M.: Proc. Lunar Sci. Conf. 2nd, **1971**, 2173–2181.
Go72	Gold, T., Bilson, E., Yerbury, M.: Proc. Lunar Sci. Conf. 3rd, **1972**, 3187–3193.
Go73	Gold, T., Bilson, E., Yerbury, M.: Proc. Lunar Sci. Conf. 4th, **1973**, 3093–3100.
Go77	Gold, T., Bilson, E., Baron, R.L.: Proc. Lunar Sci. Conf. 8th, **1977**, 1271–1275.
He73	Hemingway, B.S., Robie, R.A., Wilson, W.H.: Proc. Lunar Sci. Conf. 4th, **1973**, 2481–2487.
He74	Herminghaus, Ch., Berckhemer, H.: Proc. Lunar Sci. Conf., 5th, **1974**, 2939–2943.
Ho72	Horai, K., Fujii, N.: The Moon **4** (1972) 379–407.
Ho74	Houston, W.N., Mitchell, J.K., Carrier, W.D.: Proc. Lunar Sci. Conf. 5th, **1974**, 2361–2364.
Ho75	Horai, K., Winkler, J.L.: Proc. Lunar Sci. Conf. 6th, **1975**, 3207–3215.
Ho76	Horai, K., Winkler, J.L.: Proc. Lunar Sci. Conf. 7th, **1976**, 3183–3204.
Ho77	Horai, K., Winkler, J.L., Keihm, S.J., Langseth Jr., M.G., Fountain, J.A., West, E.A.: Lunar Science **VIII** (1977) 455–457.
Ho78	Horai, K., Winkler, J.L.: Lunar Planet. Sci. **IX** (1978) 534–536.
Ka68	Kanamori, H., Fujii, N., Mizutani, H.: J. Geophys. Res. **73** (1968) 595–605.
Ka69	Kanamori, H., Mizutani, H., Fujii, N.: J. Phys. Earth **17** (1969) 43–53.
Ka70	Kanamori, H., Nur, A., Chung, D.H., Simmons, G.: Proc. Apollo 11 Lunar Sci. Conf., **1970**, 2289–2293.
Ka71	Kanamori, H., Mizutani, H., Hamano, Y.: Proc. Lunar Sci. Conf. 2nd, **1971**, 2323–2326.
Ka71a	Katsube, T.J., Collett, L.S.: Proc. Lunar Sci. Conf. 2nd, **1971**, 2367–2379.
Ka73	Katsube, T.J., Collett, L.S.: Proc. Lunar Sci. Conf. 4th, **1973**, 3101–3110.
Ka73a	Katsube, T.J., Collett, L.S.: Proc. Lunar Sci. Conf. 4th, **1973**, 3111–3131.
Kn64	Knopoff, L.: Rev. Geophys. **2** (1964) 625–660.
La70	Latham, G.V., Ewing, M., Dorman, J., Press, F., Toksöz, N., Sutton, G., Meissner, R., Duennebier, F., Nakamura, Y., Kovach, R., Yates, M.: Science **170** (1970) 620–626.
Mi72	Mizutani, H., Fujii, N., Hamano, Y., Osako, M.: Proc. Lunar Sci. Conf. 3rd, **1972**, 2557–2564.
Mi73	Mizutani, H., Newbigging, D.F.: Proc. Lunar Sci. Conf. 4th, **1973**, 2601–2609.
Mi74	Mizutani, H., Osako, M.: Proc. Lunar Sci. Conf. 5th, **1974**, 2891–2901.
Na70	Nagata, T., Ishikawa, Y., Kinoshita, H., Kono, M., Syono, Y., Fisher, R.M.: Proc. Apollo 11 Lunar Sci. Conf., **1970**, 2325–2340.
Na72	Nagata, T., Fisher, R.M., Schwerer, F.C.: The Moon **4** (1972) 160–186.
Na75	Nagata, T., Fisher, R.M., Schwerer, F.C., Fuller, M.D., Dunn, J.R.: Proc. Lunar Sci. Conf. 6th, **1975**, 3111–3122.
Nu70	Nur, A., Simmons, G.: Int. J. Rock Mech. Min. Sci. **7** (1970) 307–314.
O'K72	O'Kelley, G.D., Eldridge, J.S., Northcutt, K.J.: Proc. Lunar Sci. Conf. 3rd, **1972**, 1659–1670.
Ol73	Olhoeft, G.R., Strangway, D.W., Frisillo, A.L.: Proc. Lunar Sci. Conf. 4th, **1973**, 3133–3149.
Ol73a	Olhoeft, G.R., Strangway, D.W., Pearce, G.W., Frisillo, A.L., Gose, W.A.: Trans. AGU **54** (1973) 601.
Ol74	Olhoeft, G.R., Frisillo, A.L., Strangway, D.W.: J. Geophys. Res. **79** (1974) 1599–1604.
Ol74a	Olhoeft, G.R., Frisillo, A.L., Strangway, D.W., Sharpe, H.: The Moon **9** (1974) 79–87.
Ol75	Olhoeft, G.R., Strangway, D.W.: Earth Planet. Sci. Lett. **24** (1975) 394–404.
Ol75a	Olhoeft, G.R., Strangway, D.W., Pearce, G.W.: Proc. Lunar Sci. Conf. 6th, **1975**, 3333–3342.
Ro71	Robie, R.A., Hemingway, B.S.: Proc. Lunar Sci. Conf. 2nd, **1971**, 2361–2365.
Ro70	Robie, R.A., Hemingway, B.S., Wilson, W.H.: Proc. Apollo 11 Lunar Sci. Conf., **1970**, 2361–2367.
Ry73	Rybach, L.: Beitr. Geol. Schweiz, Geotechn. Ser. Liefg. 51, **1973**.
Sch71	Schwerer, F.C., Nagata, T., Fisher, R.M.: The Moon **2** (1971) 408–422.
Sch73	Schwerer, F.C., Huffmann, G.P., Fisher, R.M., Nagata, T.: Proc. Lunar Sci. Conf. 4th, **1973**, 3151–3166.
Sch74	Schwerer, F.C., Huffmann, G.P., Fisher, R.M., Nagata, T.: Proc. Lunar Sci. Conf. 5th, **1974**, 2673–2687.
Sch77	Schreiber, E.: Proc. Lunar Sci. Conf. 8th, **1977**, 1201–1208.
Si65	Simmons, G., Brace, W.: J. Geophys. Res. **70** (1965) 5649–5656.

Si73	Sill, W.R., Hansen, W., Ward, S.H., Katsube, T.J., Collett, L.S.: Geophys. Geochem. Explor. Moon and Planets, Lunar Science Institute, Houston, **1973**
Si75	Simmons, G., Siegfried, R., Richter, D.: Proc. Lunar Sci. Conf. 6th, **1975**, 3227–3254.
So79	Sondergeld, C.H., Granryd, L.A., Spetzler, H.A.: Proc. Lunar Planet. Sci. Conf. 10th, **1979**, 2147–2154.
Sp80	Spetzler, H.A., Getting, I.C., Swanson, P.L.: Proc. Lunar Planet. Sci. Conf. 11th, **1980**, 1825–1835.
St70	Stephens, D.R., Lilley, E.M.: Proc. Apollo 11 Lunar Sci. Conf., **1970**, 2427–2434.
St71	Stephens, D.R., Lilley, E.M.: Proc. Lunar Sci. Conf. 2nd, **1971**, 2165–2172.
St72	Strangway, D.W., Olhoeft, G.R., Chapman, W.B., Carnes, J.: Earth Planet. Sci. Lett **16** (1972) 275–281.
St77	Stesky, R.M., Renton, B.: Proc. Lunar Sci. Conf. 8th, **1977**, 1225–1233.
St77a	Strangway, D.W., Olhoeft, G.R.: Phil. Trans. R. Soc. London Ser. A **285** (1977) 441–450.
St77b	Strangway, D.W., Pearce, G.W., Olhoeft, G.R.: Soviet-American Conf. Cosmochemistry Moon and Planets, NASA SP-370, **1977**, 417–431.
St78	Stesky, R.M.: Proc. Lunar Planet. Sci. Conf. 9th, **1978**, 3637–3649.
Ta73	Talwani, P., Nur, A., Kovach, R.L.: J. Geophys. Res. **78** (1973) 6899–6909.
Ta74	Talwani, P., Nur, A., Kovach, R.L.: Proc. Lunar Sci. Conf. 5th, **1974**, 2919–2926.
Ta75	Taylor, S.R.: Lunar Science: A Post-Apollo View, New York: Pergamon Press, **1975**, 372 p.
Ta78	Taylor, S.R.: Proc. Lunar Planet. Sci. Conf. 9th, **1978**, 15–23.
Ta79	Taylor, S.R.: Proc. Lunar Planet. Sci. Conf. 10th, **1979**, 2017–2030.
Ti72	Tittmann, B.R., Abdel-Gawad, M., Housley, R.M.: Proc. Lunar Sci. Conf. 3rd, **1972**, 2565–2575.
Ti73	Tittmann, B.R., Housley, R.M., Cirlin, E.H.: Proc. Lunar Sci. Conf. 4th, **1973**, 2631–2637.
Ti74	Tittmann, B.R., Housley, R.M., Alers, G.A., Cirlin, E.H.: Proc. Lunar Sci. Conf. 5th, **1974**, 2913–2918.
Ti75	Tittmann, B.R., Curnow, J.M., Housley, R.M.: Proc. Lunar Sci. Conf. 6th, **1975**, 3217–3226.
Ti76	Tittmann, B.R., Ahlberg, L., Curnow, J.: Proc. Lunar Sci. Conf. 7th, **1976**, 3123–3132.
Ti77	Tittmann, B.R., Ahlberg, L., Nadler, H., Curnow, J., Smith, T., Cohen, B.R.: Proc. Lunar Sci. Conf. 8th, **1977**, 1209–1224.
Ti78	Tittmann, B.R., Nadler, H., Richardson, J.M., Ahlberg, L.: Proc. Lunar Planet. Sci. Conf. 9th, **1978**, 3627–3635.
Ti79	Tittmann, B.R., Nadler, H., Clark, V., Coombe, L.: Proc. Lunar Planet. Sci. Conf. 10th, **1979**, 2131–2145.
Ti80	Tittmann, B.R., Clark, V.A., Spencer, T.W.: Proc. Lunar Planet. Sci. Conf. 11th, **1980**, 1815–1823.
To72	Todd, T., Wang, H., Baldridge, W.S., Simmons, G.: Proc. Lunar Sci. Conf. 3rd, **1972**, 2577–2586.
To73	Todd, T., Richter, D.A., Simmons, G., Wang, H.: Proc. Lunar Sci. Conf. 4th, **1973**, 2639–2662.
Tr74	Trice, R., Warren, N., Anderson, O.L.: Proc. Lunar Sci. Conf. 5th, **1974**, 2903–2911.
Wa65	Walsh, J.B.: J. Geophys. Res. **70** (1965) 381–389.
Wa71	Wang, H., Todd, T., Weidner, D., Simmons, G.: Proc. Lunar Sci. Conf. 2nd, **1971**, 2327–2336.
Wa71a	Warren, N., Schreiber, E., Scholz, C., Morrison, J.A., Norton, P.R., Kumazawa, M., Anderson, O.L.: Proc. Lunar Sci. Conf. 2nd, **1971**, 2345–2360.
Wa72	Warren, N., Anderson, O.L., Soga, N.: Proc. Lunar Sci. Conf. 3rd, **1972**, 2587–2598.
Wa73	Wang, H., Todd, T., Richter, D., Simmons, G.: Proc. Lunar Sci. Conf. 4th, **1973**, 2663–2671.
Wa73a	Warren, N., Trice, R., Soga, N., Anderson, O.L.: Proc. Lunar Sci. Conf. 4th, **1973**, 2611–2629.
Wa73b	Warren, N., Anderson, O.L.: J. Geophys. Res. **78** (1973) 6911–6925.
Wa74	Warren, N., Trice, R., Stephens, J.: Proc. Lunar Sci. Conf. 5th, **1974**, 2927–2938.
Wa75	Warren, N., Trice, R.: Proc. Lunar Sci. Conf. 6th, **1975**, 3255–3268.
Wa75a	Wasilewski, P.J., Fuller, M.D.: The Moon **14** (1975) 79–101.
We65	Wechsler, A.E., Glaser, P.E.: Icarus **4** (1965) 335–352.
We72	Wechsler, A.E., Glaser, P.E., Fountain, J.A., in: Thermal characteristics of the Moon, ed.: J. Lucas, 215–241, Cambridge: MIT Press, **1972**.
Wo70	Wood, J.A., Dickey, J.S., Marvin, U.B., Powell, B.N.: Proc. Apollo 11 Lunar Sci. Conf., **1970**, 965–988.

7.2 The age of rocks — Das Alter der Gesteine

7.2.1 Introduction — Einleitung

The age of rocks and minerals can be determined by physical methods utilizing the spontaneous decay of unstable atomic nuclei. The following nuclear transmutations play an important role in this respect:

1. Decay of radioactive isotopes (alpha, negatron and positron decay)
2. Electron capture (ec) from the K-shell
3. Spontaneous fission (sf)

The determination of the age of rocks by physical methods is often summarized as "radiometric dating". However, strictly speaking, this is only true for methods which determine the isotopic concentrations by radioactivity measurements. Most of the more important methods are based on the measurement of the abundances of parent and daughter isotopes by mass-spectrometric isotope analysis. In addition, the visual counting of tracks of spontaneous fission is used. In particular cases, the element concentrations can be determined by means of flame photometry, X-ray fluorescence analysis, neutron activation and particle accelerators.

Die Möglichkeit der Altersbestimmung von Gesteinen und Mineralen mit physikalischen Methoden beruht auf der spontanen Umwandlung instabiler Atomkerne. Folgende Arten der Kernumwandlung sind hierfür von Bedeutung:

1. Zerfall radioaktiver Isotope (Alpha-, Negatronen- und Positronen-Zerfall)
2. Elektroneneinfang aus der K-Schale (ec)
3. Spontanspaltung (sf)

Die Bestimmung des Alters der Gesteine mit physikalischen Methoden wird häufig zusammenfassend als „radiometrische Altersbestimmung" bezeichnet. Dies ist jedoch genau genommen nur für die Methoden zutreffend, bei denen man die Konzentrationen der Isotope über die Messung der Radioaktivität erhält. Bei der Mehrzahl der wichtigeren Methoden wird aber die Konzentration der Mutter- und Tochter-Isotope mit Hilfe massenspektrometrischer Isotopenanalyse bestimmt. Außerdem findet die visuelle Auszählung von Spuren der Spontanspaltung Anwendung. Für die Bestimmung von Elementkonzentrationen werden in besonderen Fällen auch Flammenphotometrie, Atomabsorptionsspektroskopie, Röntgenfluoreszenzanalyse, Neutronenaktivierung und Teilchenbeschleuniger eingesetzt.

7.2.1.1 Nuclides used — Verwendete Nuklide

Considering the origin and formation of the radioactive isotopes used, the classification of nuclides can be made as follows (Table 1):

A. Long-lived, still existing nuclides (^{40}K, ^{87}Rb, ^{147}Sm, ^{176}Lu, ^{187}Re, ^{232}Th, ^{235}U, ^{238}U).

B. Extinct nuclides (^{129}I, ^{244}Pu). In this case, the secondary products can be used for determination.

C. Stable decay products of primordial nuclides.

D. Nuclides of the uranium and thorium decay series with various lifetimes (e.g. ^{230}Th, ^{231}Pa, ^{210}Pb).

E. Cosmic-ray induced, spallogenic nuclides – most of them with short lifetime – (e.g. ^{10}Be, ^{14}C, ^{26}Al, ^{36}Cl).

Hinsichtlich Herkunft und Entstehung der verwendeten radioaktiven Isotope läßt sich folgende Unterteilung vornehmen (Tab. 1):

A. Langlebige, noch existierende Nuklide (^{40}K, ^{87}Rb, ^{147}Sm, ^{176}Lu, ^{187}Re, ^{232}Th, ^{235}U, ^{238}U).

B. Ausgestorbene Nuklide (^{129}J, ^{244}Pu). Die Bestimmung erfolgt in diesem Fall über die Folgeprodukte.

C. Stabile Zerfalls-Produkte primordialer Nuklide.

D. Nuklide der Uran- und Thorium-Zerfallsreihen mit unterschiedlicher Lebensdauer (z.B. ^{230}Th, ^{231}Pa, ^{210}Pb).

E. Durch die Einwirkung kosmischer Strahlung gebildete, spallogene Nuklide von meist kurzer Lebens- (z.B. ^{10}Be, ^{14}C, ^{26}Al, ^{36}Cl).

7.2.1.2 Occurrence of nuclides — Vorkommen der Nuklide

Suitable radioactive nuclides occur in the minerals of magmatic, metamorphic and sedimentary rocks as isotopes of main elements and of trace elements. Considering the number of applicable methods and the number of adequate minerals, the radioactive nuclides occur mainly as trace elements. Some elements exist in their specific minerals: potassium in sylvite, potassium feldspar, biotite, muscovite; uranium in

Geeignete radioaktive Nuklide kommen in Mineralen magmatischer, metamorpher und sedimentärer Gesteine als Isotope von Haupt- und Spurenelementen vor. Geht man von der Zahl der Methoden und der Zahl der verwendbaren Minerale aus, so überwiegt jedoch der Einbau als Spurenelement. Eigene Minerale besitzen die Elemente Kalium (Sylvin, Kalifeldspat, Biotit, Muscovit) Uran (Uraninit) und Thorium (Thorit),

uraninite and thorium in thorite, while the important elements rubidium and samarium and the other primordial isotopes occur only as trace elements. Cosmic radiation produces nuclides only in very low concentrations. In meteorites and on the surface of the moon, they still exist more or less *in situ*. Nuclei produced by cosmic radiation in the atmosphere (as ^{10}Be, ^{14}C, ^{32}Si) are transported by precipitation into the hydrosphere and biosphere (^{14}C) where they are incorporated into authigenic minerals and bound to detritus by adsorption.

^{210}Pb, which originates from terrestrial radon decaying in the atmosphere, is also transported into the hydrosphere by precipitation. Together with the ^{210}Pb originating from the ^{226}Ra of seawater, it enters the sediments. Besides ^{210}Pb, other nuclides of the uranium decay series (^{238}U, ^{234}U, ^{230}Th, ^{226}Ra, ^{235}U, ^{231}Pa) and of the thorium series (^{232}Th, ^{228}Ra, ^{228}Th) are incorporated or adsorbed as trace elements into marine and lacustrine sediments.

während die wichtigen Elemente Rubidium und Samarium und die anderen primordialen Isotope ausschließlich als Spurenelemente auftreten. Spallogene Nuklide werden nur in sehr geringer Konzentration erzeugt. In Meteoriten und Gesteinen der Mondoberfläche finden sie sich noch mehr oder weniger am Ort ihrer Bildung. Durch die Wirkung der kosmischen Strahlung in der Atmosphäre erzeugte Kerne (^{10}Be, ^{14}C, ^{32}Si) gelangen mit den Niederschlägen in die Hydrosphäre und Biosphäre (^{14}C) und werden dort als Spurenelemente in authigene Minerale eingebaut und adsorptiv an Detritus gebunden.

Durch Niederschläge wird auch ^{210}Pb, das durch Zerfall von terrestrischem Radon in der Atmosphäre gebildet wird, in die Hydrosphäre überführt. In mariner Umgebung gelangt es zusammen mit ^{210}Pb, das aus ^{226}Ra des Meerwassers entsteht, in die Sedimente. Neben ^{210}Pb werden auch die übrigen für Altersbestimmungen geeigneten Nuklide der Zerfallsreihen des Urans (^{238}U, ^{234}U, ^{230}Th, ^{226}Ra, ^{235}U, ^{231}Pa) und Thoriums (^{232}Th, ^{228}Ra, ^{228}Th) als Spuren durch Einbau und Adsorption in marine und limnische Sedimente aufgenommen.

7.2.1.3 Equations of age — Die Altersgleichungen

The decrease in the abundance of unstable isotopes with time corresponds to the decay equation:

Die Abnahme der Menge eines instabilen Isotops mit der Zeit erfolgt entsprechend der Zerfallsgleichung:

$$N = N_0 e^{-\lambda t} \tag{1}$$

N is the number of atomic nuclei after the time t, N_0 is the number of atomic nuclei at the beginning of the decay process and λ is the decay constant. The half-life $t_{1/2}$ is the time required for disintegration of one half of the atomic nuclei:

N ist die Anzahl der Atomkerne nach Ablauf der Zeit t, N_0 die Anzahl der Atomkerne zu Beginn des Zerfalls und λ die Zerfallskonstante. Die Zeit, nach der die Hälfte der Atomkerne zerfallen ist, wird als Halbwertszeit $t_{1/2}$ bezeichnet:

$$t_{1/2} = \frac{1}{\lambda} \ln 2 \tag{2}$$

Instead of N and N_0, the corresponding activities may be used in (1):

Anstatt N und N_0 kann man in (1) auch die entsprechenden Aktivitäten verwenden:

$$A = A_0 e^{-\lambda t} \tag{3}$$

The time required for the decrease of the activity from A_0 to A is then:

Die Zeit, die seit dem Abklingen der Aktivität von A_0 auf A verflossen ist, ist dann

$$t = \frac{1}{\lambda}(\ln A_0 - \ln A) \tag{4}$$

If $^i D_r$ is the concentration [atoms per unit weight] of a stable isotope of an element D and $^i D_r$ is directly originated by decay from its radioactive parent isotope with concentration $^k N$ then:

Wenn man mit $^i D_r$ die Konzentration in [Atome pro Gewichtseinheit] eines stabilen Isotops des Elements D bezeichnet und $^i D_r$ durch Zerfall direkt aus seinem radioaktiven Mutterisotop der Konzentration $^k N$ entstanden ist, so gilt:

$$^k N_0 = {}^k N + {}^i D_r \tag{5}$$

Here $^k N_0$ is the concentration at the beginning of the decay and $^k N$ and $^i D_r$ are the concentrations after time t.

Hierbei ist $^k N_0$ die Konzentration zu Beginn des Zerfalls und $^k N$ und $^i D_r$ sind die Konzentrationen nach der Zeit t.

Combining (1) and (5) yields:

Kombination von (1) und (5) ergibt

$$^i D_r = {}^k N(e^{\lambda t} - 1) \tag{6}$$

If it is possible to determine the present concentration of the parent isotope and of the radiogenic daughter isotope, the age can be calculated after transformation of (6):

Kann die heutige Konzentration des Mutter- und radiogenen Tochterisotops bestimmt werden, läßt sich nach Umformung von (6) das Alter berechnen:

$$t = \frac{1}{\lambda} \ln \left(\frac{{}^iD_r}{{}^kN} + 1 \right) \tag{7}$$

In most cases, ${}^iD_r$ cannot be directly determined because the measured concentration of the daughter isotope ${}^iD$ always contains a component ${}^iD_0$ already existing before the system was closed, in addition to the radiogenic component ${}^iD_r$:

${}^iD_r$ kann meist nicht direkt bestimmt werden, weil die gemessene Konzentration des Tochterisotops ${}^iD$ neben der radiogenen Komponente ${}^iD_r$ im Realfall stets eine schon bei der Schließung des Systems vorhandene Komponente ${}^iD_0$ enthält:

$$^iD = {}^iD_r + {}^iD_0 \tag{8}$$

Combining (6) and (8) yields:

Kombination von (6) und (8) ergibt:

$$^iD - {}^iD_0 = {}^kN(e^{\lambda t} - 1) \tag{9}$$

Dividing Eq. (9) by the concentration of a second non-radiogenic and stable isotope ${}^jD$ of the daughter-element yields:

Dividiert man Gl. (9) durch die Konzentration eines zweiten, nichtradiogenen und stabilen Isotops ${}^jD$ des Tochterelements, so erhält man:

$$\left(\frac{{}^iD}{{}^jD} \right) - \left(\frac{{}^iD}{{}^jD} \right)_0 = \frac{{}^kN}{{}^jD} (e^{\lambda t} - 1) \tag{10}$$

This is the equation of a straight line with the slope $(e^{\lambda t} - 1)$ and the intercept $({}^iD/{}^jD)_0$ (Fig. 1). $({}^iD/{}^jD)_0$ is called initial isotope ratio. It can be determined when at least 2 systems of the same age with the same values for $({}^iD/{}^jD)_0$ but with different ${}^kN/{}^jD$ ratios are available. The straight line defined by Eq. (10) is called an *isochron*. The age calculated from its slope is called *isochron age*. In cogenetic systems of one and the same sample, e.g. the different mineral phases of a polymineralic magmatic rock sample, this age is called an "internal" isochron age. In this case, the line represents a *mineral isochron*. If different cogenetic whole-rock samples are used for the age determination, the line is a *whole-rock isochron*. If only the analysis of a single sample and of an assumed value for $({}^iD/{}^jD)_0$ is taken as basis for the age calculation according to Eq. (10), the result is called a *model age*.

Dies ist die Gleichung einer Geraden mit der Steigung $(e^{\lambda t} - 1)$ und dem Ordinatenabschnitt $({}^iD/{}^jD)_0$ (Fig. 1). $({}^iD/{}^jD)_0$ wird als Anfangsisotopenverhältnis oder initiales Isotopenverhältnis bezeichnet. Es läßt sich bestimmen, wenn wenigstens zwei gleich alte Systeme mit gleichen Werten für $({}^iD/{}^jD)_0$ aber unterschiedlichen ${}^kN/{}^jD$-Verhältnissen zur Verfügung stehen. Die durch Gl. (10) definierte Gerade wird als *Isochrone* bezeichnet; das aus ihrer Steigung berechnete Alter nennt man *Isochronen-Alter*. Handelt es sich um kogenetische Systeme ein und derselben Probe, wie es beispielsweise für die verschiedenen Mineralphasen einer polymineralischen magmatischen Gesteinsprobe zutrifft, so spricht man von einem „internen" Isochronen-Alter; die Gerade ist in diesem Fall eine *Mineral-Isochrone*. Werden für die Altersbestimmung hingegen verschiedene kogenetische Gesamtgesteins-proben verwendet, so nennt man die Gerade eine *Gesamtgesteins-Isochrone*. Wenn einer Altersberechnung nach Gl. (10) nur die Analyse einer einzigen Probe und ein angenommener Wert für $({}^iD/{}^jD)_0$ zugrunde liegen, so spricht man von einem *Modellalter*.

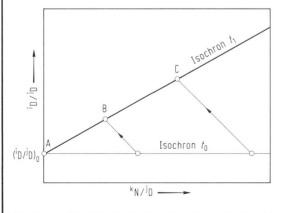

Fig. 1. Isochron diagram showing the time-dependent evolution of the isotopic ratio ${}^iD/{}^jD$ in three cogenetic systems A, B and C which represent mineral or whole-rock samples. ${}^kN$ is an instable isotope decaying to ${}^iD$. ${}^jD$ is a stable, non-radiogenic isotope of the element D. The straight line through points A to C is termed isochron. The age t of the systems at time t_1 is proportional to the slope of the isochron (see Eq. (11)). $({}^iD/{}^jD)_0$ is the initial ratio in all three systems at time t_0 when isotopic homogenization among the systems ended.

Since in the most cases $\lambda t \ll 1$, Eq. (10) can be replaced with fair approximation by

Da in den meisten Fällen $\lambda t \ll 1$ ist, läßt sich Gl. (10) in guter Annäherung durch

$$\left(\frac{{}^iD}{{}^jD}\right) - \left(\frac{{}^iD}{{}^jD}\right)_0 \simeq \frac{{}^kN}{{}^jD}\lambda t \qquad (11)$$

Choosing ${}^iD/{}^jD$ and t as coordinates yields the so-called *evolution diagram* shown in Fig. 2 which is particularly applied to the Rb−Sr and Sm−Nd methods.

ersetzen. Bei Wahl von ${}^iD/{}^jD$ und t als Koordinaten gelangt man zu der in Fig. 2 gezeigten Darstellung, die als *Entwicklungsdiagramm* bezeichnet wird und besonders in Zusammenhang mit der Rb−Sr- und Sm−Nd-Methode zur Anwendung gelangt.

The condition for the application of Eqs. (4), (7) and (10) is that the considered parent-daughter systems have been completely closed since the start of the "radiometric clock".

Für die Anwendung der Gleichungen (4), (7) und (10) gilt die Voraussetzung, daß nach dem Start der „radiometrischen Uhr" die betrachteten Systeme für das Mutter- und Tochterelement vollkommen geschlossen waren.

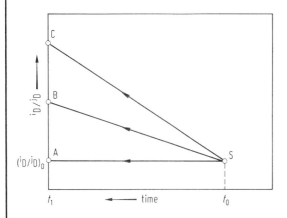

Fig. 2. Evolution diagram. Points A, B and C correspond to the systems A, B and C in Fig. 1. The diagram is the representation of Eq. (11). The slopes of the evolution lines are proportional to the ${}^kN/{}^jD$ ratios in the systems.

7.2.2 The methods for age determination of minerals and rocks— Die Methoden der Altersbestimmung von Mineralen und Gesteinen

Table 1. Methods of dating in geochronology and cosmochronology.

A. Methods using primordial nuclides and their stable decay products.

No.	Method	Radio-active primordial nuclide	Isotopic abundance (atom %)	Mode of decay [1]	Decay pro-duct(s)	Decay constant(s) [2] λ $[a^{-1}]$	Half-life $t_{1/2}$ $[10^9\,a]$	Ref. $(\lambda, t_{1/2})$
1	K−Ca	${}^{40}K$	0.0119 [Nie50]	β^- (88.8%)	${}^{40}Ca$	$\lambda_{\beta^-} = 4.72 \cdot 10^{-10}$ $\lambda_{ec} + \lambda_{ec'} = 0.585 \cdot 10^{-10}$ $\lambda = 5.305 \cdot 10^{-10}$	1.31	Ald58
2	K−Ar			ec ec' $\}$ (11.16%)	${}^{40}Ar$	$\lambda_{\beta^-} = 4.905\,(9) \cdot 10^{-10}$ $\lambda_{ec} = 0.5660\,(35) \cdot 10^{-10}$ $\lambda_{ec'} = 8.67\,(147) \cdot 10^{-13}$ $\lambda = 5.480\,(10) \cdot 10^{-10}$	1.265 (2)	Bec69
3	${}^{40}Ar/{}^{39}Ar$		0.01167 [Gar76] [Ste77]	β^+ (0.001%)		$*\lambda_{\beta^-} = 4.962 \cdot 10^{-10}$ $*\lambda_{ec} + \lambda_{ec'} = 0.581 \cdot 10^{-10}$ $*\lambda = 5.543 \cdot 10^{-10}$	1.250	Gar76, Ste77

[1] β^- negatron decay; β^+ positron decay; ec electron capture; ec' electron capture to ground state.
[2] Values marked by an asterisk recommended by the IUGS Subcommission on Geochronology 1976 [Ste77].

Table 1 (continued)

No.	Method	Radio-active primordial nuclide	Isotopic abundance (atom %)	Mode of decay [1]	Decay pro-duct(s)	Decay constant(s) [2] λ [a^{-1}]	Half-life $t_{1/2}$ [10^9 a]	Ref ($\lambda, t_{1/2}$)
4	U – Pb	^{235}U	0.72	$7\alpha + 4\beta^-$	^{207}Pb	$\lambda_{235} = 9.7216 \cdot 10^{-10}$	0.713 (16)	Fle52
		^{238}U	99.28 [Ste77]	$8\alpha + 6\beta^-$	^{206}Pb	$\lambda_{238} = 1.5369 \cdot 10^{-10}$	4.51 (1)	
5	U – He	^{235}U	0.72	$7\alpha + 4\beta^-$	^{4}He	$^*\lambda_{235} = 9.8485 \cdot 10^{-10}$	0.7038	Jaf71
		^{238}U	99.28 [Ste77]	$8\alpha + 6\beta^-$		$^*\lambda_{238} = 1.55125 \cdot 10^{-10}$	4.468	
6	Th – Pb	^{232}Th	100	$6\alpha + 4\beta^-$	^{208}Pb	$4.8813 \cdot 10^{-11}$	14.2 (7)	Sen56
						* $4.9475 \cdot 10^{-11}$	14.010	LeR63
7	Th – He	^{232}Th	100	$6\alpha + 4\beta^-$	^{4}He			
8	Lu – Hf	^{176}Lu	2.6 [See74]	β^-	^{176}Hf	$1.962 (81) \cdot 10^{-11}$	35.3 (14)	Pat80
9	Re – Os	^{187}Re	62.60 [See74]	β^-	^{187}Os	$1.61 \cdot 10^{-11}$	43 (5)	Hir63
10	Rb – Sr	^{87}Rb	27.8346 [Cat69] [Ste77]	β^-	^{87}Sr	$1.39 \cdot 10^{-11}$	50 (2)	Ald56
						$1.47 \cdot 10^{-11}$	47.0 (5)	Fly59
						* $1.42 \cdot 10^{-11}$	$48.8\,^{+0.6}_{-1.0}$	Neu74, Ste77
11	Sm – Nd	^{147}Sm	14.996 [Was81]	α	^{143}Nd	$6.54 \cdot 10^{-12}$	106.0 (8)	Lug78

B. Methods using decay products of extinct primordial nuclides.

No.	Method	Extinct radio-active nuclide	Isotopic abundance at the end of nucleo-synthesis (atom%)	Mode of decay [1]	Decay product(s)	Decay constant λ [a^{-1}]	Half-life $t_{1/2}$ [10^6 a]	Ref. ($\lambda, t_{1/2}$)
12	I – Xe	^{129}I	$> 10^{-6}$	β^-	^{129}Xe	$4.1 \cdot 10^{-8}$	17	Wea76
13	Pu – Xe	^{244}Pu		α sf	^{232}Th $^{131-136}$Xe	$8.39 \cdot 10^{-9}$	82.6	See74

C. Methods using stable decay products of primordial nuclides.

No.	Method	Parent nuclide(s)	Decay product(s)	Decay constant	Isotopic ratios used in equations
14	Common Sr	^{87}Rb	^{87}Sr	see No. 10	$\dfrac{^{87}\text{Sr}}{^{86}\text{Sr}}$
15	Pb – Pb	^{235}U ^{238}U	^{207}Pb ^{206}Pb	see No. 4	$\dfrac{^{206}\text{Pb}}{^{204}\text{Pb}}, \dfrac{^{207}\text{Pb}}{^{204}\text{Pb}}$
16	Common Pb	^{235}U ^{238}U ^{232}Th	^{207}Pb ^{206}Pb ^{208}Th	see No. 4 see No. 6	$\dfrac{^{206}\text{Pb}}{^{204}\text{Pb}}, \dfrac{^{207}\text{Pb}}{^{204}\text{Pb}}, \dfrac{^{208}\text{Pb}}{^{204}\text{Pb}}, \dfrac{^{238}\text{U}}{^{204}\text{Pb}}, \dfrac{^{232}\text{Th}}{^{204}\text{Pb}}$ For symbols of ratios in common use, see Table 2.

[1] β^- negatron decay; β^+ positron decay; α alpha decay; sf spontaneous fission.

[2] Values marked by an asterisk recommended by the IUGS Subcommission on Geochronology 1976 [Ste77].

Table 1 (continued)

D. Methods using radionuclides of the ^{230}U, ^{235}U and ^{232}Th decay series.

No.	Method	Source nuclide of decay serie	Radio nuclide	Half-life $t_{1/2}$ [a]	Principle of method	Ref. $(t_{1/2})$
17	^{228}Th/^{232}Th	^{232}Th	^{228}Th	1.913	Decay of excess ^{228}Th	See 74
18	^{228}Ra/^{226}Ra	^{232}Th ^{238}U	^{228}Ra ^{226}Ra	5.75 1622	Decrease of the ^{228}Ra/^{226}Ra ratio of unsupported Ra due to the faster decay of ^{228}Ra relative to ^{226}Ra	See 74, fau 77
19	^{210}Pb	^{238}U	^{210}Pb	22.26	Decay of excess ^{210}Pb	fau 77
20	^{231}Pa	^{235}U	^{231}Pa	32480	Decay of ^{231}Pa	fau 77
21	^{231}Pa/^{235}U	^{235}U	^{231}Pa	32480	Growth of ^{231}Pa toward equilibrium with ^{235}U	fau 77
22	^{231}Pa/^{230}Th	^{235}U ^{238}U	^{231}Pa ^{230}Th	32480 75200	Decrease of the ^{231}Pa/^{230}Th ratio due to the faster decay of (excess) ^{231}Pa relative to ^{230}Th	fau 77
23	^{230}Th/^{232}Th (Ionium method)	^{238}U	^{230}Th	75200	Decrease of the ^{230}Th/^{232}Th ratio due to the faster decay of (excess) ^{230}Th relative to ^{232}Th	fau 77
24	^{230}Th/^{238}U	^{238}U	^{230}Th	75200	Growth of ^{230}Th by decay of ^{234}U toward equilibrium with ^{238}U	fau 77
25	^{230}Th/^{234}U	^{238}U ^{238}U	^{230}Th ^{234}U	75200 $2.48 \cdot 10^5$	Growth of ^{230}Th by decay of ^{234}U toward equilibrium with ^{234}U	fau 77, Str 58
26	^{234}U/^{238}U	^{238}U	^{234}U	$2.48 \cdot 10^5$	Decay of (excess) ^{234}U from disequilibrium toward equilibrium with ^{238}U	Str 58

E. Methods using cosmic-ray-induced radionuclides.

No.	Method	Radio nuclide	Mode of decay [1])	Half life $t_{1/2}$ [a]	Ref. $(t_{1/2})$
27	^{3}H (tritium)	^{3}He	β^-	12.346	See 74
28	^{3}He/^{3}H	^{3}He	β^-	12.346	See 74
29	^{10}Be	^{10}Be	β^-	$1.6 \cdot 10^6$	See 74
30	^{14}C (radiocarbon)	^{14}C	β^-	5568 (30) 5730 (40)	God 62, Hug 64 Value adopted at the Fifth Radiocarbon Dating Conference Cambridge **1962**
31	^{22}Na	^{22}Na	β^+	2.60	See 74
32	^{22}Ne/^{22}Na	^{22}Na	β^+	2.60	See 74
33	^{26}Al	^{26}Al	β^+	$7.4 \cdot 10^5$	Wea 76
34	^{21}Ne/^{26}Al	^{26}Al	β^+	$7.4 \cdot 10^5$	Wea 76
35	^{36}Cl	^{36}Cl	β^-	$3.1 \cdot 10^5$	Wea 76

[1]) β^- negatron decay; β^+ positron decay.

Table 1 (continued)

No.	Method	Radio nuclide	Mode of decay [1])	Half life $t_{1/2}$ [a]	Ref. $(t_{1/2})$	
36	$^{36}Ar/^{36}Cl$	^{36}Cl	β^-	$3.1 \cdot 10^5$	Wea 76	
37	^{39}Ar	^{39}Ar	β^-	269	See 74	
38	$^{38}Ar/^{39}Ar$	^{39}Ar	β^-	269	See 74	
39	^{32}Si	^{32}Si	β^-	280	See 74	
40	$^{41}K/^{40}K$	^{40}K	β^-, ec, β^+	$1.250 \cdot 10^9$	Gar 76	see method No. 1 and 2
41	^{53}Mn	^{53}Mn	ec	$3.7 \cdot 10^6$	See 74	
42	$^{81}Kr/^{83}Kr$	^{81}Kr	ec	$2.1 \cdot 10^5$	See 74	

F. Methods based on radiation damage in solids.

No.	Method	Radioactive nuclide(s)	Mode of decay [1]) that contributes to the radiation damage	Kind of damage	Decay constant λ [a^{-1}]	Ref.
43	Fission-tracks	^{238}U	sf	Trails of damage produced by the fission fragments	$6.85\,(20) \cdot 10^{-17}$ $8.46\,(6) \cdot 10^{-17}$	Fle 64 Gal 70, Wag 75
44	Thermo-luminescence	^{238}U, ^{235}U, ^{232}Th ^{40}K	$\alpha, \beta^-, \beta^+, \gamma$	Accumulating production of metastable electrons		

[1]) β^- negatron decay; β^+ positron decay; ec electron capture; ec' electron capture to ground state; α alpha decay; sf spontaneous fission; γ gamma radiation.

7.2.2.1 K−Ca method — K−Ca-Methode
(Table 1, No. 1)

This method has not become very important because of its limited applicability. This is explained by the fact that the daughter isotope ^{40}Ca is also the most abundant Ca-isotope with 96.94% and because only very few minerals show a sufficiently high K/Ca ratio.

The equation for the isochron (Fig. 1) is:

Die Methode hat wegen ihrer sehr begrenzten Anwendbarkeit keine nennenswerte Bedeutung erlangt. Der Grund dafür ist, daß das Tochterisotop ^{40}Ca mit 96,94% gleichzeitig das häufigste Calcium-Isotop ist und weil außerdem nur sehr wenige Minerale ein ausreichend hohes K/Ca-Verhältnis aufweisen.

Die Gleichung für die Isochrone (Fig. 1) lautet:

$$\left(\frac{^{40}Ca}{^{44}Ca}\right) - \left(\frac{^{40}Ca}{^{44}Ca}\right)_0 = \frac{\lambda_{\beta^-}}{\lambda} \frac{^{40}K}{^{44}Ca}(e^{\lambda t} - 1) \tag{12}$$

$$\lambda = \lambda_{ec} + \lambda_{\beta^-}$$

Compared with Eq. (10) the factor $\lambda_{\beta^-}/\lambda$ has to be considered here because ^{40}Ca is only produced by negatron-decay of ^{40}K.

Im Vergleich mit Gl. (10) tritt hier noch der Faktor $\lambda_{\beta^-}/\lambda$ auf, da nur der Negatronen-Zerfall von ^{40}K zum ^{40}Ca führt.

7.2.2.2 K−Ar method — K−Ar-Methode
(Table 1, No. 2)

In Eq. (10) for the isochron (Fig. 1), a factor, λ_{ec}/λ, must be considered similar to the K−Ca method because ^{40}Ar can only be formed by electron-capture (the positron-decay is negligible)

Die Gl. (10) für die Isochrone (Fig. 1) muß, ähnlich wie bei der K−Ca-Methode noch um einen Faktor, λ_{ec}/λ, ergänzt werden, da nur der Elektroneneinfang (der Positionen-Zerfall ist vernachlässigbar klein) zur Bildung von ^{40}Ar beiträgt.

$$\left(\frac{^{40}\mathrm{Ar}}{^{36}\mathrm{Ar}}\right) - \left(\frac{^{40}\mathrm{Ar}}{^{36}\mathrm{Ar}}\right)_0 = \frac{\lambda_{ec}}{\lambda} \frac{^{40}\mathrm{K}}{^{36}\mathrm{Ar}}(e^{\lambda t}-1) \tag{13}$$

$$\lambda = \lambda_{ec} + \lambda_{\beta^-}$$

Most experiments have shown the value of the initial isotopic ratio $(^{40}\mathrm{Ar}/^{36}\mathrm{Ar})_0$ to be 295.5 because of the incorporation of atmospheric argon during the formation of the minerals or during the closure of the system. However, in minerals from metamorphic rocks, the $(^{40}\mathrm{Ar}/^{36}\mathrm{Ar})_0$ ratio can be many times higher because of the possibility of higher amounts of radiogenic argon in the gasphase. The K−Ar method is frequently applied for the age determination because potassium is the main component of many minerals and because parent and daughter elements are well separated during their incorporation into the mineral. Nevertheless, this method is more and more being superseded by the $^{40}\mathrm{Ar}/^{39}\mathrm{Ar}$ method.

Für das initiale Isotopenverhältnis $(^{40}\mathrm{Ar}/^{36}\mathrm{Ar})_0$ findet man wegen des Einbaus von Argon atmosphärischer Zusammensetzung bei der Bildung der Minerale oder bei der Schließung der Systeme meist Werte von 295,5. In Mineralen metamorpher Gesteine kann das $(^{40}\mathrm{Ar}/^{36}\mathrm{Ar})_0$-Verhältnis wegen der Möglichkeit erhöhter Anteile an radiogenem Argon in der Gasphase jedoch um ein Vielfaches größer sein. Die K−Ar-Methode hat, da Kalium Hauptbestandteil vieler Minerale ist und da beim Einbau gleichzeitig eine gute Trennung von Mutter- und Tochter-Element erfolgt, vielfach Anwendung bei der Altersbestimmung gefunden. Sie wird allerdings heute mehr und mehr durch die $^{40}\mathrm{Ar}-^{39}\mathrm{Ar}$-Methode ersetzt.

7.2.2.3 $^{40}\mathrm{Ar}/^{39}\mathrm{Ar}$ method — $^{40}\mathrm{Ar}/^{39}\mathrm{Ar}$-Methode

(Table 1, No. 3)

It is an analytic version of the K−Ar method [Tor69, Dal79]. In contrast to the K−Ar method, it allows one to ascertain whether a mineral has lost a part of its argon after its formation or whether an extraneous argon with an isotopic composition differing from atmospheric argon has been included later. The determination of $^{40}\mathrm{K}$ is replaced by the determination of radioactive $^{39}\mathrm{Ar}$ ($t_{1/2}=269$ a), which is produced by irradiation of $^{39}\mathrm{K}$ with fast neutrons through the reaction $^{39}\mathrm{K}(n,p)^{39}\mathrm{Ar}$. The $^{39}\mathrm{Ar}$ produced from $^{39}\mathrm{K}$ and the radiogenic $^{40}\mathrm{Ar}$ are released together in fractions by stepwise heating. The $^{40}\mathrm{Ar}/^{39}\mathrm{Ar}$ ratios of the different fractions are measured with a mass spectrometer after each step. By assuming that the $^{39}\mathrm{Ar}$ comes from the same lattice sites as the radiogenic $^{40}\mathrm{Ar}_r$, the apparent age of each fraction can be calculated as follows:

Sie stellt eine analytische Variante der K−Ar-Methode dar [Tur69, Dal79]. Im Gegensatz zur K−Ar-Methode gestattet sie es zu erkennen, ob ein Mineral nach seiner Bildung einen teilweisen Argonverlust erlitten hat oder ob später Fremdargon mit einer von Luftargon abweichenden Isotopenzusammensetzung aufgenommen wurde. Die Bestimmung des $^{40}\mathrm{K}$ erfolgt über radioaktives $^{39}\mathrm{Ar}$ ($t_{1/2}=269$ a), das durch Bestrahlung von $^{39}\mathrm{K}$ mit schnellen Neutronen über die Reaktion $^{39}\mathrm{K}(n,p)^{39}\mathrm{Ar}$ erzeugt wird. Durch stufenweises Entgasen bei schrittweise erhöhten Temperaturen wird das aus $^{39}\mathrm{K}$ gebildete $^{39}\mathrm{Ar}$ zusammen mit dem radiogenen $^{40}\mathrm{Ar}$ freigesetzt und das $^{40}\mathrm{Ar}/^{39}\mathrm{Ar}$-Verhältnis der einzelnen Fraktionen mit einem Massenspektrometer gemessen. Setzt man voraus, daß das $^{39}\mathrm{Ar}$ den gleichen Gitterplätzen wie das radiogene $^{40}\mathrm{Ar}_r$ entstammt, so kann für jede Fraktion ein „Alter" berechnet werden:

$$t = \frac{1}{\lambda} \ln\left(\frac{^{40}\mathrm{Ar}_r}{^{39}\mathrm{Ar}}J+1\right) \tag{14}$$

$$\lambda = \lambda_{ec} + \lambda_{\beta^-}$$

The factor J considers the neutron flux, the cross section and the duration of irradiation of the sample and is generally determined by simultaneous irradiation of a monitor sample.

Der Faktor J berücksichtigt Neutronenflußdichte, Wirkungsquerschnitt und Bestrahlungsdauer der Probe und wird im allgemeinen durch gleichzeitige Bestrahlung einer Monitor-Probe bestimmt.

$$J = \frac{e^{\lambda t_m}-1}{(^{40}\mathrm{Ar}_r/^{39}\mathrm{Ar})_m} \tag{15}$$

t_m age of the monitor sample
$(^{40}\mathrm{Ar}_r/^{39}\mathrm{Ar})_m$ ratio of the monitor sample.

For the determination of the $^{40}\mathrm{Ar}_r/^{39}\mathrm{Ar}$ ratio, corrections are necessary for the atmospheric argon and for the argon amounts produced from Ca, K, Ar and Cl during the irradiation.

t_m Alter der Monitor-Probe.
$(^{40}\mathrm{Ar}_r/^{39}\mathrm{Ar})_m$ Verhältnis in der Monitor-Probe.

Für die Bestimmung des $^{40}\mathrm{Ar}_r/^{39}\mathrm{Ar}$-Verhältnisses sind Korrekturen für Luftargon und für die Argonanteile notwendig, die bei der Bestrahlung aus Ca, K, Ar und Cl gebildet werden.

Fig. 3 shows the common presentation of the results, in the case of an undisturbed system (Fig. 3a), in the case of partial loss of argon (Fig. 3b) and in the case of later absorption of radiogenic argon (Fig. 3c). If all fractions or several neighbouring ones have the same apparent age within analytical uncertainty, this age is called *plateau-age*.

Fig. 3 zeigt die gebräuchliche Art der Darstellung der Ergebnisse und zwar für den Fall eines ungestörten Systems (Fig. 3a), für den Fall teilweisen Argonverlusts (Fig. 3b) und für den Fall einer späteren Aufnahme radiogenen Argons (Fig. 3c). Ergeben alle oder mehrere benachbarte Fraktionen innerhalb ihres Analysenfehlers das gleiche Alter, so spricht man von einem *Plateaualter*.

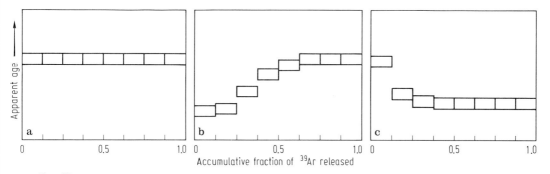

Fig. 3. $^{40}Ar/^{39}Ar$ method. The schematic diagrams show three typical patterns of apparent ages which can be obtained by fractional release of Ar from minerals during an incremental heating procedure. Each data bar represents one release fraction beginning with low temperatures at the left and ending with fusion of the samples and total degassing at the right end. Real samples show variable lengths and hights (error in age) of the bars. The apparent ages are calculated from the $^{40}Ar/^{39}Ar$ ratios of the individual fractions (Eq. (14)). a) closed isotopic system with undisturbed plateau age, b) partial Ar loss but with high-temperature fractions still indicating a plateau, c) with components of extraneous Ar.

7.2.2.4 U, Th − Pb methods and Pb − Pb method — U, Th − Pb-Methoden und Pb − Pb-Methode

(Table 1, Nos. 4, 6 and 15)

A presentation of the results of the U − Pb methods in isochron diagrams according to Fig. 1 is rarely appropriate, because the systems have often lost uranium through weathering, while Th and Pb generally remain quantitatively unchanged in the rock. The non-radiogenic, "stable" ^{204}Pb is used as a reference isotope. The half-life of ^{204}Pb of $1.4 \cdot 10^{17}$ years is so long that ^{204}Pb can be considered as stable.

Die Darstellung von Ergebnissen der U − Pb-Methode in Isochronen-Diagrammen von der Art der Fig. 1, ist nur selten sinnvoll, da die Systeme durch Verwitterungseinflüsse häufig Uran verloren haben, während Th und Pb im allgemeinen quantitativ im Gestein verbleiben. Als Bezugsisotop wird das nicht-radiogene, „stabile" ^{204}Pb verwendet. Die Halbwertszeit von ^{204}Pb ist mit $1,4 \times 10^{17}$ Jahren so lang, daß es als stabil angesehen werden kann.

As the value of the $^{238}U/^{235}U$ ratio is the same world-wide (except for some particular cases, such as the natural reactor of Oklo, Gabun), the division of the isochronic Eq. (10) for the $^{235}U − ^{207}Pb$ system by the corresponding equation for the $^{238}U − ^{206}Pb$ system yields a new equation in which the concentration of uranium no longer appears:

Da das $^{238}U/^{235}U$-Verhältnis, von besonderen Fällen abgesehen (natürlicher Reaktor von Oklo, Gabun), weltweit den gleichen Wert besitzt, erhält man durch Division der Isochronen-Gleichung (10) für das $^{235}U − ^{207}Pb$-System durch die entsprechende Gleichung für das $^{238}U − ^{206}Pb$-System eine neue Gleichung, in der die Konzentration von Uran nicht mehr erscheint:

$$\frac{(^{207}Pb/^{204}Pb) - (^{207}Pb/^{204}Pb)_0}{(^{206}Pb/^{204}Pb) - (^{206}Pb/^{204}Pb)_0} = \frac{1}{137.88}\,\frac{e^{\lambda_{235}t} - 1}{e^{\lambda_{238}t} - 1} \tag{16}$$

137.88 is the present value of the $^{238}U/^{235}U$ ratio. Fig. 4 shows the graphical presentation. In the diagram, cogenetic systems having different $^{238}U/^{204}Pb$ ratios define a straight line which is termed a *Pb − Pb isochron*. The isochron age can be determined by iterative insertion of different values for t into Eq. (16). Therefore, only the measurement of isotopic ratios of lead is required for the age determination.

137,88 ist das heutige $^{238}U/^{235}U$-Verhältnis. Fig. 4 zeigt die graphische Darstellung. Kogenetische Systeme mit unterschiedlichen $^{238}U/^{204}Pb$-Verhältnissen definieren im Diagramm eine Gerade, die *Pb − Pb-Isochrone*. Das Isochronen-Alter läßt sich mit Hilfe der Gl. (16) durch versuchsweises Einsetzen von Werten für t ermitteln. Für die Altersbestimmung ist demnach lediglich die Messung von Isotopenverhältnissen des Bleis erforderlich.

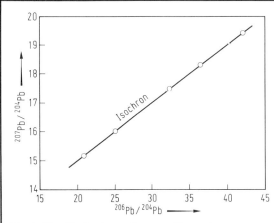

Fig. 4. Pb – Pb isochron diagram. The points represent cogenetic samples (whole rocks or minerals) with different $^{238}U/^{204}Pb$ and $^{235}U/^{204}Pb$ ratios but common initial $^{206}Pb/^{204}Pb$ and $^{207}Pb/^{204}Pb$ ratios. The points define an isochron. The time elapsed since the systems began to follow individual lead evolution can be calculated by means of Eq. (16). The slope of the isochron decreases with time because of the higher production of ^{206}Pb from uranium compared with ^{207}Pb. The initial ratios $(^{206}Pb/^{204}Pb)_0$ and $(^{207}Pb/^{204}Pb)_0$ cannot be determined without uranium analysis.

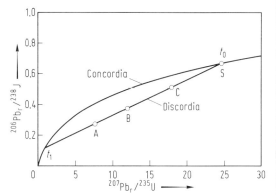

Fig. 5. Concordia diagram. Concordia is the locus of all points of systems for which the $^{206}Pb/^{238}U$ age equals the $^{207}Pb/^{235}U$ age. The curve results from the different decay constants for ^{238}U and ^{235}U (Eq. (17)). The subscript r indicates that only the radiogenic lead produced since the closure of the systems is considered. The three systems A, B and C (e.g. zircon samples) have discordant ages but fall on a straight line termed discordia. Possible interpretations of such a pattern are: a) The three samples are cogenetic and crystallized at time t_0. Lead loss and (or) uranium gain at time t_1 shifted the points from S to their present positions. b) The samples are composites of older components that were formed at t_0 with younger ones that were formed at t_1. In both cases linear extrapolations of the line to concordia yield meaningful intercept ages.

For minerals with high U/Pb ratios and a correspondingly high amounts of radiogenic lead as for instance in zircon, another kind of presentation which is called *concordia-diagram* is used (Fig. 5). The equation of the concordia curve is obtained by dividing the Eq. (6) for the $^{238}U-^{206}Pb$ system by that for the $^{235}U-^{207}Pb$ system. Rearranging then yields:

Für Minerale mit hohen U/Pb-Verhältnissen und dementsprechend hohem Anteil an radiogenem Blei, wie z. B. beim Zirkon, ist eine andere Art der Darstellung in Gebrauch, welche als *Concordia-Diagramm* bezeichnet wird (Fig. 5). Die Gleichung für die Concordia-Kurve erhält man, wenn man die entsprechenden Gln. (6) für die Zerfallssysteme $^{238}U-^{206}Pb$ und $^{235}U-^{207}Pb$ durcheinander dividiert und umformt:

$$\frac{^{206}Pb_r}{^{238}U}=\frac{^{207}Pb_r}{^{235}U}\frac{e^{\lambda_{238}t}-1}{e^{\lambda_{235}t}-1} \tag{17}$$

The index r indicates that only the radiogenic portion should be considered. The correction for the so-called *common lead* which already existed at the formation of the mineral is made as follows:

Der Index r gibt an, daß nur der radiogene Anteil berücksichtigt werden darf. Die Korrektur für schon bei der Bildung des Minerals vorhandenes, sogenanntes *gewöhnliches Blei* geschieht wie folgt:

$$^{206}Pb_r=[(^{206}Pb/^{204}Pb)-(^{206}Pb/^{204}Pb)_0]^{204}Pb \tag{18}$$
$$^{207}Pb_r=[(^{207}Pb/^{204}Pb)-(^{207}Pb/^{204}Pb)_0]^{204}Pb$$

Frequently the isotopic composition of the common lead can only be estimated. However, the resulting error for *t* is negligible or tolerable most of the time.

Die Isotopenzusammensetzung des gewöhnlichen Bleis kann allerdings häufig nur abgeschätzt werden; doch ist der dadurch bedingte Fehler für *t* meist vernachlässigbar oder tolerierbar.

Sustitution of the present value of 137.88 for the $^{238}U/^{235}U$ ratio into Eq. (17) allows, as in Eq. (16), the calculation of *t* solely from the $^{207}Pb_r/^{206}Pb_r$ ratio. This age is therefore named the $^{207}Pb-^{206}Pb$ *age*. The data points of all systems the $^{206}Pb/^{238}U$ ages of which equal their $^{207}Pb/^{235}U$ ages are located

Einsetzen des heutigen Wertes von 137,88 für das $^{238}U/^{235}U$-Verhältnis in Gl. (17) gestattet ähnlich wie bei Gl. (16), das Alter *t* lediglich aus dem $^{207}Pb_r/^{206}Pb_r$-Verhältnis zu berechnen. Dieses Alter wird deshalb als $^{207}Pb-^{206}Pb$-*Alter* bezeichnet. Alle Systeme, deren $^{206}Pb/^{238}U$-Alter gleich ihrem $^{207}Pb/^{235}U$-Alter ist,

on the concordia curve and are termed as concordant. Short opening of cogenetic concordant systems with lead loss and (or) uranium gain shifts the points in the diagram along straight lines which are named *discordias* (Fig. 5). The same pattern can result also from mixing of two concordant systems of different age. The extrapolation of such a discordia yields an upper and a lower intercept age which, in the ideal case, corresponds to the ages of the primary system formation and of the later disturbing event, respectively. However, linear pattern should be interpreted with caution because a continuous loss of lead can also lead to a linear (or in a long range linear) arrangement of the data points.

liegen mit ihren Datenpunkten auf der Concordia-Kurve und werden als konkordant bezeichnet. Kurzzeitiges Öffnen kogenetischer, konkordanter Systeme mit Bleiverlust und (oder) Urangewinn bewirkt eine Verschiebung der Punkte im Diagramm entlang von Geraden, die man als *Discordia* bezeichnet (Fig. 5). Die gleiche Anordnung kann auch durch Mischen verschieden alter Systeme erreicht werden. Extrapolation der Dicordia liefert ein oberes und ein unteres Schnittpunkts-Alter, welche im Idealfall den Altern der primären Bildung der Systeme bzw. dem der späteren Störung entsprechen. Vorsicht ist allerdings bei der Interpretation geboten, weil kontinuierlich diffusiver Bleiverlust ebenfalls zu linearer oder über größere Erstreckung linearer Anordnung der Punkte führen kann.

7.2.2.5 U, Th−He methods — U, Th−He-Methoden

(Table 1, Nos. 5 and 7)

These methods are of minor importance. They have been applied for dating undisturbed meteorites and fossil aragonite.

Diese Methoden sind von untergeordneter Bedeutung. Sie wurden für die Datierung von ungestörten Meteoriten sowie von Aragonitschalen von Fossilien verwendet.

7.2.2.6 Lu−Hf and Re−Os methods — Lu−Hf- und Re−Os-Methoden

(Table 1, Nos. 8 and 9)

Of the two methods the Lu − Hf method has become increasingly important since 1980 [Pat80]. As in the Sm − Nd method, the initial $^{176}Hf/^{177}Hf$ ratios of rocks and minerals (zircon) are of particular interest with respect to the geochemical investigation of the Earth's mantle and crust. ^{177}Hf and ^{126}Os are used as the stable reference isotopes.

Von den beiden Methoden hat die Lu − Hf-Methode seit 1980 zunehmende Bedeutung erlangt [Pat80]. Ähnlich wie bei der Sm − Nd-Methode sind die initialen $^{176}Hf/^{177}Hf$-Verhältnisse von Gesteinen und Mineralen (Zirkon) für Untersuchungen zur geochemischen Entwicklung des Erdmantels und der Erdkruste von besonderem Interesse. Als stabile Bezugsisotope werden ^{177}Hf und ^{126}Os verwendet.

7.2.2.7 Rb−Sr method — Rb−Sr-Methode

(Table 1, No. 10)

The equation for the isochron is:

Die Gleichung für die Isochrone lautet:

$$\left(\frac{^{87}Sr}{^{86}Sr}\right) - \left(\frac{^{87}Sr}{^{86}Sr}\right)_0 = \frac{^{87}Rb}{^{86}Sr}\left(e^{\lambda_{87}t} - 1\right) \tag{19}$$

The isochron diagram (Fig. 1) and, less frequently, the $^{87}Sr/^{86}Sr$ evolution diagram (Fig. 2) are used for the presentation of the results. The isochron diagram illustrates mainly the results of age determination while the evolution diagram in particular presents the initial strontium-isotope ratios in conjunction with geochemical and cosmochemical problems.

Für die Darstellung von Ergebnissen wird das Isochronen-Diagramm (Fig. 1) und, wenn auch seltener, das $^{87}Sr/^{86}Sr$-Entwicklungsdiagramm (Fig. 2) verwendet. Das Isochronen-Diagram dient besonders der Veranschaulichung von Ergebnissen der Altersbestimmung, während das Entwicklungsdiagramm mehr für die Darstellung initialer Strontium-Isotopenverhältnisse in Verbindung mit geochemischen und kosmochemischen Fragestellungen zur Anwendung gelangt.

Three facts have particularly contributed to the successful application of the Rb − Sr method:

1. Rubidium occurs as a typical trace element in all potassium bearing minerals and is therefore wide-spread.

Zum erfolgreichen Einsatz der Rb−Sr-Methode haben besonders drei Gegebenheiten beigetragen:

1. Rubidium kommt als typisches Spurenelement in allen Kalium führenden Mineralen vor und besitzt demnach eine weite Verbreitung.

2. Minerals, especially those of the SiO_2-rich magmatites, and many bodies of igneous rocks vary considerably in their Rb/Sr ratio. This often allows the determination of the isochron age with small error.

3. Since, in contrast to ^{40}Ar, radiogenic ^{87}Sr generally remains in the rocks during metamorphic events, a completely new distribution of Sr-isotopes (Fig. 6) can occur in the size range of mineral grains and also in more extended ranges. Therefore, several events which have formed the respective rock can be dated occasionally by investigation of representative samples of various volumes. In such cases, the mineral isochrons yield lower ages than the whole-rock isochrons (Fig. 6).

When analyzing minerals with high Rb/Sr ratios such as biotite, phengite and muscovite, the choice of the $(^{87}Sr/^{86}Sr)_0$ ratio usually remains without considerable influence on the model age. If young metamorphites from much older parent rocks are investigated, the initial ratios should be determined with much care, for they could be considerably increased in such cases $((^{87}Sr/^{86}Sr)_m$ in Figs 6 and 7).

2. Die Minerale, vor allem die der SiO_2-reichen Magmatite, sowie viele magmatischer Körper zeigen eine beträchtliche Variation im Rb/Sr-Verhältnis. Dies ermöglicht deshalb häufig die Bestimmung von Isochronenaltern mit kleinem Fehler.

3. Da radiogenes ^{87}Sr im Gegensatz zu ^{40}Ar bei der metamorphen Überprägung von Gesteinen im allgemeinen in diesen verbleibt, kommt es im Größenbereich der Mineralkörner, zum Teil aber auch über größere Erstreckung, zu einer vollständigen Neuverteilung der Sr-Isotope (Fig. 6). Durch Untersuchungen an repräsentativen Proben unterschiedlich großer Gesteinsvolumina können gegebenenfalls mehrere Ereignisse, welche das jeweilige Gestein geprägt haben, datiert werden. Mineral-Isochronen liefern gegenüber Gesamtgesteins-Isochronen in derartigen Fällen das niedrigere Alter (Fig. 6).

Bei der Untersuchung von Mineralen mit hohen Rb/Sr-Verhältnissen, wie Biotit, Phengit und Muscovit, bleibt die Wahl des $(^{87}Sr/^{86}Sr)$-Verhältnisses meist ohne nennenswerten Einfluß auf das Modellalter. Vorsicht ist allerdings geboten, wenn es sich um junge Metamorphite sehr viel älterer Ausgangsgesteine handelt, da in derartigen Fällen das initiale Isotopenverhältnis beträchtlich erhöht sein kann $((^{87}Sr/^{86}Sr)_m$ in Figs. 6 und 7).

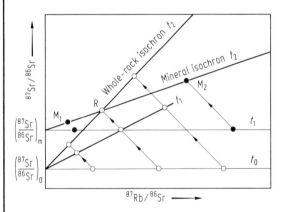

Fig. 6. Rb−Sr isochron diagram showing a case of different Sr evolution in minerals (closed symbols) compared to whole rocks (open symbols) due to an episode of thermal metamorphism. The whole-rock systems remained undisturbed from the initial stage at t_0 through t_1 until t_2. Therefore the initial ratio for the whole-rock isochron remained the same all the time. The minerals M_1 and M_2, e.g. feldspar and biotite, respectively, of the whole-rock sample R equilibrated their $^{87}Sr/^{86}Sr$ ratios during the metamorphism at t_1. Thus the initial ratio of the mineral isochron was raised from $(^{87}Sr/^{86}Sr)_0$ to $(^{87}Sr/^{86}Sr)_m$. The slope of the whole-rock isochron yields the age of crystallization of the rock whereas the mineral isochron corresponds to the age of the metamorphism.

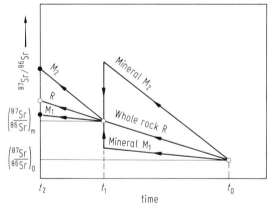

Fig. 7. $^{87}Sr/^{86}Sr$-evolution diagram for the Minerals M_1 and M_2 of the whole-rock sample R of Fig. 6. At time t_0 the rock crystallized. At t_1 an episodic thermal metamorphism opened the Rb−Sr systems and equilibrated the isotopic ratios among the individual minerals. The initial ratio of the mineral isochron was raised from $(^{87}Sr/^{86}Sr)_0$ to $(^{87}Sr/^{86}Sr)_m$, the ratio of the whole-rock system. After t_1 the minerals again were closed systems.

7.2.2.8 Sm−Nd method — Sm−Nd-Methode

(Table 1, No. 11)

The ages are calculated by the isochron method according to Eq. (10):

Die Berechnung des Alters erfolgt nach der Iso-chronen-Methode entsprechend Gl. (10):

$$\left(\frac{^{143}\text{Nd}}{^{144}\text{Nd}}\right)-\left(\frac{^{143}\text{Nd}}{^{144}\text{Nd}}\right)_0=\frac{^{147}\text{Sm}}{^{144}\text{Nd}}\,(e^{\lambda_{147}t}-1) \qquad (20)$$

The method is important for the dating of basic rocks when the Rb−Sr method fails. Furthermore, the investigation of the initial Nd-isotope ratios is of great interest in clarifying the geochemical de-velopment of the earth's mantle. As the decay constant is considerably smaller than that of ^{87}Rb and the variation of the Sm/Nd ratios in nature is relatively small, extreme accuracy of measurement is required in order to apply this method successfully [Lug75].

Die Methode ist für die Datierung basischer Ge-steine, bei der die Rb−Sr-Methode versagt, von Be-deutung. Darüber hinaus ist die Untersuchung der initialen Nd-Isotopenverhältnisse von großem Interesse für die Aufklärung der geochemischen Entwicklung des Erdmantels. Die gegenüber ^{87}Rb wesentlich kleinere Zerfallskonstante und die vergleichsweise geringere Variationsbreite des Sm/Nd-Verhältnisses in der Natur verlangt für die erfolgreiche Anwendung der Methode besondere Anstrengungen im Hinblick auf die Meß-genauigkeit [Lug75].

7.2.2.9 I−Xe and Pu−Xe methods — I−Xe- and Pu−Xe-Methoden

(Table 1, Nos. 12 and 13)

Primitive meteorites contain radiogenic ^{129}Xe from ^{129}I and fission $^{131-136}$Xe from the decay of ^{244}Pu. After cooling below the blocking temperature the Xe subsequently produced in the meteorites has been trapped. Thus, the ^{129}I/^{127}I and the ^{244}Pu/^{238}U ratios in undisturbed bodies at the time of the meteorite formation can be obtained by determination of the Xe concentration and isotopic composition. ^{127}I is a stable isotope, and the former ^{238}U concentration can be calculated from the presently measured U con-centration [Pod70, Pod70a].

Primitive Meteorite enthalten radiogenes ^{129}Xe aus ^{129}I und $^{131-136}$Xe, das durch Spontanspaltung von ^{244}Pu entstanden ist. Nach Abkühlung der Meteorite im Anschluß an ihre Entstehung wurde das von da ab gebildete Xe festgehalten. Über die Bestimmung der Konzentration und Isotopenzusammensetzung des Xe in ungestörten Meteoriten lassen sich somit die ^{129}I/^{127}I- und ^{244}Pu/^{238}U-Verhältnisse zur Zeit der Bildung erhalten. ^{127}I ist ein stabiles Isotop, und die Konzentration von ^{238}U zur Zeit der Bildung läßt sich aus den heute gemessenen Werten errechnen [Pod70, Pod70a].

If one assumes a homogeneous distribution of the ^{129}I/^{127}I and ^{244}Pu/^{238}U ratios in the solar nebula at the time of meteorite condensation, small time differences (≈ 0.5 Ma) between the formation of the individual meteorites can be determined because the half-lifes of ^{129}I and ^{244}Pu are relatively short. In case of the I−Xe method the age difference Δt between two meteorites A and B is obtained by means of the equation:

Wegen der vergleichsweise niedrigen Halbwerts-zeiten von ^{129}I und ^{244}Pu kann man bei Annahme einer homogenen Verteilung des ^{129}I/^{127}I- bzw. des ^{244}Pu/^{238}U-Verhältnisses im Sonnennebel zur Zeit der Meteoritenentstehung kleine Zeitunterschiede ($\approx 0,5$ Ma) zwischen der Entstehung der einzelnen Körper bestimmen. Den Altersunterschied Δt zwischen zwei Meteoriten A und B erhält man bei der I−Xe-Methode mit Hilfe der Gleichung:

$$\Delta t=\frac{1}{\lambda_{129}}\ln\left(\frac{(^{129}\text{I}/^{127}\text{I})_\text{A}}{(^{129}\text{I}/^{127}\text{I})_\text{B}}\right) \qquad (21)$$

Usually the values for Δt are indicated relative to the chondritic meteorite Bjurböle.

Es ist üblich, Δt auf den chondritischen Meteoriten Bjurböle zu beziehen.

7.2.2.10 Age determination with common Sr — Altersbestimmung mit gewöhnlichem Strontium

(Table 1, No. 14)

Dating of rocks from differences in the initial ratios (common Sr) is not possible. The reason is the variable chemical fractionation of the elements Rb and Sr during the formation and differentiation of magmatic melts and during formation of sedimentary rocks. If the mean Rb/Sr ratios of the rocks under

Eine Datierung von Gesteinen aufgrund von Unter-schieden in den initialen ^{87}Sr/^{86}Sr-Verhältnissen (ge-wöhnliches Sr) ist nicht möglich. Der Grund hierfür ist die unterschiedliche chemische Fraktionierung der Elemente Rb und Sr bei der Entstehung und Differen-tiation magmatischer Schmelzen und bei der Bildung

consideration are known or can be estimated, Eq. (19) can be used for the calculation of model ages for their chemical separation from a uniform "reservoir" like the Earth's mantle or the solar nebula. Assuming a homogeneous solar nebula, differences of several Ma have been obtained for the time of meteorite condensation. The initial $^{87}Sr/^{86}Sr$ of basaltic achondrites (BABI) is used as reference value [Gra73].

von Sedimentgesteinen. Wenn das mittlere Rb/Sr-Verhältnis der interessierenden Gesteine bekannt ist oder abgeschätzt werden kann, lassen sich mit Gl. (19) Modellalter für die chemische Abscheidung aus einem in der Isotopenzusammensetzung homogenen „Reservoir" wie z. B. dem Erdmantel und dem Sonnennebel berechnen. Unter Annahme eines homogenen Sonnennebels wurden auf diese Weise Unterschiede für das relative Kondensationsalter von Meteoriten erhalten, wobei man als Bezugswert das initiale $^{87}Sr/^{86}Sr$-Verhältnis basaltischer Achondrite (BABI = basaltic achondrites best initial) verwendet [Gra73].

Pb − Pb method — Pb − Pb-Methode

(see 7.2.2.4 U, Th − Pb methods)

7.2.2.11 Age determination with common lead — Altersbestimmung mit gewöhnlichem Blei

(Table 1, No. 16; Table 2)

Table 2a. Symbols and values of ratios in use with the common-lead method.

Meaning of ratios	Ratios used in equations	Equivalent symbols		Values of ratios	Ref.
		Canadian convention	Swiss convention		
Ratios at time T when the Earth formed (primordial or primeval lead)	$\left(\dfrac{^{206}Pb}{^{204}Pb}\right)_T$	a_0	α_0	9.56 9.307 (6) 9.310	Mur 62 Tat 73 Til 73
	$\left(\dfrac{^{207}Pb}{^{204}Pb}\right)_T$	b_0	β_0	10.42 10.294 (6) 10.296	Mur 62 Tat 73 Til 73
	$\left(\dfrac{^{208}Pb}{^{204}Pb}\right)_T$	c_0	γ_0	29.71 29.476 (18) 29.44	Mur 62 Tat 73 Til 73
Ratios at geologic time t before present	$\left(\dfrac{^{206}Pb}{^{204}Pb}\right)_t$	X	α		
	$\left(\dfrac{^{207}Pb}{^{204}Pb}\right)_t$	Y	β		
	$\left(\dfrac{^{208}Pb}{^{204}Pb}\right)_t$	Z	γ		
Ratios today	$\left(\dfrac{^{238}U}{^{235}U}\right)_n$			137.88	Ste 77
	$\left(\dfrac{^{238}U}{^{204}Pb}\right)_n$	$137.88 \cdot V$	μ		
	$\left(\dfrac{^{235}U}{^{204}Pb}\right)_n$	V			
	$\left(\dfrac{^{232}Th}{^{204}Pb}\right)_n$	W	$\mu \cdot k$		
	$\left(\dfrac{^{232}Th}{^{238}U}\right)_n$	$\dfrac{W}{137.88 \cdot V}$	k		

(continued)

Table 2b. Equations in use for calculation of common-lead ages [doe70].

1. Primary growth equations

$$\left(\frac{{}^{206}\text{Pb}}{{}^{204}\text{Pb}}\right)_t = \left(\frac{{}^{206}\text{Pb}}{{}^{204}\text{Pb}}\right)_T + \left(\frac{{}^{238}\text{U}}{{}^{204}\text{Pb}}\right)_n \cdot (e^{\lambda\,{}^{238}\text{U}\cdot T} - e^{\lambda\,{}^{238}\text{U}\cdot t})$$

$$\left(\frac{{}^{207}\text{Pb}}{{}^{204}\text{Pb}}\right)_t = \left(\frac{{}^{207}\text{Pb}}{{}^{204}\text{Pb}}\right)_T + \left(\frac{{}^{235}\text{U}}{{}^{204}\text{Pb}}\right)_n \cdot (e^{\lambda\,{}^{235}\text{U}\cdot T} - e^{\lambda\,{}^{235}\text{U}\cdot t})$$

$$\left(\frac{{}^{208}\text{Pb}}{{}^{204}\text{Pb}}\right)_t = \left(\frac{{}^{208}\text{Pb}}{{}^{204}\text{Pb}}\right)_T + \left(\frac{{}^{232}\text{Th}}{{}^{204}\text{Pb}}\right)_n \cdot (e^{\lambda\,{}^{232}\text{Th}\cdot T} - e^{\lambda\,{}^{232}\text{Th}\cdot t})$$

2. Primary isochron equation

$$\frac{\left(\frac{{}^{207}\text{Pb}}{{}^{204}\text{Pb}}\right)_t - \left(\frac{{}^{207}\text{Pb}}{{}^{204}\text{Pb}}\right)_T}{\left(\frac{{}^{206}\text{Pb}}{{}^{204}\text{Pb}}\right)_t - \left(\frac{{}^{206}\text{Pb}}{{}^{204}\text{Pb}}\right)_T} = \frac{1}{\left(\frac{{}^{238}\text{U}}{{}^{235}\text{U}}\right)_n} \cdot \left(\frac{e^{\lambda\,{}^{235}\text{U}\cdot T} - e^{\lambda\,{}^{235}\text{U}\cdot t}}{e^{\lambda\,{}^{238}\text{U}\cdot T} - e^{\lambda\,{}^{238}\text{U}\cdot t}}\right)$$

3. Secondary growth equations *)

$$\left(\frac{{}^{206}\text{Pb}}{{}^{204}\text{Pb}}\right)_{t'} = \left(\frac{{}^{206}\text{Pb}}{{}^{204}\text{Pb}}\right)_T \quad \left(\frac{{}^{238}\text{U}}{{}^{204}\text{Pb}}\right)_{n'} \cdot (e^{\lambda\,{}^{238}\text{U}\cdot T} - e^{\lambda\,{}^{238}\text{U}\cdot t}) + \left(\frac{{}^{238}\text{U}}{{}^{204}\text{Pb}}\right)_{n''} \cdot (e^{\lambda\,{}^{238}\text{U}\cdot t} - e^{\lambda\,{}^{238}\text{U}\cdot t'})$$

$$\left(\frac{{}^{207}\text{Pb}}{{}^{204}\text{Pb}}\right)_{t'} = \left(\frac{{}^{207}\text{Pb}}{{}^{204}\text{Pb}}\right)_T + \left(\frac{{}^{235}\text{U}}{{}^{204}\text{Pb}}\right)_{n'} \cdot (e^{\lambda\,{}^{235}\text{U}\cdot T} - e^{\lambda\,{}^{235}\text{U}\cdot t}) + \left(\frac{{}^{235}\text{U}}{{}^{204}\text{Pb}}\right)_{n''} \cdot (e^{\lambda\,{}^{235}\text{U}\cdot t} - e^{\lambda\,{}^{235}\text{U}\cdot t'})$$

$$\left(\frac{{}^{208}\text{Pb}}{{}^{204}\text{Pb}}\right)_{t'} = \left(\frac{{}^{208}\text{Pb}}{{}^{204}\text{Pb}}\right)_T + \left(\frac{{}^{232}\text{Th}}{{}^{204}\text{Pb}}\right)_{n'} \cdot (e^{\lambda\,{}^{232}\text{Th}\cdot T} - e^{\lambda\,{}^{232}\text{Th}\cdot t}) + \left(\frac{{}^{232}\text{Th}}{{}^{204}\text{Pb}}\right)_{n''} \cdot (e^{\lambda\,{}^{232}\text{Th}\cdot t} - e^{\lambda\,{}^{232}\text{Th}\cdot t'})$$

4. Secondary isochron equation

$$\frac{\left(\frac{{}^{207}\text{Pb}}{{}^{204}\text{Pb}}\right)_{t'} - \left(\frac{{}^{207}\text{Pb}}{{}^{204}\text{Pb}}\right)_t}{\left(\frac{{}^{206}\text{Pb}}{{}^{204}\text{Pb}}\right)_{t'} - \left(\frac{{}^{206}\text{Pb}}{{}^{204}\text{Pb}}\right)_t} = \frac{1}{\left(\frac{{}^{238}\text{U}}{{}^{235}\text{U}}\right)_n} \cdot \left(\frac{e^{\lambda\,{}^{235}\text{U}\cdot t} - e^{\lambda\,{}^{235}\text{U}\cdot t'}}{e^{\lambda\,{}^{238}\text{U}\cdot t} - e^{\lambda\,{}^{238}\text{U}\cdot t'}}\right)$$

*) n' indicates the "milieu" prior to time t before present, expressed in present day ratios.
   n'' indicates the "milieu" between t and t' ($t' < t$), expressed in present day ratios.

"*Common lead*" is lead existing in rocks and minerals with very low values for U/Pb and (or) Th/Pb and consequently insignificant proportions of radiogenic lead. The method has great application in the dating of ore deposits where, strictly speaking, the time of separation of the lead from uranium or thorium-bearing environment is determined. Table 2a shows the symbols in use with the common-lead method and also more recent values for primordial lead, the lead at the time when the Earth formed. The isotopic ratios in the primordial lead have been obtained by measuring uranium and thorium-free meteorites.

Before its separation lead may have evolved in an environment with constant or variable ${}^{238}\text{U}/{}^{204}\text{Pb}$ and ${}^{232}\text{Th}/{}^{204}\text{Pb}$ ratios, the change due to natural decay of U and Th not being considered. The isotopic ratios which characterized the environment in the

Als „*gewöhnliches Blei*" bezeichnet man Blei in Gesteinen und Mineralen mit sehr niedrigen U/Pb- und Th/Pb-Verhältnissen und dementsprechend vernachlässigbaren Anteilen an radiogenem Blei. Die Methode hat breite Anwendung bei der Datierung von Erzlagerstätten gefunden, wobei genau genommen der Zeitpunkt der Abtrennung des Bleis aus einem uran- bzw. thoriumhaltigen Milieu bestimmt wird. In Tab. 2a sind die gebräuchlichen Symbole sowie neuere Werte für die Blei-Isotopenzusammensetzung zum Zeitpunkt der Entstehung der Erde (primordiales Blei) aufgeführt. Die primordialen Isotopenverhältnisse wurden durch Messungen an uran- und thoriumfreien Meteoriten erhalten.

Die Entwicklung des Bleis bis zu seiner Abscheidung kann sich in einem Milieu mit konstanten oder sich ändernden ${}^{238}\text{U}/{}^{204}\text{Pb}$- und ${}^{232}\text{Th}/{}^{204}\text{Pb}$-Verhältnissen vollzogen haben. Die Veränderung durch den natürlichen Zerfall von U und Th sei hier nicht berück-

past are indicated by their corresponding present-day values (μ and μk). The growth equations for lead with single and two stage development are compiled in Table 2b.

The ages obtained for the separation of lead and the formation of lead ores are model ages because in the case of a single-stage evolution the age T of the Earth and the composition of the primodial lead has to be known. In case of a two-stage evolution, calculation requires knowledge of the age t for the beginning of the second stage and the lead isotopic composition at that time. Contemporaneous leads having the same evolutionary history but each derived from an environment with a different μ-value plot in a $^{207}Pb/^{204}Pb$ versus $^{206}Pb/^{204}Pb$ diagram as straight lines which are isochrons (Figs. 8 and 9). The isochron equations for single and two-stage development are also given in Table 2b. Here again the age calculation for the lead separation requires the knowledge of the age of the Earth or the age of the beginning of the second stage.

sichtigt. Die das Milieu in der Vergangenheit bestimmenden Isotopenverhältnisse werden durch ihre entsprechenden heutigen Werte (μ und μk) angegeben. In Tab. 2b sind die Wachstumsgleichungen für die Bleiisotopen-Verhältnisse bei ein- und zweistufiger Entwicklung zusammengestellt.

Bei den Altern, die man für die Bleiabscheidung erhält, handelt es sich um Modellalter, weil bei einstufiger Entwicklung für die Berechnung das Alter T der Erde und die Zusammensetzung des primordialen Bleis und bei zweistufiger Entwicklung das Alter t des Beginns der zweiten Stufe und die Isotopenzusammensetzung des Bleis zu diesem Zeitpunkt bekannt sein müssen. Gleichalte Bleie mit gleicher Vorgeschichte, jedoch mit unterschiedlichen μ-Werten ihrer jeweiligen Ausgangsmilieus liegen in einem $^{207}Pb/^{204}Pb - ^{206}Pb/^{204}Pb$-Diagramm (Bleientwicklungs-Diagramm, Fig. 8 und 9) auf Geraden (Isochronen). Die Isochronen-Gleichungen für ein- und zweistufige Bleientwicklung sind in Tab. 2b wiedergegeben. Zur Berechnung des Alters der Bleiabscheidung ist jedoch auch hier die Kenntnis des Alters der Erde bzw. des Beginns der zweiten Stufe notwendig.

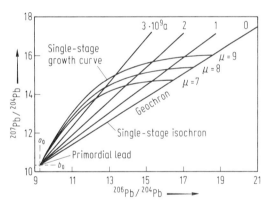

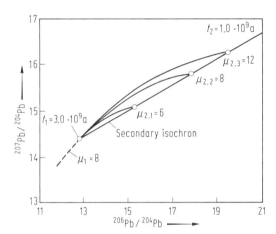

Fig. 8. Lead-evolution diagram showing single-stage model growths curves for systems having μ-values (= present day $^{238}U/^{204}Pb$ ratios) of 7, 8 and 9. The curves started at the point for primordial lead after the Earth had formed. Systems which remained undisturbed until present fall on the zero isochron, also termed geochron. Common leads that separated from the uranium-bearing systems, e.g. 3 Ga ago, fall on the 3 Ga isochron. For the growth and isochron equations, see Table 2b.

Fig. 9. Lead evolution diagram showing examples of second-stage growth curves for systems that evolved 3 Ga ago from a uniform primary system with $\mu_1 = 8$. The leads that separated from the three subsystems ($\mu_2 = 6$, 8 and 12) 1 Ga before present define a secondary isochron. For the growth and isochron equations, see Table 2b.

Age determination of terrestrial lead ores under application of the single-stage evolution model has frequently yielded too high or too low values for mineralization processes of known age which has led to their characterization as "anomalous lead". This can be explained by the fact that in reality all leads are the results of multi-stage evolutions.

Die Altersbestimmung von terrestrischen Bleierzen unter Anwendung des Einstufen-Modells hat bei Vererzungen bekannten Alters häufig zu hohe oder zu niedrige Werte ergeben, was zu ihrer Bezeichnung als „anomales Blei" geführt hat. Der Grund dafür ist, daß es sich bei den Vorkommen in Wahrheit um Blei aus einer mehrstufigen Entwicklung handelt.

7.2.2.12 Methods in which isotopes of the U and Th decay series are used — Methoden, in denen Isotope der Zerfallsreihen von Uran und Thorium verwendet werden

(Table 1, Nos. 17···26)

The methods are based on disturbed equilibria among the isotopes of the radioactive decay chains of U and Th. They are, therefore, also called "disequilibrium methods". The disequilibria are due to the difference in geochemical properties of the various elements. Chemical weathering, precipitation from aqueous solutions, biological processes, adsorption on clay minerals and fractionation during differentiation of magmatic melts can isolate or enrich certain elements of the decay series.

For dating of rocks the following observations are considered:

a) Changes in the activity A of a single isolated isotope (^{210}Pb, ^{231}Pa)

b) Changes in the activity ratio R of a relatively short-lived isotope with respect to a longer-lived isotope of *another* decay chain (^{228}Th/^{226}Ra, ^{231}Pa/^{230}Th, ^{230}Th/^{232}Th).

c) Changes of the activity ratio R of a relatively short-lived isotope compared to a longer-lived isotope of the *same* decay chain (^{228}Th/^{232}Th, ^{231}Pa/^{235}U, ^{230}Th/^{238}U, ^{230}Th/^{234}U, ^{234}U/^{238}U).

The change proceeds always from a stage of disequilibrium ($R_0 \neq 1$) toward secular equilibrium ($R = 1$).

The determination of the activities is generally performed by direct measurement of the radiation of the nuclide under consideration or indirectly by measurement of a short-lived daughter nuclide if radioactive equilibrium is sufficiently approached.

1. *Dating with a single isotope* (Table 1, Nos. 19 and 20): If the initial activity A_0 at the time of formation of the rock is known, the age can be calculated by means of Eq. (4). For the determination of the rate of deposition in marine and lacustrine sediments, the activities of samples from various depths h are measured, and the results are plotted on a diagram as shown in Fig. 10. If it is possible to draw a straight

Die Methoden beruhen auf der Störung des radioaktiven Gleichgewichts zwischen den Isotopen der Zerfallsreihen von U und Th. Sie werden deshalb auch als „Ungleichgewichts-Methoden" bezeichnet. Die Störungen sind auf das unterschiedliche geochemische Verhalten der verschiedenen Elemente zurückzuführen. Chemische Verwitterung, Fällung aus wässerigen Lösungen, biologische Prozesse, Adsorption an Tonminerale und Fraktionierung bei der Kristallisation magmatischer Schmelzen können zur Abtrennung oder Anreicherung eines Elements der Zerfallsreihen führen.

Für die Datierung von Gesteinen nutzt man:

a) die Änderung der Aktivität A eines isolierten Isotops für sich allein (^{210}Pb, ^{231}Pa)

b) die Änderung des Verhältnisses R der Aktivität eines vergleichsweise kurzlebigen Isotops gegenüber der Aktivität eines längerlebigen Isotops einer *anderen* Zerfallsreihe (^{228}Ra/^{226}Ra, ^{231}Pa/^{230}Th, ^{230}Th/^{232}Th)

c) die Änderung des Verhältnisses R der Aktivität eines vergleichsweise kurzlebigen Isotops gegenüber der Aktivität eines längerlebigen Isotops der *gleichen* Zerfallsreihe (^{228}Th/^{232}Th, ^{231}Pa/^{235}U, ^{230}Th/^{238}U, ^{230}Th/^{234}U, ^{234}U/^{238}U).

Die Änderung verläuft vom Ungleichgewicht ($R_0 \neq 1$) in Richtung auf ein säkulares Gleichgewicht ($R = 1$).

Die Bestimmung der Aktivitäten geschieht im allgemeinen durch direkte Messung des interessierenden Nuklids oder über die Messung eines kurzlebigen Folgeprodukts, sobald sich für dieses das radioaktive Gleichgewicht hinreichend eingestellt hat.

1. *Datierung mit einem einzelnen Isotop* (Tab. 1, Nr. 19 und 20): Ist die Anfangsaktivität A_0 zur Zeit der Bildung des Gesteins bekannt, so läßt sich das Alter mit Hilfe von Gl. (4) berechnen. Für die Bestimmung der Ablagerungsrate mariner oder limnischer Sedimente mißt man die Aktivität von Proben aus unterschiedlicher Tiefe h und trägt die Ergebnisse wie in Fig. 10 gezeigt auf. Läßt sich durch die Datenpunkte für einen

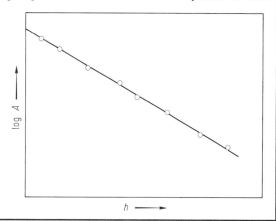

Fig. 10. Plot of logarithm of activity A, e.g. of ^{210}Pb or ^{230}Th, vs. the depth h of samples in a core of marine or lacustrine sediment. Instead of A also activity ratios R, e.g. ^{230}Th/^{232}Th, can be used. From the slope of the regression line the rate of deposition can be calculated with Eq. (22).

line through the points of a certain depth range the rate of deposition for this range can be determined by means of Eq. (22) which has been derived from Eq. (3):

bestimmten Tiefenbereich eine Gerade legen, so kann mit Hilfe der aus Gl. (3) abgeleiteten Gl. (22) die Sedimentationsrate für diesen Tiefenbereich ermittelt werden:

$$\log A - \log A_0 = -\frac{\lambda}{2.303 \, s} h \tag{22}$$

The rate of deposition s is obtained from the slope $-\lambda/2.303 \, s$ of the regression line. If Eq. (22) is applied, only the excess activities (unsupported activities) may be used. The portion of the isotope produced by decay of parent isotopes contained in the sediment has to be eliminated by appropriate sample preparation or by adequate corrections in the calculation.

Die Sedimentationsrate s erhält man aus der Steigung $-\lambda/2{,}303 \, s$ der Ausgleichsgeraden. Bei Anwendung von Gl. (22) ist zu beachten, daß nur die Überschuß-Aktivitäten (nichtgestützter Anteil) des jeweiligen Isotops eingesetzt werden dürfen. Der Anteil des Isotops, der durch den Zerfall von im Sediment enthaltenen Mutterisotopen stammt, muß durch Anwendung geeigneter Präparationsverfahren beseitigt oder bei der Berechnung berücksichtigt werden.

 2. *Dating by measuring activity ratios of isotopes of different decay chains* (Table 1, Nos. 18, 22 and 23): The activity ratio R of the isotopes i and j ($R = A_i/A_j$) at time t after formation of a rock is given by Eq. (23):

 2. *Datierung durch Messung des Verhältnisses der Aktivitäten von Isotopen aus verschiedenen Zerfallsreihen* (Tab. 1, Nr. 18, 22, 23): Das Verhältnis R der Aktivitäten der Isotope i und j ($R = A_i/A_j$) zur Zeit t nach der Bildung eines Gesteins ist durch Gl. (23) gegeben:

$$R = R_0 \frac{e^{-\lambda_i t}}{e^{-\lambda_j t}} \tag{23}$$

R is the activity ratio when the rock was formed. If $\lambda_i \ll \lambda_j$, which is the case for the ^{230}Th/^{232}Th method, Eq. (23) is simplified to:

R_0 ist das Verhältnis der Aktivitäten bei der Bildung des Gesteins. Ist $\lambda_i \ll \lambda_j$, wie bei der ^{230}Th/^{232}Th-Methode der Fall, so vereinfacht sich Gl. (23) zu:

$$R = R_0 \, e^{-\lambda_i t} \tag{24}$$

For the determination of deposition rates, the activity ratios in sediments can be used in the same way as the activities of a single isotope in Eq. (22) and Fig. 10. The choice of a reference isotope of the same element, as e.g. in the ^{230}Th/^{232}Th method, has the advantage that disturbing influences due to fluctuations in the primary concentration are reduced.

Die Verhältnisse der Aktivitäten in Sedimenten können wie die Aktivitäten eines einzelnen Isotops entsprechend Gl. (22) und Fig. 10 zur Ermittlung von Sedimentationsraten verwendet werden. Die Wahl eines Bezugsisotops desselben Elements, wie z.B. bei der ^{230}Th/^{232}Th-Methode, bietet dabei den Vorteil, daß sich störende Einflüsse primärer Konzentrationsschwankungen auf die Bestimmung der Sedimentationsrate verringern.

 An extension of Eq. (24) allows the correction for ^{230}Th supported by ^{238}U present in the rock:

 Eine Korrektur für ^{230}Th, das durch Zerfall von gleichzeitig vorhandenem ^{238}U entsteht (gestütztes ^{230}Th), ist durch Erweiterung von Gl. (24) möglich:

$$R = R_0 \, e^{-\lambda_{230} t} + \frac{A_{238}}{A_{232}} (1 - e^{-\lambda_{230} t}) \tag{25}$$

R is the activity ratio of ^{230}Th to ^{232}Th, A_{238} and A_{232} are the activities of ^{238}U and ^{232}Th, respectively.

R ist das Verhältnis der Aktivitäten von ^{230}Th zu ^{232}Th, A_{238} ist die Aktivität von ^{238}U und A_{232} die Aktivität von ^{232}Th.

 The presence of unsupported ^{230}Th in young volcanic rocks can be used for dating of the minerals and the volcanic glass (Fig. 11). The equation of the isochrons in Fig. 11 is identical with Eq. (25). From the slope m of the isochrons the ages can be calculated:

 Die Anwesenheit von ungestütztem ^{230}Th in jungen vulkanischen Gesteinen läßt sich zur Datierung der Minerale und des vulkanischen Glases verwenden (Fig. 11). Die Gleichung für die Isochronen in Fig. 11 ist identisch mit Gl. (25). Aus der Steigung m der Isochronen können die Alter berechnet werden:

$$t \simeq \frac{1}{\lambda_{230}} \ln (m+1) \tag{26}$$

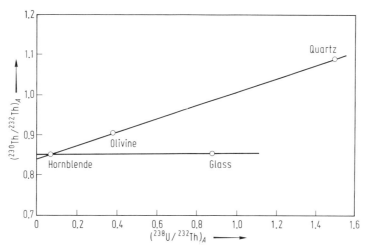

Fig. 11. Application of the ^{230}Th/^{232}Th method in dating young volcanic rocks (rhyolite from Mono Crater, California). The slopes of the isochrons are proportional to the time elapsed since crystallization of the minerals in depth (20000 a) and the solidification of the glass after eruption, respectively. The age of 20000 a was calculated with Eq. (26). After [Tad67] from [fau77]. Index A indicates activity ratios.

3. *Dating by measurement of activity ratios of isotopes of the same decay series* (Table 1, Nos. 17, 21 24···26): In seawater, ^{238}U and ^{234}U are not in radioactive equilibrium; the activity ratio A_{234}/A_{238} has a fairly constant value of 1.15. When an authigenic mineral forms, traces of U from the water are incorporated. After measuring the activity ratio R, the age of the mineral can be determined by means of Eq. (27), where $R_0 = 1.15$:

3. *Datierung durch Messung der Aktivitätsverhältnisse von Isotopen aus der gleichen Zerfallsreihe* (Tab. 1, Nr. 17, 21, 24···26): ^{238}U und ^{234}U befinden sich im Meerwasser nicht im radioaktiven Gleichgewicht; das Aktivitätsverhältnis A_{234}/A_{238} hat einen ziemlich konstanten Wert von 1,15. Bei der Bildung eines authigenen Minerals werden Spuren von Uran aus dem Wasser eingebaut. Durch Messung des Aktivitätsverhältnisses R kann das Alter des Minerals mit Hilfe der folgenden Gl. (27) bestimmt werden, wobei $R_0 = 1,15$ ist:

$$R = 1 + (R_0 - 1)\, e^{-\lambda_{234}t} \tag{27}$$

This equation is valid also for the ^{231}Pa/^{235}U method which is based on the separation of ^{235}U from its daughter when a mineral forms in the seawater. The initial ratio R_0 in the mineral is much smaller than 1, in this case. In Eq. (27) λ_{234} has to be replaced by the decay constant for ^{231}Pa.

Gl. (27) ist auch für die ^{231}Pa/^{235}U-Methode gültig, welche auf der Trennung des ^{235}U von seinem Folgeprodukt ^{231}Pa bei der Bildung eines Minerals im Meerwasser beruht. Das Anfangsverhältnis R_0 im Mineral ist in diesem Fall sehr viel kleiner als 1. Anstelle von λ_{234} ist in Gl. (27) die Zerfallskonstante für ^{231}Pa einzusetzen.

For the ^{230}Th/^{238}U method the activity ratio A_{230}/A_{238} is given by Eq. (28), provided that the initial value for A_{230}/A_{238} is nearly zero.

Bei der ^{230}Th/^{238}U-Methode ist das Aktivitätsverhältnis A_{230}/A_{238} durch die folgende Gl. (28) gegeben, vorausgesetzt daß der Anfangswert für A_{230}/A_{238} nahezu null ist:

$$\frac{A_{230}}{A_{238}} = (1 - e^{-\lambda_{230}t}) + \frac{\lambda_{230}}{\lambda_{230} + \lambda_{234}} \left[\left(\frac{A_{234}}{A_{238}} \right)_0 - 1 \right] (e^{-\lambda_{234}t} - e^{-\lambda_{238}t}) \tag{28}$$

$(A_{234}/A_{238})_0$ is the initial activity ratio of ^{234}U to ^{238}U in the sample. If right from the beginning ^{234}U is in secular equilibrium with ^{238}U then $(A_{234}/A_{238})_0$ is zero and Eq. (28) is reduced to:

$(A_{234}/A_{238})_0$ ist das initiale Aktivitätsverhältnis von ^{234}U zu ^{238}U in der Probe. Wenn sich ^{234}U von Anfang an im säkularen Gleichgewicht mit ^{238}U befindet, ist $(A_{234}/A_{238})_0$ gleich null und die Gl. (28) vereinfacht sich zu:

$$\frac{A_{230}}{A_{234}} = 1 - e^{-\lambda_{230}t} \tag{29}$$

This is the equation for the ^{230}Th/^{234}U method. Eqs. (28) and (29) are only valid if the sample contains no excess ^{234}U unsupported by ^{238}U.

The application of the ^{228}Th/^{230}Th method in dating young marine sediments is more complicated because the observed excess ^{228}Th can be derived from three independent sources: 1. From the sea-water, 2. from authigenic ^{232}Th of the sediment and 3. from authigenic, non-supported ^{226}Ra of the sediment [gol 74].

Dies ist die Gleichung der ^{230}Th/^{234}U-Methode. Die Gln. (28) und (29) gelten nur, wenn sich in der Probe kein zusätzliches, nicht durch ^{238}U gestütztes ^{234}U befindet.

Die Anwendung der ^{228}Th/^{230}Th-Methode zur Datierung junger mariner Sedimente ist komplizierter, weil das beobachtete überschüssige ^{228}Th aus drei voneinander unabhängigen Quellen stammen kann: 1. aus dem Meerwasser, 2. aus authigenem ^{232}Th des Sediments und 3. aus authigenem, nichtgestütztem ^{226}Ra des Sediments [gol74].

7.2.2.13 Methods in which cosmic-ray induced nuclides are used — Methoden, in welchen durch kosmische Strahlung erzeugte Nuklide verwendet werden

(Table 1, Nos. 27···42)

The mostly relatively short-lived nuclides are used for dating of young marine and limnic sediments, soils, fossils and ice as well as for the determination of exposure ages and terrestrial ages of meteorites.

Of great importance is the ^{14}C method (radiocarbon method, Table 1, No. 30). The age can be calculated with Eq. (4). For ^{14}C which is in equilibrium with the atmosphere an activity of $(13.56 + 0.07)$ dpm g^{-1} has been determined. This value is to be inserted as the initial activity A_0. From measurements on wood samples with known ages, it was possible to find long-term variations of A_0 in the atmosphere. These must be considered when calculating the age. Moreover corrections are necessary when the ^{13}C/^{12}C ratios in the samples show a larger fractionation. However, this can only be detected by additional analysis with a mass spectrometer.

The other methods are partly afflicted with larger errors due to uncertainties in the estimates for the initial concentrations. For the determination of the concentrations anti-coincidence counters are used. In recent years the application of particle accelerators has considerably increased the sensitivity.

To determine the *cosmic-ray exposure age* t_e of meteorites pairs of nuclides are used. Each pair consists of a stable nuclide (^iD_s) and an unstable one (^jA_r) with small mass difference.

Die größtenteils vergleichsweise kurzlebigen Nuklide werden zur Datierung von jungen marinen und limnischen Sedimenten, Böden, Fossilien und Eis sowie für die Bestimmung von Bestrahlungs-Altern und terrestrischen Altern von Meteoriten verwendet.

Größte Bedeutung hat die ^{14}C-*Methode* (Radiokohlenstoff-Methode, Tab. 1, Nr. 30) erlangt. Die Berechnung des Alters geschieht mit Hilfe der Gl. (4). Für ^{14}C, das sich im Gleichgewicht mit der Atmosphäre befindet, wurde eine Aktivität von $(13,56 \pm 0,07)$ dpm g^{-1} bestimmt, die als Anfangsaktivität A_0 einzusetzen ist. Aufgrund der Messungen an Holzproben bekannten Alters ließen sich langzeitige Schwankungen von A_0 in der Atmosphäre nachweisen, die bei der Altersberechnung berücksichtigt werden müssen. Korrekturen sind außerdem erforderlich, wenn das ^{13}C/^{12}C-Verhältnis in den Proben eine größere Fraktionierung aufweist, was jedoch nur durch getrennte Isotopenanalyse mit dem Massenspektrometer nachgewiesen werden kann.

Bei den anderen Methoden ergeben sich zum Teil größere Fehler durch die Unsicherheit bei der Abschätzung der Anfangskonzentration. Für die Konzentrations-Bestimmung werden Antikoinzidenz-Zähler und in neuerer Zeit mit bedeutend höherer Empfindlichkeit Teilchenbeschleuniger verwendet.

Für die Bestimmung des *Bestrahlungs-Alters* t_e von Meteoriten verwendet man Paare aus einem stabilen Nuklid (^iD_s) und einem instabilen Nuklid (^jA_r) mit möglichst geringem Massenunterschied:

$$t_e = \frac{^iD_s}{^jA_r} \frac{^j\bar{\sigma}_r}{^i\bar{\sigma}_s} \tag{30}$$

$^j\bar{\sigma}_r/^i\bar{\sigma}_s$ is the effective cross-section ratio for the energy spectrum of the radiation and the target of given composition and orientation. Its value is approximated by simulated radiation of the same material or by other experimentally obtained data. The nuclide pairs in use for the determination of t_e are listed in Table 3A.

$^j\bar{\sigma}_r/^i\bar{\sigma}_s$ ist das Verhältnis der effektiven Wirkungsquerschnitte für das Energiespektrum der Strahlung und gegebener Anordnung und Zusammensetzung des Targets. Man bestimmt das Verhältnis näherungsweise durch simulierte Bestrahlung des gleichen Materials oder mit Hilfe auf andere Weise experimentell ermittelter Werte. Die für die Bestimmung von t_e verwendeten Nuklid-Paare sind in Tab. 3A aufgeführt.

The *terrestrial age* can be calculated with Eq. (1), if N_0 is replaced by the saturation concentration of a suitable spallogenic radioactive nuclide (see Table 3A).

Das *terrestrische Alter* kann man mit Gl. (1) berechnen, wenn man anstelle der Anfangskonzentration N_0 die Sättigungs-Konzentration eines geeigneten spallogenen radioaktiven Nuklids (Tab. 3A) einsetzt.

7.2.2.14 Fission-track dating — Altersbestimmung mit Hilfe von Spaltspuren

(Table 1, No. 43)

Latent damages produced by the charged fission products of spontaneous fission of ^{238}U are made evident in rock glasses and minerals by etching so that they can be detected and counted under an optical microscope. The uranium concentration can be determined in a similar way. After the density of the fossil fission tracks has been determined, these are annealed by tempering the sample. Then, fission of ^{235}U can be induced by irradiation with thermal neutrons. The density of these artificially-induced fission tracks is determined in the same way as the fossil tracks. The time t, indicating how long the natural fission tracks have remained in the sample, can be calculated with the following equation:

Bei dieser Methode werden die bei der Spontanspaltung von ^{238}U durch geladene Spaltprodukte in Gesteinsgläsern und Mineralen erzeugten latenten Schäden durch Anätzen sichtbar gemacht, so daß sie unter einem Lichtmikroskop erkannt und ausgezählt werden können. Auf ähnliche Weise wird auch die Urankonzentration ermittelt. Nachdem die Dichte der fossilen Spaltspuren bestimmt worden ist, werden diese durch Tempern der Probe ausgeheilt. Danach wird durch Bestrahlung mit thermischen Neutronen eine Kernspaltung des ^{235}U induziert. Die Dichte der künstlichen Spaltspuren wird dann wie die der fossilen Spuren ermittelt. Die Zeit t, seit der die natürlichen Spaltspuren in der Probe erhalten geblieben sind, berechnet man mit der folgenden Gleichung:

$$t = \frac{1}{\lambda_{238}} \ln \left[1 + \frac{\varrho_s}{\varrho_i} \frac{\lambda_{238}}{\lambda_{sf}} \frac{1}{137.88} \varphi \, \sigma \right] \tag{31}$$

λ_{238} decay constant for α-decay of ^{238}U

λ_{sf} decay constant for the spontaneous fission of ^{238}U

ϱ_s density of the fossil fission tracks

ϱ_i density of the induced fission tracks

φ neutron dosis

σ cross section of ^{235}U for thermic neutrons

λ_{238} Zerfallskonstante für den Alpha-Zerfall von ^{238}U

λ_{sf} Zerfallskonstante für die Spontanspaltung von ^{238}U

ϱ_s Dichte der fossilen Spaltspuren

ϱ_i Dichte der induzierten Spaltspuren

φ Neutronendosis

σ Wirkungsquerschnitt des ^{235}U für thermische Neutronen

The product $\varphi\sigma$ is generally determined by the simultaneous irradiation of a monitor sample with known concentration of uranium.

The temperatures at which the fission tracks are completely annealed are known. Corrections to the calculation of t are needed if the cooling rates during the geological processes are low [Wag79, Wag77].

Das Produkt $\varphi\sigma$ wird im allgemeinen durch gleichzeitige Bestrahlung einer Monitor-Probe mit bekannter Urankonzentration bestimmt.

Die Temperaturen, bei denen die Spaltspuren vollständig ausheilen, sind bekannt. Werden während geologischer Vorgänge die Temperaturen nur langsam unterschritten, so werden entsprechende Korrekturen für die Berechnung von t erforderlich [Wag79, Wag77].

7.2.2.15 Thermoluminescence method — Thermolumineszenz-Methode

(Table 1, No. 44)

With this method we date the last heating of the sample. It is an important method for dating of ceramics in archeometry but it is only of minor importance for the age determination of rocks. Uncertainties result from disturbing influences (mechanical stress, humidity) during the geological history as well as from the fact that in young rocks the isotopes of the uranium and thorium decay series are often not in radioactive equilibrium.

Datiert wird die letzte Aufheizung der Probe. Die für die Bestimmung von Keramikfunden in der Archäometrie wichtige Methode ist für die Altersbestimmung von Gesteinen nur von geringer Bedeutung. Unsicherheiten ergeben sich aus dem Einfluß der geologischen Geschichte (mechanische Beanspruchung, Feuchtigkeit) sowie der Tatsache, daß sich in jungen Gesteinen die Isotope der Uran- und Thorium-Zerfallsreihen nicht im radioaktiven Gleichgewicht befinden.

7.2.2.16 Dating of cosmic and geologic events — Datierung kosmischer und geologischer Ereignisse

Table 3. Dating of cosmic and geologic events

	Method	Age type	Ref.
A. Meteorites [kir 78, Kir 81]			
a) Formation age of meteorites and their parent bodies			
Individual meteorites	$U-Pb$, $Th-Pb$	internal isochron age	Ado76
	$Rb-Sr$	internal isochron age	Com65, San70
	$Sm-Nd$	internal isochron age	Lug75
	$Pb-Pb$	internal isochron age	Che76
	$U-Pb$	concordia upper intercept age	Tat73, Che76
	$^{40}Ar/^{39}Ar$	plateau age	Jes80
Groups of (cogenetic) meteorites	$Lu-Hf$	whole-rock isochron age	Pat80
	$Rb-Sr$	whole-rock isochron age	Mur65, Gop70
	$Pb-Pb$	whole-rock isochron age	Tat73, Til73
b) Subsequent meteorite evolution			
Re-heating or melting by shock metamorphism due to collision, re-heating or melting by internal heat generation (igneous processes)	$K-Ar$	gas retention age	Kir63
	$^{40}Ar/^{39}Ar$	gas retention age/ plateau age	Tur69
	$U/Th-He$	gas retention age	Kir63, Tay69
	$U-Pb$	concordia intercept age	Che76
	$Rb-Sr$	internal isochron age	Gal75
Cosmic-ray exposure ages (life time of the meteorite as a small object after break-up of the meteorite parent body, measured backwards from its fall on earth)	$^{3}He/^{3}H$, $^{22}Ne/^{22}Na$ $^{21}Ne/^{26}Al$, $^{36}Ar/^{36}Cl$ $^{38}Ar/^{39}Ar$, $^{41}K/^{40}K$ $^{81}Kr/^{83}Kr$		Kir71, Cre76
Time after fall on earth (terrestrial age of the meteorite)	Range of applicability in parenthesis $^{3}H(<150a)$, $^{10}B(<8 \cdot 10^{6}a)$ $^{14}C(<5 \cdot 10^{4}a)$, $^{22}Na(<25a)$ $^{26}Al(<3.5 \cdot 10^{6}a)$, $^{36}Cl(<1.5 \cdot 10^{6}a)$ $^{39}Ar(<1500a)$, $^{53}Mn(<15 \cdot 10^{6}a)$		Koh63, Sam63
c) Measurement of small	$I-Xe$		Rey60, Pod70, Kir81
differences with respect to meteorite formation	$Pu-Xe$ common Sr	differences in initial $^{87}Sr/^{86}Sr$ ratios	Pod70a, Gra73, Kir78, Kir81
B. Earth and moon			
Formation (accretion) age of earth and moon	Indirect dating with formation ages of undisturbed meteorites (see above)		kir78
	Indirect dating with model ages for the isotopic evolution of terrestrial lead, based on isotope data of primordial lead of meteorites and of conformable strata-bound lead ores. Two-stage lead evolution models		Sta75, Cum75
Ages of terrestrial and lunar rocks	see $C-E$ below		

(continued)

Table 3 (continued)

	Method	Age type	Material	Ref.
C. Magmatic rocks				
Chemical formation or differentiation of magmatic melts from a source rock (reservoir)	Rb – Sr Sm – Nd	model ages deduced from initial $^{87}Sr/^{86}Sr$ $(^{143}Nd/^{144}Nd)$ ratios of whole-rock iso-chrons and from average Rb/Sr (Sm/Nd) ratios	magmatic rocks of variable composition but from an iso-topically uniform reservoir	McC78
Age of the crustal source rock(s) of palingenetic melts	U – Pb	concordia upper intercept ages	inherited accessory zircon	Gra73a
Intrusion and extru-sion ages ($\approx$ age of crystallization)	U – Pb Th – Pb	whole-rock isochron age (rarely appli-cable because of low U(Th)/Pb ratios and loss of U and Th)	granitic rocks	doe70, Ros73, Pid78b
	Rb – Sr	whole-rock isochron age	rocks of granitic to granodioritic com-position	fau72, Sch76, Moo72
	Sm – Nd	whole-rock isochron age	basic rocks	Jac79
	Pb – Pb	whole-rock isochron age	rocks of variable composition	doe70, Ros73, Bla71
	U – Pb	concordia upper and lower intercept ages	accessory zircon, monazite, sphene, apatite	Che76, doe70, Pid78b
Intrusion and extrusion with quick subsequent cooling	K – Ar	whole-rock model age } whole-rock isochron age }	volcanic rocks	fau77
		mineral model age } internal isochron age }	biotite, muscovite feldspars, hornblende	fau77
	$^{40}Ar/^{39}Ar$	plateau age	see K – Ar	fau77, Tur77
	Rb – Sr	mineral model age	biotite, muscovite K-feldspar	fau77, fau72
		internal isochron age	biotite, muscovite, feldspars and addi-tional minerals with low Rb/Sr ratios	fau77, fau72
	Sm – Nd	internal isochron age	plagioclase, pyroxene, ilmenite	Jac79
	fission tracks		volcanic glass, zircon, apatite, epidote, sphene	Nae79, Wag79
	$^{230}Th/^{232}Th$	internal isochron	olivine, hornblende, quartz and glass of young volcanic rocks ($< 3 \cdot 10^5$ a)	Tad67, All68a
D. Metamorphic rocks				
Contact metamorphism	K – Ar $^{40}Ar/^{39}Ar$ Rb – Sr fission tracks	model age plateau age model age	biotite, K-feldspar, feldspar biotite apatite, sphene	Har64 Ber75 Har64 Cal73
				(continued)

Table 3 (continued)

	Method	Age type	Material	Ref.

Regional meta-
morphism
	K – Ar	mineral model age
		  (gas retention age)
		internal isochron age
			biotite, muscovite,
			  phengite illite,
			  feldspars, amphiboles
				Hun79

	$^{40}Ar/^{39}Ar$	plateau age	see K – Ar	Dal79
	Rb – Sr	whole-rock isochron age	metasediments, ortho- gneisses	Hun70

		internal isochron age	biotite, muscovite,
			  phengite, feld-
			  spars, apatite
				Jäg79

	U – Pb	concordia lower
		  intercept ages
			accessory zircon
			  and monazite of
			  metasediments
			  and orthogneisses
				Geb79

Cooling age	K – Ar	gas retention age	biotite, amphibole	Hun79, Wag77, Dod79
	Rb – Sr	mineral model age internal isochron age	biotit, phengite, muscovite	Jäg79, Wag77, Dad79

	fission
	  tracks
			apatite, zircon,
			  vesuvian garnet,
			  sphene
				Wag77, Wag79

Method	Range of applica- bility [a]	Material	Ref.

E. Sediments and fossils

Age of deposition [rate of accumulation]

Method	Range of applica- bility [a]	Material	Ref.

Rb – Sr isochron
  method
	No limit	clay mineral fraction free of detrital
		  components,
			Cla79

		glauconite	Obr68
K – Ar	$\geqq 10^5$	glauconite, synsedimentary volcanic rocks (whole-rock samples and minerals)	Hun79
^{10}Be	$0.5 \cdot 10^6 \cdots 10^7$	pelagic sediments, manganese nodules	Ami66
U – He	No limit	corals, oolites	Fan65
$^{234}U/^{238}U$	$5 \cdot 10^4 \cdots 10^6$	corals, oolites,	Vec66

		calcium carbonate of biogenic and
		  inorganic origin deposited in marine and
		  nonmarine environments
			Ku65, Ku67

$^{230}Th/^{232}Th$	$2 \cdot 10^4 \cdots 3 \cdot 10^5$	pelagic sediments, manganese nodules	Gol68
^{231}Pa	$10^4 \cdots 1.4 \cdot 10^5$	pelagic sediments, manganese nodules	Ku67, Ku69
$^{231}Pa/^{230}Th$	$10^4 \cdots 1.4 \cdot 10^5$	pelagic sediments, manganese nodules	Ros61, Ku69
$^{230}Th/^{234}U$	$< 2 \cdot 10^5$	corals, oolites	Kau71
$^{231}Pa/^{235}U$	$< 10^5$	corals, oolites	Ku68
^{14}C	$10^3 \cdots 4 \cdot 10^4$	calcareous pelagic sediments, near-shore and lacustrine sediments with substantial organics or carbonates	Bro56, Eme62
^{210}Pb	$10 \cdots 100$	rapidly accumulating lacustrine and marine sediments	Koi73
$^{228}Ra/^{226}Ra$	$5 \cdots 30$	rates of coral growth	Moo72a
$^{228}Th/^{232}Th$	$1 \cdots 10$	rapidly accumulating near-shore sediments, rate of coral growth	Koi73

7.2.3 Geological time tables

Table 4. **Planetary and early terrestrial ages**

A. Meteorites [kir78; Kir81].

	Age [10^9 a] based on decay constants recommended by the IUGS subcommission on Geochronology [Ste77]	Method(s)	Age type	Ref.
Primitive meteorites which remained essentially undisturbed after their formation				
Example for ages obtained on an individual meteorite (C3 carbonaceous chondrite Allende)	4.52 (3)	$^{40}Ar/^{39}Ar$	plateau age	Jes80
	4.57 (2)	U – Pb	concordia upper intercept age	Til73
	4.565 (4)	Pb – Pb	internal isochron age	Til73
	4.553 (4)	Pb – Pb	internal isochron age	Tat76
Range of ages obtained for individual meteorites	4.4···4.6	K – Ar	model age	
		$^{40}Ar/^{39}Ar$	plateau age	
		U – Pb ⎱ Th – Pb ⎰	concordia intercept ages	
		U – Th – He	model age, gas retention age	
		Sm – Nd	internal isochron age	
		Pb – Pb	internal isochron age	
Weighted mean of individual isochron ages	4.52 (2)			Kir81
Range of ages obtained for different meteorite groups (chondrites, eucrites, iron meteorites)	4.45···4.6	Rb – Sr ⎱ Lu – Hf ⎟ Re – Os ⎟ Pb – Pb ⎰	whole-rock isochrons	
Weighted mean of whole-rock isochron ages for different meteorite groups	4.54 (3)			Kir81

B. Moon [kir78; Kir81; Ado76].

Rock type	Age [10^9 a] based on decay constants recommended by the IUGS subcommission on Geochronology [Ste77]	Method	Age type	Ref.
Oldest individual rock				
Dunite 72417	4.46 (10)	Rb – Sr	internal isochron	Pap75
Troctolite 76535	4.51 (9)	Rb – Sr	internal isochron	Pap76
Most ancient reservoirs				
Soils 10084	≈4.6	U – Pb	concordia upper intercept age	Tat70
Anorthosites	4.51 (15)	Rb – Sr	whole-rock isochron	Sch76
Troctolite and mare basalts	4.53 (10)	Sm – Nd	whole-rock isochron	Lug76

(continued)

Table 4 (continued)
C. Archean rocks of the earth.

Region Geologic/petrographic unit	Area/locality	Rock type	Age 10^6 a	Method and type of age IA isochron age WR whole rocks M minerals UI concordia upper intercept age	Constants used indicated only if not the same as those recommended by the IUGS Subcommission on Geochronology [Ste77]	Ref.
West Greenland						
'Granitic' gneisses						
Amîtsoq gneiss	Godthaab area	quartzo-feldspathic gneiss	3650 (50) } 3646 (85) }	U–Th–Pb, UI-zircon	$\lambda^{235}U = 9.72 \cdot 10^{-10}\,a^{-1}$ $\lambda^{238}U = 1.54 \cdot 10^{-10}\,a^{-1}$ $\lambda^{232}Th = 4.99 \cdot 10^{-11}\,a^{-1}$ $^{238}U/^{235}U = 137.8$	Baa73
			3550 (220)	Lu–Hf, IA-WR	$\lambda^{176}Lu = 1.96 \cdot 10^{-11}\,a^{-1}$	Pet81
			3750 (50)	Rb–Sr, IA-WR	$\lambda^{87}Rb = 1.39 \cdot 10^{-11}\,a^{-1}$	Moo72
			3620 (100)	Pb–Pb, IA-WR	$\lambda^{235}U = 9.7216 \cdot 10^{-10}\,a^{-1}$ $\lambda^{238}U = 1.5369 \cdot 10^{-10}\,a^{-1}$	Bla71
	Isua area	quartzo-feldspathic gneiss	3780 (130)	Rb–Sr, IA-WR	$\lambda^{87}Rb = 1.39 \cdot 10^{-11}\,a^{-1}$	Moo75
			3800 (120)	Pb–Pb, IA-WR	$\lambda^{235}U = 9.72 \cdot 10^{-10}\,a^{-1}$ $\lambda^{238}U = 1.537 \cdot 10^{-10}\,a^{-1}$ $^{238}U/^{235}U = 137.8$	Moo75
Nûk gneiss	Godthaab townsite and Bjøneon	dioritic to tonalitic gneiss	2890···3065	U–Th–Pb, UI-zircon		Baa81
	Bjøneon	dioritic to tonalitic gneiss	3076 (27)	Rb–Sr, IA-WR	not indicated	Baa81
	Godthaab area	quartzo-feldspathic gneiss	2780···2930	Rb–Sr, IA-WR	$\lambda^{87}Rb = 1.39 \cdot 10^{-11}\,a^{-1}$	Moo76a; Pan73
	Frederikshaab area	streaky gneiss and massive tonalite	2662 (116)	Rb–Sr, IA-WR	$\lambda^{87}Rb = 1.39 \cdot 10^{-11}\,a^{-1}$	Pid75
Isua supracrustals	Isua area	garnet-bearing quartzite, matrix of rhyolitic conglomerate	≈3800	U–Pb, UI-zircon		Baa76

(continued)

Table 4 (continued)

Region Geologic/petrographic unit	Area/locality	Rock type	Age 10^6 a	Method and type of age IA isochron age WR whole rocks M minerals UI concordia upper intercept age	Constants used indicated only if not the same as those recommended by the IUGS Subcommission on Geochronology [Ste77]	Ref.
West Greenland *Isua supracrustal* (continued)	Isua area (continued)	acid boulders from conglomerate	3769^{+11}_{-8}	U–Pb, UI-zircon		Mic77
		boulders from metamorphic volcanogenic sediments	3710^{+0}_{-90}	Rb–Sr, IA-WR	$\lambda^{87}\text{Rb}=1.39\cdot10^{-11}\,\text{a}^{-1}$	Moo75
		basic metavolcanics, pebbles and matrix from acid conglomerate	3770 (42)	Sm–Nd, IA-WR	$\lambda^{147}\text{Sm}=6.54\cdot10^{-12}\,\text{a}^{-1}$	Ham78
		banded ironstones	3760 (70)	Pb–Pb, IA-WR	$\lambda^{235}\text{U}=9.72\cdot10^{-10}\,\text{a}^{-1}$ $\lambda^{238}\text{U}=1.537\cdot10^{-10}\,\text{a}^{-1}$ $^{238}\text{U}/^{235}\text{U}=137.8$	Moo73
East Greenland *'Granitic' gneiss*	Kangerdlugssuag area	heterogeneous, layered metaigneous gneisses	2980 (60)	Pb–Pb, IA-WR		Lee76
North Norway Vikan gneisses	Vesterålen islands	migmatic dioritic to granitic gneisses	3460 (70)	Pb–Pb, IA-WR	$\lambda^{235}\text{U}=9.722\cdot10^{-10}\,\text{a}^{-1}$ $\lambda^{238}\text{U}=1.537\cdot10^{-10}\,\text{a}^{-1}$	Tay75
Scotland Lewisien gneisses	Harris Island, Outer Hebrides	quartzo-feldspathic gneiss	2770 (10)	U–Pb, UI-zircon	$\lambda^{235}\text{U}=9.72\cdot10^{-10}\,\text{a}^{-1}$ $\lambda^{238}\text{U}=1.537\cdot10^{-10}\,\text{a}^{-1}$ $^{238}\text{U}/^{235}\text{U}=137.8$	Pid72
'Grey Gneiss' complex	Islands of North Uist, Benbecula, South Uist and Barra, Outer Hebrides	homogeneous and banded gneisses	2690 (60)	Rb–Sr, IA-WR	$\lambda^{87}\text{Rb}=1.39\cdot10^{-10}\,\text{a}^{-1}$	Moo75a

(continued)

Table 4 (continued)

Region Geologic/petrographic unit	Area/locality	Rock type	Age 10^6 a	Method and type of age IA isochron age WR whole rocks M minerals UI concordia upper intercept age	Constants used indicated only if not the same as those recommended by the IUGS Subcommission on Geochronology [Ste77]	Ref.
Scotland (continued) '*Uamhaig*' gneisses Scourie gneisses	NW Scotland	quartzo-feldspathic gneisses	2680 (60)	Pb–Pb, IA-WR	$\lambda^{235}U = 9.85 \cdot 10^{-10}\,a^{-1}$ $\lambda^{238}U = 1.551 \cdot 10^{-10}\,a^{-1}$	Cha77
Labrador Uviac gneiss	Saglek Bay	quartzo-feldspathic gneiss	3622 (72)	Rb–Sr, IA-WR	$\lambda^{87}Rb = 1.39 \cdot 10^{-11}\,a^{-1}$	Hur75
Hebron gneiss	Hebron Fiord	granodioritic gneiss	3618 (218)	Rb–Sr, IA-WR	$\lambda^{87}Rb = 1.39 \cdot 10^{-11}\,a^{-1}$	Bar75
Minnesota (Canadian Shield) '*Granitic*' gneisses Morton and Montevideo gneiss	Minnesota River Valley	tonalitic to quartz-monzonitic gneisses	≈ 3800	Rb–Sr, IA-WR	$\lambda^{87}Rb = 1.39 \cdot 10^{-11}\,a^{-1}$	Gol74
Greenstone-granite terrane Northern Light Gneiss	Vermilion District	leucotonalite, leuco-granodiorite, amphibolite	2704 (123)	Rb–Sr, IA-WR	$\lambda^{87}Rb = 1.39 \cdot 10^{-11}\,a^{-1}$	Han71; Jah75
Ely Greenstone	Vermilion District	basaltic to dacitic greenstones	2690 (80)	Rb–Sr, IA-WR	$\lambda^{87}Rb = 1.39 \cdot 10^{-11}\,a^{-1}$	Jah75
Saganaga Tonalite	Vermilion District	tonalite, granodiorite	2721 (147)	Rb–Sr, IA-WR	$\lambda^{87}Rb = 1.39 \cdot 10^{-11}\,a^{-1}$	Han71; Jah75
Icarus Pluton	Vermilion District	syenodiorite, tonalite, granodiorite	2690 (21)	Rb–Sr, IA-WR	$\lambda^{87}Rb = 1.39 \cdot 10^{-11}\,a^{-1}$	Han71; Jah75
Giants Range Granite	Vermilion District	granite	2670 (18)	Rb–Sr, IA-WR	$\lambda^{87}Rb = 1.39 \cdot 10^{-11}\,a^{-1}$	Jah75
Vermilion Granite	Vermilion District	granite, leucogranodiorite	2700 (50)	Rb–Sr, IA-WR	$\lambda^{87}Rb = 1.39 \cdot 10^{-11}\,a^{-1}$	Jah75; Pet72
Newton Lake Formation	Vermilion District	mafic to acidic metavolcanics	2650 (110)	Rb–Sr, IA-WR	$\lambda^{87}Rb = 1.39 \cdot 10^{-11}\,a^{-1}$	Jah75

(continued)

Grauert

Table 4 (continued)

Region Geologic/petrographic unit	Area/locality	Rock type	Age 10^6 a	Method and type of age IA isochron age WR whole rocks M minerals UI concordia upper intercept age	Constants used indicated only if not the same as those recommended by the IUGS Subcommission on Geochronology [Ste77]	Ref.
Northern Michigan (Canadian Shield)	Watersmeet	tonalitic augen gneiss	3410 ≈3600	U–Th–Pb Sm–Nd, WR (model age)	not indicated	Pet80 McC80
Guayana (Guayana Shield) Imataca Series	Guri dam, Caroni River	feldspathic granulites	≈3200 …3700	Rb–Sr, IA-WR	λ^{87}Rb$=1.39 \cdot 10^{-11}$a^{-1}	Hur76
	La Ceiba	migmatites and granitoids	≈2720	Rb–Sr, IA-WR	λ^{87}Rb$=1.39 \cdot 10^{-11}$a^{-1}	Hur76
Zimbabwe (Rhodesian Craton) Pre-Bulawayan basement	Mashaba area	banded tonalitic, quartzdioritic and granodioritic gneisses	3580 (200)	Rb–Sr, IA-WR	λ^{87}Rb$=1.39 \cdot 10^{-11}$a^{-1}	Haw75
Greenstone belts Sebakwian Mont d'or Granite	Selukwe	granite	3420 (60)	Rb–Sr, IA-WR	λ^{87}Rb$=1.39 \cdot 10^{-11}$a^{-1}	Moo76
Bulawayan Maliyami Fm.	Que Que area	andesites and basaltic volcanics	2720 (70)	Rb–Sr, IA-WR	λ^{87}Rb$=1.39 \cdot 10^{-11}$a^{-1}	Haw75
	Bulawayo area	basaltic volcanics	2540 (90)	Rb–Sr, IA-WR	λ^{87}Rb$=1.39 \cdot 10^{-11}$a^{-1}	Haw75
Sesombi Tonalite Pluton	Que Que area	intrusive tonalite	2690 (70)	Rb–Sr, IA-WR	λ^{87}Rb$=1.39 \cdot 10^{-11}$a^{-1}	Haw75
Gwenoro Gneisses	Selukwe	tonalitic to granodioritic migmatites	2780 (30)	Rb–Sr, IA-WR	λ^{87}Rb$=1.39 \cdot 10^{-11}$a^{-1}	Haw75

(continued)

Table 4 (continued)

Region Geologic/petrographic unit	Area/locality	Rock type	Age 10^6 a	Method and type of age IA isochron age WR whole rocks M minerals UI concordia upper intercept age	Constants used indicated only if not the same as those recommended by the IUGS Subcommission on Geochronology [Ste77]	Ref.
Zimbabwe/Transvaal						
Limpopo Mobile Belt						
Sand River Formation	Messina	gneisses	3858 (116)	Rb–Sr, IA-WR	$\lambda^{87}\text{Rb}=1.39\cdot10^{-11}\,\text{a}^{-1}$	Bar77
		intrusive basic dykes	3643 (102)	Rb–Sr, IA-WR	$\lambda^{87}\text{Rb}=1.39\cdot10^{-11}\,\text{a}^{-1}$	Bar77
			3128 (84)	Rb–Sr, IA-WR	$\lambda^{87}\text{Rb}=1.39\cdot10^{-11}\,\text{a}^{-1}$	Bar77
Mushandike Granit	Fort Victoria	massive granodiorite	3520 (260)	Rb–Sr, IA–WR	$\lambda^{87}\text{Rb}=1.39\cdot10^{-11}\,\text{a}^{-1}$	Hic74
Transvaal/Swaziland (Kaapvaal Craton)						
Greenstone belts (Swaziland Supergroup)						
Onverwacht Group Komati Formation	Barberton Moutain Land	basaltic komatiite	3500 (200)	Rb–Sr, IA-M	$\lambda^{87}\text{Rb}=1.39\cdot10^{-11}\,\text{a}^{-1}$	Jah74
Middle Marker Horizon	Barberton Moutain Land	banded shales	3355 (70)	Rb–Sr, IA-WR	$\lambda^{87}\text{Rb}=1.39\cdot10^{-11}\,\text{a}^{-1}$	Hur72
Fig Tree Group	Barberton Moutain Land	shales and graywackes	2980 (40)	Rb–Sr, IA-WR	$\lambda^{87}\text{Rb}=1.39\cdot10^{-11}\,\text{a}^{-1}$	All68
Surrounding granites and gneisses						
Ancient gneiss complex	Barberton Moutain Land	tonalite gneiss	3440 (300)	Rb–Sr, IA-WR	$\lambda^{87}\text{Rb}=1.39\cdot10^{-11}\,\text{a}^{-1}$	All62
Nelspruit gneisses	Barberton Moutain Land	granite gneisses and migmatites	3000···3400	Rb–Sr, IA-WR	$\lambda^{87}\text{Rb}=1.39\cdot10^{-11}\,\text{a}^{-1}$	All68

(continued)

Table 4 (continued)

Region Geologic/petrographic unit	Area/locality	Rock type	Age 10⁶ a	Method and type of age: IA isochron age; WR whole rocks; M minerals; UI concordia upper intercept age	Constants used indicated only if not the same as those recommended by the IUGS Subcommission on Geochronology [Ste77]	Ref.
Western Australia (Pilbara Block) *Granite-greenstone terrane (Pilbara Supergroup)*						
North Star Basalt (Waarawoona Group)	NE of Marble Bar	dacite	3570 (180)	Rb−Sr, IA-WR		Jah81
			3556 (542)	Sm−Nd, IA-WR	$\lambda^{147}Sm = 6.54 \cdot 10^{-12}\,a^{-1}$	Jah81
Duffer Formation	Glen Herring area	columnar dacite	3452 (16)	U−Pb, UI-zircon		Pid78a
		silicic lava	3230 (280)	Rb−Sr, IA-WR		Jah81

Table 5. **Geologic time scale**

Eon	Era	Period		Epoch	Age	Age [10⁶ a] 1)	2)	3)	4)
Phanerozoic	Cenozoic	Quarternary		Holocene Pleistocene		1.5···2			
		Tertiary	Neogene	Pliocene Miocene		7 26			
			Paleogene	Oligocene Eocene Paleocene		37···38 53···54 65	65	66	
	Mesozoic	Cretaceous		Upper	Maastrichtian Campanian Santonian Coniacian Turonian Cenomanian	70 76 82 88 94 100	72 83 88 90 93 ≧96		73 90 95
				Lower	Albian Aptian Barremian Hauterivian Valanginian Berriasian	106 112 118 124 130 136	117 123 126 130 136 143	132	107 112 117 129
		Jurassic		Upper (Malm)	Portlandian Kimmeridgian Oxfordian	146 151 157	149 158 162		
				Middle (Dogger)	Callovian Bathonian Bajocian Aalenian	162 167 172	166 177		
				Lower (Lias)	Taorcian Pliensbachian Sinemurian Hettangian	178 183 188 190···195	187 198 212	185	
		Triassic		Upper	Rhaetian Norian Carnian	205?	220 228 233		
				Middle	Ladinian Anisian	215?	238 242		
				Lower	Skythian	225	≧247	235	

(continued)

1) Harland et al. (1964) [Har64].
2) Armstrong (1978) [Arm78] } Ages calculated or recalculated with decay constants
3) Afanas'yev and Zykov (1975) [Afa75] } recommended by IUGS Subcommission on Geochronology [Ste77].
4) Odin (1978) [Odi78].
5) Canadian shield [Eys78].
6) Ukrainian shield [Sem68].

Table 5 (continued)

Eon	Era	Period	Epoch	Age	Age [10⁶ a]			
					^{1})	^{2})	^{3})	^{4})
Phanero-zoic (con-tinued)	Paleo-zoic	Permian	Upper	Tatarian Kazanian	225 230 240	≧ 247 252 258	235	
			Lower	Kungurian Artinskian Sakmarian	255···258 265···268 280	269 278 289	280	
		Carboniferous *Pennsylvanian* ——— *Mississippian*	Upper	Stephanian Westphalian Namurian	290···295 310···315 325	308 330 341		
			Lower	Visean Tournaisian	335···340 345	355 367	345	
		Devonian	Upper	Famennian Frasnian	353 359	379 385		
			Middle	Givetian Couvinian	370	391 395		
			Lower	Emsian Siegenian Gedinnian	374 390 395	400 405 416	400	
		Silurian	Upper	Ludlovian Wenlockian		432 440		
			Lower	Llandoverian	430···440		435	
		Ordovician	Upper	Ashgilian Caradocian	445	455 465		
			Lower	Llandeilian Llanvirnian Arenigian Tremadocian	≈ 500	476 491 ≧ 509	490	
		Cambrian	Upper	Shidertinian Tuorian	515	525		
			Middle	Mayan Amgan Lenan	540	545		
			Lower	Aldanian	≈ 575 570	≈ 575	570	

Eon	Era	Period	Epoch	Age		Age [10⁶ a]			
Protero-zoic	Pre-cambrian	Upper	Algonkian	Hadrynian	^{5}) ≈ 1000	Precambrian V	1200	^{6})
				Helikian		Precambrian IV	1700	
		Middle		Aphebian ≈ 1700	≈ 1800	Precambrian III	2000	
						Precambrian II	2700	
Archaeo-zoic Azoic		Lower	Archean	≈ 2700	≈ 2700	Precambrian I	3500	

7.2.4 References for 7.2 — Literatur zu 7.2

a) Books, reviews and geologic time tables

coh78   Cohee, G.V., Glaessner, M.F., Hedberg, H.-D. (eds.): Contributions to the geologic time scale. Studies in Geology **6**, Am. Ass. Petrol. Geol., Tulsa, Oklahoma, **1978**, 388 p.

doe70   Doe, B.R.: Lead isotopes. New York-Heidelberg-Berlin: Springer, **1970**, 137 p.

eys78   Van Eysinga, F.W.B.: Geological time table, 3rd. ed., Amsterdam: Elsevier, **1978**.

fau72   Faure, G., Powell, J.L.: Strontium isotope geology. New York-Heidelberg-Berlin: Springer **1972**, 188 p.

fau77   Faure, G.: Principles of isotope geology. New York: John Wiley and Sons, **1977**, 464 p.

gol74   Goldberg, E.D., Bruland, K.: Radioactive geochronologies. In: The sea, Vol. 5: Marine chemistry, E.D. Goldberg (ed.), New York: John Wiley and Sons, **1974**, 451–489.

har64   Harland, W.B., Francis, E.H. (eds.): The Phanerozoic time-scale; a supplement: Geol. Soc. London Geol. Soc. London Quart. J. **120** suppl. (1964) 458 p.

har71   Harland, W.B., Francis, E.H. (eds.): The Phanerozoic time-scale; a supplement: Geol. Soc. London Spec. Publ. **5** (1971) 1–120.

jäg79   Jäger, E., Hunziker, J.C. (eds.): Lectures in isotope geology. Berlin-Heidelberg-New York: Springer, **1979**.

kir78   Kirsten, T.: Time and the solar system. In: The origin of the solar system, S.F. Dermott (ed.), Chichester: John Wiley and Sons, **1978**, 267–346.

b) Special references

Ado76   Adorables, E.: Progress by the consorts of Angra dos Reis. Lunar Sci. **VII** (1976) 443–445.

Afa75   Afanas'yev, G.D., Zykov, S.I.: Phanerozoic geochronological time scale in the light of significantly new decay constants. Moscow: Nauka, **1975**, 100 p.

Ald56   Aldrich, L.T., Wetherill, G.W., Tilton, G.R., Davis, G.L.: The half life of ^{87}Rb. Phys. Rev. **104** (1956) 1045–1047.

Ald58   Aldrich, L.T., Wetherill, G.W.: Geochronology by radioactive decay. Ann. Rev. Nucl. Sci. **8** (1958) 257–298.

All62   Allsopp, H.L., Roberts, H.R., Schreiner, G.D.L., Hunter, D.R.: Rb−Sr age measurements on various Swaziland granites. J. Geophys. Res. **67** (1962) 5307–5313.

All68   Allsopp, H.L., Ulrych, T.J., Nicolaysen, L.O.: Dating some significant events in the history of the Swaziland system by the Rb−Sr method. Can. J. Earth Sci. **5** (1968) 605–619.

All68a   Allègre, C.J.: ^{230}Th dating of volcanic rocks: A comment. Earth Planet. Sci. Lett. **5** (1968) 209–210.

Ami66   Amin, B.S., Kharkar, D.P., Lal, D.: Cosmogenic ^{10}Be and ^{26}Al in marine sediments. Deep-Sea Res. **13** (1966) 805.

Arm78   Armstrong, R.L.: Pre-Cenozoic Phanerozoic time scale. Computer file of critical dates and consequences of new and in-progress decay-constant revisions. In: Contributions to the geologic time scale, Cohee, G.H., Glaessner, M.F., Hedberg, H.D. (eds.). Studies in Geology **6**, Am. Ass. Petrol. Geol., Tulsa, Oklahoma, **1978**, 73–91.

Baa73   Baadsgaard, H.: U−Th−Pb dates on zircons from the early Precambrian Amîtsoq gneisses, Godthaab district, West Greenland. Earth Planet. Sci. Lett. **19** (1973) 22–28.

Baa76   Baadsgaard, H.: Further U−Pb dates on zircons from the early Precambrian rocks of the Godthåbsfjord area, West Greenland. Earth Planet. Sci. Lett. **33** (1976) 261–267.

Baa81   Baadsgaard, H., McGregor, V.R.: The U−Th−Pb systematics of zircons from the type Nûk gneisses, Godthaabsfjord, West Greenland. Geochim. Cosmochim. Acta **45** (1981) 1099–1109.

Bar75   Barton, J.M., jr.: Rb−Sr isotopic characteristics and chemistry of the 3.6-b.y. Hebron gneiss, Labrador. Earth Planet. Sci. Lett. **27** (1975) 427–435.

Bar77   Barton, J.M., Fripp, R.E.P., Ryan, B.: Rb/Sr ages and geological setting of ancient dykes in the Sand River area, Limpopo mobile belt, South Africa. Nature **267** (1977) 487–490.

Bec69   Beckinsale, R.D., Gale, N.H.: A reappraisal of the decay constants and branching ratio of ^{40}K. Earth Planet. Sci. Lett. **6** (1969) 289–294.

Ber75   Berger, G.W.: ^{40}Ar/^{39}Ar step heating of thermally overprinted biotite, hornblende and potassium feldspar from Eldora, Colorado. Earth Planet. Sci. Lett. **26** (1975) 387–408.

Bla71   Black, L.P., Gale, N.H., Moorbath, S., Pankhurst, R.J., McGregor, V.R.: Isotopic dating of very early Precambrian amphibolite facies gneisses from the Godthaab district, West Greenland. Earth Planet. Sci. Lett. **12** (1971) 245–259.

Bog67	Bogard, D., Burnett, D., Eberhardt, P., Wasserburg, J.: $^{40}Ar-^{40}K$ ages of silicate inclusions in iron meteorites. Earth Planet. Sci. Lett. **3** (1967) 275–283.
Bro56	Broecker, W.S., Kulp, J.L.: The radiocarbon method of age determination. Am. Antiq. **22** (1956) 1–11.
Cal73	Calk, L.C., Naeser, C.W.: The thermal effect of a basalt intrusion on fission tracks in quartz monzonite. J. Geol. **81** (1973) 189–198.
Cat69	Catanzaro, E.J., Murphy, T.J., Garner, E.L., Shields, W.R.: Absolute isotopic abundance ratio and atomic weight of terrestrial rubidium. J. Res. U.S. Nat. Bur. Std., Sect. A **73A** (1969) 511–516.
Cha77	Chapman, H., Moorbath, S.: Lead isotope measurements from the oldest recognized Lewisian gneisses of northwest Scotland. Nature **268** (1977) 41–42.
Che76	Chen, J.H., Tilton, G.R.: Isotopic lead investigations on the Allende carbonaceous chondrite. Geochim. Cosmochim. Acta **40** (1976) 635–643.
Cla79	Clauer, N.: A new approach to Rb–Sr dating of sedimentary rocks. In: [jäg79], p. 30–51.
Com65	Compston, W., Lovering, J.F., Vernon, M.J.: The rubidium-strontium age of the Bishopville aubrite and its component enstatite and feldspar. Geochim. Cosmochim. Acta **29** (1965) 1085–1099.
Cre76	Cressey, P.J., Bogard, D.D.: On the calculation of cosmic-ray exposure ages of stone meteorites. Geochim. Cosmochim. Acta **40** (1976) 749–762.
Cum75	Cumming, G.L., Richard, J.R.: Ore lead isotope ratios in a continuously changing earth. Earth Planet. Sci. Lett. **28** (1975) 155–171.
Dal79	Dalmeyer, R.D.: $^{40}Ar/^{39}Ar$ dating: Principles, techniques, and application in orogenic terranes. In: [jäg79], p. 77–104.
Dod79	Dodson, M.H.: Theory of cooling ages. In: [jäg79], p. 194–202.
Eme62	Emery, K.O., Bray, E.E.: Radiocarbon dating of California basin sediments. Am. Ass. Petrol. Geol. Bull. **46** (1962) 1839–1856.
Fan65	Fanale, F.P., Schaeffer, O.A.: The helium-uranium method for dating marine carbonates. Science **149** (1965) 312–317.
Fle52	Fleming, E.H., Ghiroso, A., Cunningham, B.B.: The specific alpha-activities and half-lifes of ^{234}U, ^{235}U and ^{236}U. Phys. Rev. **88** (1952) 642–652.
Fle64	Fleischer, R.L., Price, P.B: Decay constant for spontaneous fission of ^{238}U. Phys. Rev. **133** (1964) 1363–1364.
Fly59	Flynn, K.F., Glendenin, L.E.: Half-life and beta spectrum of ^{87}Rb. Phys. Rev. **116** (1959) 744–748.
Gal70	Galliker, D., Hugentobler, E., Hahn, B.: Spontane Kernspaltung von ^{238}U und ^{241}Am. Helv. Phys. Acta **43** (1970) 593.
Gal75	Gale, N., Arden, J., Hutchinson, R.: The chronology of the Nakhla achondritic meteorite. Earth Planet. Sci. Lett. **26** (1975) 195–206.
Gar76	Garner, E.L., Murphy, T.J., Gramlich, J.W., Paulsen, P.J., Barnes, I.L.: Absolute isotopic abundance ratios and atomic weight of a reference sample of potassium. J. Res. U.S. Nat. Bur. Std., Sect. A **79A** (1976) 713–725.
Geb79	Gebauer, D., Grünenfelder, M.: U–Th–Pb dating of minerals. In [jäg79], p. 105–131.
God62	Godwin, H.: Half life of radiocarbon. Nature **195** (1962) 984.
Gol68	Goldberg, E.D.: Ionium/thorium geochronologies. Earth Planet. Sci. Lett. **4** (1968) 17–21.
Gol74	Goldich, S.S., Hedge, C.E.: 3800 Myr granitic gneiss in south-western Minnesota. Nature **252** (1974) 467–468.
Gop70	Gopalan, K., Wetherill, G.W.: Rubidium-strontium studies on enstatite chondrites: Whole meteorite and mineral isochrons. J. Geophys. Res. **75** (1970) 3457–3467.
Gra73	Gray, C.M., Papanastassiou, D.A., Wasserburg, G.J.: The identification of early condensates from the solar nebula. Icarus **20** (1973) 213–239.
Gra73a	Grauert, B., Hofmann, A.: Old radiogenic lead components in zircons from the Idaho batholith and its metasedimentary aureole. Carnegie Inst. Wash., Yearbook **72** (1973) 297–299.
Ham78	Hamilton, P.J., O'Nions, H.K., Evensen, N.M., Bridgwater, D., Allaart, J.H.: Sm–Nd isotopic investigations of Isua supracrustals and implications for mantle evolution. Nature **272** (1978) 41–43.
Han71	Hansen, G.N., Goldich, S.S., Arth, J.G., Yardley, D.H.: Age of the early Precambrian rocks of the Saganaga Lake, Northern Light Lake area, Ontario-Minnesota. Can. J. Earth Sci. **8** (1971) 1110–1124.

Har64	Hart, S.R.: The petrology and isotopic-mineral age relations of a contact zone in the Front Range, Colorado. J. Geol. **72** (1964) 493–525.
Haw75	Hawkesworth, C.J., Moorbath, S., O'Nions, R.K., Wilson, J.F.: Age relationships between green-stone belts and "granites" in the Rhodesian Archean craton. Earth Planet. Sci. Lett. **25** (1975) 251–262.
Hic74	Hickman, M.H.: 3500-Myr-old granite in southern Africa. Nature **251** (1974) 295–296.
Hir63	Hirt, B., Tilton, G.R., Herr, W., Hoffmeister, W.: The half-life of ^{187}Re. In: Earth Science and Meteorites; J. Geiss, E.D. Goldberg (eds.). Amsterdam: North Holland **1963**, p. 273–280.
Hug64	Hughes, E.E., Mann, W.B.: The half-life of carbon-14, comments on the mass-spectrometric method. Int. J. Appl. Rad. Isotop. **15** (1964) 97–100.
Hun70	Hunziker, J.C.: Polymetamorphism in the Monte Rosa, Western Alps. Eclogae. Geol. Helv. **63** (1970) 151–161.
Hun79	Hunziker, J.C.: Potassium argon dating. In: [jäg79], p. 52–76.
Hur72	Hurley, P.M., Pinson, W.H., jr., Nagy, B., Teska, T.M.: Ancient age of the Middle Marker Horizon: Onverwacht Group, Swaziland Sequences, South Africa. Earth Planet. Sci. Lett. **14** (1972) 360–366.
Hur75	Hurst, R.W., Bridgwater, D., Collerson, K.D., Wetherill, G.W.: 3600 m.y. Rb−Sr ages from very early Archean gneisses from Skaglek bay, Labrador. Earth Planet. Sci. Lett. **27** (1975) 393–403.
Hur76	Hurley, P.M., Fairbairn, H.W., Gaudette, H.E.: Progress report on early Archean rocks in Liberia, Sierra Leone, and Guayana, and their general stratigraphic setting. In: The early history of the earth, B.F. Windley (ed.), London: Wiley **1976**, p. 511–521.
Jac79	Jacobsen, S.B., Wasserburg, G.J.: Nd and Sr isotopic study of the Bay of Island ophiolite complex and the evolution of the source of midocean ridge basalts. J. Geophys. Res. **84** (1979) 7429–7445.
Jäg79	Jäger, E.: The Rb−Sr method. In: [jag79], p. 13–26.
Jaf71	Jaffey, A.H., Flynn, K.F., Glendenin, L.E., Bentley, W.C., Essling, A.M.: Precision measurements of half-lifes and specific activities of ^{235}U and ^{238}U. Phys. Rev. C **4** (1971) 1889–1906.
Jah74	Jahn, B.-M., Shih, C.-Y.: On the age of the Onverwacht group, Swaziland sequence, South Africa. Geochim. Cosmochim. Acta **38** (1974) 873–885.
Jah75	Jahn, B.-M., Murthy, V.R.: Rb−Sr ages of the Archean rocks from the Vermilion district, north-eastern Minnesota. Geochim. Cosmochim. Acta **39** (1975) 1679–1689.
Jah81	Jahn, B.-M., Glikson, A.Y., Peucat, J.J., Hickman, A.H.: REE geochemistry and isotopic data of Archean silicic volcanics and granitoids from the Pilbara Block, western Australia: implications for the early crustal evolution. Geochim. Cosmochim. Acta **45** (1981) 1633–1652.
Jes80	Jessberger, E.K., Dominik, B., Staudacher, Th., Herzog, G.F.: ^{40}Ar−^{39}Ar ages of Allende. Icarus **42** (1980) 380.
Kau71	Kaufman, A., Broecker, W.S., Ku, T.L., Thurber, D.L.: The status of U-series methods of mollusk dating. Geochim. Cosmochim. Acta **35** (1971) 1155–1183.
Kir63	Kirsten, T., Krankowsky, D., Zähringer, J.: Edelgas- und Kalium-Bestimmungen an einer größeren Zahl von Steinmeteoriten. Geochim. Cosmochim. Acta **27** (1963) 13–43.
Kir71	Kirsten, T., Schaeffer, O.A.: High interactions in space. In: L.C. Yuan (ed.), Elementary Particles; Science, Technology and Society, New York: Academic Press, **1971**, p. 76–157.
Kir81	Kirsten, T.: Chronology of the solar system. In: Landolt-Börnstein, New Series, Vol. VI/2a, H. Voigt, K. Schaiffers (eds.), Berlin: Springer, **1981**, p. 273–285.
Koh63	Kohmann, T.P., Goel, P.S.: Terrestrial ages of meteorites from cosmogenic ^{14}C. Radioactive Dating, IAEA Wien, **1963**, p. 395–411.
Koi73	Koide, M., Bruland, K.W., Goldberg, E.D.: Th-228/Th-232 and Pb-210 geochronologies in marine and lake sediments. Geochim. Cosmochim. Acta **37** (1973) 1171–1187.
Ku65	Ku, T.L.: An evaluation of the U-234/U-238 method as a tool for dating pelagic sediments. J. Geophys. Res. **70** (1965) 3457–3474.
Ku67	Ku, T.L., Broecker, W.S.: Rates of sedimentation in the Arctic Ocean. In: Progress in Oceanography, Vol. 4, Oxford: Pergamon Press, **1967**, p. 95–104.
Ku68	Ku, T.L.: Protactinium-231 method of dating coral from Barbados Island. J. Geophys. Res. **73** (1968) 2271–2276.
Ku69	Ku, T.L., Broecker, W.S.: Radiochemical studies of manganese nodules of deep sea origin. Deep-Sea Res. **16** (1969) 625–637.
Lee76	Leeman, W.P., Dasch, E.J., Kays, M.A.: ^{207}Pb/^{206}Pb whole-rock age of gneisses from the Kangerdlugssuaq area, eastern Greenland. Nature **263** (1976) 469–471.

LeR63 Le Roux, L.J., Glendenin, L.E.: Half-life of ^{232}Th. Proc. Nat. Meet. On Nuclear Energy, Pretoria, South Africa, April **1963**, p. 83–94.

Lug75 Lugmair, G.W., Scheinin, N., Marti, K.: Search for extinct ^{146}Sm. 1. The isotopic abundance of ^{142}Nd in the Juvinas meteorite. Earth Planet. Sci. Lett. **27** (1975) 79–84.

Lug76 Lugmair, G., Kurtz, J., Marti, K., Scheinin, N.: The low-Sm/Nd region on the Moon: Evolution and history of troctolite and a KREEP basalt. Lunar Sci. **VII** (1976) 509–511.

Lug78 Lugmair, G.W., Marti, K.: Lunar initial ^{143}Nd/^{144}Nd: differential evolution of the lunar crust and mantle. Earth Planet. Sci. Lett. **39** (1978) 349–357.

McC78 McCulloch, M.T., Wasserburg, G.J.: Sm $-$ Nd and Rb $-$ Sr chronology of continental crust formation. Science **200** (1978) 1003–1011.

McC80 McCulloch, M.T., Wasserburg, G.J.: Sm $-$ Nd model ages from an early Archean tonalite gneiss of northern Michigan. Geol. Soc. Am. Spec. Papers **182** (1980) 135–138.

Mic77 Michard-Vitrac, A., Lancelot, J., Allègre, C.J., Moorbath, S.: U $-$ Pb ages on single zircons from the early Precambrian rocks of West Greenland and the Minnesota River Valley. Earth Planet. Sci. Lett. **35** (1977) 449–453.

Moo72 Moorbath, S., O'Nions, R.K., Pankhurst, R.J., Gale, N.H., McGregor, V.R.: Further rubidium-strontium age determinations on the very early Precambrian rocks of the Godthaab district, West Greenland. Nature Phys. Sci. **240** (1972) 78–82.

Moo72a Moore, W.S., Krishnaswamy, S.: Coral growth rates using Ra-228 and Pb-210. Earth Planet. Sci. Lett. **15** (1972) 187–190.

Moo73 Moorbath, S., O'Nions, R.K., Pankhurst, R.J.: Early Archean age of the Isua iron formation. Nature **245** (1973) 138–139.

Moo75 Moorbath, S., O'Nions, R.K., Pankhurst, R.J.: The evolution of early Precambrian crustal rocks of Isua, West Greenland: geochemical and isotopic evidence. Earth Planet. Sci. Lett. **27** (1975) 229–239.

Moo75a Moorbath, S., Powell, J.L., Taylor, P.N.: Isotopic evidence for the age and origin of the "grey gneiss" complex of southern Outer Hebrides, Scotland. J. Geol. Soc. London **131** (1975) 213–222.

Moo76 Moorbath, S., Wilson, J.F., Cotterill, P.: Early Archean age for the Sebakwian Group at Selukwe, Rhodesia. Nature **244** (1976) 536–538.

Moo76a Moorbath, S., Pankhurst, R.J.: Further Rb $-$ Sr evidence for the nature of the late Archean plutonic event in West Greenland. Nature **262** (1976) 124–126.

Mur62 Murthy, V.R., Patterson, C.C.: Primary isochron of zero age for meteorites and the earth. J. Geophys. Res. **67** (1962) 1161–1167.

Mur65 Murthy, V.R., Compston, W.: Rb $-$ Sr ages of chondrules and carbonaceous chondrites. J. Geophys. Res. **70** (1965) 5297–5307.

Nae79 Naeser, C.W.: Fission-track dating and geologic annealing of fission tracks. In: [jäg79], p. 154–169.

Neu74 Neumann, W., Huster. H.: The half-life of ^{87}-Rb measured as a difference between the isotopes ^{87}Rb and ^{85}Rb. Z. Physik **270** (1974) 121–127.

Nie50 Nier, A.O.: A redetermination of the relative abundance of the isotopes of carbon, nitrogen, oxygen, argon and potassium. Phys. Rev. **77** (1950) 789–793.

Obr68 Obradovich, J.D., Peterman, Z.E.: Geochronology of the Belt series, Montana. Can. J. Earth. Sci. **5** (1968) 737–747.

Odi78 Odin, G.S.: Results of dating Cretaceous, Paleogene Sediments, Europe. In: Contributions to the geologic time scale, G.V. Cohee, M.F. Glaessner, H.D. Hedberg (eds.). Studies in Geology **6**, Am. Ass. Petrol. Geol., Tulsa, Oklahoma **1978**, p. 127–141.

Pan73 Pankhurst, R.J., Moorbath, S., McGregor, V.R.: Late event in the geologic evolution of the Godthaab district, West Greenland, Nature Phys. Sci. **243** (1973) 24–26.

Pap75 Papanastassiou, D., Wasserburg, G.J.: Rb $-$ Sr study of a lunar dunite and evidence for early lunar differentiates. Proc. Lunar Sci. Conf. 6th, Vol. 2 (1975) 1467–1489.

Pap76 Papanastassiou, D., Wasserburg, G.J.: Early lunar differentiates and lunar initial ^{87}Sr/^{86}Sr. Lunar Sci. **VII** (1976) 665–667.

Pat80 Patchett, P.J., Tatsumoto, M.: Lu $-$ Hf total-rock isochron for the eucrite meteorites. Nature **288** (1980) 571–574.

Pet72 Peterman, Z.E., Goldich, S.S., Hedge, C.E., Yardley, D.H.: Geochronology of the Rainy Lake region, Minnesota-Ontario. Geol. Soc. Am. Mem. **135** (1972) 193–215.

Pet80 Peterman, Z.E., Zartman, R.E., Sims, P.K.: Tonalitic gneiss of early Archean age from northern Michigan. Geol. Soc. Am. Spec. Papers **182** (1980) 125–134.

Pet81	Pettingill, H.S., Pattchet, P.J.: Lu − Hf total-rock age for the Amîtsoq gneisses, West Greenland. Earth Planet. Sci. Lett. **55** (1981) 150–156.
Pid72	Pidgeon, R.T., Aftalion, M.: The geochronological significance of discordant U − Pb ages of oval-shaped zircons from a Lewisian gneiss from Harris, Outer Hebrides. Earth Planet. Sci. Lett. **17** (1972) 269–274.
Pid75	Pidgeon, R.T., Hopgood, A.M.: Geochronology of Archean gneisses and tonalites from north of Frederikshåbs isblink, S.W. Greenland. Geochim. Cosmochim. Acta **39** (1975) 1333–1346.
Pid78a	Pidgeon, R.T.: 3450-m.y.-old volcanics in the Archean layered greenstone succession of the Pilbara block, Western Australia. Earth Planet. Sci. Lett. **37** (1978) 421–428.
Pid78a	Pidgeon, R.T., Aftalion, M.: Congenetic and inherited zircon U − Pb systems in granites: Palaeozoic granites of Scotland. Geol. J., Spec. Issue **10** (1978) 183–220.
Pod70	Podosek, F.A.: Dating of meteorites by the high temperature release of iodine-correlated ^{129}Xe. Geochim. Cosmochim. Acta **34** (1970) 341–365.
Pod70a	Podosek, F.A.: The abundance of ^{244}Pu in the early solar system. Earth Planet. Sci. Lett. **8** (1970) 183–187.
Pod73	Podosek, F.A.: Thermal history of the nakhlites by the ^{40}Ar/^{39}Ar method. Earth Planet. Sci. Lett. **19** (1973) 135–144.
Rey60	Reynolds, J.H.: Determination of the age of the elements. Phys. Rev. Lett. **4** (1960) 8–10.
Ros61	Rosholt, J.N., Emiliani, C., Geiss, J., Koczy, F.F., Wangersky, P.J.: Absolute dating of deep-sea cores by the Pa-231/Th-230 method. J. Geol. **69** (1961) 162–185.
Ros73	Rosholt, J.N., Zartman, R.E., Nkomo, I.T.: Lead isotope systematics and uranium depletion in the Granite Mountains, Wyoming. Geol. Soc. Am. Bull. **84** (1973) 989–1002.
Sam63	Sammet, F., Herr, W.: Studies on the cosmic-ray produced nuclides ^{10}Be, ^{26}Al and ^{36}Cl in iron meteorites. Radioactive Dating, IAEA Wien, **1963**, p. 343–354.
San70	Sanz, H.G., Burnett, D.S., Wasserburg, G.J.: A precise ^{87}Rb/^{87}Sr age and initial ^{87}Sr/^{86}Sr for the Colomera iron meteorite. Geochim. Cosmochim. Acta **34** (1970) 1227–1239.
Sch76	Schonfeld, E.: Rb − Sr evolution of the lunar crust. Lunar Sci. **VII** (1976) 773–775.
See74	Seelmann-Eggebert, W., Pfennig, G., Münzel, H.: Karlsruher Nuklidkarte (4. Aufl.). Ges. Kernforsch., Karlsruhe, **1974**.
Sem68	Semenenko, N.P., Scherbak, A.P., Vinogradov, A.P., Tugarinov, A.I., Eliseeva, G.D., Cotlovskay, F.I., Demidenko, S.G.: Geochronology of the Ukrainian Precambrian. Can. J. Earth Sci. **5** (1968) 661–671.
Sen56	Senftle, F.E., Farley, T.A., Lazar, N.: Half-life of ^{232}Th and the branching ratio of ^{212}Bi. Phys. Rev. **104** (1956) 1629.
Sta75	Stacey, J., Kramers, J.: Approximation of terrestrial lead isotope evolution by a two-stage model. Earth Planet. Sci. Lett. **26** (1975) 207–221.
Ste77	Steiger, R.H., Jäger, E.: Subcommission on geochronology: Convention on the use of decay constants in geo- and cosmo-chronology. Earth Planet. Sci. Lett. **36** (1977) 359–362.
Str58	Strominger, D., Hollander, J.M., Seaborg, G.T.: Table of isotopes. Rev. Mod. Phys. **30** (1958) 585–904.
Tad67	Taddeuci, A.W., Broecker, W.S., Thurber, D.L.: ^{230}Th dating of volcanic rocks. Earth Planet. Sci. Lett. **3** (1967) 338–342.
Tat70	Tatsumoto, M.: Age of the Moon: an isotopic study of U − Th − Pb systematics of Apollo 11 lunar samples. Proc. Lunar Sci. Conf. 1st., Vol. 2 (1970) 1595–1612.
Tat73	Tatsumoto, M., Knight, R.J., Allègre, C.J.: Time differences in the formation of meteorites as determined from the ratio of lead-207 to lead-206. Science **180** (1973) 1279–1283.
Tat76	Tatsumoto, M., Unruh, D.M., Desborough, G.A.: U − Th − Pb and Rb − Sr systematics of Allende and U − Th − Pb systematics of Orgueil. Geochim. Cosmochim. Acta **40** (1976) 617–634.
Tay69	Taylor, G.J., Heymann, D.: Shock, reheating and the gas retention ages of chondrites. Earth Planet. Sci. Lett. **7** (1969) 151–161.
Tay75	Taylor, P.N.: An early Precambrian age for migmatitic gneisses from Vikan I Bø, Vesterålen, North Norway. Earth Planet. Sci. Lett. **27** (1975) 35–42.
Til73	Tilton, G.R.: Isotopic lead ages of chondritic meteorites. Earth Planet. Sci. Lett. **19** (1973) 321–329.
Tur69	Turner, G.: Thermal histories of meteorites by the ^{39}Ar/^{40}Ar method. In: Millman (ed.), Meteorite Research, Dordrecht: Reidel, **1969**, p. 407–417.
Tur77	Turner, G.: Potassium-argon chronology of the Moon. Phys. Chem. Earth **10** (1977) 145–195.
Vee66	Veeh, H.H.: Th-230/U-238 and U-234/U-238 ages of Pleistocene high sea level stand. J. Geophys. Res. **71** (1966) 3379–3386.

Grauert

Wag75	Wagner, G.A., Reimer, G.M., Carpenter, D.S., Faul, H., Van der Linden, R., Gijbels, R.: The spontaneous fission rate of U-238 and fission track dating. Geochim. Cosmochim. Acta **39** (1975) 1279–1286.
Wag77	Wagner, G.A., Reimer, G.M., Jäger, E.: Cooling ages derived by apatite fission-track, mica Rb−Sr and K−Ar dating: The uplift and cooling history of the Central Alps. Mem. Ist. Geol Min. Univ. Padova **30** (1977) 1–27.
Wag79	Wagner, G.A.: Correction and interpretation of fission track ages. In: [jäg79], p. 170–177.
Was81	Wasserburg, G.J., Jacobsen, S.B., DePaolo, D.J., McCullochs, M.T., Wen, T.: Precise determination of Sm/Nd ratios, Sm and Nd isotopic abundances in standard solutions. Geochim. Cosmochim. Acta **45** (1981) 2311–2323.
Wea76	Weast, R.C. (ed.): Handbook of chemistry and physics (57th ed.). Cleveland: Chemical Rubber Company Press, **1976**.

Two-dimensional survey of contents

The numbers in the survey are the page numbers of the respective subvolumes a and b of V/1 where the specified information is to be found.

Physical properties of rocks Physikalische Eigenschaften der Gesteine		Chemical elements, special materials Chemische Elemente, spezielle Materialien	Minerals Minerale
Nomenclature, chemical and mineralogical components of rocks Nomenklatur, chemische und mineralogische Komponenten der Gesteine		a 1, a 3	a 3
Density Dichte	normal conditions	a 119	a 66
	extreme conditions	a 129, a 131, a 180	a 135···a 149
Mean atomic weight, Mittleres Atom-Gewicht			b 34, b 79
Porosity, Porosität	normal conditions		
	extreme conditions		
Permeability, internal surfaces, capillarity Permeabilität, interne Oberflächen, Kapillarität			
Wave velocity and constants of elasticity Wellen-Geschwindigkeiten und Elastizitäts-Konstanten			
	normal conditions	b 21	
	extreme conditions		b 109
Inelasticity, Inelastizität			
Thermal conductivity, specific heat Wärme-Leitfähigkeit, spezifische Wärme			a 310
Thermal expansion, Thermische Ausdehnung			
Parameter of melting processes Parameter der Schmelzprozesse			
Radiogenic heat, Radiogene Wärme			
Electrical conductivity Elektrische Leitfähigkeit	normal conditions	b 241, b 262	b 241
	extreme conditions		b 291
Dielectric constant, Dielektrizitäts-Konstante		b 254	b 254
Magnetic properties, Magnetische Eigenschaften		b 308	b 308
Optical properties, Optische Eigenschaften			
Radioactivity, Radioaktivität		b 444	b 467
Absolute ages, Absolute Alter			

Zwei-dimensionale Inhaltsübersicht

Die Ziffern in der Übersicht sind die Seitenzahlen des jeweiligen Teilbandes a und b von V/1, in dem die spezielle Information zu finden ist.

Igneous rocks Magmatische Gesteine	Metamorphic rocks Metamorphe Gesteine	Sediments Sedimente	Ice Eis	Moon Mond
a 3, a 157, b 101	a 26, a 169, b 101	a 44, a 174, b 101		a 158, b 508
a 114, a 115, a 116, b 1, b 17	a 116, b 1, b 17	a 118, a 184, a 189, b 1, b 17	b 484	b 513
a 157	a 169	a 174		
b 101	b 101			
	a 184	a 184, b 107		b 513
		a 188		
		a 267, a 278		
b 8	b 8	b 8	b 496	b 514, b 525
b 35, b 99	b 35, b 99	b 35, b 99		b 514, b 525
b 141	b 141	b 141	b 501	b 523
a 315	a 321	a 325, a 344	b 484	b 529
b 111	b 111		b 485	b 536
a 21, a 345, b 137			b 484	
a 353, b 433	a 353, b 433	a 353, b 433		
b 245	b 247	b 248, b 276	b 487	b 538
b 291				
b 257	b 260	b 258	b 487	b 538
b 366	b 366	b 366		b 548
			b 493	
b 436, b 446	b 441, b 460	b 443, b 462		
b 561, b 585	b 561, b 585	b 561, b 585		b 585

Subject index of Vol. V/1 a, b and of other LB volumes

Subject	Object or substance	LB, NS *) Vol., first page	Year of publication
Abundances			
of elements	chondrites	VI/2a, p. 263	1981
	lunar surface	VI/2a, p. 258	1981
	meteorites	VI/2a, p. 258	1981
	solar photosphere	VI/2a, p. 263	1981
	solar system	VI/2a, p. 257	1981
of isotopes	solar system	VI/2a, p. 268	1981
Acoustic properties → elastic properties			
Age	chondrites	V/1b, p. 561	1982
		VI/2a, p. 277	1981
	earth	V/1b, p. 561	1982
		VI/2a, p. 283	1981
	Galaxy (milky way)	VI/2a, p. 284	1981
	lunar rocks	VI/2a, p. 277	1981
	meteorites	V/1b, p. 561	1982
		VI/2a, p. 276	1981
	moon	VI/2a, p. 282	1981
	solar system	VI/2a, p. 273	1981
	terrestrial rocks	V/1b, p. 561	1982
	universe	VI/2a, p. 284	1981
		VI/2c, sect. 9.7	1982
Composition, chemical	earth	V/1a, p. 2	1982
	minerals	V/1a, p. 6	1982
	rocks, lunar	V/1b, p. 508	1982
	terrestrial	V/1a, p. 13	1982
mineral	earth's crust	V/1a, p. 4	1982
	rocks	V/1a, p. 9	1982
		V/1b, p. 101	1982
Crystallographic properties	actinide compounds	III/12c, p. 406	1982
	elements	III/6, p. 1	1971
	ferroelectric compounds	III/16a, b	1981
	garnets	III/12a, p. 1	1978
	hexagonal ferrites	III/12c, p. 1	1982
	ice	III/7b1, p. 1	1975
	inorganic compounds	III/7a⋯f	1973⋯1983
	minerals	III/7a⋯f	1973⋯1983
	perovskites	III/12a, p. 373	1978
	rare earth compounds	III/12c, p. 372	1982
	spinels	III/12b, p. 54	1980
Density	elements	III/6, p. 1	1971
		6th ed., II/1, p. 384	1971
	ice	V/1b, p. 484	1982
		III/7b1, p. 1	1975
	inorganic compounds	III/7a⋯f	1973⋯1983
		6th ed., II/1, p. 449	1971

(continued)

*) New Series, if not stated otherwise.

Subject index (continued)

Subject	Object or substance	LB, NS *) Vol., first page	Year of publication
Density	minerals	V/1 a, p. 67	1982
	under shock compression	V/1 a, p. 131	1982
	rocks, lunar	V/1 b, p. 513	1982
	rocks, terrestrial	V/1 a, p. 114	1982
	under shock compression	V/1 a, p. 157	1982
Dielectric properties	ferroelectric substances	III/16a, b	1981
	ice	V/1 b, p. 487	1982
	inorganic compounds	6th ed., II/6, p. 452	1959
	lunar rocks	V/1 b, p. 538	1982
	minerals	V/1 b, p. 254	1982
	rocks	V/1 b, p. 257	1982
Elastic properties	elements	III/11, p. 9	1979
	ice	V/1 b, p. 496	1982
	inorganic compounds	III/11, p. 22	1979
	minerals	V/1 b, p. 22	1982
	rocks, lunar	V/1 b, p. 487, p. 493	1982
	rocks, terrestrial	V/1 b, p. 8	1982
Electric conductivity	ferroelectric substances	III/16a, b	1981
	ice	V/1 b, p. 487	1982
	lunar rocks	V/1 b, p. 538	1982
	metals	III/15a, p. 1	1982
	minerals	V/1 b, p. 241	1982
	rocks	V/1 b, p. 245	1982
Elastooptic properties	inorganic compounds	III/11, p. 495	1979
Electrooptic properties	inorganic compounds	III/11, p. 495	1979
Electrostrictive properties	inorganic compounds	III/11, p. 495	1979
Ferroelectric properties	inorganic compounds	III/16a, b	1981
Heat capacity (specific heat)	Fe and other oxides,	III/4a, b	1970
	garnets, hexagonal ferrites, perovskites, spinels	III/12a, b, c	1978, 1980, 1982
	ice	V/1 b, p. 484	1982
	inorganic substances	6th ed., II/4, p. 180	1961
	lunar rocks	V/1 b, p. 529	1982
	minerals	V/1 a, p. 313	1982
	rocks	V/1 a, p. 315	1982
Inelasticity	terrestrial rocks	V/1 b, p. 141	1982
Lattice constants → crystallographic properties			
Magnetic properties	Fe and other oxides,	III/4a, b	1970
	garnets, hexagonal ferrites, perovskites, spinels	III/12a, b, c	1978, 1980, 1982
	lunar rocks	V/1 b, p. 547	1982
	minerals	V/1 b, p. 308	1982
	rocks	V/1 b, p. 366	1982
			(continued)

*) New Series, if not stated otherwise.

Subject index (continued)

Subject	Object or substance	LB, NS *) Vol., first page	Year of publication
Optical properties	ferroelectric substances	III/16	1981
	Fe and other oxides,	III/4a, b	1970
	garnets, hexagonal ferrites, perovskites, spinels	III/12a, b, c	1978, 1980, 1982
	ice	V/1 b, p. 493	1982
	inorganic compounds	6th ed., II/8, p. **2-44**	1962
	minerals		
	→ inorganic compounds		
	rocks		
	→ inorganic compounds		
Phase transitions → crystallographic properties			
Piezoelectric, piezooptic constants	inorganic compounds	III/11, p. 287	1979
Porosity	lunar rocks	V/1 b, p. 513	1982
	rocks	V/1 a, p. 184	1982
Pyroelectric constants	inorganic compounds	III/11	1979
Radioactivity	terrestrial rocks	V/1 b, p. 433	1982
Thermal expansion	ice	V/1 b, p. 484	1982
	inorganic compounds	III/7a···f	1973···1983
		6th ed., II/1, p. 385	1971
	lunar rocks	V/1 b, p. 529	1982
	terrestrial rocks	V/1 b, p. 111	1982

*) New Series, if not stated otherwise.